FUNDAMENTALS OF APPLIED ENTOMOLOGY

THIRD EDITION

edited by **ROBERT E. PFADT**

Professor of Entomology, University of Wyoming

MACMILLAN PUBLISHING CO., INC.
New York
COLLIER MACMILLAN PUBLISHERS
London

Printed in the United States of America

Macmillan Publishing Co., Inc.
866 Third Avenue, New York, New York 10022

Collier Macmillan Canada, Ltd.

Library of Congress Cataloging in Publication Data

Pfadt, Robert E ed.
Fundamentals of applied entomology.

Includes bibliographies and index.
1. Insects, Injurious and beneficial. 2. Insect control. I. Brown, Leland R. II. Title.
SB931.P45 1971 632'.7 77-8256
ISBN 0-02-395110-9

Printing: 3 4 5 6 7 8 Year: 0 1 2 3 4

PREFACE

The multitude and rapidity of changes in applied entomology demonstrate the great value of the science for society and agriculture. A host of entomologists working in universities, state and federal government, and industry have devoted themselves to the solution of the many insect problems that affect our health and economy, bringing about improvements both in pest control and in the cultivation and protection of beneficial species.

For a variety of reasons insect control is becoming more complex. State and federal laws restrict the use of insecticides; federal government withdraws registration of many widely recommended insecticides; industry markets new insecticides but with fewer approved uses; relatively few registrations of insecticides are available for the protection of minor crops; numbers of insecticide-resistant species of insects continue to increase; and intricacies of pest management baffle most farmers.

Despite all these difficulties—or perhaps because of them—the profession of entomology is increasingly needed and esteemed in today's world. Professionalism has been the exclusive club of the professorial group in universities, but the club is now being joined by applied entomologists from other sections. Both the American Registry of Professional Entomologists and the amended Federal Insecticide, Fungicide, and Rodenticide Act have made this new professionalism possible.

Happily the science and practice of insect control are growing and progressing. We have come to realize that increase of insect populations to injurious numbers is a complex phenomenon that cannot be managed by simple methods of control. We have been forced to look at and study all phases of insect population dynamics and ways to retard population growth in order to keep insect numbers below economic injury levels. In this development the subject of applied entomology continues to expand and become more voluminous.

The third edition of this text attempts to present to students a comprehensive yet reasonable summary of the discipline of applied entomology. We have updated information, incorporated recent research, and included concepts of pest management whenever applicable. Understandably, pest management has been developed more for some crops than others. It is also true that insect control practices developed in past years for products like stored grain were not specifically called pest management but can very well be included as early examples.

We have continued the same organization of this book as begun in the first edition, but have added one chapter, "Insect Pests of Ornamental Shrubs, Shade Trees, and Turf." We have deleted a few illustrations and have added others. Because of the current emphasis placed on pest management, the variety of new insecticides, the many new discoveries in insect biology, and the drastic changes in laws governing use of pesticides, this revision has not been an easy one. We have been encouraged by the kind help of many colleagues who have given freely of their time and knowledge in reviewing chapters and in supplying information and illustrations.

We wish especially to acknowledge the professional assistance of the following colleagues.

Chapter 1. *Insects and Man*: D. Lyle Goleman, E. F. Knipling, M. D. Levin, C. D. F. Miller, John V. Osmun, M. J. Ramsay, Charles W. Rutschky, Joseph F. Spears, and Douglas Sutherland.

Chapter 2. *Insect Structure and Function*: Roger D. Akre and Carl Johansen.

Chapter 3. *Insect Growth*: Roger D. Akre and Carl Johansen.

Chapter 4. *Classification of Insects and Their Relatives*: Roger D. Akre and M. T. James.

Chapter 5. *Insects and Their Environment*: A. C. Hodson.

Chapter 6. *Principles of Insect Control*: Roger D. Akre.

Chapter 7. *Chemical Control*: C. C. Burkhardt, Carlo Ignoffo, and John E. Lloyd.

Chapter 8. *Insecticide Application Equipment*: C. C. Burkhardt.

Chapter 9. *Insect Pests of Small Grains*: R. H. Burrage, Robert L. Gallun, and E. A. Wood, Jr.

Chapter 11. *Insects of Legumes*: Donald W. Davis, Carl Johansen, and William P. Nye.

Chapter 12. *Insect Pests of Cotton*: George A. Slater and David F. Young, Jr.

Chapter 16. *Insect and Other Pests of Floricultural Crops*: Richard K. Lindquist.

Chapter 19. *Household Insects*: Gary W. Bennett and James F. Dill.

Chapter 20. *Livestock Insects and Related Pests*: Jerry F. Butler and John E. Lloyd.

Chapter 22. *Insects of Medical Importance*: Deane P. Furman and Robert F. Harwood.

R. E. P.

CONTENTS

Note: Officially approved uses of specific insecticides may change from year to year. Consult current recommendations put out by your state agricultural college and follow label directions when you want to control infestations of a pest. Do not use discussions of insecticide recommendations in this textbook as the basis for control operations.

FUNDAMENTALS OF APPLIED ENTOMOLOGY

chapter 1 / **ROBERT E. PFADT**

INSECTS AND MAN

The history of man is the record of a hungry creature in search of food, van Loon concluded in *The Story of Mankind*. In America, a land of plenty, too few people appreciate the vital importance of food in maintaining the health and life of man. Today around a half billion of the world's population are hungry and malnourished, and with current high birth rates and the prolongation of life by the humane application of medical science, the problem of raising sufficient food to fill all human wants appears to be increasing. This problem is not new; history is filled with evidence of man's precarious position in the face of famine and disease. Causing much of this trouble have been the insects, those small six-legged animals that devour man's crops, feed on his livestock, suck his blood, and infect him with disease.

Insects inhabited the earth long before man appeared. Fossils indicate that insects have been around for 350 million years; by Permian times, 200 million years ago, insects had extended their world and become numerous. When man arrived, approximately 1 million years ago, insects had evolved into nearly all of their present diversity. Although we have little proof, primitive man presumably was harassed by blood-sucking insects that not only fed on him but also spread pestilence. There is evidence that human lice, vectors of typhus and other diseases, infested ancient peoples; the dried-up remains of these pests have been found in the hair of prehistoric mummies from many parts of the world.

Yet we should not conclude that all insects, or even many, were enemies of ancient man. Just as primitive tribes do today, it is likely that ancient man ate many kinds of insects, including caterpillars, grubs, and grasshoppers. There is evidence that he utilized honey as long ago as the Stone Age. On the wall of a Spanish cave a Mesolithic artist depicted around the year 7000 B.C. a man removing honey from a natural hole in a cliff. People of ancient Egypt raised honey bees. They kept these insects in pipe hives that consisted of long clay tubes piled horizontally. The Egyptians used the honey as food and the wax for making candles, writing tablets, sealing and waterproofing substances, mummification mixtures, and medicines. A hieroglyphic honey bee is inscribed on a sarcophagus dating back to 3633 B.C., and a bas-relief describing the process of extracting honey is depicted in the temple of Ne-user-re, built around 2600 B.C. Cuneiform inscriptions record the introduction of bees into Assyria for the production of honey and wax, and clear representations of flies and preserved locusts that were used as food appear in the seals and sculptures of ancient Assyria and Babylon. The ancient Chinese practiced silk culture; fossil cocoons unearthed in late Stone Age remains indicate that oriental man cultivated the silkworm before 4700 B.C.

Ancient man not only used insects to satisfy his physical wants; he also brought them into his rituals and made them symbols of worship. To the Egyptians, dung beetles [*Scarabaeus sacer* Linnaeus (Fig. 1:1) and several other species] were sacred and symbolized eternal life. The image, carved in jade, emerald, and other stones, symbolized their sun-god Khepera, who was the "Creator" and the "Father of the Gods." These images, known as scarabs, were used widely as sacred ornaments and as symbols of resurrection in burials. When a man died, his heart was removed and a scarab inserted in its place. Scarabs, along with figures of gods, were placed on mummies lying in their sarcophagi.

Why did the Egyptians come to revere dung beetles—a reverence that seems strange at first thought? One hypothesis is that the beetles were held

Figure 1:1. The sacred scarab beetle, *Scarabaeus sacer* L., of the ancient Egyptians. *Courtesy USDA.*

sacred because the Egyptians associated these insects with their kings. Interring their deceased monarchs in tunnels cut into cliffs, they concealed the tombs from robbers with piles of refuse. The decomposing organic matter served as breeding places for the beetles, which the early Egyptians related mistakenly to their kings rather than to the refuse.

We find vivid accounts of the ravages of insect enemies in the literature of the ancient Egyptians, Hebrews, and Greeks. The Bible mentions them many times, and 11 insects have been identified specifically in the Bible by the late F. S. Bodenheimer, an Israeli entomologist. Among these are the body louse, *Pediculus humanus* L., the human flea, *Pulex irritans* L., the webbing clothes moth, *Tineola bisselliella* (Hummel), the honey bee, *Apis mellifera* L., and the desert locust, *Schistocerca gregaria* Forskål.

Joel 2:3 describes the destructiveness of locusts in these words:

> the land is as the garden of Eden before them, and behind them a desolate wilderness; yea, and nothing shall escape them.

Of the ten plagues visited upon Egypt preceding the Exodus, insects caused three and were involved in two or three others. The eighth plague was one of locusts:

> For they covered the face of the whole earth, so that the land was darkened; and they did eat every herb of the land, and all the fruit of the trees which the hail had left: and there remained not any green thing in the trees, or in the herbs of the field, through the land of Egypt.

Even in present times the peoples of Africa and the Middle East have grave trouble with periodic outbreaks of this migratory grasshopper, known commonly as the desert locust (Fig. 1:2).

During the Middle Ages insects as vectors of disease were responsible for some of the worst epidemics that have struck the human race. The Black Death, primarily bubonic plague, is one of the major calamities of history. In the middle of the fourteenth century the disease claimed the life of one quarter of the population in Europe—at least 25 million victims. Although bubonic plague is essentially a bacterial disease of rodents transmitted by the oriental rat flea, *Xenopsylla cheopis* (Rothschild), at times it causes serious epidemics among human beings. When rats die of the disease, the fleas will leave the dead animals and migrate to other hosts, including man. The flea feeds on blood and directly introduces the infection with its bite. This disease remains with us still, but it is no longer the dreaded human killer of former years.

We could continue relating the histories of fateful human diseases carried by insects—epidemic typhus by the human louse, malaria by the anopheline mosquitoes, yellow fever by the mosquito, *Aedes aegypti* L., sleeping sickness by tsetse flies, and still others—but this would take us deeply into the field of medical entomology and divert us from our main objective, the general study of applied entomology. It is instructive now to take a look at applied entomology in North America beginning with the period of colonization.

Figure 1:2. A swarm of desert locust, *Schistocerca gregaria* Forskål, in Ethiopia. *Courtesy Anti-Locust Research Centre, photograph by C. Ashall.*

INSECTS IN COLONIAL AMERICA

Fortunately not nearly as many insect pests attacked the crops of the early colonists in America as attack crops nowadays. The flood of plant materials from the Old World and of their insect enemies, which inevitably accompanied them, had not yet begun. But there were native insect species and a few introduced ones that caused much trouble. Agriculture in America during colonial times consisted of growing a few grains, fruits, and vegetables that the English had been accustomed to producing at home; corn and tobacco adopted from the Indians; and in southern colonies, where the climate was favorable, cotton, rice and indigo.

As early as 1632 armyworms caused serious damage to the settlers' corn. Fruit insects were injurious too, for a Mr. John Hull recorded in 1661, "the cankerworm hath for four years devoured most of the apples in Boston, so that the trees look in June as if it were the ninth month." In 1740 grasshoppers seriously attacked the crops of the Massachusetts Colony. The colonists, applying the best control they knew, armed themselves with bundles of brush and drove millions into the ocean. In 1743 an outbreak of armyworms caused extensive damage to grains throughout the region of the North Atlantic states.

Insects, principally mosquitoes, caused much trouble for the colonists by transmitting serious human diseases. Because of malaria the death rate among slaves working the rice swamps was high; still more slaves were imported as replacements. Yellow fever felled thousands of America's pioneers. In 1699 it killed one sixth of the population of Philadelphia, and in 1793, when the population numbered 50,000, 11,000 persons contracted the disease, and 4000 died.

Besides the native insects, colonists had to contend with those that they brought with them. Stored grain insects infested grain, and animal parasites annoyed and sucked the blood of livestock. The colonists transported to their homes clothes moths, bed bugs, and cockroaches. Peter Kalm, a Finnish naturalist, noted in 1748 the great havoc of the pea weevil. He stated that peas were no longer cultivated in Pennsylvania and to only a limited extent in New Jersey and New York because of the damage of this insect. In 1779 the Hessian fly, supposedly introduced from Europe in the straw bedding of Lord Howe's Hessian troops, caused extensive destruction to the wheat crop on Long Island.

The colonists introduced honey bees into New England early, at least by 1640, to provide themselves with honey and wax. After 1670 bee culture declined in the colonies, presumably because of American foulbrood, a devastating bee disease. The Indians were unfamiliar with honey bees and called them the "white man's fly."

VOLUNTEER ENTOMOLOGY

As America grew and agriculture expanded, insects increased their injurious attacks and forced recognition of the problem upon the populace. During these early times there were no agricultural experiment stations or federal laboratories to study any of the distressing plant and animal problems confronting both the farmer and the townsman. A keenly observant group of men—farmers, clergymen, physicians, and teachers—became interested in insects and their economic importance and began to study and write about them. One of the most prominent of the early workers was **William D. Peck**, a Massachusetts naturalist who in 1795 wrote *The Description and History of the Cankerworm*. In 1805 he became the first professor of natural history at Harvard University and in this position was able to pursue his interest and study of insects. Born in Boston in 1763, Peck became known as America's first native entomologist.

Figure 1:3. Thaddeus William Harris (1795–1856), pioneer economic entomologist, published the first comprehensive report on destructive insects in America in 1841. *From thè National Archives.*

Thaddeus William Harris, a student of Professor Peck, became a distinguished early economic entomologist (Fig. 1:3). Harris received a medical degree in 1820 and took up the practice of medicine for a short while. His interest in insects probably arose early in life: both of his parents were fascinated by natural history. In 1831 he became librarian at Harvard; while a member of a commission to study the geology and botany of Massachusetts he wrote what is now regarded as a classic in applied entomology, *A Report on the Insects of Massachusetts Injurious to Vegetation*, first published in full in 1841. The third edition (1862) is beautifully illustrated with wood engravings of insects and contains seven chapters with over 600 pages.

PROFESSIONAL ENTOMOLOGY

There were other studies and other publications on insects during the volunteer period, but we shall turn to the year 1854, a significant milestone in entomology, for it was in this year that the profession of entomology is considered to have begun in the United States. Recognition of the need for insect control led to the appointment of two entomologists to government positions. Townend Glover accepted a position with the new U.S. Agricultural Division of the Patent Office, and Asa Fitch went to work for the State of New York.

Townend Glover began his duties of collecting "information on seeds, fruits, and insects of the United States" on June 14, 1854. Though Glover was interested chiefly in his drawings of insects and in his agricultural museum, field activities took him into the South, where he studied insect pests of crops, particularly those of the orange and of cotton, and into the West, where he studied grasshoppers and the Colorado potato beetle.

Asa Fitch, a medical doctor, gave up his practice in 1838 and began to study insects in 1840 (Fig. 1:4). In 1854 the New York State Agricultural Society, authorized by the state legislature, appointed him to investigate the injurious insects of New York, thus making him the first state entomologist. Fitch made thorough studies of the life histories of many insect pests of New York and described many of the important economic species of America. He described, for example, the grape phylloxera, *Daktulosphaira vitifoliae* (Fitch), the corn leaf aphid, *Rhopalosiphum maidis* (Fitch), and the wheat jointworm, *Harmolita tritici* (Fitch). Much of his work was published in the *Transactions of the New York State Agricultural Society* in the form of 14 annual reports.

From these modest beginnings entomology in America, through the research and study of many diligent entomologists, has risen gradually to the position of an important and respected science. The developments came in two streams that often intermingled. One consisted of the advances made by the federal organization; the other, of contributions from the state experiment stations and universities. In recent years a third stream, consisting of the teams of entomologists and of other scientists in industry and commerce, has risen to great importance.

Because a full account of entomological progress in the twentieth century would fill a volume in itself, we shall consider only a few of the highlights, first taking up the contributions made in the federal service by the chiefs of entomology.

Entomology in Federal Service

Upon the retirement of Townend Glover in 1878, **Charles Valentine Riley** (Fig. 1:5) accepted the post. Born in London and educated in France and Germany, Riley immigrated to the United States at the age of 17. When he accepted the federal position, Riley was already a well-known entomologist;

Figure 1:4. Asa Fitch (1809–79), first state entomologist, investigated the injurious insects of New York and described many of the economic species of America. *Courtesy New York State Museum and Science Service.*

he had been the state entomologist of Missouri for ten years. His nine annual reports were sound pieces of research illustrated by admirable woodcuts of insects. Riley was a man with a driving but difficult temperament. He failed to give due credit to the assistants who aided him in conducting research and writing reports. If a young entomologist stood up to explain an interesting discovery that he had made on the biology of an insect, Riley immediately rose and began to expound his own findings beginning with the caustic remark, "I have long suspected that to be the case."

Riley was troubled with insomnia, and though he found it hard to sleep in his bed at home, he could sleep on a long railway journey or in a barber's chair. In addition to his Missouri reports, he is remembered for four other important accomplishments.

Figure 1:5. Charles Valentine Riley (1843–95), a dynamic investigator and organizer of agricultural entomology in America. *Courtesy University of Missouri.*

1. He was largely responsible for the founding of the **U.S. Entomological Commission** in 1877. The job of the commission was to study how to control the hordes of the Rocky Mountain grasshopper, *Melanoplus spretus* (Walsh), which had developed in the eastern foothills of the Rocky Mountains, flown eastward in 1874, and descended upon the farmlands of the Midwest, creating a national disaster. Growing crops were devoured, farms were abandoned, and the trek of settlers toward Kansas and neighboring states was halted. Three entomologists, Riley, A. S. Packard, and Cyrus Thomas, made up the commission. The results of their studies were published in five reports, the first two on the Rocky Mountain grasshopper and the other three on a variety of insects. In June 1881 the activities of the commission ceased.

2. He began the building of the organization that grew into the Bureau of Entomology.

3. He received a gold medal from the French government for his suggestion—based on his research in Missouri—that American resistant rootstocks be used to control the ravages of the grape phylloxera in France.

4. He conceived and directed the successful biological control of the cottony-cushion scale, a serious pest of citrus in California, by introducing a predaceous enemy of the scale, the lady beetle, *Rodolia cardinalis* (Muls.), from Australia.

Although Riley's accomplishments are now recognized universally by entomologists, his career was not without disappointments. In his eagerness to promote work in entomology, he went directly to Congress for obtaining increased funds rather than through regular channels. This lack of bureaucratic ethics inflamed Riley's superior, the Commissioner of Agriculture. In the heat of the argument Riley resigned, and J. H. Comstock of Cornell University was appointed federal entomologist, a position Comstock held for two years, 1879 to 1881. In 1881, after the election of President Garfield, Riley's friends were able to persuade the new administration to reappoint him to his former position.

Riley retired in 1894, and his assistant, **Dr. Leland O. Howard**, stepped in as chief (Fig. 1:6). This was a happy selection: Howard was a brilliant scientist, a leader, and a man of action. He made lasting contributions to medical entomology, biological control, insect taxonomy, and—perhaps most important of all—to the tremendous growth, efficient operation, and sound research of the Bureau of Entomology. In 1911, with Fiske as collaborator, Howard published a classic study of the gypsy moth and browntail moth, in which appeared the first clear exposition of insect population dynamics. And in 1912 to 1917, with Dyar and Knab as collaborators, he published a large four-volume work, *The Mosquitoes of North America, Central America, and the West Indies.*

In 1927 **Charles L. Marlatt** succeeded Howard as chief of the Bureau of Entomology. Marlatt had been associate chief to Howard and before that associate entomologist at Kansas State College. An expert on sawflies and scale insects, he traveled to the Orient; in China he discovered the native home of the San Jose scale. This insect had been inadvertently introduced into California around 1870 and became a serious pest of fruit and ornamental trees in America. Howard labored long and perseveringly to have Congress pass federal quarantine laws to put an end to the thoughtless introduction of foreign pests. Although the United States had suffered most from stowaway insects, it was one of the last countries to pass a satisfactory quarantine law. In 1912 the Plant Quarantine Act finally became a reality.

Under **Lee A. Strong**, chief from 1933 to 1941, the quarantine work of the federal government became more highly organized and more efficient. He was largely instrumental in the merger of two bureaus in 1934 that formed the Bureau of Entomology and Plant Quarantine.

Upon the sudden death of Strong in 1941, Dr. **P. N. Annand** became chief of the bureau and held the office through the war years. Annand's great

Figure 1:6. Leland O. Howard (1857–1950), a brilliant promoter of entomology in America and world authority in medical entomology. Howard was chief of the Bureau of Entomology from 1894 to 1927. *Courtesy USDA.*

contribution was organizing and channeling the resources of the bureau into insect control studies to help increase food and fiber and to provide the military with new methods of controlling the insect vectors of such diseases as typhus, malaria, and yellow fever. Upon Annand's death in office in 1950, **Avery S. Hoyt** succeeded him and remained chief until January 2, 1954, when the Bureau of Entomology and Plant Quarantine was abolished.

A surge of reorganization within USDA was begun in 1952 and is still continuing. In 1953 USDA established the Agricultural Research Service (ARS) and under it a number of divisions including the Entomology Research Division, Plant Protection Division, Plant Quarantine Division, and Pesticides Regulation Division.

From 1953 to 1970 **Edward F. Knipling** (Fig. 1:7) directed the Entomology Research Division. During this period, both by his example and by

Figure 1:7. Edward F. Knipling (1909–), former Director of the Entomology Research Division of the United States and world renowned for studies on insect pests of man and animals. *Courtesy USDA.*

his writing and lecturing, he set new goals and broadened the horizons of research on insect control. One of his capital ideas was the control of the screwworm and certain other major insect pests by the sterile male technique. On January 1, 1971, Dr. Knipling became Science Adviser to the Administrator of the Agricultural Research Service. He held this position until his retirement from government service, December 31, 1973. Presently he is continuing entomological research as collaborating entomologist with ARS.

During its existence from 1953 to 1972, the **Entomology Research Division** was divided both functionally and administratively into various branches and special laboratories. As finally organized, it included eight branches: Apiculture; Cotton Insects; Fruit Insects; Grain and Forage Insects; Insects Affecting Man and Animals; Insect Identification and Parasite Introduction; Pesticide Chemicals; and lastly Vegetable, Ornamental, and Specialty Crops Insects. The administration of each branch was centered in Beltsville, Maryland. In addition to the central facilities, the branches administered around 100 laboratories strategically located in various sections of the United States and three laboratories in foreign countries. Approximately 500 scientists and 800 support personnel staffed these facilities.

The Entomology Research Division also established five special research laboratories: two at Beltsville, Maryland, in 1958, Insect Pathology and Insect Physiology; one in Fargo, North Dakota, in 1964, Insect Metabolism and Radiation; one in Gainesville, Florida, in 1969, Insect Attractants, Behavior, and Basic Biology; and one in Stoneville, Mississippi, in 1970, Bioenvironmental Insect Control.

On July 1, 1972, a new organizational structure of ARS became effective, replacing 15 divisions and over 50 branches. The change was primarily administrative; no essential changes took place at the research level. Each ARS laboratory continues to have a Research Leader or Laboratory Director, who now reports to an Area Director rather than a Branch Chief.

For organizational purposes ARS has divided the United States into four regions, the Northeastern, the North Central, the Western, and the Southern. Each region is further divided into areas headed by Area Directors.

Research objectives remain much the same, but new terminology has been developed to designate them. The USDA now has 11 "missions," 40 "operating goals," and 287 "programs." The programs are further divided into "research activities." A National Program Staff oversees these activities, reviewing their scientific and practical value for the nation. In the field of entomology there are presently 31 research activities supervised by nine members of the program staff.

In order to show the nature and categories of entomological research in ARS, we have outlined below 14 of the 31 research activities, along with the names of the supervising staff scientists.

1. Insect pests of small grains, R. D. Jackson
2. Corn insect control, R. D. Jackson
3. Grass and legume insect control, R. D. Jackson
4. Cotton insect control, R. L. Ridgway
5. Tobacco insect control, R. L. Ridgway
6. Bees and other insects for crop pollination and honey production, E. C. Martin
7. Fruit and vegetable insects, M. E. Cleveland
8. Insect pests of forest and shade trees, W. Klassen
9. Control of insects, mites, and ticks on livestock and poultry, D. A. Lindquist

10. Insects affecting man, L. S. Henderson
11. Development of selective, nonpersistent insecticides, P. Schwartz
12. Biological agents to control insect pests, D. E. Bryan
13. Identification and classification of insects of agricultural importance, D. E. Bryan
14. Physiology, reproduction, and growth of insects of agricultural importance, W. Klassen

Not all entomological research of the federal government is conducted by the Agricultural Research Service. In addition, the government sponsors research by entomologists of the Forest Service (USDA) on forest insects, by entomologists of the Agricultural Marketing Service (USDA) on insect pests of stored products, by entomologists of the Public Health Service (USDHEW) on insects of medical importance, and by entomologists of the Department of Defense on insects of military importance.

Animal and Plant Health Inspection Service

The regulatory agencies within the Agricultural Research Service, namely the Agricultural Quarantine Inspection Division, the Plant Protection Division, Animal Health Division, and Veterinary Biologics Division, were separated from ARS and combined on October 31, 1971, into a new regulatory agency—the Animal and Plant Health Service. On April 2, 1972, meat and poultry inspection were transferred to the new service, which then became known as the Animal and Plant Health Inspection Service (APHIS).

The Animal Health Division and the Veterinary Biologics Division were merged into a single unit—Veterinary Services of APHIS. The functions of the Agricultural Quarantine Inspection Division and the Plant Protection Division were merged on September 3, 1972, into a single operation—**Plant Protection and Quarantine Programs (PPQ)**. An administrative subdivision of APHIS, PPQ retains the responsibilities of the two former divisions. It has the task of guarding the United States against entry of foreign plant and animal diseases and pests; of controlling, eradicating, or preventing the spread of the foreign plant diseases or pests that do slip through these defenses and become established; and of suppressing any outbreaks of endemic pests too widespread for farmers and ranchers to handle themselves.

Headquartered in Washington, D.C., and in Hyattsville, Maryland, PPQ has the following organization:

Plant Protection and Quarantine Programs

Office of the Deputy Administrator

Deputy Administrator: J. O. Lee, Jr.
Assistant Deputy Administrator: T. G. Darling
Assistant to Deputy Administrator: J. W. Gentry

Executive Office
Executive Officer: C. R. Ranek

Professional Development Staff
Chief Staff Officer: E. A. Thomas

National Program Planning Staff
Director: G. G. Rohwer
Assistant Director: J. F. Spears
Assistant to the Director: M. T. Pender
Environmental Evaluation Staff: R. L. Williamson, Chief Staff Officer
Methods Development Staff: C. M. Amyx, Acting Chief Staff Officer
New Pest Detection and Survey Staff: H. V. Autry, Chief Staff Officer
Pest Program Development Staff and Aircraft Operations: W. F. Helms, Chief Staff Officer
Plant Importation and Technical Support Staff: H. S. Shirakawa, Chief Staff Officer
Port Operations Development Staff: M. Slobodnik, Chief Staff Officer
Regulatory Support Staff: H. I. Rainwater, Chief Staff Officer

Pest control operations of PPQ are cooperative with affected states and the Republic of Mexico. Programs are jointly planned, financed, and executed. This cooperative arrangement is spelled out in memoranda of understanding with each of the states and with Mexico. State departments of agriculture are the principle cooperators, although in most cases state experiment stations and the state extension services are parties to the agreements. Affected cities, counties, industries, and farmers or ranchers likewise cooperate on certain programs.

Federal laws provide the U.S. Department of Agriculture with authority to develop and carry out programs to eradicate, control, or prevent spread of destructive plant pests. Programs are conducted in Mexico principally to prevent the spread of destructive plant pests from that country into the United States. Cooperative relations are established with Canada on program pests of mutual interest.

In protecting the United States against the inroads of pests, PPQ in cooperation with the states carries out manifold assignments:

1. Inspection of agricultural commodities at international ports of entry as authorized by federal quarantine laws and regulations.
2. Detection of pests new to the country, determination of the spread of introduced pests into new areas, and determination of levels of infestation of established agricultural insect pests through systematic surveys.

3. Prevention of the spread of introduced plant pests to noninfested areas through the enforcement of federal or state cooperative quarantines.
4. Eradication or application of control measures to prevent the spread of dangerous, introduced insects, nematodes, and plant diseases while they are still confined to a small area (Fig. 1:8).
5. Suppression of infestations of certain native pests such as grasshoppers and Mormon crickets on rangelands of western states to prevent their spread into valuable croplands and to protect the range cover.
6. Monitoring levels of insecticide residue in the areas of cooperative pest control.
7. Support of these activities through methods development investigations designed to adapt basic research findings to specific program needs.

Figure 1:8. Fumigating a grain storage building and sheds on a cattle feed yard at Bakersfield, California as part of a cooperative khapra beetle eradication program between a state invaded by a foreign pest and APHIS. Buildings are covered completely with gas-tight tarpaulins and fumigated with methyl bromide. *Courtesy USDA.*

8. Supervision of aerial surveys and of aerial application of pesticides on cooperative programs.
9. Initiation of emergency programs to control serious outbreaks of pests and assistance in the prevention of impending catastrophes of animal and human diseases.
10. Providing technical assistance in pest control to other agencies and individuals in this country and abroad.

Although the majority of cooperative programs are concerned with insect pests, there are several that embrace plant nematodes and plant diseases. Control programs are aided by domestic plant quarantines and uniform state quarantines enforced jointly by PPQ and state agencies. These programs deal with the following quarantines:

Burrowing nematode
Black stem rust
Citrus blackfly
Golden nematode
Gypsy moth and browntail moth
Imported fire ant
Japanese beetle
Mexican fruit fly
Pink bollworm
Unshu orange
West Indian sugarcane root borer
Whitefringed beetle
Witchweed

Other cooperative programs include control of the boll weevil, cereal leaf beetle, oat cyst nematode, several species of grasshoppers, and the Mormon cricket. Personnel of PPQ also make special surveys for beet leafhopper, khapra beetle, hibernating boll weevils, several species of fruit flies, and several insect pests in foreign countries.

An important function of PPQ is its work in coordinating the national cooperative economic insect survey and detection system involving federal, state, and private workers interested in general insect control. This program serves as an "intelligence" system for the defense of the nation's crops and forests against destructive insects. Field observers make observations and send them to a state coordinator. The state officer may then warn county agents, local control agencies, and other interested groups of the presence of some new insect or of the buildup of populations of insects. At the same time, he sends his report to national headquarters for inclusion in the weekly *Cooperative Plant Pest Report.* This journal, in its 27th year of publication in 1977, is prepared in headquarters and distributed to over 4,000 readers in the United States and foreign countries. It now includes information on the detection and status of weeds and plant diseases as well as insects. Thirty-two states employ entomologists who are specialists in detection, assessment, and prediction; they provide technical leadership in the survey program.

Foreign Plant Quarantines

The foreign plant quarantine services performed by PPQ were begun when Congress passed the **Plant Quarantine Act** of 1912. Originally the Federal Horticultural Board, with C. L. Marlatt as chairman (1912–1928), enforced the federal quarantine authority. In subsequent years these functions were performed by various divisions of USDA.

In the enforcement of foreign plant quarantine laws, PPQ has two primary goals. One is to protect U.S. crops, forests, ornamentals, and grasslands from the invasion and spread of foreign pests. The second is to facilitate the marketing of American agricultural products abroad by providing an export inspection and certification service to shippers to ensure that U.S. plant products meet the quarantine requirements of the importing countries.

Because the principal reason for plant quarantine laws is to guard against the entry into the United States of destructive foreign plant pests and diseases, 600 professionally trained PPQ personnel are on watch at 80 ports of entry that include major ocean, Great Lakes, air, and border ports in the continental United States, and in Hawaii, Guam, Puerto Rico, and the American Virgin Islands. The inspectors, working in cooperation with U.S. Customs officials, examine ships, airplanes, vehicles, trains, baggage, and mail to enforce the quarantine against the entry of foreign pests. They also inspect cargoes of restricted material and apply pest control methods, such as fumigation (Fig. 1:9) and other safeguards, to ensure that harmful pests are not brought in with such cargoes.

To preclear agricultural commodities destined for the United States, PPQ inspectors work also in foreign countries. At the request and expense of foreign exporters, they inspect and supervise treatment of fruits, vegetables, and flower bulbs to meet entry requirements of the United States. This service benefits U.S. agriculture by eliminating pests at the source.

EPA Office of Pesticide Programs

Upon the establishment of the **Environmental Protection Agency (EPA)** on December 2, 1970, its Office of Pesticide Programs was assigned responsibility for the registration and the supervision of manufacture, distribution, and use of insecticides and other pesticides. Four divisions carry out the many duties of this office. The Registration Division reviews labeling, registers pesticide products, and sets tolerances—the amount of a pesticide that may legally remain on an agricultural commodity after harvest. The Criteria and Evaluation Division provides scientific support for public hearings, and for other divisions of the Office of Pesticide Programs. The Operations Division attempts to improve government pesticide activities, recommends plans for state legislation and for applicator certification, and makes environmental

Figure 1:9. Nursery stock from Holland is placed in chamber and fumigated by staff of Plant Protection and Quarantine to free it from insect pests before being released to importer. *Courtesy USDA.*

impact evaluations of pesticide use. The Technical Service Division investigates the effect of pesticides on the environment and on the health of man, performs chemical analyses of pesticides, publishes *Federal Register* notices, the *EPA Compendium of Registered Pesticides*, and bulletins and periodicals on pesticides. This division also processes Freedom of Information requests and maintains files on company data supporting registrations.

In 1976 EPA established a new organizational unit, the Office of Special Pesticide Reviews. Its responsibilities are to identify pesticides that may pose health and environmental risks and to study the scientific evidence in order to arrive at valid decisions concerning the continued registration and use of products containing such chemicals.

Entomologists have contributed to the pioneering phases of EPA's pesticide programs. William M. Upholt was the first deputy administrator for pesticides and drafted early legislation. John V. Osmun, F. W. Whittemore, and James H. White authored the standards and developed the structure for the statewide applicator training and certification program.

Entomology in the States

Entomologists working in state departments and in the universities of America have contributed brilliantly to the science of entomology. Besides Asa Fitch hired in 1854 by the state of New York, other early state entomologists were **B. W. Walsh**, hired in 1866 by Illinois, and C. V. Riley, hired by Missouri in 1868. Walsh was born in England and went to Cambridge University, where he worked with Charles Darwin. He came to America at the age of 30, but not until after a varied career as a farmer and lumberman did he begin the earnest study of insects, when he was 50 years old! Though his career in entomology lasted only 11 years, he greatly influenced the standards of research and writing of a young science. Riley greatly admired the older Walsh, and learned and profited much from their association. State entomologists of this early period headed research and educational organizations, whereas present-day state entomologists are principally regulatory officials.

The need for education and research in agriculture, not just in entomology, was recognized by agricultural societies, state boards of agriculture, and numerous individuals throughout the country. From 1840 on, there was continuous activity to obtain public support for agricultural colleges. At the national level this movement culminated in the **Morrill Land-Grant College Act** of 1862, which donated to each state 30,000 acres of federal land for each of its U.S. Senators and Representatives, for the endowment, support, and maintenance of at least one college, where the leading subjects would be related to agriculture and the mechanic arts, and provided for professorships in botany and entomology. This act, together with the **Hatch Act** of 1887, which supported state agricultural experiment stations throughout the nation, became the *point d'appui* for the phenomenal growth and development of entomological research and teaching by the states.

The teaching of entomology in America, so vital for the training of new entomologists, was begun by Peck and Harris in an informal way at Harvard University. The first regular teacher of entomology was **Hermann Hagen**, who was brought to Harvard in 1867 from Germany through the invitation of Louis Agassiz, Director of the Harvard Zoological Museum. Hagen was an enthusiastic teacher and had a kindly, helpful way with fellow workers. Many students came to him for instruction, including Herbert Osborn and J. H. Comstock. At Harvard he founded the first entomological museum in this country.

The teaching of entomology got its real impetus from the newly formed agricultural colleges. Among the prominent early teachers in these institutions were B. F. Mudge at Kansas State University, A. J. Cook at Michigan State University, T. J. Burrill and S. A. Forbes at the University of Illinois, C. H. Fernald at the University of Maine and University of Massachusetts, and J. H. Comstock at Cornell University.

Figure 1:10. John Henry Comstock (1849–1931), one of America's great, early teachers of entomology, *From the National Archives.*

J. H. Comstock (Fig. 1:10), who became one of the leading teachers of entomology in the world, greatly influenced the science not so much by his excellent research on the morphology of insects as by the many fine students he turned out. He graduated from Cornell University (B.S., 1874) and for forty years thereafter worked and taught at his alma mater. There he founded the first university department of entomology in the world. The list of Comstock's students who attained distinction is a long one. He inspired sound scholarly work. Comstock repeatedly warned his students, "Be sure you are right, and then look again."

The student of entomology today is greatly indebted to these pioneer teachers; through their efforts and dedicated lives we now have so many fine teachers in so many splendid departments of entomology in America.

Entomological research, particularly the applied phases, in the states is conducted principally by the agricultural experiment stations. Because these are usually located in the agricultural college, both teaching and research are performed by one and the same department of entomology. Generally the faculty's teaching load is light, so that most entomologists spend the larger share of their time doing research and publishing their discoveries.

A significant development for agricultural entomology was the creation of the **Federal-State Extension Service** by the Smith-Lever Act of 1914. The first extension entomologist was **Thaddeus H. Parks** (Fig. 1:11), who was hired by the University of Idaho in 1913, one year prior to passage of the Act. The circumstances that led to his employment were the presence of the alfalfa weevil in Idaho and the concern of hay growers over quarantine laws enacted by California and Montana against them.

Born February 14, 1887, at Ashville, Ohio, Parks attended Ohio State University, where in 1909 he completed the B.S. degree in agriculture. In 1925 he earned the M.S. degree in entomology from the University of Illinois.

Figure 1:11. Thaddeus Hedges Parks (1887–1971), first extension entomologist in America, ably served farmers and agriculture for 47 years. *Courtesy Ohio State University.*

Parks served three states as extension entomologist: Idaho from 1913 to 1915, Kansas from 1916 to 1918, and Ohio from 1918 to 1956. On returning to Ohio State University, he became Ohio's first extension entomologist. In 1951 USDA presented him its Superior Service Award for his leadership in the development of extension entomology, his contribution in developing methods to control the Hessian fly, and his promotion of a spray service for Ohio fruit growers. In 1957, a year after his retirement, Ohio State University awarded him the honorary Doctor of Science degree.

How well the extension entomologists of the nation have performed their function of taking the discoveries made by the entomologists of the experiment stations and quickly making the information available to county agents and farmers is indicated by their number in 1976—300 extension entomologists employed in 50 states.

In April 1973 the Department of Entomology of Pennsylvania State University established a **Center for the History of Entomology**. Undoubtedly an active program to acquire historical documents of American entomology will save many pieces from unintentional destruction and make these materials available to students at a central location. Dr. Charles W. Rutschky is the Center's first archivist.

Commercial Entomology

Just as the appointments of Glover and of Fitch to federal and state positions in 1854 are considered the beginning of professional entomology in the United States, so the appointment of Dr. **Otto H. Swezey** (Fig. 1:12) in 1904, fifty years later, as assistant entomologist in the experiment station of the Hawaiian Sugar Planters Association is considered the beginning of commercial entomology. Dr. Swezey came from Ohio State University and continued in active service for thirty years. Soon other growers' organizations began to recognize the need for specialists on insect control and began to hire entomologists. Though their achievements in applied entomology have been many, we have time to mention only a few of the outstanding workers. In 1910 **Asa C. Maxon** was made director of the experiment station of the Great Western Sugar Company at Longmont, Colorado. He carried on research there for 35 years and became the leading authority on insects affecting sugar beets. In the citrus industry, Sunkist Growers since 1920 have had a progressive group of entomologists who have greatly improved insect control measures on the organization's extensive citrus plantings in Arizona and California. One of the most brilliant records in agricultural entomology has been made by the entomologists of the Pineapple Producers Association in Hawaii. In 1930 Dr. **R. N. Chapman**, the eminent insect ecologist, left the University of Minnesota to become director of the association. In the same year Dr. **Walter Carter** became head of the department of entomology of the Pineapple Research Institute. For many years the United Fruit Company has carried on investigations in tropical agriculture and on the insects that affect its extensive plantings.

Figure 1:12. Otto Herman Swezey (1869–1959), first commercial entomologist in America and a devoted student of insects. *Courtesy Hawaiian Sugar Planters Association.*

The biggest employer of commercial entomologists today is the insecticide industry, which comprises more than 50 basic manufacturers and more than 500 formulators and processors. The industry employs hundreds of entomologists to work in laboratories, on experimental farms, and in company organizations as administrators, technical representatives, and salesmen. A recent estimate of entomologists in industry puts the number at about 500.

Other areas in which the numbers of commercial entomologists have grown can be seen by the following categories of employment.

Beekeeping and associated supplies industry
Urban and industrial pest control operators
Supervised pest control services
Commercial crop spraying and dusting
Forest products industry
Grain and milling trade
Seed growers
Canning companies
Tree experts and arborists
Nurserymen
Dairy industry
Biological supply houses
Consulting entomologists

ENTOMOLOGY IN CANADA

Our historical review of applied entomology in America would be incomplete if we did not call attention to the history of the science in Canada. Before the 1850s, as in the United States, knowledge of insects depended on the observations of a few people who were interested in these animals as a hobby. They noticed the beautiful butterflies, the tormenting mosquitoes, and the devastating grasshoppers. Just as they had done in the United States, insects forced the development of entomology in Canada by their encroachments on the nation's economy.

The beginnings of Canadian agriculture stressed the exportation of lumber and wheat, and when in 1856 the Hessian fly and wheat midge caused crop destruction of over two million dollars, the people became greatly alarmed. Because the pest problem intensified as agriculture expanded and increased, insects made the entomologist one of the most important scientists in Canadian society. Famous Canadian entomologists included **James Fletcher**, the first Dominion Entomologist, and his successors, **Charles Gordon Hewitt** and **Arthur Gibson**. Entomologists who have since filled this position are H. G. Crawford, Robert Glen, and B. N. Smallman.

In 1961 a reorganization of the Department of Agriculture along "crop" lines rather than disciplines abolished the position of Dominion Entomologist. At present the Research Branch of the Department of Agriculture has many research stations all across Canada, and at most of these, groups of entomologists (sections) perform basic and applied research on insects and report to the local station director. In addition two institutes devote themselves to specialized studies in entomology: the Research Institute at London, Ontario with pesticide investigations, and the Biosystematics Research Institute at Ottawa, Ontario with insect taxonomy. The institute at Ottawa, employing 35 taxonomists and allied scientists, is one of the largest entomological museums in the world.

About 300 entomologists work in Canada, mainly in the Department of Agriculture and in the Forestry Service. The remainder work in provincial departments or teach and do research in Canadian universities. There are comparatively few entomologists working for chemical companies in Canada. And ten major Canadian universities with distinct entomology departments offer programs leading to the master's or the Ph.D. degree in entomology.

ENTOMOLOGICAL SOCIETIES

Entomological societies have played an important part in the development of the science by bringing people of common interests together, by creating a contagious enthusiasm for the study of insects, by providing a strong organization to enlighten public opinion and to sway governmental decisions and actions, and by sponsoring journals in which members can publish their research. Entomological societies started in England with the founding of the **Aurelian Society** in London about 1745. The society did not last long; the fire that burned its library and insect collections also caused its dissolution in 1748. A succession of societies followed, until finally the present and distinguished **Royal Entomological Society of London** was formed in 1833. The society publishes three scientific journals: *Ecological Entomology*, *Physiological Entomology*, and *Systematic Entomology*; and a quarterly news and reviews bulletin, *Antenna*.

In America the first society of this kind was the short-lived Entomological Society of Pennsylvania, founded at York in 1842; it disbanded around 1844. The oldest existing entomological society in America is the **American Entomological Society**, founded in Philadelphia in 1859 and called the Entomological Society of Philadelphia until 1867. It is a strong organization with an excellent insect collection and several periodical publications, notably the *Transactions*, *Entomological News*, and *Memoirs*.

Another important early society was the **Entomological Society of Canada**, founded at Toronto in 1863. This association immediately led to an increase in the number of articles on insects, and in 1868 the society began its now famous journal, *The Canadian Entomologist*. The Society also issues *Memoirs* for the publication of longer scientific treatises and the *Bulletin* for publication of news, business, and problems of interest to entomologists. It has a membership of approximately 1000, drawn from all over the world.

The federal acts of 1862 and 1887, which established the state agricultural colleges and experiment stations, opened the door for the employment of large numbers of entomologists who shortly felt the need for an association. In July 1889 James Fletcher, the Dominion Entomologist of Canada, came to Washington where he and L. O. Howard drafted a constitution for an Association of Economic Entomologists. The next month the Association was organized at Toronto. In 1908 the organization began publishing the *Journal of Economic Entomology* and in the following year changed its name to the **American Association of Economic Entomologists**. As entomology grew in America, those interested primarily in the basic rather than the applied phases of the science organized the **Entomological Society of America** in 1906. This society began publishing its annals in 1908. Because many entomologists in the United States belonged to both organizations and because there was great need for interaction among individuals involved in research, control, taxonomy, teaching, extension, and industry, the members of both organizations decided in 1953 to merge into one strong, united society and to retain the name Entomological Society of America. The Society's first president, **Charles E. Palm**, and the president-elect, **Herbert H. Ross**, were instrumental in effecting the merger. Today the society has over 7700 members with worldwide representation, an annual budget of well over $600,000, and its own headquarters building at College Park, Maryland. The society now publishes or underwrites eleven periodical publications.

Annals of the Entomological Society of America
Journal of Economic Entomology
Environmental Entomology
Insecticide and Acaricide Tests
Index of American Economic Entomology
Monographs of the Thomas Say Foundation
The Annual Review of Entomology
Miscellaneous Publications of the Entomological Society of America
Bulletin of the Entomological Society of America
Entoma—Pest Control Directory
Opportunities in Professional Entomology

An **American Registry of Professional Entomologists (ARPE)** was established in 1970 by the Entomological Society of America to identify qualified individuals who provide technical service to the public in entomologically related activities. Eighteen categories of service are presently available, including agricultural entomology, medical and veterinary entomology, pest management, regulatory entomology, urban and industrial entomology, and apiculture. Through the Registry the society is able to provide evidence of the professional competence of individuals who perform services affecting the health and economy of people and the quality of the environment. To obtain registration an individual must make application to the Registry and provide evidence of his educational and occupational qualifications. In 1976 membership in the Registry stood at 1,307. ARPE was granted self-governance in 1976; the members elected Perry L. Adkisson as its first president and John V. Osmun, the original developer of the Registry, president-elect.

IMPACT OF LAWS AND REGULATIONS ON APPLIED ENTOMOLOGY

Federal and state laws regulating pesticides in the environment and on food are crucial these days. Because these laws are important and because they are a lively topic not only among farmers and agriculturalists but also among housewives whose menus may change as a result of them, we shall digress momentarily to discuss the topic of laws and insecticides.

The first federal law concerning insecticides was passed in 1910 to protect the farmer and other users of these chemicals. Before this date many worthless preparations were advertised and sold to farmers. Even effectual ones like paris green were often adulterated. In the 1860s Benjamin D. Walsh carried on a crusade against these malpractices and addressed himself to one of these dishonest purveyors in the following language:

> We fear greatly that, instead of being a decently good entomologist, tolerably well acquainted with the noxious insects of the United States, you are a mere entomological quack; and that, instead of talking good, common, horse sense to us, you are uttering all the time nothing but bosh.

The Federal Insecticide Act of 1910 served its purpose well; it quickly put an end to nearly all the fraud in the sale of insecticides. The act set standards for existent insecticides and fungicides, forbade false claims on labels, and prohibited the inclusion of substances that would injure vegetation. There were several state laws covering insecticides before the federal act became law. The first pesticide law was adopted by New York State in 1898 to regulate the sale of paris green. Similar legislation was enacted in 1899 by Oregon and Texas and in 1901 by California, Louisiana, and Washington.

In 1947 the **Federal Insecticide, Fungicide, and Rodenticide Act** (FIFRA) supplanted the act of 1910. This new law, which regulated the marketing of rodenticides, in addition to insecticides and fungicides, required proof of efficacy and safety as well as registration with the U.S. Department of Agriculture for any economic poison marketed in interstate commerce. Because of government reorganization these responsibilities of USDA were transferred in December 1970 to the newly created Environmental Protection Agency.

In order to exercise greater federal control over the use of pesticides in the United States, FIFRA was extensively amended and signed into law by President Nixon on October 21, 1972, and again amended and signed by President Ford on November 28, 1975. The essentially new law places primary emphasis on protection of human health and the environment while retaining provisions for registration and quality of pesticides.

The provisions of FIFRA, as amended, have become stricter. Pesticides shipped intrastate as well as interstate must be registered with EPA. The law now prohibits the use of any registered pesticide in a manner inconsistent with label instructions, thus making the label a legal document. Further, the law requires the classification of pesticides into either general or restricted use depending on their hazards to the applicator or to the environment. General-use pesticides are the less hazardous ones and may be applied by the general public. Restricted-use pesticides are the more hazardous ones and may be applied only by certified applicators. Decisions on the classification of pesticides with intermediate hazards become more difficult, and some uses of a pesticide may be classified for general use for certain applications and restricted use for others.

Because it provides the user with essential information, the **label** fixed to the outside of the pesticide container is an important document. Anyone who applies a pesticide should read, study, and understand the label and then follow the directions exactly. By law the label must provide the following information:

1. Statement of ingredients and type of formulation.
2. Directions for use, including target pests.
3. Use classification—general or restricted.
4. Precautions to protect human health, livestock, wild life, and environment.

5. First aid instructions in case of poisoning.
6. The signal words *danger* and *poison*, together with skull and crossbones, for substances highly toxic to man; *warning* for moderately toxic pesticides; and *caution* for pesticides of low toxicity; there is no signal word for pesticides with the lowest category of toxicity.
7. Net weight or measure of contents.
8. Brand name, chemical name, and usually common name of the pesticide.
9. Name and address of the registrant.
10. EPA registration number of pesticide (numbers to left of hyphen identify registrant and numbers to right the individual product of registrant).
11. EPA establishment number identifying the manufacturing plant.
12. Directions for storage and disposal.

A pesticide is designated "misbranded" if any one of the requirements is false, misleading, or absent from the label. Companies may put out supplemental literature describing the uses and properties of a pesticide. All information of this sort, as well as the label, are called **labeling**.

Because of critical needs for certain pesticide uses not specified on the label, EPA has provided for relaxed enforcement by publication of **Pesticide Enforcement Policy Statements (PEPS).** By the end of 1976 six PEPS were published covering these topics:

1. Use of pesticides at less than label dosages.
2. Control of unnamed target pests in structural pest control.
3. Enforcement during reregistration of pesticides.
4. Preventive treatments in pest control.
5. Control of unanticipated pests in agriculture and other situations of nonstructural pest control (that is, pests not named on the label).
6. Use and labeling of service containers for the transportation or temporary storage of pesticides.

The rewriting and updating of labels and amendments in the law will eventually make the issuance of PEPS unnecessary for the full utilization of pesticides.

Registration of pesticides is an effective method used by EPA in carrying out its duty of supervising the marketing of these products. In 1974 more than 34,000 pesticide products, made from one or more of 900 chemical compounds, were registered. Products must be reregistered every five years.

Before EPA will register a pesticide the manufacturer must submit data showing that the product when used as directed is effective against the pests listed on the label; will not injure man, domestic animals, crops, or wild life, or damage the environment; and will not result in illegal residues on food or feed. Other required data include pesticide reaction in the environment such as rate and form of decomposition after application; degree of transport and persistence in soil, air or water; effect of light and rain on the pesticide; and acute and chronic toxicity of pesticides to animals and plants.

Certification to allow pesticide applicators to employ restricted use pesticides is an important new provision of the amended FIFRA. The intent is to have only trained persons apply the more hazardous pesticides. In providing for certification, congress has given formal endorsement to the principle that education is a reliable and workable ingredient in the regulatory process. To become certified an applicator must demonstrate competence in handling and using pesticides. Following minimum standards set by EPA, the individual states have the primary responsibility for certifying applicators. The acts of examining and certifying applicators are carried out by each state lead agency (usually the state Department of Agriculture). Training programs are conducted or endorsed by the cooperative extension service in each state. Dozens of entomologists are involved and will be from now on, for continuing education will be the basis for applicator recertification. FIFRA has divided certified applicators into private (usually farmers) and commercial. The **private applicator** who produces an agricultural commodity must show that he possesses a practical knowledge of the pests and control practices associated with his crops and livestock. He must be able to recognize the common pests and their damage, to read and understand the label, to apply pesticides according to label instructions, and to recognize symptoms of poisoning.

A **commercial applicator** is a certified applicator who either sells his service of controlling pests to farmers, home owners, and other people in need of pest control or else uses restricted-use pesticides as part of his business (e.g., a food manufacturer). Because of the great variety of pests and methods of controlling them, EPA has designated categories for commercial applicators. A commercial applicator must have specialized knowledge and skills to qualify for a particular category. EPA recognizes ten categories of pest control: (1) agricultural—plant and animal; (2) forest; (3) ornamental and turf; (4) seed treatment; (5) aquatic; (6) right-of-way; (7) industrial, institutional, structural, and health-related; (8) public health; (9) regulatory; and (10) demonstration and research. All commercial applicators must have a knowledge of the general principles and practices of pest control and demonstrate competency in the following areas:

1. Comprehension of the label and labeling.
2. Knowledge of pesticide safety. Commercial applicators must know the degree of toxicity of pesticides, the hazards of pesticides to man, and the precautions necessary to guard against injury to themselves and other people in or near treated areas. They must know the symptoms of pesticide poisoning and first aid treatments. They must recognize the importance of proper identification of pesticides, know how to

store, transport, handle, and mix pesticides safely, and know the proper methods of disposal of pesticides and pesticide containers.

3. A practical knowledge of the hazards of pesticides to the environment. They must have an awareness of the effect of pesticides on nontarget organisms such as birds, fish, game, and other beneficial species, including insects and other invertebrates.
4. A knowledge of pests; their appearance, biology, and damage.
5. A knowledge of pesticides; the different kinds, formulations, and characteristics.
6. A knowledge of application equipment; the different kinds, uses, maintenance and calibration.
7. A knowledge of application techniques; the methods of applying various formulations such as liquid emulsions, dusts, and gases, the ways to prevent drift and pesticide loss into the environment.
8. A knowledge of state and federal pesticide laws and regulations.

EPA's enforcement strategy of FIFRA has stressed industry compliance with registration requirements of pesticides and user compliance with label directions. To accomplish these goals EPA engages in four types of activities: (1) inspection of producer's establishments, (2) sampling of pesticides, (3) analysis of pesticides, and (4) surveillance of the use of pesticides.

In addition to FIFRA as amended, a section of another law, the **Federal Food, Drug, and Cosmetic Act**, as amended August 1972, indirectly regulates the use of insecticides and other pesticide chemicals. The purpose of the law, which is administered jointly by EPA and the Food and Drug Administration of the U.S. Department of Health, Education, and Welfare, is to make sure that the nation's food supply is safe for human consumption.

Besides having one of the richest diets and the most varied menus on earth, Americans daily consume an assortment of around 400 chemicals added to foods as preservatives, mold inhibitors, antioxidants, coloring agents, bleaches, thickeners, thinners, emulsifiers, moisteners, and—the substances of main interest to us—pesticides. Unfortunately, the last, applied to growing crops and to livestock for control of destructive insects and diseases, may leave residues of the toxicant. The problem that faces us, and especially the Food and Drug Administration, is what residues are safe and how much. This is not an easy problem to solve, for prohibiting use of pesticides would lower food quality and could well lead to famine in America. Yet certainly no one is willing to advise the consumption of dangerous amounts of poison. Since no sharp line exists between poisonous and nonpoisonous chemicals—for common salt can be a poison in excess, and arsenic can be a lifesaver—a compromise solution has been reached by applying the old Paracelsus law. The Swiss alchemist-physician Paracelsus put it this way:

"Poison is in everything, and no thing is without poison. The dosage makes it either a poison or a remedy."

The method by which the federal government guards our food from harmful amounts of residues is to establish "tolerances." **A tolerance** is the amount of a pesticide and its toxic alteration products that may remain as a residue on a food or feed crop or in meat or fat. First, scientists consider available data on safety and on amount of residue that remains when the insecticide is used according to directions on the label. Then a finite tolerance is set that will be safe and can be met if the grower follows label directions.

Although the majority of insecticides used presently bear finite tolerances, there are some, such as pyrethrum, *Bacillus thuringiensis*, and petroleum oils, that are exempt from this requirement because, applied to growing crops in a proper manner, they present no hazard. There is a third group of pesticides recognized by the FFDC Act. These chemicals, including lime and lime sulfur, are recognized as being safe. No tolerance or exemption from tolerance requirement is needed to support their use as pesticides.

In the 1970 government reorganization the job of setting pesticide tolerances, which was formerly held by FDA, was assigned to EPA. The FDA, however, continues to carry out field surveillance and enforcement activities. Teams of FDA chemists and inspectors check on the residues of toxicant on fresh produce, eggs, milk, and other agricultural products rather than on packaged pesticides. APHIS of USDA is responsible for monitoring residues in meat and poultry.

When excessive residues from a particular area are suspected, samples are examined from outgoing shipments of the product. If unlawful residues exist in commodities in interstate commerce, a federal court order removes the shipment from the market. Random samples of agricultural products are also taken and analyzed. The law provides for action against persons and business firms responsible for violations. Fortunately such action is a rarity. From the standpoint of health, the quantity of pesticides in the American diet is insignificant. The FDA takes regular market basket samples of food from five geographic regions representative of the entire United States. Residues of pesticides found in the samples are generally less than 1 per cent of the allowable levels.

Because the agricultural industry has a responsibility for producing healthful, wholesome food and for producing this food within the law, agricultural leaders have emphasized the importance of following official recommendations and label directions in controlling insect pests. A good rule for every grower to follow is "Keep accurate records of pesticides used on each crop, including the amounts and dates applied." State cooperative extension services publish insect control recommendations each year that not only conform with federal and state laws but also explain local variations in agriculture and environment which must be reckoned with to achieve the best results.

WHY THIS BOOK?

As one may surmise from what has been said, textbooks are not the best places for anyone to seek the latest in insect control recommendations. On the other hand, texts do serve an important role in providing both a background of information on insects—their habits, life histories, and ecology—and an appreciation of the insect problems on various crops and how we go about solving these problems.

An attempt has been made to develop principles in this book so that once they are learned, the student can handle specific insect problems himself or will know where to get the information he needs. As many students, principally "ag" students, will not have had a previous course in entomology, we begin this book with modern treatments of insect structure and function, insect growth, insect classification, and insect ecology. In order to provide the student with prerequisite information for intelligent study of the host chapters, these topics are followed by short chapters on the principles of insect control, insecticides, and application equipment.

Host chapters have been written to present a panorama of the insect problems that affect specific crops. First, we discuss the number and importance of the pests of the particular crop, or commodity, second, how these pests cause damage, and then, in general, how they are controlled. After these introductory sections we consider a limited number of pests of the crop, usually four or five, in detail. The choice of pests may not include all insect species that a student should like to study for a particular area, but he may add other species to those treated and with the help of his instructor dig out the necessary information, a task that is both edifying and rewarding.

SELECTED REFERENCES

Cloudsley-Thompson, J. L., *Insects and History* (New York: St. Martin's Press, 1976).
Essig, E. O., *A History of Entomology* (New York: Macmillan, 1931).
Howard, L. O., *A History of Applied Entomology*, Smithsonian Misc. Coll., Vol. 84, 1930.
Leonard, M. D., "The Development of Commercial Entomology in the United States," *Proceedings Tenth International Congress of Entomology*, 3:99–106 (1958).
Mallis, A., *American Entomologists* (New Brunswick: Rutgers University Press, 1971).
Montgomery, B. E., "Arthropods and Ancient Man," *Bul. Ent. Soc. Amer.*, 5:68–70 (1959).
Pellett, F. C., *History of American Beekeeping* (Ames, Iowa: Collegiate Press, 1938).
Smith, R. F., T. E. Mittler, and C. N. Smith, eds. *History of Entomology.* (Palo Alto: Annual Reviews, 1973).
True, A. C., *A History of Agricultural Experimentation and Research in the United States*, USDA Misc. Publ. 251, 1937.
Zinsser, Hans, *Rats, Lice and History* (Boston: Little, Brown, 1935).

chapter 2 / **ROBERT F. HARWOOD**

INSECT STRUCTURE AND FUNCTION

Insects thrive in more environments than any other group of animals. They dwell essentially on the earth's surface, but they have also mastered the subterranean world and the realms of air and water. They are at home in deserts and rain forests, hot springs and snow fields, dark caves and sunlit surroundings. They eat the choicest foods on man's table—and can eat the table too! It seems obvious that organisms with abilities as great as these must have extreme structural and functional diversity. It is truly amazing that such complexities are found in animals as small as insects. Though as a group they are much smaller than mammals, their anatomy and physiology for performing many commonplace activities are as complex. The vast range of environments utilized by insects is accompanied by an equal range of structural modifications and functional capabilities.

GENERAL EXTERNAL FORM

Insects, like mammals, are **bilaterally symmetrical** animals; that is, most of their external and internal structures are divided into left and right halves that are mirror images. Only minor variations of this scheme occur; there are occasional subtle differences in the mouthparts and genital structures, and major asymmetry is possible in the digestive tract.

Both sexes are frequently alike in appearance, particularly in the immature forms. However, **sexual dimorphism** of adults may be very marked. There is a tendency for female insects to be larger than males, although males of many beetle species are bigger and are armed with large mandibles or horns. In several species of moths the antennae of the males are decidedly fringed, and in mosquitoes and related Diptera the same structures of males are plumed. Females of many insects can be recognized clearly by well-developed ovipositors, males by prominent genital claspers. Extreme over-all body differences are found in scale insects, where the females are flattened, inactive, and featureless; the males are active, with well-developed legs and wings (Fig. 14:17). More subtle differences occur, such as in higher Diptera, where the compound eyes of males are practically contiguous in front, and those of females are separated.

The progenitors of modern insects possessed a pair of appendages on each segment, a characteristic that is retained in many present-day arthropods. In insects, **segmented appendages** may now occur in a largely unmodified condition (**legs**), in structures highly modified for a particular function (**mouthparts** and **genitalia**), in a temporary condition (**abdominal appendages** of larvae—which may actually be secondary developments not homologous to true segmental appendages), or they may be lacking entirely. The primitive structure of an appendage, possibly homologous with the two-branched appendage of crustaceans, consists of a several-segmented organ allowing movement in a number of directions. The leg, derived from seven segments but with five segments commonly recognized, is an example of this highly articulated type of structure (Fig. 2:2).

Adult and many immature insects have three distinct body regions: **head**, **thorax**, and **abdomen** (Fig. 2:1). Each region results from a variable degree of

Figure 2:1. External structure of an insect. This diagram, based on the generalized structure of the common cricket, shows the body regions and their associated structures.

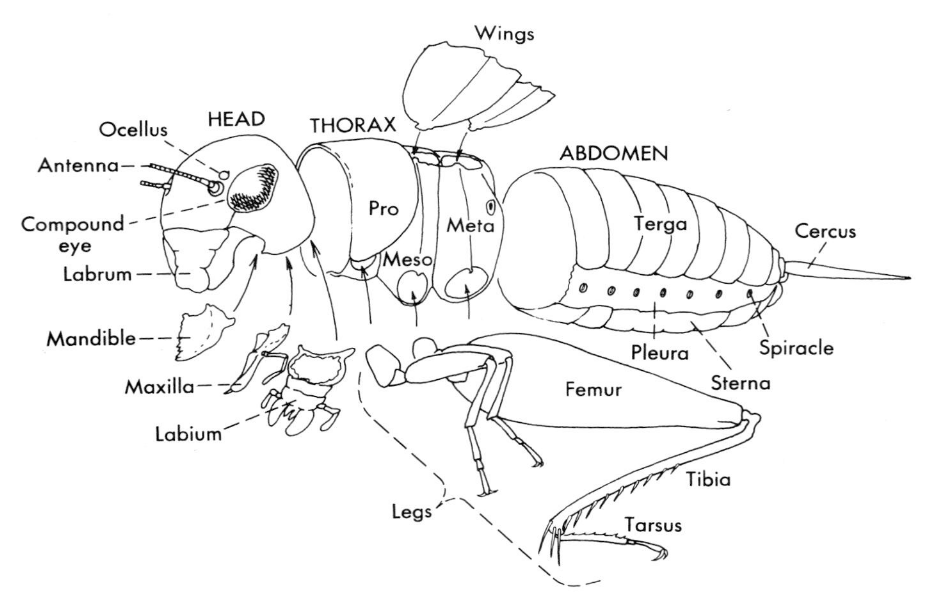

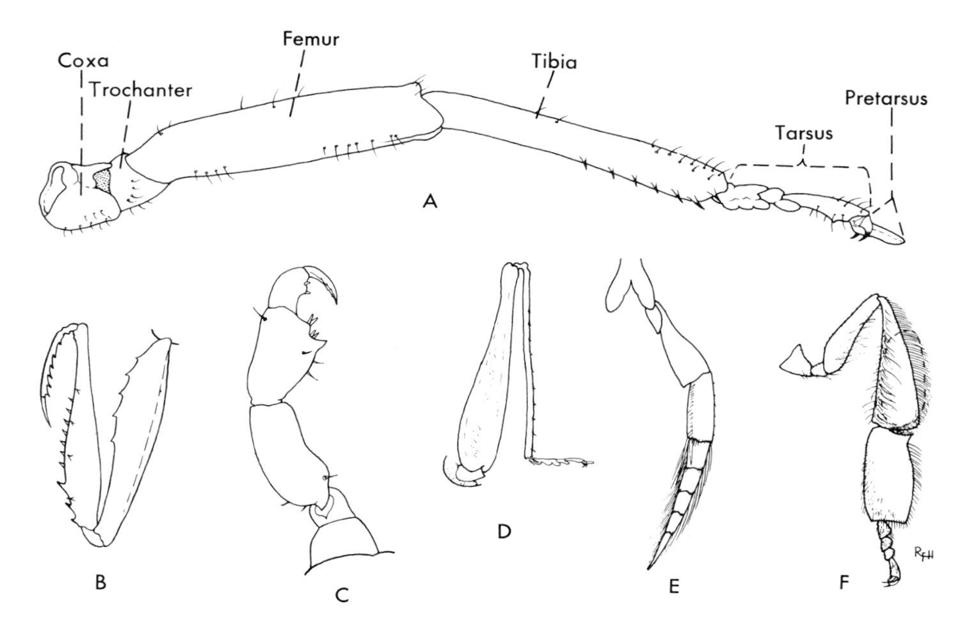

Figure 2:2. Insect legs. A, commonly observed features of an insect leg; B, the grasping and heavily spined foreleg of a mantid, which preys on other insects; C, the clinching leg of a hog louse, adapted for seizing the hairs of its host; D, the long and well-muscled hindleg of a grasshopper, adapted for jumping; E, the swimming hindleg of a diving beetle; F, the hindleg of a honey bee, adapted for carrying pollen (outlined mass is the position of the pollen ball).

fusion of segments that are distinctly separate in embryonic development. Clearly separate segments in postembryonic stages are usually evident in the abdomen, a structure which retains a certain amount of flexibility. The thorax, bearing one pair of legs per segment, is obviously a fusion of three segments. Fusion is most complete in the head region. On the basis of the presence of appendages and on embryological evidence, most investigators agree that the head is a fusion of six or seven primitive segments. Particularly in immature insects, it may be difficult to distinguish three separate body regions. A case in point is the house fly larva (Fig. 3:14G), which has no head capsule and which bears no appendages on the thorax. In specially adapted forms, such as adult female scale insects, legs may be absent, and the thorax and abdomen are more or less fused into a single unit.

Head

The insect head is a hard capsule whose form and position are generally dependent on the mouthparts and type of food ingested. A heavy head structure is found among insects with chewing mouthparts. Commonly when this type of mouthparts is present, the head is directed downward (grasshoppers) or protrudes forward as in predaceous insects which have the mouthparts developed for seizing prey in front of them; or else the mouthparts face down and backward (the cockroach). Insects with piercing-sucking mouthparts do not require as heavy a frame for the attachment of muscles as those which operate chewing mandibles, and consequently rather small heads are typical. Again head position is governed by location of the mouthparts, which may be directed forward (mosquitoes), downward (horse flies), and backward (aphids). Reasonably constant features on the adult insect head are a pair of antennae, a pair of compound eyes, and three, two, or no simple eyes (Fig. 2:1). The compound eyes are generally prominent, in some cases comprising the major area of the head.

Antennae. Antennae are sensory structures located between or just below the compound eyes. In various insects they are known to perceive odors, humidity changes, vibrations, and wind velocity and direction. Their great differences in form may be a useful tool for distinguishing families of insects. Two basal antennal segments, the **scape** and the **pedicel**, and a series of similar segments, the **flagellum**, are common (Fig. 2:3). The simplest form is the threadlike or **filiform** antenna. Variations include club-shaped or **clavate**, head-shaped or **capitate**, elbowed or **geniculate**, plate-like or **lamellate**, comb-shaped or **pectinate**, and plume-shaped or **plumose** antennae.

Mouthparts. Mouthparts of insects are believed, with the exception of the labrum-epipharynx and hypopharynx, to be derived from segmental appendages. Types of mouthparts have developed to permit the ingestion of solid foodstuffs, free liquids, plant sap, or the blood of animals. Among separate groups of insects, mouthparts quite different in structure are adapted for utilizing the same type of food. Thus, although mosquitoes, horse flies, stable flies, and fleas all feed on blood their mouthparts differ structurally.

The basic and most primitive type of mouthparts is the **chewing type**, present in grasshoppers and beetles. Here is thought to be the pattern from which much more specialized insect mouthparts were derived. In order of appearance, from anterior to posterior, chewing mouthparts consist of a single **labrum** (upper lip), a pair of **mandibles** (jaws), a pair of **maxillae** (second jaws), and a **labium** (lower lip) derived from the fusion of paired appendages (Fig. 2:1). A single structure, the **hypopharynx** (tongue-like organ), is located centrally. The inner surface of the labrum is referred to as the **epipharynx**, an area frequently membranous and inconspicuous. Sensory structures, demonstrated in various insects to function in tasting, smelling, touching, or perceiving temperature, are found abundantly on the palps of the maxillae and labium. Because they may be quite elongate, the mouthparts of adult weevils can be mistaken as piercing and sucking in nature, but actually they consist of an elongated frontal portion of the head bearing chewing mouthparts (Fig. 2:4).

There is a considerable variety of mouthparts that are not of the chewing type. In honey bees a combination of chewing and lapping structures is well

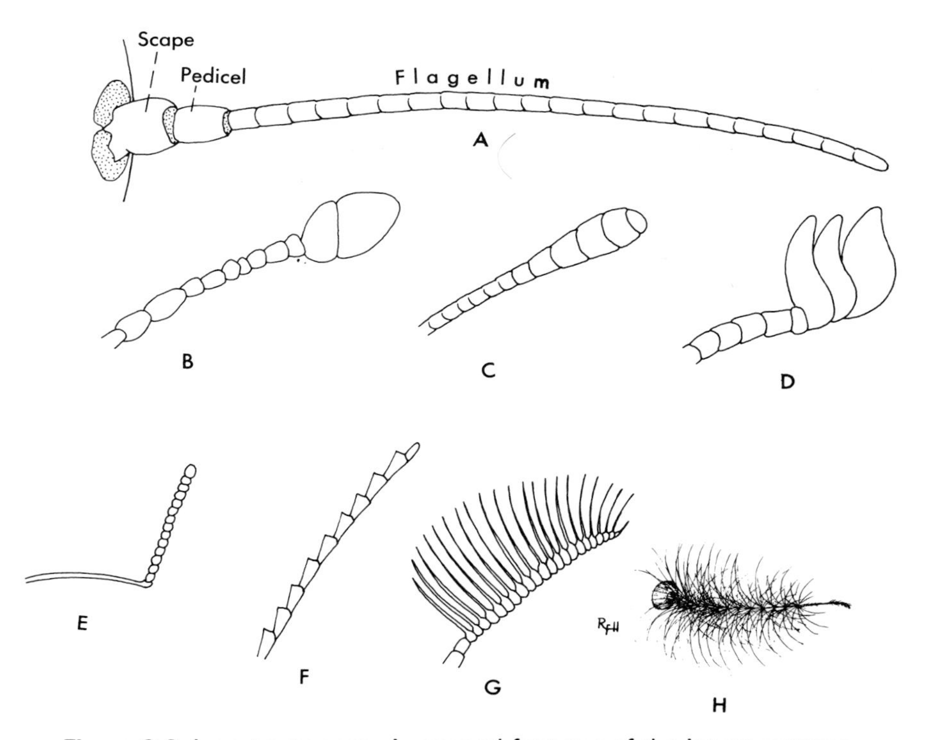

Figure 2:3. Insect antennae. A, general features of the insect antenna, represented by the filiform or filamentous type such as that of the grasshopper; B, capitate or headshaped form, such as that found on some adult clerid beetles; C, clavate or clubshaped, typical of adult tenebrionids or darkling beetles; D, lamellate or platelike, found in scarabaeid beetles such as June beetles; E, geniculate or elbowed, this sample being moniliform or necklace-like at the end, as represented by some adult weevils: F, serrate or saw-like, typical of many cerambycid beetles; G, pectinate or comb-like, especially developed in males of several moths; H, Plumose or plume-like, the antenna of a male mosquito.

Figure 2:4. A weevil. The snout of this beetle superficially resembles many piercing-sucking mouthparts, but note the enlargement of the tip showing chewing mandibles to be present.

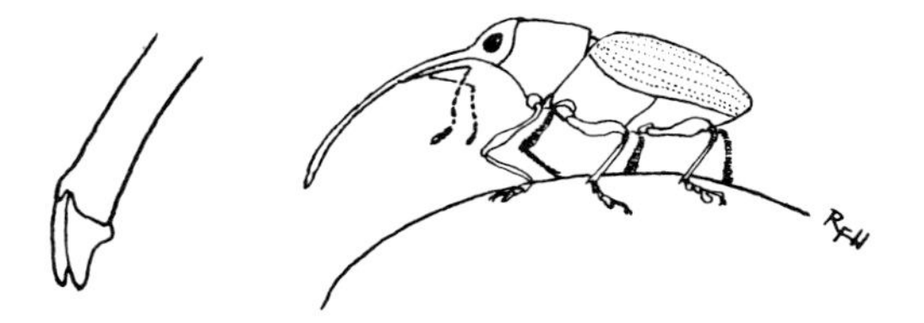

suited for manipulating solids and imbibing liquids. Thrips suck up juices that ooze from plant tissues abraded by the points of short stylets. In stable flies a small labellum on the tip of the labium, similar to the large and spongy one found on the proboscis of the house fly, contains hard, rasping teeth.

Piercing-sucking mouthparts, developed for withdrawing fluids from animals and plants, are a very important type. They are best known in Hemiptera and Homoptera and in mosquitoes and horse flies, although they are present with different modifications in several other insect groups. The parts that enter tissue consist of a group of piercing **stylets** that work as a unit called a **fascicle**. There is generally a strong pumping mechanism developed in the foregut for removal of fluids, though in some aphids it has been demonstrated that pressure from within the conductive tissues of plants is sufficient to explain the rate of removal of sap.

Fluid is imbibed by Hemiptera and Homoptera through a food channel; there is also a duct for the ejection of salivary secretions (Fig. 2:5). In some plant feeders the salivary fluids contain enzymes or toxins that can distort plant cell walls. Both the food and the salivary channels are located between a pair of stylets developed from the maxilla. These **maxillary stylets** are grooved

Figure 2:5. Piercing-sucking mouthparts of Hemiptera and Homoptera. A, sectional view of cicada head, note well-muscled pump; B, cross section of cicada mouthparts, note that the labium forms a sheath around the maxillary and mandibular stylets, which intermesh to form a fascicle; C, fascicle of milkweed bug entering tissue, the labium folding back and not piercing the surface. *A, B, redrawn from Snodgrass, Smithsonian Misc. Coll. 104, No. 7; C, original.*

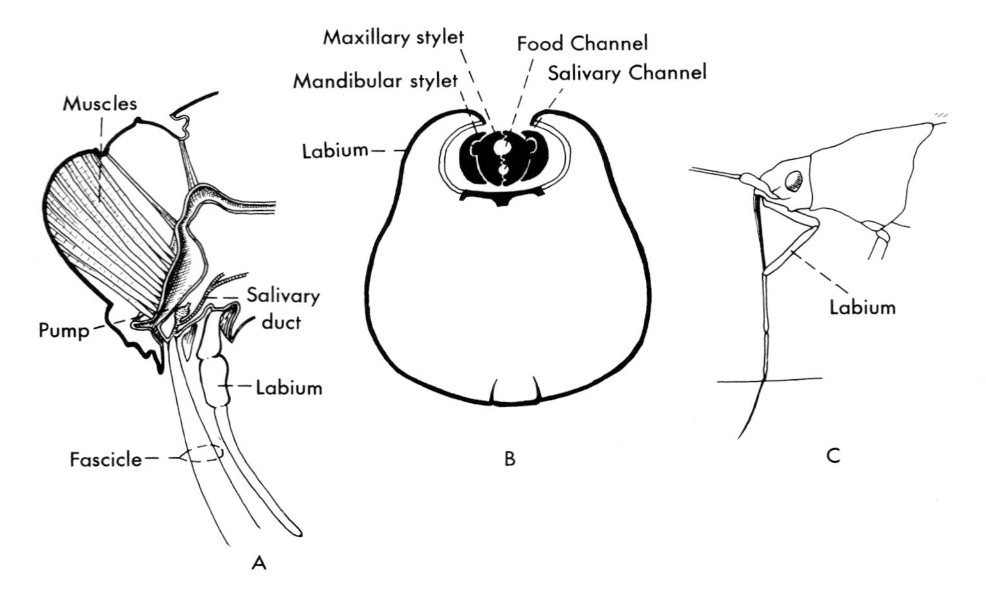

to slide on one another, and **mandibular stylets** outside of them are connected functionally by interlocking grooves; the four stylets make up the fascicle. The tips of the stylets may have minute teeth for tearing tissues. The labium is a relatively coarse and jointed enveloping sheath that holds the other mouthparts when at rest; it does not enter the tissue during feeding.

Mouthparts of mosquitoes (Fig. 2:6) likewise consist of stylets, but their component parts differ from those of Hemiptera. The labrum-epipharynx and hypopharynx form a food channel. The posterior portion of this channel is closed by the hypopharynx, which has a salivary duct. Salivary fluid contains anticoagulins that reduce the clotting of blood. The maxillae and mandibles form paired, piercing stylets that may bear fine teeth at their tips. The labium has at its tip a lobed and soft sensory structure, the **labellum**. The labium is unjointed and serves as a sheath for the remainder of the mouthparts. It flexes back during feeding and does not actually pierce the animal tissue. The

Figure 2:6. Piercing-sucking mouthparts of the mosquito. A, diagram of a mosquito head in lateral view; B, sectional view of the head showing two-well-muscled pumps for withdrawing blood; C, cross section of mouthparts, indicating the fascicle lies within a sheath formed by labium; D, diagram of fascicle entering tissue, with labium flexed backwards. *Redrawn from Snodgrass, Smithsonian Misc. Coll. 104, No. 7.*

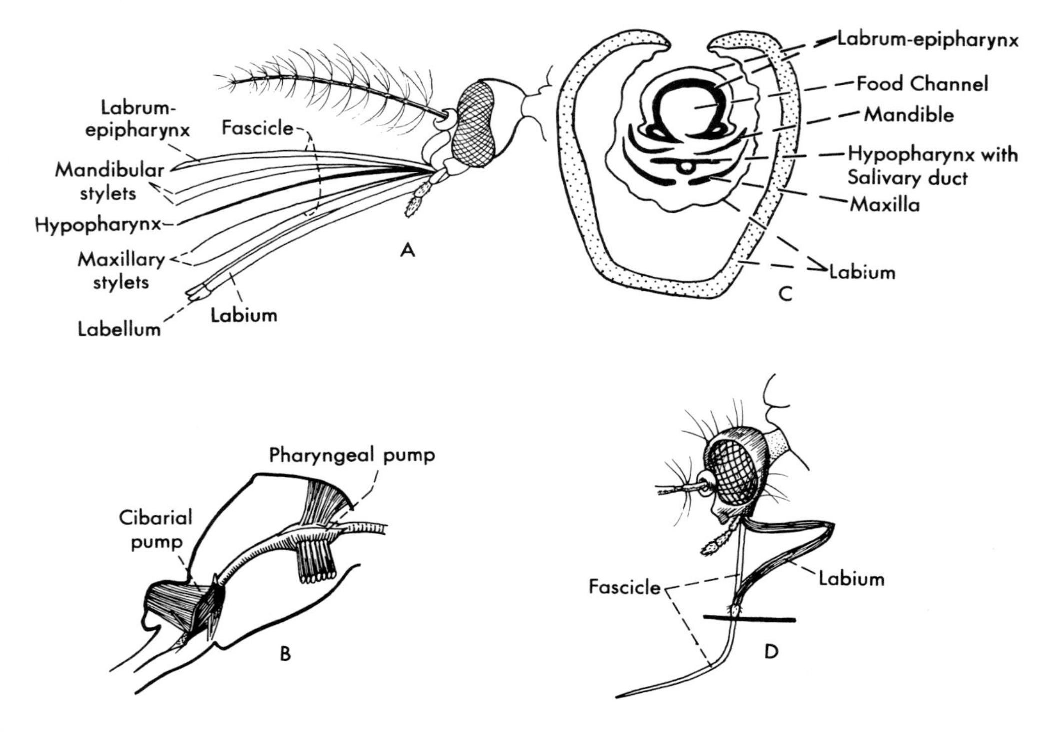

position of the mouthparts in horse flies is similar, but the stylets are relatively heavier cutting structures, and the labium is large and fleshy.

Sponging and siphoning mouthparts are highly specialized types. The sponging mouthpart is found in the house fly and its close relatives. There is a prominent, elbowed proboscis comprised of a fusion of the labium and maxillae, and at the distal end, the paired fleshy labella (Fig. 2:12D). Open trachealike channels on the surface of the labella soak up liquid by capillary action. The collected liquid is then sucked up in the food channel formed by the labrum-epipharynx and the hypopharynx, much as in the mosquito, but without these two structures being developed for piercing. Siphoning mouthparts of butterflies and moths consist of a coiled springlike structure derived from two fused portions of the maxilla (Fig. 4:18B). The insect can extend and insert this structure into flowers to withdraw nectar or into a puddle to obtain water.

Mouthparts of ticks and mites are tearing or piercing and sucking in nature (Fig. 2:7). The main cutting structures are a pair of **chelicerae**. In ticks and many mites each chelicera is a jointed appendage with cutting structures. In some mites, such as the common red spider mite, the chelicerae are simple sharp stylets that serve as piercing organs. Ticks possess a distinct median structure called the **hypostome**, which anchors in tissue by means of barbs that face downwards. The same structure is present in some mites, but not

Figure 2:7. Mouthparts of ticks. A, ventral diagram of the capitulum and mouthparts of an *Amblyomma* tick; B, section of the head of a tick, note pharyngeal pump for withdrawing blood. *A original; B, after Snodgrass, Smithsonian Misc. Coll. 110, No. 10.*

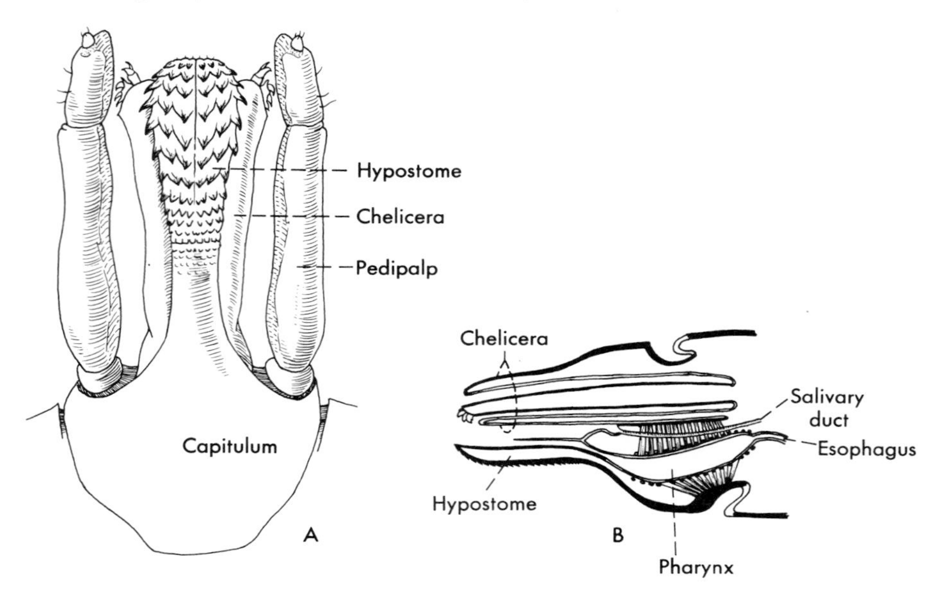

with anchoring abilities. An outer pair of segmented structures, the **pedipalps**, is also present. In ticks the pedipalps are grooved to fit protectively against the chelicerae; in mites the pedipalps may be toothed on their tips or may be modified to function as another pair of chelicerae.

Thorax and Associated Structures

The thorax serves as a center for bearing structures of transportation. It is composed of three usually fused segments. Each segment bears a pair of legs; the last two segments bear wings when both pairs are present. The thorax is often box-shaped, having a distinct dorsal surface (**nota**), a ventral surface (**sterna**), and a pair of sides (**pleura**). Internal skeletal ridges at the site of sutures brace the thorax and serve as sites of muscle attachment.

Legs. Insect legs are constructed on a basic plan derived from seven segments, with only five clearly apparent. From point of attachment to the body, these five are the **coxa**, **trochanter**, **femur**, **tibia**, and **tarsus** (Fig. 2:2). The first two segments are generally short, the next two long, and the tarsal segment small and usually subdivided into three to five subsegments. Tarsi may bear adhesive pads on their under surface; a terminal portion called the **pretarsus** usually bears claws and pads. Legs have undergone great structural changes to fit the life habits of insects. Notable examples are the swimming legs of aquatic beetles, clutching forelegs of mantids, jumping hindlegs of grasshoppers, grasping legs of lice, and pollen-carrying hindlegs of honey bees (Fig. 2:2).

Wings. Although all insects do not possess wings, these are the only organs of true flight among invertebrates. Insects that lack wings, such as the Thysanura (silverfish) and Collembola (springtails), belong to a primitive group (**Apterygota**) that never developed flight structures. Other insects had ancestors that were winged (**Pterygota**), although many present-day forms lack these structures. Loss of wings has occurred in whole orders of insects, as in fleas and lice, through the development of a parasitic habit where travel by flying is not required and where superfluous structures such as wings might even be disadvantageous for remaining on a host. In some cases, for example the fly family Hippoboscidae, wings are present on some species but are absent on others. Thus the sheep ked is without wings, whereas a related hippoboscid that attacks deer still possesses them. Some insects have lost their wings because they live in a very restricted environment, such as under bark or in caves, and do not need to fly. Aphids are remarkable in that wingless forms are commonplace, but winged forms develop to enable a change of plant hosts.

Wings are located only on the second and third thoracic segments. Although some insects had flat outgrowths on the first thoracic segment, as evidenced by fossilized remains, these outgrowths were never flight structures, and it is uncertain whether they are homologous with the wings of the succeeding two thoracic segments. In Diptera, where a single pair of wings is used for flight, the second pair has been modified into small, knobbed vibrating guidance organs, the **halters**. Beetles have the first pair of wings, called **elytra**, modified as a tough protective cover playing no part in flight (Fig. 4:14). In some instances the elytra are fused together at midline. The forewings, referred to as **tegmina**, of grasshoppers are leathery. In Hemiptera the first pair of wings are known as **hemelytra**; the basal half is leathery, and the terminal half is membranous (Fig. 4:10). Wings may be fringed (thrips, Fig. 4:8D) or bear scales (Lepidoptera and mosquitoes).

A basic pattern of **wing venation** underlies the system of veins in insect wings (Fig. 2:8). Each principal longitudinal vein is preceded by a trachea during formation of the wing in a bud or wing pad. Using this evidence and that obtained by detailed comparison of wing venation in fossil insects, it has been possible to construct a hypothetical system. Despite great structural differences between orders and families of insects, it is possible to determine homologies in wing veins correctly and to apply universally accepted names to such veins (Fig. 2:9).

Although there is much variation in wing size, shape, and structure, some generalizations are possible. The more advanced insects fold their wings close to the body, giving them the advantage of being able to explore small spaces for obtaining food and to seek shelter from a harsh environment and many of their enemies. Insects with a profusion of veins in the wings, particularly an excess of cross veins, are generally slow and unskilled fliers. The fast and highly maneuverable fliers tend to have wings that are relatively short and wide, with only a few well-developed longitudinal veins and reduced number of cross veins.

Figure 2:8. The basic plan of wing venation in insect wings. This system is based on the tracheae that precede longitudinal veins in the developing wing pads of immature insects and on a comparison of wing structure in fossil insects. In present day insects the number of longitudinal veins is generally reduced, and crossveins are common. *After Comstock, 1906,* A Manual for the Study of Insects, *by permission Cornell University Press, Ithaca, N.Y.*

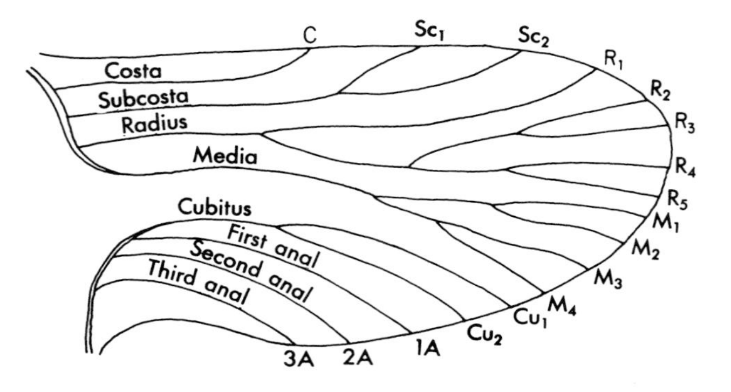

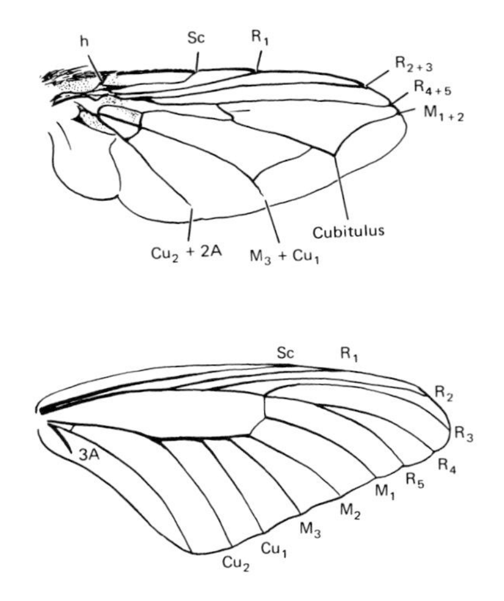

Figure 2:9. Modification of the basic plan of wing venation in two present-day insects. Above, the wing of muscoid fly; below, the forewing of a swallowtail butterfly.

Abdomen and Associated Structures

In most insects the abdomen, composed of distinct segments, has maintained a flexible condition. As many as 11 segments are found, although this number is commonly reduced to about 8 or fewer that are clearly recognizable. Large portions of the digestive tract, the reproductive system, and other vital organs are located internally. Quite possibly the supple nature of the abdomen is required for expansion as eggs enlarge in the female; certainly flexibility is a requirement for copulating, ovipositing, and stinging.

With the exception of genital structures, segmental appendages of the abdomen are seldom retained in adult insects. Larvae may possess prominent abdominal appendages; transitory and vestigial structures of such a nature may occur during embryonic development. Thysanura have simple abdominal appendages called **styli** (Fig. 2:10), abdominal **prolegs** (locomotory

Figure 2:10. A thysanuran in which abdominal appendages called styli are evident. *From Snodgrass, Smithsonian Misc. Coll. 122, No. 9.*

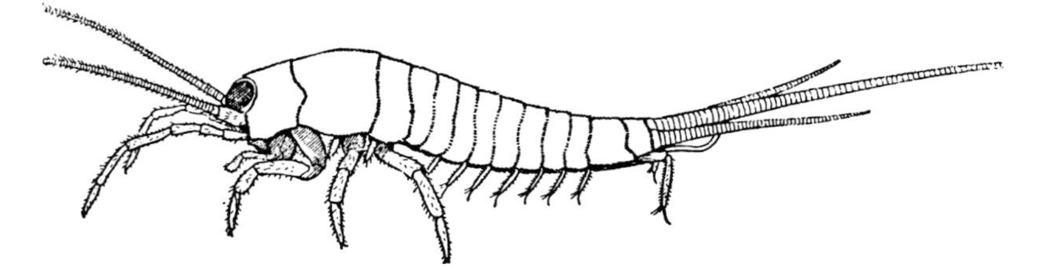

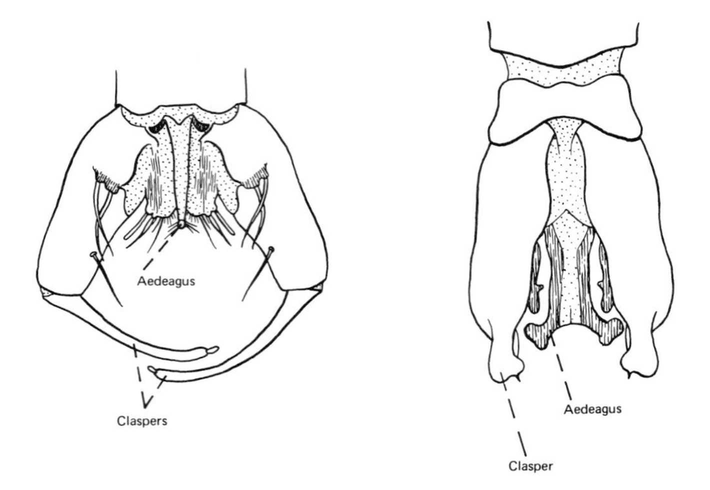

Figure 2:11. The external genitalia of male insects generally consist of an aedeagus, claspers, and various accessory structures. They are often characteristic enough to permit their use in identification to species. Left, genitalia of a male mosquito; right, genitalia of a male hornet. *After Snodgrass, Smithsonian Misc. Coll. 135, No. 6.*

structures) are present in caterpillars (Fig. 3:14C) and sawfly larvae, and abdominal appendages may be prominent respiratory structures in aquatic insects such as mayfly nymphs and aquatic Neuroptera (Fig. 3:14A).

External genitalia. The external reproductive structures of insects are often highly specialized and complicated. This is particularly true of male insects, where special sclerous structures and membranous areas have been found quite useful in distinguishing species that otherwise appear to be similar. Typically there is an intromittent organ (the **aedeagus**) and paired **claspers** (Fig. 2:11).

Female insects frequently have well-developed **ovipositors** for placing eggs in protected sites (Fig. 2:12). Whole families possess characteristically distinctive ovipositors. Longhorn grasshoppers have long, flat, blade-like ovipositors; crickets have a round, lance-shaped structure. Some of the parasitic wasps are endowed with an amazingly slender and flexible ovipositor that can enter tough materials such as the bark of a tree. Ovipositors are generally composed of three or fewer pairs of valves that fit together to form a single functional unit. In some insects like the muscoid flies, the end of the abdomen is developed into a narrow, extensive telescoping tube, more properly called the **pseudovipositor**.

INTEGUMENT AND DERIVATIVES

The outer integument or **exoskeleton** of insects is a distinctive feature (Fig. 2:15A) that helps relate them to the crustaceans, spiders, millipedes, and

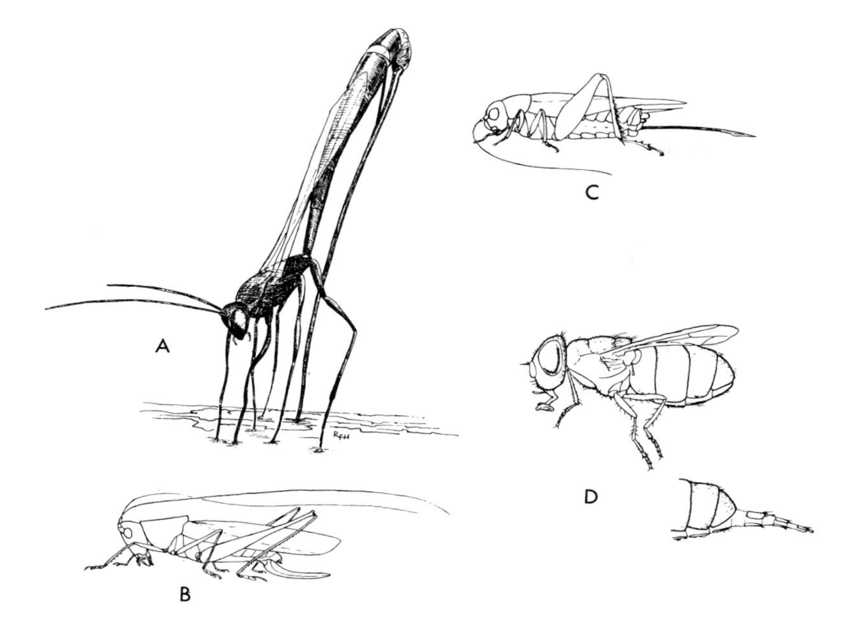

Figure 2:12. Ovipositors of insects. A, a parasitic ichneumonid wasp starting to pierce the bark of a tree with its ovipositor, in order to reach a wood-boring larva; B, the blade-like ovipositor as found in longhorn grasshoppers (Tettigoniidae); C, the lance-shaped ovipositor of crickets (Gryllidae); D, the pseudovipositor of a muscoid fly, in which the terminal abdominal segments are narrow and normally retracted.

other arthropods. It may serve as armor, skeletal support for muscles and other internal organs, and a barrier retarding loss of water from the body. Knowledge of the structure and composition of integument provides an understanding of how it serves its many vital functions.

The integument is a sheet of living epithelial cells covered by outer layers of cuticle (Fig. 2:13). This layer of epithelial cells, the **epidermis**, rests on a noncellular **basement membrane**. Occasional wax-secreting cells are found in the epidermis, as are cells that secrete a molting fluid that softens the old cuticle at time of molting.

Cuticle lying just above the epidermis makes up the major structural portion of the integument. It is separable into a relatively thin outer layer or **exocuticle** with a thick **endocuticle** beneath it. The exocuticle contains protein and a durable, though flexible, substance called **chitin**; here **pigments**, if present, are found. This portion is thinner but generally harder and less flexible than the endocuticle. The endocuticle also contains chitin and protein but lacks pigment.

The outermost layer of integument, the **epicuticle**, is much thinner than exo- and endocuticle, but it may contain several distinctly different layers. In

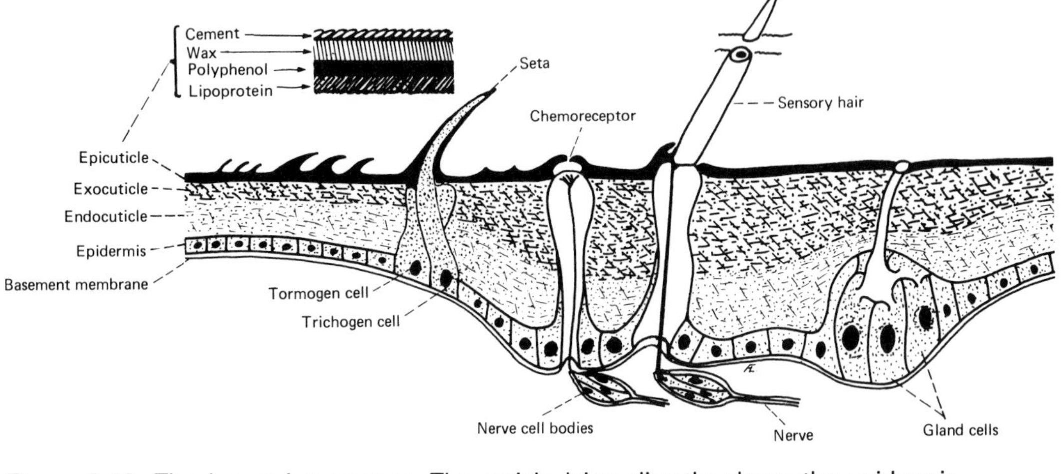

Figure 2:13. The insect integument. The cuticle lying directly above the epidermis consists of the unpigmented endocuticle, and above it a thinner, pigmented exocuticle. The external surface consists of the thin multilayered epicuticle. *Original drawing by Fred A. Lawson, University of Wyoming, Laramie.*

insects possessing a highly specialized epicuticle there is an outermost **cement** layer or **tectocuticle** which serves as a protective covering for the very thin **wax layer**, an important barrier against water loss. A **polyphenol layer** and an innermost **lipoprotein layer** composed of fatty substance in chemical combination with protein may be present as part of the epicuticle.

The three regions above the epidermal cells are deposited in stratified layers. The vertical passages through these strata are channels for sensory nerve fibers and ducts of wax glands. In addition, in some insects, plentiful and extremely fine spiral **pore canals** of uncertain function traverse the cuticle vertically.

The complex chemical nature of the integument imparts some important physical characteristics. This structure acts selectively toward the passage of liquids and gases; consequently some organic solvents and contact insecticides pass more readily through the cuticle than others. Some exchange of respiratory gases takes place through the integument in terrestrial forms, but this capability is more important in aquatic insects that have the cuticle formed into thin-walled gills (Fig. 2:14).

Prevention of water loss is particularly important to terrestrial arthropods. In soil inhabitants the wax layer is continually abraded in passing through soil particles. If soil insects are placed in a dry atmosphere, they soon undergo fatal loss of water. It is believed that the early Egyptians discovered that dusts protect stored grain from the depredations of insects. In a more modern application very fine particulate materials that both abrade and adsorb the cuticular waxes of insects, such as silica aerogel, have been found capable of controlling household pests such as cockroaches and structural pests such as termites.

Because of the relative rigidity of the insect exoskeleton, growth is accomplished by shedding, or **molting**, old cuticle and expanding into a new and larger cuticle. A definite series of changes takes place in the integument prior to molting. The epidermal cells detach from the layers of cuticle, generally increase in number by cell division, and become folded under the restricting outer layers. **Molting fluid**, containing enzymes that attack chitin and protein, digests away the thick endocuticle. The digested substances are resorbed and most likely are used in forming new exocuticle. Because the

Figure 2:14. The nymph of a damselfly (Odonata). During its growth stages this insect lives in water; respiration is aided by three leaf-shaped tracheal gills at the end of the abdomen.

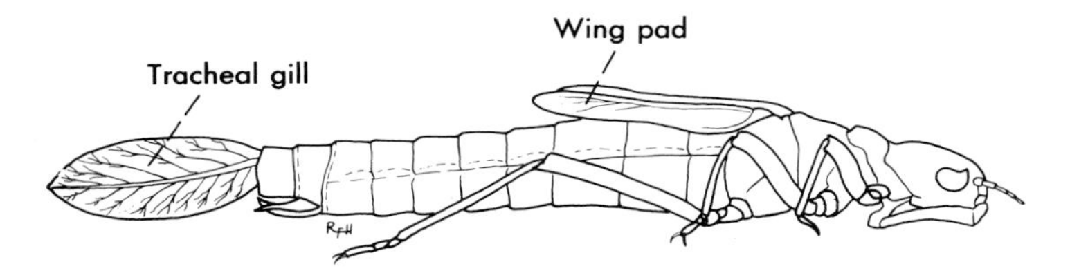

exocuticle and epicuticle remain virtually intact, there may be little or no externally visible change in the integument even though the endocuticle has disappeared just prior to molt.

Internal pressure splits the old cuticle along the back by means of muscular contractions, accompanied in terrestrial insects by the swallowing of air and in aquatic forms by the swallowing of water. A structurally weaker middorsal line is the usual site of separation of the cuticle. The lining of foregut, hindgut, and tracheae is cuticular in nature and is found attached to the inside of the shed cuticle.

Molting generally takes place in a rather secluded and protected site because the newly molted insect is rather immobile and defenseless and is subject to desiccation. The new integument is white or clear, but in insects with a dark cuticle the darkening takes place very quickly by a **tanning process** and by the deposition of pigment. The outermost protective cement layer of the epicuticle is deposited soon after molt, and the thick endocuticle develops over a longer period. Expansion of the new covering takes place by the insect's swallowing air or water.

Hardening of the cuticle, called **sclerotization**, is caused largely by the interaction of proteins with tanning agents which are polyphenols or quinones. It was originally believed that hardening was caused by the presence of the highly polymerized polysaccharide chitin, but it is now known that soft intersegmental membranes may actually have a higher proportion of chitin than hardened sclerous plates. The hard portions of the integument may be divided into distinct platelike areas called **sclerites**, separated from each other by flexible cuticle or by indented lines called **sutures.** Sutures are generally the external evidence of an internal ridge that serves as a point of attachment for muscles; or in some instances, as in the head, sutural lines may simply indicate fusion between what were originally separate and distinct segments.

There may be a number of cuticular modifications in the form of hairs, spines, and spurs. The hairs, called **setae**, are formed in a particular manner (Fig. 2:13). A **trichogen cell** secretes the seta, and a **tormogen cell** develops into the socket. One or more nerve cells may be associated closely with the seta (see section on the nervous system in this chapter), and in certain cases—for example the stinging hairs of caterpillars—the seta is hollow and serves as a channel for products produced by glandular cells. Generally the seta is a simple hairlike structure, but it may be plume-shaped in forms such as the honey bee; in Lepidoptera and mosquitoes it may be modified as a scale. **Spines** are composed of heavy inflexible outgrowths of the integument derived from several epidermal cells. **Spurs** are socketed outgrowths formed by several epidermal cells and are connected to the body wall by a joint. The outer surface of the cuticle may possess very fine structures, **microtrichia**, which bear no direct relationship to the number of epithelial cells. In some insects a loose, powderlike substance or bloom may occur on the surface of the cuticle.

RESPIRATORY SYSTEM

Insects, for the most part, are small, active animals requiring a large amount of oxygen. A butterfly at rest requires about three times as much oxygen per equivalent weight of tissue as a man under similar conditions. A butterfly in flight increases its oxygen consumption about 160 times, approximately 25 times the amount on an equivalent-weight basis required by a man running. It is obvious that insects, particularly flying insects, must have an effective respiratory system to provide the oxygen required during periods of high activity. Because carbon dioxide diffuses through tissue about 25 times as fast as oxygen, the essential problem is to supply enough oxygen to tissues; the removal of carbon dioxide is solved by the same system more or less automatically.

Oxygen is taken up directly by insect tissues without the aid of respiratory pigments such as hemoglobin (see section on the circulatory system in this chapter for exceptions). The distribution system consists of main trunks called **tracheae**. These subdivide into fine **tracheoles** that ramify among the tissues, terminating in a **tracheal end cell**. The system also includes large air sacs, especially characteristic of flying insects (Fig. 2:15D). The tracheal network is derived embryonically from the integument and, consequently, has an epithelium-like cuticular epidermis that secretes an inner lining (**intima**) similar in chemical and structural constitution to the cuticle. The tracheal tubes are braced against collapse by the presence of spiral thickenings of the intimal lining called **taenidia** (Fig. 2:16B).

Air enters the tracheal system through specialized openings, **spiracles**, on the sides of the thorax and abdomen. These spiracles are usually able to close, thus conserving water when air is not being taken up. Basically there is a total of ten pairs of spiracles, eight on the abdomen and two on the thorax. However, there is much variation in number and location of spiracles in various insects; the mosquito larva has a single pair on an air tube near the tip of the abdomen, the pupa has a single pair on the thorax and a single pair on the abdomen, and, in the adult, the spiracles are located on the thorax and abdomen.

Despite the proliferative nature of tracheae in insect tissue, an over-all plan can be discerned for the respiratory system (Fig. 2:16A). The spiracles are located along each side of the body. Short branches lead in from these spiracles to a principal **lateral longitudinal trunk**. Branches connect the lateral trunk of each side to a major **dorsal longitudinal trunk**, a **ventral trunk**, and a **visceral trunk** closely associated with the digestive and reproductive systems. Rather small tracheal commissures make cross connections in the dorsal and ventral body regions. This basic tracheal pattern can be modified; in the larvae of many Diptera, for example, the dorsal longitudinal trunk is the main functional trunk.

A large insect, such as a grasshopper, can be seen making rhythmic **respiratory movements** with the abdomen. In Coleoptera and Hemiptera

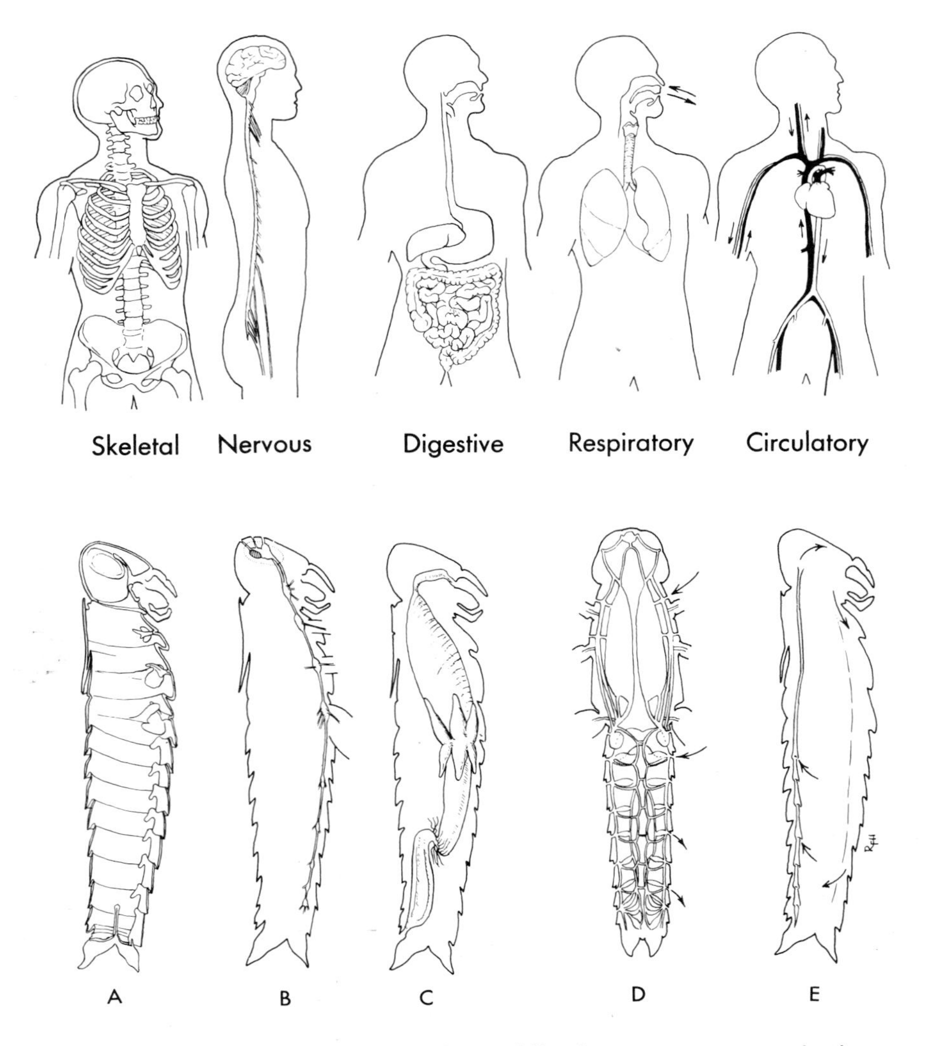

Figure 2:15. A structural comparison of five important systems in the grasshopper and man. A, skeletal. Though the insect skeleton is largely external, note internal ridges and processes for the attachment of muscles. B, nervous. The nerve cord is located dorsally in man, ventrally in insects. The brain is proportionately larger in man. C, digestive. The gut is more complex in many insects. D, respiratory. In man the respiratory system is relatively small, with gases transported by the circulatory system; insects have tracheae and air sacs that ramify throughout the tissue to exchange gases directly with the tissues. Arrows indicate direction of flow of respiratory gases. E, circulatory. The insect circulatory system is much simpler than that of man, because it is not required to carry the respiratory gases. Arrows indicate direction of blood flow.

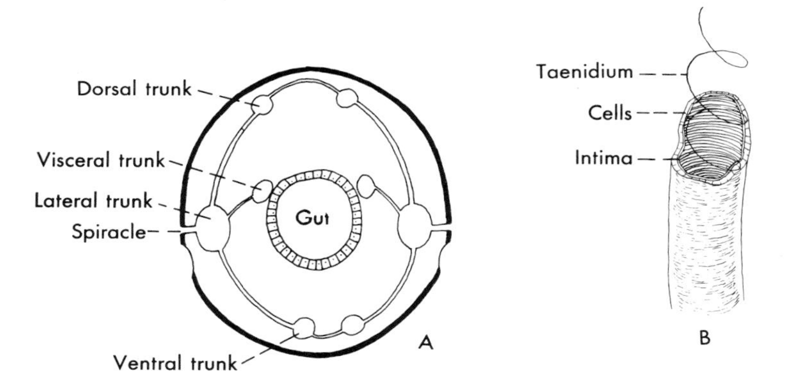

Figure 2:16. The respiratory system. A, diagrammatic cross section of the insect body to show the location of spiracles and main longitudinal trunks; B, structure of the trachea showing the spiral thickenings called taenidia which brace the inside of the tube.

these movements consist of a raising and lowering of the upper wall of the abdomen; in Odonata, Diptera, and Hymenoptera the whole abdomen lengthens and shortens slightly; in Lepidoptera and Trichoptera the upper, lower, and side surfaces of the abdomen move in and out. Close observation of the spiracles during respiration shows that they do not open and close simultaneously. Experiments have demonstrated that in certain insects respiratory movements place pressure on the air sacs resulting in a directional flow of gas caused by the control of the spiracular opening. Thus in the grasshopper air enters the two pairs of thoracic and first two pairs of abdominal spiracles and leaves through the six posterior pairs of abdominal spiracles (Fig. 2:15D). In flight, or other extreme activity, all spiracles remain open in order to allow maximum uptake of oxygen. Rhythmic opening and closing of spiracles is controlled by the central nervous system, but isolated spiracles can open in response to an increased local concentration of carbon dioxide.

The outstanding advantage of the tracheal respiratory system of insects is that it is well suited for such diverse environments as air, soil, internal environment of hosts, and water. Obviously a number of special adaptations are necessary to permit adequate oxygen consumption under these circumstances. In very small insects inhabiting wet environments, such as some Collembola and some parasitic larvae, spiracles are lacking entirely and sufficient gas exchange takes place by **diffusion** through the general body surface. Aquatic insects obtain atmospheric oxygen directly or by the diffusion of oxygen dissolved in water through a gill. A gill may have an abundance of tracheae (**tracheal gills** of damselfly nymphs, Fig. 2:14) or may be filled with body fluids (**blood gills** of Chironomid larvae). Often gills appear to be utilized only during high activity and stress, with sufficient gas diffusion occurring normally through the general body surface. Oxygen may be obtained through a **physical gill** consisting of an air bubble held at spiracular openings. The bubble supplies oxygen in greater quantity than is entrapped originally because the nitrogen present in the initial air bubble (about 80 per cent) diffuses more slowly than the oxygen. Consequently, oxygen is replenished in the bubble by diffusion of dissolved oxygen from water into the oxygen-depleted bubble. Ultimately sufficient nitrogen diffuses from the bubble into the water to reduce the size of the bubble, thus requiring its replacement. Notonectid bugs and diving beetles are typical examples of insects that breathe with the aid of a gas bubble.

It is evident that a very efficient aquatic respiratory system might be developed if a perpetual gas film could be maintained around an aquatic insect for diffusion of dissolved oxygen from the surrounding water. A few aquatic Coleoptera and Hemiptera have achieved this ideal condition through structural developments of the epicuticle. These structures consist of extremely fine hairs or scales on the integument surface which permanently trap a thin film of gas to the body. Gaseous connection is maintained between this trapped film and the spiracles, providing a system functionally resembling a tracheal gill. This type of trapped surface film breathing is called **plastron respiration**.

A further adaptation to aquatic respiration consists of obtaining oxygen from air-filled cells of aquatic plants. A well-known example occurs among mosquito larvae in the genus *Mansonia*: the air tube at the terminal end of the body is modified to form a piercing organ that enters plant air cells. In similar manner some internal parasites of insects maintain a connection with a trachea of their host.

DIGESTIVE SYSTEM

In its simplest form the digestive system of insects consists of a tube that is rather poorly differentiated into foregut, midgut, and hindgut (Figs. 2:15C and 2:17). Close examination reveals that there is generally a valve, the **stomodaeal** or **cardiac valve**, that separates foregut and midgut. The midgut and hindgut are separated by a **proctodaeal valve**. This system tends to be structurally simple in larvae and more complicated in adult insects.

There often are marked modifications of major sections of the digestive tract. Many insects that imbibe liquids have one or more reservoirs, frequently connected with the foregut by only a narrow duct, for storage of food. The midgut, particularly in some sucking insects, can consist of as many as four structurally distinct regions and may possess a large number of pockets or chambers (Fig. 2:18). The hindgut is usually simple with the terminal portion enlarged. In dragonfly nymphs the hindgut is large and complicated, serving for respiration and propulsion along with its usual functions. In many Homoptera the hindgut loops back alongside of the midgut to form a special filter chamber, where liquids and excess food components of plant sap directly bypass the absorptive areas of the midgut.

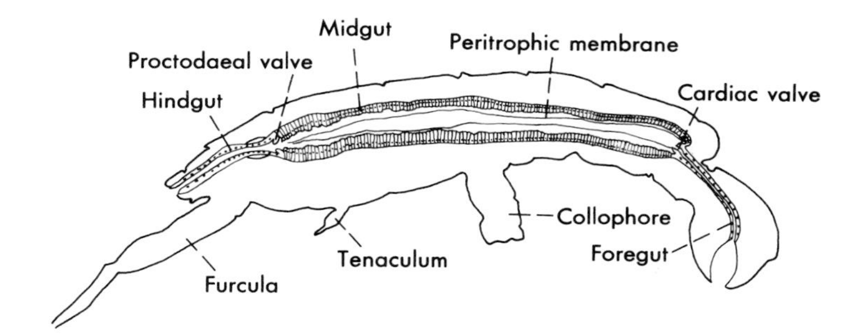

Figure 2:17. The simple digestive system of a collembolan or springtail. Gastric caeca or other outpocketings of the gut are lacking, and simple valves separate the three gut regions. The peritrophic membrane is clearly visible. Malpighian tubes, found in most insects at the junction of mid- and hindgut, are lacking in this insect. *After Folsom and Welles.*

The **foregut** is primarily for storage but may perform limited grinding and mixing of food. A distinct and narrow **pharynx** and **esophagus** may be present at the anterior end, followed by the main storage area or **crop**; the crop may be followed by a muscular, gizzardlike area called the **proventriculus**. The proventriculus can possess a fine screen of hairs to prevent the return of solid food particles (honey bees), or there may be heavy tooth-like structures that serve in screening and grinding solids as in many chewing insects. The lining

Figure 2:18. Digestive tract of a hemipteran, the squash bug. There are several outpocketings of the gut, and the midgut can be seen to consist of three distinct segments. *After Breakey.*

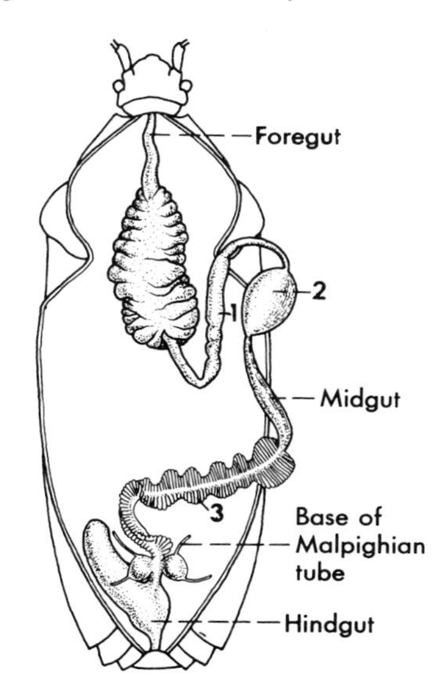

of the foregut (**intima**) is cuticular in origin and often bears setae, spines, or roughened ridges. No digestive enzymes are produced in the foregut, although limited digestion takes place in some cases from carbohydrate-splitting enzymes already mixed with the food and originating in the salivary glands, or by the regurgitation of enzymes from the midgut. Most nutrients are not absorbed through the wall of the foregut, although apparently fats can be absorbed in this manner.

The **midgut** is the main site of digestion and absorption. At its anterior end, just behind the cardiac valve, it characteristically possesses a series of outpocketings, the **gastric caeca** (Fig. 2:19), that are areas of high digestive activity. Similar outpocketings occur in other locations in many insects, but although they are also referred to as caeca, they appear to be chambers harboring microorganisms that aid in digestion. Digestive enzymes are richly present throughout the midgut but, as has been shown for the larvae of flesh flies, these enzymes may originate from specific zones, or even from specific cells. The same zoning characteristic is typical of areas of absorption. The contents of the midgut are generally more alkaline than the contents of the foregut. The acidic or basic fluids of importance in the function of digestive enzymes may also be limited to quite specific regions.

There is no cuticlelike armature or lining connected with the cells of the midgut. Especially in those insects partaking of solid food, a semipermeable structure, the **peritrophic membrane**, lies between the food and the surface of the epithelial cells (Fig. 2:17). This membrane is replaced, passing on with the food into the hindgut. It may be formed as a continuous tube, like a sausage

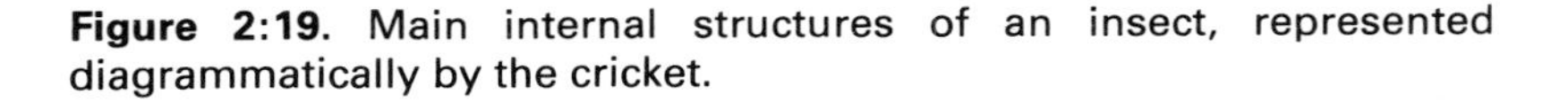

Figure 2:19. Main internal structures of an insect, represented diagrammatically by the cricket.

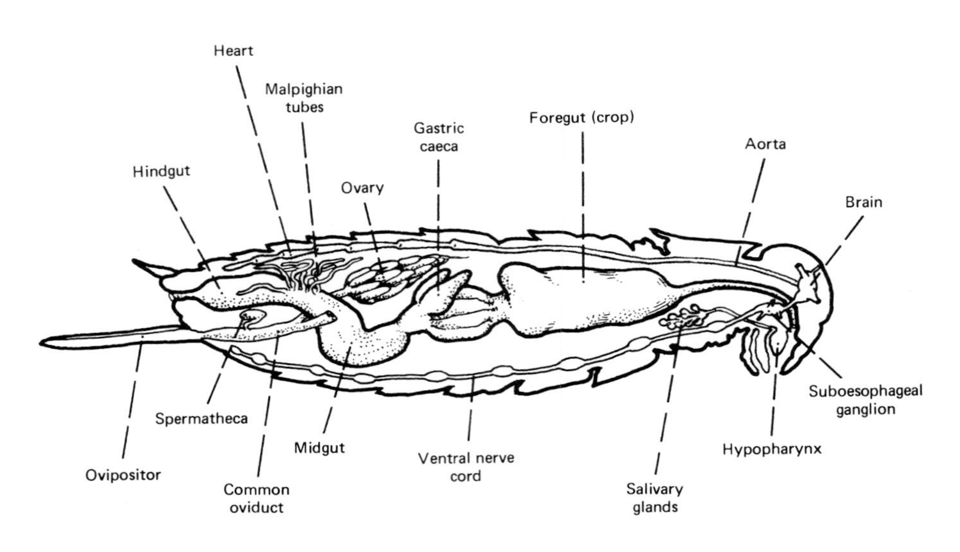

casing, by specialized cells in the region of the stomodaeal valve (fly and mosquito larvae), or in successive layers by cells that make up the entire midgut (grasshoppers). The peritrophic membrane, extremely thin and composed of chitin and protein like the cuticle, permits passage of liquids and solutes while blocking larger fragments.

The **hindgut** is generally a simple tube that enlarges to form the rectum at its terminal end. At its anterior end, near the juncture of the proctodaeal valve, the **Malpighian tubes** empty their excretory products. A thin cuticlelike lining (intima) occurs in the hindgut, generally without spines or other conspicuous armature. Often there are structures, called **rectal pads** or **papillae**, that conserve water and ions by extracting them from the gut contents and returning them to the hemolymph.

Insects possess a wide variety of digestive enzymes. These enzymes split large molecules of foodstuffs into smaller molecules which can then pass through the gut wall to be utilized. As in most animals, there are enzymes for the common foods, namely **proteases** for protein, **carbohydrases** for carbohydrates, and **lipases** for fats. In insects that undergo complete metamorphosis, differences in enzymes characteristic of larvae and adults reflect differences in the diets of these stages. Species that eat food containing a large proportion of nutrients in the form of solutes, such as plant sap or blood, generally have a smaller variety of enzymes than those species that partake of solids, which require greater enzymatic breakdown.

Symbiotic microorganisms in the insect often aid in breaking down otherwise indigestible foodstuffs. These **symbionts** benefit from their life in the insect by having an ideal environment for their development; the host provides ample food and ideal conditions of moisture and oxygen tension. The symbionts benefit their insect host in the process of breaking down largely indigestible food materials into simpler molecules by providing usable nutrients that are in excess of their own requirements or are by-products of their metabolism. Striking examples are symbiotic Protozoa that utilize wood cellulose in the gut of termites, and bacteria that aid in the digestion of beeswax ingested by larvae of the wax moth *Galleria*. The widespread possession of symbionts by insects as a group indicates a long-standing cooperative association.

NUTRITION

Detailed studies of insect nutritional requirements have primarily concerned insects feeding on grain and other stored products. In recent years further information has been obtained for insects that feed on blood, plant sap, leaves, and wood. In many cases where information is lacking, accurate analyses have not been possible for want of synthetic diets with nutrient elements that are acceptable in taste and consistency to the insect in question.

In general nutritional needs of insects resemble those of other animals. They require the same ten **amino acids** essential for protein synthesis in mammals. **Carbohydrates** are a requisite for energy and usually can be supplied in the form of sucrose or glucose. However, unlike mammals, the only vitamins needed are the water-soluble ones of the **vitamin B complex** and, for some plant-feeding species, **ascorbic acid**. **Sterols** are essential in small amounts, and although fats and oils are not generally an absolute requirement, certain unsaturated **lipids** seem beneficial in synthetic diets prepared for plant feeders. The **minerals** required by insects can be supplied by several salt mixtures found satisfactory for the diet of vertebrates.

There is little doubt that those insects which thrive on foodstuffs lacking in essential nutrients are aided by the presence of symbiotic microorganisms. These symbionts may be in the lumen of the digestive tract (see the section on digestion in this chapter), or in specialized cellular structures called **mycetomes**, believed to have been derived originally from the gut. Symbionts are often yeasts or yeast-like organisms that provide B-complex vitamins and sterol. Typically they are found in blood-, wood-, and sap-feeding insects. Symbionts in the fat body of cockroaches have been studied in detail, and symbiotic yeasts from some beetles have been cultured on artificial media.

Most plant-feeding species of insects confine their attacks to a single plant or to a few closely related plants. Chemical analyses of leaves from a variety of plants indicate that all contain rather similar nutrients. It is now evident that particular plants may be attacked by a limited group of insects because of the presence of substances that are attractive, or the absence of substances that are repellent, and not because such plants are uniquely suited from a nutritional standpoint. Apparently these **attractants** and **repellents** have developed in plants during their close and prolonged evolutionary association with insects and mites. Undoubtedly plants and insects have influenced each other greatly in this manner to evolve into the present complexity of plant-insect relationships. There is speculation that plants developed repellent or toxic substances which prevent excessive attack by insects, but also that the chemicals peculiar to a group of plants indicate to an insect its satisfactory range of hosts. For example, the mustard oils of cabbage and other cruciferous plants have been demonstrated to be the key attractive factors for the diamondback moth and some other pests of crucifers.

EXCRETION

The organs of excretion in insects, which functionally resemble kidneys in vertebrates, are the **Malpighian tubes** (Fig. 2:19). These tubes attach to the digestive system at the juncture of the midgut and hindgut. The closed end of each tube generally lies free in the body cavity, but in some insects, notably certain Coleoptera and Lepidoptera, this distal end is attached to the surface of the rectum. The tubes may lie passively in the body cavity, or they may undergo writhing movements caused by the presence of muscle strands. There is great variation in the number of tubes present in various groups of insects. Aphids, Collembola (Fig. 2:17), and some Thysanura lack these structures.

Between two and six tubes are typical of the majority of insects, and the honey bee and some Orthoptera possess as many as 150. The presence of an exceptionally large number of tubes is brought about by the branching of a much lesser number, so that only a few ducts actually connect directly with the digestive tract.

Malpighian tubes filter out water and solutes from the insect blood or hemolymph. Usually the soluble nitrogenous wastes are converted into highly insoluble **uric acid** (frequently in association with inorganic salts); this is discharged into the hindgut. The excretion of uric acid is a water-conserving mechanism; the water returns to the hemolymph through the wall of the hindgut, and in some instances through the tube near its base. The uric acid mechanism of conserving water is typical of some other animals such as reptiles and birds; by way of contrast, however, man and other mammals excrete urea largely in solution in water. Insects that do not need to conserve water because they inhabit a wet environment excrete their nitrogenous wastes in soluble form. Thus blow fly maggots and many aquatic insects excrete **ammonia** directly, and insects feeding on moist foods generally excrete **urea**.

Initially the nitrogenous waste products are formed away from the excreting organs, having been transported to these organs in soluble form by the blood. It has been shown that the enzymes necessary for formation of uric acid and other nitrogenous waste products are found in high concentration within the **fat body** of several insects. Then we have a situation that operates in a manner similar to nitrogenous excretion in vertebrates. Vertebrates produce urea and other waste products in the liver, whereas insects produce uric acid in the fat body. In both vertebrates and insects the waste products are released to the blood; in vertebrates the kidneys filter out the wastes, whereas in insects the Malpighian tubes perform this function. Water and ions are returned to the blood in the basal portion of the kidney tubules of vertebrates, and these same substances may be recovered in the basal portion of the Malpighian tubes and hindgut in insects.

A certain amount of **storage excretion** takes place quite commonly in insects. In some cases uric acid in solid concretions is packed in specialized **urate cells** in the fat body (cockroaches, Collembola). Blow fly maggots have a pair of posterior-directed Malpighian tubes for normal excretion; they also possess an anterior pair which may become packed with **granules** of calcium carbonate and calcium phosphate.

CIRCULATORY SYSTEM

The circulatory system provides a means of chemical exchange between organs. In insects it is composed of (1) fluid and blood cells, collectively called **hemolymph**; (2) a dorsal pulsatile tube, the **dorsal vessel** (Fig. 2:19), which functions to circulate the blood; (3) **accessory pulsating structures**, which aid circulation.

The insect circulatory system is called an **open system**. In man the circulatory system is closed: the blood is channeled completely from heart to arteries, to capillaries, to veins, and returns to the heart. In insects the dorsal vessel, consisting of the posterior **heart** and the anterior **aorta**, is the only real blood vessel (Fig. 2:19). Blood is projected forward toward the region of the head and passes more or less freely through the body, returning to the heart through openings along its side, called **ostia**. **Dorsal** and **ventral diaphragms** may direct the blood along the upper and lower body surfaces, but frequently these structures are missing or poorly defined. Accessory pulsating organs may be present at the base of legs, wings, or antennae, and a membrane may divide tubular appendages to distribute blood in one side and out of the other (Fig. 2:20).

The blood of insects is not an effective carrier of oxygen to tissues. Respiratory pigments are lacking for the most part, although larvae of some chironomid midges and horse bots (both Diptera) and some aquatic true bugs contain hemoglobin. In insects this hemoglobin is not contained in specialized cells such as the red blood cells or erythrocytes of vertebrates; rather, when present, it is found in the fluid portion of blood or in special fat cells. In insects that possess hemoglobin, this substance aids in oxygen transport during a shortage of oxygen in the surrounding medium rather than as the normal means of respiration.

Blood cells contain a nucleus and are so variable that they are difficult to classify into groups. As many as ten classes of blood cells divided into 32 types have been described for the larva of a noctuid moth, but most authorities recognize only five or six classes. Many of the blood cells are phagocytic and may be particularly useful in engulfing disintegrating tissue fragments during metamorphosis. Blood cells appear to play some part in growth and metamorphosis because the ratio of cell types changes between molting

Figure 2:20. Cross section of a femur, leg of the pomace fly *Drosophila*. The dividing membrane does not continue all the way to the tip of the appendage, thus permitting the outward flow of blood on one side and its return on the other. *From a photomicrograph by Miller, in Demerec, 1950,* Biology of Drosophila, *by permission John Wiley & Sons, Inc., New York.*

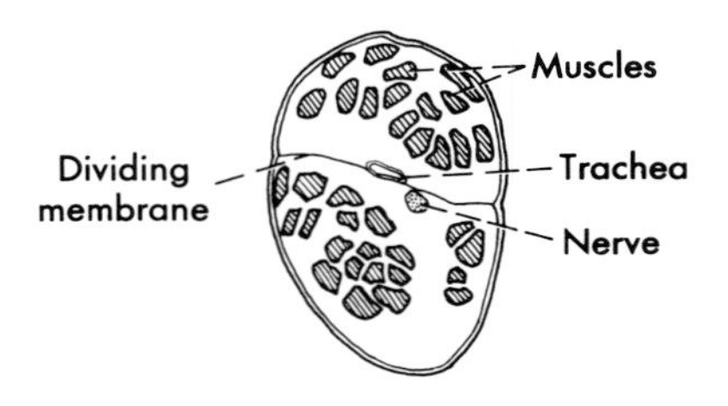

intervals. The cells may function in clotting at wound sites, serving as the center of formation of the clot or disintegrating to release clotting factors.

It is important that blood maintain an osmotic pressure optimal for the body cells. The correct ratio of sodium and potassium is also necessary for proper nerve and muscle function. The **osmotic pressure** of insect blood is due largely to amino acids, in contrast to mammals, where inorganic chlorides are of greater importance. In insect blood there is also a higher concentration of phosphates than in that of mammals. In vertebrates a relatively high ratio of sodium to potassium in blood serum and a high ratio of potassium to sodium in body cells provide proper ion balance for nerve impulse propagation and muscle activity. This same relationship holds for carnivorous insects, but in many insects whose ancestors were plant feeders, or who are themselves plant feeders, the amount of blood sodium relative to potassium is very low. A low blood-sodium-to-potassium ratio is found in many Lepidoptera, Hymenoptera, Phasmidae (walking sticks), and plant-feeding Coleoptera. Even in these cases it is notable that the body cells still maintain a high concentration of potassium. Evidently membranes and cells surrounding the nerves maintain a sodium and potassium ionic differential similar to that required for vertebrate nerve function.

MUSCULAR SYSTEM

Although the muscular system of insects is complex, practically all the body muscles of adults, nymphs, and more primitive larvae can be homologized throughout the various orders. However, the musculature of higher holometabolous larvae has yet to be compared successfully in detail with that of other insects. Within a species, the muscle groups in each segment are often similar. Origin and insertion of muscles are modified by the motion requirements of the insect in question. Differences in muscular patterns are largely the result of loss of muscles from an overall "original" arrangement. The total number of muscles is often very large, and a moth larva may have roughly three times as many muscles as are found in man.

All insect muscles are **striated** (Fig. 2:21). They may be divided into a skeletal and a visceral group, according to whether they move the body wall and its appendages or are associated with internal organs. The **skeletal muscles** are clearly recognizable as separate bundles of fibers with a definite origin and insertion. **Visceral muscles** consist of the circular and longitudinal muscles of the gut and the sheaths of interconnecting muscle fibers associated with the reproductive system and ventral diaphragm.

Four subdivisions of insect muscle type are possible, based chiefly on the location and nature of nuclei and muscle sheaths. Particularly noteworthy are the **giant mitochondria** or **sarcosomes** observable in the indirect flight muscles (Fig. 2:21). These structures are sites of high enzymatic activity required for energy transformations in this very active group of muscles.

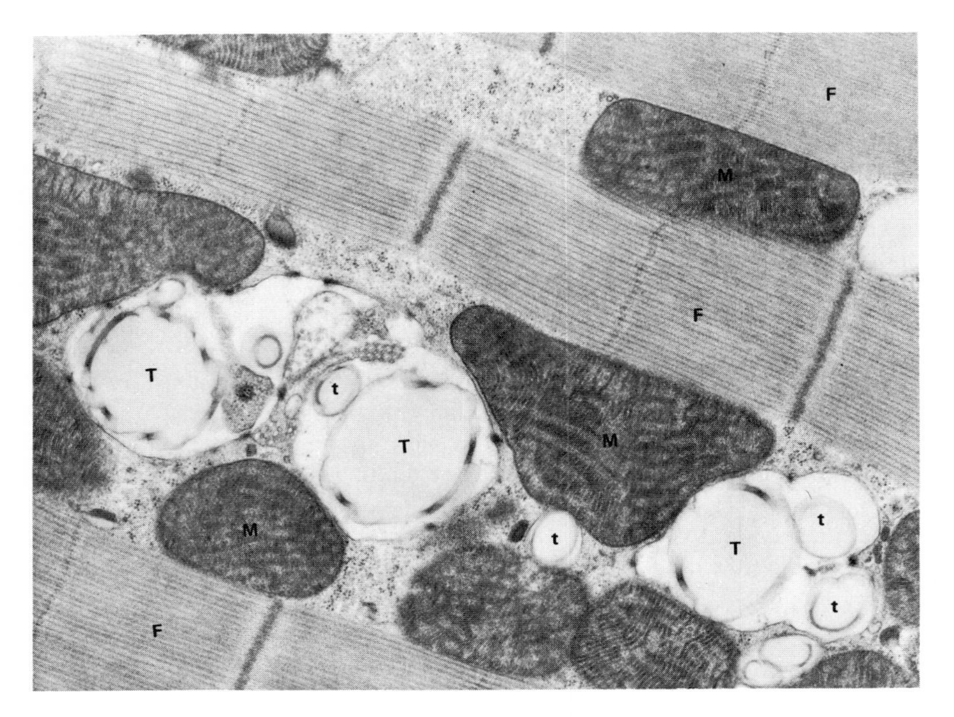

Figure 2:21. Longitudinal section of indirect flight muscle of the blow fly *Phormia*. Striated muscle fibrils (F), mitochondria (M), transversally sectioned tracheae (T), and tracheoles (t). Magnification ×28,000. *Electronmicrograph* by David S. Smith, *University of Miami, School of Medicine.*

REPRODUCTIVE SYSTEM

Insects are **bisexual**; that is, there are males and females. Reproduction without benefit of males does occur (**parthenogenesis**), but usually the combination of sperm and egg is required to produce a new individual. The reproductive system consists of paired gonads, with the common duct from each combining into a median duct that leads to the outside of the body. In conjunction with these ducts, specialized glands and chambers are found for storage of spermatozoa.

Each **ovary** consists of a group of egg tubes, or **ovarioles** (Fig. 2:22). The ovarioles are grouped together loosely and are held in position by a terminal ligament. The number of ovarioles varies from one in some aphids to several hundred in certain insects. Each ovariole is a tube of epithelial cells containing eggs in varying stages of development. This tube is divided into chambers, with primordial germ cells or **oogonia**, often in a chamber called the **germarium** at the narrow end, and eggs in various stages of development in

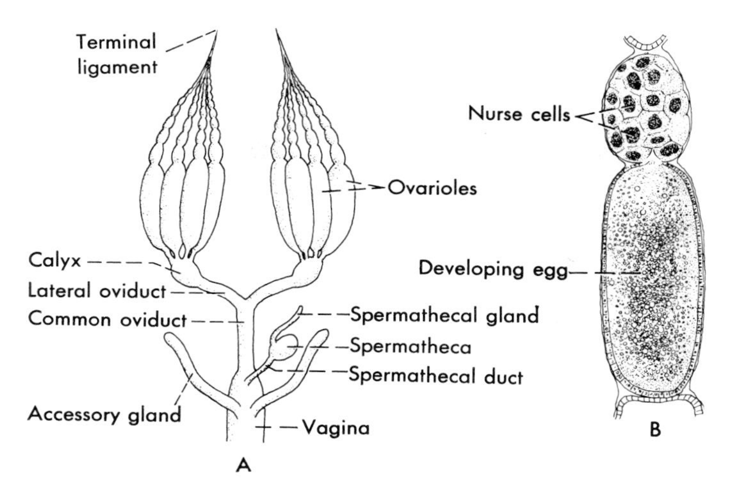

Figure 2:22. The female reproductive system. A, diagram of the whole system; B, section of an egg follicle in the honey bee, showing nurse cells associated with the developing egg. *A, after several investigators; B, original.*

succeeding chambers. Enlargement of the eggs is caused primarily by the formation of granular yolk, which is a food reserve for embryonic development. In many insects, **nurse cells** associated with the eggs serve a nutritive function (Fig. 2:22). By the time the eggs are fully developed, the epithelial cells of the ovariole have deposited a shell or **chorion**. It is the imprint of these cells that leaves the reticulated pattern readily visible on the chorion after oviposition (Fig. 2:23). At one end of the egg an opening through the chorion, the **micropyle**, permits the entrance of sperm.

Generally fertilization takes place after an egg leaves the ovary. The egg passes down the **lateral oviduct** and enters the **common oviduct**. As it passes the duct of the sperm storage organ or **spermatheca**, fertilization takes place. In some insects the release of spermatozoa for fertilization can be controlled

Figure 2:23. Diagram of an insect egg.

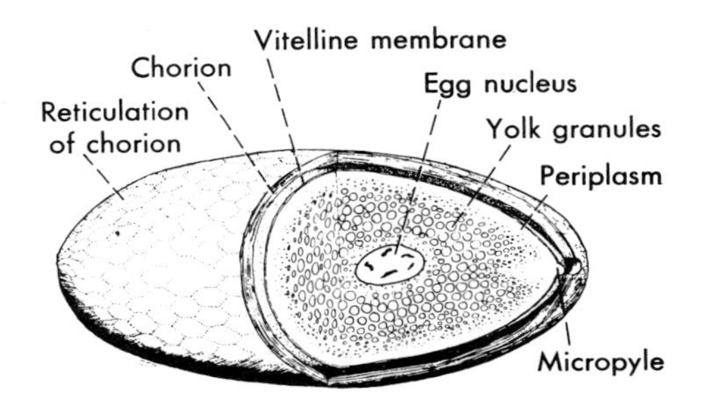

by the female. For example in the honey bee, unfertilized eggs develop into drones, and fertilized eggs develop into workers or queens. Spermatozoa can be maintained for very long periods in the spermatheca of many female insects. A small gland in conjunction with the spermatheca may produce secretions responsible for sperm longevity (Fig. 2:22A).

Generally during copulation the male intromittent organ is placed in that portion of the median oviduct called the **vagina**. In many insects, however, a special chamber for deposition of spermatozoa, the **bursa copulatrix**, is developed to one side of the median oviduct. There is particularly clear separation in the higher Lepidoptera, where the external opening of the bursa is on the segment in front of that bearing the median oviduct. In this case a **seminal duct** is present to permit spermatozoa to enter the true reproductive system of the female and ultimately to reach the spermatheca.

There are usually **accessory glands** that discharge their contents into the median oviduct. These are known as **colleterial glands** if they produce a sticky substance that glues eggs to objects. The accessory glands of a female cockroach produce protein and a tanning substance which combine to form the hard brown cover of the **ootheca**, an egg packet containing several eggs (Figs. 3:1D and 19:10). In grasshoppers the glands that form a hard foamy matrix around the eggs are extensions of the upper portion of each oviduct and should be referred to as **oviducal glands**.

Each testis of the male consists of a few to several tubular **testicular follicles** (Fig. 2:24A). These follicles are enclosed in a sheath, and frequently

Figure 2:24. The male reproductive system. A, diagram of the whole system; B, section of the testicular follicle in a grasshopper showing the growth and differentiation of germ cells into mature flagellated spermatozoa. Note that groups of developing cells are contained in packets. *A, after several investigators*; *B, original.*

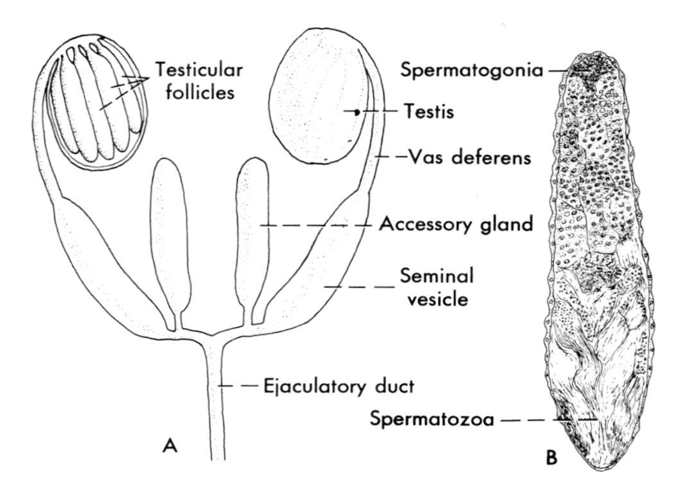

both **testes** are enclosed in this sheath to appear as a single unit. Each follicle is a tube of epithelial cells containing male germ cells in varying stages of development. At the upper end are the primordial germ cells, or **spermatogonia**, followed by packets of sperm in succeedingly advanced stages of development. The **spermatozoa**, with each sperm possessing a head and flagellum, are located at the lower end of the follicle (Fig. 2:24B).

Ducts lead from the testes to outside the body wall (Fig. 2:24A). From each testis a duct, the **vas deferens**, unites to form a median **ejaculatory duct**. Frequently **seminal vesicles**, enlargements of each vas deferens, serve as areas to store spermatozoa. **Accessory glands** are usually present, and although their function is uncertain, it seems likely that they produce seminal fluid for additional volume and for the nutriment of spermatozoa.

Physiological aspects of the reproductive system in insects are not well known. Most evidence suggests that this system does not produce sex hormones affecting secondary sexual characteristics, such as those found in vertebrates. However, cells associated with the testes in male firefly larvae are believed to secrete substances that produce male characteristics, and it may well be that a similar situation, overlooked up to now, occurs in most insects during embryonic development. Hormones controlling growth are apparently responsible for initiating the development of eggs, and in the honey bee, substances secreted by the queen will inhibit reproduction in workers. Insects as a group also have some means of controlling numbers of eggs, and excess eggs undergoing development can be resorbed if sufficient food reserves are lacking. Reserves for egg formation may be derived from nutrients consumed recently or from fats and proteins already present within the body.

NERVOUS SYSTEM AND SPECIALIZED RECEPTORS

The central nervous system of insects differs in location from that of mammals. The main **nerve cord** is located in the ventral body region beneath the gut, and the **brain** is situated in the head above the digestive tract (Figs. 2:15B and 2:19). It is quite apparent that the ganglionic centers of this system have fused to varying degrees. Both the brain and the **subesophageal ganglion** consist of three fused ganglia. Generally there are three thoracic ganglia and as many as eight abdominal ganglia; yet in many insects there is a further migration and fusion of ganglia toward the front of the body. In Hemiptera and muscoid Diptera (Fig. 2:25), for example, this condition is characterized by a large thoracic ganglionic mass that innervates the thorax and abdominal segments. Segmental nerves branch from each ganglion into the muscles and other body structures of their segment. Even when a ganglion has migrated anteriorly and has fused, it is possible to determine its segmental derivation by tracing the peripheral nerves that lead out from the ganglionic center to terminate in the segment from which it originated.

The **stomatogastric system** is a system associated with the digestive tract and is located dorsally on the foregut (Fig. 2:26). From the third or tri-

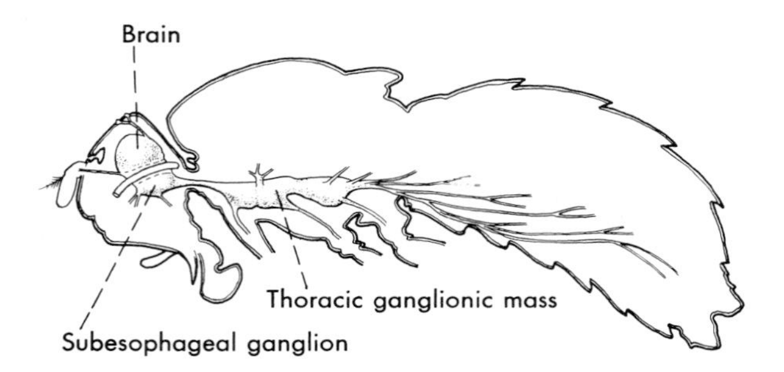

Figure 2:25. Central nervous system of a fly, *Drosophila.* Though the thoracic and abdominal ganglia consist of a fused mass, their separate origin is indicated by the segmental nerves passing to the segments of thorax and abdomen. *After Miller, in Demerec, 1950,* Biology of Drosophila, *by permission John Wiley & Sons, Inc., New York.*

tocerebral portion of the brain a pair of connectives attaches to the **frontal ganglion** on the esophagus. A short and relatively thick nerve, the **recurrent nerve,** runs under the brain from the frontal ganglion to connect with the **occipital** or **hypocerebral ganglion** in back of the brain. The occipital ganglion has a single nerve or pair of nerves connecting with a single or paired **ingluvial ganglion** on the upper wall of the foregut. Finer nervous connections are made between the ganglia of this stomatogastric system and the gut, certain mouthparts, salivary glands, and the aorta.

A **sympathetic nervous system** is frequently present in the form of **median nerves**; these nerves are particularly well developed in caterpillars. They may connect directly between adjacent ganglia of the nerve cord or may branch in each succeeding segment to the lateral body areas. Where median nerves are present, they connect with the spiracles, although spiracular innervation is supplied by lateral nerves from ganglia of the nerve cord in forms lacking median nerves. The terminal ganglion in the abdomen innervates the posterior portions of the gut and the reproductive system.

Figure 2:26. The brain and associated endocrine structures of a cockroach. The termination of the aorta and the anterior portion of the stomatogastric nervous system are also indicated. *After Cazal.*

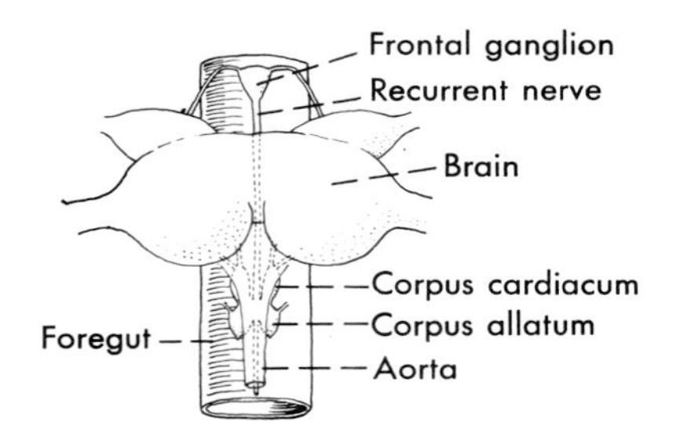

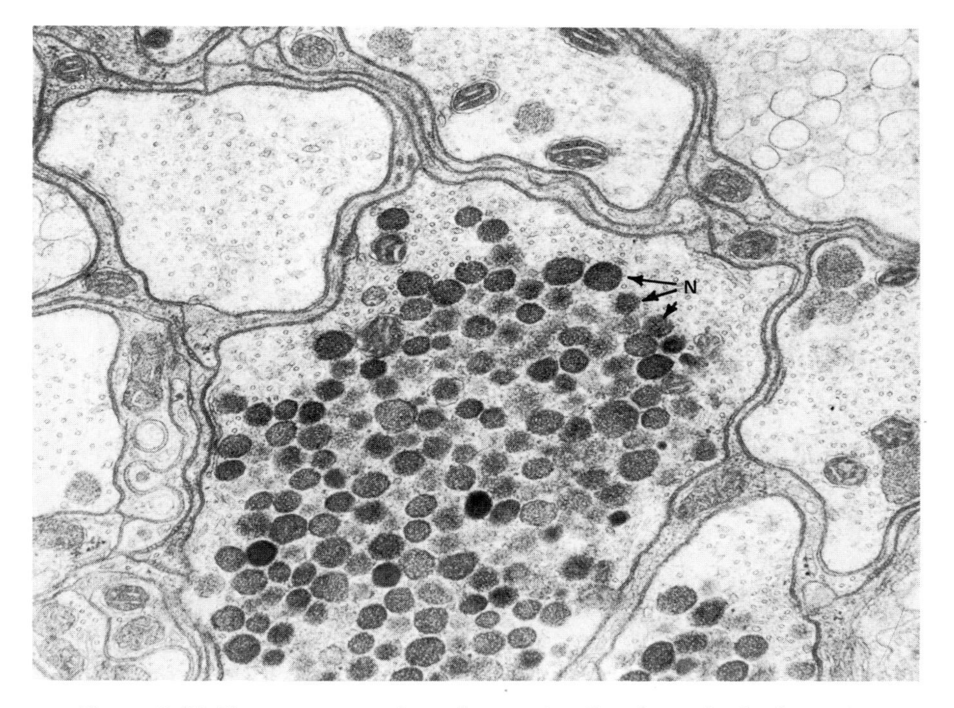

Figure 2:27. Transverse section of nerve leading from the brain to the corpus cardiacum of the stick insect *Carausius*. Axons are the larger spaces with supporting cell processes around their boundaries. Neurosecretory granules (*N*) are particularly abundant in one axon. Magnification ×35,000. *Electronmicrograph by U. Smith and D. S. Smith, University of Miami, School of Medicine.*

The enzyme and ion interchanges occurring in insect nerve transmission resemble those in vertebrates. **Cholinesterase** and **acetylcholine** are considered necessary for transmission across the synapse, which may be inhibited by drugs and poisons affecting vertebrates. **Sodium** ions within the axon and **potassium** ions outside it are likewise of primary importance in propagating an electrical charge, although the specific amounts or ratios may differ from vertebrates. Where drugs or toxicants do not affect the nerve transmission of insects to the same degree as in vertebrates, it is likely that differences in the **nerve sheath** or cells surrounding it are responsible for permitting or preventing these substances from reaching the site of activity. Ganglia and main nerves are surrounded by a noncellular sheath, and by sheathlike layers of cells resembling the myelin sheath of vertebrates (Fig. 2:27).

Sensory Receptors

A wide variety of specialized sensory structures inform an insect of the conditions surrounding it and of its internal conditions. **Sensory hairs** of the

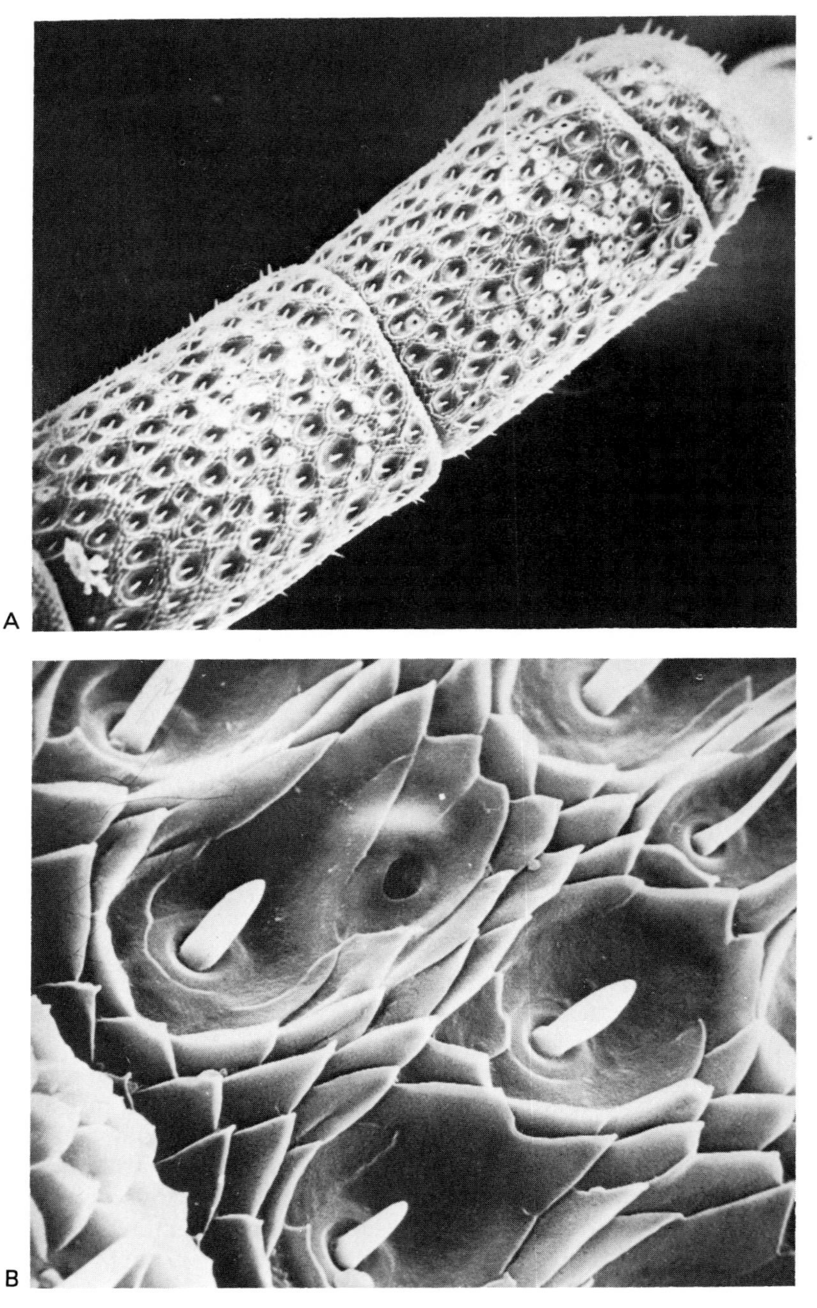

Figure 2:28. The tip of the antenna of a male desert locust as seen with a scanning electron microscope. The pits with a central peg are the sense organs concerned with smell and taste. A, at 120 magnification; B, at 1200 magnification. *Courtesy Anti-Locust Centre, photographs by Marion Kendall.*

body surface are similar in structure to ordinary setae, with the addition of one or more **sensory neurones** attaching to some point on their inner surface (Fig. 2:13). Other sensory structures on the body surface resemble sensory hairs in internal structure, but in place of setae are found thin-walled pegs, domes, or pits (Fig. 2:28). Such receptors detect stimuli of taste, smell, touch, motion, temperature, or humidity. Frequently these sensory receptors are concentrated on organs in close contact with the external environment such as palpi, tarsi, antennae, or cerci.

Insect Eyes

The eyes of insects merit special mention, not only because of their elegant structure, but also because the visual ability of some insects is second only to that of advanced vertebrates. In addition to the high degree of visual acuity that may be present, other characteristics make insects' eyes particularly well suited for vision during flight. In discussing the visual organs both simple eyes (ocelli) and compound eyes must be considered.

As many as three dorsal **simple eyes** or **ocelli** (Fig. 2:29A) are present in adult insects, although these structures are frequently reduced to two or none. Structural details suggest that these are not organs of particularly acute vision. There may be a few or several receptor units behind a single corneal area,

Figure 2:29. The eyes of adult insects, as seen in a dragonfly. A, overall position and size of ocelli and compound eye; B, enlargement diagrammatically showing how each receptor unit or ommatidium of compound eye is located beneath a hexagonal facet. Many cells within the ommatidium are deleted for simplification; C, pigment in the light-adapted position permits light, indicated by arrows, to reach only the receptor units or rhabdoms in direct line with the ommatidium.

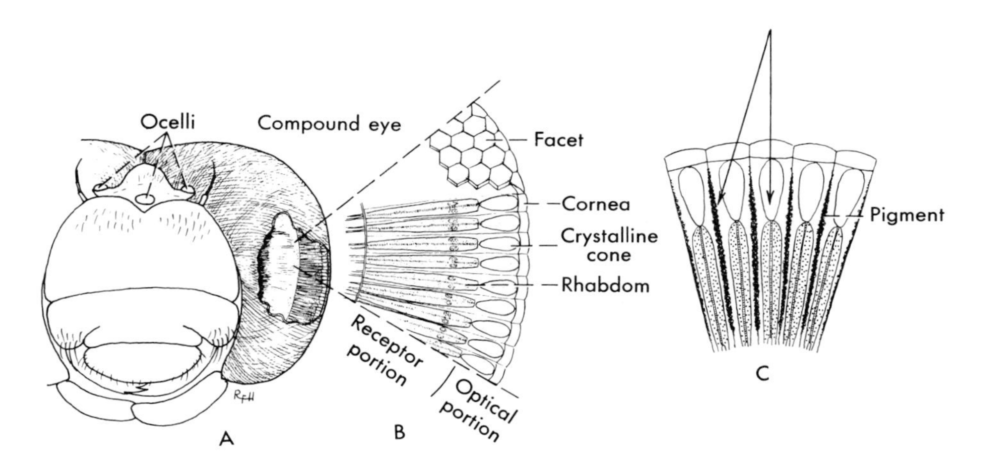

with no clear separation of such units to permit division of the image received. Thus it appears that the ocelli determine levels of light intensity, and in fact it has been shown that these structures control general activity of the insect in response to the amount of light present. Lateral simple eyes, or **stemmata**, are found in many larvae (Fig. 3:14C). These also are considered incapable of acute vision. Some insects without eyes—fly maggots, for example—distinguish between high and low light intensity by means of crude light receptors situated below the body surface.

A pair of **compound eyes** is characteristic of adult insects. As the name implies, they are made up of a large number of separate receptor units (Fig. 2:29). Each unit, an **ommatidium**, is quite a complex structure. The ommatidium can be divided into the outermost optical elements consisting of a clear **cornea** and a light-gathering device, the **crystalline cone**. In back of these the **retinula cells** which are sensory neurons, form a rod-shaped structure possessing an internal **rhabdom** or light receptor unit. Photochemical reactions take place within the retinula.

The compound eyes of many nocturnally active insects adapt to conditions of high and low light intensity. **Adaptation** takes place through the migration of pigment granules within cells that form a sheath around each ommatidium. In dim light the pigment migrates toward the outermost part of the eye, leaving the retinula receptor elements uncovered. Under these circumstances a maximum quantity of light can reach the retinula because rays of light entering the optic portion at an angle impinge on the bared receptors of adjacent ommatidia. Pigment ensheaths the receptors during periods of high light intensity, and therefore the retinula cells receive only that light entering the optic portion directly in front of the ommatidium in which they are located (Fig. 2:29C). Although the dark-adapted eye has increased sensitivity because of maximal light activation of each receptor unit, it seems obvious that under such circumstances vision must be less acute.

A great ability to **perceive movement** is found in some insects. This ability is particularly advantageous in the flying and predaceous species. Like man, many insects distinguish a maximum of about 50 flickers of light per second as separate events, yet the honey bee can distinguish a maximum of 300 flickers per second. In flying insects the **facets** (cornea) are arranged much closer together in portions of the eye most necessary for vision in flight. By this means more receptor units are stimulated per unit of time during movement, permitting fine distinctions of movement.

Behavioral experiments, particularly with honey bees, have demonstrated that insects have **color vision**, and they can make some distinctions in the **shape of objects**. In general insects perceive in a shorter wave length portion of the color spectrum than man, so that they see the near ultraviolet into the blue-green range quite well, but they are unable to distinguish red from black or gray. Electrophysiological studies have demonstrated that in certain species specific retinula cells have specific characteristic color responses, similar to the cone cells in the retina of man.

FLIGHT

Flight is a unique characteristic of insects; they are the only invertebrates endowed with this ability. Not only can they fly, but the best-flying insects fly as well as or better than birds. Some insects, such as the house fly and other higher Diptera, have extreme maneuverability in flight; others, such as locusts, can sustain flight for hundreds of miles.

A series of events occurs in an insect for the **initiation** and **maintenance of flight**. Loss of tarsal contact, as when an insect jumps up from a surface, has been demonstrated to initiate reflexive flight movement of the wings in a great many species. Once the insect is in flight certain signals indicate that flight can be economically continued; an airstream must continuously hit receptors on the front part of the head in locusts, and various flies will sustain flight for longer periods in such an airstream. When the velocity of the airstream causes pressures suggesting a loss of forward motion, wing movements cease. The mosquito relies on a moving pattern received by the bottom portion of the eye, indicating forward movement, in order to maintain flight. Under experimental conditions flight can be maintained in flies and bees until the energy reserves, in the form of sugars or glycogen, are depleted; flight can be started again by feeding sugar. It has been shown that prolonged flights of locusts and butterflies can be maintained by the utilization of fats rather than sugars.

Figure 2:30. Diagram of indirect flight muscles in a horse fly, in cross section and as seen on the median plane. Main wing movement is aided by the elastic nature of the body wall springing up and down in response to contraction of the dorsoventral muscles to cause the upstroke (A), and contraction of the longitudinal muscles to cause the downstroke (B); C, relationship of large flight muscles in the right half of the thorax. Much smaller direct flight muscles attach near the articulation of the wing bases to modify the orientation of each wing during up and down movement, thus affecting rate of climb, hovering, and turning.

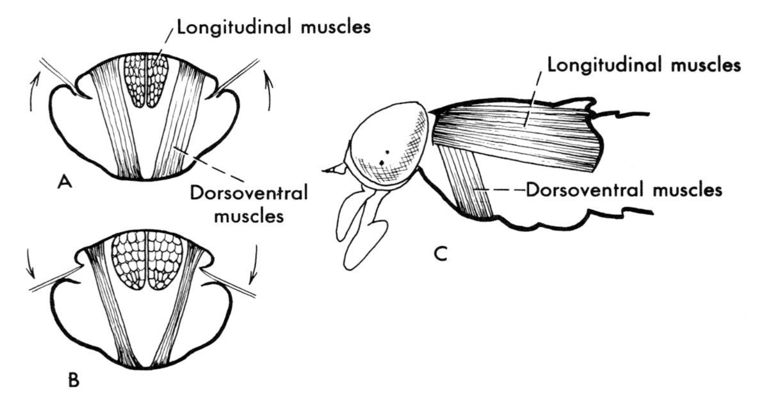

Flight direction is controlled by varying the angle of the wings. The basic up-and-down movement of wings is provided by muscles extending between the top and bottom of the thorax, thereby depressing it and providing the upstroke, and by longitudinal muscles (indirect flight muscles) that compress the thorax lengthwise to provide the down or power stroke (Fig. 2:30). Other muscles, attached to small plates in the wing base, cause changes in the angle of the stroke. In flies the second pair of wings is modified into knobbed **halters** (club-shaped structures), which serve as vibrating gyroscopic guidance organs. Pressure changes detected by groups of receptors on the halters are transmitted to the central nervous system and integrated to cause necessary adjustments in the wing movement.

The **frequency of wing beat** is most remarkable, ranging from about 5 beats per second in butterflies to over 1000 beats per second in a ceratopogonid fly. Muscle contracts in response to an impulse from the stimulating nerve, and a ratio of one contraction of flight muscle per impulse can readily be demonstrated in insects with a relatively slow wing beat. In fast-flying insects, however, the flight muscles may contract as many as 40 times per nerve impulse, thus making very fast beats possible. This extremely fast wing beat is typical of Diptera, Hymenoptera, Coleoptera, and some Homoptera.

CHEMICAL COMMUNICATION

The number and variety of methods insects use to communicate among themselves are amazing for such unreasoning creatures. Although many insects commonly communicate by visual signals or by calling through their chirps, clicks, songs, and sounds of wing beat, many others communicate by release of chemical messengers, special compounds produced by insects. A receiving insect may make a behavioral response upon contact or at very close range with the released chemical, or in cases of sex attraction, the response may be elicited by following a stream of attractive vapor for a mile or more. Such responses to chemical clues are divided basically into two types: (1) **pheromones**, the groups of chemicals that provide messages between individuals of the same species (intraspecific), and (2) **allelochemicals**, the substances that affect individuals or populations of a species different from the source (interspecific).

Two types of pheromones have been discovered among insects, the fast-acting **releasers** and the slow-acting primers. An important group of releasers are the **sex pheromones** that are secreted by several orders of insects, notably moths, beetles, flies, and the Hymenoptera. Much is now known about the sex pheromone of the codling moth. The female of this species secretes the pheromone in a gland (Fig. 2:31) that opens dorsally in the intersegmental fold between the last two segments of the abdomen. Normally these two segments lie telescoped within the abdomen, but when a virgin female seeks a mate she will protrude them and release the pheromone.

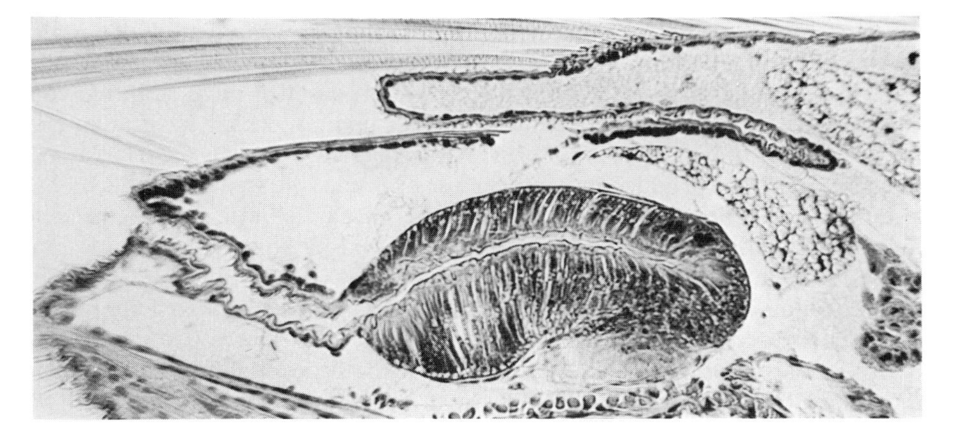

Figure 2:31. Longitudinal section of the pheromone gland and duct of the codling moth. *Courtesy Martin M. Barnes, University of California, Riverside.*

Codling moth males of the right age detect the pheromone in the air, become sexually excited, and respond immediately by flying to the source and mating with the 'calling'' female. The sex pheromone of the codling moth has been identified as (*E*, *E*)-8, 10-dodecadien-1-ol.

A second group of pheromones brings about aggregation or assembly of individuals of a species, a type of behavior that has survival value. Assembly may be temporary, as in the overwintering of lady bird beetles, or permanent, as for honey bees in a colony. Bark beetles produce **aggregation pheromones**. The first bark beetle that finds a susceptible tree releases a pheromone that attracts others to join in the attack.

A third group of pheromones are the **trail-marking secretions**. Ants and termites leave a chemical trail which others of the same species follow. Honey bee workers release an aerial odor that allows other workers to locate a new food source.

Alarm pheromones, a fourth group, are chemicals that alter the behavior of social insects to retreat or to attack. In the well-known behavior of the honey bee, a sting from one bee elicits aggressive behavior of other bees. This aggression of guard bees starts when the sting apparatus of one bee is torn from its body and embedded in the victim's skin. A special gland that remains attached to the sting releases the sting pheromone sensed by the other bees. Immediately it stimulates additional stinging of the invader.

The slow-acting **primer pheromones** are secreted by social insects to maintain the organization of the colony. The mandibular glands of the queen honey bee produce a chemical called queen substance that inhibits the workers, which ingest it, from producing eggs and building queen cells. Termites produce, and pass from one individual to another, primer pheromones that foster a favorable caste ratio of soldiers, workers, and reproductives.

Allelochemicals differ from pheromones in that they stimulate individuals of another species rather than those of the species producing the chemical. The allelochemicals are separated into allomones, which have adaptive advantages to the producer organism, and kairomones, which have adaptive advantages to the receiver organism. **Allomones** include the great array of defense secretions produced by insects: for example, the irritant produced by bombardier beetles, toxins in larvae of the monarch butterfly that render them unpalatable to birds, and venoms produced by stinging Hymenoptera that help them to subdue prey.

Kairomones include plant substances that warn insects of toxicity to the receiver or odors that attract insects to the right food source. Coumarin, an odorous constituent of sweetclover, attracts flying sweetclover weevils to their food plant. Specific lipids produced by the tobacco budworm provide a cue for braconid parasites (tiny wasps) to lay eggs in their host.

The study of chemical communication among insects is still young. Doubtless the division of messenger chemicals into the types described above and the categorization of pheromones into primer and releaser will some day be considered arbitrary. Many familiar elements of insect life cycles and life styles may eventually become ascribed to pheromones and hormones as yet unidentified. Moreover, the details of the olfactory processes are virtually unknown. Currently several pheromones are being evaluated as direct compounds for insect control by disrupting mating success or otherwise upsetting normal behavior. Ultimately some communication chemicals will be employed in an integrated fashion with conventional control measures.

STRUCTURE AND FUNCTION IN MITES AND TICKS

External Form

Mites and ticks belong to the order Acarina and have a body consisting of two regions, the **cephalothorax** and **abdomen**. There are variations of this basic plan, particularly in mites whose body appears to consist of a single unit, or whose abdomen is subdivided into segments. There may be a distinct headlike region, but it is not structurally similar to the head of insects. This area, bearing the feeding organs, is called the **gnathosoma** (jaw body). In ticks, the region is developed into a headlike structure called the **capitulum** (Fig. 2:7A). Many mites have a distinct demarcation of the body between the second and third pair of legs (Fig. 2:32); the anterior portion is called the **proterosoma** (preceding body), and the posterior portion is called the **hysterosoma** (behind body).

Normally there are four pairs of legs in adult acarines. The form emerging from an egg is called a larva and has only three pairs of legs. At the first molt a hindmost pair of legs is added, making a total of four pairs, and the immature form is now called a nymph. Typically, mites go through three nymphal stages, although there are many variations. The nymphs generally resemble adults in appearance. Mite families are especially variable in form. Many

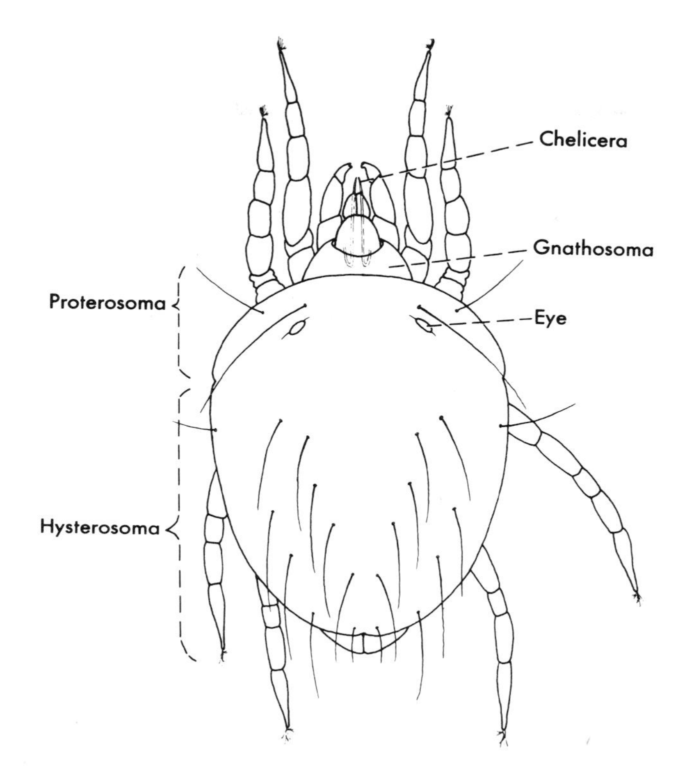

Figure 2:32. General external form of a tetranychid or spider mite. *After Malcolm.*

parasitic mites are peculiarly shaped; the plant-feeding eriophyid or blister mites are narrow and elongate with only two pairs of legs (Fig. 4:5A). Acarine legs are usually six-segmented, but subdivision to form a total of seven segments occurs, and fusion or deletion may result in as few as two segments.

The integument, like that of insects, is composed of several layers secreted by a sheet of epidermal cells. The chemical nature of these layers is believed to be similar to insect cuticle. There is often a very rich surface sculpturing of the cuticle which is distinctive enough to be of aid in classifying mites. There may be areas of the integument that are thicker, forming sclerites or plates. Setae are often limited to definite locations, and they can be so characteristic of species as to be of prime importance in classification. Special sensory setae are common, especially on the ends of appendages.

Internal Form and Function

For the sake of brevity, description of internal structures and their functions will be limited to mites. Knowledge in this area can be supplemented by

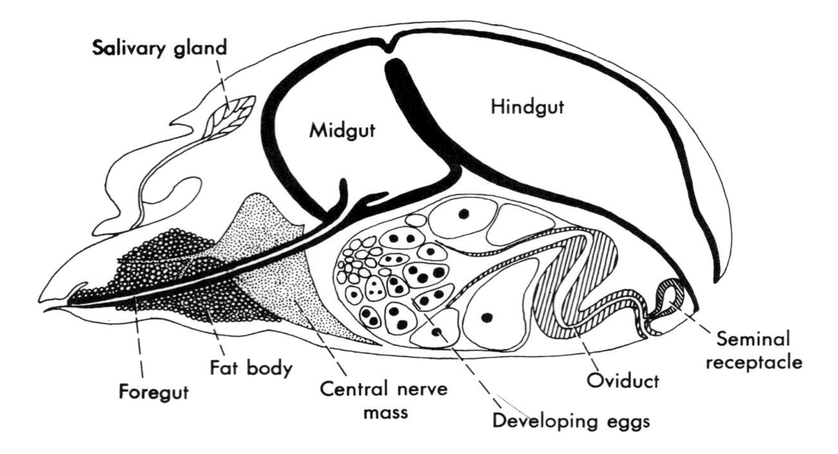

Figure 2:33. Internal anatomy of the twospotted spider mite. *Redrawn from Blauvelt.*

reference to Baker and Wharton's *An Introduction to Acarology*. The internal organization of the twospotted spider mite *Tetranychus urticae* is illustrated (Fig. 2:33) because it is sufficiently typical, and this is the most widely encountered mite of agricultural importance.

The digestive system consists of fore-, mid-, and hindgut. Each gut section may be relatively simple or further subdivided into definite regions. In the twospotted spider mite the spot characteristics are caused by the presence of dark food in pouchlike chambers on each side of the midgut, readily visible through the integument. Though detailed studies on mite digestive enzymes are lacking, limited investigations suggest similarities to the enzymes of insects, and certainly a wide variety of foodstuffs can be utilized.

Several structures have been described as having excretory functions. There are usually excretory tubes in connection with the hindgut, and certain of the digestive tract cells are excretory in function, as is a gland at the base of the leg. The main nitrogenous waste product is guanine, a compound chemically similar to uric acid.

The mite circulatory system is extremely simple. A heart is located dorsally in only a few species; in others there is free circulation of blood consisting of hemolymph and cells.

The muscles, as in insects, are striated and are located so as to perform the required movements.

Reproduction is by sexual means. Males possess a pair of testes, with ducts leading outward much as in insects; females have a pair of ovaries, or a single ovary, and ducts and glands similar to those in insects are present. Spermatozoa may be transferred in a sperm packet or **spermatophore**. There are several species of mites for which males are unknown, and parthenogenesis also occurs in many species that have males.

The nervous system seems relatively simple. Ganglia of the central nervous system have fused into a mass that surrounds the foregut. From this mass nerves radiate to the eyes, mouthparts, legs, and interior organs. Motor fibers connect with the muscles, and sensory fibers lead from the often elaborate external sensory setae to the central system. From one to five simple eyes may be present on the upper anterior portion of the body, but frequently these eyes are lacking. They probably only perceive changes in light intensity and not clear images.

The respiratory system is quite variable. Tracheal tubes with branching tracheoles may distribute gases, but often such tubes are not present. External breathing openings, the **stigmata**, are characteristic in form and location and are important structures for identification. The tracheae ramify from stigmata, or there may be a number of sacs leading from the stigmata into the tissues.

Hormonal mechanisms in ticks and mites are virtually unknown, though an analogue of the juvenile hormone of insects has been shown capable of breaking the diapause form of a tick. Because molting occurs, it is likely that this event, as with insects, is hormonally controlled. The diapause forms of some mites are known to be controlled by the length of the day, as is the case with certain insects.

SELECTED REFERENCES

Baker, E. W., and G. W. Wharton, *An Introduction to Acarology* (New York: Macmillan, 1952).
Beroza, M., ed., *Chemicals Controlling Insect Behavior* (New York: Academic Press, 1970).
Chapman, R. F., *The Insects—Structure and Function* (New York: American Elsevier, 1969).
Hughes, T. E., *Mites or the Acari* (London: Athlone Press, 1959).
Pringle, J. W., *Insect Flight* (Cambridge: Univ. Press, 1957).
Rockstein, M., ed., *The Physiology of Insecta*, Vols. I–VI (New York: Academic Press, 1973–74).
Roeder, K. D., *Insect Physiology* (New York: Wiley, 1953).
Shorey, H. H., *Animal Communication by Pheromones* (New York: Academic Press, 1976).
Snodgrass, R. E., *Principles of Insect Morphology* (New York: McGraw-Hill, 1935).
Whittaker, R. H., and P. P. Feeny, "Allelochemics: Chemical Interactions between Species," *Science* 171: 757–70 (1971).
Wigglesworth, V. B., *The Physiology of Insect Metamorphosis* (Cambridge: Univ. Press, 1954).
———, *The Principles of Insect Physiology* (New York: Dutton, 1965).
———, *The Life of Insects* (New York: New American Library, 1968).

chapter 3 / **ROBERT F. HARWOOD**

INSECT GROWTH

The growth of insects, both as individual organisms and as exploding populations, is a subject of interest and much speculation. The uninitiated observer is amazed by a sudden appearance of insects where previously they seemed lacking or unimportant. It is small wonder that insects were used as evidence for man's erstwhile belief in spontaneous generation, the development of living things from inanimate substance. Complexities in the development of insects fascinate even the professional entomologist. Differences in the body form and behavior of various stages of an insect may make it difficult for one to realize without detailed study that a single organism is being observed.

We shall discuss the growth of the individual insect as an orderly process that proceeds in a set pattern. Growth is an increase in size, but in insects this increase occurs in stepwise fashion, which may be accompanied by distinct changes in form.

THE EGG

Insects develop from eggs, which are usually deposited by the female in sites favoring survival. In assessing insect populations the potential present in their eggs is often overlooked. For the most part they are small, generally one-half millimeter or less in their greatest dimension. Furthermore, egg sites are frequently hidden from view. For example, many leafhoppers, longhorn grasshoppers, and weevils deposit eggs inside plant tissues, where they can remain moist and out of sight. Locusts, crickets, and a number of beetles deposit eggs beneath the surface of soil. Parasitic insects, particularly parasitic wasps, may lay eggs inside their host.

Eggs that are laid in water are seldom noted. Some caddisflies and damselflies oviposit in plant stems beneath the water surface. Dragonflies and mayflies may drop their eggs on the water while flying. Gelatinous egg masses are deposited in water by several midges and caddisflies.

Some insects place their eggs in conspicuous locations. Prominent egg masses of mantids glued to a twig are particularly noticeable. Lacewings attach to foliage single, small light green eggs on a slender stalk. Many true bugs lay masses of eggs with bizarre shapes and gaudy colors.

The external **appearance of insect eggs** is extremely diverse (Fig. 3:1). They range in shape from spherical to elongate; outgrowths for their attachment to objects or for respiration may be present, as well as ridges and caplike structures. They may be held in a case called an **ootheca** (cockroaches), or in a so-called **egg pod** (grasshoppers). Usually a fine reticulation of the surface of the chorion or shell can be seen at high magnification. This surface structure is the imprint of cells lining each egg chamber in the ovary, which were left there as these cells secreted the chorion.

The **number of eggs** produced by females of different species is extremely variable and is related to the chances for survival to the adult stage or—in the case of social insects—to the production of large colonies. Among parasitic forms that spend their entire life on a host, there is little danger of loss during development. Accordingly the sheep ked produces a single offspring at a time, and the total number in a lifetime is small. Parasitic flies that deposit their

Figure 3:1. Shapes of insect eggs. While insect eggs are often relatively simple in form, the egg of the water scorpion *Ranatra* (A) is located in plant tissue with respiratory filaments in the surrounding water; B, lacewings lay eggs and attach each one to surfaces by a long stalk; C, eggs of stink bugs and other Hemiptera are frequently bizarre in appearance and laid in groups; D, the cockroach lays its eggs in a packet or ootheca, each surface indentation indicating the position of an egg. *A, after Hungerford; B, C, and D, original.*

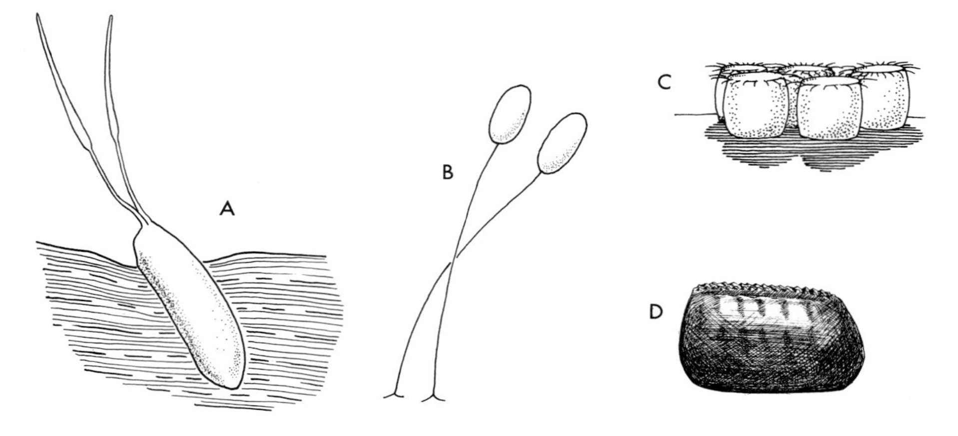

eggs on vegetation or other sites where there is only a slight chance of contact with a host may scatter over 1000 eggs. The female of sexual forms in aphids may lay only a single egg directly onto the host plant. Among the social insects the queen honey bee can lay more than 2000 eggs a day during periods of peak production, and over one million eggs during a five-year period of productivity. Perhaps 100 to 200 is the usual number of eggs produced by most insects during a life time. Some parasitic Hymenoptera, notably Braconidae and Chalcididae, solve the problem of attrition by a number of embryos developing from each egg; this phenomenon is called **polyembryony**. In these insects early cell divisions within the egg result in as many as a hundred embryos, which grow into a like number of individuals (Fig. 3:2).

When fully developed within the ovary (Fig. 2:22), the egg is ready for fertilization. The shell or **chorion** at this time is completed and consists mainly of tanned proteins as in cuticle. Further subdivisions of chorion into layers of varying chemical composition often occur. The chorion is relatively impermeable and therefore has an opening or openings for the entrance of sperm, called the **micropyle** if single, or **micropylar pores** if multiple. The yolk granules may be surrounded by a very thin **vitelline membrane**, lying beneath and next to the chorion. A limiting vitelline membrane cannot be demonstrated in some insect eggs. Yolk granules and fat globules fill the egg in a matrix of protoplasm. The granular material is usually lacking in the periphery; instead there is a poorly defined clear outer area called the **periplasm**. Prior to fertilization the nucleus of the egg may be situated in almost any position.

Among insects the act of mating usually takes place quite a while before fertilization of the eggs; spermatozoa are stored inside the female, generally entering the egg only in association with the act of egg laying (**oviposition**). Longevity of sperm within the female may be remarkable; evidence suggests a period as long as five years in queen honey bees. In such cases spermatozoa

Figure 3:2. Polyembryony. Section of an egg of *Platygaster vernalis* Myers, a wasp that parasitizes the Hessian fly. Four embryos in an early stage of development are visible. *After USDA.*

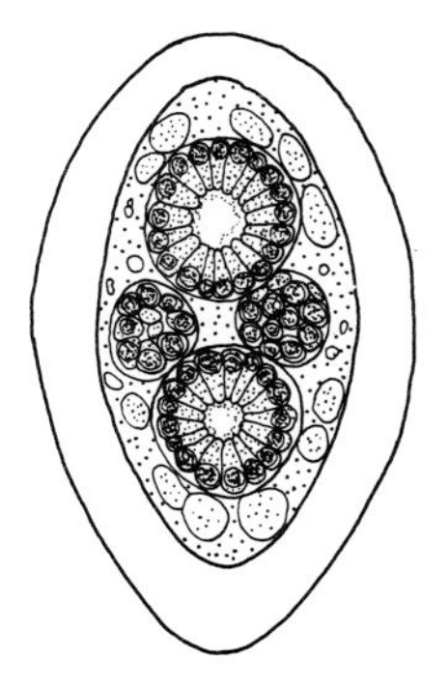

are stored in an inactive state in the spermatheca of the female and are activated as they pass through the spermathecal duct to fertilize a passing egg.

The female may be able to selectively prevent entrance of spermatozoa into the egg. Drone bees are produced from eggs that develop without the stimulus of a spermatozoan. Thus the drone has the genetic constitution of its mother, the queen; this character has been used in artificial insemination by selecting spermatozoa from drones produced by queens with particularly desirable characteristics.

FERTILIZATION OF THE EGG

Normal **bisexual reproduction** occurs in most insects, mites, and ticks, but there are a number of exceptions. Insects are known for which no males exist, and in a large number of species, although males are usually required, the females can produce offspring in their absence. This development of young without benefit of males is termed **parthenogenesis**. Several species of cockroaches can produce offspring without males, and some weevils (notably whitefringed beetles) have been found to have females only. A number of species of mites have females only, and parthenogenetic strains of ticks are known. An intermediate condition exists, such as in aphids, where summer offspring are produced in the absence of males, but normal sexual reproduction occurs in the fall. Whether we refer to the production of young with or without males, the process is sexual reproduction. The egg is an element essential to sexual reproduction, and it matters little whether a spermatozoan is required to initiate development or whether reproduction can be induced by other means.

Fertilization usually takes place as the egg traverses the common oviduct past the point where the spermatheca enters it. Following entrance of spermatozoa, the micropylar openings seal off to prevent the loss of water from this region after the egg leaves the body of the female. The **female pronucleus** usually develops after the penetration of spermatozoa. After that event the germinal vesicle migrates to the yolk periphery and undergoes reductional divisions to become the haploid female pronucleus. The female pronucleus and a haploid spermatozoan (**male pronucleus**) fuse, and the diploid **zygote nucleus** moves into the yolk mass, where it begins to divide. Several extra spermatozoa may enter the egg, but they disintegrate.

EARLY ORGANIZATION AND TOPOGRAPHY

Before fertilization, and with the first cell divisions, the insect egg often shows evidence of **presumptive organization** into the regions of the developing individual. Eggs in the oviduct are lined up in the same relative position as the mother; at this time there is a presumptive head end, right and left sides, and top and bottom. At the first few divisions following fertilization the resultant cells may be committed toward development of structures or body regions of the forming insect. Investigators have shown that limited cellular destruction during early development by such means as ultraviolet light or a needle prick will cause specific defects in the insect that emerges from the egg. In some cases early injury causes defects in the larva and not the adult, whereas later injury affects adult structures and not the larva. Thus early larval and adult organization appear to be distinctly separate.

Following fertilization, the first cell divisions typically take place near the center of the egg. Because there is a relatively large amount of yolk present, the cells divide by **meroblastic cleavage**, where the whole egg does not divide but rather divisions are limited to a local region. Meroblastic cleavage is common in animal eggs possessing much yolk, the usual situation in insects. The dividing cells migrate to the periphery to form a single layer of cells, the **blastoderm**. A region on the ventral surface consists of cells that are thicker and makes up the **ventral plate** or main area of embryonic development (Fig. 3:3A,B,C).

A second layer of cells forms beneath the surface of the cells that make up the ventral plate, a process called **gastrulation** (stomach formation). This inner layer of cells develops by (1) the infolding and separation of a trough on the midventral line (Fig. 3:3D,E), (2) by a lateral overgrowth and fusion of cells at the lateral margins of the ventral plate, or (3) by simple division of cells in the midventral area in a manner leaving an inner sheath of cells. The two layers of embryonic tissue formed are the matrix from which organs and organ systems are derived. The outer layer (**ectoderm**) will form the outer covering and derivatives in the embryo. The inner layer (**mesoderm**) will form many of the inner structures of the developing individual. (See p. 87 for discussion of endoderm.)

A subdivision into distinct body parts (**segmentation**) is apparent soon after the development of ectoderm and mesoderm. This separation is indicated by thickening of the mesoderm tissue in the middle of each segment and by its remaining thin at the margin between segments. More distinct evidence of segmentation soon follows, with growth of the head region and the appearance of appendages (Fig. 3:4). The front portion of the head forms two enlarged lobes, and just behind these lobes appendages delimiting five more segments of the head appear, making a total of six segments which compose the head. Behind the head three slightly wider segments with larger appendages form the thorax. A maximum of 12 narrow succeeding segments indicate the extent of the abdomen. Appendages may form on the abdominal segments, remaining on the individual after hatching or disappearing during further embryonic growth.

The developing embryonic area becomes covered by the lateral blastoderm layer, which forms two extraembryonic membranes. In the simplest mode of formation this occurs by a direct growth of blastoderm over the embryo. The outer membrane closest to the chorion is called the **serosa**. Within the serosa the membrane directly over the embryo is called the **amnion**.

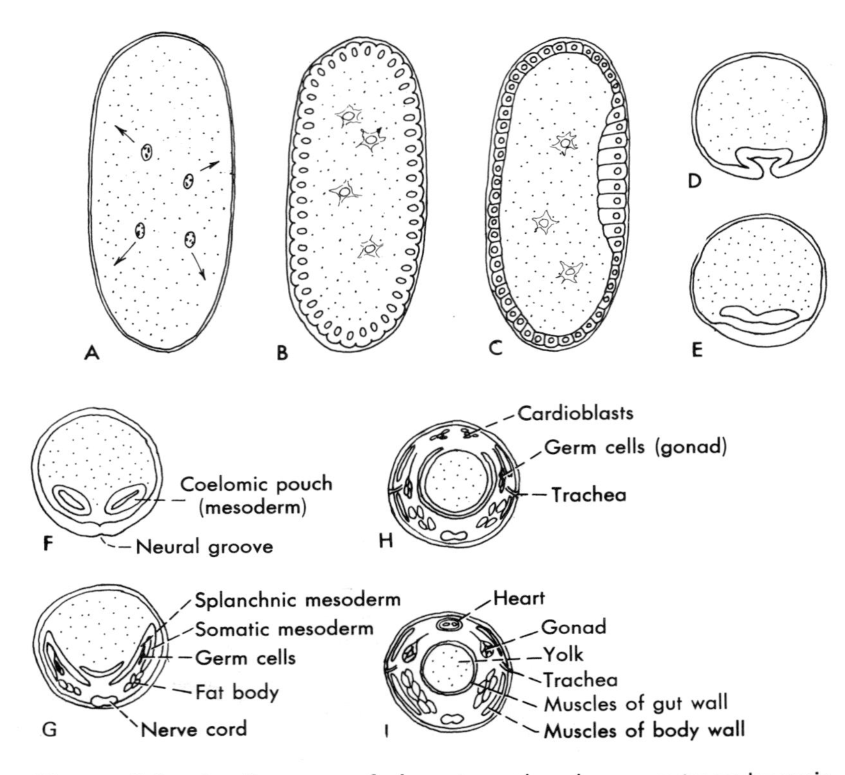

Figure 3:3. A diagram of insect embryology, extraembryonic membranes left out. A, migration of the first cleavage nuclei to the surface; B, blastoderm with vitellophag cells remaining in the yolk; C, formation of the ventral plate; D, E, and F, formation of mesoderm; G, H, and I, development of internal organs. *Redrawn from Hagan, 1951,* Embryology of the Viviparous Insects, *by permission The Ronald Press, Company, New York.*

In most insects a migration of the embryo into the yolk forms the amnion and serosa (Fig. 3:5). During later development the embryo returns to the surface. The process of involution of the embryo and its subsequent return to the surface is called **blastokinesis**. The function of blastokinesis is uncertain; shifts in position of this magnitude hardly seem necessary to develop the extraembryonic membranes. In some insects, notably the grasshopper, embryonic growth ceases for the duration of harsh seasonal conditions such as cold or dry spells, with the embryo remaining inside protected by both membranes. When favorable climatic conditions resume, growth recommences, and the embryo returns to the surface. In this case it appears that blastokinesis is a mechanism for protecting the embryo.

The extraembryonic membranes undergo a variable fate in different insects. In the majority of cases, as the lateral margins of the embryo expand

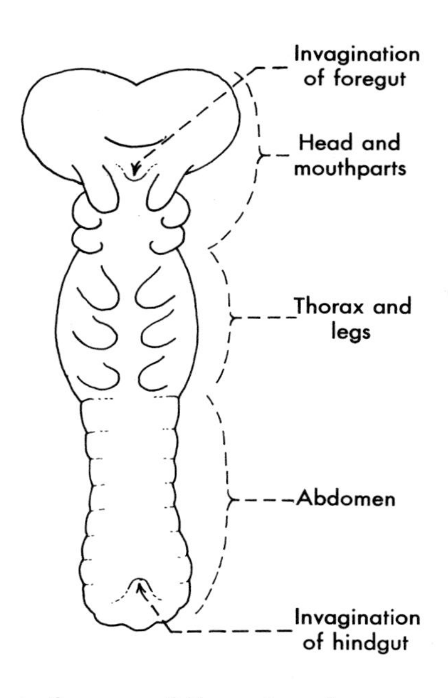

Figure 3:4. General diagram of the external appearance of an insect embryo when body segmentation is just completed.

after blastokinesis, the membranes are pushed toward the dorsal surface and are finally engulfed within the body where they disintegrate. In some instances the serosa remains next to the chorion as a fine layer similar to insect cuticle and serves as a potent barrier preventing loss of water. This serosal membrane may adhere to the chorion and be readily mistaken for a vitelline membrane, as seems to be the case in mosquito eggs. In grasshoppers the serosa is retained as a loose bag around the nymph at time of hatching. The new grasshopper nymph wriggles in this bag to the soil surface, and there sheds the bag in what amounts to a second hatching (Fig. 3:9D,E).

Figure 3:5. Diagrammatic representation of the extraembryonic membranes of an insect, formed by involution of the embryo into the yolk.

THE DEVELOPMENT OF ORGANS AND ORGAN SYSTEMS

Formation of the ectoderm and mesoderm sets the stage for **differentiation** of tissues into the structures that develop into the particular form of the insect which emerges from the egg.

Ectodermal Derivatives

The central nervous system is derived from ectoderm. Shortly after the mesodermal layer has differentiated, two **neural ridges** form parallel to one another, making a central trough, the **neural groove** (Fig. 3:3F). This groove sinks below the surface and forms a tube of tissue (the nerve cord) that separates from the surface ectoderm, which then closes back over it. The **nerve cord** remains in this relative position (the ventral surface of the developing animal) further forming its various subdivisions as the body regions become distinct. This cord eventually forms the main ganglionic complex that comprises the **brain**, a ganglionic mass that becomes the **subesophageal ganglion**, three **thoracic ganglia**, and eight **abdominal ganglia**. The nerve cord does not remain a single tube but separates into two **connectives** that span the space between ganglia along its whole length.

As we have seen, the nerve cord and ganglia are derived from a groove in the midventral region of the ectoderm. In further development **efferent nerves** extend from the segmental ganglia to the muscles and glands within their respective segments. The **afferent nerves** that receive stimuli of touch, taste, smell, light, and sound for transmission to the central nervous system also develop from ectoderm, but as specialized cells within the integument. At appropriate areas special nerve cells, such as those associated with a sensory hair or with a receptor organ such as the eye, grow inwardly to form a nerve connecting with the nearest ganglion of the central nervous system.

The **tracheal tubes** of the respiratory system also develop as invaginations of the ectoderm. These invaginations take place along the lateral areas of the animal in a segmental region. From these initial invaginations longitudinal tracheal trunks interconnect, and further ramifications of the tracheae permeate the tissues to form a complete respiratory system by the time of hatching.

Ectoderm also forms the anterior and posterior ends of the tube that forms the digestive tract. As the head and posterior end of an embryo become distinct, an invagination near each extremity develops into the **foregut** and **hindgut** (Fig. 3:4). The ectodermal derivation of these regions of the gut is apparent later by its structural resemblance to the integument, as it possesses an epithelial layer which forms an inner cuticle-like lining. Although there is some question about the formation of **Malpighian tubes**, it is generally agreed that they are formed from the anterior end of the hindgut.

Other relatively minor structures are derived from ectoderm. The terminal portion of ducts connecting the reproductive system with the outer surface of the animal are formed by invaginations of this tissue. Much glandular tissue, such as the **salivary glands**, is also ectodermal in origin.

Mesodermal Derivatives

The mesoderm develops into many important internal structures of insects. After separating from the ectoderm, this layer divides into right and left halves. Each block of cells hollows out and extends laterally, forming an inner and outer sheet of tissue on each side. The outer layer is the **somatic mesoderm**, which forms the **skeletal muscles** of the body wall and appendages. The inner layer is the **splanchnic mesoderm**, which forms the **visceral muscles** associated with the digestive tract and reproductive system (Fig. 3:3F–I).

The **circulatory system** is derived from somatic mesoderm. Advancing upper margins of the somatic layer fuse to form the tubular heart and aorta in the middorsal line (Fig. 3:3H,I). Blood cells are derived initially from cells remaining in the hollow formed between the inner and outer mesodermal layers. Further blood cells are produced by division of these primordial cells. Cells of the **fat body** are first noted in the same region as the primordial blood cells.

Most of the reproductive system develops in association with the somatic mesodermal layer. Although the primordial germ cells may have been formed in the first few cellular divisions, the gonads (ovaries and testes) that develop from these cells come to lie in the abdomen against the upper and inner surface of the somatic mesoderm (Fig. 3:3G–I). **Germ cells**, namely spermatozoa and ova, develop from these primordia, as do nurse cells in the ovary if these are present. The enveloping tissue which forms **sperm tubes** and **ovarioles**, and the first portion of the ducts leading from them, is derived from mesoderm.

Endodermal Development

Vertebrates have a well-defined inner layer of embryonic tissue, the endoderm. There is some question whether such a tissue is present in insects, but if it is present, it forms the **midgut**. At the very first divisions after fertilization a few cells scatter within the yolk. These cells aid in consuming the nutritive reserves in yolk and accordingly are called **vittelophags** (yolk eaters). In some insects, groups of cells among the vitellophags are thought to migrate at the appropriate time to connect with the advancing midgut and hindgut to form the central tubular section of midgut. In other cases cells forming the midgut are first noted to be present at the advancing ends of the foregut and hindgut and presumably have come to lie in this position after early derivation from mesodermal cells. For this reason the mesoderm tissue at time of formation can also be called the **mesentoderm** layer. At any rate the midgut appears to be of different origin from portions of gut in front and in back of it. The fully developed midgut is structurally different from digestive tract at either end because it lacks a well-defined, cuticlelike lining. The midgut forms in the embryo at about the time of completion of the lateral margins of body wall, with the remaining yolk contained in this portion of the digestive tract at the time of hatching.

HATCHING

When embryonic development is finished, various hatching mechanisms aid a new individual to emerge from the egg. For example, in Hemiptera there may be a point of structural weakness in the chorion, forming a cap or **operculum** that breaks off at a distinct line of fracture (Fig. 3:6C). In some insects, notably grasshoppers, a pair of glandular structures called **pleuropodia**, which resemble appendages of the first abdominal segment in the mature embryo, secrete enzymes that partially destroy the surrounding membranes (Fig. 3:6B). Most mature embryos increase their internal pressure by swallowing surrounding fluids and air. Such internal pressures swell the head or evert special **vesicles** which in turn fracture the chorion. In many insects a **hatching spine** is present on the head for cutting the chorion or for applying pressure at one point (Fig. 3:6A). In Lepidoptera the new larva simply chews its way out of the shell (Fig. 3:6D).

Figure 3:6. Mechanisms of hatching. A, the mosquito larva has a spined egg burster for applying pressure to the egg shell at the site of fracture; B, in grasshoppers a secretion from the glandular pleuropodia digests the inner portion of the shell just before hatching. If the egg is tied off in the middle, the portion lacking pleuropodia is not digested; C, many insect eggs have a cap portion which is popped off during hatching. An emerging nymph of the bed bug, *Cimex*, swallows air to expand the body; D, many insects, such as the larva of the European cabbage butterfly, simply chew through the egg shell. *B, after Slifer; C, after Sikes and Wigglesworth; D, after Essig, 1947,* College Entomology *by permission Marie W. Essig.*

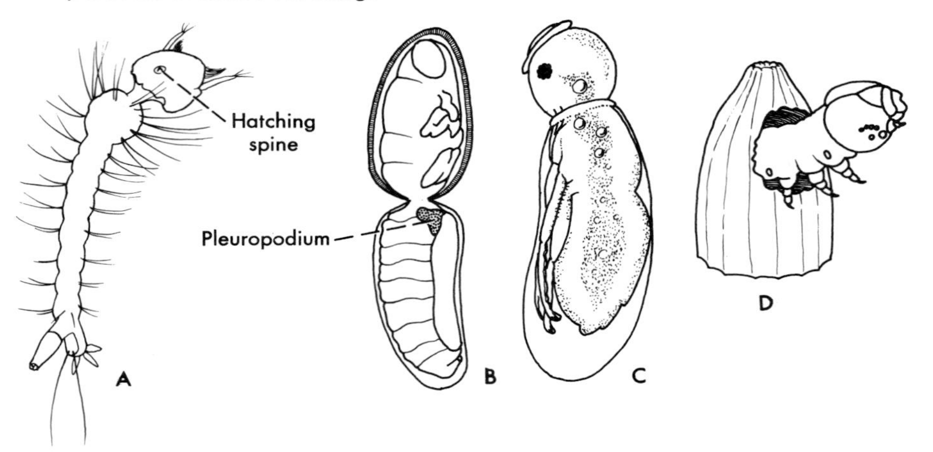

There is a wide variation in the external structures of insects at the time of hatching. Some are legless, whereas most possess well-developed organs of locomotion. Some are quite inactive, although most seek food actively. Such major differences in development and activity suggested to early investigators that insects hatching with poorly developed locomotory structures had really emerged prematurely, and subsequent growth consisted of a sort of delayed embryonic development. Active and well-developed insects were thought to have completed embryonic growth by hatching time. It was difficult, of course, to explain why most insects laid eggs, yet many flies deposited larvae, aphids bore nymphs, and the sheep ked laid larvae which formed puparial cases almost immediately. We now realize that variations in postembryonic growth are involved, because generally even legless larvae have undeveloped rudiments of larval or adult structures present within their body.

METAMORPHOSIS

Upon hatching, insects grow by a series of molts, shedding the old cuticle and expanding into a new and larger one. When the adult stage is reached, molting in most insects ceases. At each molt externally visible changes occur; this type of growth is called **metamorphosis**, a change in form or transformation. There is considerable variation in the amount of bodily change at molting, and the degree of this change is used to classify insects into three generally accepted types: no metamorphosis, simple metamorphosis, and complex metamorphosis. From a physiological standpoint, because similar hormonal mechanisms control postembryonic growth, there is a tendency to refer to all development that follows hatching as larval growth.

The role of hormones in controlling the growth of insects is being clarified rapidly. Simple but ingenious experiments have determined the endocrine structures involved and the role they play. More sophisticated chemical approaches have revealed the nature of hormones produced, particularly in those hormonal events controlling growth and metamorphosis.

The endocrine structures that operate during growth and metamorphosis are associated with the head and thorax (Figs. 2:26 and 3:7). A specific event

Figure 3:7. Corpora allata and prothoracic glands in the larva of a honey bee. *After Formigoni.*

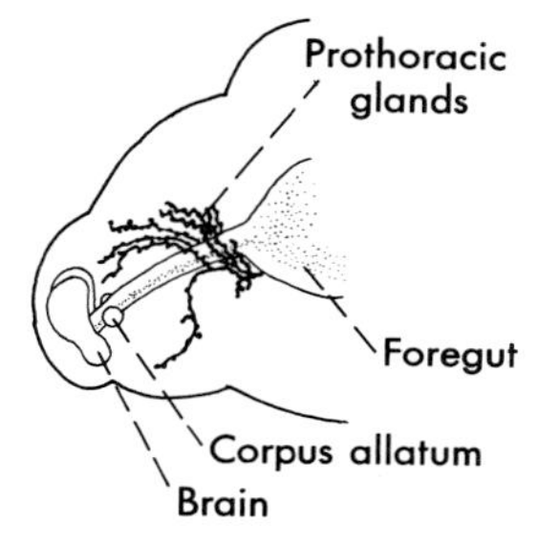

may be required to trigger growth, and this event often occurs normally in the life of an insect. For example, the event triggering nymphal growth in the blood-sucking bug *Rhodnius* is the distension of the gut after feeding; the hibernating pupa of a species of moth requires cold temperatures to restart development. These triggering stimuli are relayed to the brain where nerve cells, called **neurosecretory cells**, begin secretion of a substance that can be stained or otherwise demonstrated. Prior to molting the neurosecretory substance builds up in a storage organ in back of and beneath the brain, the **corpus cardiacum**, where it is released into the blood to affect other endocrine structures further.

Molting and growth are controlled by hormones released from the **corpora allata** near the corpus cardiacum in the head and the **prothoracic glands** in the region of the thorax. The corpora allata release the **juvenile hormone**, which juvenilizes (maintains the immature form of) an insect during early molts. The corpora allata stop secretion at the end of immature growth, resulting in an adult insect in the next stage. Miniature adults can be developed by removal of the corpora allata in early growth stages. Strangely enough, the juvenile hormone produced by the corpora allata is present again in relatively large amounts in insects after the adult condition has been achieved, and it has been shown that this hormone is necessary for egg maturation. For example the treatment of diapausing alfalfa weevils with juvenile hormone or close chemical relatives will cause resumption of egg production.

Hormones causing molting (**ecdysones**) are produced by the prothoracic glands. Differentiation into the adult is blocked in larvae and nymphs by the presence of juvenile hormone from the corpora allata; further molting during the adult stage does not occur because the prothoracic glands disappear. An exception occurs in Thysanura (silverfish) and Collembola (springtails), which retain the prothoracic glands and molt as adults.

Ecdysones and juvenile hormone have been isolated and characterized chemically; the latter hormone has been synthesized, and various chemical analogs are being tested for insect control. Promising approaches lie in treatments that force insects to remain in immature form, thus to die, or to cause the resumption of growth at a time the insect would be destroyed by unfavorable weather conditions.

In describing the events of metamorphosis several special terms, such as *stadium*, *instar*, *imago*, and *stage*, are often used. The **stadium** is the time interval between the molts of insects, and **instar** is the form assumed by an insect during a particular stadium. When an insect emerges from the egg it is said to be in its first larval or nymphal instar; at the end of this stadium the insect molts and enters the second larval or nymphal instar, and so on. The final instar is the adult insect or **imago**. **Stage** is a distinct, sharply different period of an insect's life, for example, the egg stage, larval stage, pupal stage, and adult stage.

In what is considered the most primitive type of postembryonic growth, the structural changes are almost imperceptible (Fig. 3:8). Insects exhibiting this type of development are said to undergo **no metamorphosis**, and they are referred to as **Ametabola**. Molting may continue throughout their adult life. Collembola (springtails) and Thysanura (silverfish) are the insects most commonly encountered in this group. Immature Ametabola are called nymphs.

Insects that undergo **simple metamorphosis** belong to the **Paurometabola**. In this type of growth there is a relatively gradual change in external appearance in the molting steps from egg to adult (Figs. 3:9 and 3:10). The immature forms are called nymphs and have feeding habits usually similar to those of the adult. In the final molt functional wings are developed in those insects that possess wings. Wing pads can be seen in nymphs prior to development of the functional wings of adults. Grasshoppers, termites, stoneflies, chewing and sucking lice, and true bugs are common examples of insects with simple metamorphosis. A greater deviation of metamorphosis occurs among some of this group whose immature stages are aquatic. These are sometimes called **Hemimetabola**, and their nymphs can be referred to as

Figure 3:8. No metamorphosis. Development of Ametabola, as represented by a thysanuran. Group of eggs, A; the nymphs, B to G, quite closely resemble the adult, H. There is scarcely any major difference between the external appearance of the oldest nymph and the adult. *From Metcalf, Flint, and Metcalf, 1962,* Destructive and Useful Insects, *by permission of McGraw-Hill, Inc., New York.*

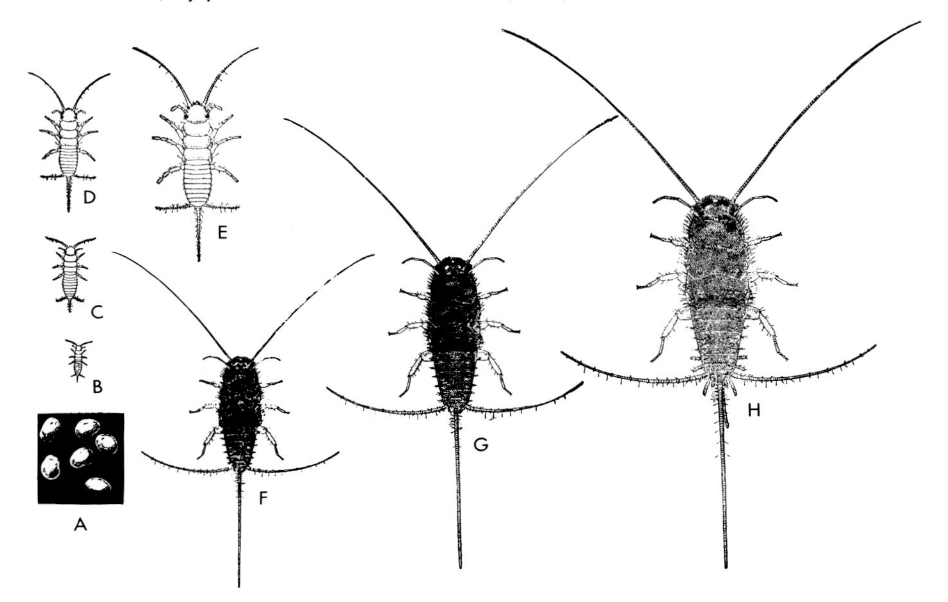

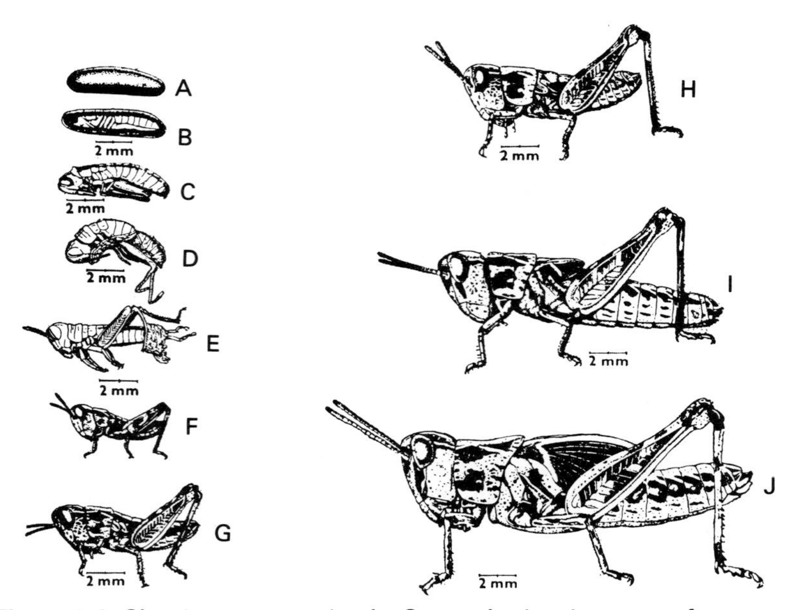

Figure 3:9. Simple metamorphosis. Stages in development of a grasshopper from egg to last nymphal instar. A, newly laid egg; B, embryo within egg; C, hatched nymph inclosed in serosa; D and E, nymph shedding serosa; F to J, first to fifth nymphal instars. *Courtesy Commonwealth Scientific and Industrial Research Organization, Australia, original drawing by Mr. D. C. Swan.*

naiads. The aquatic insects in this category are the mayflies, stoneflies, dragonflies, and damselflies.

The greatest external and internal structural changes take place among insects with **complex metamorphosis**, the **Holometabola** (Fig. 3:11). These are regarded as the most advanced insects, or the most recent in an evolutionary sense. The postembryonic form is a **larva**, which feeds actively and may have locomotory appendages developed to varying degrees. Fly larvae (maggots), for example, are completely legless, whereas many beetle larvae have well-developed legs. After several molts the inactive **pupa** is formed.

Figure 3:10. Adult grasshopper with fully developed, functional wings. *Courtesy University of Arizona, original drawing by Dr. E. R. Tinkham.*

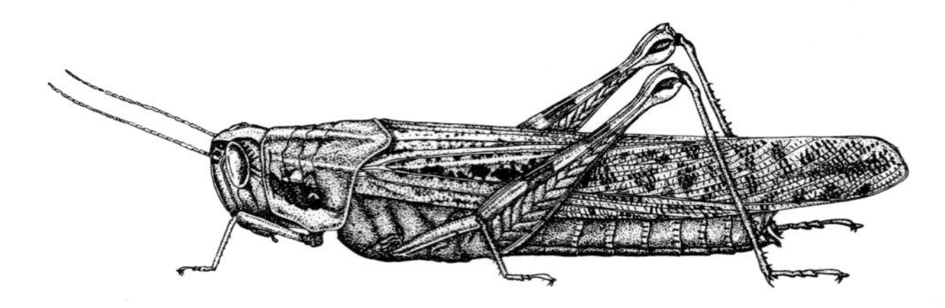

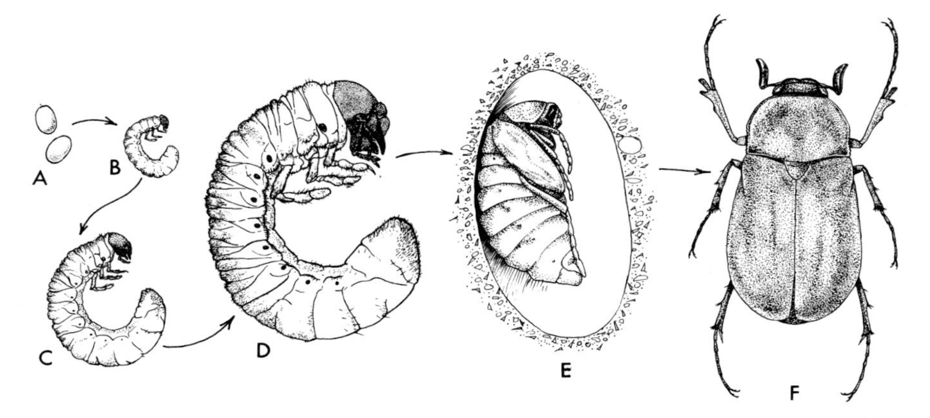

Figure 3:11. Complex metamorphosis, as found in a May beetle. From eggs (A), three larval instars develop (B, C, and D), pupation occurs in a cell in the soil (E), and finally the adult beetle (F) emerges.

The pupa does not feed and is incapable of locomotion (mosquito pupae are a notable exception), but major structural reorganizations are taking place internally. At the termination of the pupal stage a highly active adult results. These great variations in structure during the growth of holometabolous insects are often accompanied by different habits. Thus larvae and adults may eat different foods and occupy radically different environments. Beetles, flies, fleas, moths and butterflies, and bees and wasps are common representatives of the Holometabola.

There are several instances where insect transformation does not fit neatly into the ordinary classifications of metamorphosis. Thrips belong to the Paurometabola, yet the last two nymphal instars do not feed and may be inactive (Fig. 3:12), a situation resembling the pupal stage in Holometabola.

Figure 3:12. Development in a thrips, simple metamorphosis with non-feeding stages resembling pupae. A and B, first and second nymphal instars; C. propupa; D, pupa; E, adult female. *From Snodgrass, 1954, Smithsonian Misc. Coll. 122, No. 9.*

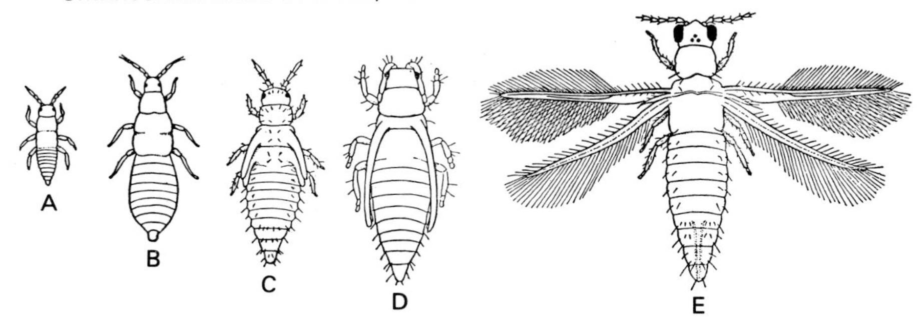

Some insects, among which the Meloidae or blister beetles provide an example, undergo what is called **hypermetamorphosis**. In this case growth is by complex metamorphosis, but there are distinct changes in external form and habits at each successive larval molt.

The pupae of insects, being inactive and defenseless, frequently have some means of protection and concealment (Fig. 3:13). Such protection is sought by the last larval instar. In fact this last larval form does not feed just prior to pupation, often has a distinct appearance, and is known as the **prepupa**. In many insects, particularly Neuroptera, caddisflies, fleas, many Lepidoptera, and Hymenoptera, a **cocoon** of silk is woven by the larva. Pupation takes place within the cocoon. Insects which pupate in soil often do so in an **earthen cell**. Trichoptera larvae build cases by gluing together pebbles, sticks, or other fragments in which they live and subsequently pupate. Some moths cover the silken cocoon with body hairs.

LARVAL FORMS

The larval form is highly variable in insects, but there are sufficient similarities in overall structure to permit a classification of type that is useful in describing their appearance. (1) **Campodeiform** (Fig. 3:14A) larvae have a flattened body with long legs, usually with filaments on the end of the abdomen. Larvae of diving beetles in the family Dytiscidae and many other beetles, Neuroptera, and Trichoptera are typical examples. (2) **Carabiform** larvae (Fig. 3:14B) are similar to the campodeiform type, but the legs are shorter and filaments are lacking on the end of the body. They receive their name from the larvae of carabid beetles. Chrysomelid beetle larvae are of the same type. (3) **Eruciform** larvae (Fig. 3:14C) are cylindrical; they have a

Figure 3:13. Various types of insect cocoons. The cocoons of many moths are encased in leaves of the host plant (A), debris sticks to the cocoon of a flea (B), the alfalfa weevil spins its lacy cocoon among leaves of the host plant (C), and many caddisflies simply seal off the case of cemented particles in which the larva developed (D). *C, courtesy USDA; A, B, and D, original.*

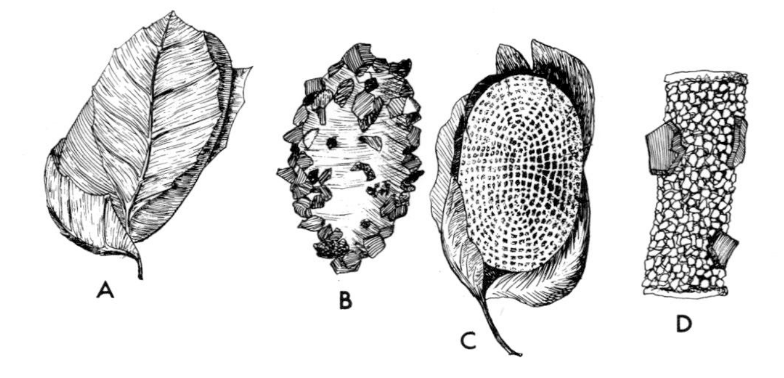

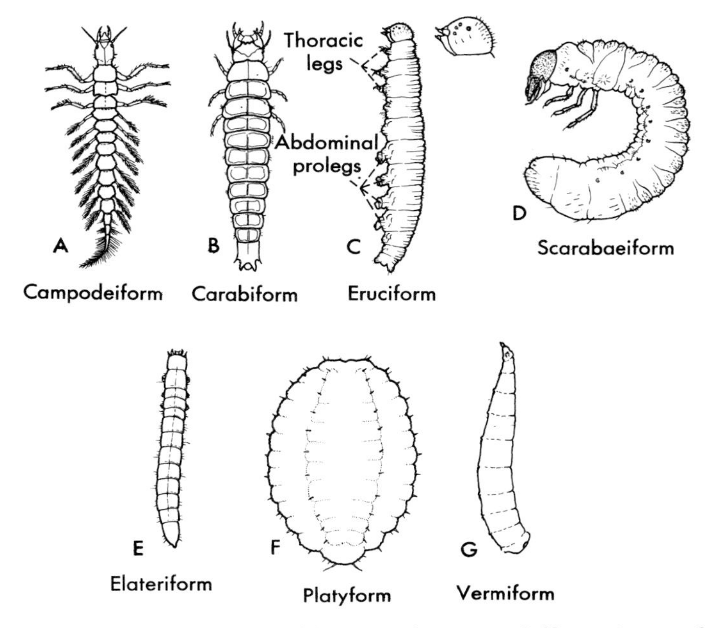

Figure 3:14. The larval forms of insects. A, campodeiform, larva of an aquatic neuropteran, *Sialis*; B, carabiform larva of a carabid or ground beetle; C, eruciform caterpillar, the larva of a butterfly—enlargement of head shows lateral simple eyes or stemmata; D, scarabaeiform, white grub, the larva of a June beetle; E, elateriform, wireworm, larva of a click beetle; F, platyform, flattened larva of an uncommon moth; G, vermiform, a fly maggot. *Redrawn from Peterson,* Larvae of Insects, *1951, by permission Alvah Peterson.*

well-formed head, thoracic legs, and abdominal prolegs. Larvae of Lepidoptera and sawflies are typical examples. (4) **Scarabaeiform** larvae (Fig. 3:14D) are C-shaped, have a well-developed head, and usually possess thoracic legs but lack prolegs. The type is named after larvae of scarabaeid beetles (dung beetles, May beetles) and is also represented by the larvae of weevils and furniture beetles. (5) **Elateriform** larvae (Fig. 3:14E) are cylindrical, smooth, and relatively tough-skinned larvae with short legs. They are named after larvae of elaterid or click beetles (wireworms). Larvae in the beetle family Tenebrionidae also have this appearance. (6) **Platyform** larvae (Fig. 3:14F) are broad and flat with legs short or absent. This type is not common, but examples are found among larvae of some syrphid flies, certain caterpillars, and blister beetles. (7) **Vermiform** or wormlike larvae (Fig. 3:14G) are cylindrical and elongate, without appendages of locomotion. This form is represented by the larvae of higher Diptera (maggots), some woodboring beetles, fleas, and higher Hymenoptera, including the honey bee.

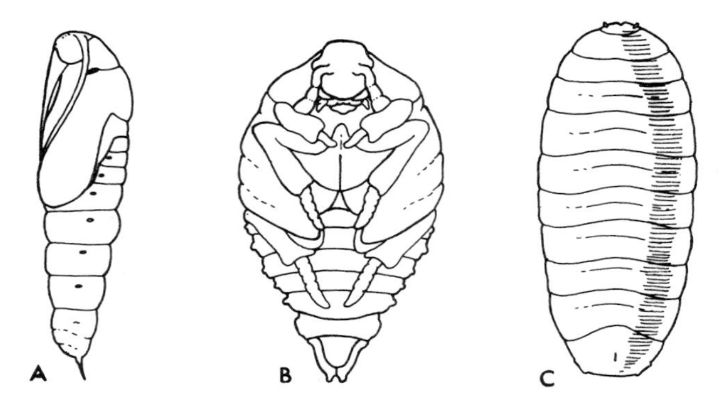

Figure 3:15. Pupal form of insects. A, the obtect pupa of a moth; B, the exarate pupa of a scarabaeid beetle; C, the coarctate pupa of a fly. *Redrawn from Peterson, 1951,* Larvae of Insects, *by permission Alvah Peterson.*

PUPAL FORMS

Pupal forms can be classified by means of the appearance of developing appendages. (1) In **obtect** pupae the appendages are close to the body, held there by a tight-fitting outer envelope (Fig. 3:15A). Pupae of this type are found in Lepidoptera and many Coleoptera, as well as the less advanced Diptera. The obtect pupa of butterflies is often angular, may be metallic colored, and is called a **chrysalis**. (2) When the appendages develop free of the body the pupa is called **exarate** (Fig. 3:15B). This type of pupa is characteristic of Neuroptera, Trichoptera, and most Coleoptera. (3) If the visible pupal case is smooth with no appendages apparent, it is referred to as **coarctate** (Fig. 3:15C). This condition is typical of the more advanced or cyclorrhaphous Diptera. The pupal case in reality is the hardened and expanded skin of the third larval instar; within this shell the fourth larval instar molts to form an exarate pupa.

SELECTED REFERENCES

Elzinga, R. J., *Fundamentals of Entomology* (Englewood Cliffs, N.J.: Prentice-Hall, 1978).
Hagan, H. R., *Embryology of the Viviparous Insects* (New York: Ronald, 1951).
Horn, D. J., *Biology of Insects* (Philadelphia: Saunders, 1976).
Johannsen, O. A., and F. H. Butt, *Embryology of Insects and Myriapods* (New York: McGraw-Hill, 1941).
Romoser, W. S., *The Science of Entomology* (New York: Macmillan, 1973).

chapter 4 / **CARL JOHANSEN**

CLASSIFICATION OF INSECTS AND THEIR RELATIVES

The animal kingdom is divided into a number of large groups known as **phyla**. The following is a list of the ten most important phyla of animals, including a few examples of each:

Invertebrates

1. Protozoa—single-celled animals, as amoeba
2. Porifera—sponges
3. Coelenterata—jellyfishes, corals
4. Platyhelminthes—flatworms, flukes, tapeworms
5. Aschelminthes—roundworms, trichina
6. Mollusca—snails, slugs, clams
7. Echinodermata—starfish, sea cucumbers
8. Annelida—segmented worms, earthworms, leeches
9. Arthropoda—insects, spiders, crayfish, millipedes

Vertebrates

10. Chordata—fishes, amphibians, reptiles, birds, mammals

In this book, we are concerned mainly with Arthropoda, the largest group of animals; more than three fourths of the total number of species belong to this phylum. Other invertebrate phyla which contain agricultural pests are the Aschelminthes, Platyhelminthes, and Mollusca. These include, respectively, plant- and animal-infesting roundworms, flukes and tapeworms that parasitize higher animals, and slugs and snails that are crop pests.

ARTHROPODA

Arthropods are quite variable in structure. They include such diverse animals as millipedes, spiders, crayfish, and insects. All arthropods, however, have the following characteristics in common:

1. Series of ringlike segments—each animal in this group has a body formed of a number of segments.
2. Jointed appendages—they have legs and other appendages that are made up of segments jointed together.
3. Exoskeleton—they possess an outer covering of a hardened, horny material, an external skeleton.
4. Bilateral symmetry—the arrangement of body parts is such that they can be split down the middle so as to form two equal portions.
5. Ventral nerve cord—the nerve cord (more or less equivalent to spinal cord of man) is found within the lower part of the body.
6. Dorsal heart—the heart is a simple tubelike structure found within the upper part of the body.

In order to classify animals into groups and subgroups that can be studied and understood more easily, a **system of nomenclature** has been developed. The largest groups in the animal kingdom are the phyla. Each phylum is divided into a number of classes, each class into a number of orders, each order into a number of families, and so on. Let us take the codling moth, the familiar "worm in the apple," as an example and see how this classification system works:

Kingdom—Animal
Phylum—Arthropoda
Class—Insecta
Order—Lepidoptera
Family—Olethreutidae
Genus—Laspeyresia
Species—pomonella

We derive the **scientific name** of the codling moth, *Laspeyresia pomonella* (Linnaeus), by combining the generic and specific names. Note that this scientific name is italicized in print; when written by hand it is underlined to indicate italics. Correct style also includes capitalizing the genus but not the species. You will see that sometimes a man's name, either abbreviated or in full, is listed after the scientific name of an insect. This is the person who first described the species in question. If his name is in parentheses, it indicates that the scientific name has been changed in some way since the species was first described.

Farmers and entomologists alike have a tendency to use common names for important crop and livestock insects. This practice has become so entrenched that the Entomological Society of America regularly publishes a list of approved common names. In this book we shall often refer to insects only by their common names. Yet for scientific accuracy and as a key to the literature of an insect, the scientific name is indispensable. For this reason we have included an appendix of common-scientific names which the student may consult.

Characteristics of the major classes of arthropods are presented in Table 4:1, and representatives of four of these classes are illustrated in Fig. 4:1. Only the classes Arachnida and Insecta will be discussed in greater detail in this chapter.

Recognition of the pest species involved in an agricultural problem is of utmost importance. Different control measures are often required even for two closely related kinds of insects due to variations in their life cycles or in their reactions to insecticidal materials. There are about one million species of insects and 30,000 species of mites and ticks which have been described and cataloged; many more remain to be discovered. Obviously, no one can learn to recognize more than a small portion of the total. An entomologist who has worked in one region for most of his life is not likely to be familiar with more than a few hundred insects out of the thousands of kinds that are present.

We shall discuss the groups of insects, mites, and ticks that are of major agricultural importance and take note of a limited number of families that contain almost all of the major pest and beneficial species. By studying the characteristics of these families, one can attain a practical background which may be applied to the agricultural pest species throughout the United States, indeed most of the world.

CLASS ARACHNIDA

Almost all members of the class Arachnida are air-breathing arthropods. The body is usually divided into two regions, the **cephalothorax** and the **abdomen**; however, sun spiders appear to have three regions, whereas the entire body of mites and ticks is more or less fused into one region. Antennae are absent, and eyes are simple. Adults have four pair of legs attached to the cephalothorax. Most forms breathe by means of **book lungs**—pouchlike organs containing membranous flaps resembling the pages of a book—situated in the underside of the abdomen. Most mites and ticks breathe by means of **tracheae**.

The order **Araneida** contains the spiders (Fig. 4:2). They are generally beneficial—they feed mainly upon insects. **Scorpionida** are the scorpions, which are most common in the Southwestern states (for illustration of this and other orders see the following Key to Major Orders of Arachnida). The sting of the common species is painful but not dangerous. Pseudoscorpions belong to the order **Chelonethida**. These unusual little animals are commonly found

Table 4:1. Major classes of the Phylum Arthropoda and Their Characteristics

Class and Examples	Body Regions	Pairs of Legs	Pairs of Antennae	Breathing Organs	Eyes	Agricultural Importance
Crustacea Crayfish, crabs, sowbugs, etc.	2 Cephalothorax and abdomen	5 or more	2	Gills	Compound	Sowbugs are a minor pest, especially on greenhouse crops
Arachnida Spiders, scorpions, daddy longlegs, mites, ticks, etc.	2 Cephalothorax and abdomen	4	None	Book lungs or tracheae	Simple	The mite and tick group contains major pests of both animals and plants and some beneficial predators
Diplopoda millipedes	2 Head and body (rounded form)	30 or more (2 pr. per segment)	1 (short)	Tracheae	Simple	A few species feed on roots and tubers, especially vegetable crops, mostly minor
Symphyla Symphylans	2 Head and body (flattened form)	12 (1 pr. per segment)	1 (long)	Tracheae	None	Garden symphylan is extremely damaging to all types of crops in some areas, feeds on roots
Chilopoda Centipedes	2 Head and body (flattened form)	15 or more (1 pr. per segment)	1 (long)	Tracheae	Usually simple	Feed on insects; rather insignificant agriculturally
Insecta Bugs, beetles, butterflies, bees, etc.	3 Head, thorax and abdomen	3	1	Tracheae	Compound, often possess simple eyes also	Greatest number of major agricultural pests; also many beneficial species

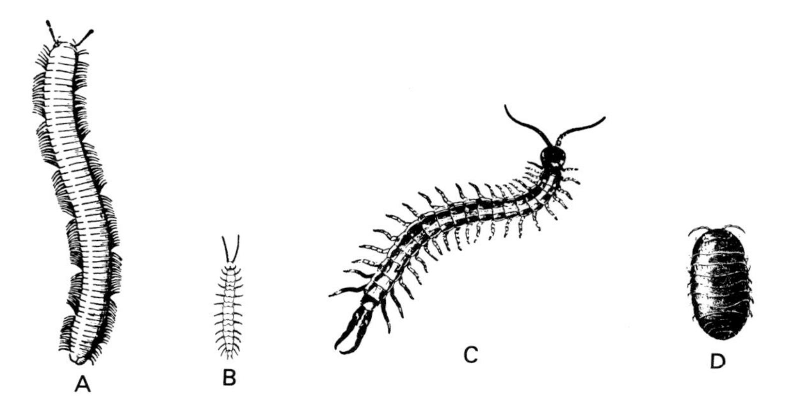

Figure 4:1. Representative arthropods (magnification approx 2×). A, millipede; B, garden symphylan; C, centipede; D, sowbug. *A, C, D, courtesy USDA; B, courtesy Waterhouse, Wash. Agr. Exp. Sta.*

under bark or stones and inside buildings. They feed upon mites and small insects. **Phalangida** are the daddy longlegs. Their food consists of plant juices, dead plant and animal material, and possibly live insects.

Whipscorpions belong to the order **Pedipalpida** and have slender whiplike tails without stings. They are nocturnal and apparently predaceous upon other small animals. **Solpugida** are the sun spiders. Their large pincer-like chelicerae (false jaws) give them a fierce appearance, but they are harmless. They prey upon spiders and insects and are found in warm, arid areas throughout the West.

Order Acarina

Acarina is by far the most important of the arachnid orders. It contains the mites, which attack both plants and animals, and the ticks, which are pests of animals. Most newly hatched acarina have six legs, whereas most later stages have eight. The cephalothorax and abdomen are fused into a single body region. Mouthparts are adapted for biting, piercing, and sucking. Acarina

Figure 4:2. The black widow spider, a poisonous species, exposing ventral side and characteristic hour-glass mark on abdomen. *Courtesy Utah Agr. Exp. Sta.*

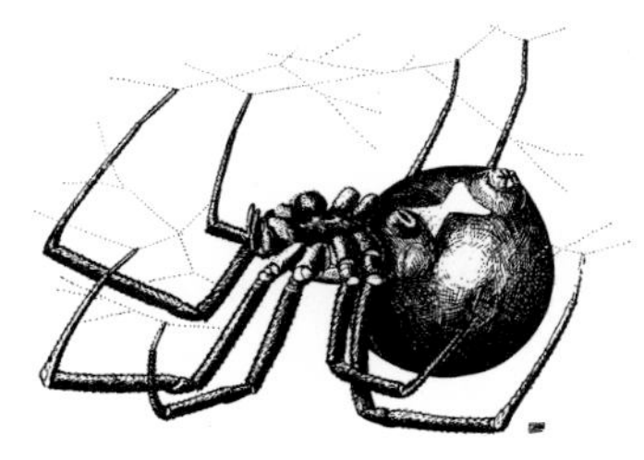

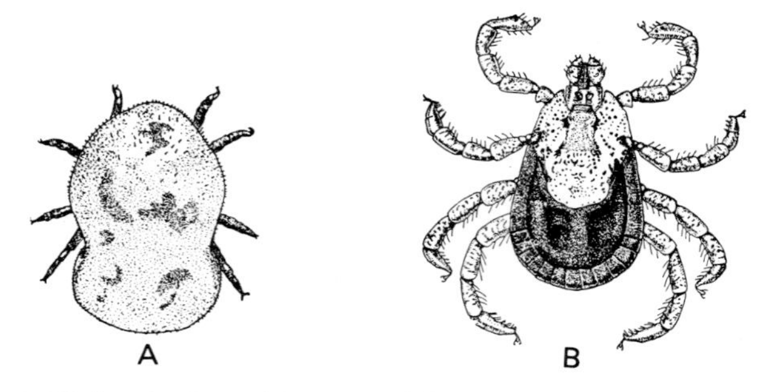

Figure 4:3. Ticks. A, **Argasidae**, ear tick; B, **Ixodidae**, Pacific Coast tick. *Courtesy USDA.*

breathe by means of tracheae or (apparently) through the body surface. Mites and ticks range in size from less than 1 mm to 15 mm long.

Ticks. Ticks (Fig. 4:3) are major agricultural pests because they suck the blood of domestic animals and also transmit disease organisms. One can distinguish ticks from mites by their larger size, leathery skin, and spiracles located behind the third or fourth pair of legs.

KEY TO THE MAJOR ORDERS OF ARACHNIDA

1. Abdomen clearly segmented 2
   Abdominal segments fused, some forms with a few obscure sclerites 7
2. Abdomen with long, taillike extension . . . 3
   Abdomen without taillike extension 4
3. Tail thick, terminating in large sting (Fig. 1) . . . . scorpions: *Scorpionida*
   Tail whiplike, without sting (Fig. 2) . . . . whipscorpions: *Pedipalpida*
4. Pedipalpi chelate (pincerlike) (Fig. 3) . . . . pseudoscorpions: *Chelonethida*
   Pedipalpi not chelate 5
5. Chelicerae (false jaws) not chelate . . . . tailless whipscorpions: *Pedipalpida*
   Chelicerae chelate 6
6. Head region broadly joined to thorax, both covered with unsegmented carapace; legs usually long (Fig. 4) . . . . .daddy longlegs: *Phalangida*
   Head region joined necklike to thorax; carapace with segmentation (Fig. 5) . . . . sun spiders: *Solpugida*

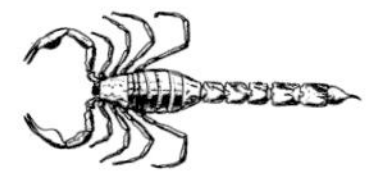

Figure 1. Scorpion

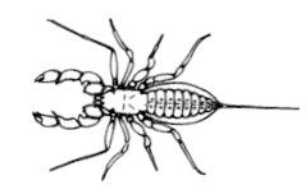

Figure 2. Whipscorpion

Figure 3. Pseudoscorpion

Figure 4. Daddy longlegs

7. Cephalothorax and abdomen narrowly joined by distinct stalk (see Fig. 4:2) . . . . spiders: *Araneida*
   Cephalothorax and abdomen broadly joined (see Figs. 4:3 and 4:4) . . . . mites and ticks: *Acarina*

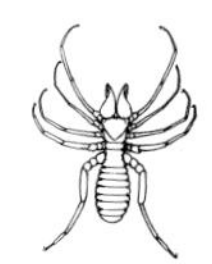

Figure 5. Sun spider

Figures 1, 2, 3, and 5 are courtesy of E. W. Baker, The Institute of Acarology; Figure 4 is from *CDC Pictorial Keys*, 1967, courtesy Communicable Disease Center, Public Health Service.

Mites. Mites, too, are major agricultural pests (Figs. 4:4, 4:5, and 4:6). Various kinds attack both plants and animals, and a few are beneficial predators upon pest species. Mites are smaller than ticks, some of them being microscopic in size. The skin is not leathery, and the tracheal openings (stigmata), when present, are near the head region or the bases of certain legs.

Figure 4:4. Mites. A, **Phytoseiidae**, a predator mite **Amblyseius cucumeris**; B, **Dermanyssidae**, chicken mite; C, **Demodicidae**, follicle mite; D, **Tetranychidae**, Banks grass mite. *A, after Cunliffe and Baker*; *B, from USDA*; *C, from Baker et al.*, Parasitic Mites, *by permission National Pest Control Association*; *D, courtesy Malcolm, Wash. Agr. Exp. Sta., Pullman.*

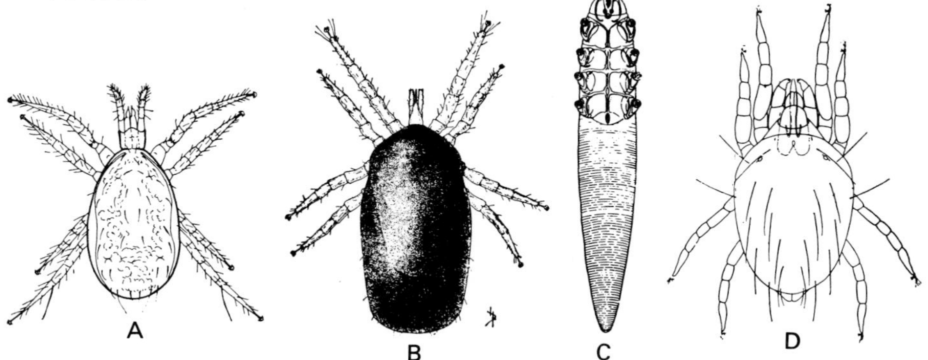

Figure 4:5. Mites. A, **Eriophyidae**, peach silver mite; B, **Tarsonemidae**, cyclamen mite; C, **Trombiculidae**, adult chigger mite, *A, courtesy Ohio Agr. Res. Dev. Center*; *B, courtesy University of California*; *C, courtesy USDA.*

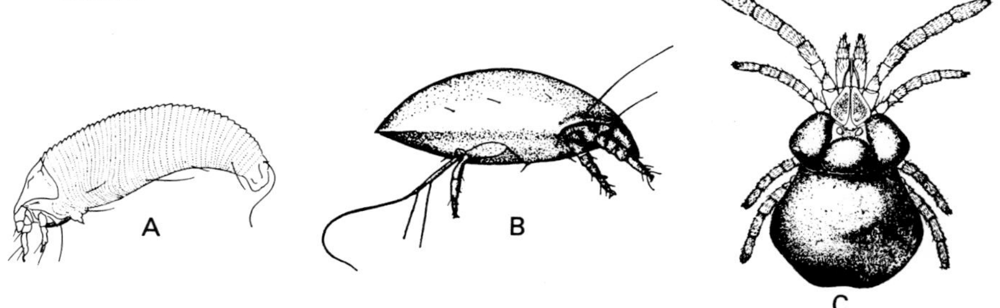

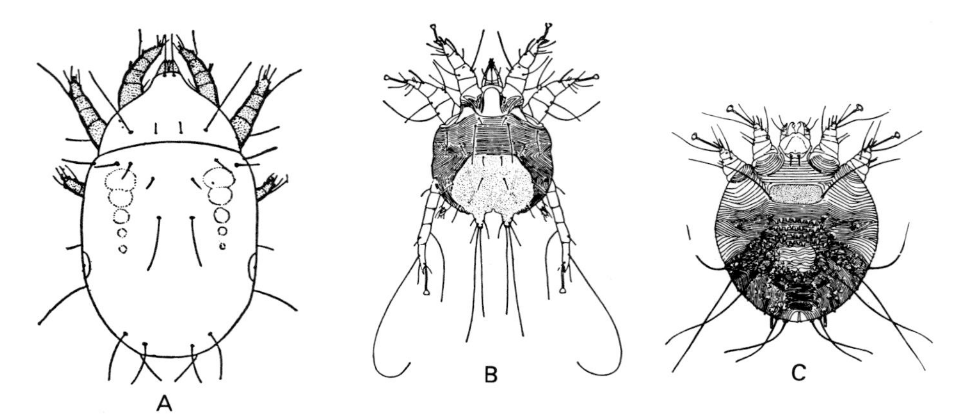

Figure 4:6. Mites. A, **Acaridae**, bulb mite; B, **Psoroptidae**, scab mite; C, **Sarcoptidae**, itch mite. *A, from Conn. Agr. Exp. Sta.; B, from Baker and Wharton, 1952, An Introduction to Acarology, by permission Macmillan Publishing Co., Inc.; C, from Baker et al., 1956.* Parasitic Mites, *by permission National Pest Control Association.*

KEY TO THE MAJOR ECONOMIC FAMILIES OF ACARINA

1. Large, leathery parasites of vertebrates; hypostome with recurved teeth (Fig. 1 and Fig. 2:7A) (ticks) 2
   Small, usually soft-bodied forms with various food habits; hypostome without recurved teeth (mites) 3
2. Scutum (dorsal shield) present (Fig. 4:3B) .... hardbacked ticks: *Ixodidae*
   Scutum absent (Fig. 4:3A) .... softbacked ticks: *Argasidae*
3. Pedipalpi with tined claws (Fig. 2); stigmata (spiracles) lateral to coxae (Fig. 3) . 4
   Pedipalpi simple; stigmata (when present) not lateral to coxae 5
4. Ventral and anal plates contiguous (Fig. 4); not with fixed digit developed or chelicerae long, whiplike; mostly predators of phytophagous species .... predator mites: *Phytoseiidae*
   Ventral and anal plates not contiguous; fixed digit developed or chelicerae long, whiplike (Fig. 5); parasites of vertebrates .... chicken mites: *Dermanyssidae*

Figure 1

Figure 2

Figure 3

Figure 4

Figure 5

5. Tracheal system present, except in Eriophyidae and male Tarsonemidae; with styletlike chelicerae 6
   Tracheal system absent; chelicerae chelate (pincerlike) (Fig. 6) 10
6. Body elongate, wormlike, annulate (ringed) 7
   Body rounded, not wormlike or annulate 8
7. With two pairs of legs on anterior end (Fig. 7); phytophagous ....
   .... rust mites: *Eriophyidae*

   With four pairs of stumpy legs on anterior end (Fig. 8); parasites of vertebrates .... follicle mites: *Demodicidae*
8. Fourth pair of legs highly modified with terminal whiplike setae in female (Fig. 9), as flanged claspers in male (Fig. 10) ...... thread-footed mites: *Tarsonemidae*
   Fourth pair of legs not highly modified 9
9. Chelicerae form long, recurved stylets (Fig. 11); body not figure eight-shaped or densely clothed with setae; phytophagous .... spider mites: *Tetranychidae*
   Chelicerae short, strong; body of adult usually figure eight-shaped and densely clothed with setae (Fig. 12); later stages predaceous .... chiggers: *Trombiculidae*
10. With two pairs well-developed genital discs (Fig. 13); with normal walking legs; feed on stored products, fungi, and decaying organic matter ....
    ....cereal mites: *Acaridae*
    Genital discs greatly reduced or absent; legs modified for parastic life, usually terminating in caruncles (bell-like membranous extensions of tarsi) (Fig. 14) ... 11
11. Female genital opening an inverted "U" with apodemes (Fig. 15); mostly parasites of mammals .... scab mites: *Psoroptidae*
    Female genital opening a single transverse slit without apodemes (Fig. 16); parasites of vertebrates ... itch mites: *Sarcoptidae*

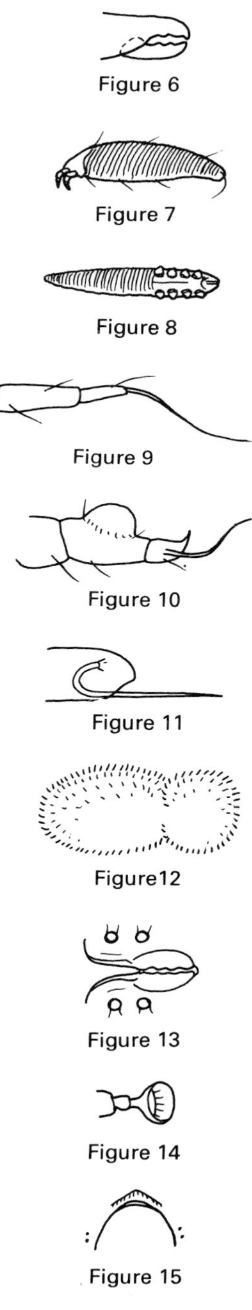

Figure 6
Figure 7
Figure 8
Figure 9
Figure 10
Figure 11
Figure12
Figure 13
Figure 14
Figure 15
Figure 16

CLASS INSECTA

Insects represent the largest and most diverse class of animals. Adult insects differ from other arthropods in usually having two pairs of wings, three distinct body regions (head, thorax, and abdomen), and three pairs of legs. The antennae are variable in form, and most adult insects possess both compound and simple eyes. Insects breathe by means of an air tube system (tracheae). Table 4:2 presents a brief résumé of the 26 orders of insects.

Table 4:2. Characteristics of Insect Orders[1]

Order	Appearance	Common names	Metamorphosis
1. Protura	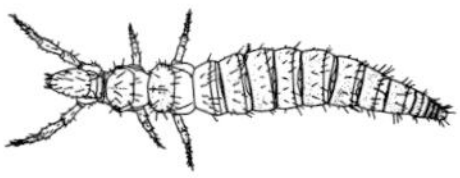	Telsontails	None
2. Thysanura	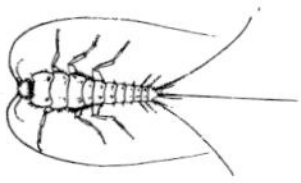	Silverfish and allies	None
3. Collembola	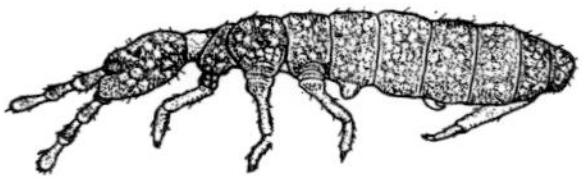	Springtails	None
4. Ephemeroptera		Mayflies	Simple
5. Odonata	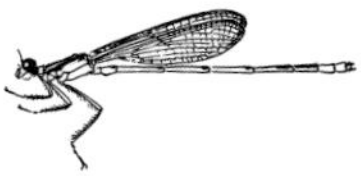	Dragonflies, damselflies	Simple
6. Plecoptera		Stoneflies	Simple
7. Orthoptera	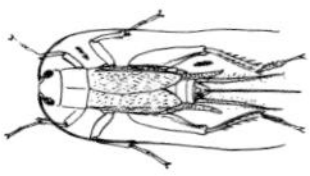	Grasshoppers, crickets, cockroaches, etc.	Simple

Insects are divided into two subclasses, the **Apterygota**, wingless insects, and the **Pterygota**, winged insects. The Apterygota contain small, primitive insects which grow and develop with little or no metamorphosis. They include three orders: Protura, Thysanura, and Collembola.

The Pterygota contain the other 23 orders of insects. Some may lack wings, but the condition is a secondary one, for their thoracic structure definitely places them in the winged subclass.

Wings	Mouth Parts	Distinctive Features	Habitat	Agricultural Importance[2]	
				Beneficial	Injurious
None	Sucking	Without antennae	Terrestrial and semi-aquatic	–	–
None	Chewing	Fine scales on body; 3 tails end of abdomen	Terrestrial and semi-aquatic	–	–
None	Chewing	Furcula (spring) on bottom of abdomen	Terrestrial and semi-aquatic	–	+
1–2 pr. membranous	Chewing; vestigial in adult	Forewings larger than hindwings; 2 or 3 tails end of abdomen	Aquatic as nymphs	–	–
2 pr. membranous	Chewing	Large insects with unfolded, nearly alike wings; eyes big; no cerci	Aquatic as nymphs	+	–
2 pr. membranous	Chewing	Two long cerci	Aquatic as nymphs	–	+
Tegmina and membranous hindwings	Chewing	Forewings leathery, hindwings membranous	Terrestrial	–	++++

Table 4:2 (continued)

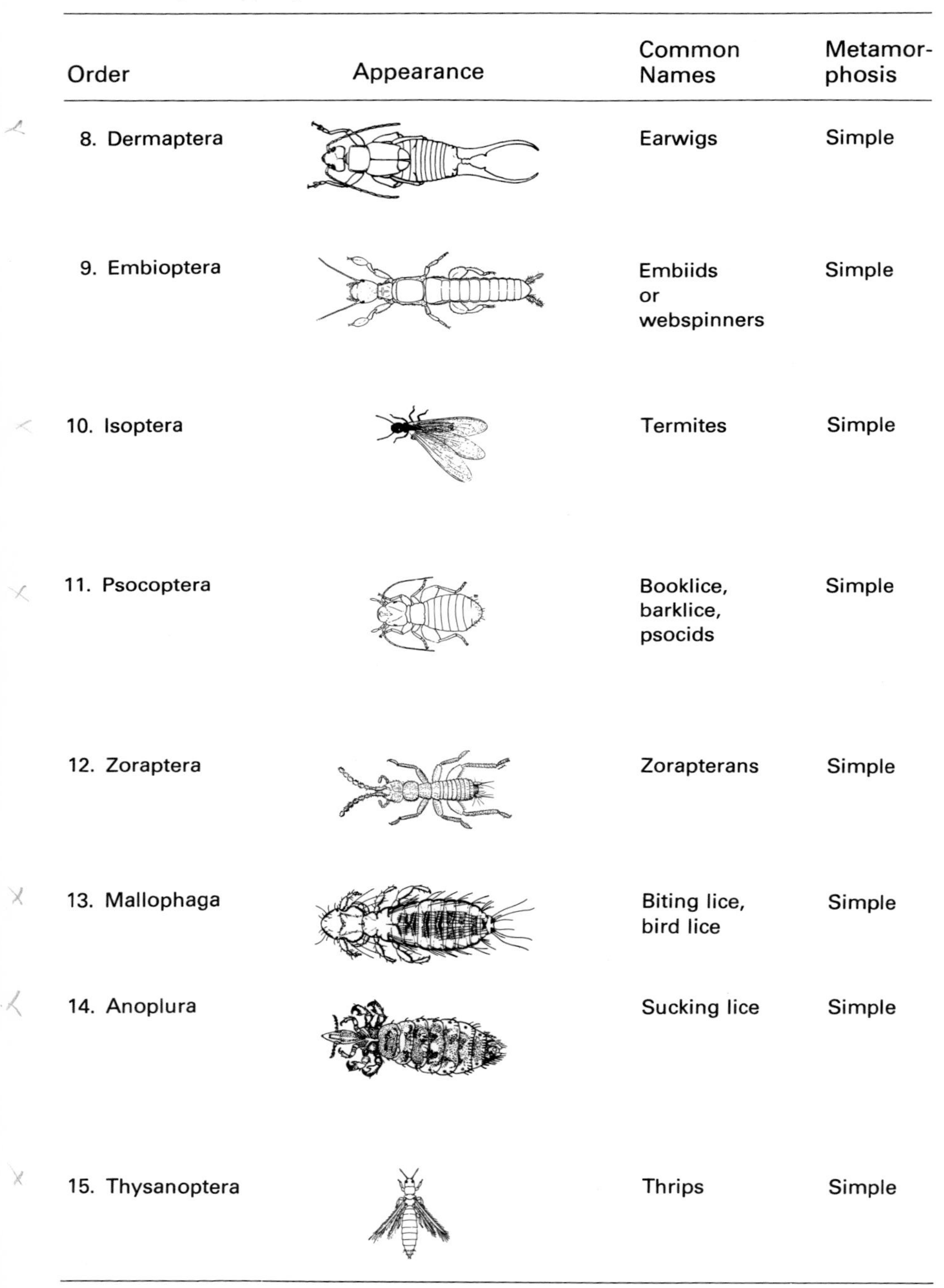

Order	Appearance	Common Names	Metamorphosis
8. Dermaptera		Earwigs	Simple
9. Embioptera		Embiids or webspinners	Simple
10. Isoptera		Termites	Simple
11. Psocoptera		Booklice, barklice, psocids	Simple
12. Zoraptera		Zorapterans	Simple
13. Mallophaga		Biting lice, bird lice	Simple
14. Anoplura		Sucking lice	Simple
15. Thysanoptera		Thrips	Simple

Wings	Mouth Parts	Distinctive Features	Habitat	Agricultural Importance	
				Beneficial	Injurious
Short, hard tegmina and membranous hindwings	Chewing	Forceps at end of abdomen	Terrestrial	–	++
2 pr. membranous or none	Chewing	Long, flat body; tarsi of forelegs enlarged for spinning silk	Terrestrial and semi-aquatic	–	–
2 pr. membranous or none	Chewing	Sexual forms pigmented, others pale, without pigment	Terrestrial	–	++
2 pr. membranous or none	Chewing	Small insects with roofed wings or none; mouthparts with maxillary "pick"	Terrestrial	–	–
2 pr. membranous or none	Chewing	Minute insects with long bead-like antennae	Terrestrial and semi-aquatic	–	–
None	Chewing	Minute, flat insects with wide head	Terrestrial, ecto-parasitic	–	++
None	Piercing-sucking	Minute, flat insects with narrow head; mouthparts retractile into head	Terrestrial, ecto-parasitic	–	+++
2 pr. slender, fringed with hair	Rasping-sucking	Tarsi bladder-like at tip	Terrestrial	+	+++

Table 4:2 (continued)

Order	Appearance	Common Names	Metamorphosis
16. Hemiptera	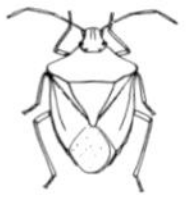	Bugs	Simple
17. Homoptera		Aphids, scales, leafhoppers	Simple
18. Neuroptera	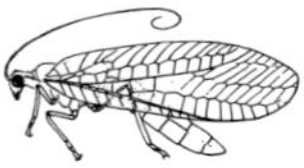	Lacewings and allies	Complex
19. Coleoptera		Beetles, weevils	Complex
20. Strepsiptera		Twisted-winged insects, stylopids	Complex
21. Mecoptera		Scorpionflies	Complex
22. Trichoptera		Caddisflies	Complex
23. Lepidoptera		Butterflies, moths, skippers	Complex

Wings	Mouth Parts	Distinctive Features	Habitat	Agricultural Importance	
				Beneficial	Injurious
Hemelytra and membranous hindwings or none	Piercing-sucking	Beak arises from front of head; meso-thorax with triangular scutellum	Terrestrial and aquatic	+++	++++
2 pr. membranous or none	Piercing-sucking	Beak arises from back of head	Terrestrial	–	+++++
2 pr. membranous	Chewing	Wings net-veined and roofed; cerci absent	Terrestrial and aquatic	++	–
Elytra and membranous hindwings	Chewing	Forewings hard and veinless; prothorax large	Terrestrial and aquatic	++++	+++++
Front reduced, hind membranous, none in female	Chewing	Female maggotlike with head and thorax fused	Terrestrial, most parasitic in insects	–	–
2 pr. membranous or none	Chewing	Head elongated ventrally forming a beak	Terrestrial	–	–
2 pr. membranous with hairs or scales	Chewing	Wings covered with hairs and held rooflike	Aquatic as larvae and pupae	–	–
2 pr. membranous with scales	Siphoning in adults, chewing in larvae	Proboscis long and coiled, extensible	Terrestrial	–	+++++

Table 4:2 (continued)

Order	Appearance	Common Names	Metamorphosis
24. Hymenoptera		Ants, bees, wasps, etc.	Complex
25. Diptera		Flies	Complex
26. Siphonaptera		Fleas	Complex

[1] Figures 3, 11, and 17 from Essig, *Insects and Mites of Western North America, 1958,* by permission Macmillan Publishing Co., Inc.; Figures 1, 9, 12, 20, 21 from Essig, *College Entomology, 1947,* by permission Marie W. Essig; Figure 19 courtesy Ohio Agr. Exp. Sta.; other figures courtesy USDA.

[2] Dash indicates minor agricultural importance; plus indicates extent of major agricultural importance.

The Pterygota are further divided into the **Paleoptera**, or ancient, winged insects, whose wings are held permanently at right angles to the body, and the **Neoptera**, or modern winged insects, whose wings can be folded and held close to the body. Paleopterans were abundant in ancient times and included many orders now extinct. Today they are represented by two orders, the Ephemeroptera and the Odonata. The Neopterans, being able to fold their wings, have an advantage in their ability to escape from enemies not only by flying away but also by running fast and by hiding in crevices. It is an interesting fact that although wingless species occur in all of the Neopteran orders, none of the living or extinct Paleoptera is wingless.

Subgroups which separate the Pterygota according to type of metamorphosis are **Exopterygota** and **Endopterygota**. Literally these terms indicate external and internal development of the wings. The last nine orders which

Wings	Mouth Parts	Distinctive Features	Habitat	Agricultural Importance	
				Beneficial	Injurious
2 pr. membranous or none	Chewing or chewing-lapping	Membranous wings with few veins; base of abdomen usually constricted	Terrestrial	+++++	++
Membranous and halters or none	Piercing-sucking or sponging; chewing in larvae	Forewings usually present; hind-wings reduced to halters	Terrestrial, many aquatic larvae and pupae	++++	+++++
None	Piercing-sucking; chewing in larvae	Small insects with laterally compressed bodies	Terrestrial, ecto-parasitic as adults	–	++

show the greatest specialization in feeding habits and development are included in the Endopterygota.

Because immature insects, larvae and nymphs, are often the major active feeding stages, knowledge of them is especially important in agricultural entomology. In the early days of the development of insect classification only the adult insects were studied. Even today, the adults are much better known than the nymphs or larvae. Characteristics of the immature stages of the insect orders of major agricultural importance are presented in Table 4:3.

Anyone who can learn three "key" characteristics of an insect order plus another distinctive feature will be able to separate them fairly well. The three basic characteristics are type of mouthparts, type of wings, and type of metamorphosis. In this chapter we shall take special note of 14 insect orders and 89 families of major economic importance.

Table 4:3. Characteristics of Immature Insects in Agriculturally Important Orders[1]

Order	Type of Young	Appearance	Number of Instars	Wing Pads	Mouth-parts	Thoracic Legs	Abdominal Prolegs	Eyes
Orthoptera	Nymph		Usually 5 or 6	Usually in late instars	Chewing	Well developed	None	Compound and simple
Dermaptera	Nymph		4–5	Usually present	Chewing	Well developed	None	Compound and simple
Isoptera	Nymph		6	In late instars of sexual forms	Chewing	Well developed	None	Compound and simple; or none
Mallophaga	Nymph		3	None	Chewing	Special form	None	Compound, poorly developed
Anoplura	Nymph		3	None	Piercing-sucking	Special form	None	Compound, small when present

Thysanoptera	Nymph	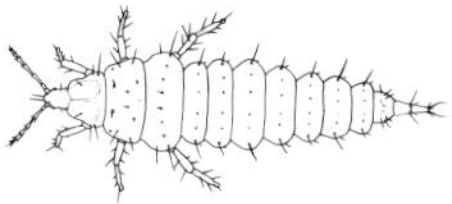	4	Usually present	Rasping-sucking	Short	None	Compound, sometimes simple
Hemiptera	Nymph	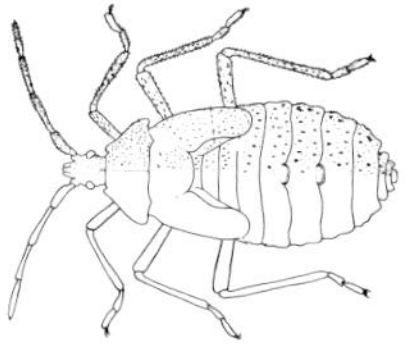	Usually 5	Usually present	Piercing-sucking	Well developed	None	Compound, sometimes simple
Homoptera	Nymph	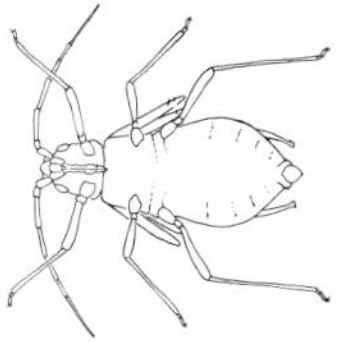	3–5	Often present	Piercing-sucking	Well developed	None	Compound, sometimes simple
Neuroptera	Larva (antlion, aphidlion, others)	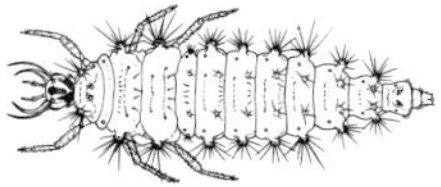	3 or more	None	Chewing, jaws modified for sucking	Well developed	1 caudal pr. in some aquatics	Usually 5–7 pr. simple
Coleoptera	Larva (grub, wireworm, others)	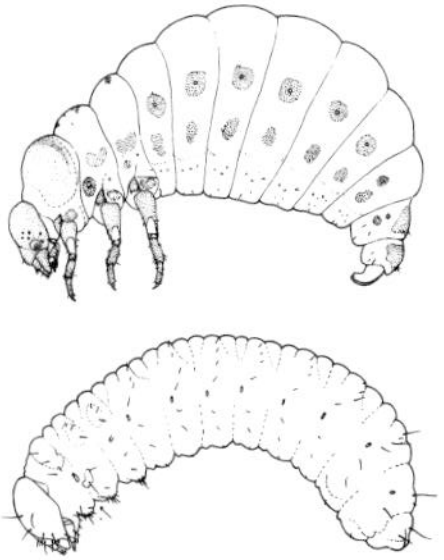	Usually 5 (3–25)	None	Chewing	None to well developed	None	1–6 pr. simple

Table 4:3 (continued)

Order	Type of Young	Appearance	Number of Instars	Wing Pads	Mouth-parts	Thoracic Legs	Abdominal Prolegs	Eyes
Lepidoptera	Larva (caterpillar)		Usually 5–6	None	Chewing	None to well developed	5 pr. with crochets, (also 2–4)	1–8 pr. simple
Hymenoptera	Larva		Usually 4–5	None	Chewing	None to well developed	None or 8 pr. without crochets (also 6–7)	1 pr. simple or none
Diptera	Larva (maggot, wriggler, others)		Usually 4–6	None	Chewing, maggot has mouth hooks	None	None (several have false legs)	None, sometimes simple (eye spots)
Siphonaptera	Larva		3	None	Chewing	None	None	None

[1] Figures of isopteran and mallophagan from Essig, *College Entomology, 1947,* by permission Marie W. Essig; figure of anopluran original; all other figures from Alvah Peterson, *Larvae of Insects*, Part I, 1948, Part II, 1951, by permission Helen H. Peterson.

KEY TO THE ORDERS OF INSECTA[1]

1. Winged insects 2
   Wingless or vestigial-winged insects (wingless forms of Zoraptera, Embioptera and Mecoptera not included in key, but indicated in figures) 23

Figure 1

2. With only one pair of wings 3
   With two pairs of wings 5

3. Wings net-veined; halters absent .... certain mayflies: *Ephemeroptera*
   Wings not net-veined; halters present 4

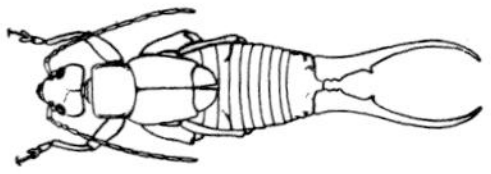

4. Wings with highly reduced venation; caudal filaments usually present; minute, delicate insects (see Fig. 14: 17G) ..... male mealybugs and scale insects: .. *Homoptera*
   Wings with longitudinal and a few cross veins; no caudal filaments (Fig. 1) .... flies, mosquitoes, etc.: *Diptera*

Figure 2

5. Forewings horny, without veins, meeting in a straight line over middle of body and usually concealing membranous hindwings (certain forms have the hindwings vestigial or absent) 6
   Forewings not as above 7

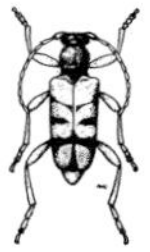

Figure 3

6. Abdomen provided with forcepslike cerci at posterior end (Fig. 2) .... earwigs: *Dermaptera*
   Abdomen without forcepslike cerci (Fig. 3) .... beetles, weevils: *Coleoptera*

7. Two pairs of wings unlike in structure ... 8
   Two pairs of wings similar in structure .. 10

8. Forewings reduced to slender club-shaped appendages; hindwings folded fanlike at rest (Fig. 4) ....twistedwinged insects: *Strepsiptera*
   Forewings not as above 9

Figure 4

[1] Figures 1, 3, 5, 12, and 24 are from Essig, *Insects and Mites of Western North America, 1958*, by permission Macmillan Publishing Co., Inc.; Figures 4, 8, 14, 18, 19, 20, 21, and 25 are from Essig, *College Entomology, 1947*, by permission Marie W. Essig; Figures 6 and 9 are original; Figures 10 and 15 are courtesy Illinois Natural History Survey; Figure 11 is after Needham; Figure 17 is courtesy California Insect Survey; other figures are courtesy USDA.

9. Forewings thick and leathery at base and membranous at tip; mouthparts form a sucking beak (Fig. 5) . .bugs: *Hemiptera*
   Forewings leathery throughout; chewing mouthparts (Fig. 6) ....
   grasshoppers, cockroaches, etc.: *Orthoptera*

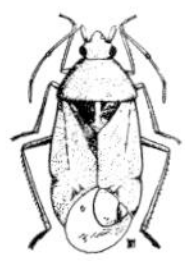

Figure 5

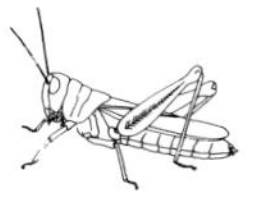

10. Wings partially or more often entirely covered with microscopic scales (Fig. 7) .... moths, butterflies: *Lepidoptera*
    Wings transparent or covered with fine hair 11

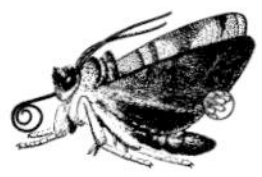

Figure 6

11. Wings very narrow and fringed with long hairs, small slender-bodied insects (Fig. 8) .... thrips: *Thysanoptera*
    Wings not as above 12

Figure 7

12. Mouthparts a piercing-sucking beak arising from the rear of the head near the first pair of legs (Fig. 9) ....
    aphids, leafhoppers, etc.: *Homoptera*
    Mouthparts not a piercing-sucking beak and normally situated at the front of the head 13

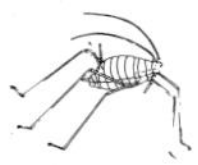

Figure 8

Figure 9

13. Antennae small and bristle-like 14
    Antennae conspicuous and of many forms 15

14. Fore and hindwings nearly equal in size; tip of abdomen without terminal filaments (Fig. 10) ....
    .... dragonflies, damselflies: *Odonata*
    Forewings much larger than hindwings; tip of abdomen with 2–3 long terminal filaments (Fig. 11) . mayflies: *Ephemeroptera*

Figure 10

15. Wings with many veins and cross veins 16
    Wings with few veins and cross veins ... 20

Figure 11

16. Hind tarsi with fewer than five segments . 17
    Hind tarsi with five segments 18

Figure 12

17. Tarsi three-segmented; hindwings as large as or larger (wider) than forewings (Fig. 12) .... stoneflies: *Plecoptera*
    Tarsi four-segmented; forewings and hindwings of equal size (Fig. 13) ....
    .... most termites: *Isoptera*

Figure 13

18. Head prolonged into a beak (Fig. 14) .... scorpionflies: *Mecoptera*
    Head not prolonged into a beak 19
19. Wings covered with fine hair (Fig. 15) .... caddisflies: *Trichoptera*
    Wings transparent, not covered with hair (Fig. 16) .... lacewings: *Neuroptera*
20. Tarsi two- or three-segmented; wings approximately equal in size 21
    Tarsi usually five-segmented; forewings larger than hindwings (Fig. 17) .... ants, bees, wasps: *Hymenoptera*
21. Basal tarsal segment of foreleg greatly enlarged (Fig. 18) .... webspinners: *Embioptera*
    Basal tarsal segment not enlarged 22
22. Cerci present; body less than 3 mm long (Fig. 19) .... zorapterans: *Zoraptera*
    Cerci absent; body 3 mm long or longer .... barklice: *Psocoptera*
23. Abdomen composed of six or fewer segments with ventral spring apparatus (Fig. 20) .... springtails: *Collembola*
    Abdomen with more than six segments and no spring 24
24. Abdominal segments 1–3 each with a pair of small ventral appendages; antennae, eyes, and cerci absent; minute and rare (Fig. 21) .... telsontails: *Protura*
    Abdomen and appendages not as above . 25
25. Abdomen with 2–3 long terminal appendages or pair of forcepslike cerci, segments 2–7 may each have a pair of small ventral leglike appendages (Fig. 22) .... bristletails: *Thysanura*
    Abdomen without terminal filaments or ventral appendages 26
26. Mouthparts fitted for chewing 27
    Mouthparts fitted for piercing, lapping, or sucking, sometimes concealed 31
27. Louselike insects 28
    Insects not louselike, various forms 29

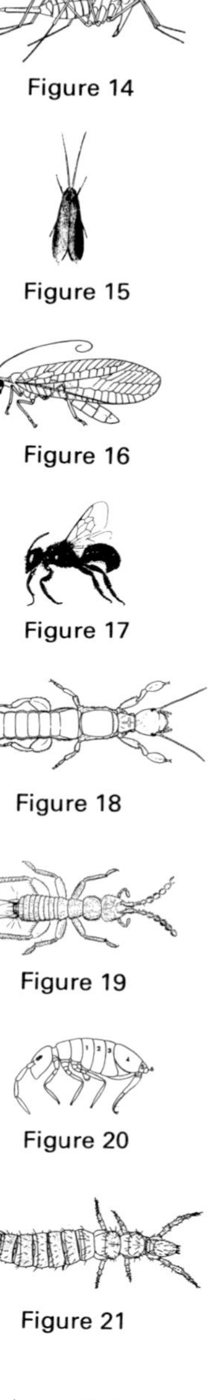

Figure 14

Figure 15

Figure 16

Figure 17

Figure 18

Figure 19

Figure 20

Figure 21

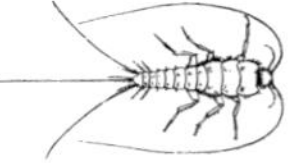

Figure 22

28. Antennae with five or less segments (Fig. 23) .... chewing lice: *Mallophaga*
    Antennae with more than five segments (Fig. 24) .... booklice: *Psocoptera*

29. Abdomen constricted at the base (see Fig. 4:28) .... ants, wasps: *Hymenoptera*
    Abdomen not constricted at the base ... 30

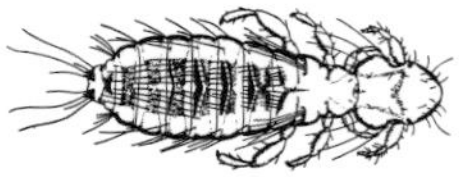

Figure 23

30. Body very slender and linear, hindlegs fitted for jumping, or body oval and flattened (see Fig. 4:7) ..... walkingsticks, grasshoppers, cockroaches: *Orthoptera*
    Body or legs not as above; body antlike, but abdomen broadly joined to the thorax (see Fig. 19:19) ...... termites: *Isoptera*

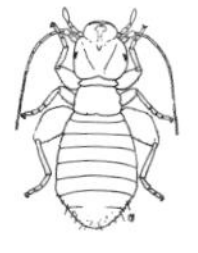

Figure 24

31. Tarsi with five segments 32
    Tarsi with fewer than five segments 34

32. Body strongly compressed laterally (Fig. 25) .... fleas: *Siphonaptera*
    Body not strongly compressed laterally .. 33

33. Abdomen not distinctly segmented, covered with hairs (see Fig. 20:13A) .... sheep ked, other flies: *Diptera*
    Abdomen distinctly segmented, covered with scales (see Fig. 4:22A) .... females of some bagworms, measuring worms, and tussock moths: *Lepidoptera*

Figure 25

34. Last tarsal segment a bladderlike organ, without well-developed claws .... thrips: *Thysanoptera*
    Last tarsal segment with one or two claws 35

35. Louselike form; sucking beak not evident (Fig. 26) .... sucking lice: *Anoplura*
    Insect not louselike; sucking beak evident 36

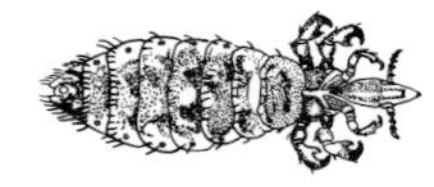

Figure 26

36. Beak arising from front of head (Fig. 4:9) .... bedbugs, water striders, etc.: *Hemiptera*
    Beak arising from rear of head near the first pair of legs (Fig. 9) .... aphids, scale insects, etc.: *Homoptera*

Order Orthoptera: Grasshoppers, Katydids, Crickets, Cockroaches, Mantids, and Walkingsticks (Fig. 4:7)

Mouthparts—typical chewing mouthparts.

Wings—forewings elongate and narrow, modified into somewhat hardened tegmina; hindwings membranous with extensive folded area.

Metamorphosis—simple.

Additional features—wingless and short-winged forms common; mainly terrestrial with good powers of running or jumping.

Orthopterans are a diverse group of insects. Many, such as the grasshoppers, crickets, and katydids, have large jumping legs. The cockroaches, mantids, and walkingsticks, however, are quite different in structure. Grasshoppers and crickets produce chirping or clacking sounds by rubbing one body

Figure 4:7. Orthoptera. A, **Blattidae,** American cockroach; B, **Tettigoniidae,** broad-winged katydid; C, **Gryllidae,** field cricket; D, **Acrididae,** migratory grasshopper. *A, B, and C, from Essig, 1958,* Insects and Mites of Western North America, *by permission Macmillan Publishing Co., Inc., New York; D, courtesy Ariz. Agr. Exp. Sta., original drawing by Dr. E. R. Tinkham.*

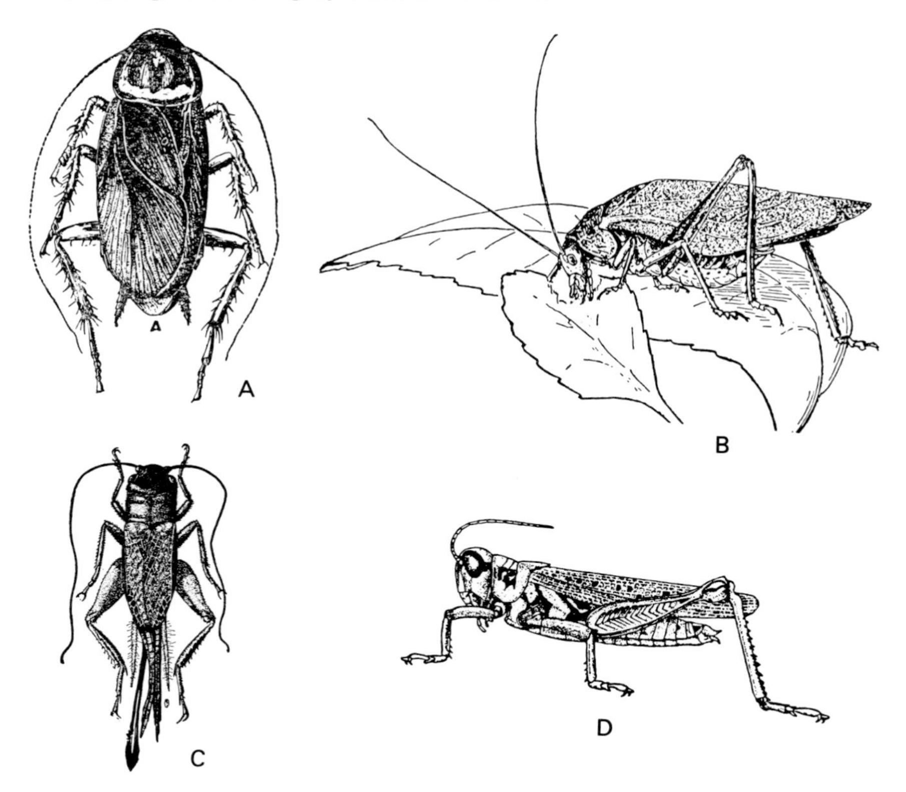

part against another. Most are plant feeders, and some are very destructive to vegetation. Two species of locusts (Old-World migratory grasshoppers) periodically cause severe damage throughout the Mediterranean region, North Africa, the Middle East, Pakistan, and India.

KEY TO THE MAJOR ECONOMIC FAMILIES OF ORTHOPTERA

1. Hindlegs adapted for jumping; body not flattened and oval; head not covered by pronotum 2
   Hindlegs not adapted for jumping; body flattened and oval; head covered by pronotum (see Fig. 4:7A) .... cockroaches: *Blattidae*

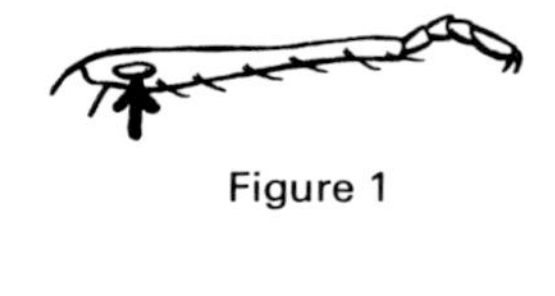

Figure 1

2. Antennae much longer than pronotum; tympanum in front tibiae (Fig. 1) 3
   Antennae about as long as pronotum; tympanum present on first abdominal tergum (shorthorn grasshoppers) (Fig. 2) .... grasshoppers: *Acrididae*

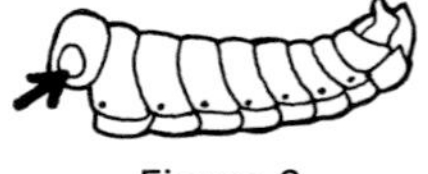

Figure 2

3. Tarsi four-segmented; ovipositor usually swordlike or sicklelike (longhorn grasshoppers) (Fig. 3) .... katydids and others: *Tettigoniidae*
   Tarsi three-segmented; ovipositor spear-shaped or awl-shaped (Fig. 4) .... crickets: *Gryllidae*

Figure 3

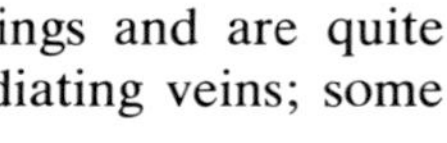

Figure 4

Order Dermaptera: Earwigs (Fig. 4:8A)

Mouthparts—typical chewing mouthparts.

Wings—forewings are hardened covers over hindwings and are quite short; hindwings are membranous and rounded with radiating veins; some lack wings.

Metamorphosis—simple.

Additional feature—large forcepslike cerci at end of abdomen.

Earwigs are a familiar group, mainly because of the introduced European earwig, an all-too-common household and garden pest. The female has an interesting habit of brooding and protecting her eggs and young almost like a hen. Abdominal glands secrete a foul-smelling liquid.

Order Isoptera: Termites (Fig. 19:19)

Mouthparts—chewing.

Wings—two pairs membranous of equal size and shape; both winged and wingless forms occur in a colony.

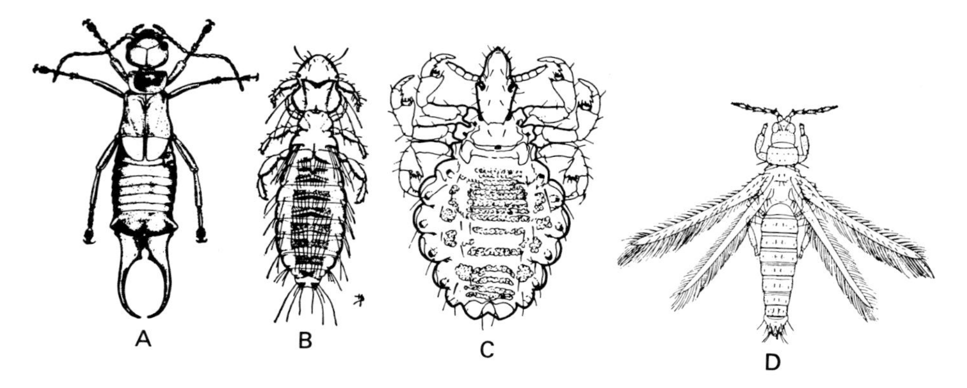

Figure 4:8. Dermaptera, Mallophaga, Anoplura, and Thysanoptera. A, **Forficulidae**, European earwig; B, **Philopteridae**, chicken head louse; C, **Haematopinidae**, hog louse; D, **Thripidae**, pear thrips. *A, courtesy Wash. Agr. Ext. Serv.; B, courtesy USDA; C, after Ferris; D, courtesy Calif. Agr. Exp. Sta.*

Metamorphosis—simple.

Additional features—termites are social insects with two or more forms or castes occurring in each species (reproductives, supplementary reproductives, soldiers, workers, and nasuti); termites are sometimes confused with ants and are called "white ants"; they are easily distinguished from ants in that they are light colored and soft bodied, with the abdomen broadly joined to thorax and with the antennae not elbowed.

Order Mallophaga: Chewing Lice (Fig. 4:8B)

Mouthparts—chewing.

Wings—wingless.

Metamorphosis—simple.

Additional features—small or minute, flat, broad-headed external parasites; compound eyes not well developed; no ocelli.

Chewing lice are mainly parasitic on birds and include a number of species that infest domestic poultry. Each species attacks a certain kind of poultry or domestic mammal. Heavy infestations do not necessarily kill the host, but cause debilitation, usually leading to death from disease.

Order Anoplura: Sucking Lice (Fig. 4:8C)

Mouthparts—piercing-sucking.

Wings—wingless.

Metamorphosis—simple.

Additional features—small, external parasites of mammals; compound eyes reduced or absent; no ocelli.

Both the adults and the nymphs of sucking lice bear a superficial resemblance to a crab, because of the flattened body and clawed, crablike legs. The clasping tarsi, with movable claw and opposing thumblike structure, are uniquely adapted to attaching to the hairs of the host. All of the pest species found on hogs, sheep, horses, and cattle are in one family, known as the wrinkled sucking lice.

Order Thysanoptera: Thrips (Fig. 4:8D)

Mouthparts—rasping-sucking.

Wings—similar in form, long, very narrow, and fringed with long hairs; wings sometimes absent.

Metamorphosis—intermediate between simple and complex.

Additional features—cerci absent; small slender-bodied insects.

The destructive species of thrips are tiny, only about 1 to 1.5 mm in length. They are unique among insects because of their bristle wings, bladderlike adhesive structures on the tarsi, and cone-shaped rasping-sucking mouthparts. *Thrips* is both singular and plural: there is no such thing as a thrip. These miniature pests can cause severe damage to crops by destroying buds or blossoms, whitening or curling foliage, and deforming or scarring fruits.

KEY TO THE MAJOR ECONOMIC FAMILIES OF MALLOPHAGA

1. Antennae lying in grooves on sides of head (Fig. 1); head broadly triangular, expanded behind eyes .... poultry body lice: *Menoponidae*
   Antennae exposed; head not broadly triangular and expanded behind eyes 2
2. Antennae five-segmented (Fig. 2); tarsi with two claws: .... feather chewing lice: *Philopteridae*
   Antennae three-segmented (Fig. 3); tarsi with one claw; .... mammal chewing lice: *Trichodectidae*

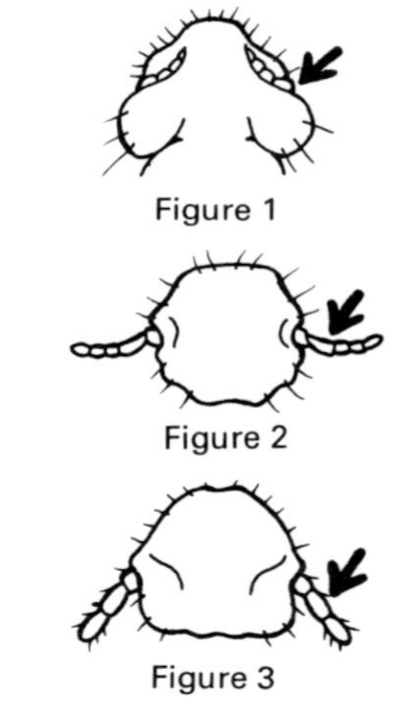

Figure 1

Figure 2

Figure 3

Order Hemiptera: Bugs (Figs. 4:9 and 4:10)

Mouthparts—piercing-sucking.

Wings—overlapping on abdomen; forewings are hemelytra (with hardened basal portions and membranous tips); hindwings membranous.

Metamorphosis—simple.

Additional feature—base of rostrum usually not touching anterior coxae ("beak" arises at front of underside of head).

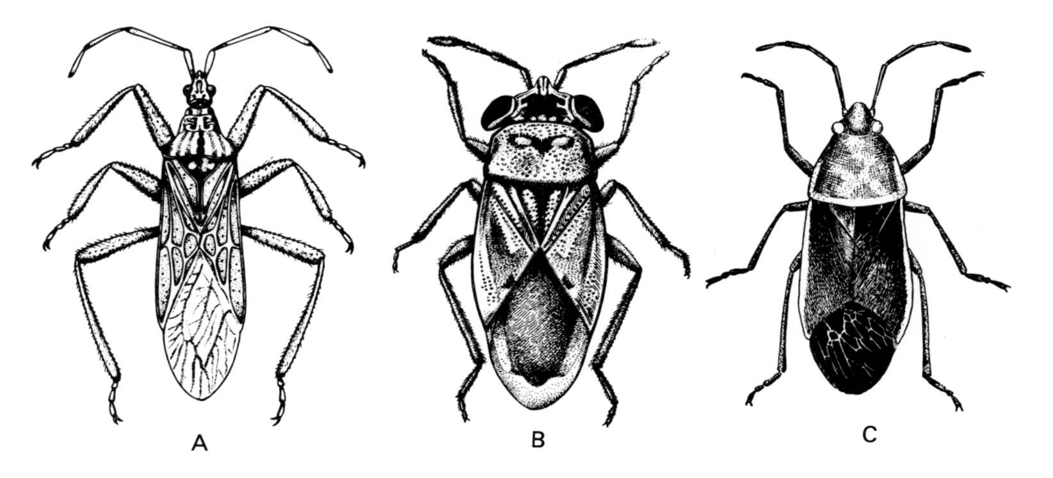

Figure 4:9. Hemiptera, A, **Nabidae,** damsel bug; B, **Lygaeidae,** a bigeyed bug; C, **Pyrrhocoridae,** a bordered plant bug. *A and B, courtesy University of California,* Berkeley, Calif. *C, from Essig,* 1958, Insects and Mites of Western North America, *by permission Macmillan Publishing Co., Inc.*

Bugs form a large and varied group. They are usually classified as short-horned (aquatic bugs with very short, concealed antennae) and long-horned (mostly terrestrial with long antennae). The latter group contains both crop pests such as the plant bugs, chinch bugs, squash bugs and most stink bugs, and beneficial predators such as the damsel bugs, pirate bugs, bigeyed bugs, certain assassin bugs and certain stink bugs. It also contains the

Figure 4:10. Hemiptera. A, **Coreidae,** squash bug; B, **Pentatomidae,** harlequin bug; C, **Miridae,** tarnished plant bug. *A and B, courtesy USDA; C, courtesy Illinois Natural History Survey.*

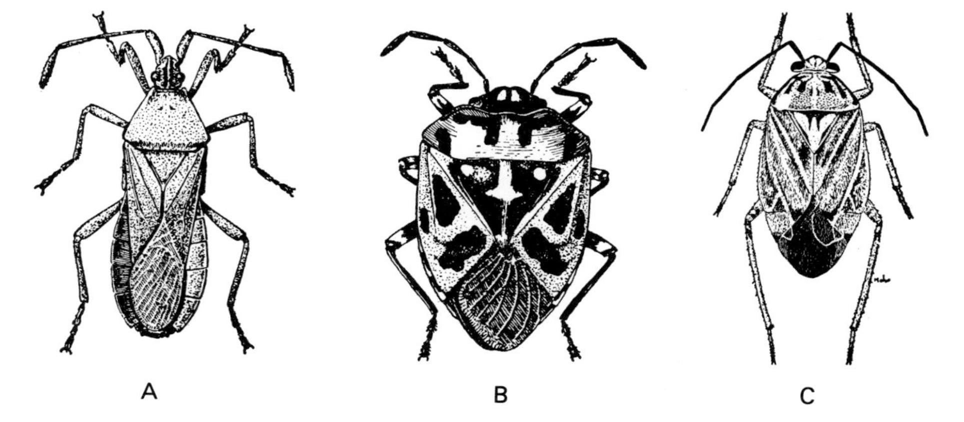

bloodsucking conenose bugs, which transmit the dread Chagas disease of man in South and Central America, and the bed bugs, which are annoying pests of man and animals.

KEY TO THE MAJOR ECONOMIC FAMILIES OF HEMIPTERA

1. Antennae shorter than head; concealed in grooves beneath the eyes (short-horned bugs: most of the aquatic bugs) ....
   Antennae longer than head and plainly visible from above (long-horned bugs: terrestrial bugs and water striders) 2

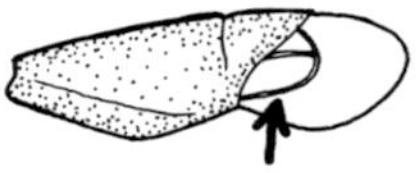

Figure 1

2. Forewings reduced to short pads; body flattened and modified for ectoparasitic habit (see Fig. 19:1E) .... bed bugs: *Cimicidae*
   Forewings not reduced to short pads; body not flattened and modified for ectoparasitic habit 3

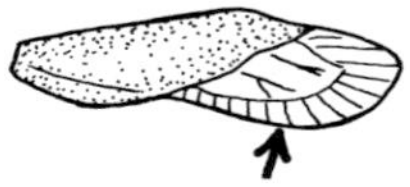

Figure 2

3. Membranous portion of forewing with two closed cells (Fig. 1) .. plant bugs: *Miridae*
   Membranous portion of forewing without two closed cells 4

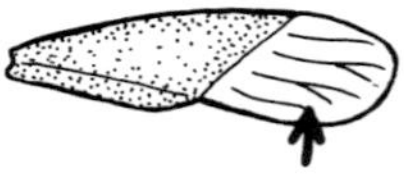

Figure 3

4. Membrane with row of small cells around margin (Fig. 2) .... damsel bugs: *Nabidae*
   Membrane without row of small cells around margin 5

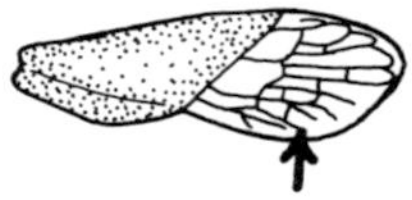

Figure 4

5. Membrane with 4–5 open veins (Fig. 3) .... chinch bugs: *Lygaeidae*
   Membrane without 4–5 open veins (may be many) 6

Figure 5

6. Membrane with many branched veins and cells but without numerous longitudinal veins (Fig. 4) .... red bugs: *Pyrrhocoridae*
   Membrane without many branched veins and cells but with numerous longitudinal veins (Fig. 5) 7

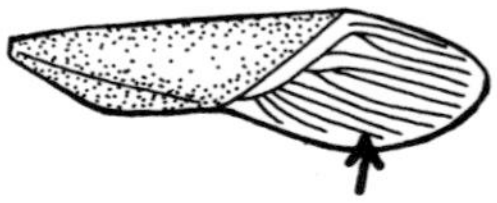

7. Antennae four-segmented, scutellum usually not large ..... squash bugs: *Coreidae*
   Antennae five-segmented; scutellum very large (Fig. 6) .. stink bugs: *Pentatomidae*

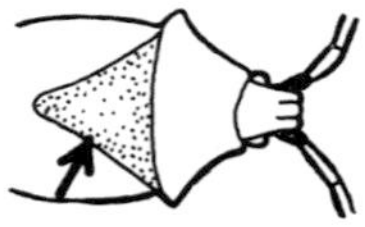

Figure 6

Order Homoptera: Cicadas, Treehoppers, Spittlebugs, Leafhoppers, Aphids, Whiteflies, Scale Insects, and Others (Figs. 4:11 and 4:12)

Figure 4:11. Homoptera. A, **Membracidae**, buffalo treehopper; B, **Cicadellidae**, apple leafhopper; C, **Psyllidae**, pear psylla. *A, courtesy USDA; B, from Essig, 1958,* Insects and Mites of Western North America, *by permission Macmillan Publishing Co., Inc.; C, courtesy Conn. Agr. Exp. Sta.*

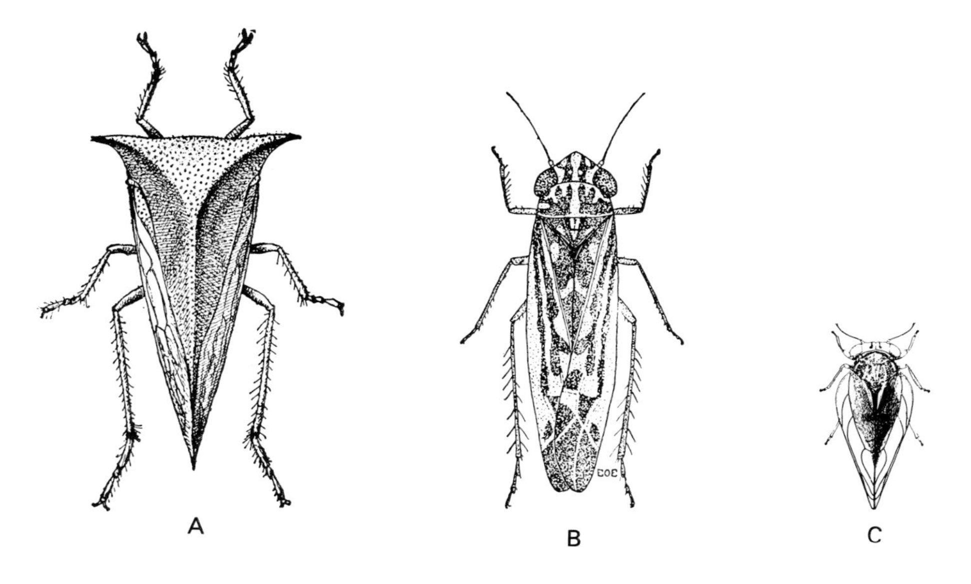

Figure 4:12. Homoptera. A, **Aphididae**, turnip aphid; B, **Phylloxeridae**, grape phylloxera; C, **Pseudococcidae**, grape mealybug; D, **Coccidae**, cottonycushion scale. *A, B, and C, courtesy USDA; C, original.*

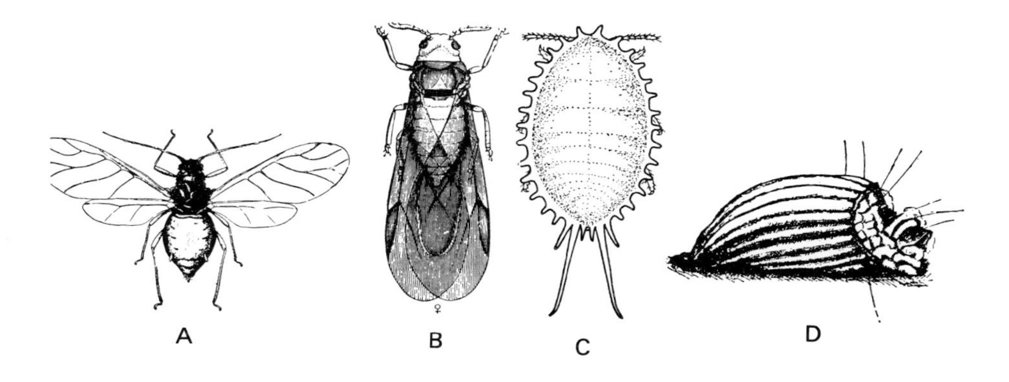

Mouthparts—piercing-sucking.

Wings—usually sloping over sides of body at rest; forewings uniform texture throughout; some forms wingless.

Metamorphosis—simple.

Additional feature—base of sucking beak extending between anterior coxae ("beak" arises at rear of underside of head).

Order *Homoptera* is a diverse group, closely related to the bugs. Those with bristlelike antennae include cicadas, leafhoppers, and spittlebugs. Those with threadlike antennae include psyllids, whiteflies, aphids, scale insects and mealybugs. Many have quite complex life cycles. For example, aphids typically pass through a number of distinct female forms: stem mothers that develop from winter eggs, spring migrants, summer females, fall migrants, and egg-laying females. Winged males that develop in late fall mate with the wingless oviparous forms to complete the cycle. Most of the important vectors of plant viruses are in this order.

KEY TO THE MAJOR ECONOMIC FAMILIES OF HOMOPTERA

1. Antennae setaceous (bristlelike) (Fig. 1) . . . 2
   Antennae filiform (threadlike) (Fig. 2) or rudimentary 4

2. Pronotum extending backward over abdomen (Fig. 3) . . treehoppers: *Membracidae*
   Pronotum not extending backward over abdomen 3

3. Hind tibiae with one or more rows of small spines (Fig. 4) . leafhoppers: *Cicadellidae*
   Hind tibiae with one or two stout spines and usually a circlet of spines at apex (Fig. 5) . . . . spittlebugs: *Cercopidae*

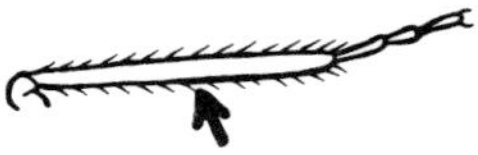

4. Tarsi two-segmented and with two claws . 5
   Tarsi one-segmented and with single claw (Coccoidea) 8

5. Hind femora enlarged for jumping (Fig. 6); antennae with 5–10 (usually 10) segments . . . . psyllids: *Psyllidae*
   Hind femora not enlarged for jumping; antennae with 3–7 segments 6

6. Wings opaque, usually covered with white powdery wax (Fig. 7) . . . . whiteflies: *Aleyrodidae*
   Wings transparent when present 7

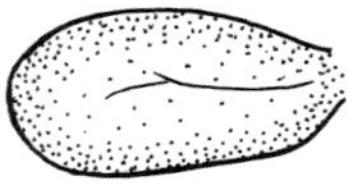

Figure 1

Figure 2

Figure 3

Figure 4

Figure 5

Figure 6

Figure 7

7. Cornicles (pair of tubules on top of rear of abdomen) usually present and conspicuous (Fig. 8); wing venation not highly reduced .... aphids: *Aphididae*
   Cornicles not present; wing venation highly reduced (Fig. 9) ....
   ....bark aphids, gall aphid, and phylloxerans: *Phylloxeridae*
8. Body usually hidden by waxy or scalelike covering; sessile during most of life 9
   Body covered with powdery wax; mobile throughout life (see Fig. 4:12C) ....
   .... mealybugs: *Pseudococcidae*
9. Body covered with hardened shell formed from wax, shed skins, and fibrous material easily removable (see Fig. 14:17E); females without posterior end cleft ....
   .... armored scales: *Diaspididae*
   Body covered with soft wax not easily removable (see Fig. 4:12D); if not covered with soft wax, then females with hard, smooth, often greatly convex, exoskeleton with posterior end cleft (see Fig. 14:2) ...
   .... soft scales: *Coccidae*

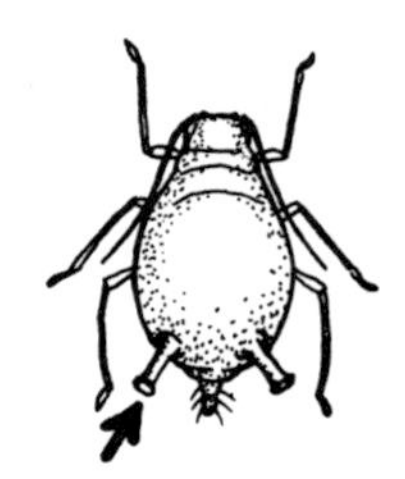

Figure 8

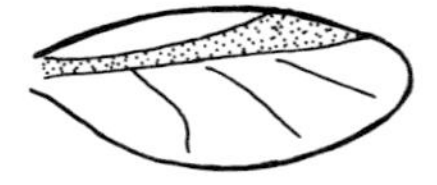

Figure 9

Order Neuroptera: Lacewings, Dobsonflies, Antlions, and Others (Fig. 4:13)

Figure 4:13. Neuroptera. A, green lacewing adult; B, green lacewing larva. *A, courtesy Illinois Natural History Survey; B, courtesy Cornell University.*

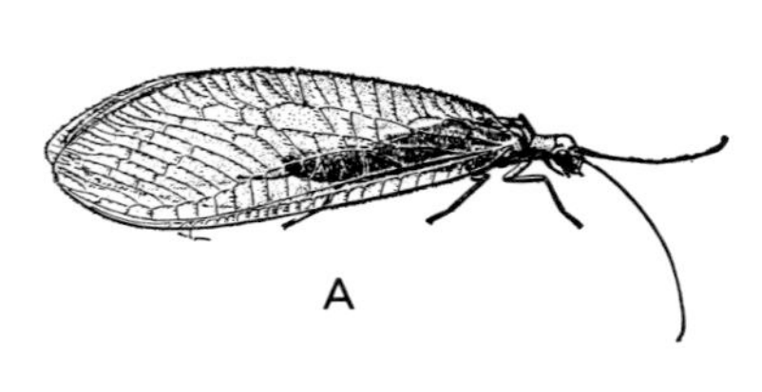

A

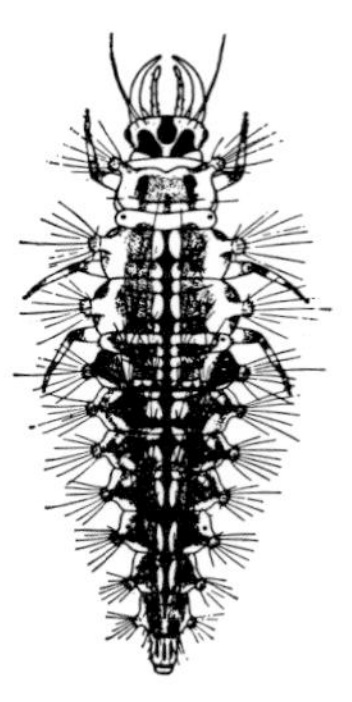

B

Mouthparts—chewing.

Wings—two pairs large membranous wings with many cross veins and longitudinal veins, held rooflike over abdomen at rest.

Metamorphosis—complex.

Additional features—antennae usually long and many-segmented; cerci absent.

Most neuropterans are predaceous. The larvae have a flattened, wedge-shaped body and long sicklelike jaws, giving them the appearance of a miniature alligator. They suck the juices of aphids, scale insects, and other prey through their grooved jaws.

Order Coleoptera: Beetles and Weevils (Figs. 4:14, 4:15, and 4:16)

Figure 4:14. Coleoptera. A, **Carabidae**, ground beetle; B, **Elateridae**, click beetle; C, **Buprestidae**, metallic wood borer. *A, courtesy Kansas State University; B, courtesy USDA; C, courtesy J. N. Knull.*

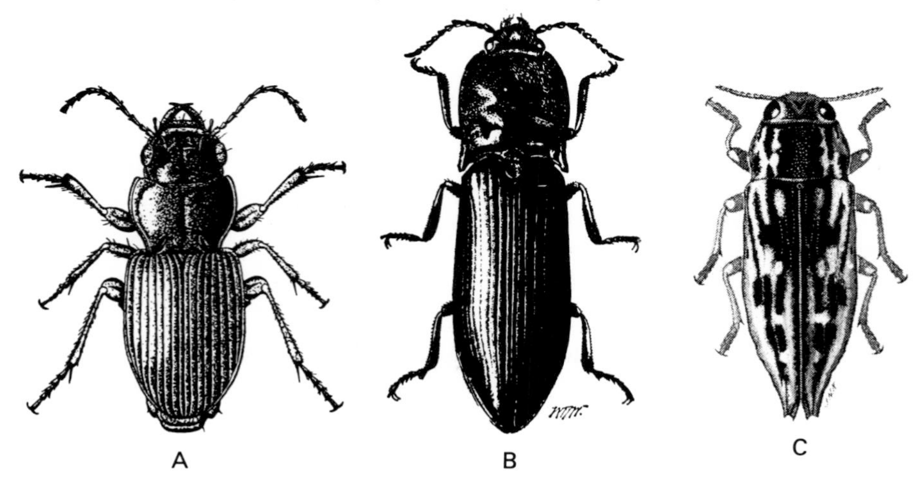

Figure 4:15. Coleoptera. A, **Coccinellidae**, lady beetle; B, **Scarabaeidae**, June beetle; C, **Cerambycidae**, locust borer. *A, from Essig, 1958,* Insects and Mites of Western North America, *by permission Macmillan Publishing Co., Inc.; B, C from USDA.*

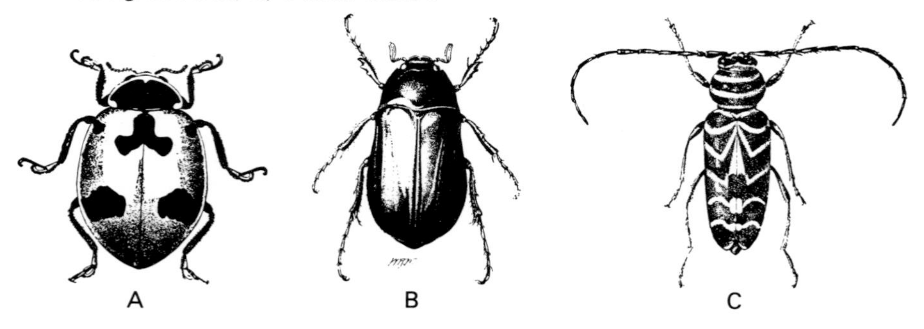

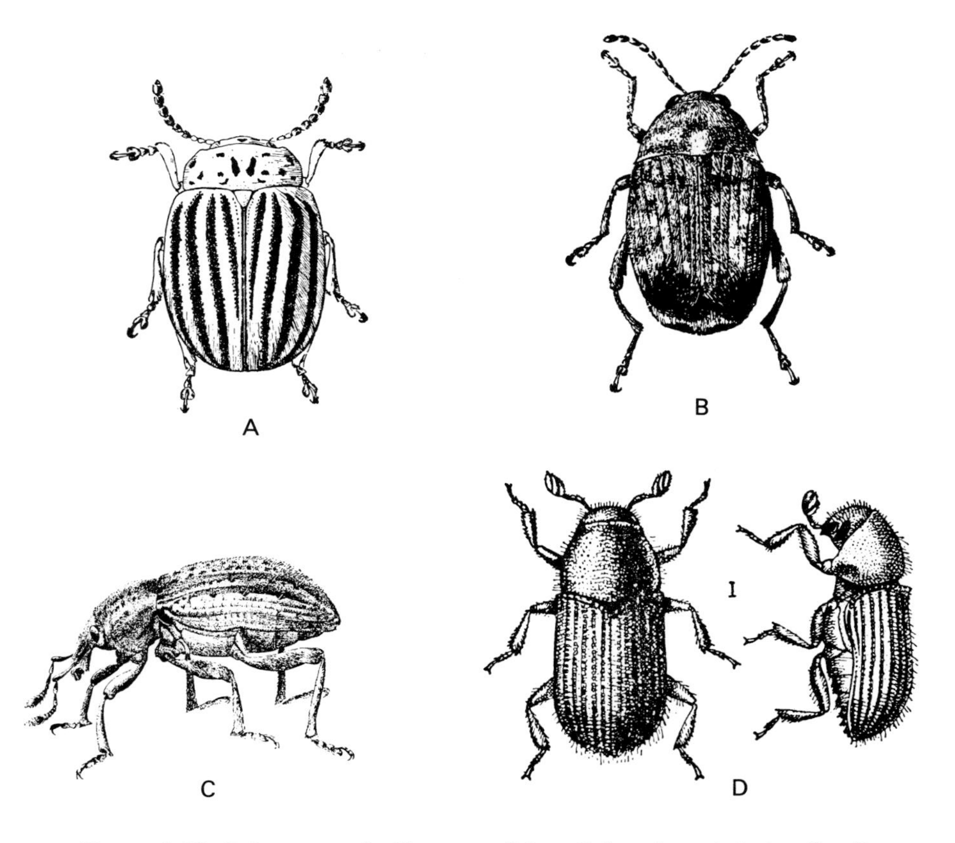

Figure 4:16. Coleoptera. A, **Chrysomelidae,** Colorado potato beetle; B, **Bruchidae,** pea weevil; C, **Curculionidae,** clover leaf weevil; D, **Scolytidae,** peach bark beetle. *Courtesy USDA.*

Mouthparts—chewing.

Wings—forewings are elytra (hard, shell-like covers) which meet in a straight line over the abdomen when not in flight; hindwings membranous.

Metamorphosis—complex.

Additional features—prothorax large and mobile; mesothorax much reduced.

Beetles are the largest order of insects; within the order, the weevils are the largest family, numbering at least 70,000 species. Beetles feed on all types of plants and plant parts and on stored products. Ground beetles, which are beneficial predators, are in the divided sternum group, whereas most pest types, wireworms, white grubs, leaf beetles, and flour beetles are in the undivided sternum group. Beetles are an exception to the rule among insects with complex metamorphosis in that the larvae and the adults often feed on the same material and have somewhat similar habits.

KEY TO THE MAJOR ECONOMIC FAMILIES OF COLEOPTERA

1. Head not prolonged into a snout; gular sutures (on underside of head) double .. 2
   Head usually prolonged into a snout; gular sutures fused or lacking 13

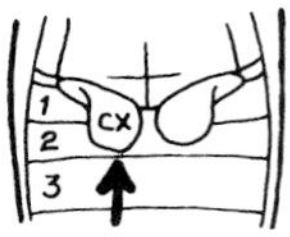

2. First abdominal sternum divided by hind coxae (Fig. 1) ....
   .... ground beetles: *Carabidae*
   First abdominal sternum not divided by hind coxae 3

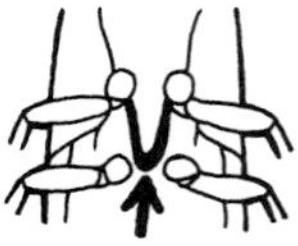

3. Click mechanism (prosternal spine fitting into groove in mesosternum) present (Fig. 2) .. click beetles, wireworms: *Elateridae*
   Click mechanism not present 4

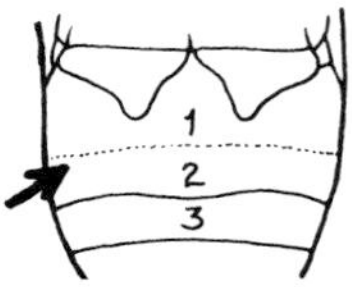

4. First two abdominal segments fused (Fig. 3); body usually metallic ....
   .... flatheaded wood borers: *Buprestidae*
   First two abdominal segments not fused; body not usually metallic 5

5. Hind coxae dilated and grooved for reception of femora; often small hairy or scaly beetles 6
   Hind coxae not dilated and grooved; not hairy or scaly beetles 7

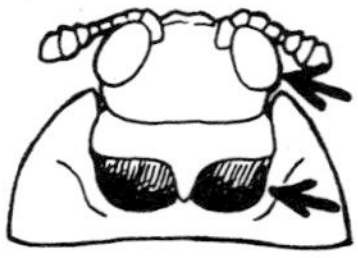

6. Head concealed from above; front coxal cavities open behind (Fig. 4) ....
   .... skin beetles: *Dermestidae*
   Head not concealed from above; front coxal cavities closed behind (Fig. 5) ....
   .... fruitworm beetles: *Byturidae*

7. Tarsi apparently three-segmented (third segment minute and fused to base of fourth) (Fig. 6); body almost hemispherical ....
   .... lady beetles: *Coccinellidae*
   Tarsi not apparently three-segmented; body not almost hemispherical 8

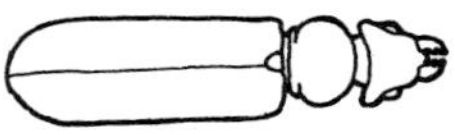

8. Body highly flattened and narrow (Fig. 7) flat bark beetles, grain beetles: *Cucujidae*
   Body not highly flattened and narrow ... 9

Figure 1

Figure 2

Figure 3

Figure 4

Figure 5

Figure 6

Figure 7

9. Tarsal formula 5–5–4 (Fig. 8) .... darkling beetles, flour beetles: *Tenebrionidae*
   Tarsal formula not 5–5–4 10
10. Antennae with club formed from movable plates (lamellate) (Fig. 9) .... white grubs: *Scarabaeidae*
    Antennae without club formed from movable plates 11
11. Elytra short, exposing tip of abdomen; pronotum greatly narrowed anteriorly (Fig. 10) ...... seed beetles: *Bruchidae*
    Elytra not short; pronotum not generally narrowed anteriorly 12
12. Antennae usually less than $\frac{1}{2}$ length of body; usually small, rounded, and brightly colored (see Fig. 4:16A) .... leaf beetles: *Chrysomelidae*
    Antennae at least $\frac{1}{2}$ length of body, often longer; small to large, elongate, and with or without color patterns (see Fig. 4:15C) roundheaded wood borers: *Cerambycidae*
13. Head prolonged into a definite snout (Fig. 11); small to large beetles of varying shapes .... weevils: *Curculionidae*
    Head slightly prolonged into an obscure snout (Fig. 12); small, cylindrical beetles (usually less than $\frac{1}{3}$ inch in length) .... bark beetles: *Scolytidae*

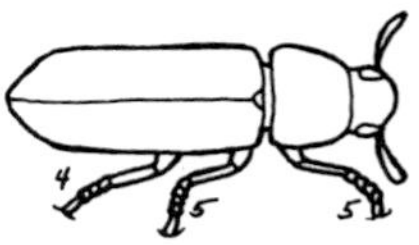

Figure 8

Figure 9

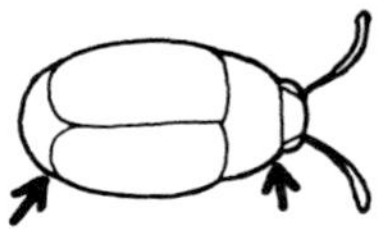

Figure 10

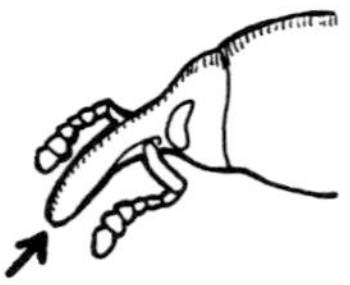

Figure 11

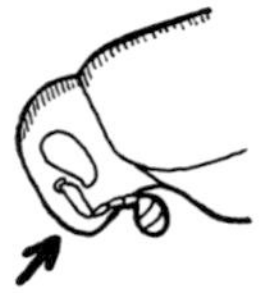

Figure 12

Order Lepidoptera: Butterflies and Moths (Figs. 4:17 through 4:23)

Mouthparts—chewing in larvae; siphoning in adults.

Wings—two pairs of membranous wings covered with a layer of minute scales.

Metamorphosis—complex.

Additional features—legs relatively small; caterpillars (larvae) have cluster of lateral ocelli on each side of head (as contrasted to one on each side in the caterpillarlike larvae of some Hymenoptera).

Moths and butterflies are easily distinguished: moths have various forms of antennae (not knobbed), relatively large bodies, usually are night-fliers, and pupate in silken cocoons; butterflies, on the other hand, have knobbed antennae, slender bodies, are day-fliers, and have naked pupae (chrysalids). Many of the destructive pest species are moths. The larvae, called caterpillars, feed voraciously on plants or plant materials with their chewing mouthparts.

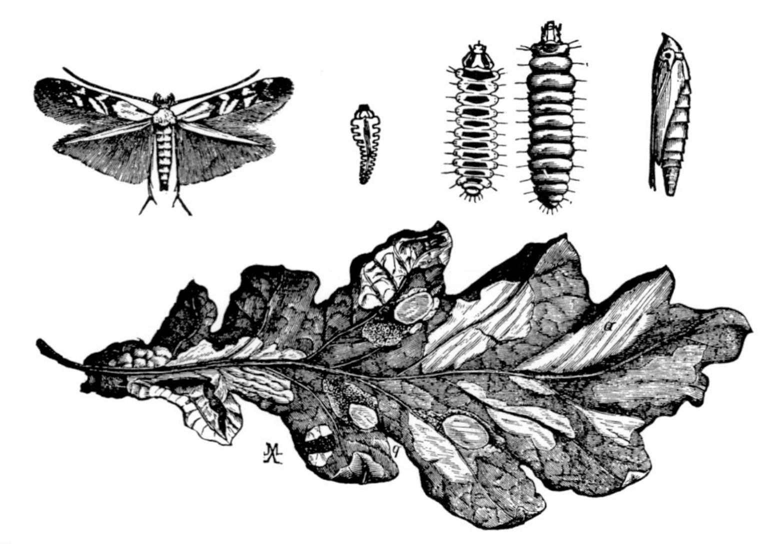

Figure 4:17. Lepidoptera. **Gracillariidae**, Solitary oak leafminer. Adult moth, three larval instars, pupa, and oak leaf showing mines of larvae. *From Comstock, 1940,* An Introduction to Entomology, *by permission Cornell University Press, Ithaca, N.Y.*

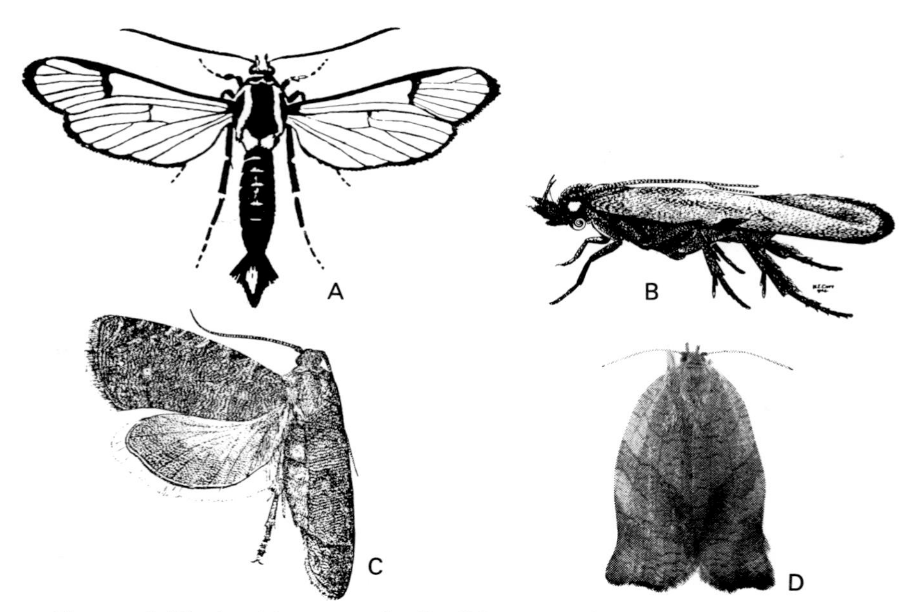

Figure 4:18. Lepidoptera. A, **Sesiidae,** peachtree borer adult; B, **Gelechiidae**, peach twig borer adult; C, **Olethreutidae**, oriental fruit moth; D, **Tortricidae**, obliquebanded leafroller adult. *A and D, courtesy USDA; B, courtesy Calif. Agr. Exp. Sta.; C, courtesy Wood and Selkregg.*

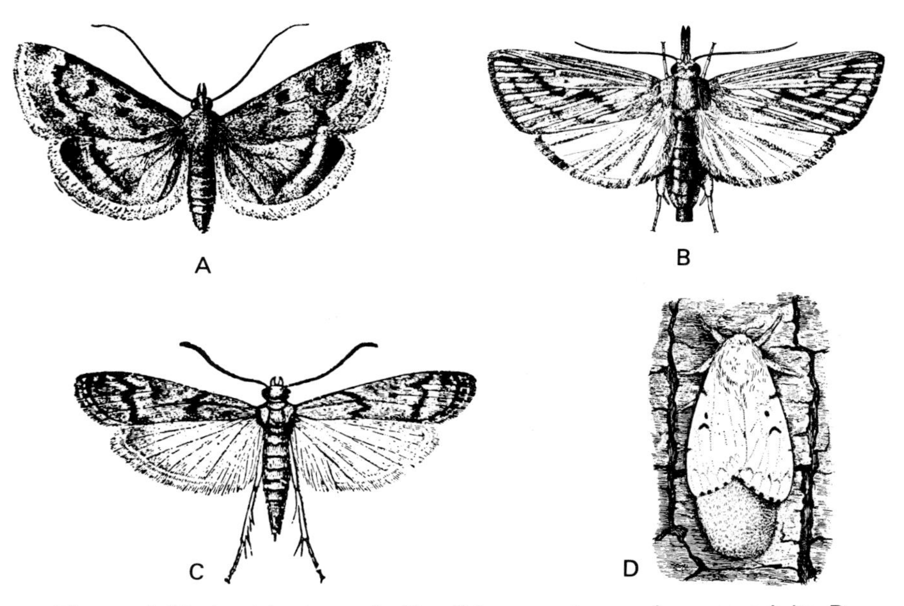

Figure 4:19. Lepidoptera. A, **Pyralidae**, garden webworm adult; B, **Pyralidae**, sugarcane borer adult; C, **Pyralidae**, Mediterranean flour moth; D, **Lymantriidae**, Gypsy moth. *A, B, and C, courtesy USDA*; *D, courtesy Conn. Agr. Exp. Sta.*

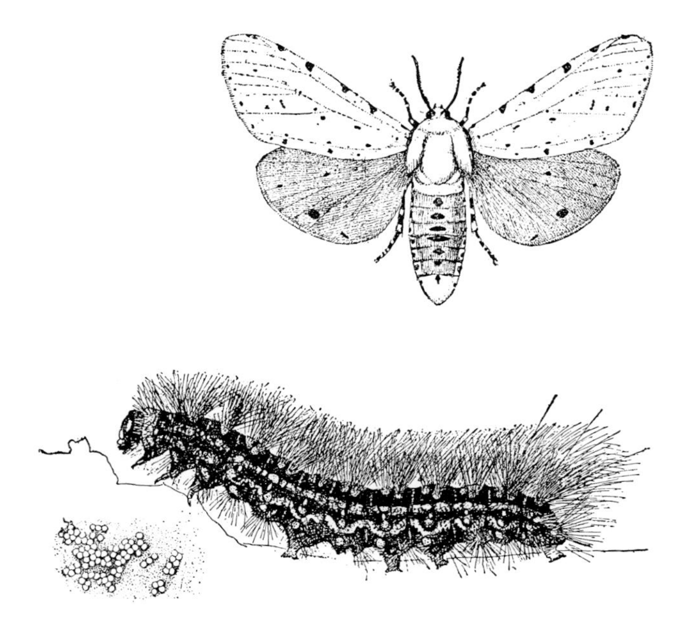

Figure 4:20. Lepidoptera. **Arctiidae**, Saltmarsh caterpillar. Adult moth, caterpillar and eggs. *Courtesy USDA.*

Figure 4:21. Lepidoptera. A, **Sphingidae**, tobacco hornworm adult; B, **Pieridae**, alfalfa caterpillar adult. *Courtesy USDA*.

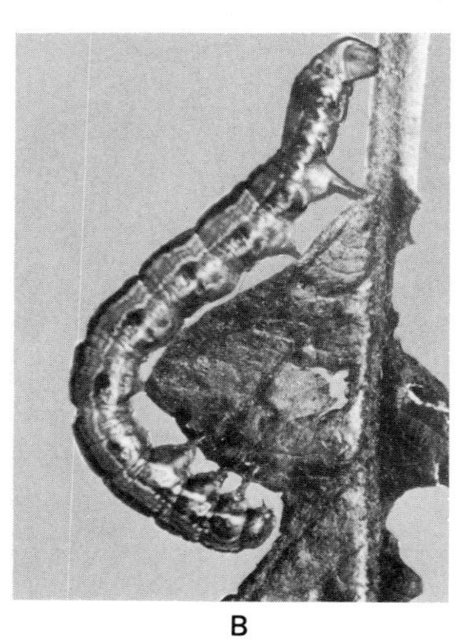

Figure 4:22. Lepidoptera. **Geometridae**, fall cankerworm. A, female moth laying eggs, note that it is wingless; B, larva or cankerworm, feeding; C, pupa inside cocoon in soil; D, the winged male moth. *Courtesy USDA*.

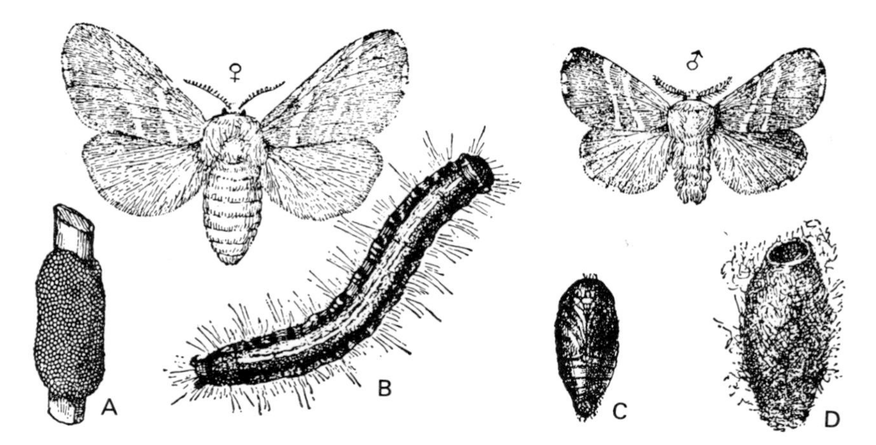

Figure 4:23. Lepidoptera. **Lasiocampidae,** eastern tent caterpillar. Top, female and male moth; A, eggs; B, caterpillar; C, pupa; D, cocoon. *Courtesy USDA.*

The cutworm family is the largest group, with about 5,000 species in North Temperate regions. They are known by such names as armyworms, bollworms, earworms and loopers and comprise many of the most directly destructive crop pests in the world. Another family, the pyralid moths, include eighteen rice stem borers, which are the major pests of this staple crop of two thirds of the world's humans.

KEY TO THE MAJOR ECONOMIC FAMILIES OF LEPIDOPTERA

1. Antennae threadlike and knobbed at tip (Fig. 1); no frenulum (spine or spines on leading edge of hindwing which unite fore- and hindwings in flight); ocelli absent (butterflies) 2

   Antennae of various forms, but usually not knobbed (Fig. 2); if antennae clubbed, then frenulum present (Fig. 3); ocelli often present (moths) 3

2. Small to medium-sized butterflies; white, yellow, or orange wings often marked with black (see Fig. 4:21B) .... white butterflies: *Pieridae*

   Small butterflies; metallic blue, green, copper, or bronze wings, sometimes with bright color markings and/or tiny tail-like projections (Fig. 4) .... blues, coppers, and hairstreaks: *Lycaenidae*

Figure 1

Figure 2

Figure 3

Figure 4

3. Hindwings with soft scales and small anal regions; palpi not unusually large and forming a snoutlike projection 4
   Hindwings with firm, fine scales and ample anal regions (Fig. 5); palpi often large and forming snout-like projection (Fig. 6) . . . . . flour moths, grass moths, etc.: *Pyralidae*

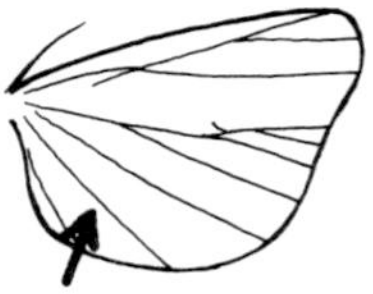

Figure 5

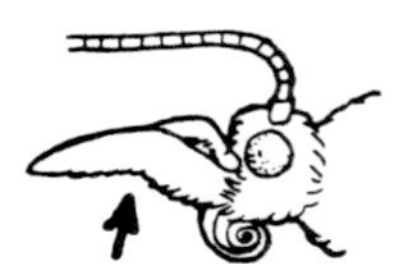

Figure 6

4. Forewings usually long, narrow, and pointed; hindwings usually short; body stout and tapered at both ends (see Fig. 4:21A) . . . . sphinx moths: *Sphingidae*
   Forewings not long, narrow, and pointed; hindwings often nearly as large as forewings; body not usually stout and tapered 5

Figure 7

5. Labial palpi with third segment long and slender, usually tapering, the palpi upturned (Fig. 7) . . . . gelechiid moths: *Gelechiidae*
   Labial palpi without third segment long and slender, usually tapering, or the palpi upturned 6

Figure 8

6. A large part of the wings, especially the hindwings, devoid of scales except on veins and margins (Fig. 8) . . . . clearwing moths: *Sesiidae*
   Wings normally scaled throughout and without extensive transparent areas 7

Figure 9

7. Hindwings with second and third veins[1] usually stalked (fused basally to form single vein) (Fig. 9); hindwings approximately equal in expanse to forewings; small to medium-sized moths 8
   Hindwings without second and third veins stalked; hindwings usually distinctly smaller than forewings; small to large moths . 9

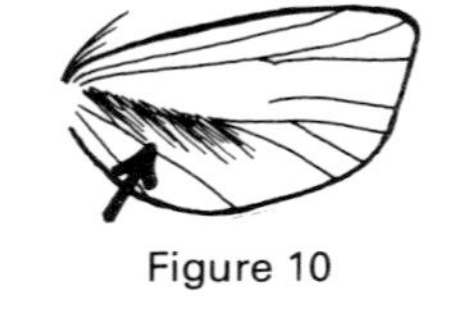

Figure 10

8. Hindwings usually with fringe of long hairs on upper side of basal part of vein 6 (Fig. 10); if fringe lacking, then veins 7 and 8 in front wing close together at tip (Fig. 11); outline of moth not typically bell-shaped at rest . . . . fruit moths: *Olethreutidae*

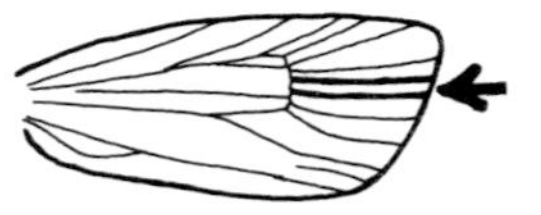

Figure 11

[1] This simplified designation of veins is not a standard system.

Hindwings usually without fringe of long hairs on vein 6; if fringe present, then veins 5 and 6 in front wing stalked (Fig. 12); outline of moth typically bell-shaped at rest (Fig. 13) .... leafroller moths: *Tortricidae*

9. Very small moths with narrow, pointed wings; hind margins of wings with wide fringes of scales 10
   Medium to large-sized moths without narrow, pointed wings; hind margins of wings without wide fringes of scales 11

10. Forewing usually without accessory cell; hindwing often has hump along leading edge near base (Fig. 14) .... leafblotch miner moths: *Gracillariidae*
    Forewing usually with accessory cell (Fig. 15); hindwing tapers smoothly to apex .... clothes moths: *Tineidae*

11. Proboscis rarely absent; body slender; legs slender, with few or no hairs; forewings marked with wavy parallel bands (Fig. 16) .... measuringworm moths: *Geometridae*
    Proboscis reduced or absent; body robust; legs well developed, either very hairy or spiny; forewings without wavy parallel bands 12

12. Ocelli absent; tympanum not developed on metepimeron (just below base of hindwing); frenulum absent; base of leading edge of hindwing greatly expanded (Fig. 17) .... tent caterpillar moths: *Lasiocampidae*
    Ocelli usually present (except *Lymantriidae*); tympanum developed on metepimeron (Fig. 18); frenulum present; base of leading edge of hindwing not greatly expanded 13

13. Forewings brightly marked in contrasting colors, sometimes plain white or yellow; vein below discal cell in hindwing appears to be four-branched (Fig. 19) .... tiger moths: *Arctiidae*

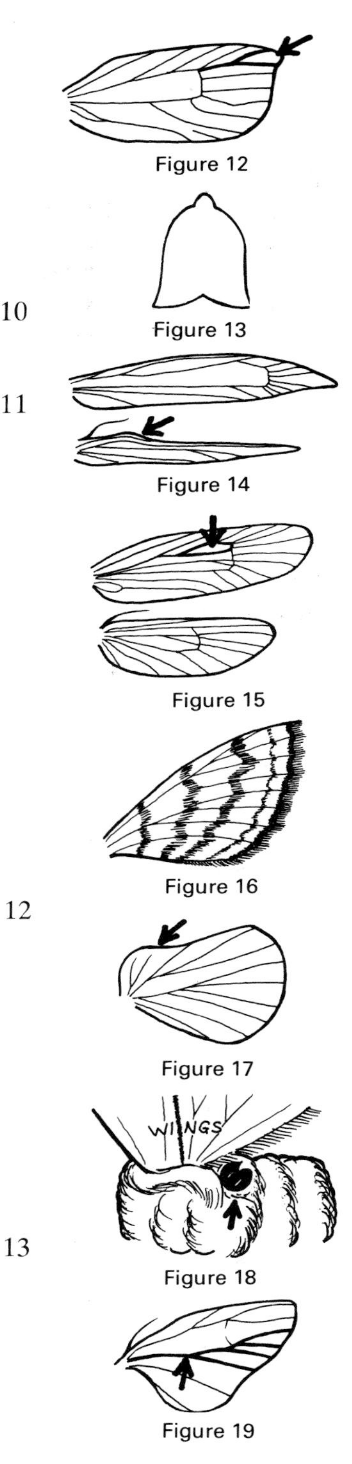

Figure 12
Figure 13
Figure 14
Figure 15
Figure 16
Figure 17
Figure 18
Figure 19

Forewings usually dull gray or brown; vein below discal cell in hindwing often appears three-branched (Fig. 20) 14

14. Antennae usually threadlike; ocelli usually present; leading edges of wings quite straight (Fig. 21) .... cutworm moths: *Noctuidae*
Antennae feathery; ocelli absent; leading edges of wings rounded (Fig. 22) .... tussock moths: *Lymantriidae*

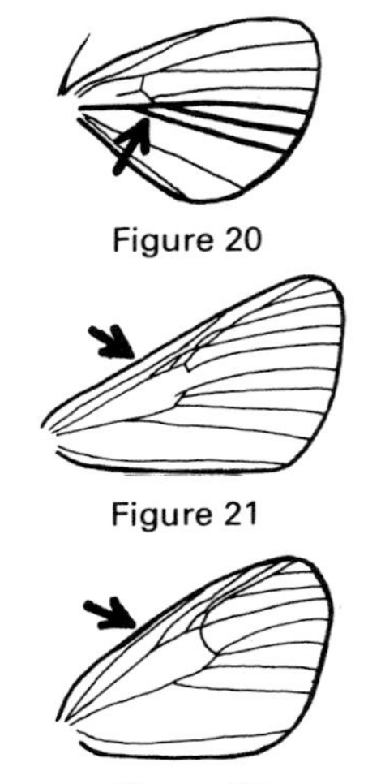
Figure 20

Figure 21

Figure 22

Order Hymenoptera: Sawflies, Ants, Wasps, and Bees (Figs. 4:24 through 4:28)

Figure 4:24. Hymenoptera. **Tenthredinidae**, A, cherry fruit sawfly; B, **Cephidae**, wheat stem sawfly. *Courtesy USDA.*

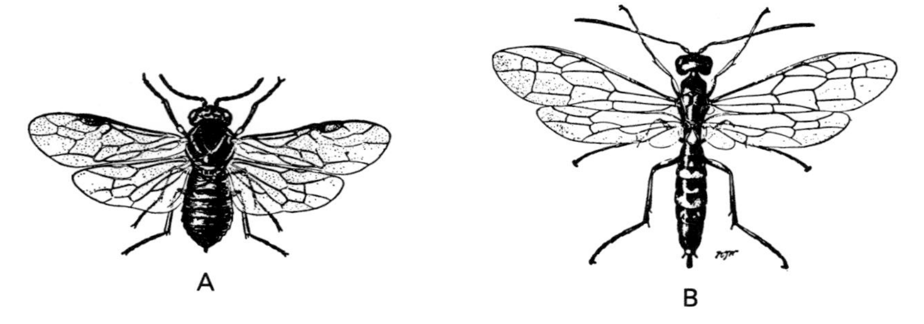

Figure 4:25. Hymenoptera. A, **Braconidae**, *Apanteles thompsoni*, a parasite of European corn borer; B, **Eurytomidae**, wheat jointworm, a chalcid; C, **Eulophidae**, *Aphelinus mali*, a parasite of woolly apple aphid, a chalcid. *Courtesy USDA.*

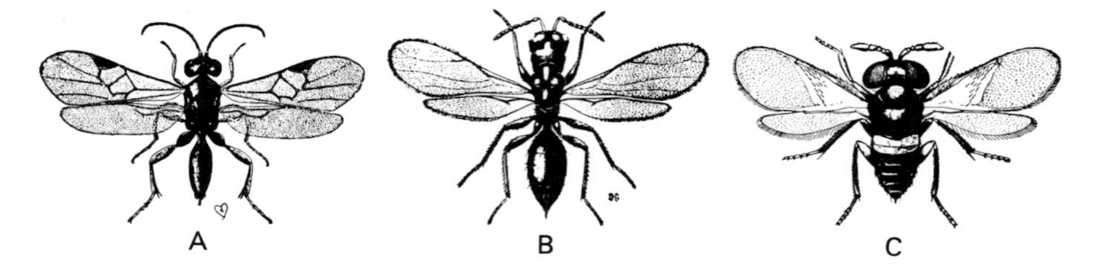

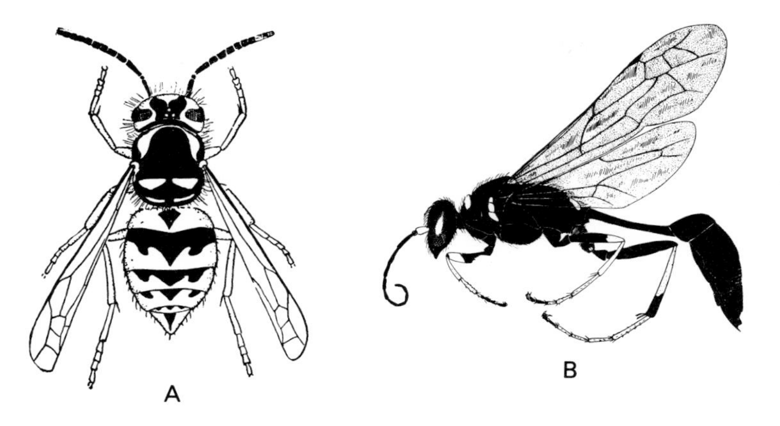

Figure 4:26. Hymenoptera. A, **Vespidae**, western yellowjacket; B, **Sphecidae**, common mud dauber. *A, from Essig, 1958,* Insects and Mites of Western North America, *by permission Macmillan Publishing Co., Inc.; B, from CDC Pictorial Keys, 1967, courtesy Communicable Disease Center, Public Health Service.*

Figure 4:27. Hymenoptera. A, **Halictidae**, alkali bee; B, **Megachilidae**, alfalfa leafcutting bee. These two species are being managed on a commercial scale for production of alfalfa seed. *A, photograph by Jack D. Eves; B, photograph by Roger D. Akre, both, Entomology Department, Washington State University, Pullman.*

Mouthparts—chewing in larvae; chewing or chewing-lapping in adults.

Wings—usually two pairs of membranous wings; the hind pair often attached to the fore pair with minute hooks.

Metamorphosis—complex.

Additional features—abdomen often with slender waist; female with ovipositor often specialized as a piercing, sawing, or stinging organ.

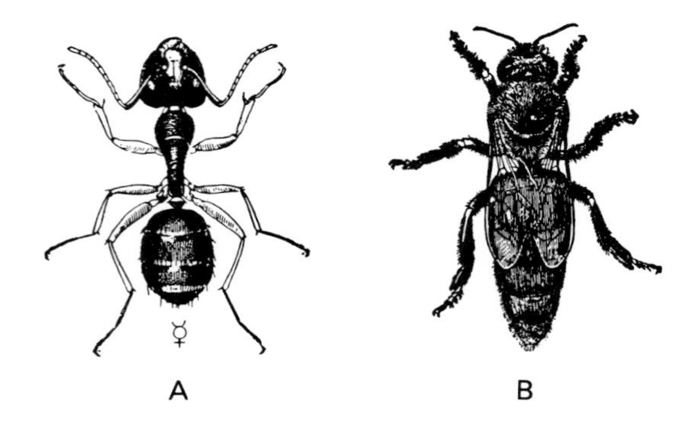

Figure 4:28. Hymenoptera. A, **Formicidae**, black carpenter ant; B, **Apidae**, honey bee queen. *Courtesy USDA*.

Hymenopterans are separated into two groups: the nonconstricted abdomen group, including sawflies and horntails, and the constricted abdomen group, including parasitic wasps, social wasps, threadwaisted wasps, ants, and bees. Most of the group are beneficial parasites, predators, or pollinators. The few species of crop pests are found mainly among the seed chalcids and sawflies. Ants, social wasps, and social bees are some of the most interesting insects because of their immense colonies, development of castes, division of labor, and advanced methods of communication.

KEY TO THE MAJOR ECONOMIC FAMILIES OF HYMENOPTERA

1. Abdomen broadly joined to thorax (nonconstricted abdomen group) 2
   Abdomen joined to thorax by a slender petiole or waist (constricted abdomen group) 3

Figure 1

2. Body robust; abdomen not compressed laterally (see Fig. 4:24A); fore tibiae with two apical spurs .... sawflies: *Tenthredinidae*
   Body slender; abdomen compressed laterally (see Fig. 4:24B); fore tibiae with one apical spur .... stem sawflies: *Cephidae*

Figure 2

3. Posterior trochanter consisting of two segments (Fig. 1) 4
   Posterior trochanter consisting of a single segment 6

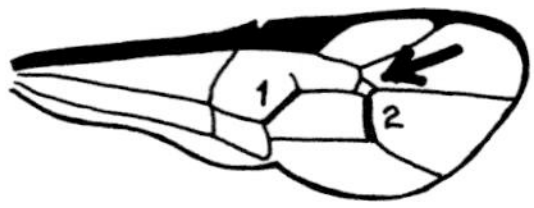

Figure 3

4. Forewings with a conspicuous, dark, thickened spot (stigma) about midway on leading edge; venation not highly reduced 5
Forewings without a conspicuous, dark, thickened spot on leading edge; venation highly reduced (Fig. 2) .... chalcids (19 families): *Chalcidoidea*
5. Two recurrent veins in forewing; small submarginal cell (areolet) often present (Fig. 3) ... ichneumons: *Ichneumonidae*
One recurrent vein in forewing; small submarginal cell not present (Fig. 4) .... braconids: *Braconidae*
6. Antennae geniculate (elbowed); first 1–2 abdominal segments often with dorsal hump (Fig. 5) .... ants: *Formicidae*
Antennae not geniculate; first 1–2 abdominal segments without hump 7
7. First discoidal cell very long (Fig. 6); wings usually folded lengthwise at rest .... social wasps, potter wasps: *Vespidae*
First discoidal cell not unusually long; wings not folded lengthwise at rest 8
8. Body hairs unbranched; abdomen often petiolate (stalked); posterior angle of pronotum lobelike (Fig. 7) .... mud daubers, threadwaisted wasps: *Sphecidae*
Body hairs, especially on thorax, branched (Fig. 8); abdomen not petiolate; posterior angle of pronotum not lobelike (bees) 9
9. Basal vein strongly arched (Fig. 9); often with metallic colors .... sweat bees, mining bees: *Halictidae*
Basal vein not strongly arched; without metallic colors 10
10. With two submarginal cells in forewing; abdomen typically boat-shaped (Fig. 10) .... leafcutting bees: *Megachilidae*
Usually with three submarginal cells in forewing (Fig. 11); abdomen not typically boat-shaped .... honey bees, bumble bees, robust mining bees: *Apidae*

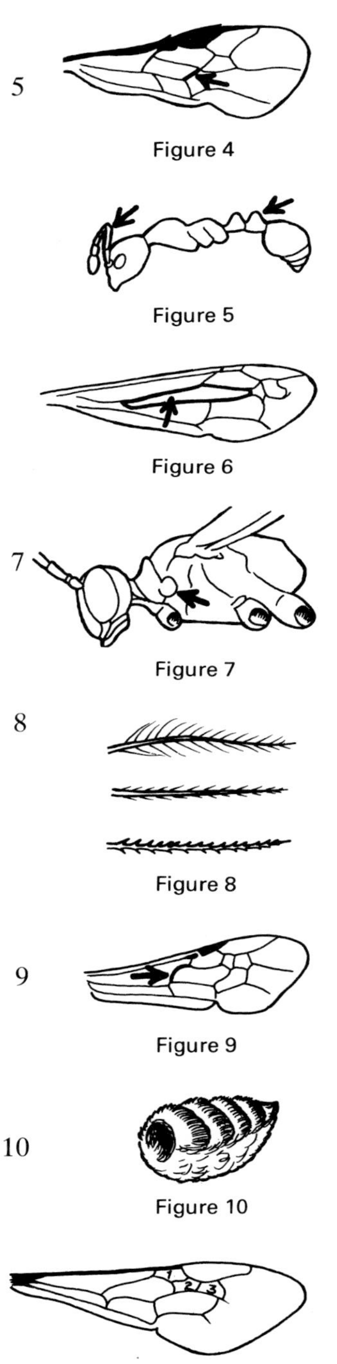

Figure 4
Figure 5
Figure 6
Figure 7
Figure 8
Figure 9
Figure 10
Figure 11

Order Diptera: Flies (Figs. 4:29 through 4:35)

Mouthparts—chewing or reduced in larvae; piercing-sucking, sponging, or vestigial in adults.

Wings—one pair of membranous forewings; hindwings modified as halters.

Metamorphosis—complex.

Additional features—head, thorax, and abdomen quite distinct; compound eyes usually large.

Figure 4:29. Diptera. A, **Psychodidae,** pappataci sand fly; B, **Ceratopogonidae,** a salt marsh biting midge. *A, from* Fauna of U.S.S.R.—Diptera, *1968, courtesy U.S. National Science Foundation*: *B, from Essig, 1942,* College Entomology, *by permission Marie W. Essig.*

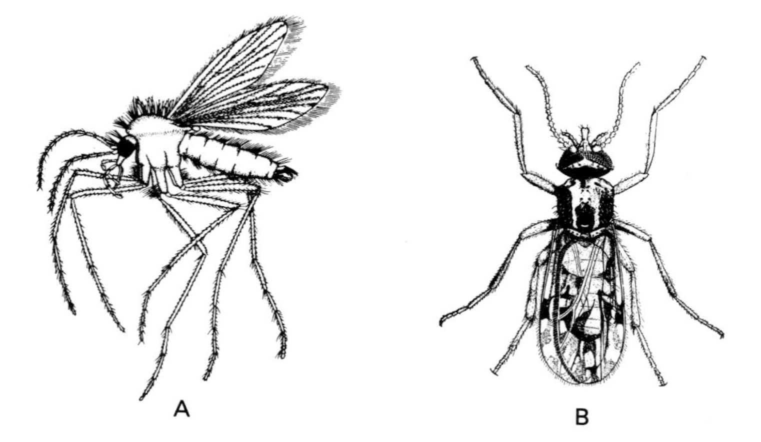

Figure 4:30. Diptera. A, **Culicidae,** yellowfever mosquito; B, **Cecidomyiidae** *Aphidoletes meridionalis,* an aphid predator. *Courtesy USDA.*

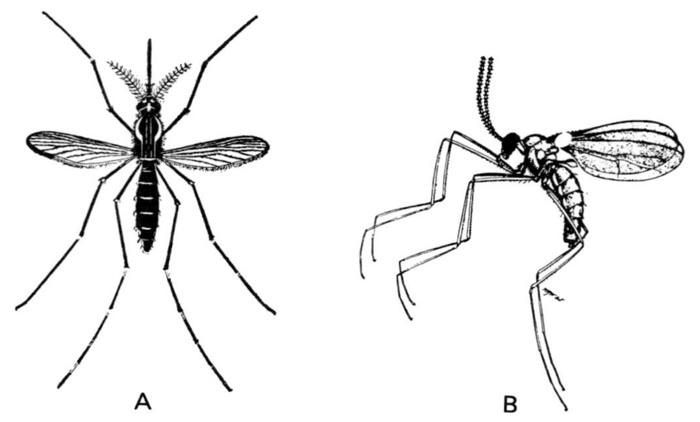

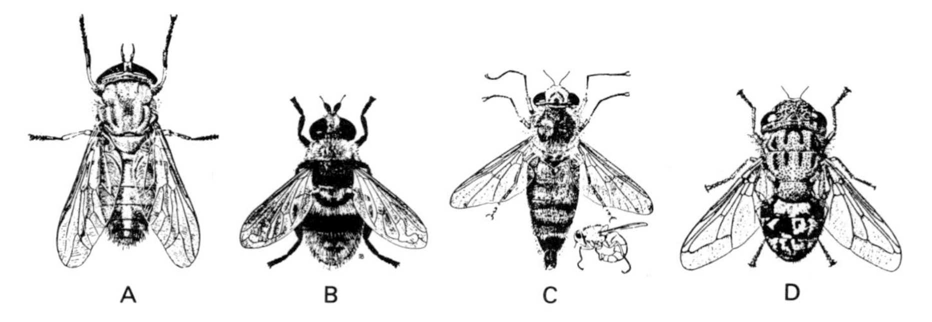

Figure 4:31. Diptera. A, **Tabanidae**, striped horse fly, B, **Syrphidae**, narcissus bulb fly; C, **Gasterophilidae**, horse bot fly; D, **Oestridae**, sheep bot fly. *Courtesy USDA*.

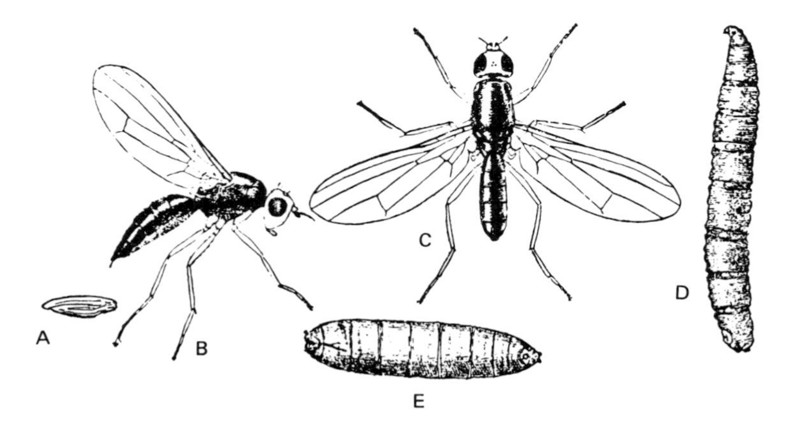

Figure 4:32. Diptera. **Psilidae**, carrot rust fly. A, egg; B, female; C, male; D, larva; E, pupa. *Courtesy Agricultural Extension, British Columbia*.

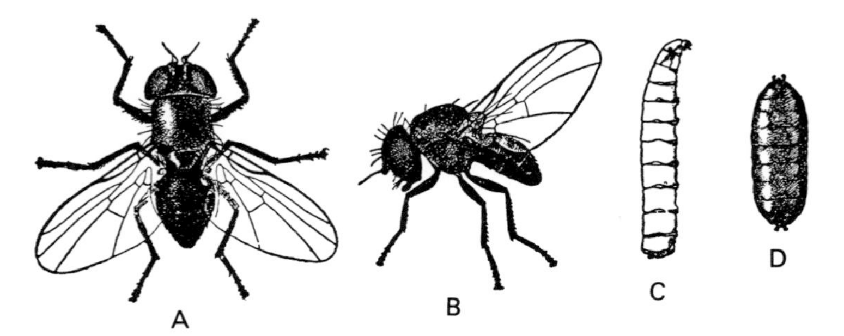

Figure 4:33. Diptera. **Agromyzidae**, Asparagus miner. A and B, adults; C, larva or miner; D, pupa. *Courtesy USDA*.

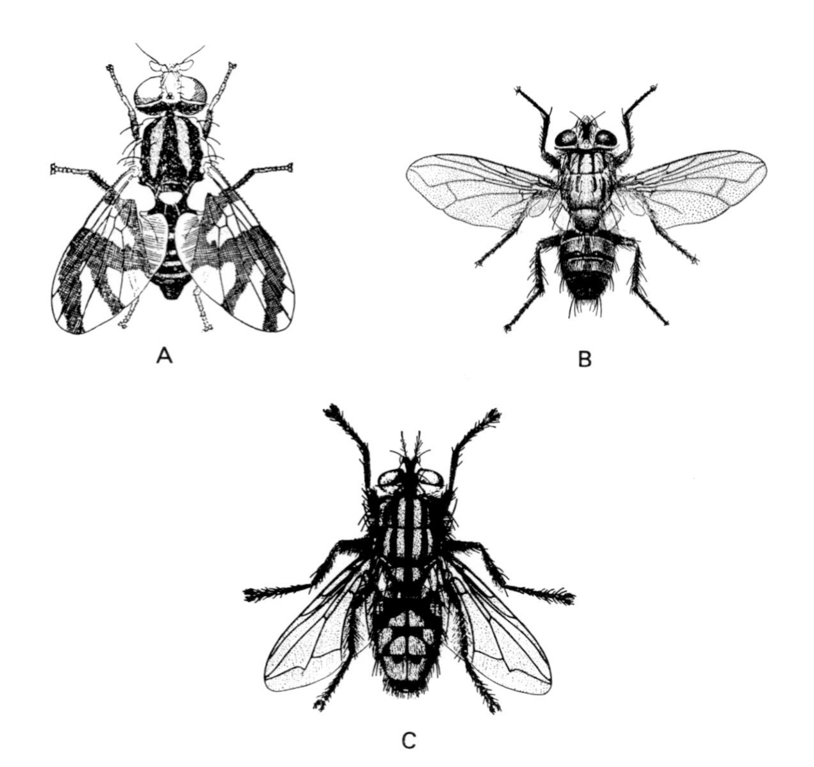

Figure 4:34. Diptera. A, **Tephritidae**, apple maggot adult; B, **Tachinidae**, *Compsilura concinnata*, parasite of satin moth, gypsy moth, browntail moth, and tent caterpillars; C, **Sarcophagidae**, redtailed flesh fly. *A, from Snodgrass,* 1954, Insect Metamorphosis, *B, and C, courtesy USDA*.

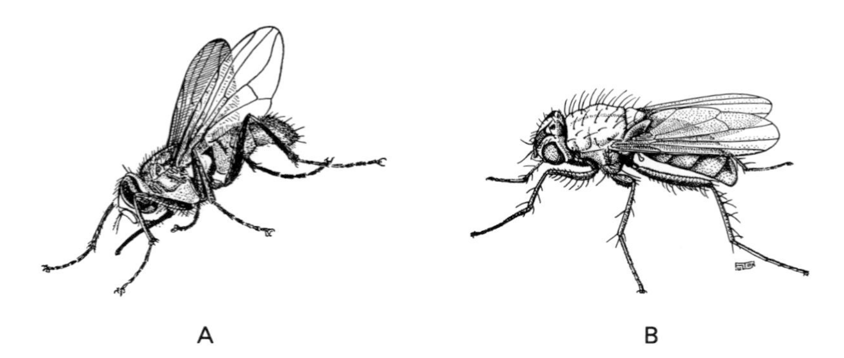

Figure 4:35. Diptera. A, **Muscidae**, stable fly, B, **Anthomyiidae** cabbage maggot adult. *A, courtesy USDA; B, courtesy No. Car. Agr. Exp. Sta.*

Primitive types of flies with long antennae include the sand flies, mosquitoes, biting midges, and gall midges. Higher flies, typically with three-segmented antennae, are separated into those that emerge from the puparium through a T-shaped opening and those that emerge through a circular opening. Horse flies are the only agriculturally important family in the T-group. Flies contain most of the major livestock pests: horse bots, cattle grubs, sheep bots, sheep keds and disease-transmitting mosquitoes, biting midges, horse flies, stable flies, blow flies and house flies. Crop pest groups include fruit flies, rust flies, and root maggots. Major beneficials are the parasitic tachina flies and the predaceous syrphid or flower flies.

KEY TO THE MAJOR ECONOMIC FAMILIES OF DIPTERA

1. Antennae many-segmented (long-horned flies) 2
   Antennae with five or fewer segments, usually three (higher Diptera) 5

2. Wings with scales or dense hair on veins . 3
   Wings without scales or dense hair on veins 4

3. Wings very hairy, giving an opaque moth-like appearance (Fig. 1) .... sand flies: *Psychodidae*
   Wings with scales along the veins (Fig. 2) .... mosquitoes: *Culicidae*

4. Tibiae with spurs; legs not unusually long and slender; costa (vein of leading edge of wing) not continuing around wing; with piercing-sucking mouthparts (Fig. 3) .... biting midges: *Ceratopogonidae*
   Tibiae without spurs; legs unusually long and slender; costa continuing around wing (Fig. 4); without piercing-sucking mouthparts ...... gall midges: *Cecidomyiidae*

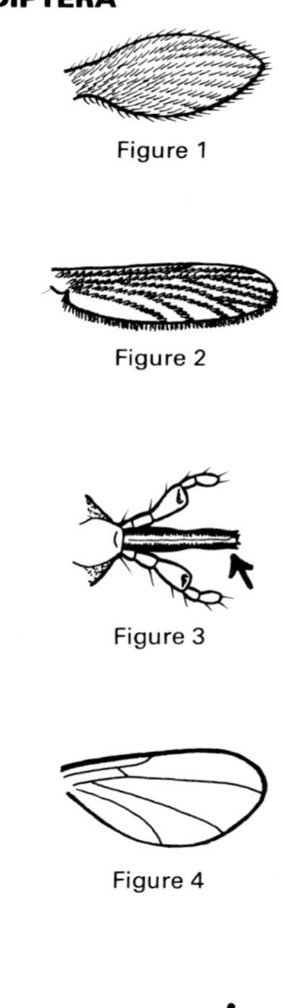

Figure 1

Figure 2

Figure 3

Figure 4

5. Head broad, thin, and convex; antennae with third segment often toothed and distal portion annulate (ringed) (Fig. 5) .... horse flies: *Tabanidae*
   Head not broad, thin, and convex; antennae without third segment toothed or distal portion annulate 6

Figure 5

6. Wing with spurious (false) vein between third and fourth veins (Fig. 6) . . . . flower flies: *Syrphidae*
   Wing without spurious vein 7

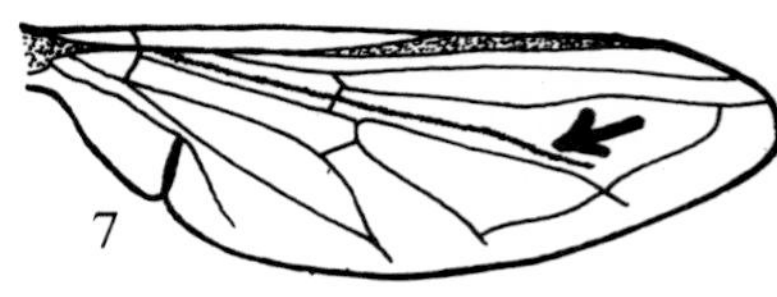

Figure 6

7. Second vein short, bends at right angle towards leading edge of wing; wings usually with bands of color (Fig. 7) . . . . fruit flies: *Tephritidae*
   Second vein not short and with right-angle bend; wings not usually with bands of color 8

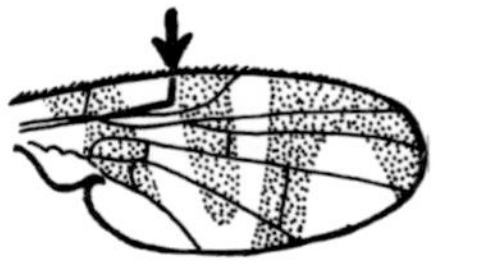

Figure 7

8. Mesonotum without bristles except above wings; characteristic weakened area across basal third of wing (Fig. 8) . . . . rust flies: *Psilidae*
   Mesonotum with more or less complete bristle pattern; no weakened area across basal third of wing 9

Figure 8

9. Postocellar bristles on top of head divergent (Fig. 9), if absent, the arista (large bristle in apical antennal segment) absent . . . . leafminer flies: *Agromyzidae*
   Postocellar bristles not divergent; arista always present 10

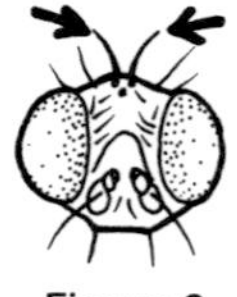

Figure 9

10. Hypopleural bristles absent (Fig. 10); arista plumose (with branches) 11
    Hypopleural bristles present (Fig. 11); arista bare (except in some tachinids) . . . 12

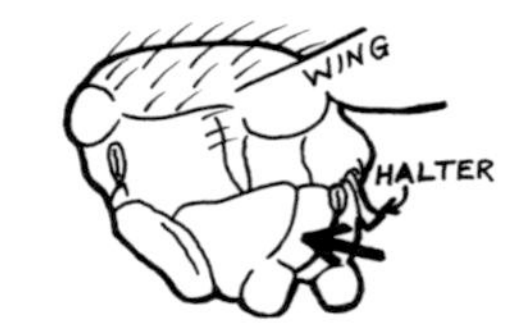

Figure 10

11. Sixth vein reaching wing margin, at least as a fold (Fig. 12) . . . . rootmaggot flies: *Anthomyiidae*
    Sixth vein never reaching wing margin (Fig. 13) . . . . house flies: *Muscidae*

Figure 11

12. Mouthparts vestigial 13
    Mouthparts normally developed 14

13. Third and fourth veins diverge distally (Fig. 14); light yellow-brown in color . . . . horse bots: *Gasterophilidae*
    Third and fourth veins do not diverge distally (Fig. 15); dark brown in color . . . . bot and warble flies: *Oestridae*

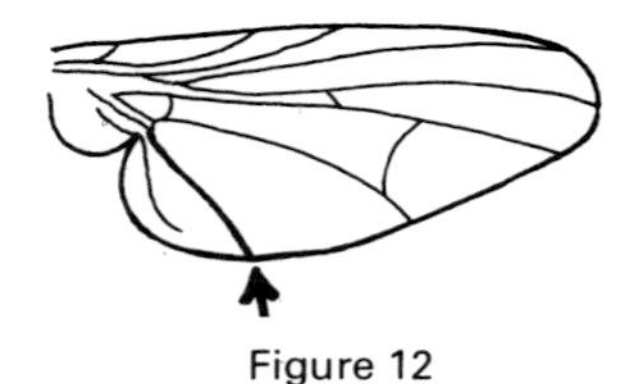

Figure 12

14. Head fitting into emargination of thorax (Fig. 16); usually wingless ....
.... louse flies: *Hippoboscidae*
Head not fitting into emargination of thorax; winged 15

15. Postscutellum (area of thorax immediately below scutellum) developed (Fig. 17); arista usually bare; abdomen usually bristly .... tachina flies: *Tachinidae*
Postscutellum not developed; arista usually plumose; abdomen not bristly 16

16. Thorax typically gray with three black lines; abdomen typically gray checkered, never metallic (see Fig. 4:34C) ....
.... flesh flies: *Sarcophagidae*
Thorax not gray and without black lines; abdomen, with few exceptions, strongly metallic blue or green (see Fig. 20:12)
.... blow flies: *Calliphoridae*

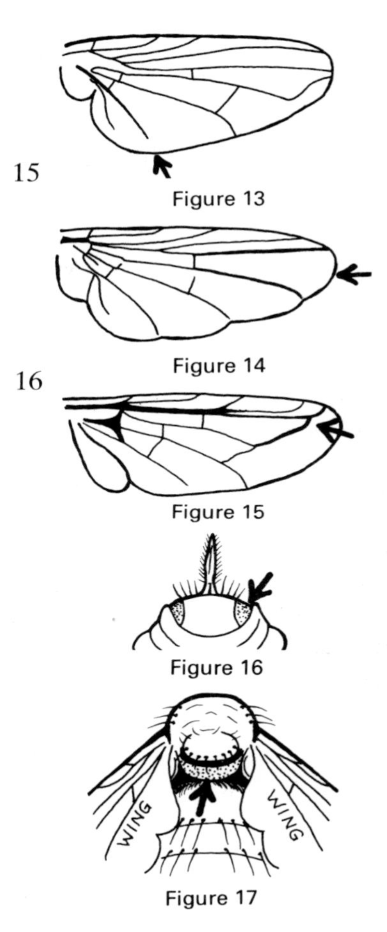

Figure 13
Figure 14
Figure 15
Figure 16
Figure 17

Order Siphonaptera: Fleas (Figs. 21:9 and 21:10)

Mouthparts—piercing-sucking in adult; chewing in larva.
Wings—wingless.
Metamorphosis—complex.
Additional features—body strongly flattened laterally and with many backward-projecting spines; long legs fitted for jumping; short antennae lie in grooves in head.

SELECTED REFERENCES

Baker, E. W., T. M. Evans, D. J. Gould, W. B. Hull, and H. L. Keegan, *A Manual of Parasitic Mites of Medical or Economic Importance* (Elizabeth, N.J.: Nat. Pest Control Assoc., Inc., 1956).

Blackwelder, R. E., *Taxonomy—A Text and Reference Book* (New York: Wiley, 1967).

Borror, D. J., D. M. DeLong, and C. A. Triplehorn, *An Introduction to the Study of Insects*, 4th ed. (New York: Rinehart, 1976).
Borror, D. J., and R. E. White, *A Field Guide to the Insects*, (Boston: Mifflin Co., 1970).
Comstock, J. H., *An Introduction to Entomology* (Ithaca, N.Y.: Comstock, 1940).
CSIRO, *The Insects of Australia* (Carlton, Victoria: Melbourne Univ. Press, 1970).
Essig, E. O., *Insects and Mites of Western North America* (New York: Macmillan, 1958).
Imms, A. D., *A General Textbook of Entomology* (New York: Dutton, 1964).
Jaques, H. E., *How to Know the Insects* (Dubuque, Iowa: Wm. C. Brown Co., 1947).
Jeppson, L. R., H. H. Keifer, and E. W. Baker, *Mites Injurious to Economic Plants* (Berkeley: Univ. Calif. Press, 1975).
Krantz, G. W., *A Manual of Acarology* (Corvallis, Ore.: OSU Bookstores, 1971).
Peterson, A., *Larvae of Insects, Part I: Lepidoptera and Hymenoptera*, 5th ed. (Ann Arbor, Mich.: Edwards Brothers, 1965).
———, *Larvae of Insects, Part II: Coleoptera, Diptera, Neuroptera, Siphonaptera, Mecoptera, Trichoptera*, 4th ed. (Ann Arbor, Mich.: Edwards Brothers, 1960).
Pritchard, A. E., and E. W. Baker, *A Revision of the Spider Mite Family Tetranychidae*, Pac. Coast Ent. Soc. Memoirs Series, Vol. 2, 1955.
Ross, H. H., *A Textbook of Entomology*, 3rd ed., (New York: Wiley, 1967).
Swain, R. B., *The Insect Guide* (New York: Doubleday, 1948).
Usinger, R. L., ed., *Aquatic Insects of California* (Berkeley: Univ. Calif. Press, 1956).

chapter 5 / **H. C. CHIANG**

INSECTS AND THEIR ENVIRONMENT

DEFINITION

The study of the interrelationships between organisms and their environment is called **ecology**. The term is used today in much the same sense as it was used one century ago by the German zoologist Ernst Haeckel (1834—1919), who coined it. In such a definition of ecology, there are three terms that require clarification.

First, **organisms** may be defined as (1) single individuals—one unit, (2) a number of individuals of a given species—a population, or (3) a number of individuals of more than one species—a community or a multispecific population.

Second, **environment** is the space and conditions surrounding an organism. It may be as small as a thin layer of air around an individual or as large as a continent or the entire world.

Third, **interrelationships** describe how organisms are affected by numerous physical factors in the environment—temperature, humidity, wind, evaporation, and light. Organisms are also affected by biotic factors such as the actions of individuals of other species; perhaps even more, they are affected by the actions of other individuals of the same species. For insects, the individuals of other species could be their food (plants or animals), their natural enemies (predators or parasites), or their competitors for food and physical resources. The concept of interrelationships also consists of a converse relationship between organisms and their environment. That is, organisms act upon the environment by reducing or depleting vegetation through their feeding or by changing the environment through their metabolic discharges. These interrelationships are shown in Fig. 5:1.

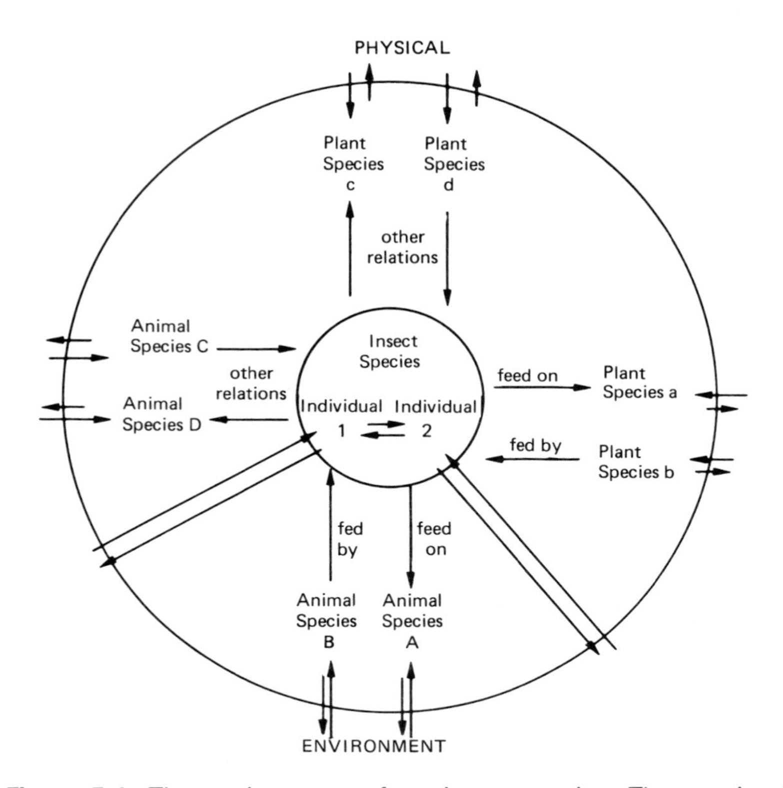

Figure 5:1. The environment of an insect species. The species is affected by the physical environment, and it also affects the physical environment. The individuals of this species also interact with each other. The species may have trophic relations with other animals (A and B), and plants (a, and b); they have other types of relationships with other animals (C and D) and plants (c and d). All of these animal and plant species interact with the physical environment.

Chapter 1 stated that this text is designed to provide background information in applied entomology and that ecology is an important part of such background information. Insects, like other animals, depend on vegetation or animal material for nutrition. If this material happens to be important to human welfare and human economy, the particular insect species is regarded as a pest. If the species happens to feed on pest insects, it is considered beneficial. Thus from man's point of view the status of an insect depends upon its ecological relationships to other species of animals and plants.

In management of pest insects or beneficial insects, we must know how they respond to these natural environmental factors. This management process is essentially a device to exploit ecological relationships to our advantage.

SPECIES DIVERSITY AND NUMERICAL ABUNDANCE OF INSECTS

It is a matter of point of view whether insects live in man's world or whether men live in the insect's world. The fact is that men and insects share the same world. Therefore, there is a constant interaction between man and insects, and the intensity of this interaction is contingent upon the size of the human population and upon the numbers of insects in terms of both individuals and species.

Insects have the largest number of species in the animal kingdom. There are between two and three times as many species of insects as all other animals combined. In fact, there are more species of Diptera (flies) than there are vertebrates, and there are more Coleoptera (beetles and weevils) than all invertebrates, other than insects. Such a multitude of species has enabled insects to evolve a great diversity of habits and to invade a wide range of habitats. Insects are found on land, in fresh water, in the air, in the soil, in and on animals and plants, and in human and animal habitations.

These small creatures also possess a great diversity of structures and life cycles. The three most significant features of insects are (1) their small size, which allows them to live in sites too restricted for many other animals and which permits them to flourish on small quantities of food, (2) their short life, which makes it possible for them to complete a life cycle in sites that are benign for only a short time, and (3) their great mobility, particularly their ability to fly, which enables them to travel to favorable sites and avoid adverse conditions (Fig. 5:2).

ENVIRONMENTAL CONDITIONS

Physicochemical Factors

The life processes of an insect are affected by the physicochemical conditions of the place in which it lives. These factors may be separated into two types: (1) energy and forces and (2) molecules. Some are important to all insects, such as energy in the form of temperature and light. Others are important only to certain species. Factors such as air pressure and moisture are important only to terrestrial forms, whereas water currents and dissolved chemicals are important only to aquatic forms. These factors are summarized as a general survey in Table 5:1. Several less common factors are included because in recent years research has demonstrated their effects upon insects. It will be well for us to keep our minds open as to what constitutes the physicochemical environment of insects. We certainly cannot define this environment on the basis of our own sensory perception.

On earth, almost all energy originates from solar radiation. Qualitatively, radiation energy is characterized by its wave length. It travels in waves, with each oscillation covering a certain distance. Its entire spectrum ranges from

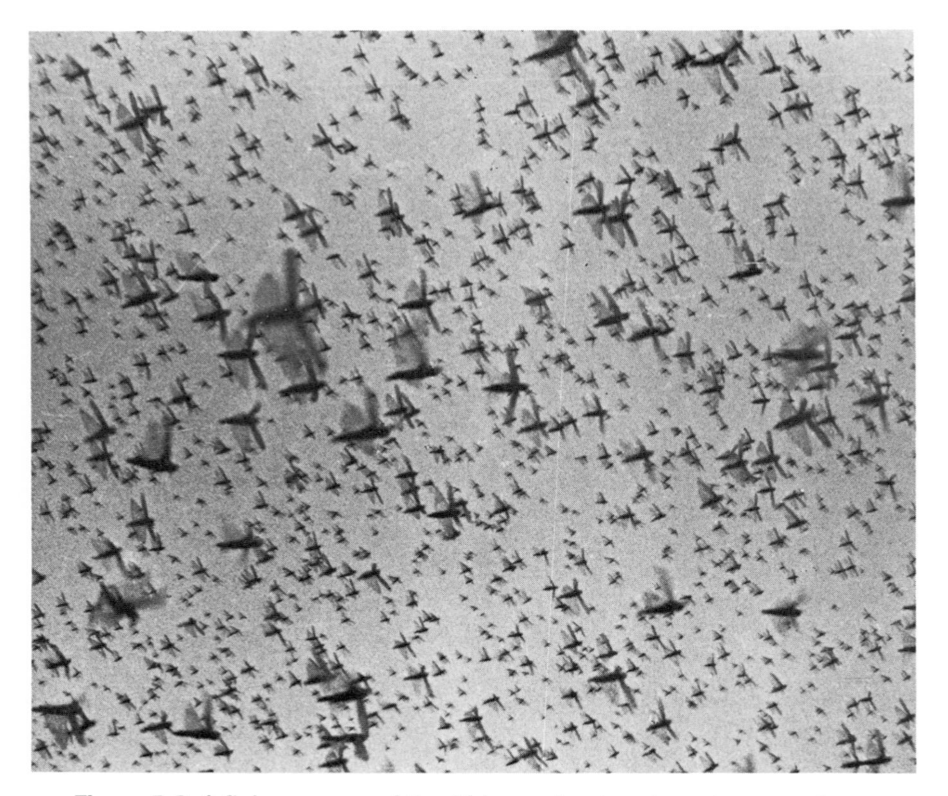

Figure 5:2. A flying swarm of the African migratory locust as seen from below. These insects migrate with the prevailing winds; in Africa they travel northward from February to September and then southward from October to January in response to the latitudinal displacement of the intertropical front and its associated wind field. *Courtesy Anti-Locust Research Centre, photographed by H. A. Dade.*

the short gamma ray, with wave lengths measured in angstroms, to the long radio waves measured in meters. Starting from the short end and going to the long end of the spectrum, we note that radioactivity may cause ionization of molecules and thus damage insect cells. Ultraviolet light may denature proteins and thus also damage the cells, yet some insects are known to orient themselves to the ultraviolet light reflected by flowers. Visible light affects the development and behavior of insects. The long end of the visible spectrum and infrared radiation produce heat. The infrared radiation at the far end affects insect behavior.

Quantitatively, solar radiation travels the expanse of space, and an average amount of about 2 cal/cm^2/min reaches the top of the earth's atmosphere. This is called the **solar constant**. Approximately one third of this amount is reflected back into space and lost; one third is absorbed by gas and other molecules in the air; and one third reaches the earth's surface. Much of the last amount is absorbed by the earth and reradiated back to the air. In this process, the temperatures of the substratum—be it soil, rock, lake, or ocean and the air above it—are increased. Only a small portion of the energy that reaches the earth will be absorbed by vegetation. Of that which plants absorb, a certain amount is converted to heat through plant respiration; the remainder is converted to plant protoplasm through photosynthesis. Protoplasm, which represents energy stored in chemical bonds, supplies energy directly to phytophagous insects and indirectly to other insects.

The effects of physicochemical factors upon insects will be discussed later in this chapter in the section entitled "Life Processes of Insects". Let us now turn our attention to the important biotic factors affecting insects.

Biotic Factors

Living components in the environment constitute the biotic factors affecting insects. These include food, natural enemies, and competitors. The latter group includes individuals of the same species as well as individuals of other species. In fact, competition among individuals of the same species is usually more intense because their requirements are almost identical.

Food factors. **Phytophagous** insects utilize plants, the primary producers in the community, whereas **entomophagous** species feed on other insects. In general, phytophagous insects are usually considered injurious, and entomophagous insects are considered beneficial.

Insects vary in the flexibility of their feeding habits. They are polyphagous, oligophagous, or monophagous if they feed on many, a few, or a single species of plant or animal, respectively. Examples are the migratory grasshopper (polyphagous), the imported cabbageworm (oligophagous), and the monarch butterfly (monophagous). For certain insects, a plant is always acceptable unless it contains a repellent substance. For certain other insects, a plant is always rejected unless it contains a substance that stimulates feeding. Most insects, however, are intermediate in their ability to discriminate and to select host plants.

The quality of the food plant that affects the growth and survival of insects may vary even within the same species of plant. The water content of cotton affects the incidence of diapause of cotton insects. Larvae of the European corn borer survive better on the upper parts rather than the lower parts of the corn plant. Some varieties of a given species of host plant are more suitable than others for development and survival of insects. In fact, the growing of crop varieties least suitable for insects (resistant varieties) is one effective method of discouraging infestations (Fig. 5:12).

Quantity and distribution of food are also crucial factors. The monoculture system, with one crop dominating large areas, is responsible for some of our insect problems.

Intraspecific factors: other individuals of the same species. Other individuals of the same species are part of the environment of a given

Table 5:1. Physicochemical Factors of Ecological Significance and Their Measurements

	Factor	Unit	Explanation
A.	Energy and forces		
	1. Radiation		
	Quantity	Solar constant	2 calories per cm^2 per minute
	Wave length	Angstrom (Å)	1×10^{-8} cm
	Frequency	Hertz (Hz)	1 Hertz = 1 cycle per second
	2. Heat		
	Measurements	Calorie	Amount of heat required to raise temperature of 1 ml water by 1°C (specifically from 15 to 16°C)
	Heat of crystalization	79.67 cal	Heat released by 1 gm of water at 0°C to 1 gm of ice at 0°C
	Heat of vaporization	539.55 cal	Heat absorbed by 1 gm of water at 100°C to 1 gm of vapor at 100°C
	3. Temperature	°C	A state of molecular excitation equivalent to 1/100 of the entire range from freezing to boiling of water (with freezing point at 0°C)
		°F	A state of molecular excitation equivalent to 5/9 of 1°C (with freezing point of water at 32°F)
	4. Visible light		
	Intensity	Foot-candle	The light from a point source of 1 lumen falling on 1 ft^2
		Lux	The light from a point source of 1 lumen falling on 1 m^2
		Lumen	The light emitted by 1 international candle
	Color		Visible lights of different wave lengths from 4000 Å (violet) to 7000 Å (red)
	Polarization		Lights of a given plane of vibration
	Photoperiod		The hours of light in sequence with hours of darkness
	5. Invisible light		
	Infrared (IR)		0.022 cm to 7000 Å
	Ultraviolet (UV)		4000 Å to 1600 Å
	6. Radioactivity		
	Emission	Curie	The amount of material in which 3.7×10^{10} atoms disintegrate each second
	Absorption	Rad	The absorbed dose of 100 ergs of energy per gram of tissue
	Specific activity	mc/gm	Microcurie of radioactive isotope per gram of stable isotope

Factor	Unit	Explanation
7. Sound		
Intensity	Decibel	The logarithm of the ratio of the intensity of the sound to the intensity of a standard
Frequency	Hertz (Hz)	1 Hertz = 1 cycle per second
8. Gravitation	G	An acceleration of 32 ft/sec/sec or 9.8 cm/sec/sec (the average gravitation of the earth)
9. Magnetism	Oe (Oersted)	The magnetic intensity that can repel another pole of the same polarity placed 1 cm away in vacuum with a force of 1 dyne
10. Electric field		
Potential	Volt	The potential that will cause a current of 1 ampere in a conductor whose resistance is 1 ohm
11. Pressure		
Air	mm Mercury (Hg)	At sea level the air pressure is 760 mm Hg or 14.7 lbs/in^2, which is also called one atmosphere (atm)
Water		One atm of pressure per 10.07 m of water depth
12. Current, air or water speed	cm/sec mph	Centimeters per second Miles per hour
B. Molecules		
1. Water		
Density	1	1 gm/1 ml, at 4°C
Specific heat	1	1 cal needed to raise temperature of 1 gm of water by 1°C
Precipitation	cm or in	Centimeters or inches
Moisture:		
Absolute humidity	gm water/ L air	Gram of water per liter of air
Relative humidity	%	Per cent of saturation
Vapor pressure	mm Hg	Partial pressure caused by moisture
Saturation deficit	mm Hg	Partial pressure of water needed to reach saturation
2. Chemicals		
Acidity	pH	Indication of concentration of hydrogen ions (H^+)
Concentration of chemicals	ppm ppb	Parts per million of solvent Parts per billion of solvent

individual. One example is that among most insect species an individual of the opposite sex is essential for reproduction. In some species, one sex may produce a pheromone that acts as a chemical attractant to the other sex, a behavioral response that facilitates mating even when only a few individuals are present in a large area.

Certain species have the habit of aggregating, particularly when larvae hatch from the same egg mass. Tent caterpillars jointly spin their silk threads into a tent and stay within it during the day. This structure is helpful because it provides protection under a variety of circumstances. The ultimate example of aggregation is found among social insects. In this group the individuals are polymorphic and become so dependent upon each other that no individual can live outside the colony.

The presence of other individuals of the same species may also constitute a disadvantage. For one thing, the competition among individuals for food can become very intense. If a given amount of food is sufficient for a certain number of insects to complete their development, a slightly larger number of insects may result in total destruction of the group simply because food is exhausted before any of the individuals reach maturity. Individuals, however, usually exhibit biological variability. Those that feed faster, develop faster, or require less food to complete development may survive at the expense of the slower individuals. Thus, intraspecific competition brings about the selection of adapted individuals.

The effect of competition applies to other resources as well, such as suitable habitat. In Australia a method has been devised to force the indigenous populations into areas less suitable by introducing large numbers of sterile individuals into the habitat which compete for food and space but do not multiply. Too many individuals of a species may get in each other's way, causing a decrease in mating and oviposition. Laboratory cultures of *Drosophila* (fruit flies) are often reduced in this way.

Another aspect of the effect of other individuals of the same species is pollution of the environment. Pollution is not just a problem with human populations. Insects such as flour beetles produce metabolic wastes which, in high concentrations, render their habitat unsuitable for further occupancy. The more individuals there are in the population, the sooner this critical level of pollution is reached. Pollution sometimes causes insects to disperse to a more favorable habitat.

Interspecific factors: other animals. Other animals may be natural enemies of an insect species. Usually a balance has been developed through natural selection, so that a predator does not always get its prey and a parasite does not search out the last host (Fig. 5:3). In fact, it has been found that the reaction times of the preying mantis and its prey are about the same. This means that half the time the mantis gets its prey, and half the time the prey escapes. A parasite population may kill most of the host individuals in a short time and then have difficulty finding the last few host individuals. This allows time for the host to build up its population again. Had the predators and

A B

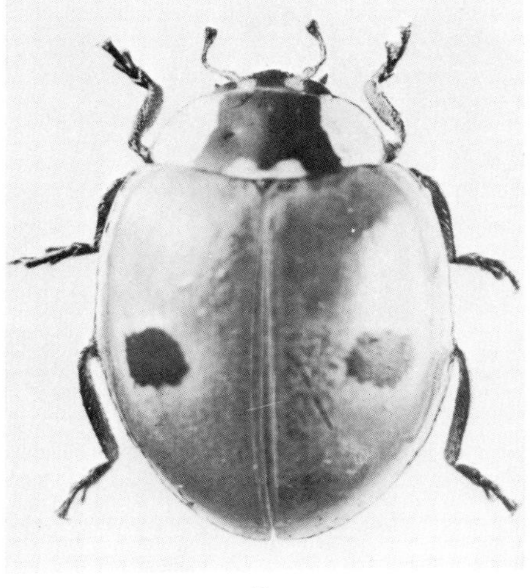

C

Figure 5:3. Life stages of the twospotted lady beetle. A, eggs attached to leaf; B, larva feeding on aphids; it consumes up to 400 aphids before entering the pupal stage (not shown); C, adult lady beetle. *Courtesy USDA.*

parasites been extremely efficient, they would have eliminated all prey and hosts; then they themselves would have eventually perished. A reservoir of host individuals is usually essential for a biological control program of lasting effectiveness (Fig. 5:4).

Another animal may be an insect's competitor. Competition may be for food or for other resources. A small species, which needs less food to complete development, usually has an advantage over a larger species. A species that completes development in a short time usually will have an advantage over a species that requires a longer time. A species that can withstand longer starvation and more weight loss has the better chance of survival when food is temporarily short; a polyphagous species will be more fortunate than a monophagous species because it can move to a different food source.

Competing species do not have to be on the host at the same time to compete with one another. Species *A* feeding on a tree could affect the quantity and quality of food available to species *B*, which feeds on it at a later date. And species *B* could affect the food of the next generation of species *A*.

Figure 5:4. Larva of the fall *Tiphia*, a wasp parasite of the Japanese beetle. This parasite was introduced into the United States from Japan where it keeps the beetle from becoming a serious pest. The *Tiphia* larva (center) feeds only on grubs of the Japanese beetle (larger larva). *Courtesy USDA.*

Some insects may affect the host in a manner such that subsequent generations of their own species can no longer live there, and at the same time create conditions favorable to insects of other species. Trees weakened by defoliating insects may be attacked by secondary invaders such as bark beetles and borers. This is the basic principle involved in **community succession**. Other species may be beneficial or even essential for the survival of a given insect. Corn field ants and corn root aphids have a mutually beneficial relationship, and protozoan symbionts are essential for the nutrition of termites.

Environmental Conditions in a Crop Field

A crop field is a habitat; as such it supports a community, and the whole may be considered an **agroecosystem**. Like any stand of vegetation, the crop plants alter the physical conditions of the habitat. First, they prevent some light from reaching the ground. This, in turn, prevents soil temperature from rising. The reduction is directly related to the density and height of plants, and also to the size of leaves. The shielding effect is caused by both reflection and absorption of light by the plants.

Second, the plants reduce wind through friction. This reduction of wind, like that of light, is related to the density, height, and leaf size of plants. Reduction in wind, in turn, results in a decrease in evaporation from the ground surface as well as from plant surfaces. Reduced evaporation results in greater retention of moisture in soil and vegetation. It also tends to minimize the heat loss usually associated with evaporation.

The surface temperature of plant tissues is usually higher than that of the surrounding air when the plants are exposed to the sun, but lower during the night. Thus insects staying on such exposed tissues are actually subjected to a greater temperature fluctuation than that shown by the usual climatological records. On the other hand, insects sheltered by plant foliage may be subjected to less extreme fluctuations.

The temperature of the soil is usually more moderate than that of the air. Soil temperature at 60 cm below the surface fluctuates very little each day, whereas at 200 cm below, it fluctuates very little during an entire year. Most insects are located above these depths. White grubs and other soil insects are known to migrate vertically in the winter in response to soil temperature. Vegetation tends to moderate the soil temperature even further. Soil temperature at open spots in a crop field, however, can be raised to a level higher than soil insects can tolerate. The reason for this high temperature is that light and heat rays reach the open spot, and the surrounding plants reduce wind and evaporation; thus the heat can be retained.

The presence of crops in the soil creates a gradient of chemicals. There is evidence that some roots release oxygen, thus establishing a higher oxygen concentration near the roots. The ways in which this may affect insects is not known as yet. Plant roots also release secretions that stimulate nematode cysts to hatch. A similar relationship may exist in the hatching of insect eggs.

Many agronomic practices greatly affect the insects in a field. Tillage may clear the field surface and deprive insects of their food plants. Growers of wheat obtain effective control of grasshoppers by spring tillage that destroys all weeds and volunteer growth just before or soon after these insects begin to hatch. Irrigation has encouraged some insects, such as the western corn rootworm, to establish in new sites. Inorganic fertilizers are known to affect plant chemistry, which affects insect populations indirectly. High nitrogen is known to favor insects, whereas overabundance of certain other elements may have adverse effects. An organic fertilizer such as manure on the one hand increases soil fertility and provides conditions favorable for a predatory fauna that may reduce pest insects in the soil, but on the other hand it encourages certain pests such as seedcorn maggots and millipedes.

A productive crop field should be free of other plants. The presence of weeds may affect the infestations of pest insects in both a positive and a negative way. In a field of barley, the corn leaf aphid is usually established first on foxtail grass; here a weed assists a pest insect species. Contrarily, the presence of grasses and ragweed near a corn field may harbor the common stalk borer, an early host of *Lydella grisescens*, a parasitic fly. This fly later parasitizes the European corn borer. Here the presence of a weed serves to reduce a pest insect.

As a general rule, where the vegetation stand is purer, the community is simpler and hence the populations of insect pests fluctuate more. Preservation of hedge rows or other uncultivated areas which tend to provide a more complex community has been suggested as a way of stabilizing populations of pest insects at low densities.

LIFE PROCESSES OF INSECTS

In spite of their small size and short lives, insects have complex structures, intricate activity patterns, and varied life cycles. They are sensitive to adverse environmental conditions, yet retain a strong potential to produce large populations. In this section we shall discuss briefly the major aspects of insect life processes and the effects of environmental factors upon these processes and upon the success of the individual species. The discussions will treat (1) insect development and how it is affected by environmental conditions, particularly temperature; (2) variations in insect life cycles and their causes; (3) insect life span; (4) insect tolerance of adverse conditions; (5) insect fecundity; (6) insect mobility; and finally (7) insect behavior. All of these features of insect life are important for our understanding of the growth of insect populations to injurious numbers.

Insect Development

Insects, like all organisms, live within a relatively narrow range of environmental conditions. The upper and lower limits of an environmental factor adversely affect their survival; within these limits the environment influences their rate of development. The development of insects is particularly sensitive to temperature because these animals are ectothermic. There is a general relationship between environmental temperature and insect activity and survival (Fig. 5:5).

The life stages—egg, larva, pupa, and nymph—which lead to the adult were described in Chapter 3. If we take one of the immature life stages, the eggs, as an example, and expose a number of them to a gradation of temperature, we find that they take various lengths of time to complete development. Within limits, the relationship follows a hyperbola (Fig. 5:6A). If the results are expressed in different terms, namely, the rate of development, the relationship then follows a slightly curved line (Fig. 5:6B). The developmental rate is the reciprocal of the developmental time. For example, if an insect egg takes ten days to complete its total development, it completes 1/10, or 10 per cent, of the development each day. The former (ten days) is the time, and the latter (10 per cent) is the rate.

This **temperature curve** shows three significant features. (1) The middle section (a) is approximately a straight line; within this temperature range, the rate of development is directly proportional to temperature. This is the **optimal range**. (2) The descending section (b) of the curve suggests that development is slowing down with further increase in temperature. With still

Figure 5:5. The relationship between temperature and activity level of insects. The graph shows the lower and upper lethal points which are the limits of tolerance to extreme temperatures. Above the lower lethal point, insect metabolism and activity increase with increase in temperature up to a point and then decrease with further increase until the upper lethal point is reached.

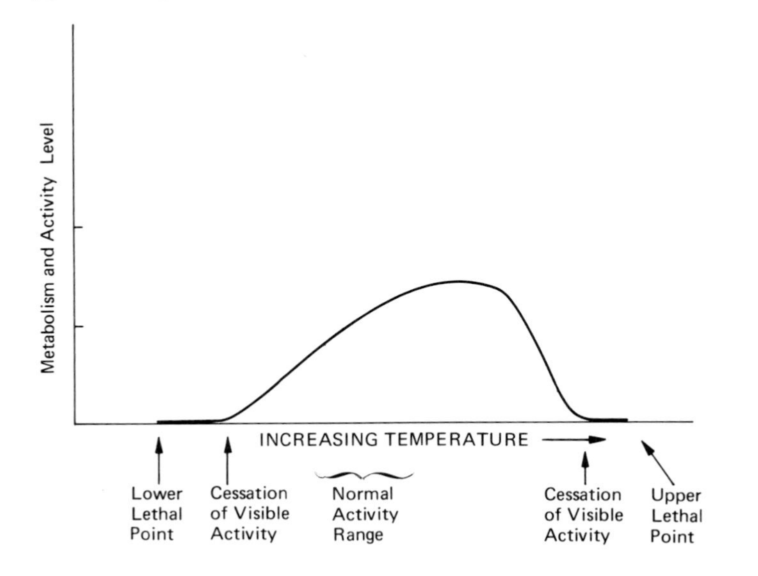

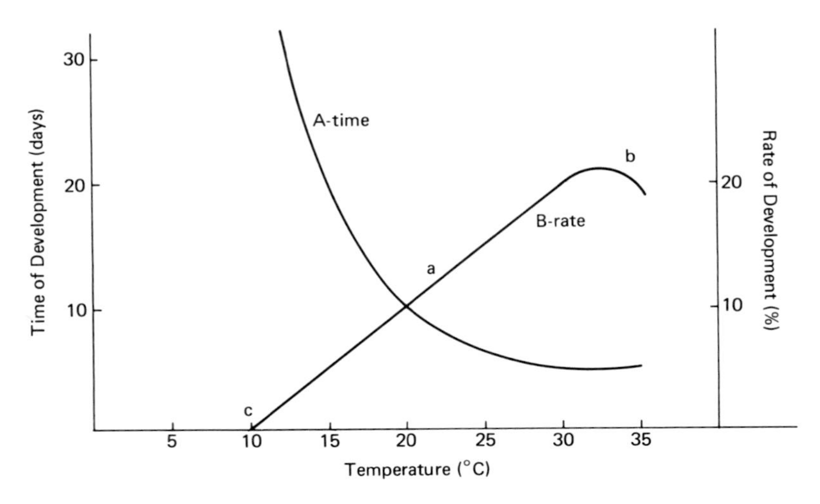

Figure 5:6. The relationship between time and rate of insect development with temperature. See text for explanation.

further increases the insect will succumb, indicating that the **upper lethal temperature** has been reached. (3) The curve intersects the abscissa; at this point of intersect (c) the rate of development is zero. The particular temperature indicated is called the **threshold temperature**.

The presence of the threshold temperature implies that only the part of the environmental temperature above it, the **effective temperature**, promotes development. For example, if the eggs of an insect have a threshold of 50°F and if they are kept in a temperature chamber at 68°F, the effective temperature will be the difference of the two, 18 F degrees.

At a given temperature a life stage requires a certain length of time to complete its development. During this time the insect is exposed to the amount of heat required for development. For practical purposes, the amount of heat is expressed not in calories but in heat units of **day-degrees** above threshold. If the eggs completed their development at 68°F in ten days, they require $180(18 \times 10)$ day-degrees above the threshold of 50°F. Some researchers express this as 180 dd/50. At other temperatures within the optimal range, the total heat units will always be 180 whether it is accumulated in 10 days at 18 units a day, 20 days at 9 units a day, or any other combination. This total amount is called the **thermal constant**.

The concept of thermal constant is used to predict insect development from field temperatures. Because temperatures fluctuate throughout the day, the daily mean temperature $\left(\frac{\text{daily maximum} + \text{daily minimum}}{2}\right)$ is used. For example, if a given day has a maximum of 72°F and a minimum of 56°F, it would have a mean of 64°F, and a heat unit accumulation of 14 day-degrees.

Over a period of several days, the next day may be colder with a mean of 60°F and a heat unit accumulation of 10, and the day following may be warmer with a mean of 70°F and a heat unit accumulation of 20. The expectation is that whenever the sum of the heat units $(14+10+20+\cdots)$ reaches 180, the eggs will hatch.

It is clear that the above method needs some refinements. The following are three conditions which require closer inspection. (1) The temperature condition on plant surfaces may be different from that of the air. Thus, the predicted date based on air temperature will be too early if plant surfaces are cooler than the air and too late if plant surfaces are warmer than the air. (2) On cooler days and in the morning hours of the day, some insects bask in the sun and absorb heat above the level indicated by the air temperature. When this happens, the date predicted on the basis of air temperature will be too late. (3) If a day has a maximum of 58°F and a minimum of 42°F, the daily mean will be 50°F, which means that there will be no heat unit accumulated for that day. Yet there would be several hours during the day when the temperature is actually above 50°F and some development would have taken place. The predicted date will be too late if many such days occur.

For insects that live in the soil, it is possible to predict their development by the air temperature records, if the correlation of the temperatures of the air and the soil is known. If the correlation is established on the basis of bare soil, the actual completion of development will be later if the field is covered with vegetation or plant debris, because the ground cover tends to slow down the warming of the soil.

In this discussion we used eggs as an example, but the same general relationship applies to nymphs, larvae, and pupae. We have discussed only the effects of temperature because it is the most important factor affecting development. Light, particularly photoperiod light, is important to metamorphosis and diapause and will be discussed in the next section under "multivoltine species." Humidity is important to survival but has relatively little effect on the rate of development. The effects of gravitation and electromagnetism on *Drosophila* studied recently show that these insects completed two generations under 10G with normal rate of development, but they exhibited morphological abnormalities in a high magnetic field. Once the insect has reached the adult stage, the environmental factors will have their greatest effect upon survival, reproductive capacity, and behavioral activities.

Insect Life Cycles

The development from egg to adult and again to egg represents a life cycle. Most insect species complete one life cycle or one generation each year and are considered **univoltine**. Some insects complete more than one life cycle each year and are considered **multivoltine**. Others complete one life cycle in a longer period than one year and are considered **perennial**.

The stage that lives through the winter season is called the **overwintering stage**. It may be any one of the life stages: egg stage of the grasshopper, partially grown larva of the armyworm, mature larva of the codling moth, pupa of the imported cabbageworm, nymph of some scale insects, and adult of the alfalfa weevil.

Univoltine. In most of the univoltine species, the overwintering stage is usually cold hardy and is in **diapause**, a state of very low metabolism.

Multivoltine. Multivoltine species are divided into two groups, homodynamic and heterodynamic. The **homodynamic** species are those that can carry on generations throughout the year without interruption if environmental conditions are favorable. This group includes many of the stored grain and household insects. The **heterodynamic** species are those in which one of the generations during the year goes into diapause. Examples are numerous, but we shall discuss only aphids and the European corn borer.

Aphids may have many generations during the season, most of which are reproduced parthenogenetically and ovoviviparously by summer females. Reproduction of the so-called sexual generation is induced by the progressively shortening photoperiods in the fall. Some species of aphids respond to lengthening days in spring and produce a winged generation, which migrates from the winter host to the summer host.

In North America, the European corn borer (Figs. 10:12 and 10:13) has a partial second generation in northern areas of its distribution. In midsummer, many of the larvae reach maturity and continue their development, which results in the production of a second generation. This is because long days encourage continuous development. On the other hand, short days encourage diapause. Those larvae reaching maturity late enter diapause. The temperature during the season plays an indirect role. If the temperature is high, a relatively high proportion of the population will reach maturity in the early part of the season when days are long. Thus the proportion of the population carrying on a second generation depends directly upon the photoperiod and indirectly upon the seasonal temperature.

Recent studies show that the critical length of the photoperiod which triggers pupation varies with the source of the population. Compare Minnesota and Missouri as habitats of the borer: Minnesota, in the north, has longer summer days than has Missouri. Thus, the borers in Minnesota are adapted to longer days than are borers in Missouri. When Minnesota borers were allowed to grow in a corn field in Missouri, they responded to the short days, and a greater proportion entered diapause than Minnesota borers that remained in Minnesota. This happened despite the fact that there were more warm days for second-generation individuals to reach maturity in Missouri than in Minnesota. Conversely, when Missouri borers were allowed to grow in Minnesota, more of them continued their development and pupated than did the Missouri borers in Missouri, even though there were fewer warm days for development in Minnesota than in Missouri. These results suggest that, once a population is adapted to local photoperiod conditions, it will not alter its diapause or pupation to changed conditions of temperature during a single season.

It may be concluded that for the heterodynamic species, the number of generations in a year is affected by photoperiod and temperature and that the relationship results from adaptation to conditions of the habitat.

Perennial. The perennial species require more than one year to complete one life cycle. The reasons for this prolongation vary among the different species. It may be due to nutrition, intrinsic factors, temperature, or a variable life cycle.

Nutrition. Many wood-boring forms spend more than one year in the larval stage because of the low nutrient level of their foods. Experiments have demonstrated that the development of some cerambycids is accelerated if the larvae feed upon wood impregnated with peptone or diastase. Another example of the nutritional factor prolonging the life cycle is found in a buprestid, *Melanophila california*. Within a healthy tree, the larvae may not develop at all for a period of four years and will die during the fifth year. If adverse conditions seriously weaken the host tree at any time during the first four years, the larvae will commence development and complete metamorphosis in one summer. In this case, nutritional factors essential for larval development are associated with a dying tree.

Intrinsic factors. The perennial life cycle of some species seems to be intrinsic. Different species of periodical cicadas spend from 13 to 17 years as nymphs, whereas white grubs and mayflies have life cycles of from 2 to 3 years. The life cycle of cicadas is so rigid that there seems little doubt that it is fixed genetically. This fact is supported by some theoretical considerations based upon behavioral isolation. Because experiments such as those conducted on wood-boring forms are lacking for white grubs and mayflies, nutritional factors may still be involved to a degree in these insects.

Temperature. A species may be univoltine in a warm climate but perennial in a cooler climate. Such is the case with two species of walkingsticks, one that inhabits Australia and one that inhabits the United States. Eggs of the walkingstick in northern Australia (warm climate) tend to hatch in less than one year, whereas those in the south (cooler climate) tend to hatch in the second spring, particularly if the summer is cool.

The walkingstick of the United States oviposits too late in fall for the eggs to undergo much development in northern regions of its distribution. The following spring in such regions, eggs resume development and continue for the entire season but do not hatch. They overwinter again and are ready to hatch within a short time, early in the third summer. Enjoying favorable weather, the nymphs develop rapidly, and the adults appear in late summer and oviposit. In the southern areas, because of the warmer temperatures and longer growing season, eggs develop partially after being laid and hatch during the second season. But only the early nymphs complete development and lay eggs during that season. Those hatching late become partially grown and perish in late fall or in winter. Thus, because of their response to temperature conditions, the walkingstick maintains high populations with a two-year life cycle in the north but low populations with a one-year life cycle in the south.

Variable life cycle. Some species have a prolonged dormancy, so that a generation may last more than one year. Corn rootworm eggs are laid in the fall and usually hatch during the subsequent summer, but a small percentage may hatch without overwintering, and another small percentage do not hatch until the second summer. Thus a portion of the population has a two-year life cycle. Similarly, the flaxseed stage of Hessian fly may remain dormant until the following year. A remarkable example of variable life cycle is that of the orange wheat midge in England. Adult emergence from a soil sample kept in an insectary was observed to continue for 12 years, indicating that the dormant period varied from 1 to 12 seasons.

Insect Life Span

In the previous section we discussed the length of time required by insects to develop and complete their life cycle. Now we shall discuss how long they live during the immature stage and how long during the adult stage.

During the life span of an insect, three periods may be recognized with reference to reproduction: the prereproductive, reproductive, and postreproductive periods. Naturally the immature stages belong to the prereproductive period, excepting the rare forms that carry on **paedogenesis**, or reproduction by immature individuals. The prereproductive period continues into the beginning part of the adult stage. Adults require varying lengths of time to reach sexual maturity, mate, and reproduce; this takes from a few hours in some moths and mayflies to a much longer time for others as in beetles and grasshoppers.

Likewise, the length of the reproductive period varies among species. Some moths lay only one batch of eggs in one sitting (whitemarked tussock), whereas the house fly lays numerous batches of eggs over several weeks, and flour beetles produce eggs over many months. The length of the postreproductive period is variable but lies within a narrower range than the other periods. Usually it lasts a few days, although in mealworm adults it is as long as the reproductive period.

A knowledge of the length of the three periods of the life span has considerable significance in applied entomology. For example, the length of the prereproductive period is crucial in cases where control is aimed at the adults before and during oviposition, such as with the plum curculio. In the case of the European corn borer, the length of reproductive period has practical importance because larvae of this insect, which first feed on leaves and later enter sheaths and stalks, are vulnerable to insecticide only when feeding on leaves. Because only one insecticide application is feasible economically, the application is timed to coincide with the peak of egg hatching to prevent the majority of larvae from entering plant tissues. Catching insects in traps is used to time the application of control measures. If the species has a long and active postreproductive period, with many adults being caught after oviposition, the value of the trap catch as an aid in timing control will be greatly reduced.

Insect Tolerance of Adverse Conditions

Not all individuals of a population live through the full life span. In fact, most individuals die young because they have limited tolerance to environmental conditions, which are rarely optimal. Three categories of tolerance with regard to the limits may be recognized. A species may have (1) a minimum limit relative to some factors, (2) a maximum limit to other factors, and (3) both minimum and maximum limits to still other factors.

The food factor applies very well to the first category: there is a minimum amount the individual needs in order to survive. Overabundance is usually not a problem, because the individual regulates how much it takes by internal mechanisms. The factor of space is also included in the first category. An individual needs a minimum amount of space, particularly for oviposition and overwintering. Again, overabundance of such space has no affect. It is interesting to note that, in a culture bottle, *Drosophila* keep a certain minimum distance from each other, and that the amount of this "elbow room" varies with the genetic lines of the flies. Another example is the oxygen concentration in an aquatic environment. It may drop below the minimum tolerance level for an aquatic insect, but there cannot be too high a concentration because the amount of oxygen that water will contain is limited by the solubility of oxygen in water.

In the second category, factors having a maximum limit are those that involve certain biotic relationships. Insects are often attacked by pathogenic microbes. There is a maximum level of infection the host can tolerate, but there is no such criterion as a minimum infection. Similarly, competition among species for resources can reach maximum tolerance level, but it is difficult to conceive a minimum tolerance level of competition.

Many factors belong to the third category—temperature, humidity, light (in terms of both intensity and number of hours daily), and pH. Figure 5:6 illustrates the relationship of metabolism and activities with temperature. Beyond the optimal range, activity slows down, and finally death occurs at both the minimum and the maximum limits. This graph can be applied, with modifications, to the other factors.

Certain biotic factors also fall in this category, such as population density, which may be defined as the number of individuals per unit space or area. Two individuals of opposite sexes constitute a minimum density for most insects. For practical purposes, more than two are needed in any sizeable area for effective population increase. There is also a maximum tolerable density beyond which individuals would exhaust their food and other resources. For a variety of reasons, the maximum density that can be supported is usually reached before insects exhaust their resources.

The tolerance of an organism to an environmental factor is affected not only by the intensity of the factor, but also by its duration. In general, the length of time an individual can tolerate an adverse factor is inversely proportional to its intensity. Organisms can withstand starvation, crowding, heat, cold, drying, and many other extreme conditions for a short period, but they suffer from prolonged exposures.

The response of *Drosophila* to low temperatures illustrates this point. Figure 5:7 shows the effect of intensity and duration of cold upon the adult fly. At any one of the low temperatures, mortality increases with an increase in exposure and follows an S-shaped curve. With the lowering of the temperature, a greater mortality occurs at a given duration. The slope of the curves shows that mortality increases much more rapidly at lower temperatures.

The fact that mortality follows a sigmoid curve means that some individuals are more hardy and can withstand longer exposures than others. It is a general biological rule that individuals vary in their degree of tolerance to any environmental factor. A more gentle slope indicates greater variability, whereas a steep slope indicates more uniform response. The fact that the mortality curve is steeper at a lower temperature means that as the conditions become more severe, the response becomes more uniform. It is not difficult to imagine that if a group of insects were exposed to dry ice or another extreme, boiling water, all would die instantly, simply because the harshness of the condition is beyond the tolerance of even the most hardy individuals.

Short of the dry ice temperature, some insects can tolerate for brief periods temperatures much below the optimum range. When the insect is placed in a low-temperature chamber, one might expect the temperature of the insect body to gradually drop to that of the chamber. What happens, however, is that at about −15 to −30°C, the temperature of the insect suddenly bounces up a few degrees and stays there for a short time, then drops again and eventually comes to equilibrium with the chamber temperature. The low point just before the rebound is called the **supercooling point**. The rebound in body temperature is caused by production of heat of crystallization which is released when freezing occurs. If the insect is removed from the chamber soon after the rebound, it will be frozen and will shatter

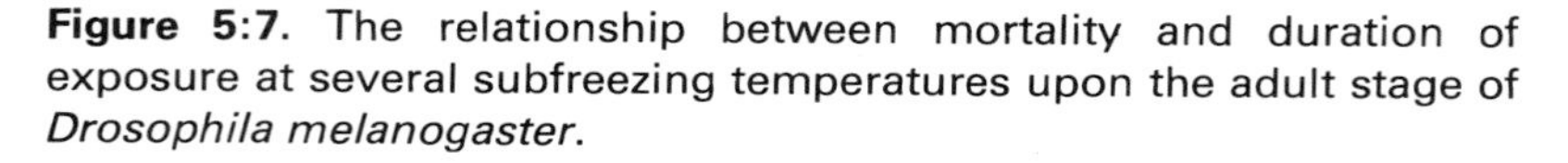

Figure 5:7. The relationship between mortality and duration of exposure at several subfreezing temperatures upon the adult stage of *Drosophila melanogaster*.

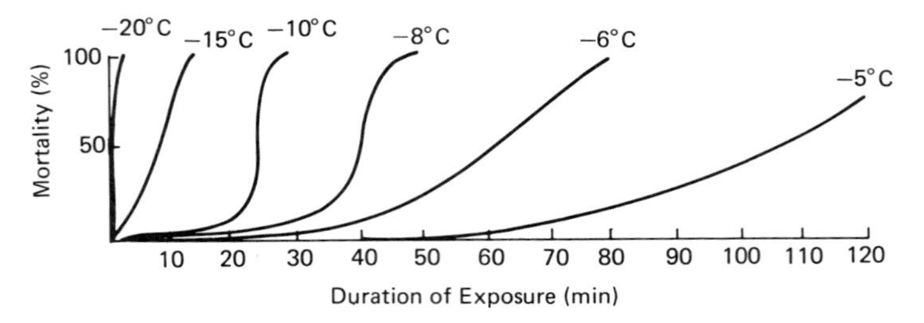

upon impact with a hard surface. If allowed to thaw out from such a frozen condition, however, individuals of some species will walk away. But, if these same insects are kept in the chamber for an extended period, they succumb to the low temperature.

We mentioned earlier the electrostatic field as an environmental factor affecting *Drosophila*. It has been found that supercooled insects freeze at higher temperatures when they are kept in a high-intensity field, indicating that the response of insects to low temperature is influenced by the electric condition of the environment.

From the point of view of applied entomology, a knowledge of the tolerance level of insects to various factors is fundamental to the design of effective and economical control measures. After all, killing of pests by whatever means is done by creating a condition beyond the tolerance of insects.

Insect Fecundity

The fecundity of insects is the level of reproduction, expressed in terms of the number of eggs per female, or of young in ovoviviparous forms. In spite of their vulnerability to adverse conditions, insects are successful, as shown by their abundance in both species and individuals. One of the reasons for this success is their high fecundity. A classical example is the queen termite, which lays six thousand eggs a day for 15 or more years. Another classical example on the other end of the scale is the oviparous rosy apple aphid, which lays only one egg per female. The species maintains itself in spite of this low rate of oviposition because the high viviparous reproductive capacity through many generations during summer builds up large numbers of oviparous females. Fecundity is generally fixed for a species, but its expression may be influenced by many ecological factors.

Biotic factors. The amount of food available to the immature stage may affect the size of adults, and in some species adult size influences fecundity. The quality of larval food is also important. Adults of the cotton bollworm have greater fecundity when as larvae they feed on corn rather than on cotton or beans. The amount and quality of food available during the adult stage also affects fecundity, as shown in *Drosophila* and in the migratory grasshopper. In some parasitic wasps, the adult punctures a host pupa and licks the fluid that comes out of the wound. Wasps feeding on young host pupae have a higher fecundity than those feeding on older pupae.

Another biotic factor that may affect fecundity is disease. Although most insect diseases cause mortality before the insect can reach the adult stage (Fig. 5:8), some diseases, such as those caused by the microsporidia, *Nosema*, reduce fecundity of adults. Nematode infections likewise reduce the fecundity of adults.

Physical factors. A variety of physical factors, particularly temperature, humidity, and light, are known to affect insect fecundity. The following examples illustrate this influence. The greenbug, *Schizaphis graminum*, has a

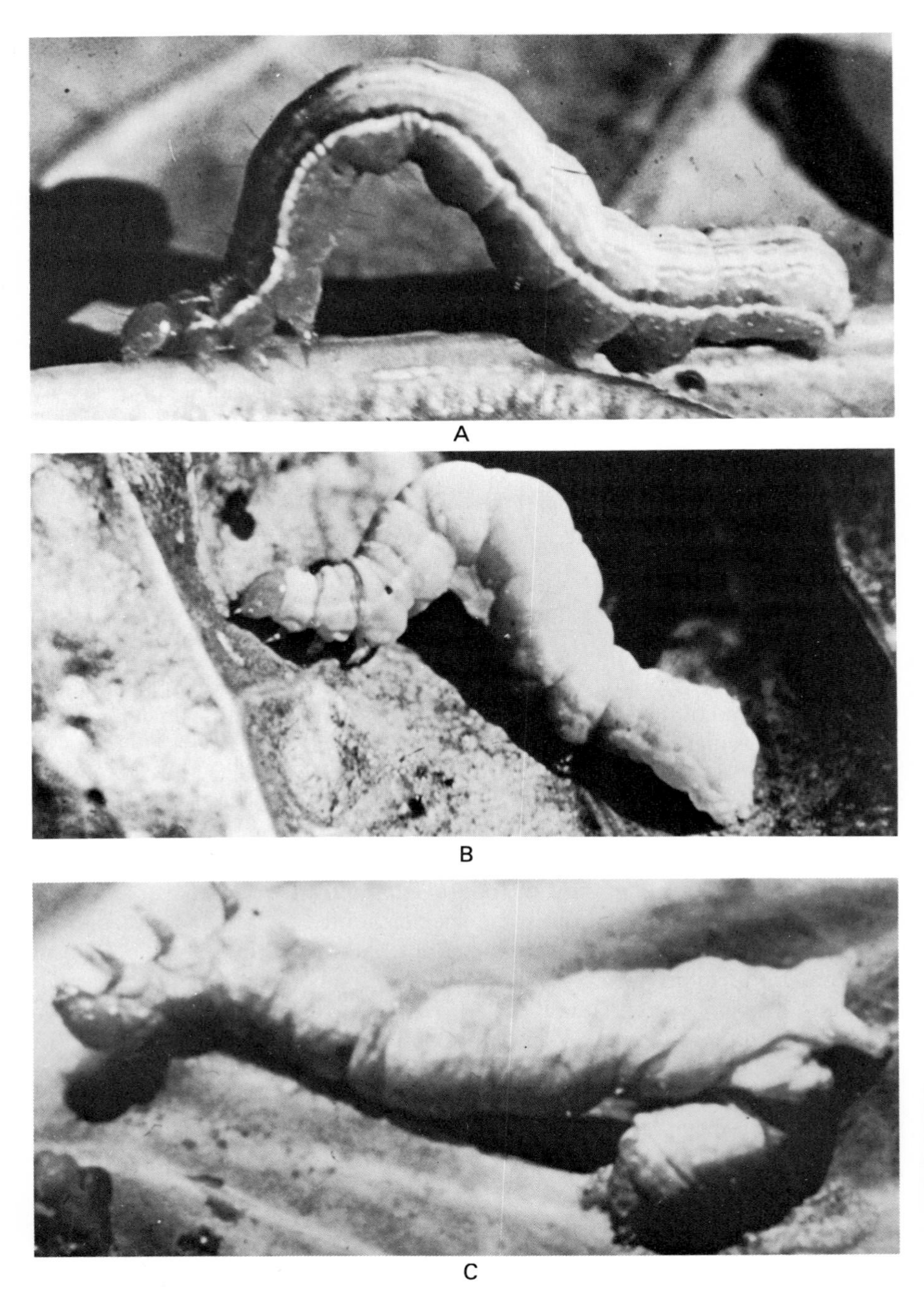

Figure 5:8. A polyhedrosis virus disease of the cabbage looper. A, healthy cabbage looper; B, cabbage looper in advanced stage of disease; C, cabbage looper dead from the disease. *Courtesy USDA.*

peak fecundity when held at a temperature of 25°C; the African migratory locust has a peak fecundity when kept at a relative humidity of 70 per cent; the fruit fly, *Dacus tryoni*, produces more eggs under high than under low illuminance.

Recent experiments of exposing insects to subfreezing temperature have yielded surprising results. When adults of the milkweed bug and flour beetle were kept at subfreezing temperatures for a period of time, a certain proportion were killed. In subsequent oviposition tests it was found that the survivors had a higher fecundity than members of the control series that were not exposed to low temperature. Evidently when low temperature selected the more cold-hardy individuals, it also selected the more fecund ones. In applied entomology we customarily think that after the cold of winter kills a certain percentage of hibernating individuals, the population will be reduced proportionally the following spring. Yet the results cited here imply that the population could be higher than that predicted on the basis of winter kill.

Insect Mobility

The high degree of mobility of insects is one of the chief factors contributing to their success. Mobility may be active or passive. The active form involves walking, jumping, and flying. Flies walk on vertical surfaces and even upside down, and ants walk on difficult terrain with heavy loads. Special structures such as claws and pulvilli contribute to their ability to grasp, whereas the exoskeleton contributes to their capacity to handle heavy weights.

For traversing long distances, flying is of greater importance than walking. Strong fliers include migratory locusts and many species of butterflies. Records indicate that African migratory locusts have flown to England from the Black Sea area, spanning an air distance of 1600 miles.

In passive mobility insects are carried by water currents, animals, manmade vehicles, or wind. The ability to survive a journey on water depends on the tolerance of an insect to submergence and its strength to stay on floating objects. A ride on another animal is facilitated by morphological anchoring adaptations. As for manmade vehicles, all means of transportation that serve man also serve insects quite well. As the speed of these machines increases, the chance for survival of the hitchhikers also increases. Successful transport in luggage compartments and other recesses in high-altitude aircraft depends on the ability of insects to survive extremely low temperatures.

Riding the wind is responsible for the long-distance movement of many insects, including some weak fliers (e.g., aphids and leafhoppers). Three steps are involved: the takeoff, the ride itself, and the landing. The ride itself is mainly passive, but the two other processes require positive action on the part of insects. The takeoff may involve flying into the air as do adult insects, or it may involve simply letting go of the substratum, as do many larval forms. The latter are usually aided by body hairs or a ballooning thread. Landing takes place when wind speed decreases and the insect simply drops or when wind passes through vegetation which combs out the insects. Insects may also land by actively disengaging themselves from the wind. The ride itself involves little energy expenditure except that used to stay aloft. The insect's ability to withstand desiccation is also a key factor for healthy arrival.

The above remarks point out the importance of mobility to dispersal of insects, but mobility has yet another aspect. There is evidence that in various insects body exercise is essential for sexual maturity, mating, and oviposition. From these facts, one may even suggest that mobility is required to bring about normal reproduction, and that dispersal is only indirectly involved.

Flight activity is affected by factors such as temperature and humidity. Recent laboratory studies have shown that flight in negatively ionized air is longer, faster, and steadier than flight in positively ionized air or in laboratory air. Knowledge of weather conditions that favor dispersal is helpful in predicting infestations of wind-borne insects, and information on the mode of dispersal is important for designing and evaluating quarantine procedures.

Insect Behavior

The behavior of insects varies in complexity depending on the species, from random movements, to oriented movements, to an instinctive chain of activities, to intraspecific communication. Random movements can be observed now and then in any insect. During the processes of feeding, mating, and oviposition, however, the insect shows definite orientations to physicochemical factors.

Orientation. An orientation is a movement relative to a stimulus. The stimulus may be temperature, humidity, gravity, contact surfaces, water, salt, chemicals, and so forth. The prefixes indicating these factors are *thermo-*, *hygro-*, *geo-*, *thigmo-*, *hydro-*, *salino-*, and *chemo-*, respectively. As far as insects are concerned orientation in its simplest form can be separated into two categories: (1) **taxis**, merely turning and moving straight toward or away from a source of stimulus, and (2) **kinesis**, random movements initiated by the stimulus.

An orientation can be positive or negative. The insect making a taxis movement will end up respectively nearer to or further away from the stimulus. In the kinesis response, however, a positive photokinesis means that the insect is stimulated by light to take up a random movement. This movement will cease when the insect by chance goes into a shaded area. Thus, an insect with a positive photokinesis would eventually rest in the shade.

Besides reacting to light as a discrete source, insects respond to several other qualities of light. Some insects are selectively sensitive to lights of different colors. The attraction of yellow light for aphids is utilized in the design of field traps to detect their presence. It is known that midges are attracted to diffused light but are repelled by a concentrated source. Similarly, aphids are repelled by shiny surfaces. On the basis of this response, aluminum foil is used to protect plants from aphid establishment. Sawfly and lepidopterous larvae are known to move at definite angles with the plane of polarized light, so that they do not double their own tracks. An even higher degree of sophistication is exhibited by the honey bee in its orientation with polarized light. Recently investigators have demonstrated that certain buprestid beetles are sensitive to infrared radiation from forest fires, and that cockchafers are sensitive to electromagnetic fields.

Orientation patterns may be affected by many physiological and ecological factors. A sawfly larva may show a negative phototaxis when hungry but a positive phototaxis when fed. Cutworms, however, show a reversed pattern. Adult syrphids and coccinellids show a positive phototaxis in cool air but a negative phototaxis in warm air. The pattern may also change from one life stage to the next. In fact, there are considerable changes during the first 24 hours of the adult life of *Drosophila*.

Instinctive activity patterns. Certain insects (e.g., ants, social bees, and digger wasps) exhibit elaborate activity patterns in their nest building and their caring for young. The digger wasp preparing for oviposition goes through a chain of reactions. When the female is ready to oviposit, she locates a spider, a caterpillar, a cicada, or a grub, stings it, digs a hole in the ground, places it in the hole, attaches one egg to its abdomen, and covers the hole with soil (Fig. 5:9). Each action involves orientation to particular chemical and physical stimuli.

Figure 5:9. A cicada killer wasp carrying prey to her nest, which she makes in the ground by digging. *Courtesy William P. Nye, USDA.*

Intraspecific communication. Communication between individuals of the same species represents behavior of considerable complexity. In the honey bee there are several categories: (1) Self-informing—a bee moves around in the hive, inspects the situation, decides what needs to be done, and does it. (2) Flexibility—although bees change their tasks as they grow older, they will do the work normally done by younger or older bees when needed. (3) Feedback—the bees change the type of activity on cue from other bees. When field bees return with nectar, house bees will unload it. Under usual conditions, the most concentrated nectar is unloaded first. If the hive is hot, however, house bees, which cool it by fanning and evaporating water, unload the most watery nectar first. When this occurs, the field bees take the cue and bring back water on the next trip. When the hive is sufficiently cooled off, a reverse transfer of information takes place. (4) Dancing—a field bee upon returning to the hive performs a dance to indicate the direction, distance, quantity, and quality of nectar source. The interpretation of the dance differs slightly among honey bees of different races. Further studies show that the code indicating distance is dictated by the energy expended by the field bee on its way out. For example, if the energy expenditure is raised when heading into the wind, the dance would indicate a greater distance than the actual one. Recent research has indicated that odor is also an important stimulus in transferring information.

POPULATION DYNAMICS

Population dynamics is the study of population growth and its causes. Population growth may be positive or negative; this results, respectively, in an increase or a decrease in the number of individuals. In order to follow the changes, it is necessary first to determine the number of individuals in a population. Because of this necessity the discussion of population growth will be preceded by a brief description of sampling methods.

Sampling Methods

With most field populations it is not feasible to count all individuals in a natural habitat of any size. Thus in most cases, only a number of small areas in the habitat are examined. These units are called **samples**. The samples may be selected within the habitat in several ways, as illustrated in Fig. 5:10.

Systematic sampling. Samples are distributed with uniform spacing between them. This procedure is used where the habitat is fairly homogeneous.

Random sampling. Samples are distributed in a random fashion. This procedure is more useful in determining the general density in a heterogeneous habitat.

Selective sampling. Samples are located in special areas in the habitat. This method is used if it is known that certain sites are favored by the insect, such as on a southern slope, in low land, or under vegetation. Selection of these sites increases the efficiency of the sampling. Selective samples are also

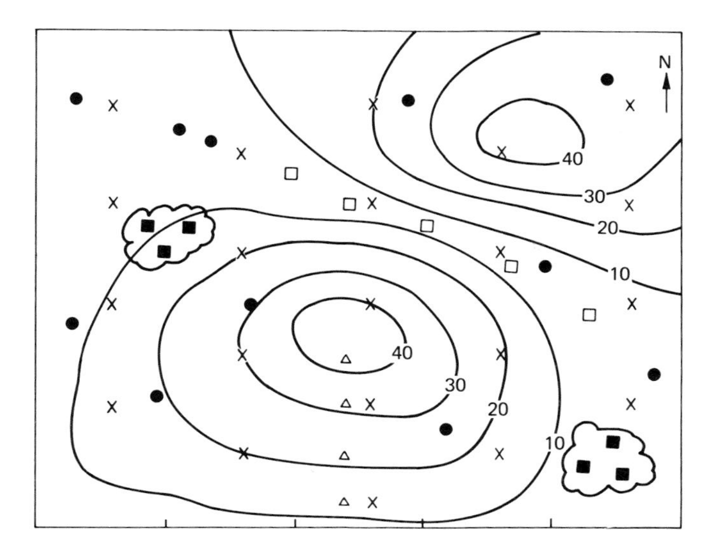

Figure 5:10. Several insect-sampling methods. ●—random samples, ×—systematic samples, △—selective samples on a southern exposure, □—selective samples on low land, and ■—selective samples under vegetation.

chosen to assess the influence of various ecological factors. For example, samples could be taken and compared on high land and on low land, or on different exposures of a hill.

After the individuals in the samples are counted, the number of insects per acre can be computed because the size of the sample area is known. This method is called **actual count**. Another method is the **relative count**: sometimes the information required is simply whether site *A* has more or fewer insects than site *B*. The sites may be different under natural conditions, such as the elevation or exposure mentioned earlier, or under experimental conditions, such as varieties of host plants or applications of several insecticides. The sampling procedure could consist of sweeping plants with a net in a standardized manner or collecting insects for a given period of time in both sites. When insects are too numerous to be counted, it is common to rate the density. The lowest abundance is given the rating of 1, and the highest 3, 5, or 9.

For certain insects, it is possible to estimate the population indirectly by checking signs of insect activity rather than the insects themselves. Entomologists have used the degree of tree defoliation caused by insect infestation, the number of injured plants in a field crop, and the number of dropped fruits in an orchard as measures of population levels. Similarly, they have used the number of cast skins, fecal pellets, and empty cocoons.

Population Growth and Natural Balance

If female insects achieve full fecundity of the species and if all the young survive to maturity, the number of individuals present after a few generations will be enormous. For example, if each female of a species produces 4 eggs, 50 per cent of which are females, the offspring will exceed 1 million in 20 generations; if the female produces 200 eggs, also with 50 per cent females, the 1 million mark will be reached in 3 generations. The relationship is a simple exponential one (Fig. 5:11A) with a base of 2 and a power of 20 in the first example, and a base of 100 with a power of 3 in the second example.

In reality, the full expression of fecundity and survival is approached only in very rare cases, when the environmental conditions are exceptionally favorable. The condition seldom lasts more than one or two generations. An exponential increase is possible only if the conditions remain fairly favorable for several consecutive generations. Even in these cases, the population density itself, when reaching high levels, produces adverse conditions, such as increased competition or degradation of the habitat, which stop further increase.

Unfavorable environmental conditions may reduce fecundity and cause mortality in different stages of the life cycle. Referring back to the two examples, if 2 of the 4 individuals, or 98 of the 100 individuals of the second

Figure 5:11. Population growth. A, exponential increase; B, population changes responding to many environmental factors; C, population changes responding to a few key factors.

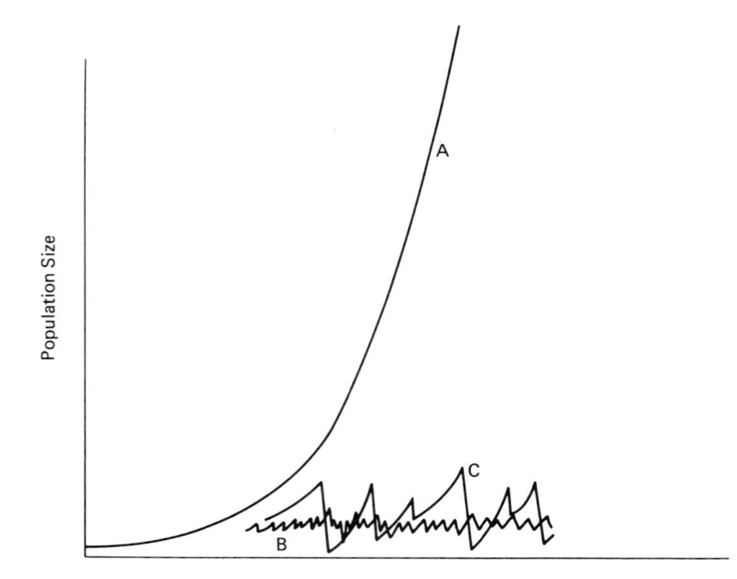

generation, die during the egg or subsequent stages due to unfavorable environmental factors, the population will be maintained at the same level as the parent generation. The mortality ratios of 0.50 (2/4) and 0.98 (98/100) are those that balance the populations of two successve generations. This ratio of balance varies with species and is directly proportional to fecundity. When a mortality ratio higher than that for balance occurs, a reduction in population will result.

In most insects, the number of individuals in one year may be higher or lower than in the preceding year, but over a period of years the level of population always seems to return to normal. This relationship means that the mortality may be higher or lower than the ratio of balance in different years, but over an extended period, the long-term mortality ratio is at the balance level. This phenomenon is called the **balance of nature**. The balance between the responses of a species and its environment is similar to the balance between a predatory species and its prey, or a parasitic species and its host. It is the result of natural selection and organic evolution.

To understand the causes of fluctuations of an insect population requires detailed studies of the ecology of the species—not only the effects of individual environmental factors but also the effects of these factors collectively. Some species are sensitive to a larger number of environmental factors. They respond to changes in any of these factors, and as a consequence exhibit a high frequency of ups and downs. The pattern of fluctuation may appear irregular, almost haphazard, and is difficult to predict (Fig. 5:11B). Other species are tolerant to most factors and sensitive to only one or a few factors. Their population fluctuations can be predicted on the basis of the degree of favorability of such **key factors** (Fig. 5:11C). We may conclude that the frequency of fluctuations is related to the number of environmental factors to which the species is sensitive.

Besides the intrinsic sensitivity of the species, the characteristics of the environment play a part in causing the fluctuations of populations. A species tends to expand its distribution until it is halted by unfavorable conditions. The environment at the fringe tends to be more variable and less optimal than in the main areas of distribution. Consequently, populations in fringe areas show a greater frequency of fluctuations than those in more favorable areas.

Another aspect of fluctuation is magnitude. We mentioned that species sensitive to many factors show a high frequency of fluctuation. At the same time, because of their sensitivity, these populations seldom have the opportunity to build up for any length of time, and thus they tend to fluctuate with oscillations of small magnitude. In contrast, species that are tolerant to most factors have a greater chance for a continuous buildup. Such steady increases lead inexorably to severe food shortages, epidemics of disease, and high rates of parasitism which cause populations to crash. The great buildups followed by the crashes result in fluctuations of great magnitude. We see that the causes of population crashes are often related to high population densities. Populations in the main areas of distribution are subject more to density-dependent biotic factors and fluctuate with greater magnitude than do fringe populations.

Although all insects maintain a balance with the environment, the density at which the balance is maintained varies greatly with different species. Generally speaking, most species in a community are represented by a few individuals, whereas a relatively few species make up most of the biomass. The following examples suggest that the factors of food, natural enemies, habitat, and genetic change are responsible, singly or in combination, for the level of balance. We should not conclude, however, that the factors discussed could be the only ones responsible for maintaining the population of a species at a specific level of balance.

Food factor. Prior to 1859 populations of the Colorado potato beetle fed on buffalobur, *Solanum rostratum* Dunal, on the plains east of the Rocky Mountains and maintained themselves at low densities. After potatoes were introduced to the area, they began to feed on this plant instead of buffalobur. With more and better food available the beetles increased tremendously in number, widened their distribution, and became a major pest. A tenebrionid beetle, *Cynaeus angustus*, found at the base of yucca plants in California, was considered rare prior to 1938. When the insects infiltrated grain storage bins where food was ample and conditions favorable, they underwent a population explosion and have remained at high levels ever since. The Hessian fly provides a different example of the influence of the food factor on population size (Fig. 5:12). This insect, which is a severe pest of wheat, was very abundant in California and in Kansas prior to 1944. After extensive planting

Figure 5:12. Differential fall injury by Hessian fly to plants of several wheat varieties. Varieties still standing and growing are resistant; few or no insects develop from just as many eggs laid on them as susceptible varieties. *Courtesy Kansas State University, Manhattan.*

in these states of wheat varieties resistant to the insect, the populations dropped to noneconomic levels. When Kansas growers began planting high yielding but susceptible varieties again, populations responded by increasing to damaging levels.

Habitat factor. Several species of wireworms in the Pacific Northwest were restricted to wetlands and usually maintained low population levels. After irrigation of crop lands began, these wireworms increased in numbers and became important pests. Tipulid larvae used to be very abundant in prairie land and damaged grain in newly broken ground. Populations have dropped to low noneconomic levels after repeated cultivation. The eutrophication of rivers and lake has reduced the populations of high–oxygen-demanding forms such as mayflies and led to an increase in low–oxygen-demanding forms such as chironomids.

Natural enemies factor. Many introduced species, if established in suitable habitats, tend to increase at near exponential rates until limited by certain environmental factors. Often the populations stay at very high levels. It is generally believed that lack of natural enemies is responsible for the high level of balance. This possibility is substantiated whenever the natural enemies imported from the native home of the pest actually reduce the pest population. A notable example is the control of the cottonycushion scale infesting citrus in California by the vedalia lady beetle imported from Australia.

Genetic factor. The alfalfa weevil was rarely found in northern Montana and southeastern Alberta prior to 1954. Gradual selection by the northern climate resulted in a change in the genetics of a strain of the weevil, and now the insect is common in the north. The northern strain responds slowly to rising temperature in early spring. The slow activation from cold prevents the insects from leaving their hibernation sites too quickly and being caught by a late cold spell, which often occurs, and from drinking water, which raises their supercooling point.

The examples above have been selected to illustrate the effects of specific factors. If a single factor is limiting the population while all other factors are favorable, the improvement of this one factor would markedly increase the population, as the addition of soil moisture did for wireworms and the availability of stored grain for tenebrionids. Conversely, a factor may be degraded to the extent that it starts to limit the population even though all other factors remain favorable; examples are the influence of resistant wheat varieties on Hessian fly and the release of vedalia beetles on cottonycushion scales. We want to emphasize that for most species the mechanisms probably involve a greater number of factors and are more complex.

Operation of Mortality Factors

We pointed out that insects suffer mortality when physical conditions exceed their tolerance, when natural enemies take their toll, and when resources become scarce. Mortality factors may be characterized in different ways, on the basis of their operation. They have different effects on population dynamics and different values for pest management.

Consistency of operation. Mortality factors differ in their consistency. Our **first theorem** is that in their effects on an insect population, the factors that are variable are more crucial than those that are constant. An example may be the reaction of a population to seasonal factors. An insect may suffer a consistent level of mortality every year because of winter conditions, and the surviving individuals may suffer from spring conditions, which are highly variable from year to year. The spring conditions rather than the winter conditions will determine the year's initial population size, even though the winter mortality may be much higher than the spring mortality.

The following example explains consistency of population reaction to the host factor. An insect may suffer a given large mortality on all varieties of the host plant because of the insect's intrinsic weakness. The surviving individuals will be subject to varied additional mortality depending upon the resistance (or susceptibility) level of the variety of the host plant. Thus, the host factor is more crucial than the intrinsic factor in determining the size of the final population, even though the intrinsic mortality may be much higher than the host-induced mortality.

Specificity of operation. Mortality factors, notably the natural enemies, differ in their specificity. A parasite may affect one single host species or many species. Our **second theorem** is that a higher mortality will occur to the host if the general parasite attacks before the specific parasite. The following example will illustrate this relationship.

Let us assume that two host species, *A* and *B*, are present and that species *A* is of economic importance, so that the reduction of its population is a matter of concern. Again assume that two parasitic species are present—that one is general to both hosts, and the other is specific to host *A*. The total effect on host *A* will differ according to the sequence of parasite attacks (I and II),

Table 5:2. Importance of the Specificity of Operation of Mortality Factors

	Host A	Host B
Initial population	100	100
I. General parasite kills a total of 50 hosts	25	25
Remaining	75	75
Specific parasite kills a total of 50 hosts (all A)	50	0
Remaining	25	75
Total mortality %	75	25
II. Specific parasite kills a total of 50 (all A)	50	0
Remaining	50	100
General parasite kills a total of 50	17	33
Remaining	33	67
Total mortality %	67	33

as shown in Table 5:2. It is clear that sequence I produced a higher total mortality on host *A* (75 per cent) than did sequence II (67 per cent).

Density dependency of operation. Mortality factors may differ in their density relationship. Density-dependent factors are those that produce increasingly higher mortality with increasing insect density, whereas density-independent factors exert a constant level of effect regardless of popualtion density. Our **third theorem** is that a higher mortality will occur to the population if the density-dependent factors exert their effects before density-independent factors do.

Let us be concerned with just one density-dependent and one density-independent factor. The quantitative relationships between population density and percentage of mortality with the two factors are shown in Fig. 5:13. Two examples are then given in Table 5:3 with two different initial populations, 100 and 50. In sequence I, the density-dependent factor operated first, but in sequence II the density-independent factor operated first. Table 5:3 shows that sequence I produced a higher mortality at both initial population levels.

Granted, in the examples given for this theorem as well as for the second theorem, the differences were not very large. But within critical ranges, a slight additional percentage of difference in mortality could reduce the population to a noneconomic level.

Target of operation. Mortality factors may affect different stages of insect species. Our **fourth theorem** is that from the standpoint of population

Table 5:3. Importance of Density Dependency of Operation of Mortality Factors

Initial population of insect (2 examples)	100	50
I. Density-dependent factors kill off:		
%	50	30
Number	50	15
Remaining	50	35
Density-independent factors kill off:		
%	20	20
Number	10	7
Remaining	40	28
Total mortality %	60	44
II. Density-independent factors kill off:		
%	20	20
Number	20	10
Remaining	80	40
Density-dependent factors kill off:		
%	40	25
Number	32	10
Remaining	48	30
Total mortality %	52	40

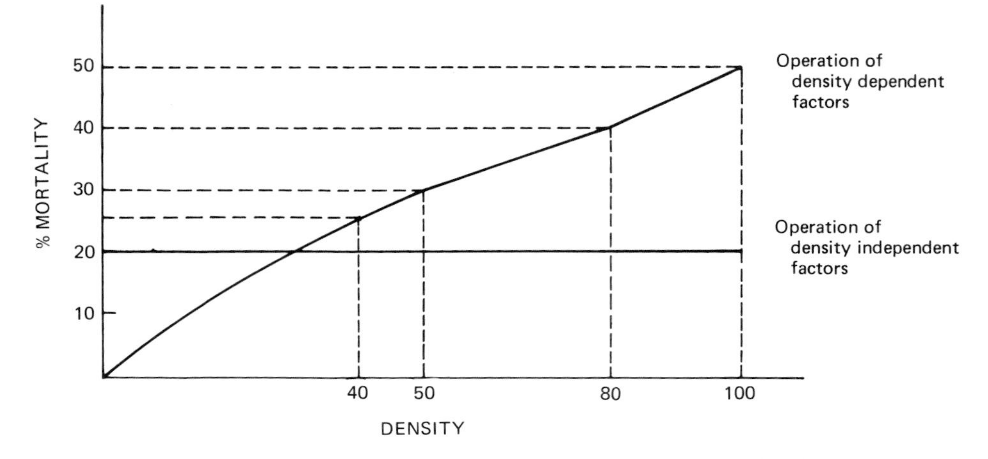

Figure 5:13. Quantitative relationships between population density and percent mortality due to density dependent and density independent factors.

dynamics the factors affecting the adult stage prior to reproduction are more crucial than those affecting the immature stages. The adult represents the final stage of development, which survived continual attrition during successive immature stages. In a hypothetically stable population with one female laying 100 eggs of which 80 hatch, 20 pupate, and 2 reach the adult stage, destruction of one adult, retrospectively, is equivalent to the destruction of 50 eggs, 40 larvae, or 10 pupae. Prospectively, the elimination of one female adult prior to reproduction is equivalent to the elimination of 100 eggs of the subsequent generation. It should be clear that this theorem does not apply when damage to crops is considered, as damage is done also if not more often by immature forms.

ECOLOGICAL CONSIDERATION IN APPLIED ENTOMOLOGY

In the introduction to this chapter, we pointed out that the process of managing insect populations is essentially a device to exploit to our advantage the ecological relationships of insects with their environment. The immediate objective of control, however, is the protection of human welfare and commodities from the adverse effects of a given species of insect in a given location and at a given time. Yet we know that manipulations of environmental factors may have far-reaching effects on other species, in other locations, and at later times. Thus we cannot overemphasize the fact that the design of management programs must consider the total interrelationships of biotic and abiotic components of the community over an extended period of time. In other words, the program must have an ecosystem approach.

The possible untoward side effects of a control procedure, particularly one that employs insecticides, on the community and environment must be considered in any scheme of pest management. As expected, the insecticide reaches the target population. With the immediate objective of crop protection, this face of the insecticidal action is often the sole consideration. However, when the ecosystem, of which the target species is only a small part, is considered, the presence of the insecticide among other components becomes important. Insecticides reach several categories of nontarget species: the plant and animal hosts of the target species, the natural enemies of the target species living in the community, and other animals, including vertebrates, living in or near the community. In addition, some of the insecticide may stay in the abiotic components of the environment, that is, air, soil, and water.

Effect of Insecticides on the Target Species

It is not uncommon to find the simplistic viewpoint that 1 per cent of the insect population killed is 1 per cent of the potential loss saved. This presupposes that crop loss follows a simple arithmetical relationship with the size of an insect population. This is not always true. A wormy apple is a wormy apple whether it has one, two, or more worms. A cabbage plant may produce a good head in spite of cabbageworms if only the outside leaves are damaged. Even ignoring the complex relationships between insect population and crop loss, the presupposition still may not be true because killing the target species may have complicating results. In the long run, the species may become resistant to the chemicals simply because of continued selection pressure. This response to selection is a basic biological characteristic of all living things. With an insecticidal control procedure of high intensity carried out year after year, selection is more severe and persistent than natural selection pressure; hence resistance may show up in a relatively shorter period of time. Development of resistant populations is not a response to chemicals alone. Continuous planting of resistant crop varieties has produced insect strains that have overcome host resistance and survive on such varieties.

When an insecticide kills the target species successfully, there still may be some negative effects, such as causing other pest species to increase. Two examples may be cited. (1) Application of soil insecticides in bean fields to control wireworms may cause an increase in seedcorn maggots, because although wireworms are mainly phytophagous, they prey considerably on seedcorn maggots. Under natural conditions, this predation is sufficiently effective to suppress populations of the seedcorn maggot. (2) In cotton fields, aphids and mites are preyed upon by lacewing larvae and bugs such as the *Nabid* and *Orius*. Later these predators feed on cotton bollworms and *Lygus* bugs. Applications of insecticide, which eliminate aphids and mites successfully, deprive the predators of food, or poison them, causing them either to die or to leave the cotton field. Later when cotton bollworms and *Lygus* bugs arrive, these pests survive in higher numbers because of reduced pressure from predators.

Effect of Insecticides on Nontarget Species

The most obvious of the nontarget species are the plant and animal hosts on which the target species feed. The insecticide absorbed by such plants and animals may affect them physiologically and, in turn, affect the animals or humans who consume them. Much of the problem in this area has been solved by prescribing a minimum period of time between the last insecticidal application and the time of harvest or slaughter of livestock, and by subsequent cleaning.

Because many natural enemies of the target species live in the same community, they represent the next category immediately affected by the insecticide. This category includes other arthropods and vertebrates. Most important are those arthropods that serve as natural enemies of the target species and other pest insects and mites. The reduction of such natural enemies may cause a resurgence of the target species and may unleash other pests which are normally controlled by natural enemies. Once chemicals have been used on some crops, higher dosages and greater frequencies become necessary in subsequent seasons because of the lack of natural enemies. As for the unleashing of other pests, it is now known that mite infestations in orchards follow the control of the codling moth, and heavy bollworm infestations in cotton fields follow control of the boll weevil. Another category of nontarget species often affected adversely by insecticide treatments is the insect pollinators of crop plants.

Vertebrates and other wildlife may receive pesticides through direct exposure and through feeding on contaminated food. This exposure may cause mortality, reducing the population of such wildlife. The reduction of insectivorous species means a lessening of predation pressure on insects. Another aspect of the residue problem is the concentration of persistent chemicals, increasing within organisms through different trophic levels, namely from insects to insectivores and then to higher predators. The vertebrates, because of their great mobility, may be responsible for some of the movement of chemicals from the site of initial application to distant communities.

Effect of Insecticides in the Abiotic Components of the Environment

Much of the insecticide applied in a crop field does not reach the plants or the insects. Instead, it falls on the soil or stays in the air. Some of the insecticide striking the soil may enter surface and underground water. Insecticides in water and air then may be carried from the site of application to other locations. Thus the presence of insecticides may be expanded in space in a pattern beyond the control of the operators. Uncontrolled expansion of persistent chemicals may produce unexpected and unwanted effects on the quality of the environment, human welfare, and wildlife.

The preceding discussion shows that the design of a management program should, whenever possible, consider all the ramifications of its effects on the ecosystem. Such consideration will not only lead to assessment and avoidance of the repercussions, but will also expand the horizon of research for new programs. An interesting example from recent studies involves the control of the green peach aphid. This species uses the peach as its winter host and infests other plants, such as sugarbeets and potatoes, in summer. Investigators discovered that applying herbicides on peach trees in the late season partially defoliated the trees and caused the aphids to move onto the few remaining leaves. The concentration of prey increased the efficiency of predators, which reduced the overwintering aphid population to a much lower level. As a consequence, infestations of sugarbeets and potatoes by spring generations were reduced. In this management program, the chemical used was a herbicide rather than an insecticide, and it was applied remotely both in space and time from the crops to be protected. Development of this particular program was possible only with detailed knowledge of the ecology of the insect including life cycle, dispersal, food habits, and natural enemies.

SELECTED REFERENCES

Allee, W. C., E. Emerson, O. Park, Th. Park, and K. L. Schmidt, *Principles of Animal Ecology* (Philadelphia: Saunders, 1949).

Andrewartha, H. G., and L. C. Birch, *The Distribution and Abundance of Animals* (Chicago: Univ. Chicago Press, 1954).

Barnes, H. F., *Gall Midges of Economic Importance*, Vol. VII (London: Crosby Lockwood and Son, 1956).

Barnothy, M. F., ed., *Biological Effects of Magnetic Field* (New York: Plenum, 1964).

Chavin, Remy, *The World of an Insect* (New York: McGraw-Hill, 1967).

Clark, L. R., R. W. Geier, R. D. Hughes, and R. F. Morris, *The Ecology of Insect Populations in Theory and Practice* (London: Methuen, 1967).

Dempster, J. P., *Animal Population Ecology* (New York: Academic Press, 1975).

Dice, L. R., *Natural Communities* (Ann Arbor: Univ. Mich. Press, 1952).

Elton, C. S., *The Patterns of Animal Communities* (London: Methuen, 1966).

Evans, W. G., "Perception of Infrared Radiation from Forest Fires by *Melanophila acuminata* (Buprestidae, Coleop.)," *Ecology*, 47:1061–5 (1966).

Fraenkel, G. S., and D. L. Gunn, *The Orientation of Animals* (New York: Dover, 1961).

Johnson, C. G., *Migration and Dispersal of Insects by Flight* (London: Methuen, 1969).

Lindauer, M., *Communication Among Social Bees* (Cambridge, Mass.: Harvard Univ. Press, 1961).

Linsley, E. G., "Ecology of Cerambycidae," *Ann. Rev. Entomol.*, 4:99–138 (1959).

Maw, M. G., "Effects of Air Ions on Duration and Rate of Sustained Flight of the Blowfly, *Phaenicia sericata* Meigen," *Canad. Entomol.*, 97:552–6 (1965).

Painter, R. H., "Resistance of Plants to Insects," *Ann. Rev. Entomol.*, 3:267–90 (1958).

Price, P. W., *Insect Ecology* (New York: Wiley, 1975).

Rabb, R. L., and F. R. Guthrie, *Concepts of Pest Management* (Raleigh: North Carolina State University, 1970).
Saini, R. S., and H. C. Chiang, "The Effects of Sub-freezing Temperatures on the Longevity and Fecundity of *Oncopeltus fasciatus* (Dallas)," *Ecology*, 47:473–7 (1966).
Salt, R. W., "Effect of Electrostatic Field on Freezing of Supercooled Water and Insects," *Science*, 133:458–9 (1961).
Schneider, F., "Systematische Variationen in der elektrischen, magnetischen und geographisch-ultraoptischen Orientierung des Maikafers," *Vierteljahrsschrift der Naturforschenden Gessellschaft in Zurich*, 108:373–416 (1963).
Southwood, T. R. E., *Ecological Methods* (London: Methuen, 1966).
Varley, G. C., G. R. Gradwell, M. P. Hassell, *Insect Population Ecology* (Berkeley: Univ. Calif. Press, 1974).

chapter 6 / **CARL JOHANSEN**

PRINCIPLES OF INSECT CONTROL

Insects are the only animals giving man a real battle for supremacy. They have been upon this earth for about 350 million years and have developed special adaptations to live under many environmental conditions. Insects become "pests" of man when their existence conflicts with his profit, convenience, or welfare.

Man has been on the earth for a much shorter period of time and is actually the "intruder." He is the most upsetting factor in the balance of nature. Development of crop agriculture led to concentrations of host plants that literally spread the table for pest insects. Domestication and large-scale production of livestock also provided ideal conditions for the multiplication of parasitic insects. Storage of foodstuffs led to other pest problems. Increases in human population were followed by increases in human lice, mosquitoes, and insect-borne diseases.

Present-day insect problems, created or aggravated by the concentrations of host plants and animals, are diverse and complex, with no simple solutions. Farmers and ranchers must follow the instructions and advice of competent fieldmen carefully in order to cope effectively with damaging pest populations. Specialists who disseminate control information must be able to adapt the lowest cost program which is not hazardous to man and domestic animals and which does not cause undesirable side effects in the environment. In the following pages we shall discuss the principles and methods that have proved useful in man's war against his small but powerful foes.

PRELIMINARY CONSIDERATIONS

Insect Classification and Life History

Knowledge of insect classification, growth and development, and life cycles is requisite to the conduct of control programs. Accurate determinations of insect species, keys to the published work of others, are valuable tools in developing control measures. Life cycle data are essential in the timing of controls. One of the most familiar principles of insect control is that of the "weakest link." Only through a thorough knowledge of a pest's life cycle can one hope to aim control measures effectively at its most vulnerable stage. Even closely related insects vary in their life histories and in their reactions to control measures.

Numbers of Insects

Not only are insects numerous in species, at least 1 million already described in the world, but also they tend to occur in very large numbers in many kinds of habitats. Each acre of farmland may contain several hundred million insects and other arthropods. Insects have relatively short life cycles and produce large numbers of progeny per female. The occurrence of tremendous populations of insects and their high reproductive potentials tend to reduce the effectiveness of control programs.

Introduced Pests

Many of the most destructive insect pests in the United States were introduced from foreign countries. About 50 per cent have been transported accidentally, and "new" species from other continents are being discovered here each year. In addition, another 10 to 20 per cent of the pest insects in an area have spread from other parts of the United States. As we shall see in a later section in this chapter entitled "Biological Control," enemies of introduced pests form the basis for almost all projects in biological control.

Crop Values

The unit value of an agricultural crop is an important consideration. Control of pest insects is usually justifiable when the increase in marketable yield produced is worth more than the cost of the control. However, insect damage to pastures or range land may not be in terms of marketable value but may represent a considerable loss, as would be seen, for example, in the interruption of the seeding cycle of grasses. In the case of low-unit-value crops, such as certain forage crops, the feasibility of controlling pests is difficult to determine. Costly controls can be applied more logically to florist crops and fruit crops than to field crops and cereals.

Consumer Pressure

Consumer demands also have an effect upon insect control. Certain fruits and vegetables may be rejected by the processor if one or more insects are found in samples taken from a truckload. Not even the slightest blemish or "sting" is allowed on certain top-grade fruits. These high standards often force the use of chemical controls, whether they represent the most biologically sound method or not.

Survey and Detection

Competent personnel must detect low-level infestations of pest species before they become damaging. Sound control programs are based on accurate knowledge of the distribution and abundance of the pest insects. Presence of parasites and predators often permits modification of standard chemical control programs and reduction of undesirable side effects of insecticides. Detection is also important in recognizing incipient outbreaks of "new" pests which may be eradicated feasibly if recognized soon enough.

Preventive Control

Preventive control measures can be applied when one knows through experience that a certain pest or pests will develop to a damaging degree in a given area year after year. It is often true that applications early in the season are more efficient than later ones. Early treatments tend to control a pest species before it has reached its maximum rate of development and reproduction and before the crop foliage has grown difficult to penetrate with sprays and dusts. Tree fruit pests are often controlled in this way, because consumer demand and unit value provide the impetus for making applications when only a few insects are present. In contrast, one should wait until a certain pest population level is reached before treating field crops. Some seasons may not favor the development of damaging numbers of the insect, or natural enemies may keep it under control.

Community Projects

Community projects are required in some pest control situations. Control of a virus disease by controlling its insect vector, control of pests of domestic animals and man, and control of insects that readily migrate from one area or crop to another are examples where the cooperative effort of everyone in a sizable locality may be essential. Supervised programs have been organized in some areas in which groups of growers or individual growers hire qualified persons to check fields and provide control recommendations.

As modern agriculture has become a highly organized and intensified operation, effective insect control must be managed on an even more knowledgeable basis. Large-scale community projects, often involving the creation of Pest Control Districts, have proved effective on cotton in the South, on tree fruit in the Pacific Northwest, and in many other situations.

New Methods and Materials

Development of many effective insecticides in recent years has changed the insect control picture in many ways. The numbers of applications previously required with the older kinds of insecticides made the cost of control often prohibitive. These older materials could only be afforded on high-value fruit and vegetable crops. Now it is often feasible to treat field and range crops and even forests because modern insecticides can be used at extremely low dosages. The eradication of introduced pests has also become a more definite possibility. Development of new insecticide formulations and types of application equipment has gone hand in hand with the appearance of new insecticides.

Professional Advice and Assistance

The Agricultural Extension Service forms an important part of all pest control programs. The grower cannot benefit from insect control research conducted by USDA or state agricultural experiment station personnel unless their findings are somehow brought to his attention. It is the job of extension personnel, particularly the county agent, to disseminate such information to the farmers. Fieldmen working for agricultural chemical companies, processors, or groups of growers also perform a vital function in carrying control recommendations to the individual farmers.

Causes of Insect Outbreaks

Outbreaks or epidemics of insect pests are usually caused by one or more of the following:

1. Large-scale culture of a single crop or stock animal by man.
2. Introduction of a pest into a favorable new area without its natural enemies.
3. Favorable weather conditions for rapid development and multiplication of a pest; these conditions may also be unfavorable to natural enemies.
4. Use of insecticides which kill the natural enemies of a pest, exert other effects favorable to a pest, or reduce the competing species of a pest while allowing it to multiply unmolested or only partially controlled.
5. Use of poor cultural practices which encourage buildup of pest infestations.
6. Destruction of natural biotic communities which otherwise provide regulation of insect population levels.

Note that all but the third cause are usually the direct result of man's activities. In the following discussion of kinds of insect controls, you will see that these factors are being counteracted in various ways.

BIOLOGICAL CONTROL

Biological control is the reduction of insect populations by means of living organisms encouraged by man. The essential difference between this control method and natural control, discussed in Chapter 5, is simply that in biological control natural enemies are encouraged and disseminated by man.

The basis for most biological control work lies in the fact that half of the major injurious insects in the United States were introduced from foreign countries. Usually when foreign insects are transported into this country accidentally, they arrive without the parasites and predators that attack them in their native land. Furthermore, native American parasites tend to be specific and not adapted to attack an introduced species.

Almost all biological control programs involve searching for parasites and predators of an introduced pest in the native land of the pest. Therefore, most projects are developed by central governments. The United States maintains biological control laboratories in various foreign countries. Teams are composed of several units: "foreign explorers," who look for potential control agents; "quarantine," where parasites or predators are checked to make sure they will be entirely beneficial; "insectary production," for increasing the numbers of the control agents (Fig. 6:1); and "release and recovery," to establish the parasites or predators in the United States. Other countries conduct biological control projects in a similar manner.

Advantages of Biological Control

Only a limited number of introduced pests have been exterminated successfully with chemicals. It would seem wiser in many instances to attempt a

Figure 6:1. Mass culturing of the braconid wasp, *Macrocentrus ancylivorus* Roch., a parasite of the oriental fruit moth. Stacked trays of potatoes are infested with potato tuberworms which serve as host for culture of the parasite. *Courtesy University of California, Berkeley.*

biological control program, which might reduce pest populations below the level that causes economic damage. If successful, biological control has the tremendous advantage that it becomes self-sustaining and integrated into the normal environment of the control area.

If pest populations are reduced to the point where economic damage is negligible, the control program is a success. Biological controls tend to be particularly useful on low-unit-value crops where complete control may not be required or where chemicals are not recommended. Pests of field and forage crops, forests, and range may be controlled economically by biological methods; chemical controls against such pests may be impractical.

A biological control destined to be fully effective will be easily and quickly established. If an imported parasite or predator is not established within three years of careful releases under good conditions for its development, the program may be justifiably discontinued.

Problems of Biological Control

One of the greatest underlying problems in biological control work is the enormous size of insect populations. Even though thousands of an imported parasite may be released, it takes time for the parasite to reproduce sufficiently to bring the pest under control. A farmer often feels that he cannot wait for the natural enemy to do the job; he needs a marketable crop each year.

When a parasite has become well established in a new area, it may control 90 per cent of the injurious species. However, this degree of control has not been obtained often, and even the most successful campaigns are likely to result in fluctuating populations of the pest. Such results are not good enough, especially for the grower who raises a high-value crop for human consumption. The consumer demands a "perfect," unblemished product. Naturally, the farmer usually resorts to chemical control measures, and the biological control agent is likely to be killed in the process. The irony of such cases (and they have been numerous) is that often the chemical control is unsatisfactory or will cause greater problems.

Other technical difficulties involve such questions as which parasites or predators to introduce, whether to use more than one parasitic species at a time, how to eliminate secondary parasites that prey on the beneficial form, and whether a program of continuous liberations may be feasible.

Biological Agents

Many types of animals have been utilized as biological control agents: top minnows, toads, birds, mites, spiders, snails, and insects. Most of the successful cases to date have involved the use of parasitic or predaceous insects. Ninety-five imported species were established in the United States during the sixty-year period ending in 1956 (81 parasites and 14 predators). A greater number, 390 imported species, did not become established during this same time. A recent survey shows 225 (59 in the United States) successful examples of biological control on a world-wide basis. Sixty-six of these were evaluated as completely effective, 88 as substantially effective, and 71 as partially effective.

The major kinds of parasitic insects that have been utilized are chalcidoids, braconids, ichneumonids (Hymenoptera), and tachinids (Diptera); the major predators have been the coccinellids (Coleoptera). Two basic kinds of programs are (1) introduction of exotic parasites and predators and (2) augmentation of established parasites and predators (mass culture and periodic release).

Use of disease organisms, "microbial control," has developed into a more important part of biological control in comparatively recent years. To date, more than 915 microorganisms are known to be pathogenic to insects, including about 25 species of bacteria, 420 species of viruses and rickettsiae, 255 species of fungi, 200 species of protozoa, and 15 species of nematodes.

Until 1970 only certain bacterial spores, the causal agents of milky diseases in the Japanese beetle, and *Bacillus thuringiensis*, which kills many kinds of caterpillars, were marketed commercially as insecticidal materials. *B. thuringiensis* is of particular interest because it produces a toxin that may serve as a model for a new class of insecticides. Two types of virus diseases, polyhedrosis and granulosis, have shown considerable promise in controlling insect pests. The first commercial virus pesticide, *Heliothis* NPV (nuclear polyhedrosis virus), was federally registered and granted permanent exemption from the requirement of tolerance in 1973.

Microbial controls have a number of interesting advantages: there is no chemical residue problem; they are usually quite specific and are not a hazard to beneficial insects or to man; commercial preparations could be made quite inexpensively, and low dosages would suffice in control work; they are compatible with insecticidal chemicals and quite versatile; few insects have developed resistance to microbial organisms to date. Disadvantages of this method include timing of applications, which is very critical because of dependence on weather conditions and lack of residual action in many cases; specificity, which is a disadvantage where several major pests must be controlled on a single crop; some types of microorganisms that are difficult to produce in quantity and cheaply; fungi, especially, which require high atmospheric moisture for successful development. We shall refer to the commercially available insect pathogens in Chapter 7 as microbial insecticides.

Weeds

The biological control of weeds is a related topic. Because this method involves the introduction of insects to attack weeds, the same branch of the USDA investigates the biological control of both insects and of weeds. The outstanding success in the United States has been the use of certain leaf beetles, *Chrysolina quadrigemina* Suffr. and *C. hyperici* (Forst.), to control the Klamathweed throughout most of the West. These beetles have completely eradicated Klamathweed from thousands of acres of rangeland. *C.*

quadrigemina has been the more effective of the two Klamathweed beetles. Its life cycle and habits are better synchronized with the development of the host weed than are those of *C. hyperici.*

Another weed control campaign that shows great promise involves the alligator weed, *Alternanthera philoxeroides.* This weed covered an estimated 97,000 acres of waterways and marshes in the South in 1963. A flea beetle, *Agasicles hygrophila,* was first released in 1964. The addition of a pyralid stemborer, *Vogtia malloi,* in 1971 is expected to expand permanent control of alligator weed to a wide range of conditions. At present, the weed is under biological control in Florida, and satisfactory to spectacular results have been reported from South Carolina, Georgia, Alabama, and Texas.

CULTURAL CONTROL

Cultural control is the reduction of insect populations by the utilization of agricultural practices. It has also been defined as "making environments unfavorable for pests."

The method more or less associated with agricultural production usually involves certain changes in the normal farming practices rather than the addition of special procedures. It sometimes consists of avoidance of factors that favor an increase of injurious insects.

Knowledge of the life history or bionomics of a pest species is essential to the effective use of cultural control methods. The principle of the "weakest link" or most vulnerable part of the life cycle usually applies. The environment is changed by altering farming practices at the correct time so as to kill the pests or to slow down their multiplication. In this way, the method is aimed more at prevention than at cure. If the environment is unfavorable, the pest may not reach a population level that will cause serious damage.

Cultural controls are often used when chemical or biological methods have not yet been devised for an injurious species. Cleanup of the sources of infestation and changes in the planting or harvesting time are particularly important when no effective method of killing the pest is known. However, these methods are also used in combination with other controls.

Sometimes cultural control practices are inconvenient to the farmer and are dropped quickly after other methods have been developed. At other times, the fact that insect control was the primary reason for establishing certain practices may be forgotten with the passage of time. Methods such as certain crop rotations may become so well established that they are simply conducted as standard practices. Because cultural methods are usually economical, they are especially useful against pests of low-unit-value crops. Practices that reduce the chances of buildup of pest populations may hold them below the level that will cause economic damage. Such methods are particularly applicable to field crops and forests.

Rotation

Certain kinds of crop rotations may aid in the control of pests. Insects that are reduced effectively by rotations usually have a long life cycle and a limited host range and are relatively immobile in some stages of their development. Changing crops in a rotation system isolates such pests from their food supply. Long-cycle insects such as wireworms and white grubs are good examples.

Planting row crops following sod is a poor type of rotation. Wireworms, cutworms, or webworms are concentrated on the row crop in this way and may cause serious damage. Planting two similar crops in succession also tends to increase insect problems.

Usually rotations are applied more effectively to field crops than to truck crops because of the larger size of plantings and because isolation of the relatively immobile stage of an insect is more complete. Rotation has also been used to some extent as a livestock pest control technique. The best-known example is maintenance of pastures completely free of animals for an extended period of time to kill ticks by starvation.

Location

Careful choice of crops to be planted adjacent to each other may help reduce insect damage. Similar crops may be attacked by the same pests, or one crop may attract a pest which later moves on to the adjacent field. Mixed crops may also deter the infestation of an injurious insect. Plantings containing mixtures of two or more species of forest trees have been particularly effective in preventing damage by certain forest pests.

Trap Crop

Small plantings of a susceptible or preferred crop may be established near a major crop to act as a "trap." After the pest insect has been attracted to the trap crop, it is usually treated with insecticides, plowed under, or both.

Tillage

The use of tillage operations to reduce populations of soil-inhabiting insects may work in several ways: it may change physical condition of soil, bury a stage of the pest, expose a stage of the insect, mechanically damage some stage of the insect, eliminate host plants of the pest, or hasten the growth or increase the vigor of the crop.

Clean Culture

Removal of crop residues, disposal of volunteer plants, and burning of chaff stacks are measures commonly applied against vegetable and field crop insects. This method is especially useful for control of caterpillars or beetles that hibernate in plant debris. Burning weeds in drainage ditches has been used as a novel way to suppress overwintering populations of the summer form of the green peach aphid. This program reduced the incidence of beet western yellows virus disease in subsequent seasons. Communitywide practices of stalk shredding followed by plowing under of cotton residues in the fall have provided control of the pink bollworm in some localities.

Timing

Changes in planting time or harvesting time are used to keep the infesting stage of a pest separated from the susceptible stage of the host. Timing is effective as a control measure when the crop concerned may be infested or injured only during a brief period, or when the infesting stage in the life of the insect is very brief.

Resistant Plant and Animal Varieties

The sources of resistance to insects in crops have been classified as nonpreference, antibiosis, and tolerance. Insect preference for a certain host plant is related to color, light reflection, physical structure of the surface, and chemical stimuli such as taste and odor. A resistant variety may have the quality of **nonpreference** by lacking one or more of the preferred factors or characteristics of the host plant.

Antibiosis is defined as an adverse effect of the plant upon the insect, either because of the deleterious effect of a specific chemical or because of the lack of a specific nutrient requirement.

Tolerance is the term applied to the general vigor of certain plants, which may be able to withstand the attack of pests such as sucking insects. Tolerance also includes the ability to repair tissues and recover from an attack.

A series of regional Plant Introduction Stations (ARS, USDA) provide an essential service in finding new plant stocks and testing them under a variety of conditions. This federal organization acts as the world germ plasm depository for cultivated crops used in breeding resistant strains.

Advantages of the use of resistant varieties include a cumulative and persistent effect, which often eliminates pest damage within a few seasons; lack of dangers to man and domestic animals; low cost (once the program is established); and utility in integrated control systems. Currently, six varieties of wheat resistant to the wheat stem sawfly are being grown on several million acres in Canada and the northern states. At least twenty varieties of wheat resistant to the Hessian fly are being recommended in various states. There has also been minor use of resistance in domestic animals to such pests as the horn fly and certain ticks.

Management

Most cultural control methods could be described as management. However, cultural procedures that aid in the control of livestock pests are especially referred to as management control. Such practices as destruction of breeding places of pests—cleanup of pens, barns, and shelters—and isolation of infested animals exemplify the methods used.

MECHANICAL AND PHYSICAL CONTROL

Mechanical control is the reduction of insect populations by means of devices that affect them directly or that alter their physical environment radically. These methods are often hard to distinguish from cultural methods. However, mechanical controls involve special physical measures rather than normal farm practices. They tend to require considerable time and labor and often are impractical on a large scale.

Hand picking, shingling, and trapping are familiar mechanical methods of insect control. Screens, barriers, sticky bands, and shading devices represent other mechanical methods; hopper-dozers, drags, and entoleters are specialized equipment for collecting or smashing pests.

Physical controls include the use of electricity, sound waves, infrared rays, X rays, or light to kill insects or attract them to killing mechanisms. Probably the most common physical methods are those that employ heat or cold. Cold storage of farm produce will usually eliminate further insect activity and damage during the storage period, even though the pests may not be killed. Both heat and cold have been used to reduce insect populations in grain elevators. No insect can survive temperatures of 140 to 150°F for very long. Soaking flower bulbs in hot water is a standard method of controlling certain pests.

LEGAL CONTROL

Legal control is the lawful regulation of areas to eradicate, prevent, or control infestation or reduce damage by insects. This method involves mainly the use of quarantines and pest control procedures. Federal and state officials often work with legally established local, community, or county districts, as in mosquito or grasshopper control projects. See Table 6:1 for a summary of legal control operations against agricultural pests.

Plant and Animal Quarantines

Fundamental prerequisites of plant quarantines are that (1) the pest must offer an actual or expected threat to large interests, (2) no substitute action less disruptive of normal trade is available, (3) the objectives must be reasonably possible to attain, and (4) the economic gains from pest control must outweigh the cost of administration and the interference with normal trade activity.

Some of the problems of plant quarantine work are (1) high-speed transportation and increased travel, (2) lack of information about potential pests in foreign countries with whom we trade, and (3) incomplete knowledge of the biology of the pest. One aspect of the third problem is that insects sometimes cause serious damage in a new place. In their native land they may be of minor concern because of natural enemies, weather conditions, and existing agricultural methods. However, when they are introduced into a favorable new area accidentally, considerable injury may result.

Table 6:1. Summary of Legal Control Operations Against Agricultural Pests

Activity	Major Federal Legislation	Administering Agency
Animal quarantines	Animal quarantine and related laws of 1884 and 1930 (plus acts of 1890, 1903, 1905, and 1962)	Veterinary Services, APHIS, USDA
Export certification of domestic animals	Animal quarantine acts of 1903 and 1907 (plus acts of 1884, 1890, 1891, and 1962)	Veterinary Services, APHIS, USDA
Import of animal products and related materials	Animal import and tariff acts of 1903 and 1930	Plant Protection and Quarantine Programs, APHIS, USDA
Plant quarantines (foreign and domestic)	Plant Quarantine Act of 1912	Plant Protection and Quarantine Programs, APHIS, USDA
Terminal inspection of plant materials	Terminal Inspection Act of 1915	State authorities (in cooperation with the Secretary of Agriculture and the U.S. Post Office Dept.)
Large-scale plant pest control programs (except forest pests)	Incipient or Emergency Outbreak Resolution of 1938	Plant Protection and Quarantine Programs, APHIS, USDA
Eradication of plant pests (except forest pests)	USDA Organic Act of 1944, 102–a	Plant Protection and Quarantine Programs, APHIS, USDA
Export certification of plant products	USDA Organic Act of 1944, 102–b (enlarges authority of 1926)	Plant Protection and Quarantine Programs, APHIS, USDA, and state officials, under authority of Secretary of Agriculture[1]
Eradication of animal pests	USDA Organic Act of 1956 (and laws of 1944 and 1948)	Veterinary Services, APHIS, USDA
Movement of a living plant pest into the United States or between states	Federal Plant Pest Act of 1957 (repeals acts of 1905 and 1951, amended 1960 and 1967)	Plant Protection and Quarantine Programs, APHIS, USDA

[1] PPQ, APHIS, USDA is the agency responsible for inspection and certification when the importing country requires a federal certificate. Some inspections may be conducted by other cooperating USDA personnel or authorized state inspectors and may be used as a basis for issuance of federal certificates by federal plant quarantine inspectors. A state may inspect and issue state certificates if in accordance with the international model certificate and when acceptable by the importing country. Each country has its own import regulations. The responsibility of PPQ is to keep up to date on these regulations and to distribute information that will assist the exporter of U.S. agricultural products.

The Federal Environmental Pesticide Control Act of 1972 (modified FIFRA) affects the use of pesticides in carrying out operations under above major federal legislation. However, adequate exemptions have been provided.

Both plant and animal quarantines are administered by the Animal and Plant Health Inspection Service (APHIS). Plant quarantines are based mainly on the **Plant Quarantine Act of 1912**, the **Mexican Border Act of 1942**, and various state regulations. Federal plant quarantines are administered by the Plant Protection and Quarantine Programs, APHIS, employing about 600 inspectors at some 85 ports of entry. A number of **domestic quarantines** (within the United States and its territories) have been established. Some of the more important are directed against the following: infestations in Hawaiian fruits and vegetables, gypsy and browntail moths, Japanese beetle, pink bollworm, infestations in Puerto Rican fruits and vegetables, Mexican fruit fly, and whitefringed beetles. **Foreign quarantines** (directed against materials from other countries) include sugar cane, sweet potatoes and yams, nursery stock, corn and related plants, rice, fruits and vegetables, and cut flowers.

Quarantines for protection against infectious diseases of domestic animals are administered by the Veterinary Services, APHIS. Foreign animal quarantines are based on laws enacted in 1890, 1926, and 1962; domestic quarantines are based on laws of 1903, 1905, 1928, and 1962.

Eradication and Control

Some provision for the eradication of introduced pests was made under the Plant Quarantine Act of 1912. Later, this activity was more specifically established by the joint resolution of 1937, **Control of Incipient or Emergency Outbreaks of Insect Pests or Plant Diseases**, and by a section of the **Department of Agriculture Organic Act of 1944** as amended. An amendment in 1957 to the Organic Act provides for the large-scale control by the Plant Protection and Quarantine Programs, APHIS, in cooperation with states, of the spread of newly introduced pests such as imported fire ant, khapra beetle, and soybean cyst nematode.

Several special methods improvement laboratories have been established to develop control techniques against newly introduced pests. Use of Mirex bait for imported fire ant control is an example of a new technique that has eliminated both the hazards to fish and wildlife and the environmental contamination produced by the previous program. Ultra-low-volume application of insecticide is a system which has proven particularly effective against the grasshopper, the boll weevil, and the cereal leaf beetle.

State and federal agencies also cooperate with farmers and grower organizations on programs against pests that are capable of periodic outbreaks over large areas. Probably the best-known suppression programs are the grasshopper control operations in the West. The Environmental Protection Agency has the responsibility for pesticide monitoring in such large-scale programs. All of the following factors must be considered: effectiveness, possible hazards to applicators and public, residues in food or feed crops, hazards to fish and wildlife, danger to honey bees as well as other pollinators and other beneficial insects, effects on soil organisms, water pollution and contamination of surrounding areas, possible systemic action in plants, and persistence in soil.

Control and eradication of arthropod-borne diseases of domestic animals are provided for in the **Department of Agriculture Organic Act of 1956**, based on earlier laws of 1884, 1903, 1944, and 1948. The Veterinary Services, APHIS, is currently conducting campaigns against cattle tick fever, screwworm myiasis, bluetongue of sheep, and other maladies. A special act for the development of an eradication program for cattle grubs was passed in 1948. The Entomology Research and Animal Disease and Parasite Research units of ARS work on this problem.

Successful eradication campaigns have been conducted against the Mediterranean fruit fly (1929–30 and 1956–57), the parlatoria date scale, the citrus blackfly, and the white garden snail. The Hall scale is believed to have been eradicated from the United States; the cattle tick control program is now confined to occasional reinfestations in Texas and California; apparently the screwworm has been eradicated from the southeastern states. A campaign started in 1962 had by 1966 reduced the screwworm populations until none were self-sustaining north of the Rio Grande River. Outbreaks occurred, however, in 1968, 1972, 1973, and 1974, particularly in south Texas. Even in these years, the program has been cost-effective and has protected the majority of livestock.

The medfly campaign of 1929 required 18 months and cost 7 million dollars, whereas the campaign of 1956 took about the same length of time and cost 13 million dollars. Since 1957, an intensive trapping program has been conducted to detect new invasions as quickly as possible. New infestations detected in 1962, 1963, and 1966 were eradicated in 8 months, $2\frac{1}{2}$ months, and 44 days, respectively, indicating that constant surveillance is a good investment.

Export Certification, Terminal Inspection, and Movement of Plant Pests

Export certification of domestic plant materials according to the sanitary requirements of the foreign country to which they are being exported was included originally under the Plant Quarantine Act of 1912. Now, such activities are provided for in the **Department of Agriculture Organic Act of 1944**. Import regulations of foreign countries are translated, summarized, and made available to U.S. exporters. Prevention of the export of diseased cattle was specifically included in the act establishing the Bureau of Animal Industry in 1884. Export certification of domestic animals is based on laws enacted in 1890 and 1903 and more recent amendments.

Inspection of plants and plant products moving across state lines was established by the **Terminal Inspection Act of 1915**. When such inspections are desired by state officials of a particular state, they may request, through the Secretary of Agriculture, that the U.S. Postmaster General direct all mail packages containing designated plants or plant products to the proper state official for inspection. Importation or interstate movement of any living pest organism injurious to cultivated plants or plant products is prohibited or restricted by the **Federal Plant Pest Act of 1957**, which replaced the **Insect Pest Act of 1905**. Movement of living insect pests and plant disease organisms for scientific purposes can only be conducted by permit obtained according to the provisions of this act. Approval is based on careful review by both the federal and the state agencies involved. Eradication of insects that may be introduced by intercontinental airplane travel is also covered by the Plant Pest Act.

Although tourists are likely to be annoyed at inspection activities at state or national boundaries, they undergo only minor inconveniences when one considers the millions of dollars that quarantine and eradication programs have saved the United States over the years. Even if a serious pest is only prevented or retarded for a few years, the monetary savings to a specific industry, such as citrus growing, may be considerable. Some of the recent trends for increasing the effectiveness of plant quarantine work are preshipment inspection, treatment of airplanes and ships in foreign countries, a special training center for quarantine workers at the Port of New York, plant quarantine courses offered in several colleges, training of foreign nationals at the **Plant Quarantine Training Center**, and more effective education of tourists concerning quarantine regulations.

REPRODUCTIVE CONTROL

Reproductive control is the reduction of insect populations by means of physical treatments or substances that cause sterility, alter sexual behavior, or otherwise disrupt the normal reproduction of insects. Manipulating the natural reproduction of insects and related pests is a recent development which shows considerable promise. It involves various methods of sterilizing insects, the use of sex attractants, and genetic manipulations.

The greatest success to date with sterilization has been the eradication of the screwworm from both the Southeast and the Southwest. This project depended upon the mass rearing, irradiation, and release of the sterilized male screwworm into wild populations (sterile insect release method or SIRM).

Major prerequisites regarding the species to be controlled by radiation-induced sterility are as follows:

1. It can be reared on a mass production scale economically.
2. It will not become a nuisance or source of injury after release.
3. It is a species in which the males and females mix over a considerable area before mating occurs.
4. It is a species in which the influence of multiple matings by the females is known.

Chemicals that sterilize insects may some day prove to be much more practical and economical than radiation techniques. The basic advantage is that it would be unnecessary to mass-rear the pest to be controlled. Chemosterilants combined with suitable, specific attractants might be placed throughout an area in order to attract and sterilize wild insect populations.

A third method of sterilizing insects or rendering them innocuous to man involves mass rearing of genetically manipulated individuals to replace the pest populations. Inherited sterility, dissociation of obnoxious properties, and introduction of conditional lethal traits are the genetic tools. Pest species being studied at the present time are the Australian sheep blow fly, *Culex*, *Anopheline*, and *Aedes* mosquitoes, the house fly, a field cricket, and the twospotted spider mite. A major problem for these programs is maintenance of good adaptive ability and competitiveness in laboratory-reared strains.

Modern chemical techniques have made it possible to determine the chemical configurations of the natural sex attractants of insects. Many have been synthesized and at least 12 are available commercially, for example: muscalure (house fly), codlelure (codling moth), disparlure (gypsy moth), looplure (cabbage looper), and grandlure (boll weevil). Artificial pheromones can be used to attract male pest insects to traps, to confuse them by masking the location of females, to time control applications most efficiently, and to locate the major sources of pest infestation in an area.

CHEMICAL CONTROL

Chemical control is the reduction of insect populations or prevention of insect injury by the use of materials to poison them, attract them to other devices, or repel them from specified areas. This topic is discussed in Chapters 7 and 8.

INTEGRATED PEST MANAGEMENT

Integrated control is the management of insect populations by the utilization of all suitable techniques in a compatible manner, so that damage is kept below economic levels. It is an ecological approach that not only avoids economic damage but also minimizes adverse side effects. Principal considerations of the integrated approach to pest management are the agroecosystem, the economic threshold, and the least disruptive program.

Agroecosystem is the complex of organisms and environmental features found in an agricultural field, orchard, or pasture. It differs from a natural ecosystem in that it is relatively artificial. Typically it will contain a few highly abundant arthropod species adapted to the crop involved and many more (usually several hundred) relatively rare species. The most effective system for controlling pests can be devised only after thorough knowledge has been gained of the principal factors underlying the fluctuations of populations.

Economic injury level is the pest population level at which economic damage begins to occur. These levels must be determined for each crop and locality, in relation to specific consumer requirements. Control measures are aimed at keeping the pest below these economic levels, in the most favorable manner.

Economic threshold, a related concept, is defined as the pest density at which control measures should be applied to prevent an increasing pest population from reaching the economic injury level. This increase may be caused by expanding numbers of pests or accruing biomass (weight) of pests as individuals grow. The economic threshold is always lower than the economic injury level.

Least disruptive program is the development of adequate controls which do not upset the desirable features of the agroecosystem. Use of a selective insecticide that destroys many of the pest individuals, while leaving a reservoir of predators and parasites to eliminate those that remain, is the most common method of achieving these results. Another system might involve the use of an insect-resistant crop variety that, combined with the effects of predators and parasites, restricts the pest below economic injury level without the use of chemicals. Currently, an outstanding commercial operation involves five or more cultural practices, natural and introduced parasites and predators, and limited use of selective insecticides.

Examples

Successful integrated programs of pest control have been developed in many parts of the world and on many different crops. Examples include control of the San Jose scale in western Europe with a eulophid parasite and selective insecticides; control of the rice water weevil in Arkansas with a combination of seed treatment and removal of weed grasses; control of a complex of cotton pests in Peru with reduced planting of ratoon cotton, preparation of soil without irrigation, encouragement of natural enemies, mandatory planting and crop residue destruction dates, and carefully selected insecticides; control of the corn earworm in the Southeast with use of resistant corn hybrids and certain insecticides; control of both the grape mealybug and the citrus mealybug in South Africa by encouraging parasites and predators and by controlling ants (which attack the natural enemies) with barriers or poisons; control of the pea aphid and the spotted alfalfa aphid on alfalfa throughout much of the West with minimum dosages of systemic insecticides which allow surviving predators and parasites to control the remaining aphids; control of the cabbage aphid in England with systemic insecticides applied as soil treatments so that parasites and predators remain unharmed; and control of the omnivorous leafroller on grapes in California with field sanitation and limited chemical treatments.

A special type of integrated program, involving carefully selected chemical control of insects plus natural control of spider mites, has been developed for use on tree fruits. Predators will provide good control of spider mites, if the natural enemies are not killed by insecticides. This system has been used in western Europe, Canada, Michigan, West Virginia, California, and

Washington. In California, the predator mite, *Metaseiulus occidentalis*, controls the European red mite and the twospotted spider mite on peach, whereas three insect pests are controlled with a dormant spray plus two to three carefully selected and timed insecticide applications during the growing season. The same predator controls the McDaniel spider mite and the European red mite on apples in Washington; a modified spray program controls the codling moth.

Necessity of Integrated Approach

Sole reliance on insecticides for control of pests may lead to the following problems: (1) selection of resistance to chemicals in pest populations; (2) resurgence of treated populations; (3) outbreaks of secondary pests; (4) residues on food and forage products and legal complications; (5) destruction of beneficial predators, parasites, and pollinators; (6) hazards to applicators; domestic animals, fish, and wildlife; and (7) expense of pesticides, involving recurrent costs for equipment, labor, and material.

Insect populations most often increase to injurious levels because of man's activities; therefore it is important that disruptive actions be identified and either avoided or minimized. Making use of carefully timed selective insecticide applications that allow predators and parasites to continue to reduce pest species is one of the most effective ways this may be achieved. Use of cultural methods such as resistant varieties or crop rotations may provide additional help in counteracting the deleterious actions of man. Whatever methods are employed, they must be organized carefully into a compatible system based on precise knowledge of the agroecosystem.

Requirements

In order to develop a successful, integrated program, economic thresholds must be determined, mortality factors in the agroecosystem must be elucidated, and the least disruptive supplemental methods of control must be devised. Development requires trained personnel who can obtain the data necessary for detailed analysis of the problem. Once an integrated program has been organized and tested, operation on a commercial scale requires competent field personnel—men who can translate the research findings into practical use for the growers. Professional sample analyses (e.g., commercial mite-counting services) are an important adjunct to large-scale integrated programs.

The Emerging Concept of Pest Management

Although the term *pest management* is relatively new, the concepts on which it is based have been developed over many decades. As early as 1880, S. A. Forbes of the University of Illinois was advocating an ecological approach and combinations of control measures for insect control. The realization of the value of monitoring pest populations by sampling led to cotton scouting programs in the South by 1925 and to organized pea weevil programs in the Pacific Northwest by 1935. During the 1940s Ray F. Smith developed an outline of *supervised control* in California. Ecological studies in the apple orchards of Nova Scotia initiated in 1943 soon led to the development of *modified spray programs.* In a popular article outlining practical ways of combining chemical and biological methods, B. R. Bartlett in 1956 coined the term *integrated control.* Shortly afterward V. M. Stern, R. F. Smith, R. van den Bosch, and K. S. Hagen, published a definitive article on the *integrated control concept.* In 1961 the Australian entomologists P. W. Geier and L. R. Clark coined the phrase *pest management* for programs in which control methods fit into the biology of the pest species. They chose the term to emphasize the need not only for a broad approach, but also for recognition of ecological principles in pest control.

Currently, pest management has evolved to include all pests of a given crop—insects, weeds, plant diseases, and nematodes. It considers the interacting factors of the environment, including the effects of adjacent crops and cultural practices, upon the crop and its pests. The major difference between pest management and standard pest control programs is that in pest management sampling of pests and beneficial species and monitoring of crop development and of environmental factors are regularly done. The data not only provide a detailed picture of pest conditions but also an accurate assessment of population trends with time. Growers may then use the information to make pest management decisions. Data from several seasons of intensive sampling can also provide the basis of a computerized recall program to aid in more complex evaluations.

Special programs are being developed for urban pest control and pests of man and animals. Since 1970 dozens of pilot projects have been initiated with joint state-federal financing. Responsibility for implementing pest management programs is vested in the Cooperative State Extension Service with support from the USDA and EPA. The programs represent a high level of technical achievement and interdisciplinary cooperation.

SELECTED REFERENCES

Burges, H. D., and N. W. Hussey, eds., *Microbial Control of Insects and Mites* (New York: Academic Press, 1971).

Bushland, R. C., et al., *Sterility Principle for Insect Control or Eradication* (Vienna: Internat. Atomic Energy Agency, 1971).

Camp, A. F., "Modern Quarantine Problems," *Ann. Rev. Ent.*, 1:367–78 (1956).

Clausen, C. P., "Biological Control of Insects," *Ann. Rev. Ent.*, 3:291–310 (1958).

Croft, B. A., and A. W. A. Brown, "Responses of Arthropod Natural Enemies to Insecticides," *Ann. Rev. Ent.*, 20:285–335 (1975).

Davidson, G., *Genetic Control of Insect Pests* (New York: Academic Press, 1974).

DeBach, P., ed., *Biological Control of Insect Pests and Weeds* (New York: Van Nostrand Reinhold, 1964).

DeBach, P., *Biological Control by Natural Enemies* (London: Cambridge Univ. Press, 1974).

Huffaker, C. B., "Fundamentals of Biological Control of Weeds," *Hilgardia*, 27:101–57 (1957).

Huffaker, C. B., ed., *Biological Control* (New York: Plenum Press, 1971).

Huffaker, C. B., and B. A. Croft, "Integrated Pest Management in the U.S.: Progress and Promise," *Environmental Health Perspectives*, 14:167–83 (1976).

Huffaker, C. B., and P. S. Messenger, eds., *Theory and Practice of Biological Control* (New York: Academic Press, 1976).

Isely, D., *Methods of Insect Control*, Parts I and II (Minneapolis: Burgess, 1946 and 1947).

Kilgore, W. W., and R. L. Doutt, eds., *Pest Control* (New York: Academic Press, 1967).

LaBrecque, G. C., and C. N. Smith, eds., *Principles of Insect Chemosterilization* (New York: Appleton, 1968).

Maxwell, F. G., and F. A. Harris, ed., *Proceedings of the Summer Institute on Biological Control of Plant Insects and Diseases* (Jackson: Univ. Press of Mississippi, 1974).

McGovran, E. R., Chairman, *Insect-Pest Management and Control, In Principles of Plant and Animal Pest Control*, Vol. 3 (Washington, D.C.: Nat. Acad. Sci., 1969).

Metcalf, R. L., and W. H. Luckman, *Introduction to Insect Pest Management* (New York: Wiley, 1975).

Painter, R. H., *Insect Resistance in Crop Plants* (New York: Macmillan, 1951).

———, "Resistance of Plants to Insects," *Ann. Rev. Ent.*, 3:267–90 (1958).

Popham, W. L., and D. G. Hall, "Insect Eradication Programs," *Ann. Rev. Ent.*, 3:335–54 (1958).

Rabb, R. L., and F. E. Guthrie, eds., *Concepts of Pest Management* (Raleigh: North Carolina State Univ., 1970).

Ryan, H. J., *Plant Quarantines in California* (Univ. Calif. Div. Agr. Sci., 1969).

Smith, R. F., and W. W. Allen, "Insect Control and the Balance of Nature," *Sci. Am.*, 190(6):38–42 (1954).

Spears, J. F., *A Review of Federal Domestic Plant Quarantines* (Hyattsville, M.: Plant Protection & Quarant. Prog., APHIS, USDA, 1974).

Steinhaus, E. A., "Microbial Control—the Emergence of an Idea," *Hilgardia*, 26:107–60 (1956).

Stern, V. M., "Economic Thresholds," *Ann. Rev. Ent.*, 18:259–80 (1973).

Stern, V. M., R. F. Smith, R. van den Bosch, and K. S. Hagen, "The Integrated Control Concept," *Hilgardia*, 29:81–101 (1959).

Turnbull, A. L., and D. A. Chant, "The Practice and Theory of Biological Control of Insects in Canada," *Can. J. Zool.*, 39:697–753 (1961).

Van den Bosch, R., and V. M. Stern, "The Integration of Chemical and Biological Control of Arthropod Pests," *Ann. Rev. Ent.*, 7:367–86 (1962).

van Emden, H. F., and G. F. Williams, "Insect Stability and Diversity in Agro-ecosystems," *Ann. Rev. Ent.*, 19:455–75 (1974).

Waterhouse, D. F., Chairman, *Proceedings of the FAO Symposium on Integrated Pest Control*, Parts 1, 2, and 3 (FAO, UN, 1966).

Wellington, W. G., "The Synoptic Approach to Studies of Insects and Climate," *Ann. Rev. Ent.*, 2:143–62 (1957).

Wilson, F., "The Biological Control of Weeds," *Ann. Rev. Ent.*, 9:225–44 (1964).

Winteringham, F. P. W., "Mechanisms of Selective Insecticidal Action," *Ann. Rev. Ent.*, 14:409–42 (1969).

chapter 7 / **W. DON FRONK**

CHEMICAL CONTROL

Discoveries of new synthetic insecticides since World War II have sparked exciting advances and major breakthroughs in the control of insect enemies. Chemicals have subdued the pests that once caused national calamities—widespread crop destruction, wholesale death of domestic animals, and epidemics of insect-borne human diseases. Because modern insecticides are effective, reliable, and cheap, we in America resort to them more and more for the solution of our many problems with insects.

Adverse side effects, however, have developed under practices of frequent or routine application of persistent insecticides, and in recent years alternatives to chemical controls have been sought. A more judicious use of insecticides is being insisted upon, and methods are now being developed to integrate the use of insecticides into an over-all program that will control pests yet cause little harm to the environment.

Insecticides are not new; man very likely employed chemicals for control of insects before he learned to write. Homer, in about 1000 B.C., spoke of "pest-averting sulfur." Cato, around 200 B.C., advised boiling a mixture of bitumen (mineral pitch or asphalt) in such a way that the fumes would blow through grape leaves and thus rid them of insect pests. The Romans treated themselves with hellebore to destroy infections of human lice.

Dioscorides (A.D. 40–90), the Greek physician, knew of the toxic nature of arsenic, and the Chinese were applying arsenic sulfides to control garden pests before A.D. 900. The first record of exploitation of arsenic as an insecticide by the western world dates from 1669, when white arsenic was suggested for inclusion with honey as an ant bait. Marco Polo (1254–1323), in narrating his travels through Asia, told of the treatment of camel mange with mineral oil. By 1690 decoctions of tobacco were being applied to pear trees for control of the pear lace bug. Before 1800 the Persians utilized pyrethrum as an insecticide for control of human pests such as fleas.

The modern use of insecticides dates from 1867, when paris green was first applied to potatoes for control of the Colorado potato beetle. Until 1939 most insecticides were inorganic chemicals, plus a few insecticides derived from plants, but the discovery of the insecticidal nature of DDT in 1939 revolutionized our concept of insecticides and of insect control. In 1941–42, English and French investigators discovered the value of benzene hexachloride. During the same period, the Germans opened the field of organophosphorus insecticides, which led to development of parathion, tepp, coumaphos, demeton, and many more. Other families of chemicals are now being investigated for possible insecticidal activity. Insecticides have been found among the urethanes, sulfones, sulfonates, carbamates, and other classes of compounds.

INSECTICIDE FORMULATIONS

An insecticide, as it appears on the market, is composed of a toxicant or active ingredient, which is the poisonous substance, and one or more inert materials, which are nonpoisonous but which nevertheless have a purpose in the formulation. These may function to dissolve the poison, act as carriers, dilute the toxicant, or act as emulsifiers, dispersants, or spreader-stickers. Insecticides are seldom used full strength but are formulated in ways to dilute, and extend them, and make them easier to apply. For a variety of reasons the type and quality of a toxicant's formulation has a great influence on its effectiveness as an insecticide.

The most common formulations are dusts, granules, insecticide-fertilizer mixtures, wettable powders, emulsifiable concentrates, flowables, solutions, soluble powders, aerosols, and fumigants. Chemical companies continue to improve formulations of their pesticides and to choose confusingly descriptive names and initials for special formulations, such as emulsifiable liquid (EL), sprayable (S), spray concentrate (SC), and water miscible (WM).

Dusts (D)

Insecticides that are to be used dry are mixed with or impregnated on organic materials such as walnut shell flour or pulverized minerals such as talc, pyrophyllite, bentonite, and attapulgite. Choice of carrier is important because an appropriate one will impart desirable physical properties to the insecticide. The finished dust may be from 0.1 to 10 per cent active material. Ground to a fine size, most dust particles will pass through a 325 mesh screen and range in size from 1 to 40 microns. In general the toxicity of an insecticide increases as the particle size decreases. Because some insecticidal compounds are inactivated by alkali, the potential reaction of the diluent is an important factor in the manufacture of a dust. Also important in formulation are the catalytically active sites on dust particles which, to assure insecticide stability, must be deactivated.

Granular Formulations (G)

Granular formulations are much like dusts except for larger particle size. The range of particle size in a granular product is designated by a two-figure mesh classification. For example, $\frac{30}{60}$ means that virtually all of the granules will pass through a standard 30 mesh sieve (30 openings per linear inch), while only a negligible quantity will pass through a standard 60 mesh sieve. Some common granule sizes are $\frac{15}{30}$, $\frac{20}{40}$, $\frac{24}{48}$, and $\frac{30}{60}$.

Formulation of granules is accomplished by spraying or impregnating pellets of clay or other inert materials with a solution of toxicant. The common insecticide concentrations in granules range from 5 to 20 per cent. Insecticide granules have several advantages. They are easy to apply with accurate control of rate and placement, they drift very little; and with precautions they are safe to handle. Insecticide granules are usually used as dressings on or under the surface of the soil or to penetrate foliage and are applied with special granule applicators. As a soil insecticidal formulation, granules usually outperform liquids, especially when banded over the row. Granules of the future may consist of two or more carriers or blends of granules that will release toxicants at different rates. One attractive suggestion is a product which would release a high concentration of toxicant for a short period to obtain a quick massive kill, followed by slow release over a long period to continue the control.

Insecticide-Fertilizer Mixtures (I-F)

Insecticide-fertilizer mixtures may be formulated by spraying insecticide directly onto the fertilizer. Such mixtures are applied at the regular fertilizing time to provide both plant nutrients and control of soil insects, and above ground insects if systemic insecticides are selected. The user of insecticide-fertilizer mixtures must consider both parts of the mixture. That is, the material must be employed in such a manner that fertilizer and insecticide can perform their functions. This depends on when and how the mixture is placed in the soil and its relation to the placement of the seed or developing plant. Insecticides or herbicides may be added to liquid nitrogen fertilizers just prior to application.

Wettable Powders (WP)

Wettable powders have the appearance of dusts but are meant to be diluted and suspended in water and used as sprays. To make an insecticidal dust act in this manner, a dispersing and wetting agent is added to the formulation. Wettable powders are more concentrated than dusts, containing as high as 75 per cent toxicant. They are second in usage only to emulsifiable concentrates. Wettable powders have lower toxicity to plants and livestock, perform well over a wide variety of conditions, and are compatible with liquid fertilizer solutions.

Emulsifiable Concentrates (EC)

The most common and versatile formulation is the emulsifiable concentrate, a formulation that consists of insecticide, solvent for the insecticide, and emulsifying agent. Mixing the concentrate with water forms an emulsion of the oil-in-water type. The solvent may volatilize quickly after spraying, leaving a deposit of toxicant after the water has evaporated.

The use of an **emulsifier** serves several purposes: (1) it allows for the dilution of a water-insoluble chemical with water, (2) it reduces the surface tension of the spray, thus allowing it to spread and wet the treated surface, and (3) it enables the spray to make better contact with the insect cuticle. The types of insecticide emulsifiers include alkaline soaps, organic amines, sulfates of long-chain alcohols, sulfonated aliphatic esters and amides, mixed aliphatic-aromatic sulfonates, nonionic types (ethers, alcohols and esters of polyhydric alcohols, and long-chain fatty acids), and natural materials such as proteins and gums.

Emulsions are not stable and tend to separate into their component parts. This action, called "breaking," can be controlled by the amount of agitation and amount of emulsifier in the mixture. A fairly quick breaking mixture is preferred for plant spraying because it results in heavier deposits of toxicant by limiting runoff. Quick-breaking mixtures, however, do not wet nor spread as well as the slower-breaking mixtures.

An emulsifiable concentrate left standing will sometimes separate into its various parts because of the differences in specific gravity of the components. Simply shaking the container will return these components to the proper form.

Flowables (F)

The flowable formulation is a suspension of finely divided particles in an oil or water base. It is a compromise formulation: it offers the advantages of an emulsifiable concentrate but, because of solubility characteristics of the insecticide, it is more suitable for formulation as a wettable powder. There are two common types. The water-based flowable consists of an insoluble material, the insecticide, often on an inert carrier, ground into a water base; a suspending agent; a thickener; and an antifreeze agent. The oil-based flowable consists of the insecticide ground in a highly refined paraffinic oil, plus a suspending agent and a surface-acting agent. Flowables follow close behind wettable powders in usage.

Solutions

Not many of the present synthetic organic insecticides are soluble in water, but most are soluble in organic solvents. Some materials are dissolved in an organic solvent and used directly for insect control. Such solutions are seldom used on plants because of phytotoxic reactions, but they are used on livestock, in the household, in barns and buildings, and for spraying on the surface of water to control mosquitoes and other aquatic insects.

Soluble Powders (SP)

A few synthetic insecticides—organophosphorus, carbamates, and others—are water-soluble. These are often formulated as dry soluble powders containing 80 to 95 per cent active ingredient. The "inert ingredients" consist of adjuvants to spread and stick the insecticide on foliage.

Aerosols

Aerosols are minute particles suspended in air, such as fog or mist. Similarly, insecticides may be suspended in air as minute particles whose diameters range from 0.1 to 50 microns. The dispersion of insecticide into aerosol form may be accomplished by burning, vaporizing with heat, atomizing mechanically, or releasing through a small hole an insecticide that has been dissolved in a liquefied gas. In the last method, the released gas volatilizes rapidly and leaves small particles of the insecticide floating in air. The popular household "aerosol bomb" operates in this manner.

Fumigants

Insecticides used in the gaseous form are known as fumigants. With few exceptions, these are used where the gas can be confined (in buildings, storage bins, ship holds, and even in soil). Fumigants are most often formulated as liquids under pressure and are held in cans or tanks. When the liquid is released in open air, it changes back to a gas. Quite often fumigants are formulated as mixtures of two or more gases to increase safety and effectiveness. In some cases, the gas is made at the location to be fumigated. Hydrogen cyanide is produced, for example, by dropping calcium cyanide into earthenware crocks filled with sulfuric acid.

Miscellaneous Formulations

Special formulations may be found for specific uses. Boli (large pills) and capsules are used with a balling gun to introduce insecticides into the stomachs of animals. Concentrated oil solutions of insecticides, called pour-ons may be poured onto an animal's back for control of internal or external blood-sucking parasites. The very volatile insecticide dichlorvos, which is also a fumigant, has been formulated with waxes or plastics into solid blocks. When the blocks are hung in rooms, there is a slow release of the insecticide in vapor form, and a single block will control insects for several weeks. Similar material made into animal collars can be used to control pests such as fleas. Insecticides may be mixed in shampoos intended for use on humans or house pets. Poison baits consist of toxicants combined with a foodstuff attractive to the insect pest. Recently some insecticides have been enclosed in microcapsules. Applied in sprays, the capsules break down and release the active ingredient on the foliage.

COMPATIBILITY OF AGRICULTURAL CHEMICALS

When certain chemicals are brought together, they will react to form a compound that differs from either parent. This simple fact has been ignored at times by applicators of agricultural chemicals, with regrettable results. Before combining chemicals in a single spray mixture, one should know how or if they will react. This information is given in a compatibility chart (Fig. 7:1).

There are several forms of incompatibility: (1) **chemical incompatibility**, in which the various chemicals react to form different compounds (usually this reaction arises from use of synthetic organic compounds with an alkaline material); (2) **phytotoxic incompatibility**, in which there may be no chemical reaction, but the mixture causes injury to plants whereas the component parts used separately cause no injury; (3) **physical incompatibility**, in which the chemicals being used change their physical form to one that is unstable and hazardous for application.

FACTORS INFLUENCING THE EFFECTIVENESS OF INSECTICIDES

The route by which the chemical enters the insect affects its toxicity. The cuticle is impervious to polar chemicals applied in a polar solvent; the same compound, however, can penetrate quite rapidly if introduced into the greasy, outermost integumental layer by an organic solvent. For this reason, adding an oil to a spray mixture often increases its toxicity. Insecticides taken orally are absorbed in the midgut. Some chemicals are repellent to insects and are avoided while eating, or if eaten they may be regurgitated. Some poisons are either not absorbed in the gut or are inactivated there. Water is unable to penetrate the tracheal system because of its high surface tension. The addition of a material to lower the surface tension allows the spray to enter. Insecticides that have a rapid effect probably enter largely through the tracheae or the cuticle.

The developmental stage of an insect influences its susceptibility to insecticides. In general, larvae and nymphs are easier to kill than pupae and adults, and the early instars are often more susceptible than the later ones. Eggs are usually most susceptible just before hatching.

Environmental conditions may alter the effectiveness of insecticides. The rate at which the insecticide is absorbed into the insect body and the rate at which the insect detoxifies the insecticide inside its body are proportional to the rise in temperature. Thus the most effective temperature conditions for a successful kill are a high temperature to get the poison inside the insect

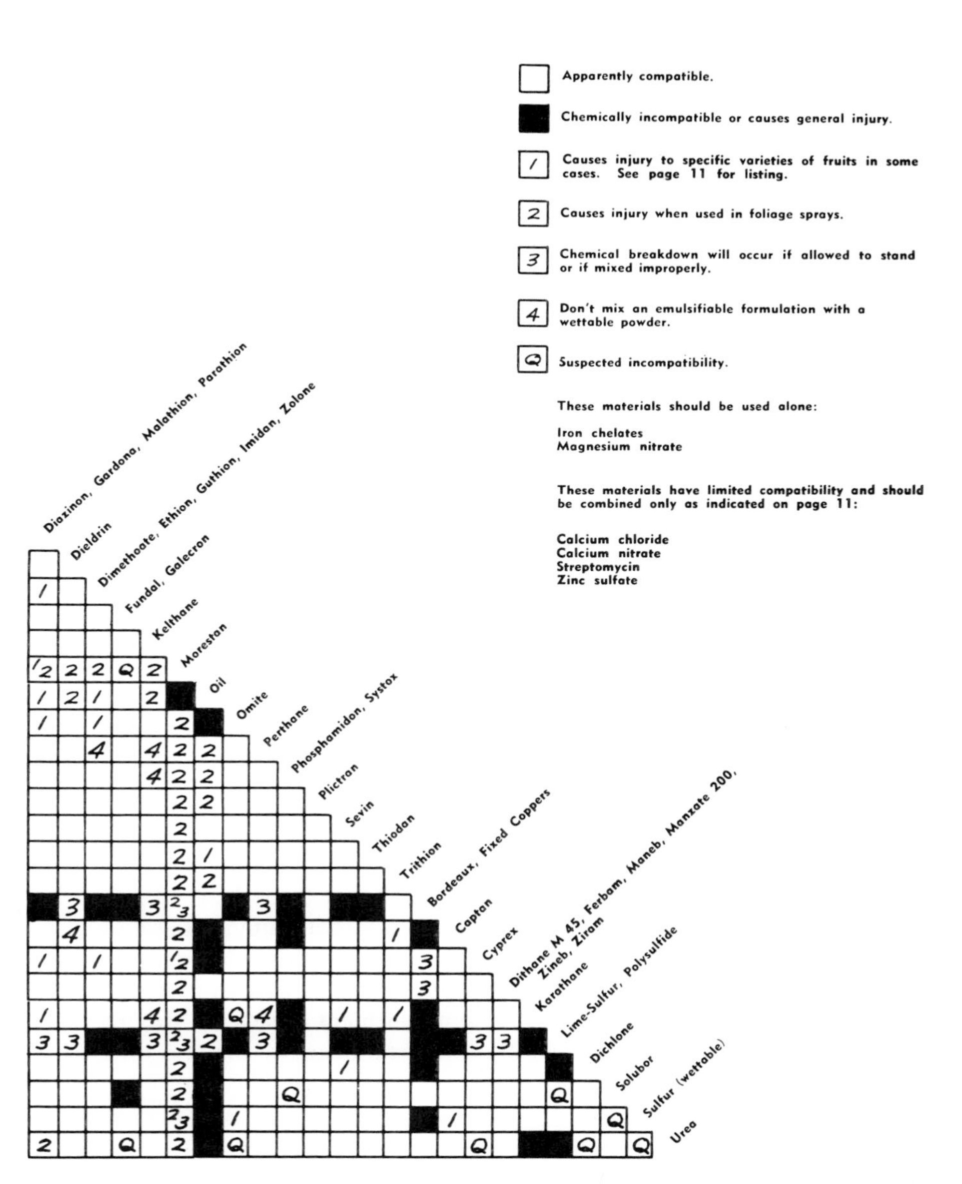

Figure 7:1. Compatibility of insecticides, fungicides, and several other classes of agricultural chemicals. (From Washington State University Extension Bulletin 419, 1976 Spray Guide for Tree Fruits in Eastern Washington.)

followed by a low temperature, which slows the rate of detoxification. Some chemicals, such as DDT and methoxychlor, are an exception to this rule in that they are most effective at continual relatively low temperatures.

Low humidities have a deleterious drying effect on the fine mists produced in concentrate spraying. Rains wash off water-soluble sprays, but most insecticides do not fall into this category. Wind and rain cause the "weathering" of the less tenacious portions of spray deposits, and sunlight, though slower in action, causes breakdown for as long as the residues remain. Air currents are of importance, as winds of greater than 6 mph carry spray and dust away from the point of application, resulting in uneven distribution. Air currents may seriously interfere with the airplane application of insecticides, and whenever strong rising currents prevail it becomes difficult to get a good deposit on the ground or target vegetation.

The condition of the plant may influence the effectiveness of an insecticide. For plants such as cabbage, cauliflower, and others with waxy leaves, spray material tends to run off the leaf and drip to the ground. The addition of a wetting agent to the spray will help to correct this condition. Heavy foliage or dense fur may prevent an insecticide from penetrating deeply into the area to be treated. Again the addition of a wetting agent will give some help, as will increasing the pressure at which the application is made. After application of insecticides the new foliage of rapidly growing plants lacks a surface residue of toxicant. An apple tree may double its foliage area in two weeks. Some cucurbit plants may grow five inches in a day.

MODE OF ACTION OF INSECTICIDES

The "mode of action of an insecticide" means the way in which the chemical acts upon the system of an insect to cause its death. Although much research has been done on this subject, there is still much to be learned before we know exactly how insecticides kill. Attempts to determine a basic mode of action are fraught with difficulties, for the reason that life processes are numerous, complex, and interdependent. Lethal action of an insecticide may be due to multiple effects involving the impairment of several life processes.

Realizing the inadequacies of present knowledge, we may for the sake of convenience classify insecticides into four groups: physical poisons, protoplasmic poisons, metabolic inhibitors, and nerve poisons. **Physical poisons** kill by some physical action, such as excluding air (e.g., mineral oils), or by abrasive or sorptive actions resulting in loss of water (e.g., certain dusts such as silica aerogel). **Protoplasmic poisons**, such as the arsenicals, kill by precipitating protein. **Metabolic inhibitors** include a variety of compounds that interfere with the normal metabolism of insects. They include (1) respiratory poisons such as hydrogen cyanide, rotenone, and dinitrophenols, which deactivate respiratory enzymes; (2) inhibitors of mixed-function oxidase, such as the pyrethrin synergists; (3) inhibitors of carbohydrate metabolism, such as sodium fluoroacetate; and (4) inhibitors of amine metabolism, such as chlordimeform. The **nerve poisons** make up the largest group of modern insecticides. They are divided into (1) anticholinesterase compounds, such as the organophosphorus and carbamate insecticides; (2) compounds that affect ion permeability of nerve membranes, such as DDT, chlordane, lindane, and pyrethrins; and (3) compounds that affect the nerve receptors or synaptic ganglia, such as nicotine.

The discovery of organophosphorus insecticides has stimulated much research on the physiology of the insect nervous system. This system is high in **acetylcholine** (ACh), a chemical known to be involved in the synaptic transmission of nerve impulses in mammals. To function normally, ACh must be removed from the synaptic region immediately after the nerve impulse is mediated. The removal is accomplished by an enzyme called **cholinesterase** (ChE). Although the exact nature and function of chemical mediators of the nerve impulse in insects are yet to be discovered, it is known that ChE is important in the insect nervous process. Organophosphorus and carbamate insecticides are active inhibitors of ChE in both mammals and insects. Inhibition results in facilitation of nerve impulses and increased excitation caused by an accumulation of acetylcholine at nerve endings.

SYNERGISTS FOR INSECTICIDES

Some chemicals have the property of greatly increasing toxicity of certain insecticides. When the increased toxicity is markedly greater than the sum of the two used separately, it is called a synergistic action. This is like saying that 2 plus 2 equals 10. For example, chemical *A* may kill 30 per cent of the flies sprayed and chemical *B* 10 per cent. Should these two be mixed and applied together you might expect at best a 40 per cent kill. However, if synergism takes place, the kill may be as high as 90 per cent or more. Most synergists have been used with pyrethrum or allethrin. First discovered by U.S. workers in 1938, insecticide synergists now include such materials as **sesamex**, **piperonyl butoxide**, **sulfoxide**, and **MGK 264**. Synergistic action is important because it provides a means for a more effective insecticide at less cost.

HAZARDS OF INSECTICIDES

Most people think the only danger in the use of insecticides is accidental poisoning. There are many other dangers, however, and anyone using insecticides should have some knowledge of their existence and possible means of avoiding them.

Because the hazards of using insecticides are closely related to their mammalian toxicities, the latter are important considerations for both health officials and agriculturalists. Most information on mammalian toxicities is obtained by conducting tests on small mammals—white mice, white rats, and rabbits. Chemicals are administered to experimental animals in several ways to provide information on acute oral toxicity, acute dermal toxicity, inhalation toxicity, and subacute or chronic feeding toxicity.

Results of acute oral and dermal toxicities are usually given in terms of LD_{50}, a designation for the dose lethal to 50 per cent of the test animals. The LD_{50} is expressed in milligrams of toxicant per kilogram of body weight of the test animal (mg/kg). For example, the acute oral LD_{50} of DDT for rats is 250 mg/kg. If one were to feed this amount of toxicant to each test rat of a large group, one could expect about 50 per cent to die and 50 per cent to survive. Having a list of LD_{50}s available gives us a handy way of comparing the toxicities of insecticides. The acute oral LD_{50}s for rats in mg/kg of some common insecticides are methoxychlor 6000, malathion 1500, lindane 125, and parathion 3. A lower figure indicates greater toxicity. From these data we can calculate that lindane is 12 times more toxic than malathion and that parathion is 42 times more toxic than lindane.

The toxicity of an insecticide varies with the route of entry into the mammalian body. Parathion has the greatest toxicity to man through inhalation, only one third as much orally, and one tenth dermally. In general acute dermal toxicity figures are usually larger than those of oral toxicity, as it takes a greater amount of insecticide placed on the skin to cause death than an ingested amount. Inhalation toxicities are often expressed as LC_{50}s (lethal concentration, μg/L) and chronic toxicities as parts per million (ppm) of the daily diet for a specified number of days.

Pesticides are classified into four categories based on acute toxicities:

1. **Highly Toxic**, oral LD_{50} 0–50 mg/kg, dermal LD_{50} 0–200 mg/kg, inhalation LC_{50} 0–2000 μg/L, DANGER, skull and crossbones, and POISON on label.
2. **Moderately Toxic**, oral LD_{50} 51–500 mg/kg, dermal LD_{50} 201–2000 mg/kg, inhalation LC_{50} 2001–20,000 μg/L, WARNING on label.
3. **Slightly Toxic**, oral LD_{50} 501–5000 mg/kg, dermal LD_{50} 2000–20,000 mg/kg, CAUTION on label.
4. **Relatively Nontoxic**, oral LD_{50} 5000+mg/kg, dermal LD_{50} 20,000+mg/kg, no signal words on label.

Despite wide differences in toxic properties, the hazards of insecticides are not proportionately different. There are several reasons for the lack of direct relationship. First, the rates at which they are employed to control insects may vary. Parathion is more toxic than malathion, but the amount used per acre is much less for parathion than for malathion. Second, the residual characteristics of insecticides vary. Parathion breaks down more quickly on foliage than DDT. Third, recommendations and precautions for use vary. Malathion may be applied to livestock for controlling a number of pests, whereas parathion is never recommended for direct application to livestock. More precautions, such as wearing a face mask, are taken in applying parathion than in applying malathion.

Another important fact concerning hazards is that different formulations of the same insecticide vary in their capacity for toxicity. Spilled on one's skin, a 98 per cent ULV formulation is more dangerous than a 25 per cent emulsifiable concentrate. There is less danger in applying granules of 2 per cent AI than in applying granules of 20 per cent AI. Sprays made with wettable powders are less toxic to plants and livestocks than sprays made with emulsifiable concentrates.

Residues on crops and in meat present a problem of chronic poisoning. The federal government and many states have laws that determine insecticide residue tolerances for human and animal food. If these tolerances are exceeded, the product may be seized and destroyed.

If used improperly, insecticides may upset the insect balance in such a manner that greater insect damage results from treatment than from no treatment at all. Some chemicals are very destructive to predators and parasites of insect pests without being particularly effective against the pest. The result is that the pest increases greatly in number. Insecticides may also kill pollinating insects with a resulting poor fruit or seed set. Wildlife may be adversely affected by unwise application of insecticides. Fish have been killed outright through direct application of insecticides. The number of such accidents is small, however; for example, in 1967 insecticides used in agriculture and in mosquito control were responsible for only 11 per cent of the reported fish kills. Large mammals are not much affected by insecticide applications.

Certain insecticides—several of the chlorinated hydrocarbons, for example—exhibit a phenomenon known as biological magnification. A portion of the chemical is stored in the fat of the animal that consumes the insecticide-contaminated organism. These consumers are in turn eaten by animals higher in the food chain. In this way sufficient insecticide may accumulate in animals near the top of the chain to interfere seriously with their normal body functions.

Insecticides may have a deleterious effect on plants, either directly by affecting growth of the plant or indirectly by tainting the edible part. Certain chemicals used on stored grain for insect control may lower or destroy seed germination. Others leave a visible residue and lower the quality of the grain. Insecticides may accumulate in the soil to such an extent that it becomes impossible to grow plants.

In treating buildings with flammable fumigants, one risks the danger of explosion and burning. Application of insecticides may involve the user in a legal battle should excessive drift occur, or should excessive residues remain on forage or food.

Last, using insecticides may lead to the development of resistance in insects. This discovery is not new; it was noted many years ago that scale insects became resistant to lime-sulfur sprays and HCN fumigation, and the codling moth to arsenical sprays. Resistance to insecticides has been discovered in 375 species of insects and acarines, such as the house fly, certain mosquitoes, the cabbage butterfly, the boll weevil, corn rootworms, and the twospotted spider mite.

SAFE USE OF INSECTICIDES

Insecticides are poisons and should be used with great care. If insecticides are used as directed, however, there is slight chance of injury. Depending on the nature of the insecticide, all or some of the following precautions should be followed: (1) read and follow all instructions on the label carefully. Have container with label available to show physician in case of accidental poisoning; (2) wear a face mask to avoid inhalation of poison; (3) wear dustproof goggles and protective clothing; (4) change clothing as soon as insecticidal treatment is completed; (5) wash off immediately, with plenty of fresh, clean water, any insecticide spilled on body; (6) bathe as soon as possible after using insecticides; (7) avoid remaining in drift of spray or dust; (8) destroy all empty containers; (9) store insecticides where children, irresponsible persons, or livestock cannot reach them; (10) call a physician immediately if any symptoms of poisoning appear in a person who is using or has recently used insecticides.

Each state maintains Poison Control Centers where persons can obtain information day and night concerning emergency treatment of accidental poisoning. Persons using pesticides frequently should keep the telephone number of the Poison Control Center near and available at all times. They should also undergo periodical medical examinations.

The importance of reading and following instructions on the label cannot be stressed enough. The label contains all the information necessary for the effective and safe use of the insecticide.

NAMES OF INSECTICIDES

Frequently the names of insecticides are confusing to the user. An insecticide will have a complicated chemical designation or name, but it may also have a code designation, a common name, and several trade names. Code designations are usually dropped early and replaced by trade names. Official common names are often assigned after an insecticide has been on the market and the public has become familiar with a trade name. In some cases the trade name is made the official common name.

In the discussion that follows, common names along with better-known trade names will be presented to permit the reader to locate additional information in the literature. The common name is not capitalized unless it begins a sentence or unless the name is a set of initials, for example, DDT or TDE. The first letter of the trade name is capitalized and in the present text will be in parentheses if it follows a common name. You will find trade and common names and descriptions of insecticides in the glossary of this text.

CLASSES OF INSECTICIDES

Insecticides may be classified in several ways. One of the most widely used systems, until the advent of the new synthetics, was based on the mode of entry into the insect—stomach, contact, and fumigant poisons. **Stomach** poisons are materials which are ingested by the insect and which kill primarily by action on or absorption from the digestive system. Usually they are limited to the control of chewing insects. **Contact poisons** are absorbed through the body wall and must come in direct contact with the insect to kill. They are usually required against sucking insects. **Fumigant poisons** enter the tracheal system in the form of a gas. Insects such as stored grain pests, living within an enclosure, are readily killed with fumigants. Soil insects may also be controlled with fumigants.

Classification by mode of entry breaks down with the newer insecticides because many of them enter the insect body in more than one way. To avoid overlapping, classification is usually based now on the chemical nature of the insecticide. The major divisions are inorganic and organic. The organic insecticides are divided further into oils, botanicals, and synthetics. Oils are petroleum products used primarily against tree fruit insects. Botanicals are of plant origin and include such insecticides as rotenone, pyrethrum, sabadilla, and ryania. The synthetic organic compounds are by far the most important and include the more recently discovered insecticides such as DDT, lindane, toxaphene, chlordane, parathion, malathion, and carbaryl.

INORGANICS

Though inorganic insecticides have been replaced largely by the more efficient organics, some still find a place in American agriculture. Lead arsenate is commonly used in the form of **acid lead arsenate** ($PbHAsO_4$) and less so as **basic lead arsenate**, a mixture of $Pb_4(PbOH)(AsO_4)_3 \cdot H_2O$ and $Pb_5(PbOH)_2(AsO_4)_4$. Acid lead arsenate is used primarily in fruit orchards, on forest and shade trees, and on shrubs to control chewing insects. Although not as toxic to insects, basic lead arsenate is safer to use on plants in foggy coastal regions than is the more soluble acid lead arsenate.

Sodium fluoride (NaF) was once a common insecticide used in cockroach and ant bait. A related compound, **sodium fluosilicate** (Na_2SiF_6), is used now in baits for ants, cockroaches, and grasshoppers. **Cryolite** (Na_3AlF_6), which is mined in Greenland as well as manufactured in the United States, has proven effective against a number of truck-crop insects and is used on plants that are generally sensitive to chemical injury.

Sulfur is the only element that has been found valuable as an insecticide. Finely ground sulfur has been applied widely as a dust for control of mites and of certain fungi. The addition of a wetting agent to sulfur makes it possible to use sulfur as a spray. Boiling together sulfur and freshly slaked or hydrated lime produces a mixture of compounds known as **liquid lime-sulfur**. The active materials in the mixture are probably the calcium polysulfides (CaS_4, CaS_5, and others). This material has been used as a fungicide, and on fruit trees against scales, aphids, and mites. Liquid lime-sulfur may be mixed with a stabilizer and evaporated to dryness to form **dry lime-sulfur**. Dry lime-sulfur is less effective than the liquid but is easier to handle.

Many other inorganic compounds are now or have been employed as insecticides. **Sodium selenate** (Na_2SeO_4), applied in the soil, acts as a systemic and was used in the past primarily to control aphids and mites on greenhouse ornamentals. Cabbage maggots were controlled with **mercurous chloride** (calomel), Hg_2Cl_2, solution. **Phosphorus** has been used in cockroach bait and **thallous sulfate** (Tl_2SO_4) in ant bait.

OILS

Oils in their natural state are highly phytotoxic, but when used in an emulsion they may, under certain conditions, be safely applied to plants. Mineral oil is a heterogeneous mixture of saturated and unsaturated chain and cyclic hydrocarbons. Certain parts of this mixture are much more useful as insecticides than others. To designate the quality of oil, data commonly supplied on the label are viscosity, boiling or distillation range, and sulfonation rating (purity or degree of refinement).

Viscosity is usually stated in terms of the time in seconds for 60 cc to flow through a standard orifice. In general oils of low viscosity are safer to use than are those of higher viscosity; however, the viscosity rating may be misleading, because oils from different parts of the country having the same viscosity rating may react quite differently when sprayed on foliage.

Boiling or **distillation range** is a more important character of oil and is an indirect indication of volatility. Phytotoxicity increases with increase in distillation range. However, it is also true that heavier spray oil fractions have a greater effectiveness against insects than do lighter oils. In both cases this probably has to do with the period of time the oils remain in contact with the plant or animal surface. The lighter oils, being more volatile, soon escape into the air, whereas the heavier oils remain in contact with the surface for a longer time.

Oils are composed of both saturated and unsaturated hydrocarbons. The unsaturated hydrocarbons are unstable and readily form compounds that are toxic to plants. Consequently if the amount of unsaturated hydrocarbons present in an oil is lower, that oil is safer for use on plants. The usual testing procedure of determining the amount of unsaturated hydrocarbons present is by means of the sulfonation test. The oil to be tested is reacted with strong sulfuric acid, and the unsaturated hydrocarbons that react with the acid sink to the bottom. The unreacted part, the **unsulfonated residue** (U.R.), is measured as a percentage and is used as a measure of purity. Dormant oils have a U.R. rating of 50 to 90 per cent and the more highly refined summer oils a rating of 90 to 96 per cent.

Oils are employed in a number of ways. They may be used as solvents or carriers for insecticides. Diesel fuel is often used as a carrier for insecticide in airplane application. Oil may also serve to carry an insecticide over water being treated to control mosquitoes.

Oils by themselves are insecticidal and, based on time of usage, are classified as either summer or dormant oils. **Summer oils**, which are highly refined and less phytotoxic, are applied to trees in foliage. For example, citrus trees may be treated with summer oils to control mites and scale insects. The **dormant oils**, which are less refined, are applied when no foliage is present. To increase plant safety as well as insecticidal action, new types of oils have been developed; these are called **superior** (or **supreme**) **spray oils**. From the time buds show green up to the time leaves are one-half inch long, apple trees may be given a delayed dormant spray with these oils to control aphids, scale insects, and mites.

Oil is not applied full strength to trees but is diluted with water and applied as an emulsion containing around 1 to 4 or more per cent oil. Several types of agricultural spray oil stocks are formulated. One type, called **emulsible oils**, contains 95 to 99 per cent oil plus emulsifier. Some formulations of these produce an emulsion instantly when poured into a tank of water, whereas others require preliminary agitation with a small amount of water. The former are often referred to as **miscible oils** and the latter as emulsible oils.

A second type of spray oil stock is the **concentrated emulsion**. These oils are preformed emulsions in a concentrated state and contain about 83 percent oil plus emulsifier and water. They have the appearance of a whitish paste; some are flowable, others are thick like mayonnaise.

A third type, **tank-mix oil**, has the oil, the emulsifier, and the water added separately to the spray tank. Violent agitation forms the emulsion. Such an emulsion is called "quick breaking" because it separates quickly into its component parts upon coming in contact with the plant surface.

The use of oils has several advantages. They are relatively cheap, have a good spreading capacity, are easy to mix, and are reasonably safe to animal life. An important consideration is the fact that insects have not developed any resistance to them. They have several disadvantages such as injury to rubber hose of sprayers, instability in storage, phytotoxicity, and low toxicity to insects. The last characteristic may be corrected by preparation of combination sprays. A highly efficient pesticidal spray is a fungicide-insecticide-oil combination that can be prepared in the spray tank just before application.

BOTANICALS

Plant products have several uses in insect control. Some act as attractants (e.g., geraniol and eugenol), some as repellents (e.g., citronella and oil of cedar), and some as solvents or extenders (e.g., cottonseed oil and walnut-shell flour). Their primary role, however, is as insect toxicants. Nicotine, pyrethrum, and rotenone are the most important.

Nicotine ($C_{10}H_{14}N_2$)

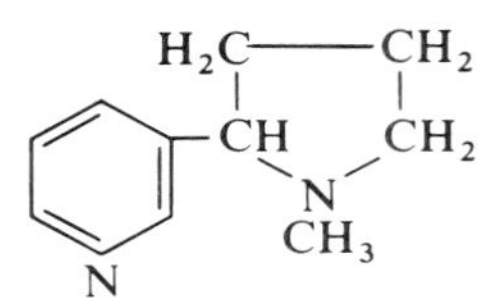

Nicotine was first used as a decoction in 1690 against the pear lace bug in France. This same form was used in the 1700s against soft-bodied sucking insects and the plum curculio. Tobacco smoke was directed onto plants by 1773 for control of aphids and other insects. In 1828 the chemical nature of nicotine was discovered and named, and the structure was established in 1893. The form most commonly used today, **nicotine sulfate**, was patented in 1908.

Nicotine comes chiefly from two plants, *Nicotiana tabacum*, which is the common tobacco plant, and *N. rustica*, which is grown primarily for its high nicotine alkaloid content. The nicotine content of this last plant varies from 2 to 20 per cent depending on the climate and condition of the soil. However, in the United States, *N. tabacum* is the only commercial source of nicotine. Nicotine is taken from leaves and stems of waste tobacco by steam distillation or by solvent extraction. The stems and leaves vary from 0.5 per cent to 3 per cent nicotine. Tobacco contains 11 other alkaloids, but only two of these, anabasine and nornicotine, have received much attention as insecticides. Pure nicotine is a colorless liquid soluble in water and in most organic solvents. It darkens and becomes viscous on exposure to air. It is an organic base which reacts with acids, producing salts which are usually water-soluble.

Nicotine is highly toxic to a great number of insects. Evidence indicates that it acts on the nerve ganglia and synapses of the insect. It is toxic when ingested, absorbed through the body wall, or taken in through the tracheae. Its mammalian toxicity is very high. It causes headache and vomiting followed by respiratory failure due to paralysis of the breathing muscles. It can be absorbed through the skin in lethal amounts. Nicotine has a very low phytotoxicity and is used on many plants where it is economically feasible.

Nicotine is formulated in several ways. The chief form is nicotine sulfate, which is relatively safe to handle and dissolves readily in water to make a useful spray. Nicotine may also be used as a dust in the form of finely ground crude tobacco, but the usual formulation consists of nicotine sulfate absorbed on an active (bentonite) or inactive (gypsum) carrier. The active carriers change the nicotine sulfate into free nicotine in the presence of moisture. When used as a fumigant dried tobacco may be burned and the smoke directed among the infested plants. Extracts may be vaporized by heat, or they may be absorbed on a combustible material which, when burned, releases the nicotine. Nicotine may also serve as a stomach poison when it is fixed or stabilized as a salt. Nicotine may be incorporated into aerosol formulations and dispensed to form a space fumigant.

Pyrethrum derives its toxicity from four esters, pyrethrin I, pyrethrin II, cinerin I, and cinerin II.

Pyrethrin I ($C_{21}H_{30}O_3$)

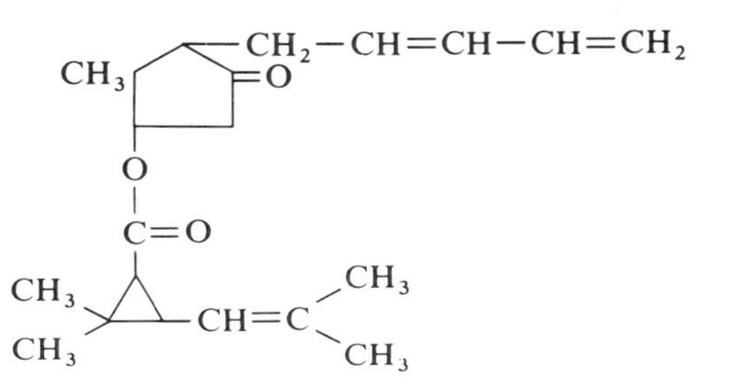

The structural formulas of the other three compounds are similar to that of pyrethrin I.

The insecticidal activity of pyrethrum was discovered in Iran (then called Persia) around 1800. Great secrecy surrounded the source of the material, which was called Persian Powder and was sold at extravagant prices for louse and flea control. The discovery of the source of the powder, according to some, was made by an Armenian merchant who had been traveling in the Caucasus Mountains. Another version of the story is that the secret was given to Russian soldiers by prisoners of war during the Russo-Persian War in 1827 or the Russo-Turkish war of 1828–29. The original powder came from the daisies *Chrysanthemum coccineum* and *C. carneum*. In 1840 it was found that *C. cinerariaefolium*, produced in Dalmatia, had a higher insecticidal activity, and this species of daisy soon replaced the previous sources.

Pyrethrum was introduced into the United States about 1858 and found wide application. From 1876 to about 1914 an attempt was made to grow pyrethrum commercially in California. The flower has also been experimentally grown around Fort Collins, Colorado. But because neither of these two ventures was a commercial success, the United States still imports all of its pyrethrum.

Pyrethrum is produced by grinding the flowers and mixing them with a dust diluent or, more commonly, by extracting the active materials with solvents and formulating the extracts into sprays and dusts. The active materials are esters that are rapidly hydrolyzed by alkali and decomposed by sunlight.

Pyrethrum is toxic to most insects with which it comes in contact. Its primary target is the ganglia of the insect central nervous system; because it acts rapidly, affected insects fall quickly. Such action is called "knockdown." Insects knocked down do not necessarily die; some will recover if allowed to do so. Pyrethrum lacks persistence in the field because of its breakdown by sunlight; thus it leaves no harmful residue. It acts almost entirely as a contact poison. Pyrethrum is relatively harmless to mammals by ingestion, although it is poisonous if injected into the blood stream, and the dust may cause allergic reactions in some people. It is harmless to plants. Two disadvantages of pyrethrum are that it breaks down rather rapidly and it is relatively expensive.

Pyrethrum is formulated as dusts, sprays, and aerosols. Usually a synergist such as sulfoxide or piperonyl butoxide is added to increase toxicity and also to reduce the amount of pyrethrum necessary for a satisfactory kill.

A material with such low mammalian toxicity would obviously find many uses. It can be applied to edible plants shortly before harvest because it leaves no harmful residue. It is effective against a wide range of insects and is one of the insecticides preferred by the backyard gardener. Pyrethrum has been used on livestock as a toxicant for external parasites and as an insect repellent. One of its largest uses is in household sprays. Pyrethrum mixed with stored grain protects the grain from insect injury.

Rotenone ($C_{23}H_{22}O_6$)

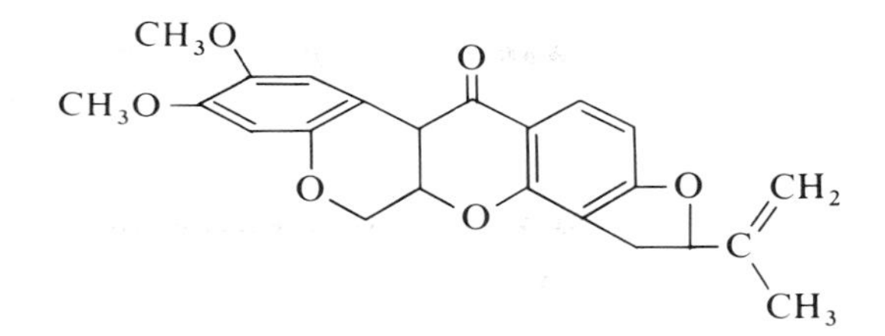

Rotenone is found in 68 species of leguminous plants. The important commercial ones are two species of *Derris*, which grow in the Far East and contain 5 to 9 per cent rotenone, and several species of *Lonchocarpus*, which grow in the Amazon Valley of South America and contain 8 to 11 per cent rotenone.

To obtain rotenone the roots of the plants are dried, powdered, and either mixed with a diluent powder to be used as a dust or extracted and the extract used in making sprays or dusts. Pure rotenone is white, crystalline, and insoluble in water but soluble in many organic solvents. It is rapidly oxidized in the presence of light or alkali.

Rotenone is a selective insecticide; it is very effective against some insects and inactive against others. It kills insects by inhibiting respiratory metabolism and blocking nerve conduction. Rotenone acts as both a stomach and a contact poison. Although mammalian oral toxicity is high—the LD_{50} ranges from 10 to 30 mg/kg—rotenone may be used with relative safety on most mammals except swine, which are highly susceptible. It causes no phytotoxicity.

The usual formulations of rotenone are dusts, which must be mixed with nonalkaline carriers to prevent breakdown; wettable powders; emulsifiable powders; emulsifiable concentrates; and aerosols. Rotenone finds many applications. It is widely used on vegetable and ornamentals and is used for control of external pests of livestock. The material has several advantages, such as safety to both mammals and plants, but it also has disadvantages, such as breakdown in storage, slow action, and low toxicity to some insects.

There are several other botanical insecticides of minor importance. **Sabadilla** comes from the seeds of a lily grown in Venezuela. It has been used against human lice and for control of Homoptera and Hemiptera. **Ryania** is obtained from stems and roots of *Ryania speciosa*, family Flacourtiaceae. It is less toxic to mammals than rotenone; it is more stable and possesses a longer residual action. It has been used primarily against the European corn borer, codling moth, citrus thrips, and some other lepidopterous insects. **Hellebore**, which comes from a lily, was in wide use many years ago for control of pests of vegetables, fruit, and livestock.

SYNTHETIC ORGANIC INSECTICIDES

Although several synthetic organic insecticides were known previous to World War II, not until the discovery of DDT did interest in these chemicals become widespread. Following the release of DDT, many companies became active in searching for new insecticides. The majority of synthetic organic insecticides now in use were developed after 1947, although several were marketed before. Dinitrocresol was sold as an insecticide in Germany as early as 1892, and by 1932 several thiocyanates were commercially available in the United States.

The development of the organophosphorus insecticides began in Germany. During World War II, Gerhard Schrader and other German chemists searched for synthetic compounds to replace insecticides that had become unavailable in Germany because of war conditions. Several organophosphorus compounds proved highly toxic to insects. From these chemicals came a number of effective and practical insecticides, such as parathion, tepp, schradan, and demeton. Rapid developments make an up-to-date classification difficult. One generally applicable list has the following headings: (1) chlorinated hydrocarbons, (2) phosphorus-containing compounds, (3) carbamates, (4) synthetic pyrethroids, (5) insect growth regulators, (6) specific miticides, (7) sulfur-containing compounds, (8) nitrophenols and derivatives, (9) thiocyanates, (10) fumigants, (11) repellents, (12) attractants, and (13) miscellaneous compounds.

Chlorinated Hydrocarbon Insecticides

The chlorinated hydrocarbon insecticides are molecules composed of chlorine, hydrogen, carbon, and occasionally oxygen or sulfur. They are also known as organochlorine insecticides.

DDT (Dichloro Diphenyl Trichloroethane)

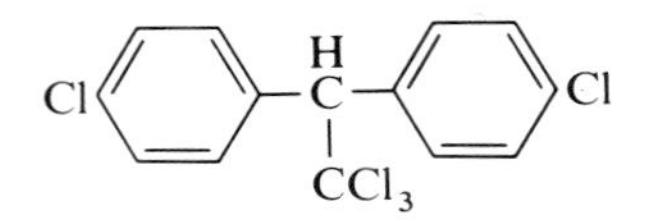

DDT is one of the most important insecticides ever devised by man. It has saved millions of lives, directly through control of insect vectors of human diseases, and indirectly through control of crop pests with resulting increase in food. Though condemned by environmentalists, it has saved millions of acres of forest from destruction by insects.

In 1932 chemists of the Geigy Company in Switzerland began a search for an improved mothproofing compound. Seven years later **Dr. Paul Müller** found a material that showed great promise. The company was given a patent on it by the Swiss government on March 7, 1940, and later it became known as DDT. DDT soon proved itself as an agricultural insecticide by stopping an outbreak of the Colorado potato beetle during 1941 in Switzerland. At the same time it was found highly effective in controlling fleas, lice, mosquitoes, and flies. In 1942 samples of the compound were sent to the British branch of Geigy Company, and two hundred pounds were sent to the United States. Production of DDT in the United States began after a pilot plant was established at Norwood, Ohio, in May 1943. All of these developments were kept secret at the time.

The first wide use of DDT came in 1944, when typhus broke out among the civilian population of Naples. Mass delousing with the new insecticide was started, and within three weeks the typhus outbreak was brought under control. Never before during a war had an epidemic of typhus been halted. In World War I, 25 per cent of the Serbian soldiers perished from typhus, and Russia lost several million to the disease.

In Naples 1.3 million civilians were treated with DDT during January 1944. It became more and more difficult to keep secret such mass medication and such spectacular control of a dreaded disease. On August 2, 1944, the British government disclosed the identity of the chemical. The world recognized the major advance made in medical science through the development of DDT, and in 1948 Dr Müller was awarded the Nobel Prize for medicine.

DDT is made by the condensation of chloral and monochlorobenzene in the presence of sulfuric acid. In its pure state it is a crystalline powder with a melting point of 108°C. The crude form is a waxy solid with a melting point near 89°C. It is insoluble in water, but soluble in most organic solvents. DDT is broken down by alkali and by organic bases, otherwise it is stable and inert. It is not attacked by ordinary acids.

DDT is effective against a great number of insect pests. However, it has little toxicity to most Orthoptera, the boll weevil, Mexican bean beetle, and most aphids. Also, it has relatively little acute toxicity to mammals. But there is concern about possible chronic effects in mammals exposed to low levels of DDT over a long period of time. Human deaths attributed to DDT have always included a solvent, and in all probability it was the solvent which caused the death and not the DDT. This chemical has been widely used in insect control, yet no increase in illness of applicators has been detected. Workers exposed to DDT in its manufacture have shown no ill effects. DDT is known to be stored up to certain levels in the fatty tissues of animals and to be excreted in the milk.

Although DDT is safe to use on most plants, the crude form causes phytotoxic symptoms on cucurbits, tomatoes, and several other plants. Pure DDT (mp 108°C) does not cause this reaction. One surprising effect of DDT is that at very low concentrations it stimulates growth in potatoes and cabbage much as plant hormones do. In fact, symptoms of plant injury by DDT are often "hormonelike" in nature.

DDT is formulated in several ways: solution, emulsifiable concentrates, dusts, wettable powders, aerosol bombs, and granules. Its greatest boon to mankind has been its effectiveness in controlling the insects that carry human diseases.

DDT is a stable product that does not break down readily and thus tends to remain in the environment. It is highly insoluble in water and does not leach to any extent from the soil. It accumulates in fat and thus becomes stored in the bodies of animals.

DDT is virtually banned in the United States, and its demise is a sad example of hasty research, biased reporting, unfounded claims and emotional response. Despite much experimentation, there still remain many unanswered questions concerning DDT. Unfortunately emotionalism has so clouded the issues that any research done now is so violently attacked or supported by the pros and cons that the factual data become lost in the fray.

Methoxychlor, another development of the Geigy Company, is closely related to DDT but is safe to use where environmental contamination is a concern. In general, it is useful against the same insect pests as DDT; however, it is more effective against plum curculio and the Mexican bean beetle and less effective against the European corn borer and corn earworm. It is only $\frac{1}{25}$ to $\frac{1}{50}$ as toxic to mammals as DDT, and accumulates less in the fat and is excreted less into milk. It is usually safer to apply on plants than DDT. Another valuable insecticide related chemically to DDT is **perthane**.

Toxaphene ($C_{10}H_{10}Cl_8$), an American development, was discovered by George Buntin of Hercules Incorporated. It is a yellow, waxy solid, insoluble in water but soluble in many organic solvents. Toxaphene is made by chlorinating camphene until it contains 69 per cent chlorine. It is broken down by strong alkali and sunlight. It is effective against a large number of insects and is about four times more toxic to mammals than DDT. It has low chronic toxicity, and accumulations in fat dissipate rather quickly in the living animal. It is safe on most plants, but is highly toxic to cucurbits. It is formulated much like DDT. It has been used against grasshoppers, cutworms, webworms, cotton insects, legume insects, and livestock insects. It is very toxic to fish, and care must be taken to see that none of it gets into a water system.

Benzene hexachloride ($C_6H_6Cl_6$) was first synthesized by Michael Faraday in 1825. BHC, as benzene hexachloride is commonly called, was found (independently) to be an active insecticide by scientists in France and in Great Britain in 1941 and 1942. BHC occurs in at least six isomers, each one of which exhibits distinctly different biological activity. The gamma isomer is much more active insecticidally than the other isomers. The purified form of BHC (99 per cent gamma isomer) is known as **lindane**. Lindane is white to colorless and lacks an odor. It is a stable compound but is broken down by alkali. Lindane is effective against a great number of insects. It acts as a stomach, contact, and fumigant poison.

Aldrin ($C_{12}H_8Cl_6$), **dieldrin** ($C_{12}H_8Cl_6O$), **chlordane** ($C_{10}H_6Cl_8$), **heptachlor** ($C_{10}H_5Cl_7$), **endrin** (isomer of dieldrin), and **endosulfan** ($C_9H_6Cl_6O_3S$) are chemically related (cyclodienes); all except the last were developed in the United States. Aldrin, dieldrin, and endrin are stable in alkalis. All are insoluble in water and soluble in most organic solvents. They are more toxic to mammals than DDT, and endrin is highly toxic, falling within the range of toxicity found among the more toxic phosphorus compounds. Most are residual insecticides, effective against a wide range of insects. Endrin is particularly effective against lepidopterous larvae, such as cutworms and webworms. Because of their persistent nature the cyclodienes have been used in the soil with success. Development of resistance by soil insects and the danger of residues on crops grown in years following soil insecticide application have reduced the value of these compounds. The EPA has greatly limited the use of cyclodienes in the United States.

Organophosphorus Insecticides

The synthetic compounds known as organophosphorus insecticides are organic molecules containing phosphorus. Thousands of these compounds have been synthesized, and many more are possible, at least in theory. The group represents the largest number (about one hundred) of synthetic insecticides under intensive experimentation and in commercial use today. Most organophosphorus insecticides have the structure

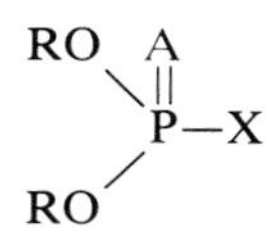

where R is an ethyl or methyl group, A is either a sulfur atom or an oxygen atom, and X is variable and more complex than the other substituents. Compounds such as malathion and parathion contain a sulfur atom in place of A in the above structural formula. Such compounds must be activated or converted to their oxygen analogue (=O substituted for =S) before they are insecticidal. This conversion actually occurs in the animal's body. The oxygen analogue of malathion is called malaoxon; that of parathion is called paraoxon.

Some organophosphorus insecticides have the ability to act systemically; that is, they are absorbed by plants, rendering the sap toxic to insects, or they are taken in by animals, rendering the blood toxic. Examples of nonsystemic or only slightly systemic compounds are parathion, tepp, malathion, and diazinon. Systemic compounds include crufomate, demeton, coumaphos, famphur, dimethoate, disulfoton, phorate, mevinphos, and trichlorfon.

Parathion ($C_{10}H_{14}NO_5PS$)

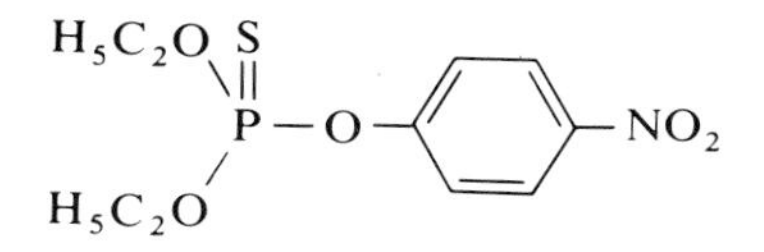

Pure parathion is a pale yellow liquid, but in its usual crude form it is a brown liquid with an odor of garlic. It is slightly soluble in water and hydrolyzes rapidly in alkali solution. It is highly toxic to both insects and mammals. Symptoms of parathion poisoning are headache, nausea, and constriction of pupils. Atropine is an antidote.

Parathion is effective against a great many insects and mites. It is noncumulative in mammals, but highly toxic. It is safe for use on most plants, except for a few varieties of plums and pear and on McIntosh apples. It is used widely in production of fruit, vegetables, ornamentals, and field crops. Its high mammalian toxicity precludes its application on livestock, in the household, or where it will come in contact with man. To avoid the danger of dermal toxicity of parathion, a method has been developed to enclose it in microcapsules. Under field conditions the capsules break down and release the parathion.

Methyl parathion, the dimethyl analogue of parathion, is less hazardous to mammals than parathion, apparently because of lower dermal toxicity. It has the same toxicity to insects. Among the organophosphorus insecticides synthesized in the United States, methyl parathion is produced in the greatest quantity. Production in 1974 reached a record high of 51.4 million pounds. Much is used to control insects on cotton and as a substitute for DDT.

Naled is a moderately toxic insecticide that can be used on numerous crops close to harvest and for control of insects that threaten public health. **Chlorpyrifos** (Lorsban) and **fenthion** (Baytex) are registered for use against mosquitoes and flies and are employed by pest control operators against household and premise pests. A formulation of fenthion (Queletox) is effective in bird control.

A number of organophosphorus insecticides have found specialized uses: the insecticide **dichlorvos** (Vapona) is only moderately toxic to humans and animals, yet highly effective in controlling insects that threaten public health and those that threaten livestock. Both dichlorvos and **crotoxyphos** (Ciodrin), another moderately toxic phosphate, provide effective control of livestock ectoparasites with no problems of residues in milk. A unique characteristic of dichlorvos is its very high vapor pressure, which allows it to act as a fumigant as well as a contact insecticide. **Temophos** (Abate) is effective against the larvae of mosquitoes, blackflies, and midges, yet it is only slightly toxic or has no effect on mammals, wildlife, and many aquatic organisms.

Animal-systemic insecticides, now commonly employed by livestock growers, belong to the organophosphorus group of insecticides. Examples of compounds effective against cattle lice and cattle grubs are **coumaphos** (Co-Ral), **crufomate** (Ruelene), **ronnel** (Korlan), **famphur** (Warbex), **phosmet** (Prolate), and **trichlorfon** (Neguvon). Trichlorfon is also registered for use against insects on forages, grains, vegetables, ornamentals, and in the home as well as in fly baits.

A number of organophosphorus insecticides are plant-systemic and are effective against certain pests of a wide variety of crops. Examples are **demeton** (Systox), **dicrotophos** (Bidrin), **dimethoate** (Cygon), **disulfoton** (Di-Syston), **mevinphos** (Phosdrin), **monocrotophos** (Azodrin), **oxydemetonmethyl** (MetaSystox-R), **phorate** (Thimet), and **phosphamidon** (Dimecron). Most of the plant-systemic phosphates are highly toxic to mammals, with the exception of oxydemetonmethyl and dimethoate, which are moderately toxic. Plant-systemic insecticides are valuable where injurious forms such as aphids and spider mites must be controlled and beneficial forms such as pollinators and predators must be protected.

Carbamate Insecticides

Carbamates in use and activity resemble organophosphorous insecticides. Some have low mammalian toxicity, whereas others are very toxic. These compounds break down readily and leave no harmful residues.

The carbamate insecticides are so named because structurally they are esters of the unstable carbamic acid:

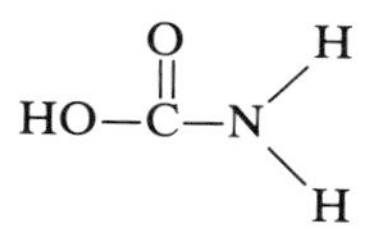

The carbamates currently in use as commercial pesticides and also the experimental compounds have the general formula

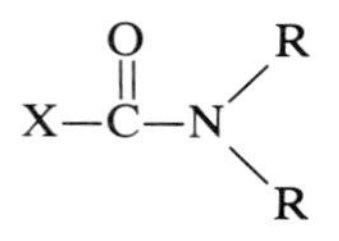

Usually one R is a methyl group and the other a hydrogen atom. The X is variable, complex, and often aromatic. Considerable variation is possible; both R's may be methyl, H, or entirely different.

Carbaryl ($C_{12}H_{11}NO_2$)

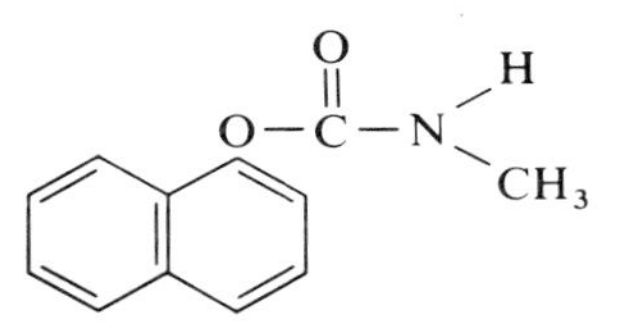

This compound is well known by its trade name Sevin. Its low mammalian acute oral toxicity—LD_{50} 307 mg/kg—makes it valuable in areas where more toxic compounds must be avoided. It is extremely toxic to honey bees and should not be applied where there is danger of killing these valuable insects. It is recommended in home gardens and on commercial crops against such insects as blister beetles, Japanese beetle, grasshoppers, scale insects, and Colorado potato beetle. It is particularly effective against the Mexican bean beetle.

Other carbamates have proved to be effective in insect control. **Carbofuran** (Furadan) is recommended for the control of soil insects and nematodes and of foliar insect pests of a variety of crops. Applied to the soil, **metalkamate** (Bux) is used in large quantities to control rootworms infesting corn. **Methomyl** (Lannate and Nudrin) controls a broad spectrum of insects in many commercial vegetable and ornamental crops, field crops, and certain fruit crops. **Propoxur** (Baygon), a moderately toxic carbamate with fast knockdown and long residual, is recommended against household and turf insects. **Aldicarb** (Temik), a highly toxic carbamate, is a systemic insecticide, acaricide, and nematicide for soil application as granules. It has good residual activity of three to ten weeks or more. Several other carbamates have systemic activity including carbofuran and methomyl. **Formetanate hydrochloride** (Carzol), both an acaricide and an insecticide, is effective against mites, thrips, and lygus bugs. It is registered for use on citrus, deciduous fruit, and alfalfa.

Miscellaneous Compounds

Insecticides have been discovered among a miscellaneous group of compounds. Two of these, **chlordimeform** and **chlordimeform hydrochloride**, have proved to be both insecticidal and acaricidal. As acaricides they control eggs, nymphs, and adults of plant-feeding mites, including strains that are resistant to organophosphorus compounds. Because of their possible carcinogenicity the manufacturer has withdrawn them from the market in 1976 as a precaution until all studies are completed on their safety.

Synthetic Pyrethroids

Synthetic pyrethroid compounds are gaining in favor for insect control. They are related to the natural pyrethrins, but are synthesized from petroleum based chemicals. The work done on the structure of the toxic constituents of pyrethrum and on the synthesis of related compounds is now regarded as classic research in chemistry. The first practical compound of the group, allethrin, was synthesized by USDA chemists who reported their success in the *Journal of the American Chemical Society* in 1949. Its characteristics of slight toxicity to mammals, high toxicity to insects, and quick knockdown are similar to those of the pyrethrins. Pest control operators use allethrin for flushing cockroaches to determine the existence or extent of an infestation.

Resmethrin, discovered by scientists at the Rothamsted Experiment Station in England, has characteristics that give it advantages over the pyrethrins. It is less toxic to mammals and more toxic to insects; it does not require synergists, has greater residual activity, and is lower priced. It is being developed in America by S. B. Penick & Co. and is presently registered for use against insects infesting households, processing plants, and greenhouses. Resmethrin is particularly effective against the greenhouse whitefly. It is also useful outdoors to control flying insects causing annoyance to people.

Several experimental synthetic pyrethroids with highly toxic properties to lepidopteran larvae and other insects are currently being developed by FMC Corporation, S. B. Penick & Co., ICI, and Shell Chemical Co.

Insect Growth Regulators

Insect growth regulators (IGRs) have been referred to as third-generation insecticides. They are compounds that alter the normal growth patterns of insects in various ways and indirectly cause their death. Several IGRs are synthetic analogues of juvenile hormones, the natural insect hormones secreted by the corpora allata, and elicit the same sort of responses. Others are dissimilar to any known juvenile hormones but may interfere with the metabolic degradation of natural hormones. Physiologically effective concentrations of IGRs are not toxic to the target insect but cause metabolic errors and asynchronous development, which eventually result in death of the pest. IGRs have several advantages as pesticides. One is their fine specificity, affecting target pests while sparing predators and parasites. Another is their low toxicity to other forms of life including man, livestock, and wildlife. A third is their effectiveness at extremely low dosages.

Only a few IGRs are presently available for insect control, but many are available for testing. **Methoprene**, an acyclic sesquiterpenoid similar to the juvenile hormone of the cecropia moth, is an IGR used to abate floodwater mosquitoes. It received full registration for this purpose in January 1975 and thus became the first IGR successfully developed for commercial use. At very low concentrations in water, 0.1 to 1 ppb, methoprene causes mosquito larvae to develop intermediate characters of larva-pupa or pupa-adult and prevents emergence of the adults. **Diflubenzuron** (Dimilin), a compound unrelated to juvenile hormones, is effective against mosquito eggs and larvae and leaf-feeding lepidopterous larvae. At present it is registered for control of the gypsy moth. It interferes with the formation of the insect's cuticle by inhibiting chitin synthesis. Other IGRs undergoing field trials are hydroprene, kinoprene, triprene, RO 20-3600, Stauffer R 20458, and MON-585.

SPECIFIC MITICIDES

Although some insecticides, such as parathion and malathion, are effective miticides, the majority of insecticides have little activity against phytophagous mites. To control this serious group of crop pests special chemicals, known as specific miticides, have been developed. They are relatively noninsecticidal but show a high degree of toxicity to mites. Chemically, they are often DDT relatives, nitrophenol derivatives, or organosulfur compounds.

Dicofol (Kelthane)

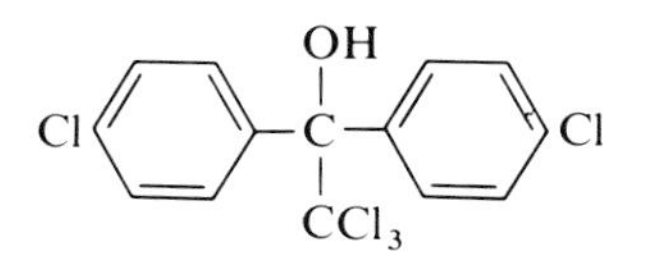

Dicofol—synthesized in 1953 by Dr. H. F. Wilson of Rohm & Haas Company—is an acaricide that provides high initial kill and long residual control of most species of agricultural mites. Related to DDT, the pure compound is a white solid; the technical product is a brown viscous oil, soluble in most organic solvents but practically insoluble in water. It has little insecticidal activity and is only moderately toxic to mammals. In 1971 it was the leading miticide produced in the United States. Other miticides related to DDT include **chlorobenzilate**, **chloropropylate**, and **bromopropylate**.

Several useful miticides are organosulfur compounds; they include **chlorbenside**, **fenson**, **oxythioquinox**, **propargite**, and **tetradifon**. Oxythioquinox controls powdery mildew as well as mites.

Nitrophenol derivatives include **dinocap** (Karathane) and **binapacryl** (Morocide), both of which act as acaricides and fungicides (powdery mildew); they are registered for use on fruit trees and several other plant groups.

Two of the newer miticides **cyhexatin** (Plictran) and **fenbutatin-oxide** (Vendex) are organotin compounds. They are effective against a wide range of plant feeding mites including strains resistant to some other miticides.

Pentac, a chlorinated aryl hydrocarbon, is particularly effective against the twospotted spider mite and is recommended for use on greenhouse floral crops.

FUMIGANTS

Fumigants are chemicals that exist as gases at required temperatures and pressures and are insecticidal at specific concentrations. They afford a practical solution for insect control within an enclosure. To be a good fumigant, a chemical should possess certain qualities. It must be volatile and penetrate deeply into stored products. The gas should be toxic or repellent to insects and mites, but should not be corrosive or interfere with seed quality if used on stored grain. The gas must desorb from treated products so that no toxic residue remains. Certain fumigants are also used in the soil weeks before planting to control insects, nematodes, and fungi. Two serious hazards of fumigation are the flammability of certain gases and the danger of accidental poisoning of man.

Hydrogen cyanide (HCN) is a colorless gas with an almond odor. It is toxic to all insects and is highly toxic to man, with dangerous amounts entering quickly through the lungs and even through the skin. It is not toxic to plants at recommended dosages, although it may injure plants bearing copper residues from previous spraying. Mills, warehouses, greenhouses, other buildings, and citrus trees under gasproof tarpaulins are often treated with HCN. It is nonflammable but is rather difficult to remove from treated materials.

Carbon disulfide (CS_2) is a colorless to yellow liquid with an unpleasant odor. It is highly flammable and is usually diluted with four parts of carbon tetrachloride to reduce fire hazard. Carbon disulfide is highly absorptive and penetrating; it has been used in buildings and in the soil. Besides its high flammability it also has the disadvantage of reacting with many surfaces, leaving yellow stains.

Methyl bromide (CH_3Br) has in recent years become a widely utilized fumigant. It is a colorless gas with a faintly sweetish odor. It is stable and nonflammable, and has high insect toxicity and some acaricidal properties. It has very high penetrating properties and is rapidly desorbed from treated materials. It is quite safe for use on dormant plants. Methyl bromide is used in the fumigation of plant products imported into this country. It is useful in mills, warehouses, and granaries. Methyl bromide is highly volatile and needs special equipment for application.

Ethylene dichloride ($C_2H_4Cl_2$) is a sweet-smelling, noncorrosive liquid. Because it is flammable, it is usually mixed with carbon tetrachloride. This mixture is used mainly for fumigation of stored products, but has also been used in emulsion form to control the peachtree borer. It has relatively low mammalian toxicity.

Carbon tetrachloride (CCl_4) is much less toxic to insects than the usual insect fumigants. It is neither flammable nor explosive and is mixed with other fumigants to reduce the hazards of fire and explosion.

Paradichlorobenzene ($C_6H_4Cl_2$) and **naphthalene** ($C_{10}H_8$) are solids which slowly give off gas. They have been used as soil fumigants and are popular in the form of moth balls or flakes for clothes moth control.

Dichloropropene ($C_3H_4Cl_2$) and **dichloropropane** ($C_3H_6Cl_2$) mixture (D-D mixture) is a common soil fumigant. The mixture is toxic to nematodes as well as to all soil insects. Because this fumigant is phytotoxic and may cause off-flavor in potatoes, the soil must be treated well in advance of planting date. Other fumigants applied to soil include **ethylene dibromide** (EDB), **dibromochloropropane** (DBCP), and **dichloropropenes mixture** (DCP).

INSECT REPELLENTS

Insect repellents are much more important in the protection of man and animals from insect attack than in the protection of plants. **Bordeaux mixture**, which acts as a fungicide, also has a repelling effect on some plant-feeding insects and has been used in potato fields for that purpose. Inactive dusts have been used on cucurbits to repel cucumber beetles.

Pyrethrum in low concentrations repels blood-sucking insects. This characteristic has been exploited in cattle sprays. Another repellent suitable for application to livestock is **butoxy polypropylene glycol**.

During World War II U.S. troops applied **dimethyl phthalate** to repel anopheles mosquitoes, the vectors of malaria. The chemical also repels other blood-sucking insects, ticks, and chiggers. **Deet** (diethyl toluamide), a discovery of the USDA, is one of the best all-purpose insect repellents developed so far. It protects the wearer against mosquitoes, chiggers, ticks, fleas, and biting flies. Its resistance to rubbing gives it a long-lasting quality, and it can be applied safely to skin and most fabrics.

Several other repellents available commercially include ethyl hexanediol, Indalone, and MGK Repellent 11.

INSECT ATTRACTANTS

Many insects take advantage of natural odors to guide themselves to vital needs of food, host plants, mates, and oviposition sites. These behavioral responses have been exploited by entomologists in various ways to control insects. Attractants are used to sample local populations by trapping, to lure insects into traps or to poisons for reducing pest populations, to offset repellent properties of certain sprays, and to lure insects away from crops.

The first insect attractants were natural products such as sugar, molasses, yeast extract, fatty acids, protein hydrolysates, geraniol, and eugenol. Sugar mixed with certain organophosphorus or carbamate insecticides makes a good bait for supplemental control of house flies. Molasses is recommended as an ingredient in bran baits for control of cutworms. **Geraniol and eugenol** mixture attracts Japanese beetles to traps (Fig. 17:7). The strong attractancy of geraniol and eugenol, both natural compounds, was discovered by USDA scientists in the mid-1920s. Only recently has the combination been superseded by an even stronger synthetic lure of **phenethyl proprionate and eugenol**.

It is interesting how a chance observation led to the discovery of **methyl eugenol**, one of the outstanding modern insect attractants. A neighbor of F. M. Howlett, an English entomologist working at the Pusa Agricultural Research Institute in India, had been troubled by some kind of fly settling on him at a time when he was using oil of citronella sprinkled on his handkerchief to repel mosquitoes. Howlett had been investigating essential oils to attract fruit flies but had not tried citronella because it in no way resembled the smell of the preferred host plants, mangoes and peaches. He decided to try the substance by wetting a handkerchief and exposing it on the ground near a peach orchard. In less than half an hour the handkerchief lying in a crumpled heap was almost hidden by a crowd of male fruit flies. Detailed research in 1912–13 led him to discover that small amounts of methyl eugenol in the mixture of compounds that compose citronella was the powerful attractant of males of the oriental fruit fly. Methyl eugenol and also eugenol are now produced synthetically.

During the fifties and sixties entomologists discovered several synthetic attractants through a tedious program of screening and testing thousands of compounds prepared by chemists. A powerful lure for the male of the melon fly was found in the compound 4-(*p*-hydroxyphenyl)-2-butanone acetate. This substance, now called **cue-lure**, is used in traps to detect infestations of the pest. Other synthetic insect attractants in current use are **medlure** and **trimedlure** for males of the Mediterranean fruit fly.

Many insects produce pheromones that attract the opposite sex. Entomologists have been able to locate the glands and to extract the attractive substances from insects, and chemists have succeeded in identifying and synthesizing several of these natural compounds. Pheromones are effective at very low concentrations, 10^{-7} to 10^{-12} μg, and can attract insects from a distance of several miles. Most are specific in their action, attracting only one sex of one species of insect. Synthetic pheromones are available for attracting the males of several moths, including **disparlure** for the gypsy moth, **gossyplure** for the pink bollworm, **codlelure** for the codling moth, **looplure** for the cabbage looper, and **virelure** for the tobacco budworm. **Grandlure**, a mixture of four synthetic compounds that copy the natural pheromone produced by the male boll weevil, acts as a sex attractant for the female and as an assembly attractant for both males and females of the species.

MICROBIAL INSECTICIDES

Microbial insecticides are safe, effective products developed from insect pathogens. They can be applied to crops to control insects in much the same way as chemical insecticides. Although thousands of insect pathogens occur among the viruses, bacteria, fungi, and protozoa, only a few have as yet been developed into microbial insecticides. Throughout the world about 15 different insect pathogens have become commercially available. Presently in the United States two bacteria and two viruses are registered by EPA for use as insecticides. An effective microbial insecticide has several advantages: it is specific against the target pest, it is neither toxic nor pathogenic to man, it spares predators and parasites, it has no deleterious effect upon the environment, and it usually adds nothing new to the environment because pathogens occur naturally.

One of the bacterial insecticides is *Bacillus popilliae* Dutky, an obligate bacterial pathogen that causes a milky disease in the larvae of the Japanese beetle and other scarab beetles. This bacterial product is manufactured in Fairfax Biological Laboratory, Clinton Corners, New York, under the trade name "Doom." When dust preparations are applied to the soil or infested turf, the larvae ingest the spores and contract the disease. The bacteria invade the hemolymph and multiply there, turning it a milky color and eventually killing the larvae. To control outbreaks of the Japanese beetle in 14 eastern states and the District of Columbia, government agencies distributed and applied spore preparations from 1939 to 1951.

The other bacterial insecticide, *Bacillus thuringiensis* Berliner, is a nonobligate pathogen. In nature it can occur free of its hosts, and it is easily cultured on artificial medium. Registered as an insecticide in 1959, *B. thuringiensis* is a more recent development than *B. popilliae.* Within its vegetative cells *B. thuringiensis* forms a protein crystal that becomes toxic upon ingestion by the insect. In some species, such as the tomato hornworm, the toxin erodes the gut wall, causing paralysis and, within several days, death. Applied as a dust or spray, *B. thuringiensis* is used to protect vegetables, field crops, fruit, shade, and forest trees from several insect pests, particularly lepidopterous larvae. Several companies sell *B. thuringiensis* under different trade names.

Over 500 viruses have been isolated and described from insects. All five of the described types—the nuclear polyhedrosis viruses, the cytoplasmic polyhedrosis viruses, the granulosis inclusion viruses, the entomopox viruses, and the noninclusion viruses—have been tested as viral insecticides.

One of the **nuclear polyhedrosis viruses**, which infects *Heliothis* species, has been the most extensively tested insect pathogen and is the world's first naturally occurring virus to be registered as a pesticide. It has been approved by EPA (1975) for use against two serious cotton pests, the bollworm and the tobacco budworm. When a field of cotton is sprayed with virus preparation, *Heliothis* larvae feeding on the cotton plants ingest the virus. The virus particles then invade cells of the midgut and multiply in the nuclei. Progeny of the virus enter the blood stream, where secondary infection of other tissues occurs. Diseased larvae generally feed less, and upon death the larvae completely disintegrate to release more infective virus. Young larvae will develop symptoms one to two days after ingesting virus and begin dying on the third day. About 95 per cent of total mortality is reached within a week. See Fig. 5:8 for appearance of polyhedrosis virus disease of the cabbage looper.

In 1976 a second nuclear polyhedrosis virus was approved by EPA for insecticidal use. It is an effective pathogen for controlling outbreaks of the Douglas-fir tussock moth. The viral insecticide acts the same way as the naturally occurring virus except that it speeds up the natural epizootic and thereby protects trees from severe defoliation.

SELECTED REFERENCES

Bailey, S. F., and L. M. Smith, *Handbook of Agricultural Pest Control* (New York: Industry Publications, Inc., 1951).

Brown, A. W. A., *Insect Control by Chemicals* (New York: Wiley, 1951).

Corbett, J. R., *The Biochemical Mode of Action of Pesticides* (New York: Academic Press, 1974).

Dethier, V. G., *Chemical Insect Attractants and Repellents* (Blakiston, 1947).

Entoma: A Directory of Insect and Plant Pest Control, Entomological Soc. of America (issued annually).

Frear, D. E. H., *Chemistry of Insecticides, Fungicides, and Herbicides* (New York: Van Nostrand Reinhold, 1948).

Hassall, K. A., *Pesticides: World Crop Protection*, Vol. 2 (Cleveland: CRC Press, 1969).

Jacobson, M., *Insect Sex Attractants* (New York: Wiley–Interscience, 1965).

Martin, H., *Guide to the Chemicals used in Crop Protection*, Canada Dept. of Agr., 1957.

Matsumura, F., *Toxicology of Insecticides* (New York: Plenum Press, 1975).

Melnikov, N. N., *Chemistry of Pesticides* (New York: Springer-Verlag, 1971).

Metcalf, R. L., *Organic Insecticides* (New York: Wiley–Interscience, 1955).

O'Brien, R. D., *Insecticides Action and Metabolism* (New York: Academic Press, 1967).

Radeleff, R. D., *Veterinary Toxicology* (Philadelphia: Lea & Febiger, 1970).

Shepard, H. H., *The Chemistry and Action of Insecticides* (New York: McGraw-Hill, 1951).

Spencer, E. Y., *Guide to the Chemicals Used in Crop Protection*, Agr. Canada Publ. 1093, 1973.

Summers, M., et al., eds., *Baculoviruses for Insect Pest Control: Safety Considerations* (Washington, D.C.: Amer. Soc. Microbiology, 1975).

USDA, *The Pesticide Review 1975*, USDA Agr. Stabilization and Conservation Service, 1968.

Van Valkenburg, W., ed., *Pesticide Formulations* (New York: Marcel Dekker, 1973).

chapter 8 / **W. DON FRONK**

INSECTICIDE APPLICATION EQUIPMENT

Recent advances in chemical control of insects have come not only from the discovery of more effective insecticides but also from the development of new and better machines for applying these chemicals. Early farm sprayers, such as those invented by John Bean in 1883, were small, hand-operated tanks that had to be carried about the field being treated. Somewhat later, bigger tanks were placed on wheels with larger but still hand-operated pumps. In 1887 the first traction sprayer was designed, and in 1894 sprayers powered by steam engines were invented. The first complete power sprayer operated by a gasoline engine was not marketed until around 1900. In 1911 the Bean Spray Pump Company perfected the modern pressure regulator, and in 1914 the Friend Sprayer Company introduced the spray gun.

Although improvements came gradually, we today have available a multitude of types and sizes of sprayers and dusters that provide the means for almost any practical application of insecticide. Insect problems, ranging from unwanted insects in the home to insects infesting thousands of acres of range or forest, are neither so small nor so big that we can not find some piece of equipment suitable for applying a chemical to help with a solution. Yet in spite of this variety, improvements are still needed in both ground and aerial equipment, and we can expect to see further changes and advances in the future.

A study in 1964 by the Economic Research Service of the USDA revealed that one half of the U.S. farmers surveyed owned power sprayers and that more than 80 per cent were operated by power takeoff, 15 per cent worked by auxiliary engines, and 4 per cent were self-propelled. The surveyors estimated that farmers in the United States owned a million power sprayers. They found that self-propelled sprayers were heavily concentrated in the Delta states, where these machines were used chiefly for control of cotton pests; power-takeoff driven sprayers were most popular in the Corn Belt states for use on corn and other grain crops; auxiliary engine sprayers were most common in the Great Lake states. The surveyors found that dusting equipment is less popular; only 6 per cent of farmers surveyed had power dusters. They estimated that farmers in the United States owned 112,000 power dusters. Ten per cent of the farmers had machine attachments and owned an estimated 172,000 of these units.

Between 1960 and 1964 farmers on the average paid $2400 for new self-propelled sprayers, $250 for power-takeoff sprayers, $215 for power-takeoff dusters, and between $131 and $231 for machine attachments.

The majority of farmers surveyed had hand-operated or other non-powered pesticide application equipment. In 1964 farmers and ranchers had an estimated 1,091,000 hand sprayers, 343,000 knapsack sprayers, 520,000 livestock rubbers, 279,000 hand dusters, and 15,000 dipping vats.

In this chapter we shall present a short summary of the common types of sprayers, dusters, and granule applicators. Special application equipment will be discussed in appropriate host chapters. Only brief mention will be made of the operation and function of the various parts of sprayers and dusters, as this subject belongs more properly to the field of agricultural engineering.

SPRAYERS

The household **intermittent and continuous sprayers** are used in homes to control insects such as house flies, mosquitoes, ants, clothes moths, and cockroaches. One type of nozzle produces a fine mist to put a space spray into the air; another type produces a coarse spray for residual deposits. **Electric household sprayers** are often used by professional pest control operators to control insect pests in restaurants and hotels. **Electric misters** (Fig. 8:1A) dispense oil solutions or emulsions of insecticide in aerosol range. They are useful for space spraying, for fogging the inside of barns, and for placing a fine coating of insecticide on cattle and horses to control various ectoparasites.

The **compressed air sprayer** (Fig. 19:5) has many applications. Gardeners use them to apply sprays to lawns, shrubs, fruit trees, and vegetables. They can be used to control insects in the home, in barns, and in poultry houses. They are used by public health workers for applying residual deposits of insecticides in homes and other buildings to control mosquitoes and other insects of medical importance. Tanks range from $1\frac{1}{2}$ to 5 gal and should be filled no more than $\frac{2}{3}$ to $\frac{3}{4}$ full with spray materials, leaving an air space above the liquid.

A convenient device for the householder to use in spraying shrubs, trees, lawn, and garden is the **hose sprayer** (Fig. 8:1B). Well-designed newer models are constructed of noncorrosive materials and have adjustable metering mechanisms and strainers.

Figure 8:1. A, an electric mister which can be clipped directly to gallon containers of ready-to-use insecticide; B, hose sprayer; insecticide and water mix in head and water pressure forces spray from orifice. *A, courtesy Samuel Halaby, Inc., Rochester, N.Y.; B, courtesy Gilmour Manufacturing Company, Somerset, Pa.*

Knapsack sprayers (Fig. 8:2), carried on the back of the operator, have tanks with a capacity of 4 to 6 gal and develop maximum pressures of 80 to 180 psi (pounds per square inch). The handle of the pumps extends over the shoulder or at hip height and must be operated by hand while spraying. Because these sprayers develop higher pressures and have greater capacity than most compressed air sprayers they are better suited for large gardens or truck farms.

Figure 8:2. Knapsack sprayer being used to apply insecticide to a vegetable garden. *From D. B. Smith & Company, Utica, N.Y.*

Bucket- and barrel-type sprayers are useful for small plantings. The bucket type may be used for spraying ornamental plantings and small trees and the barrel type for slightly larger jobs such as small orchards and vegetable gardens. These sprayers, equipped with plunger type pumps and an air chamber to maintain constant pressure, develop pressures up to 250 psi.

Slide or trombone sprayers, like the bucket sprayers, are used mostly on ornamental plantings and small trees. Pressure is provided by a telescoping pump that has a nozzle on the discharge end. The spray is usually carried in a bucket.

Wheelbarrow sprayers may be either hand- or engine-operated. To be used most efficiently, the hand-operated type requires two men, one to pump and one to spray. The wheelbarrow sprayers are the largest hand-operated units and may be used for spraying trees, orchards, greenhouses, and vegetable gardens. The tanks range in capacity from 5 to 20 gal, and may be equipped with pressure gauge and agitator. The pumps develop up to 200 psi pressure.

Traction sprayers are small, row-crop machines that derive power for the pump from the supporting traction wheels. These sprayers have tanks of about 25-gal capacity and a 4- to 6-nozzle boom; they develop maximum pressures of around 150 psi.

Power Sprayers

A common sprayer on American farms today is the **multipurpose hydraulic** type (Fig. 8:3). Equipped with spray gun and hose as well as a boom, these

Figure 8:3. Multipurpose hydraulic sprayer for general farm use employing boom to apply insecticide for insect control. *Courtesy F. E. Meyers & Bro. Co., Ashland, Ohio.*

sprayers are used for many purposes—spraying a wide variety of crops for insects and weeds, spraying livestock and farm buildings, washing machinery, and even fighting fires. They have reciprocating piston or plunger-type pumps, which develop pressures ranging from 30 to 400 or more psi. They deliver from 3 to 8 gal of spray per minute, and tanks vary in size from 50 to 200 gal. Both skid- and wheel-mounted types are available.

The **estate-type sprayer** is a small wheel-mounted sprayer popularly used on golf greens, estates, large gardens, and greenhouses. Tank sizes range from 15 to 50 gal, and pumps develop up to 300 lb pressure and discharge at rates of $1\frac{1}{2}$ to 3 gal/min.

Low-pressure, low-volume sprayers (Fig. 8:4) commonly used to control weeds and field crop insects, are inexpensive machines that can be obtained as spray kits for assembling on the farm. They operate at pressures of 15 to 125 lb and deliver 1 to 25 gal/min. A rotary-type pump is usually supplied for direct mounting on the tractor power-takeoff shaft.

High-pressure, high-volume sprayers (Fig. 8:5) are large sprayers developed to take care of the needs of fruit growers and truck farmers. They

Figure 8:4. A low-pressure, low-volume tractor-mounted sprayer. *Courtesy Farm Equipment Institute.*

Figure 8:5. High-pressure, high-volume sprayer being used on a truck farm; note suspension of nozzles which direct spray to underside of leaves. *Courtesy Oliver Corporation, South Bend, Ind.*

have plunger-type pumps that deliver from 8 to 85 gal/min at maximum pressures of 400 to 1000 psi and have tanks up to 600-gal capacity. Booms are of various designs to fit the requirements of treating vegetable and fruit crops. For many vegetable crops, the nozzles are suspended and arranged to cover the plants on the top, sides, and underside of foliage. For tree fruits the sprayers may be equipped with hand guns or with an automatic spray head.

The **self-propelled, high-clearance sprayers** (Fig. 12:3) have been developed to treat crops that grow tall, such as corn and cotton. They are usually equipped with low-pressure, low-volume pumps and with booms that are adjustable for either high- or low-growing crops.

A **stationary spray plant** is found almost exclusively in orchards. It is composed of a spiderweb of pipes throughout the orchard, connected to a tank and pump housed in a shed. Hose lines are attached to outlets in the orchard, and the hoses are dragged about the trees while spraying. Systems such as these may be used in hilly country where tractors and heavy equipment are unable to go. Spraying may be done when the ground is too wet for a tractor. This system works best where temperatures do not go so low as to freeze pipes laid on the ground or buried shallow. Pumps usually develop 600 to 1000+ psi, and tanks are about 500- to 600-gal capacity. For treatment of crops irrigated by sprinklers, center-pivot and other types, metering devices are available that attach to the system and permit accurate application of pesticides.

Air sprayers (Fig. 8:6), also called blower, concentrate, air-blast, and air-mist sprayers, are a recent development, although the idea goes back to the 1920s. The principle of the sprayer is that an air blast carries the insecticide and replaces most of the water of high-pressure, high-volume sprayers. A series of nozzles or a shear plate directs spray under low pressure into an airstream that carries the atomized liquid to the surface to be treated. Air sprayers are rated in terms of air capacity in cubic feet per minute and velocity in miles per hour. Low-volume blower sprayers deliver 250 up to 30,000 cu ft of air per minute at velocities around 150 mph, high-volume sprayers deliver 20,000 to 100,000 cu ft/min at velocities up to 100 mph.

Figure 8:6. An air sprayer being used to treat apple trees in a commercial orchard. Twin 42-in. fans deliver 95,000 cu ft/min air volume. *Courtesy F. E. Meyers & Bro. Co., Ashland, Ohio.*

Many insecticides may be used in air sprayers at two to five times or even more the concentration normally employed with dilute sprays. These machines save both time and materials and can cut the cost of orchard spraying 75 per cent. They are also efficient for treating shade trees.

Aerosol generators or **fog machines** disperse insecticide into the air in the form of small particles ranging in diameter from 0.1 to 50 microns. They are used to control mosquitoes and flies in recreational parks, resort areas, and in communities.

A major advance in insect control has been made by the adaptation and use of **aircraft** to apply insecticides. Airplane application has several advantages—quick coverage of large acreages, no mechanical damage to crops, treatment of inaccessible areas such as forests, rangeland, and swamps, and treatment of fields after rains when ground is soft. Many of the airplanes used for insecticide application are modified Stearman and N3N trainers and pleasure planes. Several companies are building specially designed agricultural airplanes (Fig. 8:7), and the helicopter is becoming very popular for

Figure 8:7. A specially designed agricultural aircraft, the Grumman Ag-Cat. *Courtesy Grumman Aircraft Engineering Corp., Bethpage, L.I., N.Y.*

aerial application (Fig. 8:8). This machine is able to treat small and odd-shaped fields and can make application close to trees, power lines, and other obstructions. Its rotors serve to drive the insecticide to the ground and penetrate into dense foliage.

Ultra-low-volume (ULV) spraying makes use of small, rotating atomizing nozzles or other devices to break the spray into small, uniform droplets. By this means as little as one ounce of chemical may be spread uniformly over one acre. Several insecticides (malathion, for example) applied undiluted in this manner may be more effective than the same chemical applied in the usual way. Because no water or other diluent is necessary, airplanes can treat a much greater area with a single load of spray.

Figure 8:8. Helicopter applying insecticide to backwaters near the Twin Cities, Minnesota in a program of mosquito abatement sponsored by the Metropolitan Mosquito Control District. Versatile application equipment is designed for snap-on installation and removal. *Courtesy American Cyanamid, Co., Princeton, N.J.*

Few farmers use aircraft themselves for treating crops. Most agricultural airplanes are the tools of professional aerial applicators, who serve the farmer in protecting crops from diseases, weeds, and insects.

Sprayer Parts

To operate a power sprayer properly, make repairs, and replace parts, one should know and understand the functions of the units that make up the sprayer: power source, pump, tank, agitator, pressure regulator, valves, pressure gauge, screens and strainers, distribution system, and nozzles (Fig. 8:9).

The **power source** is either a gasoline engine or the power-takeoff shaft of the tractor. The engines used most commonly develop 1 to 3 hp and are air-cooled. On larger machines an engine developing 30 hp or even as much as 65 hp may be required.

The **pump** is one of the most important parts of the sprayer. Gear, roller impeller, piston, and plunger are the most common types (Fig. 8:10). The piston and plunger pumps develop pressure by pushing directly on the liquid in a cylinder. Gear and roller impeller pumps develop pressure by passing the liquid through teeth or rollers, which pick up the liquid and force it through an outlet.

Figure 8:9. Diagram of a typical power sprayer. *From* Sprayer and Duster Manual, *by permission National Sprayer & Duster Association.*

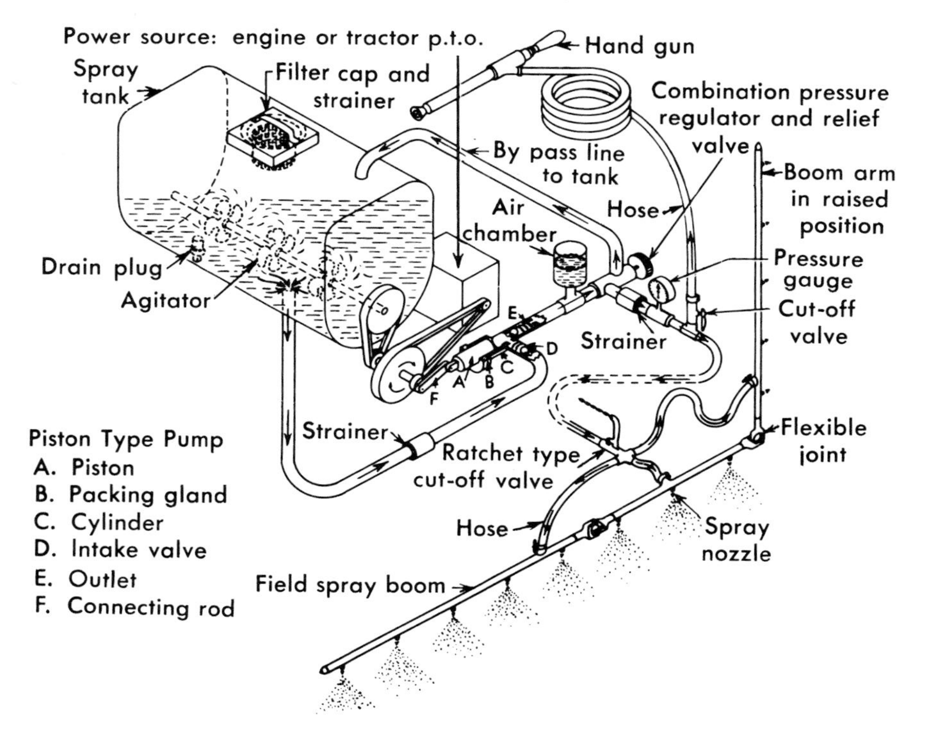

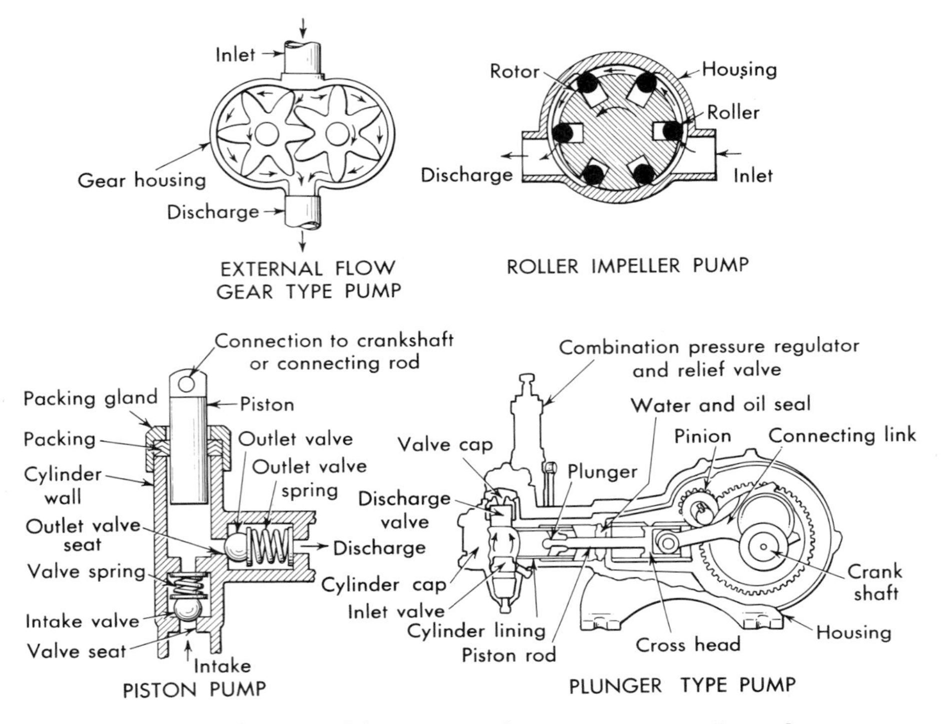

Figure 8:10. Diagram of four types of sprayer pumps. *From* Sprayer and Duster Manual, *by permission National Sprayer & Duster Association.*

Piston pumps deliver 2 to 8 gal/min and develop pressures up to 400 psi. The plunger type is used at 7 to 70 gal/min and develops pressures up to 1000 psi. Gear and some impeller pumps will, when new, develop pressures up to 300 to 400 psi. However, after a short time, wear reduces the pressure to 50 to 100 psi with a maximum output of around 20 to 30 gal/min.

Tanks vary in size from 50 to 600 gal or larger on power-operated sprayers. The tank may be either of wood or metal. It is usually equipped with a large opening and strainer at top to enable cleaning and inspection. A plug in the bottom makes it possible to drain the tank. The spray liquid is kept mixed in the tank by means of a mechanical **agitator** or by hydraulic action. With piston and plunger pumps, an **air chamber** is connected on the discharge line to level out the pulsations of the pump. Such an arrangement prevents spurts of liquid from the discharge end.

A **pressure gauge** is located on the discharge side of the pump or on the air tank to enable the operator to select the pressure he wishes. The pressure is regulated by means of a **pressure regulator**, usually of the plunger type. If the pressure in the line exceeds the tension selected on the spring of the plunger, a ball valve will lift and pypass the excess liquid back into the tank.

This regulator not only allows the operator to select his pressure but also acts as a safety valve and keeps the liquid in circulation while the discharge end is closed.

Strainers are located in the opening of the tank and in the suction line between the tank and the pump. They remove any foreign matter that might lodge in the pump, line valves, or nozzles. Strainers should be regularly removed and cleaned.

From the pump, the liquid is carried through **pipes** and **hoses** to the nozzles. Improper pipe or hose size can cause a decrease in pressure, because pressure loss varies with the square of the velocity of flow. The hose should be made to withstand about three times the regular spraying pressure.

The liquid spray emerges from a nozzle or nozzles on a hand **spray gun** or from several nozzles on a **spray boom**. Booms are usually adjustable to various heights to compensate for height of the plants. Most booms are about 21 ft long.

The **nozzle** is a vital part of a sprayer; it breaks up the liquid stream and spreads it out into the proper-sized spray droplets. The nozzle and the pressure determine the degree of atomization and the amount of liquid discharged. Nozzles are manufactured so as to give differences in rate of discharge, angle of spray, and pattern of spray. Nozzles are designed for either high or low pressure and for producing a fan-shaped, solid cone, or hollow cone spray pattern, and for spray angles ranging from a straight stream to 100°.

High-pressure nozzles are composed of a body, strainer, whirl plate, gasket, disk, and cap (Fig. 8:11). The body encloses all the parts except the cap. The whirl plate twists the liquid about at high velocity before it is discharged through the disc. This whirling motion contributes to breaking the liquid into droplets, and the resulting swirling action of the spray moves foliage and provides better coverage of all surfaces. The whirl plate may be a disc with holes cut into it at angles, a screw type, or a cylinder with slits cut at an angle through it. A hole in the center of the whirl produces a solid cone. All nozzles have a number that represents the diameter of the opening in the disc in $\frac{1}{64}$ of an inch. Sizes 2 to 4 are usually used for row crops and size 10 on

Figure 8:11. Parts of a high pressure nozzle. *Courtesy John Bean Division, FMC Corporation, Lansing, Mich.*

large-capacity hand or orchard guns. The largest size available is 20. The gasket prevents leakage and the cap holds the disc in place. Many systems have a small strainer located in the line just above the nozzle itself.

High-pressure nozzles may well be used at low pressures, but the **weed-spraying nozzle** is commonly employed on low-pressure sprayers. These nozzles usually produce a fan-shaped spray pattern. The weed-spraying nozzle is composed of a body, strainer assembly, cap, and orifice tip. The liquid passes through the strainer and is broken up in passing through the hole in the tip. By selecting tips of a certain size and with a certain shape of opening, the operator can determine spray pattern and discharge rate.

Practical foam spraying has been achieved by the development of special nozzles which aspirate and mix air with the liquid spray. **Foam nozzles**, available commercially, have several advantages: uniform suspension of fine solids, minimal drift, visible coverage, and little runoff. Disadvantages are the additional cost for nozzle hardware and for foaming agent and the fact that fewer foam droplets reach the target. More field research is needed on the effectiveness of foam sprays for insect control compared with emulsion sprays.

Sprayer Care

Any sprayer will last longer and perform more satisfactorily if it is properly cared for and maintained. Study the manufacturer's instruction manual. Proper lubrication is important, and care should be taken to check each lubrication site. With a new sprayer, it is wise to start it slowly and, using water only, check for any leaks and proper operation of valves, pump, motor, and gauge. After each use, drain the sprayer and flush with clean water. Remove and clean thoroughly all nozzles and strainers. Before placing the sprayer in storage for a period of time, add 2 to 5 gal of oil to the tank and fill with clean water. Pump this mixture out until the tank is dry. A thin coat of oil will be left throughout the system. Be sure there is no water left in any part of the system during freezing weather.

If the sprayer has been used for weed spraying, it is necessary to pay special attention to cleaning before it can be used for other spraying jobs.

Sprayer Calibration

In order to spray the correct amount of material per acre, it is necessary to calibrate the sprayer. The number of gallons per acre that a unit discharges is dependent upon the speed of the sprayer over the ground, the operating pressure, the spacing of the nozzles on the boom, and the size of nozzle tips. In an average situation for row crops, speeds of around 4 mph, pressures between 50 to 100 psi, nozzle spacings of 12 to 20 in, and nozzle sizes of 2 to 7 are practical ranges for operating the sprayer. If a grower knows the desirable application rate (gallons per acre) and the pressure, ground speed, and nozzle spacing of his equipment, he can get the correct nozzle size from the manufacturer's tables of nozzle tips.

A simple method of calibration utilizes a measured course. For convenience 40 rods or 660 ft may be chosen. The swath width is determined by boom length. Fill the tank of the sprayer with water to a known point, then operate the sprayer over the course at constant pressure and a desirable speed for spraying your crop. Next measure gallons of water to refill the tank and insert data into the following formula:

$$\frac{\text{gal to refill} \times 66}{\text{width of swath in ft}} = \text{gal per acre output}$$

Nozzles on the boom may be arranged to give complete coverage of the field (broadcast application) or arranged to give a band application over the planting row. In either case the above formula applies, and recommendations will instruct the grower to apply so many lb AI per acre or gal per acre of mixed spray in a broadcast or band treatment.

Another simple method utilizes a plastic or glass measuring jar that fastens under one nozzle and catches the spray discharged while driving over a measured distance. The device is marked on the side to indicate the number of gallons per acre being discharged by the sprayer for a number of different boom widths.

DUSTERS

Dusters are not as mechanically complicated as sprayers and are lighter in weight than comparable sprayers. Like sprayers, dusters may be either hand operated or powered by a motor. The hand dusters may be a simple **plunger or bellows** type used chiefly about the house or in a small garden. **Crank dusters** are practical for treating small acreages of row crops, such as tobacco, cotton, and vegetables. The dust is held in a hopper and discharged through an adjustable feeding device. The dust is kept stirred up in the hopper by means of a mechanical agitator. A fan is operated by a hand crank through a series of high-speed gears. The blast of air picks up the dust and carries it out through the nozzles. **Knapsack dusters** are bellows-type dusters operated by means of a handle. Each pump of the handle produces a spot application of dust. In some areas vegetable growers use lightweight, back-pack, power dusters.

Power row-crop dusters (Fig. 8:12) are commonly used in the South to treat peanuts, cotton, and tobacco. Power dusters are composed of a source of power, hopper, agitator, metering device, fan, discharge tubes, boom, and nozzles (Fig. 8:13). The **hopper** usually holds 2 to 4 cu ft of dust. The **agitator** inside the hopper keeps the dust fluffed up so it will flow evenly through the metering device. The **metering device** can be regulated so that 5 to 50 lb of dust may be delivered per acre. Common **fans** are centrifugal ones which operate at 2200 to 3400 rpm and deliver 500 to 1000 cu ft of air per minute at velocities ranging from 50 to 100 mph. The dust is carried through wire-reinforced rubber hose or flexible metal **tubes** to the nozzles. The **nozzles**,

Figure 8:12. A power row-crop duster in operation. *Courtesy John Bean Division, FMC Corporation, Lansing, Mich.*

Figure 8:13. Diagram of a typical power duster. *From* Sprayer and Duster Manual, *by permission National Sprayer & Duster Association.*

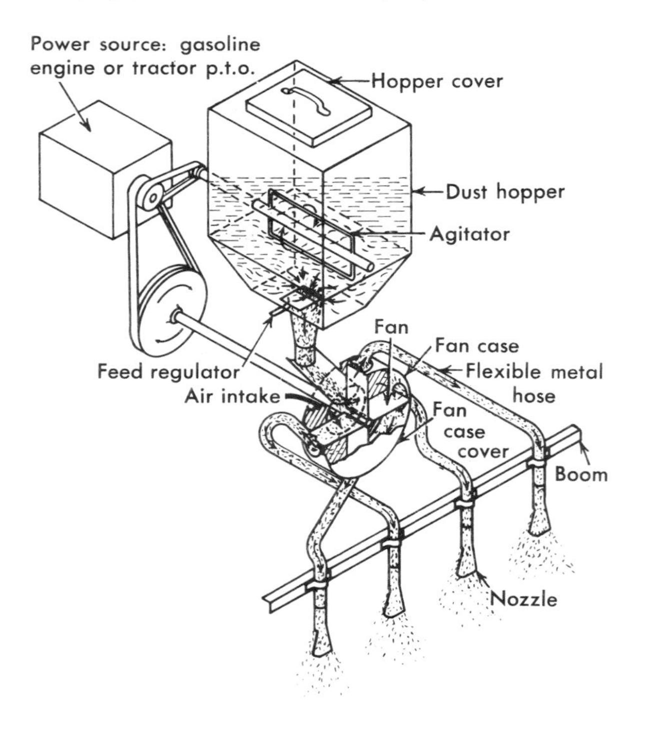

scoop-shaped metal pieces attached to the tubes, serve to direct the stream of dust. A metal **boom**, commonly 18 to 20 ft long, supports the tubes and nozzles. The **power** for the duster may come from a separate engine (from $1\frac{1}{2}$ hp up to 6 hp), or it may operate from the power takeoff of a tractor. Large fruit dusters may require engines of 15 to 25 hp or more.

As with sprayers, the operator should be thoroughly familiar with the manufacturer's operational manual. The recommended fan speed should never be exceeded. All designated spots should be lubricated carefully. After each dusting, the excess dust should be removed from the hopper; it will tend to cake if left standing too long. The duster should always be stored in a dry place.

To calibrate a duster, it is necessary to measure off a known area and measure the amount of dust delivered. More dust or less can be applied by adjusting the metering device.

GRANULE APPLICATORS

Granule applicators are designed to provide a means of accurately and uniformly dispersing granular formulations of insecticides. Because granular formulations have been found useful for controlling certain pests of row crops, much emphasis has been given to the development of row-crop applicators which attach to planters (Fig. 8:14), to cultivators, to high-clearance tractors, and to several other farm machines. Granular row-crop applicators are convenient tools for applying insecticides in the row for control of soil pests of corn, sugar beets, and several other crops. Granular formulations of systemic insecticides, applied as a band or side dressing at planting time or topically on growing plants, may give season-long control of sucking insect pests. Granules, whether systemic or nonsystemic, applied over rows of corn into the whorls also control chewing insects such as European corn borers. Trailer-type granule applicators (Fig. 10:6) are also available. These machines are quite versatile; they can apply granules broadcast or in bands.

Granule applicators should be calibrated similarly to sprayers. The amount of granules spread by applicators depends on the size of the metering opening (feeder-gate setting), speed of travel, field roughness, and the flow rate of the granules (which is affected by size, shape, texture, and density of the granules, as well as by temperature and humidity). Because of such variables, it is difficult to predict a certain rate at a certain setting. Use the manufacturer's instructions as a guide, but run a check by actually catching the granule discharge per row from a measured distance. The feeder-gate control on a granule applicator is usually a dial or lever that regulates the rate of flow of the granules. On band application equipment, raising or lowering the spreader height will change the width of the band. Adjust the height to give the desired band width.

Variation in size and density of granule particles makes a different rate setting necessary for each chemical applied. Recommendations often call for

Figure 8:14. Row-crop granule applicator attached to corn planter. The three pair of white hoppers contain insecticide granules which are directed in a 7-inch band over the planting furrow by a fan-shaped nozzle attached to the end of a malleable tube leading from each hopper. Press wheels follow application of granules and compact the insecticide into the drill row. *Courtesy Noble Manufacturing Company, Sac City, Iowa.*

a specified number of ounces of formulation, for example 10G, per 1000 ft of row. With this type of recommendation row spacing width can be disregarded. Although kept constant per 1000 linear feet, pounds of granules per acre (or lb AI/acre) will increase as row spacing decreases.

Most granules applied to the soil should be at least lightly incorporated. This is frequently accomplished by incorporation wheels, chain drags, or press wheels. Proper rpm of the rotor blades inside the granule boxes is important. Too many rpm cause attrition, resulting in "fines" that affect the drift, flowability, distribution, and ultimately efficacy of the granules.

Major Problems

Although agricultural engineers have improved application equipment over the years, major problems connected with spray drift, deposition, and effectiveness related to droplet size still require solution. Drift—that part of the spray which moves outside of the intended treatment area—can be a serious hazard. It is composed of small drops whose movement is influenced predominantly by air currents rather than gravitational forces. Particles of a coarse spray (400-micron size) will drift about nine feet while falling 10 ft in a 3 mph wind. A fine spray or a dust (20-micron size) will drift 1109 ft under the same conditions, whereas the usual dusts and aerosols (10-micron size) will drift almost 1 mile.

Attention to several details of making the application will help mitigate the danger of drift. The best time to dust or spray is early morning or late evening when there is little air movement. In no case should application be made when there is a wind greater than 8 mph. If recommendations allow, use spray application rates of at least 10 gal/acre and do not exceed 30 psi of pressure. Higher pressures produce smaller droplets with consequent greater drift. Keep the nozzles as close as possible to the area being treated.

An obvious solution to the drift problem would be the development of a sprayer that would generate sprays composed of uniform drops and a size having a low potential for drift. At the present time all sprayers force the liquid out the nozzle orifice into a thin sheet that breaks up into filaments, and then into drops of different sizes, large drops interspersed with much smaller drops. Agricultural engineers are attempting to perfect means of separating or removing the driftable component of sprays from the atomizers of existing equipment. Other ideas to solve the drift problem are to apply insecticide in a foam or with a rotary brush machine. The latter device has been used experimentally to apply systemic insecticides to the stems of cotton plants. It is a precise and highly efficient method of application, but as yet it is limited to control of insects susceptible to present systemic insecticides.

A machine has been invented which simultaneously generates and pumps foam. Reports indicate that foam suspends the insecticide evenly on crops, eliminates drift, and significantly reduces the dosage needed to control insect pests. Foam clings to the plant for up to an hour before breaking down and allowing suspended insecticide to reach plant surfaces.

The efficiency of application equipment in placing the insecticide where it is needed is rather low. In an experiment to determine the relative efficiency of low-level (5 to 20 ft) aerial applications of ULV and emulsions of insecticide, it was found that 50 per cent of the ULV insecticide but only 10 to 18 per cent of the emulsion insecticide reached the target area, which was an uncovered ground surface. Greater difficulties exist in treating foliage, as there is little control over drops to obtain a desired pattern or deposition level.

Because present sprayers deliver a range of droplet sizes, entomologists have been unable to determine the best droplet sizes for maximum effectiveness of insecticides. Sprays have been characterized by the mass median diameter (MMD) of the drops, but this statistic has little value in finding out which size is the most effective. Research done with aerial sprays dispensed from several kinds of nozzles—Micronair, minispin, Bals spinning discs, Fischer spinning screens, and the common flat fan nozzle—revealed that the minispin produced more uniform droplets, but insect kill was no better with it than with the other nozzles tested.

A new device for achieving more uniform droplet size is the hypodermic-needle atomizer (Microfoil boom). Installed on a helicopter, the Microfoil boom produces drops of 800 to 1000 microns with needles of 0.33 mm internal diameter and in an airstream of less than 60 mi per hour. Because of the larger but fewer drops the Microfoil boom has been limited to the application of systemic herbicides.

SELECTED REFERENCES

Akesson, N. B., and W. A. Harvey, *Chemical Weed Control Equipment*, Calif. Agr. Exp. Sta. Circ. 389, 1948.

Akesson, N. B., and W. E. Yates, *The Use of Aircraft in Agriculture*, FAO Agr. Development Paper 94, 1974.

Heddon, O. K., J. D. Wilson, and J. P. Sleesman, *Equipment for Applying Soil Pesticides*, USDA Agr. Handbook 297, 1966.

Hough, Walter S., and A. F. Mason, *Spraying, Dusting and Fumigating of Plants* (New York: Macmillan, 1951).

Ingerson, Howard, "One Acre to Seventy a Day," *Amer. Fruit Grower*, 79 (3):10, 11, 48 (1959).

Irons, Frank, *Hand Sprayers and Dusters*, USDA Home and Garden Bul. 63, 1967.

Jenkens, R., T. Eichers, P. Andrilenas, and A. Fox, *Pesticide Application Equipment Owned by Farmers, 48 States*, USDA Agr. Econ. Report 161, 1969.

USDA, *Power Sprayers and Dusters*, USDA Farmers' Bul. 2223, 1966.

USDA, *Aerial Application of Agricultural Chemicals*, USDA Agr. Handbook 287, 1965.

USDA, *Farm Power Sprayer*, Pacific Northwest Cooperative Extension Bul. 23, 1958.

Yeomans, A. H., R. A. Fulton, F. F. Smith, and R. L. Busbey, *Respiratory Devices for Protection Against Certain Pesticides*, USDA ARS 33-76-2, 1966.

chapter 9 / **ROBERT E. PFADT**

INSECT PESTS OF SMALL GRAINS

Many insects—well over a hundred destructive species—find fields of small grains favorable places to room and board. Fortunately, only a few of these are major pests.

THE PESTS

The most destructive insect pest of wheat in the United States has been the **Hessian fly**, a mosquitolike midge introduced from Europe (Fig. 9:13). Its larvae feed on juices of wheat stems and occasionally on the stems of barley and rye. This insect damages wheat over a wide region from the Atlantic coast to the Great Plains and from the Canadian provinces to Georgia and South Carolina. Humid Pacific Coast areas likewise suffer its attack. In recent years the growing of resistant varieties has substantially reduced populations of this insect and mitigated its damage.

The **greenbug**, a species of aphid, is the most destructive pest of small grains in central and southwestern states. In one year it can cause a loss of more than fifty million bushels of small grain. Additional destructive aphids of small grains include the **corn leaf aphid, birdcherry oat aphid, apple grain aphid**, and **English grain aphid**. The **chinch bug** (Fig. 10:8), a serious pest of corn and sorghums, damages small grains too, building up highly destructive populations in the central and eastern United States.

Grasshoppers (Fig. 9:7), particularly the **migratory grasshopper**, injured small grains in North America as long ago as early pioneer days. Today, they are especially bad pests in the more arid regions west of the Mississippi River. A western insect, the **Mormon cricket**, periodically damages small grains.

The **armyworm** (Fig. 9:11), widely injurious throughout the United States and Canada, is a dangerous pest of small grains, other cereals, and grasses. Soon after aggregations of flying adults settle in grain fields and deposit their eggs, hordes of larvae hatch and begin to devour the crop.

The **cereal leaf beetle** (Fig. 9:1), a new insect pest of small grains in the United States and Canada, has caused much concern among growers. First

Figure 9:1. Life stages of the cereal leaf beetle: the adult, a recently molted larva, a larva with the usual covering of moist fecal material, two eggs on leaf section, and pupal cell from soil. *Courtesy USDA.*

identified in 1962 from collections made in Berrien County, Michigan, the beetle has spread southward to Kentucky and Virginia and eastward to the New England States. It is of Old World origin, having been known as a pest of small grains in Europe since 1737.

Several insects bore in the stems of wheat (Fig. 9:2). The most serious are the **wheat stem sawfly**, the **wheat jointworm**, and the **wheat strawworm**. The sawfly injures crops in the northern Great Plains states and in the Great Plains provinces, the jointworm in wheat-growing states east of the Mississippi River, and the strawworm in almost all wheat-growing states and provinces. Of somewhat less importance are **wheat stem billbugs**, the **wheat stem maggot**, the **European wheat stem sawfly**, the **black grain stem sawfly**, the **lesser cornstalk borer**, and the **stalk borer**.

A number of important insect enemies of small grains lurk in the soil. They include wireworms, false wireworms, white grubs, and cutworms. **Wireworms** (Fig. 9:19) are most severe in the northern Great Plains states, the Pacific Northwest, and in western provinces; **false wireworms** in arid western states; **white grubs** in Kansas and Oklahoma; cutworms throughout the United States and Canada. The **pale western cutworm** (Fig. 9:3), **red-backed cutworm**, and **army cutworm** periodically destroy large acreages of small grains.

Figure 9:2. Three species of insects that bore in the stems of wheat. A, wheat jointworm; B, wheat strawworm within stem of wheat; C, wheat stem maggot and damage to wheat stem. *A and B, courtesy USDA; C, courtesy Kan. Agr. Exp. Sta.*

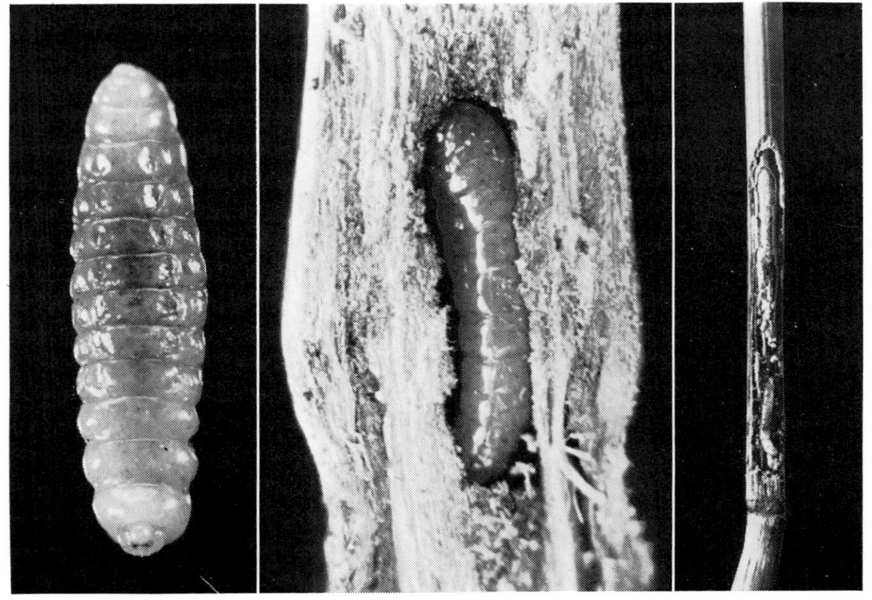

Figure 9:3. The pale western cutworm, a serious pest of small grains in western North America. *Courtesy University of Nebraska, Lincoln.*

Several species of mites attack wheat. Four of the most destructive are the **Banks grass mite** and **brown wheat mite**, which cause injury in the West, often during droughts; the **winter grain mite**, which damages small grains during cool moist weather in the Southwest; and the widely distributed **wheat curl mite**, which transmits the virus of wheat streak mosaic.

THE INJURY

Insects injure small grains in the field from the time the seed is planted until the grain is harvested. Both wireworms and false wireworms feed on the planted seed. In dry soil they even destroy a crop before rains stimulate germination and growth. After the grain germinates, these pests devour the tender sprouts just as they push out of the seeds.

Wireworms also kill seedlings by boring into and shredding the underground portion of stems. Wireworms and white grubs feed on roots and sever them from the plant. Cutworms, white grubs, and false wireworms cut off young plants near soil level (Fig. 9:4). In addition, the wounds left by soil pests allow rot pathogens to enter the plant.

Both larvae and adults of the cereal leaf beetle feed on the upper surfaces of leaves, biting lengthwise between leaf veins. Adults chew through leaf tissues completely, splitting the leaves lengthwise; the larvae seldom gnaw through the lower surface but consume greater areas than adults. In spring, winter wheat is mature enough to tolerate feeding and still yield grain, but plants of oats and barley sown in spring are often too young to withstand the injury. Severe feeding injury reduces yields up to about 25 per cent in winter wheat but may cause the complete destruction of spring-grown oats and barley. Detailed studies in Michigan and Indiana have identified the nature

Figure 9:4. Destruction of young wheat plants by pale western cutworms begins at the edges or on the knolls of fields and may continue until all plants are killed. *Courtesy University of Wyoming, Laramie.*

and extent of injury to oats by the cereal leaf beetle. Feeding of the larvae causes a reduction in yield by reducing kernel weight, kernel number, and straw weight. The last is important because straw is worth more than half the value of the grain.

Grasshoppers, Mormon crickets, and armyworms may devour young plants completely; they may strip the leaves from older plants, feed on maturing heads, or cut through the stems below the heads.

By injecting toxic secretions while feeding, Hessian fly larvae retard or kill seedlings and reduce the yields of older plants. Weakened stems of older plants are likely to cause the crop to lodge. The wheat stem sawfly, wheat jointworm, and wheat strawworm bore within culms and obstruct the flow of sap. This damage reduces the number and weight of kernels. Boring insects also cause grain to lodge.

Insects such as the chinch bug and various aphids impoverish plants by sucking juices from leaves or stems. Moreover, they produce fatal necroses by injecting toxic saliva. Small grain pests may transmit serious plant diseases, such as **wheat streak mosaic** by the wheat curl mite, **striate** by the painted leafhopper, barley yellow dwarf by several species of aphids, and aster yellows disease of barley by the aster leafhopper.

Barley yellow dwarf, a new and widespread virus disease of cereals characterized by leaves rapidly turning light green and yellow, beginning at the tips, is causing much concern in the United States and Canada. Transmission of the virus is solely by aphids. Several species have already been incriminated, including the English grain aphid, birdcherry oat aphid, corn leaf aphid, greenbug, rose grass aphid, and bluegrass aphid. Recent research has demonstrated that different strains of the virus are present and that they are usually carried by different species of aphids. Vector specificity, however, is not absolute, for in some cases one aphid species may transmit the strain usually carried by another.

In 1959 yellow dwarf losses of spring oats ranged up to more than one third of the crop in the northcentral and northwestern states and in 1960 losses ranged from 10 to 50 percent in infected fields in the northeastern states. Amount of injury depends on the stage at which the plants become infected. Young plants are frequently killed, older plants are stunted. When infection occurs in late stages of development, only the flag leaf exhibits signs of yellowing or reddening. As with other virus diseases, yield of grain shows greatest reduction when plants become infected early in their growth.

Recently plant pathologists have discovered that **aster yellows** disease of barley, caused by a mycoplasmalike organism and transmitted by the aster leafhopper, results in heavy losses of grain. The early foliage symptoms of this plant disease and of barley yellow dwarf are similar; leaves develop chlorotic blotches which intensify in color and eventually coalesce, resulting in a general yellowing. Circumstantial evidence indicates that aster yellows occurs in barley and to a lesser extent in other small grains every year. Considerable barley was probably infected in 1957, when this disease was very destructive to flax in the northcentral states and prairie provinces. Barley is more vulnerable than flax because it is a more attractive host to the aster leafhopper, which attacks this crop earlier and in greater numbers, and because the vector readily acquires and transmits aster yellows from barley to barley and other hosts, whereas it infrequently acquires and transmits the disease from flax to flax.

CULTURAL CONTROL

Because of the low acre value of small grains and because small grains can tolerate a moderate amount of insect injury with little loss of yield and quality, growers rely heavily on cultural methods to control insect pests. The methods fall into six broad categories: promotion of vigorous plants and stands; rotations; clean culture; tillage, timing of planting or harvesting; and planting early maturing and resistant varieties.

Vigorous stands more easily withstand the attacks of Hessian flies, greenbugs and other aphids, wheat jointworms, and wireworms. Against these pests, the method owes its effectiveness to the ability of strong plants to tolerate injury better than weak ones. It also operates successfully against chinch bugs, but somewhat differently. These insects prefer thin stands; they seldom congregate in heavy stands because damp, shady places are unfavorable habitats. Clover grown with the grain provides an even less favorable chinch bug environment.

Rotation of small grains with alfalfa, sweet clover, soybeans, flax, mustard, and other plants reduces the population of false wireworms, white grubs, brown wheat mites, and wheat stem sawflies. Control mainly results from substitution of a plant upon which the insect starves or refuses to oviposit. Rotation methods need not be applied regularly; growers may need only to plant substitute crops when danger of an outbreak looms. For example, they may sow sorghums in place of wheat when grasshopper populations are increasing. Crop grasshoppers relish wheat but practically refuse sorghums as food. To control wheat strawworms, plant wheat at a distance of 65 yd or more from stubble or straw because the spring form of the adult is flightless and cannot travel very far to infest young wheat.

Survival of large, injurious populations of insect pests depends on sufficient food remaining available through their entire period of growth and reproduction. When **clean culture** destroys food, pest populations fall to insignificant levels. Insect pests of small grains maintain themselves chiefly on volunteer grain or in stubble from time of harvest until the next crop appears. Plowing under stubble and volunteer grain aids greatly in keeping pest populations down. Also helpful is the destruction of weeds on which pests may feed. Clean culture aids in the control of Hessian flies, greenbugs, wheat curl mites, false wireworms, wheat strawworms, stem sawflies, and grasshoppers.

Tillage methods may bury egg, larval, or pupal stages so deeply in the soil that the insects die before they can emerge from the surface. Deep plowing is effective against the egg stage of grasshoppers and the larval and pupal stages of Hessian flies, wheat jointworms, wheat strawworms, stem sawflies, and others. Shallow tillage may destroy pests by crushing the insects or by throwing them to the surface where they are exposed to enemies and unfavorable weather.

In some areas cultural control methods that require plowing expose soil to dangers of water and wind erosion. When faced with this problem, growers may use the cultural method only when severe infestations threaten heavy losses. Plowing may also interfere temporarily with growing red clover or alfalfa in grain; however, farmers may plant other crops such as soybeans and cowpeas after grain harvest.

Timing of planting or harvesting may control small grain enemies. In the fall, Hessian flies infest early-planted winter wheat but live too briefly as adults to infest late-planted wheat. Entomologists have determined for various regions the dates on which to plant wheat to escape Hessian fly damage. These periods usually coincide with the planting times that agronomists have found result in highest yields. By harvesting early, before stems fall to the ground, one can reduce wheat stem sawfly losses. Because larvae actively girdle stems of spring wheat in August, the longer one delays harvest during this month the greater is the loss from lodging.

The planting of **resistant varieties** is one of the more promising methods of insect pest control. Varieties of hard winter wheat (such as Gage and Larned) and soft winter wheat (such as Arthur and Benhur) are resistant to Hessian flies. Several varieties of wheat have been bred with solid stems to resist the wheat stem sawfly. Spring wheats that do so are Rescue, Chinook, and Cypress, developed in Canada, and Sawtana, Fortuna, Lew, and Tioga, developed in the United States. Fortuna is the first variety to combine both sawfly and rust resistance. Rego is a hard red winter wheat with sawfly resistance. There are several small grain varieties resistant to greenbugs and to chinch bugs.

Research on resistance of wheats to the cereal leaf beetle has revealed that dense leaf pubescence is a deterrent to oviposition and exposes those eggs which are deposited to desiccation, resulting in 90 per cent mortality. This characteristic likewise allows less than 20 per cent of first instar larvae to survive.

Some varieties may escape insect injury by maturing early. Kansas growers have reduced infestations of wheat strawworm by growing such varieties as Pawnee and Triumph, which mature earlier than varieties formerly grown, such as Turkey and Blackhull.

CHEMICAL CONTROL

Cultural methods do not adequately control all pests of small grains, and sometimes they fail even against pests upon which they are usually successful. In either case we may be able to resort to a chemical method in an integrated control program.

Insecticides are effective against some enemies of small grains but are worthless against others. Although the new systemic insecticides may prove effective, the usual chemical treatments do not control important pests such as the Hessian fly, the wheat stem sawfly, the wheat jointworm, and the wheat strawworm. Surface-active insecticides, however, do control many serious pests such as grasshoppers, cutworms, wireworms, white grubs, aphids, and mites. Chemicals commonly used to control insect pests of small grains belong to two families of synthetic insecticides—the organochlorine and the organophosphorus compounds. Carbaryl, a carbamate, is also recommended for control of certain small grain insects.

Malathion or toxaphene is effective for control of grasshoppers; lindane as a seed coating is recommended to control wireworms; endrin, endosulfan, toxaphene, or trichlorfon controls cutworms; malathion or azinphosmethyl controls the cereal leaf beetle; demeton, dimethoate, disulfoton, malathion, methyl parathion, or parathion controls aphids; phorate controls Hessian flies, aphids, and grasshoppers.

Each year growers should obtain the new recommendations of their state agricultural college and extension service for controlling small grain insects. Recommendations change rapidly as research turns up new information on the effectiveness and the toxicology of insecticides. Follow label directions. Use recommended amounts; exceeding suggested dosages is not only illegal but is wasteful of chemicals and may be injurious to seeds or plants.

Treat seed that is plump and has high germination, because insecticides may injure poor seed. If diesel oil is used as diluent for insecticides applied by airplane, never put on more than one gallon per acre or apply it when temperature is above 75°F; burning of foliage may result. Diesel oil used on small grains after the boot stage may render the flowers sterile. For personal safety follow all precautions printed on the label of containers, particularly when using an organophosphorus insecticide.

CONTROL EQUIPMENT

Small grain growers commonly apply insecticides with low-pressure, boom-type sprayers; dusters are used less frequently. Power for this equipment can come from the power takeoff of a tractor or from a small gasoline engine. Trucks or tractors pull the equipment over the fields. While planting, growers may simultaneously apply insecticidal granules in the furrow by means of grass-seeder or granule applicator attachments on the seed drill. Granules of systemic insecticides provide control of several above-ground small grain pests. Because large acreages of small grains frequently need immediate treatment to stop the ravages of pests, growers often hire aerial applicators to treat the fields. Airplanes are rapid and have the advantage of not running down growing or maturing crops (Fig. 9:5).

Figure 9:5. Airplane spraying wheat field to control an infestation of insect pests. *Courtesy American Cyanamid Co., Princeton, N.J.*

Growers use seed or soil treatments to control small grain pests that live in the soil. Seed treatment is more economical and more widely practiced. Frequently, growers use an insecticide-fungicide mixture so that the treatment combats plant diseases as well as insect pests. First it is necessary to pass seed through a fanning mill to get rid of trash and light seed that would unnecessarily dilute the chemical. Insecticides may be applied as dust, slurry (a thin paste formed by addition of a little water to wettable powder), or liquid. Patented machines apply these materials efficiently. Because seed treaters are fairly expensive, growers generally have custom operators clean and treat their seed. Less efficient methods of treatment include mixing grain and dust with a shovel, mixing them in the drill box, or treating seed with a homemade machine.

REPRESENTATIVE SMALL GRAIN INSECTS

For detailed study we have chosen the following serious pests of small grains: grasshoppers, armyworms, Hessian flies, greenbugs, and wireworms.

Migratory Grasshopper. *Melanoplus sanguinipes* (Fabricius)

Differential Grasshopper. *Melanoplus differentialis* (Thomas)

Twostriped Grasshopper. *Melanoplus bivittatus* (Say)

Redlegged Grasshopper. *Melanoplus femurrubrum* (De Geer)

Clearwinged Grasshopper. *Camnula pellucida* (Scudder)

Bigheaded Grasshopper. *Aulocara elliotti* (Thomas)

[Orthoptera:Acrididae]

Although grasshoppers cause injury over all of the United States and much of Canada, they reach greatest numbers and do the severest damage in areas with average annual precipitation of 10 to 30 in. (Fig. 9:6). About six hundred native species of grasshoppers inhabit the two countries. Five species are responsible for 90 per cent of the total grasshopper damage to cultivated crops, whereas more than 25 species injure range and pasture lands. Grasshoppers especially relish small grains but readily devour other crops such as corn, alfalfa, clover, soybeans, and flax. Indeed, there are few cultivated crops that do not need protection during an outbreak of grasshoppers.

Although grasshopper damage consists primarily of the destruction of seedlings or the gradual defoliation of older plants, it may occur in other ways. Grasshoppers may feed on particular parts of plants and cause injury that far exceeds estimates based on the number of the attacking pests alone. For example, by biting through the stems of small grains, they sever the heads; by feeding on the ripening kernels, they cause extensive shattering. Grasshoppers' feeding on corn silks prevents fertilization and filling of the ears. The preference of grasshoppers for the blooms of plants causes considerable loss of legume and vegetable seed crops. Furthermore, we know

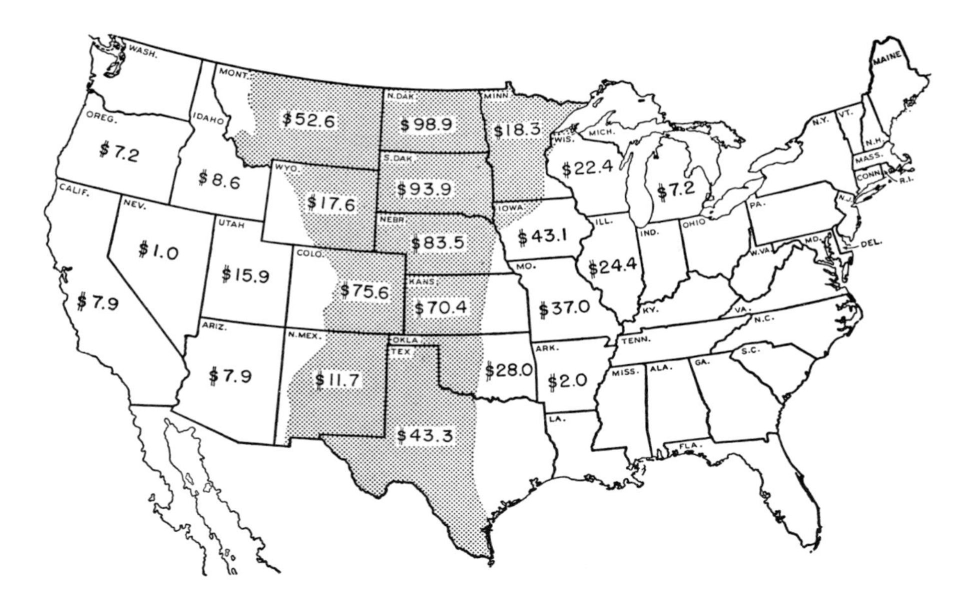

Figure 9:6. Estimated grasshopper damage to crops during the 25-year period, 1925 to 1949, by state. Heaviest losses took place in areas with average annual rainfall of 10 to 30 in (shaded region on map). *Courtesy USDA*.

that grasshoppers can transmit plant diseases such as potato spindle tuber, turnip yellow mosaic, turnip crinkle, tobacco mosaic, and tobacco ringspot.

Injury to range and pasture vegetation parallels somewhat closely the injury to crops. The main damage is the defoliation or destruction of the plant. Heavy infestations of 25 to 75 grasshoppers per square yard may completely destroy all vegetation in an area. This overgrazing leaves the soil exposed to erosion by wind and water. Some of the worst "dust bowls" have followed grasshopper outbreaks. Aggravating the damage, very small numbers of grasshoppers may destroy seedling grass, which soil conservationists plant to revegetate the range.

Of all the injurious species of grasshoppers in North America, the migratory grasshopper is the most serious (Fig. 9:7F). This pest ranges from Nova Scotia south to Florida, east to the Pacific Coast, and on into Mexico. It prefers light, well-drained soils. In the North this species has one generation annually, but in central states it may have two generations, and in southern Arizona three. It is particularly destructive to small grains and alfalfa. In winter wheat growing areas, the seedling plants of fall-seeded, small grains are especially attractive to first-generation adults and second-generation nymphs. The grasshoppers move into these green fields from stubble, weedy margins, or pasture lands and destroy row after row of plants as they penetrate deeper into the field. The migratory grasshopper is also a serious

Figure 9:7. Six species of injurious grasshoppers (not to scale). A, redlegged grasshopper; B, twostriped grasshopper; C, clearwinged grasshopper; D, differential grasshopper; E, bigheaded grasshopper; F, migratory grasshopper. *D, courtesy Shell Chemical Company; others courtesy University of Wyoming, Laramie.*

pest on weedy ranges and pastures. It is a difficult species to control because it deposits its egg pods throughout fields as well as in the margins and because it migrates extensively.

The differential grasshopper (Fig. 9:7D) is a serious pest of corn, soybeans, clover, and alfalfa, but under outbreak conditions it also destroys small grains and other cultivated crops. It prefers heavy soil and relatively rank vegetation. Depositing eggs in the crowns of grass, it concentrates them along roadsides and field margins. The species ranges widely in Mexico and in the United States, but it is found only occasionally in large numbers north of the southern borders of Minnesota, North Dakota, and Montana.

The twostriped grasshopper (Fig. 9:7B) is a common species throughout Canada and the United States. It is a serious pest of grains, forage legumes, and pastures.

The redlegged grasshopper (Fig. 9:7A), preferring low, comparatively moist ground, inhabits field margins, roadsides, pastures, and meadows. It is especially destructive to clovers and alfalfa. The species ranges from central Mexico, through all of the United States, to the interior of Canada. A subspecies, the southern redlegged grasshopper, *Melanoplus femurrubrum propinquus* Scudder, is the most abundant grasshopper in the southeastern states.

The clearwinged grasshopper (Fig. 9:7C) inhabits the northern states and the Canadian provinces from the Atlantic to the Pacific coasts. Its favorite haunts are mountain meadows, pastures, and roadsides. It is injurious chiefly to range and pasture grasses, but during outbreaks it also damages small grains.

The bigheaded grasshopper (Fig. 9:7E) is a grassland species occurring widely in western North America from the Canadian provinces south into Mexico. It is primarily a pest of rangeland grasses but occasionally damages small grains.

Description. The migratory grasshopper, like most other grasshoppers, deposits its eggs in the ground in clusters. A gluelike secretion holds the eggs together and also binds soil particles around the eggs. The eggs thus become enclosed in a small case called an egg pod (Fig. 9:8). Pods of different species

Figure 9:8. Egg pod and three loose eggs of the migratory grasshopper. *Courtesy University of Wyoming, Laramie.*

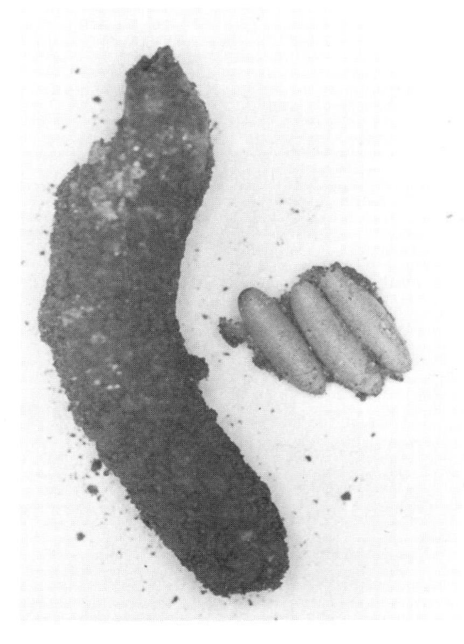

are of different sizes and shapes and contain different numbers of eggs. The egg pod of the migratory grasshopper is about 1 in. long and contains an average of 20 cream-colored eggs, each egg being about 4 mm long.

Newly hatched nymphs are pale yellow but soon turn dull yellow with dark brown markings. They are approximately 4 mm long and have somewhat the same appearance as adult grasshoppers except for smaller size and lack of wings. Adults are approximately 1 in. long and are light brown with dark markings.

The differential grasshopper is a large species of the genus *Melanoplus*; adult males are about $1\frac{1}{4}$ in. and females about $1\frac{1}{2}$ in. long. They are usually bright yellow with dark markings, though a small number in a population are melanistic. Dark chevrons on the hind femur are distinguishing marks.

The twostriped grasshopper, another large species of *Melanoplus*, is about the same size as the differential. The general color is greenish yellow or olive. Two light yellow stripes, from which this grasshopper gets its common name, run down the back from the head to the wing tips. The yellow, outside face of the hind femur has a conspicuous black dorsal band.

The redlegged grasshopper, a medium sized species of *Melanoplus*, is about $\frac{3}{4}$ in. long. The ventral surface of the body is bright yellow; the rest of the body is reddish brown. The species derives its name from the red hind tibia.

The clearwinged grasshopper is a medium-sized grasshopper; males are about $\frac{3}{4}$ in. long and females about 1 in. long. Body and front wing vary in color from yellow to brown and are marked with large, dark brown spots.

The bigheaded grasshopper is also a medium-sized grasshopper; males are about $\frac{3}{4}$ in. long and females about 1 in. long. General body color is grayish brown. The hind tibia is colored deep blue. This species is called the bigheaded grasshopper because the head appears large in relation to the size of its body.

Life history. Most species of economic grasshoppers have a single generation annually and overwinter as eggs in the soil (Fig. 9:9). The migratory grasshopper, however, completes as many as two generations a year in Kansas and three generations a year in southern Arizona. Another exception is the southern redlegged grasshopper, which has two and sometimes a partial third generation in Florida. Both species have the usual single generation each year in northern areas.

The eggs of grasshoppers, stimulated by warming soils and spring rains, hatch over a period of four to six weeks. Ordinarily, the various species hatch at different times. For example, the migratory and the twostriped grasshoppers hatch out two to three weeks earlier than the differential and redlegged grasshoppers. In addition to the built-in time clock of grasshoppers, the earliness or lateness of the season may vary the time of hatching by as much as two months. After emerging, the young nymphs feed on nearby green vegetation. Grasshoppers are selective in their choice of food plants. Unless forced by hunger, they do not eat just anything green as commonly supposed.

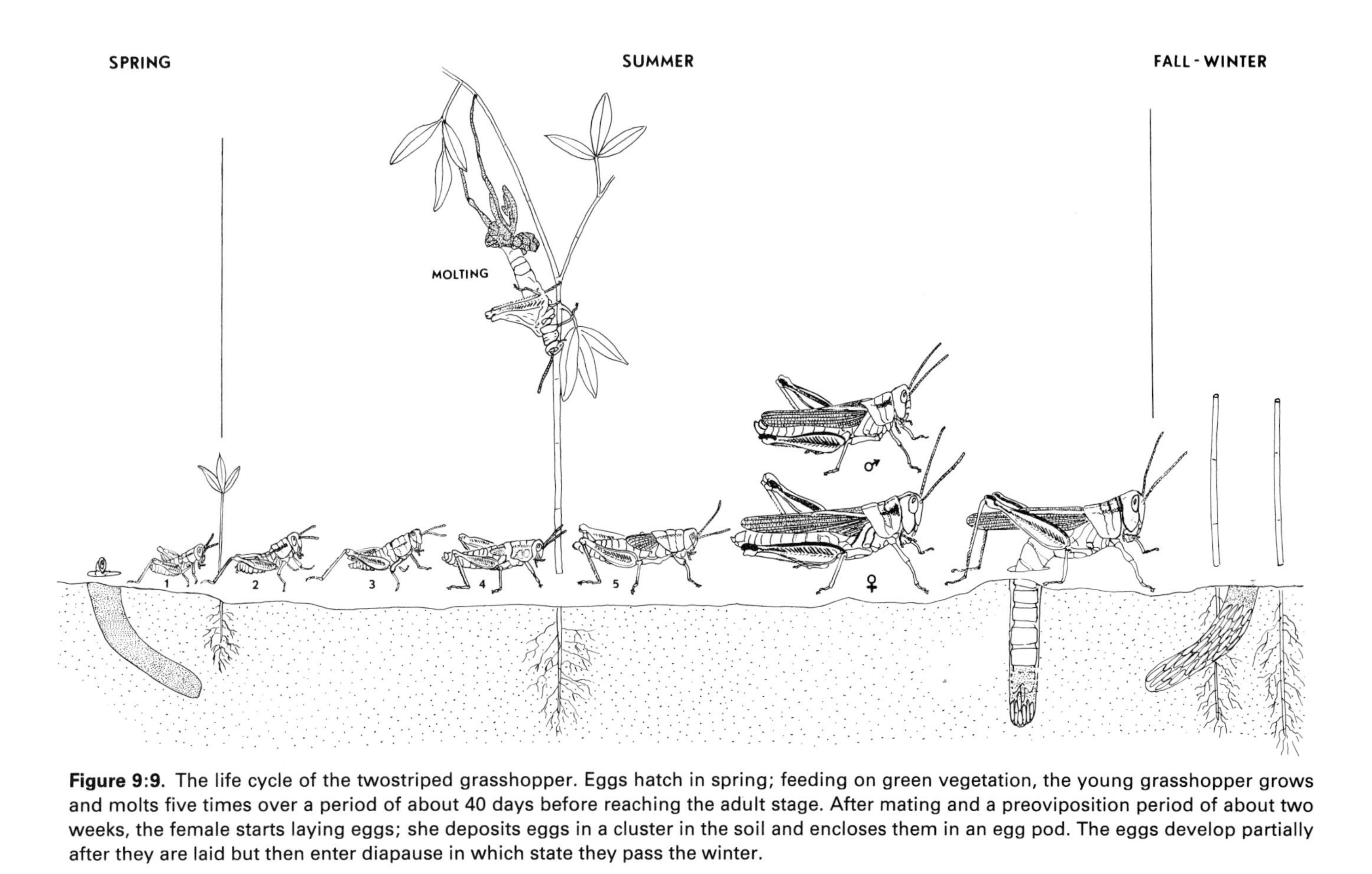

Figure 9:9. The life cycle of the twostriped grasshopper. Eggs hatch in spring; feeding on green vegetation, the young grasshopper grows and molts five times over a period of about 40 days before reaching the adult stage. After mating and a preoviposition period of about two weeks, the female starts laying eggs; she deposits eggs in a cluster in the soil and encloses them in an egg pod. The eggs develop partially after they are laid but then enter diapause in which state they pass the winter.

All six species of grasshoppers considered here seem to relish small grains, but they may differ in their preferences for other plants.

Nymphal development proceeds at a rapid rate when the weather is warm and not too wet. Grasshoppers go through five or six nymphal instars in 35 to 50 days to become full-fledged adults. After mating and passing through a preovipositional period of about two weeks, the female grasshopper produces her first cluster of eggs. She deposits them in the soil, the sites varying with the species. A female differential or twostriped grasshopper usually oviposits in sod or weedy ground bordering the crop upon which they are feeding. The female migratory grasshopper may deposit eggs over an entire field of grain stubble, alfalfa, weedy idle land, or rangeland. Female clearwinged grasshoppers may come long distances to congregate and lay eggs in sod, forming so-called egg beds. The bigheaded grasshopper deposits eggs in bare spaces between range vegetation.

The number of eggs in a cluster varies with the species. The pods of the bigheaded grasshopper contain from 6 to 10 eggs, those of the migratory and clearwinged 15 to 25, and those of the redlegged 25 to 30, whereas those of the twostriped and differential contain as many as 50 to 150. The grasshoppers that lay fewer eggs per pod usually produce more pods. Although a long-lived female migratory grasshopper may lay as many as 800 eggs, average egg production even under the most favorable conditions is much less. Based on the laboratory research of several entomologists, calculations show that it is possible for an average female migratory grasshopper to multiply herself 83 times each generation (one female produces 83 females of the next generation).

Populations of grasshoppers in both the nymphal and the adult stages do much moving and shifting about from field borders into fields and from one field into another. Mass flights of the migratory grasshopper may cover distances as great as 575 miles during a single season. The most spectacular and most injurious migrations of this species for over half a century befell the northern Great Plains states and Canadian provinces in the summers of 1938, 1939, and 1940. Flights started on clear days usually between 11:00 A.M. and 1:00 P.M. when temperatures approached 80°F, and winds came in gentle gusts. Flying with the wind, grasshoppers attained speeds of 10 or more mph. As evening drew near the swarms drifted slowly to the ground, infesting and destroying crops and range vegetation. When conditions again became favorable for flight, the hordes took to the air and moved on.

Few individual grasshoppers live long enough to realize their full reproductive potential because of the many natural enemies that constantly prey upon them. Some enemies attack the nymphs and adults; others prey upon the eggs. Robber flies, sphecid wasps, spiders, rodents, and birds feed upon the nymphs and adults, and the maggots of flesh flies, tachinid flies, and tangleveined flies parasitize these stages internally. Rodents and the larvae of bee flies, blister beetles, and ground beetles feed upon the eggs; the larva of

Scelio, a small wasp, parasitizes the egg. Other predators, parasites, and many diseases also attack grasshoppers.

Although predators and parasites help to keep grasshopper populations down and may terminate serious infestations, many entomologists believe outbreaks are caused not by any relaxation of enemies, but rather by favorable weather and an abundant supply of high-quality food during the nymphal and adult stages. Populations reach outbreak levels when the density of adult grasshoppers becomes 25 or more per square yard. In severe outbreaks numbers may reach as high as 80 or more per square yard.

The late J. R. Parker, America's famed student of grasshoppers, concluded that outbreaks develop slowly; grasshoppers double their numbers annually in a succession of three or four favorable years and then triple or quadruple themselves in an exceptionally favorable fourth or fifth year to initiate an outbreak.

Control. Good grasshopper control depends upon the employment of both cultural and chemical methods. Although recently-discovered insecticides make the protection of crops, pasture, and range much easier and more certain than in the past, they have by no means replaced cultural methods that still play an important role in grasshopper control.

Cultural methods include (1) elimination of weedy habitats, (2) planting resistant crops, (3) tilling land infested with eggs, and (4) timing of seeding.

1. Favorable habitats for the migratory, twostriped, differential, and redlegged grasshoppers are weedy field margins, fence rows, and roadsides. From these areas grasshoppers, during periods of moderate densities, work into the crop and cause material damage. Worst of all, these habitats and weedy idle land harbor the nucleus of an outbreak. We can eliminate weedy habitats by tillage, use of herbicides, and by perennial grass plantings.

2. During bad grasshopper years, we can cut losses by reducing the acreage of susceptible small grains and substituting resistant sorghums such as sorgo and kafir, which provide valuable feed for livestock. Grasshoppers rarely feed upon sorghums over 8 in. high.

3. The migratory grasshopper deposits many eggs throughout fields of small grains both before and after harvest. Plowing with a mold-board plow to a depth of 5 to 6 in., followed by packing, will cover up the eggs so that few nymphs can reach the surface. Either fall plowing or spring plowing before eggs hatch is helpful. In regions where there is danger of soil blowing, only spring plowing is advisable. Plow under egg-infested stubble land, planned for summer fallow, before eggs hatch in spring. Then maintain clean fallow through the summer to destroy host plants of grasshoppers such as volunteer wheat and various weeds and to keep the soil loose and unfavorable for egg laying.

Another method providing effective control is spring tillage that destroys all weed and volunteer growth just before or soon after grasshoppers begin to hatch. Most newly hatched grasshoppers starve to death before finding their way out of a field with no green food. The one-way disk, properly adjusted, is the most effective farm implement for this purpose.

4. Seeding spring small grains early is a recommended practice, as it gives the plants a chance to develop before grasshoppers hatch. Older plants can withstand a longer period of grasshopper feeding than younger plants. This practice allows the grower more time to obtain and to apply insecticides successfully. Furthermore, the sooner a grower plants and then harvests a crop, the less chance there is of migrating grasshoppers entering and destroying the grain.

Controlling grasshoppers with the newer insecticides is remarkably successful, provided growers discover infestations and apply treatment in time. Within three days after application 90 per cent of the grasshoppers die; most of the survivors succumb soon afterward. Recommended insecticides with dosage ranges of active ingredient to be applied per acre are the following: malathion 12 to 16 oz, toxaphene 16 to 24 oz, carbaryl 8 to 16 oz, diazinon 8 to 12 oz, mevinphos 4 to 8 oz, and naled 8 to 12 oz. In Canada dimethoate at 4 to 6 oz per acre and carbofuran at 2 oz per acre are also recommended.

These insecticides are applied most effectively as broadcast sprays with low-gallonage sprayers. Timing the treatment when grasshoppers are young, have not laid eggs, and are still confined mainly to the margins of fields, fence rows, roadsides, stubble fields, and idle land makes the job both cheaper and more effective. Unless more grasshoppers migrate into the treated area, control may last for several years.

If grasshoppers hatch in spring-seeded small grain, prompt treatment with an insecticide will prevent serious damage. When fields are soft, airplane application is necessary. In the fall of the year migrating nymphal and adult grasshoppers often menace fall-seeded small grains. These grasshoppers should be killed before the new wheat emerges. In protecting newly sprouted wheat, baits may be more effective because the plants at this time have little leaf surface to hold lethal concentrations of the insecticidal sprays. Or a grower may choose to apply a systemic insecticide to the soil at fall seeding. Granules of phorate at a rate of 1 lb AI per acre may be applied through a grass-seeder or granule applicator attachment into the seed furrow as the wheat is planted. Treating a 30 to 40 feet border of a field suffices, because in migrating later into the growing wheat, grasshoppers obtain toxic doses as they feed on the first rows of seedlings. Because growers plant wheat at different row spacings a more precise recommendation is to apply 10 per cent phorate granules at a rate of $2\frac{1}{2}$ oz per 1000 feet of row.

In the West large tracts of grassland are subject to periodical outbreaks of range grasshoppers. Economical control has been obtained through application of 8 fl oz of ULV (technical) malathion per acre by airplane. As the operation is timed precisely so that no egg laying takes place before treatment, control may last for periods exceeding five years.

Because grasshoppers are a national menace and migrate without regard to state boundaries, the federal government maintains an organization of grasshopper specialists to aid the states and individual farmers and ranchers in controlling these pests. A central office collects, analyzes, and publishes reports from fieldmen, entomologists who at one time were called "grasshopper supervisors." Almost every state in the bad grasshopper areas of the West employ grasshopper fieldmen. Their duties include making surveys of grasshopper adults in summer, eggs in fall, and nymphs in spring so that farmers and ranchers may learn of impending dangers and organize control. Because of the migratory nature of grasshoppers community action is vital for successful control of these insects.

Armyworm. *Pseudaletia unipuncta* (Haworth) [Lepidoptera:Noctuidae]

The armyworm, a widely distributed and injurious cutworm, lives in most regions of the world. In North America it destroys grass and grain crops over a wide area. Records of its destructiveness go back as far as colonial times; in 1743 a ravaging outbreak took place in the area now known as the North Atlantic States. Recent outbreaks in North America have occurred in the years 1937–39, 1953–54, 1964, and 1969; furthermore, almost every year somewhere on this continent local populations of the armyworm cause crop destruction.

An attack by armyworms is notable for its suddenness and severity. Aggregations of the adult moths, flying or wind-carried, alight in green fields of grain at night to deposit millions of eggs. Upon hatching, the larvae usually go unnoticed until the serious condition of the crop calls attention to their presence (Fig. 9:10). They may be so numerous, 30 to a square foot, as to devour a crop completely before a grower can apply control measures. As food becomes exhausted, the larvae assume a gregarious marching habit and crawl away to fresh fields.

Because armyworms prefer grasses as food they are particularly destructive to small grains, corn, and forage grasses. They may occasionally injure broadleaved crops such as alfalfa, clover, sugar beets, and tobacco. In attacking grains, they eat the succulent leaves first. Then as plants become stripped of foliage, they feed on other parts. In headed small grains, they may feed on awns and tender kernels and frequently will cut through the stem a short distance below the head.

Losses resulting from armyworm attacks reach astonishing proportions during outbreak years. In Kentucky during the 1953 outbreak, armyworms consumed around $10 million worth of crops and pastures on an estimated 1,085,000 acres. In Minnesota during 1954 armyworms caused a crop loss amounting to $12 million (Table 9:1).

Description. Female moths lay eggs in narrow bands on leaf blades or under the leaf sheaths of grasses and grains. The eggs are minute, greenish white, and globular. Young armyworms are pale green. The full-grown armyworm is a conspicuously striped, yellow to brownish green cutworm about $1\frac{1}{2}$ to 2 in. long. The adult moths are pale brown to grayish brown with a

Figure 9:10. Laying back wheat plants exposes heavy infestation of armyworms attacking the crop. Note at lower right how leaves are being consumed. *Courtesy USDA.*

Table 9:1. Crop Losses from Armyworm in Minnesota in 1954 and Savings Resulting from Control

	Estimated Loss		Estimated Savings	
Crop	Bushels	Dollar Value	Bushels	Dollar Value
Oats	7,255,000	4,570,650	15,495,000	9,761,850
Barley	3,466,000	3,569,980	7,963,000	8,201,890
Wheat	963,000	2,070,450	4,697,000	10,098,550
Corn	1,296,000	1,788,480	8,228,000	11,354,640

wingspread of about $1\frac{1}{2}$ in. In the center of each front wing is a characteristic white spot (Fig. 9:11).

Life history. From Tennessee northward as far as New York the armyworms pass the winter as partially grown larvae in the soil or under ground litter in succulent stands of grass. No prolonged arrest of growth occurs at any stage, other than that caused by the direct action of adverse environmental conditions. In Canada the armyworm is apparently unable to overwinter; infestations arise from the migration of moths that are produced farther south.

When spring arrives those larvae that have overwintered successfully resume their feeding and developing. Upon reaching full growth they cease feeding for four days, after which they enter a pupal period that lasts about 15 to 20 days. The adults emerge in May and June. Mating takes place at night, the peak sexual activity occurring during the fifth hour after sunset. Although one mating is sufficient for the female to lay fertile eggs, usually multiple matings occur. Before ovipositing, females feed for seven to ten days on sweet substances such as honeydew, nectar, or decaying fruit. They lay their eggs at night in clusters of 25 to 134 on grass or grains, usually in folded blades or under leaf sheaths. An individual female has a life span of about 17 days and may produce as many as 2000 eggs.

After six to ten days of incubation the eggs hatch into small caterpillars that soon begin to feed on the leaf surface down to the parenchyma leaving the lower leaf surface a transparent membrane. Older larvae feed from the edge of the leaf and devour all leaf tissues. Their usual feeding time is at night, from dusk to dawn, but in cloudy weather they may eat during daylight hours. During the day they ordinarily hide in the crowns of plants or under surface litter.

Figure 9:11. Life stages of the armyworm. A, adult moths; B, eggs; C, larva; D, pupa. *A and B, courtesy Canada Dept. Agr.; C, courtesy USDA; D, courtesy University of Nebraska, Lincoln.*

The armyworm has six larval instars and requires three to four weeks to complete larval development. The last or sixth instar is the longest, about seven days. This instar consumes more than 80 per cent of all foliage eaten during the entire larval period.

Full-grown larvae pupate in flimsy cocoons under litter or in earthen cells 2 to 3 inches down in the soil. Following pupation, in August and September, the second swarm of adults emerges and oviposits on lush grass. The eggs hatch, and the larvae develop partially before winter sets in.

The farther south in the United States the armyworm resides, the more generations it has each year. In the central states it has three to four generations annually, whereas in the South it has five or more generations and all stages may be present and developing throughout the winter. During summer, populations of the armyworm are low in these southern regions. Recent research has shown that the armyworm has an upper temperature limit of 31°C, above which development of eggs and larvae is retarded, and mortality becomes prevalent especially among larvae of the last three instars. Summer temperatures above 31°C are common and could account for the decline in numbers at this time of year.

Hymenopterous and dipterous parasites and various diseases take their toll of armyworms. A recent five-year study of parasites in Louisiana revealed a parasitism of 35 to 39 per cent each year. Of the more than 20 parasites found, the small wasp, *Apanteles militaris* (Walsh), was the most common. It accounted for a high and variable mortality ranging from 5 to 24 per cent in different years. Outbreaks of the armyworm have been observed to follow years of drought which are detrimental to populations of *A. militaris*. Diseases, not yet identified, have infrequently eliminated large populations in Louisiana.

The source of armyworm infestations is still somewhat uncertain. A locality may raise its own armyworms when conditions are favorable, or it may be invaded by flying moths that originate in other areas. Outbreaks in the northern United States and in Canada depend upon moth flights from more southern areas.

Control. An essential requirement for effective and profitable control is the early discovery of infestations. Shrewd growers inspect crops regularly for armyworms as well as for other insect pests. Armyworms hide under ground litter and in plant crowns during the day. A danger sign is feeding injury to foliage—chunks eaten out of the leaves or entire leaves eaten away almost completely. The armyworm prefers low, wet areas covered by rank stands of grain or grain flattened by wind or hail. These areas are the first to harbor the pests and deserve careful examination.

Insecticidal sprays applied by ground equipment or airplane are effective in controlling infestations of armyworm. Recommended insecticides include toxaphene $1\frac{1}{2}$ lb, carbaryl 1 lb, malathion $1\frac{1}{4}$ lb, trichlorfon $\frac{3}{4}$ lb, or parathion $\frac{1}{4}$ lb of AI per acre. Do not use parathion or trichlorfon on rye—these insecticides are phytotoxic to this grain—and do not apply carbaryl to small grains after the boot stage.

Hessian Fly. *Mayetiola destructor* (Say) [Diptera:Cecidomyiidae]

The Hessian fly used to be the most destructive insect enemy of wheat in the United States. Widespread outbreaks of this pest occurred at irregular intervals that lasted from one to as many as six years. Local outbreaks caused considerable crop losses nearly every year.

The Hessian fly is an introduced insect, supposedly a stowaway from Europe in straw used for bedding Hessian troops during the Revolutionary War. Regardless of how it first arrived, the Hessian fly not only crossed the Atlantic some time in the latter half of the eighteenth century but several times in the nineteenth century. It now occurs widely in the United States and in sections of Canada, although it inflicts serious injury over a somewhat lesser area.

The Hessian fly has two main broods. One attacks winter wheat in fall, and the other attacks both winter and spring wheat in spring. The larvae alone feed on the wheat and cause damage. Current studies indicate that the injury results from the larval discharge of salivary secretions into the plant, which upsets its metabolism and growth, and that the injury caused by larval extraction of sap is relatively minor. Still the number of larvae infesting a plant determines amount of injury. The presence of one or two larvae on a plant may not be serious, but during severe outbreaks numbers may average 20 to 30 per plant and run as high as 80 on some individual plants.

Wheat infested before jointing takes on a characteristic stunted appearance. The leaves of an infested plant are shorter, broader, more erect and darker green than the leaves of an uninfested plant. Larvae may kill individual tillers or the entire plant. The leaves turn yellow and finally brown. Infested plants that do survive injury in fall are more susceptible to winter killing and to diseases. Hessian flies pave the way for invasion of fungi that cause crown and basal stem rot of wheat.

Wheat infested after jointing suffers a different kind of damage. The larvae attack the stems above joints, causing injury that interferes with the transfer of food to the developing heads. Infested culms yield 25 to 30 per cent less grain by weight than uninfested. Plants with larvae-weakened stems tend to fall down or lodge (Fig. 9:12). Lodged grain is lost grain, because binders or combines cannot pick it up.

In the United States losses caused by Hessian flies have gradually lessened as resistant varieties of wheat have become available for various sections of the country. In 1915, before any resistant varieties were grown, an outbreak of Hessian flies resulted in a loss of wheat amounting to $100 million. In 1945, the last year of general distribution of susceptible varieties the loss amounted to $37 million. At present, damage to small grains, mostly

Figure 9:12. Lodging of winter wheat resulting from infestation by spring generation of Hessian fly. *Courtesy Kan. Agr. Exp. Sta., Manhattan.*

winter wheat, averages around $16 million per year. During the last decade farmers in the eastern soft wheat region have planted chiefly Hessian fly resistant varieties. In Indiana over 95 per cent of the wheat acreage has been grown to fly-resistant varieties for the past eight years, keeping fly populations much below the level that causes economic damage.

In addition to wheat, Hessian flies also injure barley, rye, emmer, and spelt. The fly does not develop readily in rye and has never been found in oats.

Description. The eggs, laid in the grooves of the upper surface of wheat leaves, are about 0.5 mm long and visible to the unaided eye. Their surface is a glossy red, which deepens with age. The newly hatched larvae are about the same size and hue as the eggs, but the color changes to white within a few days. Full-grown larvae are glistening white. A translucent green stripe runs down the middle of the back where the stomach contents show through the integument. The puparium, or "flaxseed" so called because of its resemblance to the seed of flax—is dark brown and about 3 to 5 mm long. Adults are mosquitolike in form, black, and about 2.5 mm long. The abdomen of the female appears red because of the eggs showing through the body wall (Figs. 9:13 and 9:14).

Figure 9;13. A, female Hessian fly; B, eggs on wheat leaf. *A, courtesy Kan. Agr. Exp. Sta.; B, courtesy USDA.*

Life History. In the winter wheat areas of the United States the Hessian fly has two principal broods annually, one in spring and one in fall. The species passes the winter as larvae protected in puparia on stubble, volunteer wheat, and early seeded winter wheat. When warm, wet, spring weather arrives, the larvae enter the pupal stage for two to three weeks.

The flies emerge in March in southern states and in April or May in northern states. Emergence usually takes place early in the morning and mating occurs soon afterwards, often in less than an hour. Adults do no feeding, as far as is known. Females may begin laying eggs within 15 min after mating. They prefer to deposit eggs on young wheat plants, usually in the grooves on the upper surface of the leaves. Egg capacity varies with the size of each female and ranges from 30 to 485 eggs. Oviposition is completed in a short time, as females do not usually live longer than two or three days. Eggs hatch in 3 to 15 days.

The newly emerged larvae migrate down the leaves to feeding positions underneath the leaf sheaths next to the stems. In young tillers they migrate down to the crown of the plant and feed between the leaf sheaths below soil level, but on leaves of older plants, they stop just above joints. The larvae take from 12 to 25 hours to crawl to their feeding positions, and many die

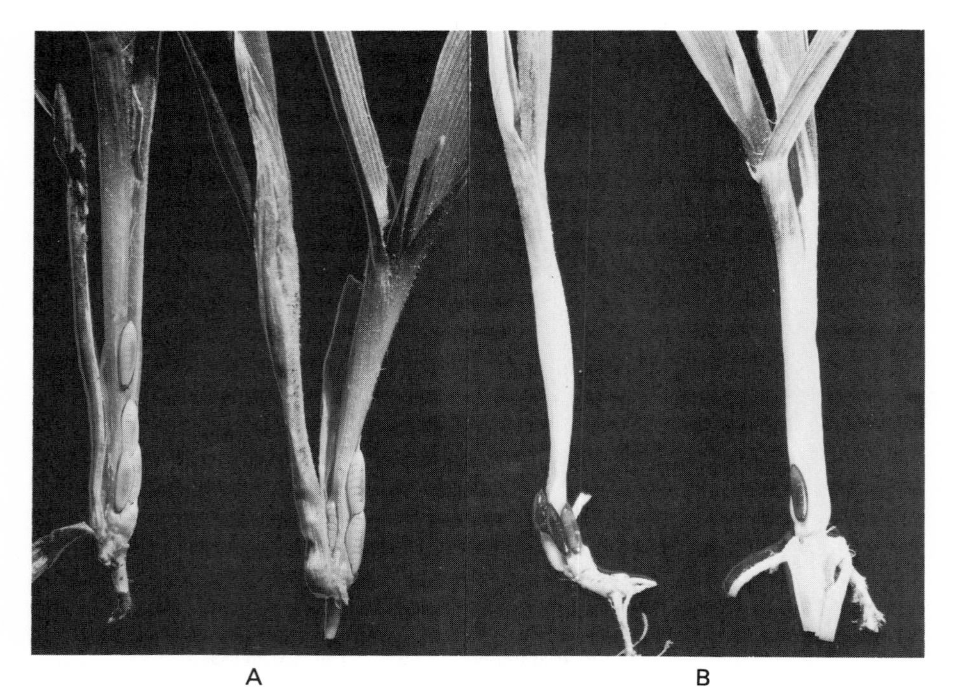

Figure 9:14. A, Hessian fly larvae feeding on stems of young wheat; B, puparia or "flaxseeds" on tillers of seedling wheat. *Courtesy Kan. Agr. Exp. Sta., Manhattan.*

before completing the migration. Conditions that cause high mortality are low relative humidity, wind, temperatures around freezing, and heavy rainfall.

The larvae feed on plant juices, which they obtain by cupping their mouthparts against the stem and sucking intermittently. The period of growth varies from two to six weeks, depending on temperature. When full grown, the larva contracts from the old larval skin, which hardens and becomes the puparium, or flaxseed. Within flaxseeds, larvae remain dormant during summer.

From the last days of August in the North to November in the South, the adults emerge from the flaxseeds and produce the eggs of the fall generation. Females lay eggs on volunteer wheat and early-planted winter wheat. Larvae hatching from these eggs migrate down the leaves to the crowns. There they develop to the flaxseed stage and pass the winter.

Not all flaxseeds of either the spring or the fall generation become adult flies during a single emergence period. Some remain dormant and emerge later with subsequent generations. For this reason we call the group of flies of approximately the same age and present at approximately the same time in a wheat field a "brood" rather than a "generation." In addition to the two main broods, favorable rainfall and temperature may foster supplementary broods. One supplementary brood may follow the spring brood, one may occur in summer, and one may follow the fall brood. During mild winters in the southeastern U.S. the Hessian fly reproduces continuous generations.

In California, in the Pacific Northwest, and in the northern spring wheat regions, the seasonal history of the Hessian fly varies from that described for the winter wheat areas of the United States. Differences in climate induce these variations. Active stages of the fly synchronize with the humid, moderately warm periods of the year. Dry, hot weather forces the species to estivate and, if prolonged, causes mortality among the flaxseeds. This combination of factors also prevents the Hessian fly from doing serious injury in the arid sections of the West and Southwest.

Investigations of resistant varieties of wheat have disclosed the existence of races or biotypes of the Hessian fly. So far eight races—the distinction is based on ability to infest different wheat varieties—have been isolated in the United States. The Great Plains race is prevalent west of central Kansas, and a race *A* is prevalent in the eastern soft wheat region. Other races that have been isolated are designated *B*, *C*, *D*, and *E*; they occur principally in the eastern soft wheat region. As varieties of wheat resistant to these biotypes are developed, their proportions change in Hessian fly populations. No doubt further search will disclose additional biotypes or strains based not only on varying ability to infest varieties of wheat but on ability to survive under local environmental conditions. Biotypes *F* and *G* have been developed in the laboratory, and a recent study has shown that Hessian flies of western Nebraska are adapted and better able to survive drier, colder, and windier conditions than flies from northeast Kansas, an area with a more moderate climate.

Thirty-five species of parasites attack the Hessian fly and play a part in controlling populations of this pest.

Control. The three most important methods of Hessian fly control consist of cultural practices: delayed planting to escape fall infestation, planting resistant varieties, and clean cultivation or good management of volunteer wheat. Wheat sown late enough will escape the egg laying of the fall brood of flies. Entomologists and agronomists have worked out average safe dates for planting wheat to avoid Hessian fly. These dates are approximate and vary from year to year depending on weather. Agricultural colleges each fall notify growers of the safe time for seeding. Modern farm machinery makes delayed planting practical because it allows farmers to plant much more quickly than was formerly possible. There are, however, disadvantages in late sowing of wheat in that the practice precludes use of the crop for grazing and for early control of soil erosion.

One should sow wheat as soon as possible after the safe date to allow the plant to make sufficient growth before winter sets in. In California adults and larvae are active only from February to May so that early planting, rather than late, reduces injury.

For various regions of North America, plant breeders have developed varieties of wheat that have both desirable agronomic characteristics and resistance to the Hessian fly. Resistant hard red winter wheat varieties adapted to the Western Plains include Gage, Larned, Ottawa, Parker, and Warrior. Resistant soft red winter wheat varieties adapted to the eastern soft wheat region include Abe, Arthur, Arthur 71, Benhur, Knox 62, Monon, Oasis, Redcoat, and Reed. Other Hessian fly–resistant varieties are available for these regions and also for less extensive wheat growing areas. The numbers of acres planted to established varieties change drastically as newer varieties, which may not always incorporate insect resistance, become available.

Although control of the Hessian fly by the planting of resistant varieties is one of the great achievements in applied entomology, the victory over this pest is not final. The insect is able through selection and possibly mutation to develop biotypes that can survive and multiply on resistant varieties. The eight biotypes that have been isolated differ from one another in their ability, or inability, to infest wheat having the H_7, H_8, H_3, or H_6 genes for resistance. One suggested way of continuing the successful control of the Hessian fly is to prolong the useful life of a resistant variety by breeding for tolerance and nonpreference in wheat as well as for antibiosis, as is being done under the present program. Another suggested way is to rear large numbers of a biotype with dominant genes for avirulence to resistant varieties and release them into the native fly population. In theory, releasing 19 Great Plains flies to every one native fly in the eastern soft wheat region may eradicate the wheat pests within five generations.

The third important cultural control method is the destruction by tillage practices of all volunteer wheat as soon as it sprouts and the destruction of other host plants, chiefly grasses that grow as weeds, in and around wheat fields. Other cultural control methods include plowing infested stubble to bury and destroy the Hessian flies, rotation of wheat with other crops, and the adoption of recommended cropping practices.

A chemical method that provides a grower with an alternative plan of management is available for controlling the Hessian fly. It consists of applying 10 per cent granular phorate or disulfoton, systemic insecticides, evenly in the furrow at time of planting at the rate of 10 lb/acre (1 lb AI). Research results of treating wheat seed with carbofuran indicate that this method shows some promise as an economical control of the fall population of Hessian fly.

Greenbug. *Schizaphis graminum* (Rondani) [Homoptera:Aphididae]

The greenbug is a widely distributed aphid—not a bug—in North and South America, Europe, Africa, and Asia. It is a serious pest of oats, barley, and wheat in this country, particularly in the central states from Texas to North Dakota and Minnesota. Infestations may also extend into Manitoba and Saskatchewan. Heavy numbers of greenbugs cause total destruction of both winter and spring grains. Entomologists now consider the greenbug the most destructive insect of small grains in the Great Plains area.

Severe greenbug outbreaks occurred for the first time on sorghum crops in the Southwest and in Kansas, Nebraska, and South Dakota in 1968. The aphids (Biotype *C*) infested several million acres of grain sorghum in all stages of plant growth. This biotype is also very destructive to small grains and has become the prevalent form in the Midwest, Southwest and Northwest.

Infestations of greenbug develop in two ways. First, in southern states as in Texas and Oklahoma, small isolated aggregations of greenbugs may begin to increase and kill out circular areas of wheat a few feet in diameter. As the grain dies, the aphids move out to healthy wheat in increasing numbers until eventually the dead areas of wheat join. **Greenbug spots** have a characteristic appearance. The inner area consists of brown, dead plants surrounded by a circle of bright yellow plants. Immediately outside of this circle the grain is green. Second, infestations may develop as an invasion of an entire field by swarms of migrating greenbugs. If this happens, all plants become injured and die at about the same time.

Light or moderate infestations of greenbugs do not cause total destruction but do cause measurable reduction in yield. In fall, greenbugs may thin stands and prevent tillering of winter small grains; in winter and in early spring, they may cause greenbug spots; in spring, they may stunt the heads by feeding inside the boot leaf (Fig. 9:15).

Greenbugs injure plants directly by injecting saliva and by sucking up juices. Powerful enzymes in the saliva alter the cells and their contents and eventually kill the living tissues. Leaf injury is evident as yellow spots with necrotic centers. Greenbugs may also act as vectors of plant diseases; they are able to transmit the viruses of sugarcane mosaic and of barley yellow dwarf.

Growers can detect greenbug infestations in sorghum by looking for reddish spots on the leaves caused by small colonies of the aphids feeding on the underside of a leaf. As colonies grow, the reddened leaf areas enlarge; eventually the leaf begins to die, turning brown from the outer edges toward the center.

The first report of greenbugs in North America came from Virginia in 1882. Since then the country has suffered 22 outbreaks. One of the most severe centered in Texas and Oklahoma in 1942. Losses in these states totaled more than 61 million bushels of grain valued at $38 million. Besides destroying grain, greenbugs reduce the pasture value of small grains. Although the greenbug is primarily a pest of oats, barley, and wheat, it also feeds on other small grains, corn, rice, sorghum, and forage grasses.

Entomologists studying the biology of the greenbug have discovered differences among populations of this insect and have designated four biotypes, *A*, *B*, *C*, and *D*. Biotype *A* is the original greenbug, against which plant breeders have developed resistance in certain lines of wheat, barley, and oats. Biotype *B* was discovered in 1961 in greenhouse cultures on barley and

Figure 9:15. Greenbug injury to wheat. Normal head on left; others stunted by feeding of greenbugs when heads were within the boot. *Courtesy Kan. Agr. Exp. Sta., Manhattan.*

became prevalent in the field in 1965. It does not differ morphologically from biotype *A*, but all wheats are susceptible to *B*. Biotype *C* is new. It appeared suddenly in 1968 as a severe pest of sorghum. It was found first in Nebraska in May, and by mid-July it was present in most of the states west of the Mississippi River. The sorghum greenbug differs from the first two biotypes in its efficient utilization of sorghum and johnsongrass as host plants, as well as small grains; by its tolerance of high temperatures up to 110°F; and by its morphology. It is light yellow in color, the cornicles are entirely yellowish green, and it has more antennal sensoria. Reared in cages, the sorghum greenbug produces 10 per cent males and a proportion of egg-laying females. The facts indicate that it is probably a recent introduction from the Mediterranean region where similar biotypes are pests of sorghum. Biotype *D* was first noticed in 1975 in Texas as a race of greenbug resistant to disulfoton and more tolerant of other insecticides. It is morphologically indistinguishable from *C*, from which it probably developed after continuous exposure of populations to insecticides.

Description. Forms of the greenbug usually seen are winged and wingless females and their young, produced parthenogenetically and viviparously (Fig. 9:16). The wingless female is 2 mm long and pale yellowish to bluish green with a darker green dorsal stripe running down the back. The winged female is slightly smaller but is similar in body color to the wingless female; it is distinguishable from other cereal aphids by the single fork of the median vein in the forewings (Fig. 9:17).

Sexual forms of this aphid do occur but are not commonly seen. The egg-laying female is wingless and similar in appearance to the wingless female described above; the male is winged, colored like the winged female but smaller, measuring about 1.3 mm long. The female lays one to ten eggs in the folds of grass leaves. The eggs are pale yellow when first laid but become shiny black by the end of the third day. They are kidney-shaped and about 0.8 mm long.

Life history. Of the four adult forms of this aphid, the wingless viviparous female occurs most abundantly and regularly. Individuals infest cereals in the green stages of growth and may produce three or four generations a month. Optimum temperature for reproduction is from 70 to 75°F, but greenbugs can reproduce at temperatures between 50 and 92°F. Under favorable

Figure 9:16. Greenbug stages. Winged (A) and wingless (B) female adults and eggs (C) attached to grass leaf. *Courtesy USDA.*

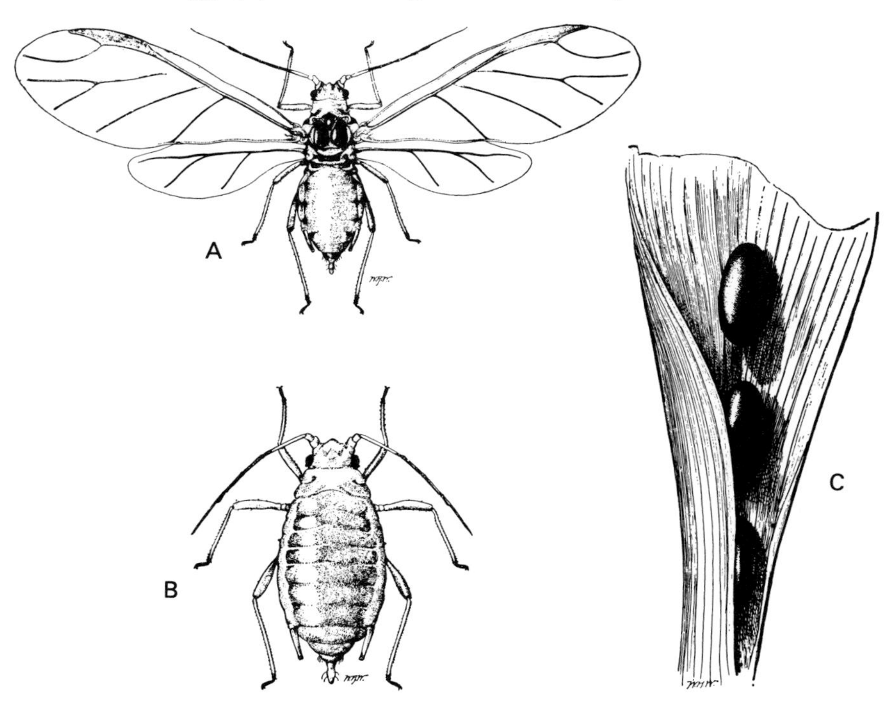

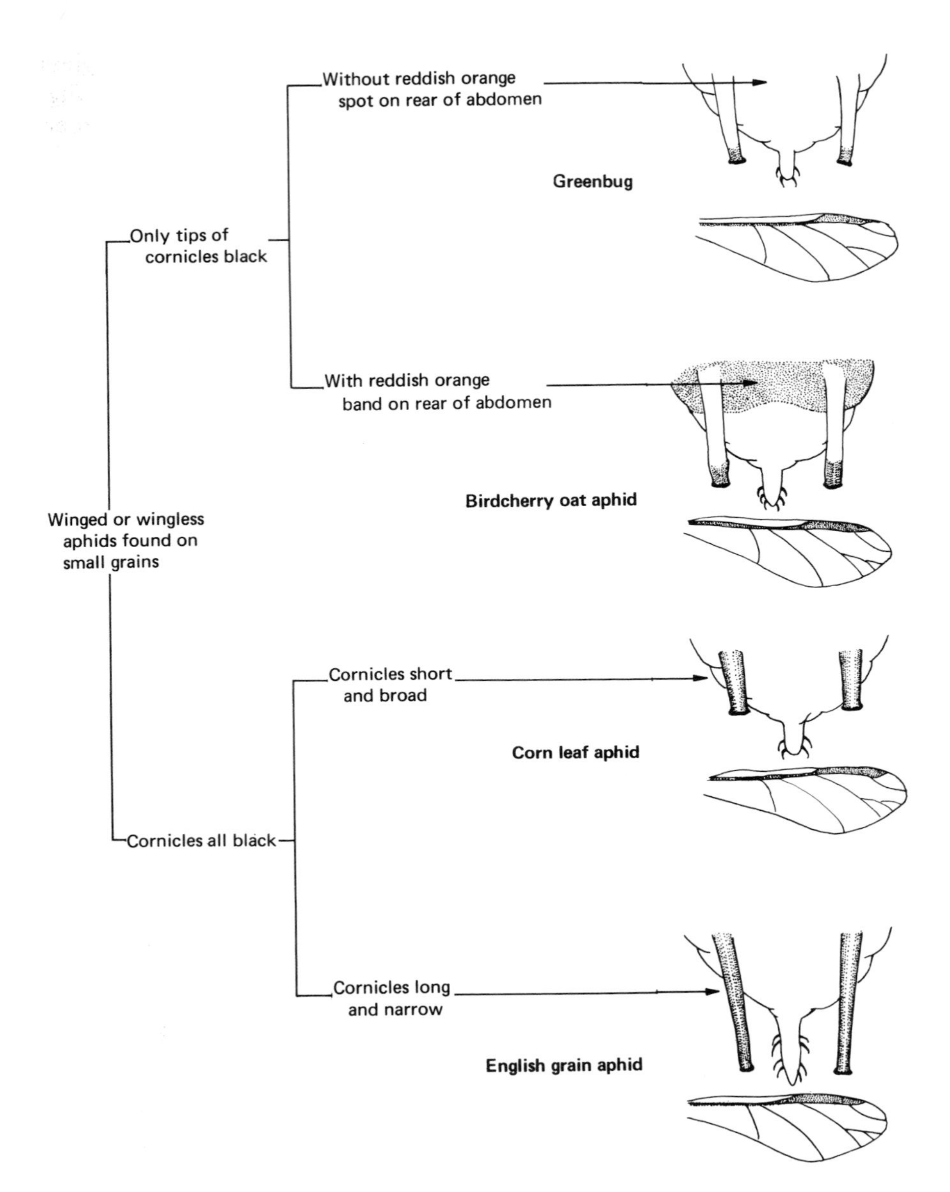

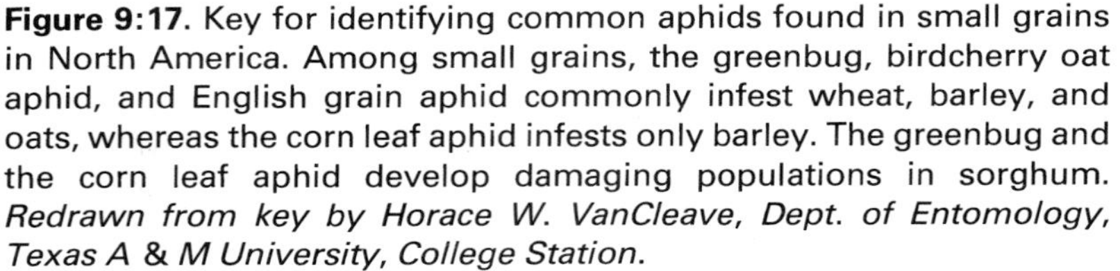

Figure 9:17. Key for identifying common aphids found in small grains in North America. Among small grains, the greenbug, birdcherry oat aphid, and English grain aphid commonly infest wheat, barley, and oats, whereas the corn leaf aphid infests only barley. The greenbug and the corn leaf aphid develop damaging populations in sorghum. *Redrawn from key by Horace W. VanCleave, Dept. of Entomology, Texas A & M University, College Station.*

temperatures the young pass through four nymphal instars in about a week. A few hours after becoming adult, wingless females start giving birth to living young which are all females. Young adults bear three or four offspring each day, but occasionally some may bear as many as ten in one day. An individual female of *A* or *B* biotype reared on a susceptible variety of barley is able to produce about 100 offspring during a reproductive period of 28 days. A biotype-*C* female under similar favorable conditions can produce 145 young.

Most progeny of wingless females are also wingless, but some are winged. Larger numbers of the winged form develop when the parent generation is subjected to poor nutritional conditions such as exist during periods of drought or in a maturing cereal crop. Other factors that may increase the proportion of winged females are variable temperatures averaging about 60°F, as in spring and fall, and day length of about $12\frac{1}{2}$ hours. The production of winged aphids is a means of dispersal and an assurance of species survival. When small grains begin to mature and dry out, large numbers of winged aphids develop and eventually migrate. They are carried long distances by the wind; the majority no doubt perish, but the lucky ones land on suitable host plants to perpetuate the species. As many as twenty generations of the viviparous females may develop during one season. Of biological interest is the discovery of paedogenesis in biotype *C*. While they are still immature (last nymphal instar), around 2 per cent of the winged females bear 1 to 4 young each.

Because of continuously low, cold winter temperatures the greenbug cannot survive in any of its stages in the northern states and in Canada. Although the sexual forms, induced by decreasing day length in fall acting upon the parent generation, occur in central states, the populations of economic importance consist of the colonies of viviparous females which survive and reproduce through the winter in southern states such as Oklahoma and Texas.

Outbreaks of greenbugs in the South correlate well with a succession of favorable weather conditions. The summer previous to an outbreak is cool and has excessive rain. Such weather delays harvest and causes the grain to lodge and shatter. The enormous volunteer grain crop that springs up provides a favorable habitat for greenbugs. This condition—followed by a mild winter, a normal spring, and a cool, wet summer—favors the production of huge numbers of aphids but retards their natural enemies.

Outbreaks in the North arise from the migration of winged aphids originating in southern grain fields. The migrants may leave wintering places in the South as early as March and April. Weather conditions in spring foster strong southerly winds that blow over maturing southern grain fields, where many winged greenbugs are active. The winds of 200 to 600 mi per day carry migrants to areas in the central states; heavy infestations may develop there. When the grain in these areas begins to mature in May, migrants are produced again in large numbers and carried by the southerly winds to northern states and Canadian provinces. Available evidence—

meteorological, phenological, suitability of host plants, and records of the presence of migrants—indicates that the invasion of northern areas takes place by a series of successive flights and generations.

Field studies have shown that winter infestations of greenbugs in Texas may arise in two ways, from populations that have lived through the summer as active aphids within the area or from populations that have migrated in fall into the area from the north.

Greenbugs have a number of natural enemies, which eventually counter increase of populations. A small wasp, *Lysiphlebus testaceipes* (Cresson), is one of the most efficient parasites. Female wasps deposit eggs within the bodies of aphids. The eggs hatch into larvae that feed on the internal organs of the hosts and kill them. The convergent lady beetle and other species of lady beetles are efficient predators in both their larval and adult stages. Other useful predators are damsel bugs and the larvae of lacewing and of syrphid flies. Unfortunately these parasites and predators are inadequate in preventing greenbug damage of cereals. During outbreaks, greenbug populations develop so rapidly that crop damage occurs before the natural enemies become effective.

Control. Both cultural and chemical methods are used to control greenbugs. Three cultural practices are feasible. One is to grow vigorous plants which, under attack of greenbugs, are injured less and yield more than weak plants. Agronomic practices that foster strong, healthy plants include preparing friable, firm seedbeds, planting good seed of suitable varieties, fertilizing where needed, regulating adequate soil moisture, and rotating crops systematically. Greenbugs injure barley, oats, and wheat sown in the fall more severely when they follow grain sorghum in rotation than when they follow soybeans, corn, or wheat. Providing more nitrogen for small grains, either by applying fertilizer or by rotating with legumes, reduces greenbugs and greenbug damage.

The second cultural practice is to grow resistant varieties of grain such as Kerr or Will barley in the Great Plains, or ERA barley in Texas. Many greenbug-resistant grain sorghums are now available commercially in a wide range of germplasm and maturities. One oat line, P.I. 186270, is resistant to all biotypes of the greenbug. The crossing of this line with Cimarron oats has yielded promising resistant progeny. There are no resistant varieties of wheat with suitable agronomic characteristics. The development of varieties that are resistant or tolerant to the greenbug is an active area of research, so that we may expect eventual relief from the ravages of this pest.

The third control practice is to destroy volunteer grains by discing or plowing. Volunteer grains provide shelter and food during the summer months for colonies of greenbugs; in this way they may constitute an important link in a chain of events that leads to an outbreak the following winter. Preventing growth of volunteer small grains is particularly advisable in Texas and Oklahoma, where serious outbreaks may originate from oversummering local populations.

When cultural and natural controls cannot cope with injurious populations, growers can protect their crops effectively with insecticides. Damage may occur in fall to seedlings of winter grain or in spring to growing plants of both winter and spring grains. For best results, spray with insecticides at the first signs of injury or follow local recommendations on threshold numbers, for example 50 aphids per foot of row on plants 4 to 6 in. tall, 200 aphids on plants 6 to 10 in., 300 aphids on plants 18 to 20 in. and 800 aphids on plants 30 in. and taller. Recommended insecticides include demeton, parathion, or methyl parathion at a rate of 4 to 8 oz per acre, or malathion at 1 lb per acre of active ingredient. Temperature at time of application of insecticide and for at least three hours afterwards should be 50°F or above.

Foliar sprays of parathion, malathion, demeton, or diazinon are effective in controlling greenbugs on sorghum. Damaging populations occur early on seedlings or later on larger plants. Growers may protect seedlings by applying granules of disulfoton, carbofuran, or phorate in a 4- to 6-inch band in front of the press wheel of drills at planting time; or treatment may be made after the seedlings have emerged by applying the granules in a narrow band over the rows.

The greenbug has begun to show insecticidal resistance in various sections of the country, particularly to dimethoate and disulfoton, two insecticides currently recommended for control of this pest.

Prairie Grain Wireworm. *Ctenicera destructor* (Brown)

Great Basin Wireworm. *Ctenicera pruinina* (Horn)

Corn Wireworm. *Melanotus communis* Gyllenhal

Wheat Wireworm. *Agriotes mancus* (Say)
[Coleoptera:Elateridae]

Wireworms, the larval stage of click beetles, comprise a large family in the order Coleoptera. Approximately 700 native species inhabit the United States and Canada. In the larval stage most feed on the roots and underground stems of grasses and forbs. A few live in decaying wood and may prey on small arthropods. Several species are important pests of small grains, corn, cotton, tobacco, potatoes, and many other crops.

The most serious wireworm injury to small grains occurs in the northern wheat-growing areas of North America. Several species are responsible for the losses. In the northern Great Plains states and in the western provinces of Canada, the prairie grain wireworm is the principal pest. This species prefers well-drained, light and medium soils of the open prairies but also inhabits irrigated sandy soil.

In the Pacific Northwest and in British Columbia, the Great Basin wireworm is the prominent soil pest of dry-land small grains. It lives only in regions that normally receive precipitation of 15 or fewer inches annually. In Canada's western provinces, *Hypolithus bicolor* Eschscholtz follows closely the prairie grain wireworm in importance. Widely distributed over open prairies and in parklands, it is more abundant in areas of heavy soil, particularly in meadows and in fields recently broken from sod.

Although wireworm damage to small grains is not as great in the central and southern wheat-growing areas of the United States, several species of wireworms do sometimes cause injury there. The wheat wireworm, an important pest of corn and potatoes, may also injure wheat. It occurs in northeastern and midwestern states and in the eastern provinces of Canada. It prefers poorly drained soil planted to meadows or sod crops. The corn wireworms, *Melanotus communis* Gyll, and others of the genus, are widely distributed in North America and may at times injure wheat. A minor wireworm pest of wheat distributed widely in the United States and Canada is *Aeolus mellillus* (Say). Many other species infest fields of small grains and may cause injury.

To learn about wireworms and their control in small grains, we shall consider in some detail the life history and control of the prairie grain wireworm. Larvae of the species feed underground on the seeds and the young seedling plants of wheat and other small grains in the spring of the year. After the plants have stooled out, wireworms no longer cause severe damage, but often by this time they have ruined entire stands.

Wireworms attack the crop as soon as the seeds are planted. In feeding, they hollow out the kernels of grain leaving only the husks. After unharmed seeds have sprouted, wireworms bore through leaf sheaths into the stems of young plants. This feeding destroys the growing point and kills the central shoot. After several days the leaves and roots also die. Stems injured by wireworms are not cut off but have a characteristic shredded appearance at the point of injury. The marks of wireworm attack in fields are bare areas of various sizes, general thinning of stand (Fig. 9:18), and reduced yields. Wireworm wounds may also allow disease organisms to enter plants and cause rots.

A survey of damage by wireworms in grain crops of Saskatchewan from 1954 to 1961 showed a decline in number of damaged wheat fields with more than 10 per cent thinning, a trend consistent with the view that insecticidal seed treatments have reduced population levels of wireworms. The proportion of fields showing no thinning, however, remained about the same—approximately 36 per cent—indicating that insecticides have not reduced the potential hazard of wireworms. Average damage was found to be greatest in wheat, less in barley, and least in oats.

Fields most seriously affected by wireworms are those planted the first year after being summer fallowed or after being in grass for several years. Wireworm control in these fields increases yields from 1 to 12 bushels of wheat per acre, with an average increase of about 5 bushels under favorable moisture conditions. Treatment of seed with insecticides has been shown to reduce wireworm numbers by approximately 70 per cent and wireworm damage by approximately 90 per cent.

Figure 9:18. Wheat in foreground shows damage by wireworms. Wheat in background shows no damage by wireworms; they were controlled by planting insecticide-treated seed. *Courtesy Rhodia Inc., Chipman Division, New Brunswick, N.J.*

Description. The prairie grain wireworm deposits eggs in clusters in the soil. They are minute, oval, and pearly white. Young wireworms are creamy white but change to a shiny yellow as they grow older. Full-grown larvae reach a length of about 1 in. They transform to fragile white pupae. The adults are dark brown to nearly black beetles $\frac{1}{4}$ to $\frac{1}{2}$ in. long (Fig. 9:19).

Life history. The life cycle of the prairie wireworm takes a minimum of two years to complete and may require as many as nine years, with most of this time in the larval stage. Adult beetles emerge from hibernation in the soil during April and May and mate almost immediately. Both sexes mate only once. The females then seek protection in the soil or under stones for about ten days while their eggs develop. Males stay on the soil surface wandering about until they succumb to predation, high July temperatures, or old age.

Adults feed very little; indeed, females can develop normal numbers of eggs without eating any food. Both sexes bear well-developed wings, but males fly very little, and it is likely that females do not fly at all. They crawl about freely but rarely move for very long in any one direction. From late

Figure 9:19. Stages of the prairie grain wireworm. A, eggs; B, larva; C, pupa; D, adult. *A, courtesy Canada Dept. Agri.; B and D, courtesy Ralph Bird; C, courtesy John F. Doane.*

May until the end of June, female beetles make frequent journeys into the soil to various depths up to 6 in. to lay clusters of eggs. They lay the majority under clods and in cracks of soil. The length of the oviposition period of individual females averages about seven days.

A female develops up to 400 eggs but probably deposits only half that number. After approximately a month of incubation, the eggs hatch into small, creamy white wireworms. The latter burrow immediately up or down, seeking food and favorable soil temperature and moisture. Wireworms prefer moist soil and temperatures between 70° and 85°F. Heat and dryness in summer and cold in winter drive wireworms down to lower levels in the soil and suppress their activity. During the winter larvae hibernate at depths of 2 to 10 in.

Young wireworms require living vegetation for food. The first two instars feed on root hairs of grass and cereals, and fungal mycelia; the next three instars feed on moist seeds, coleoptiles, and stems, as well as root hairs and fungi. At this stage of development many young wireworms perish from starvation. They cannot survive on the roots and stems of herbaceous plants. Older wireworms, not so restricted as to diet, are able to survive even on humus in the soil. Feeding is heaviest in spring and fall, when soil environment is optimum.

The mechanism by which wireworms promptly locate food in the soil has puzzled entomologists for a long time. Recent experiments with the prairie grain wireworm indicate that the CO_2 emitted by germinating seeds, roots of growing plants, and decomposing organic matter is sensed by the wireworms, which follow the gradient of this gas to its source.

Two to four years after hatching, the wireworms usually reach full growth. This stage occurs in July after the larvae have molted from eight to ten times. They crawl upwards to within 2 to 4 in. of the surface—if the soil is firm and moist—where they pupate in small cells. After a few days they molt to fragile white pupae. Within two to four weeks the pupae transform to adult beetles, which remain in their cells until the following spring.

Although predators seldom significantly reduce numbers of prairie grain wireworms in their underground hiding places, ground beetle larvae and wireworms of other species prey on them, and they may cannibalize each other. Bacterial and fungal diseases of the larvae, however, are widespread and are probably important factors in natural control. Birds feed on the adults.

Control. Since the 1920s cultural methods have been moderately successful in controlling wireworms in fields of small grains. More recently chemical control with aldrin, dieldrin, heptachlor, or lindane has provided greater and surer protection. Registrations of all these insecticides except lindane have been canceled. Fortunately lindane is effective and still available for treating seeds of small grains. This insecticide is applied economically by dressing the seeds with slurries or liquids. Automatic seed-treating machines cover each seed uniformly with the insecticide, a dye for recognition of treated seed, a sticker, and often a fungicide for control of bunt and smut. In Canada lindane is applied at $\frac{3}{8}$ oz AI per bushel of seed; in the United States it is applied in amounts ranging from $\frac{1}{2}$ to 1 oz AI per bushel.

Growers may treat seed themselves by employing metering devices that feed the dressing onto the seed as it passes through a grain auger, or they may sprinkle dust or wettable powder over the seeds in the drill box and mix well with a stick.

In treating seed, observe these precautions: (1) Use recommended dosages as printed on container label—overdoses may injure seeds and young plants. (2) Do not feed treated seed to livestock or mix with clean grain (3) Do not touch or breathe dusts.

Cultural control consists of cultivating and seeding practices that make the environment unsuitable for wireworms or hasten the growth of plants. Sow seed moderately early, when the soil has first become thoroughly warm and only when there is enough moisture at seed depth to ensure rapid germination. If moisture is insufficient, wait for rain. In dry soil, wireworms injure a larger number of seeds because lacking moisture they eat only the softer part of each grain. Sow seed as shallow as $1\frac{1}{2}$ to 2 in. when moisture is adequate. Use a press drill or follow immediately with a heavy packer. When planting untreated seed, heavier than normal seeding is advisable in order to provide an overabundance of wireworm food and allow survival of sufficient numbers of seed and young plants to provide a good stand. Using plump vigorous seed with high germination rate helps for the same reason.

Clean summer fallow will starve the newly hatched larvae, but it does not harm older wireworms. Shallow tillage during the pupal period in late July and early August crushes the soft pupae or throws them to the surface, where they perish. Deep plowing and cultivating are inadvisable at any time; these practices provide highly favorable soil environments for the wireworms.

SELECTED REFERENCES

Arnason, A. P., and W. B. Fox, *Wireworms Control in the Prairie Provinces*, Can. Dept. Agr. Processed Publ. 111, 1948.

Bruehl, G. W., *Barley Yellow Dwarf*, Amer. Phytopath. Soc. Monogr. 1, 1961.

Calkins, C. O., and V. M. Kirk, *False Wireworms of Economic Importance in South Dakota (Coleoptera: Tenebrionidae)*, South Dakota State Univ. Bul. 633, 1975.

Dahms, R. G., *Preventing Greenbug Outbreaks*, USDA Leaflet 309, 1951.

Daniels, N. E., H. L. Chada, D. Ashdown, and E. A. Cleveland, *Greenbugs and Some Other Pests of Small Grains*, Texas Agr. Exp. Sta. Bul. 845, 1956.

Davis, E. G., J. A. Callenbach, and J. A. Munro, *The Wheat Stem Sawfly and Its Control*, USDA EC-14, 1950.

Ehler, L. E., *A Review of the Spider-Mite Problem on Grain Sorghum and Corn in West Texas*, Texas Agr. Exp. Sta. Bul. 1149, 1974.

Gallun, R. L., *The Hessian Fly: How to Control It*, USDA Leaflet 533, 1965.

Gallun, R. L., and L. P. Reitz, *Wheat Cultivars Resistant to Races of Hessian Fly*, USDA Agr. Res. Serv. Prod. Res. Rpt. No. 134, 1971.

Guppy, J. C., "Life History and Behaviour of the Armyworm, *Pseudaletia unipuncta* (Haw.) (Lepidoptera: Noctuidae), in Eastern Ontario," *Can. Entomol.*, 93: 1141–53 (1961).

Lane, M. C., *The Great Basin Wireworm in the Pacific Northwest*, USDA Farmers' Bul. 1657, 1931.

Mitchener, A. V., *Field Crop Insects and Their Control in the Prairie Provinces*, Line Elevators Farm Service Bul. 8, 1956.

Painter, R. H., H. R. Bryson, and D. A. Wilbur, *Insects That Attack Wheat in Kansas*, Kan. Agr. Exp. Sta. Bul. 367, 1954.

Parker, J. R., and R. V. Connin, *Grasshoppers: Their Habits and Damage*, USDA Agr. Info. Bul. 287, 1964.

Phillips, W. J., and F. W. Poos, *The Wheat Strawworm and Its Control*, USDA Farmers' Bul. 1323, 1953.

Putnam, L. G., and R. H. Handford, *Control of Grasshoppers in the Prairie Provinces*, Can. Dept. Agr. Publ. 1036, 1958.

Somsen, H. W., and K. L. Oppenlander, *Hessian Fly Biotype Distribution, Resistant Wheat Varieties and Control Practices in Hard Red Winter Wheat*, USDA ARS-NC-34, 1975.

Starks, K. J., R. L. Burton, G. L. Teetes, and E. A. Wood, Jr., *Release of Parasitoids to Control Greenbugs on Sorghum*, USDA ARS-S-91, 1976.

USDA, *Grasshopper Control*, USDA Farmers' Bul. 2193, 1975.

USDA, *The Wheat Jointworm: How to Fight It*, USDA Leaflet 380, 1954.

Wadley, F. M., "Ecology of *Toxoptera graminum*, Especially as to Factors Affecting Importance in the Northern U.S.," *Ann. Entomol. Soc. Amer.*, 24:325–95 (1931).

Walkden, H. H., *Cutworms, Armyworms, and Related Species Attacking Cereal and Forage Crops in the Central Great Plains*, USDA Circ. 849, 1950.

Wallace, L. E., and F. H. McNeal, *Stem Sawflies of Economic Importance in Grain Crops in the United States*, USDA Tech. Bul. 1350, 1966.

Walton, W. R., and C. M. Packard, *The Armyworm and Its Control*, USDA Farmers' Bul. 1850, 1951.
Young, W. R., and G. L. Teetes, "Sorghum Entomology," *Ann. Rev. Entomol.*, 22:193–218, 1977.

chapter 10 / **C. C. BURKHARDT**

INSECT PESTS OF CORN

Corn has the highest value of any crop produced in the United States. It is raised in every state of the Union, occupies approximately 85 million acres or one fifth of the total crop land, is grown on two thirds of all farms, and is valued at almost $6 billion annually. A recent estimate indicates that field insects cause an average yearly damage to this agricultural giant of $900 million. Over 25 major destructive species and many more minor ones are responsible. Some of these pests are readily visible as they feed on the leaves, stems, or silks, but the majority hide and feed inside the stalk and roots, in the ears, in the whorl, or underground.

THE PESTS

A national survey has revealed that there are seven corn insects that can be placed in the "most destructive" category. They are the European corn borer, corn earworm, fall armyworm, corn leaf aphid, southwestern corn borer, rice weevil, and soil insects (a complex of species).

The **European corn borer**, a minor foreign pest that was accidentally introduced near Boston in about 1910, has become the single most injurious insect enemy of corn in America (Figs. 10:12 and 10:13). It now occurs in 39 states and is spreading farther south and west every year. Next in line is a native species, the **corn earworm** (Fig. 10:9). Although it ranks second among pests of the total corn crop, it is the number-one enemy of sweet corn and is a serious pest of cotton, tomatoes, and tobacco as well. The **fall armyworm**, a close relative of the corn earworm, is an important pest of corn, particularly in the northeastern and southern states.

About the time that the European corn borer was spreading into the Corn Belt, another foreign pest, the **southwestern corn borer** (Fig. 10:2), was expanding its range in the Southwest, having originally invaded the area from Mexico in 1913. Today this insect is causing much concern because of its rapid migration eastward from Oklahoma and Arkansas to central Alabama and central Tennessee, and northward to central Kansas. All or portions of 15 states were infested by 1974. This insect attacks sorghum as well as corn.

The **rice weevil**, called the "corn weevil" by many southern farmers, attacks growing corn in the South shortly after the roasting ear stage (Fig. 18:5). At harvest this insect is brought into storage where it continues to breed and to do damage as a stored grain pest. The **corn leaf aphid**, a widely distributed insect, reduces pollen, causing partial to complete barrenness.

In North America over thirty kinds of **soil insects** (Fig. 10:1) are enemies of the corn crop. This diversity of species does not exist in any one field or in any one area, but when corn in a field is beset with soil insects, the damage often originates from the activities of a complex of species. Highly destructive

Figure 10:1. Complex of soil insects destructive to corn. Left to right, Top row: cutworm, sod webworm, and billbug; Middle row: flea beetle larva, seedcorn maggot, seedcorn beetle, corn rootworm, and corn root aphid; Bottom row: white grub, wireworm, and grape colaspis. Figures not to scale. *Courtesy Illinois Natural History Survey.*

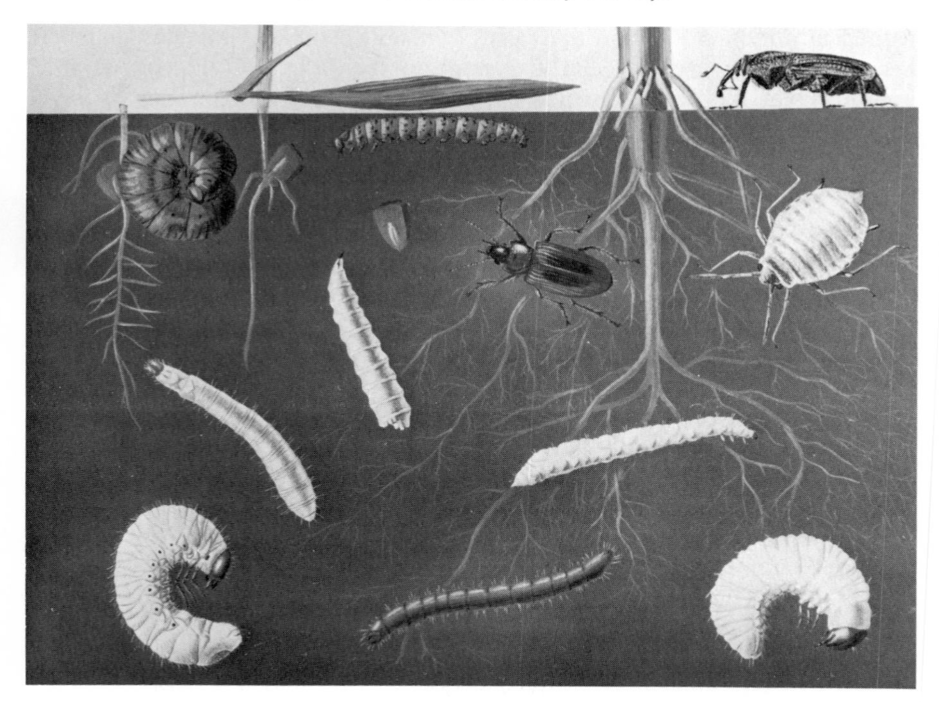

Figure 10:2. Corn stalk cut to expose the southwestern corn borer and its tunnel. *Courtesy Kan. Agr. Exp. Sta.*

soil insects include corn rootworms, cutworms, wireworms, billbugs, webworms, seedcorn beetle, and seedcorn maggot.

In addition to these major pests, there are insects of great regional importance such as grasshoppers, chinch bugs, sap beetles, and spider mites. **Grasshoppers** have caused destruction of the corn crop west of the Mississippi River. One of the chief species involved in these depredations has been the **differential grasshopper**, which in many places has earned the name of "corn hopper" (Fig. 9:7D).

In the East and South, **sap beetles**, *Carpophilus* spp., are destructive pests of the ear of sweet corn, being both primary and secondary invaders (Fig. 10:3). The **twospotted spider mite** is a serious pest of corn in the West. And the **chinch bug**, a widely distributed native insect, causes much damage during outbreak years in the Midwest and Southwest (Fig. 10:8). A relatively new but serious insect pest of corn is the western bean cutworm, which causes heavy damage in Nebraska. It has been found in eastern Wyoming, Colorado, and in Kansas. It is rapidly moving eastward across Nebraska and it may soon become a real threat to other Corn Belt states. Another often serious pest of corn in Colorado, Kansas, Nebraska, Wyoming, and regions to the south and north of these states is the Banks grass mite.

Figure 10:3. A, larvae of dusky sap beetle feeding on ear of canning sweet corn; B, sap beetle damage to sweet corn. *Courtesy University of Maryland, College Park.*

THE INJURY

Insects begin attacking corn as soon as the seed enters the ground and never let up through the entire period of plant growth to maturity. An old Indian rule for planting corn goes

> One for the bug,
> One for the crow,
> One to rot,
> And two to grow.

Insects injure corn seed severely when cool, wet weather delays germination. By eating into and hollowing out the seed, wireworms, seedcorn maggots, seedcorn beetles, thief ants, and the larvae of palestriped flea beetles seriously reduce stands.

Soil pests such as rootworms, white grubs, wireworms, corn root aphids, and other root feeders stunt plant growth and reduce yields. Corn plants injured by rootworms frequently fall over or lodge during hard rains or following irrigation and wind.

Young corn plants bear the brunt of much insect injury. Heavy feeding on leaves by armyworms, grasshoppers, or flea beetles retards growth and reduces yields. Cutworms kill plants by cutting them off at or below ground level. Corn earworms, fall armyworms, European corn borers, southern cornstalk borers, and southwestern corn borers enter the whorl and feed on the tender, folded leaves and often destroy the central bud. When injured leaves unfold they appear ragged. Growers refer to this injury as "budworm damage." Wireworms feed not only on the roots of young corn plants but also on underground stems; they may tunnel upward within them, killing the central shoot. Chinch bugs, which pierce leaves and stems and suck the sap, cause young plants to wilt and die.

As corn plants become older and stronger, they can better withstand the feeding injuries of various insect pests. Yet even during the later stages of corn growth, insects, especially those attacking the silks and ears and those tunneling the stalks, cause serious damage. Insects that relish corn silks often prevent pollination and cause poorly filled ears and nubbins. Corn earworms work their way into the developing ear through the tip, feed on the soft kernels, and contaminate the ear with excreta. Fall armyworms also damage corn ears, but they differ from corn earworms in that they burrow into the shank and ear at various places. Entrance into an ear through the side or butt end destroys the value of sweet corn for human consumption. Western bean cutworms can also cause serious damage to ears.

Corn leaf aphids may infest the leaves, tassel, and ears in such great numbers that they seriously interfere with pollen production and cause plants to be barren or to produce only nubbins. Data indicate that barrenness is not caused merely by suppression of pollination but by a physiological effect of the aphid upon the corn plant. According to recent research, even light infestations may reduce yields by as much as 10 per cent through decreased size and weight of ears.

Borers tunneling within the stalk cause injury in several ways. In young corn they stunt the plants, which subsequently yield less. In older plants they do not greatly affect yield, but they weaken stalks mechanically so that the plants break over in strong winds. Close to harvest time, the southwestern corn borer girdles the inside of the stalk about 3 to 4 in. above soil surface. This injury results in the lodging of about 75 per cent of the attacked stalks (Fig. 10:4). Borers, particularly the European corn borer, enter and weaken the shanks, causing ears to break off and fall to the ground. The lodging of plants and the breaking off of ears during picking makes harvesting slow and difficult. Corn on the ground, which often falls prey to rodents, birds, and molds, forces growers to salvage by pasturing the field with livestock or by picking it off the ground. The European corn borer and corn earworm often feed on the caps of the kernels of dent corn, resulting in a serious cleaning problem in seed corn.

In southern states stored grain insects such as the rice weevil and Angoumois grain moth infest and destroy mature grain standing in the field. The Angoumois grain moth also infests corn in the field in areas as far north as southern Indiana, particularly popcorn and hominy corn, causing serious

Figure 10:4. Lodged corn plants resulting from girdling damage by southwestern corn borers. *Courtesy Kan. Agr. Exp. Sta.*

losses. At harvest these pests are carried into storage, where they continue their destruction.

Besides directly injuring corn, insects aid in the spread and development of corn diseases. Corn seeds wounded by soil insects are highly susceptible to attack by soil fungi, the most common cause of seed rots and seedling diseases. By tunneling into the ear or boring into the stalk, insects open pathways for the entrance of ear rot and stalk rot fungi. They then facilitate infection by wounding host tissues.

Flea beetles and corn rootworms transmit bacterial wilt, known also as Stewart's disease, which is especially destructive to sweet corn. Flea beetles carry the bacteria within their bodies during winter and are ready to transmit the disease as soon as they come out of hibernation in spring.

Insects also spread several of the virus diseases of corn. The leafhopper, *Dalbulus maidis* (DeL. & W.), transmits the virus of corn stunt; the melon aphid, the virus of southern celery mosaic, which also infects corn; and the corn planthopper, the virus of corn mosaic.

CULTURAL CONTROL

Cultural and chemical methods are the chief means of controlling corn pests, although entomologists have employed biological methods against the

European corn borer by introducing parasites of this pest from foreign countries.

We may divide **cultural methods** into seven categories: (1) promotion of vigorous plants and stands, (2) rotation, (3) clean culture, (4) tillage, (5) sanitation, (6) modification in the timing or method of planting or harvesting, and (7) resistant varieties.

1. **Vigorous corn plants** tolerate the feeding of insects better than weak ones and may overcome or outgrow injury. Growers produce vigorous plants by carrying out good agronomic practices. This cultural method is useful against such injurious pests as corn earworm, southern cornstalk borer, corn rootworms, wireworms, corn root aphid, and corn leaf aphid. Because chinch bugs prefer thinner or poorer parts of grain fields, growing heavy stands helps to control these pests.

2. One of the most effective ways of avoiding some insect troubles is the proper **rotation** of corn with other crops. Allowing corn to occupy ground for only one or two years in rotation prevents large buildups of pests favored by this crop. Species against which rotation works include rootworms, wireworms, white grubs, corn root aphid, and billbugs. Planting immune crops instead of corn on newly broken sod land often averts serious damage from cutworms, wireworms, billbugs, and white grubs. Planting immune crops when a corn pest is unusually abundant is also desirable. Various legumes, and sometimes small grains or vegetables, make good substitute crops.

3. **Clean culture** or the elimination of weeds and volunteer plants stifles the development of insect populations by doing away with plants upon which the insects feed or upon which females lay their eggs. This method is useful in controlling the fall armyworm, flea beetles, southern corn rootworm, corn root aphid, grasshoppers, and stalk borer.

4. With **tillage** operations we can control insect pests in several ways. First, we can bury insects in dormant stages so deep that they are unable to emerge from the soil. This measure has been recommended for controlling the European corn borer, the southwestern corn borer, the southern cornstalk borer, and grasshoppers. Second, we can bring dormant stages to the surface and expose them to inclement weather and enemies. Insects controllable by this method include the southwestern corn borer, southern cornstalk borer, webworms, billbugs, and grasshoppers. Third, just before planting we can break up the nests of thief ants and cornfield ants and scatter corn root aphids. The operation reduces the numbers of these pests and gives the corn a chance to outgrow them.

5. **Sanitation**—cleaning up of corn refuse and stubble in fields, feed lots, and other places on the farm—cuts down the number of surviving individuals of European corn borer, southern cornstalk borer, and southwestern corn borer. Sanitation also helps in controlling field infestations of rice weevil. Because field infestations are started by weevils which have migrated from nearby infested stored grain, a cleanup of bins and cribs helps control this pest.

6. **Modifying the time or method of planting or harvesting** is sometimes useful in controlling insect pests of corn. Some insect troubles can be avoided by planting early, some by planting late, and still others by planting at the normal time. What one does depends on what kinds of pests are prevalent and how serious they are. Planting early sometimes helps to reduce injury by the southwestern corn borer, lesser cornstalk borer, southern corn rootworm in the South, corn leaf aphid, and chinch bug. Planting late and shallow to get a quick germination lessens damage to seed by pests inhabiting the soil. This practice is useful against wireworms, seedcorn maggot, seedcorn beetle, and flea beetles. Planting at the usual time for a given locality is best where both European corn borer and corn earworm are troublesome. Harvesting early before the southwestern corn borer starts girdling within the stem prevents lodging losses caused by this insect. Seeding more thickly than normal to obtain a good stand is sometimes helpful in reducing the damage of the southern corn rootworm in the lowlands of the South.

7. We can prevent or reduce the injury of a few insect pests by **planting resistant or tolerant varieties** adapted to the locality. Insects partially controllable by this method include the European corn borer (Fig. 10:5), corn earworm, corn rootworm, rice weevil, and chinch bug.

Figure 10:5. European corn borer injury to leaves of corn plant. A, resistant hybrid; B, susceptible hybrid. Note the feeding holes. *Courtesy USDA.*

A

B

CHEMICAL CONTROL

Cultural methods may not be sufficiently effective; therefore corn needs the protection provided by insecticides. Growers one to two decades ago used the chlorinated hydrocarbons more than any other group of insecticides in controlling corn pests. Even when corn crop residues are not to be used as feed for dairy animals or animals being finished for slaughter, growers can no longer apply most chlorinated hydrocarbons to control the corn earworm and European corn borer.

Because of the persistent pesticidal residues associated with many of the chlorinated hydrocarbons and the increased restrictions and limitations placed upon their use, these insecticides are being replaced rapidly by organophosphorus and carbamate insecticides. This transition is further motivated because of the development of resistance to chlorinated hydrocarbon insecticides in many insects and mites. Organophosphorus insecticides, such as diazinon, dimethoate, disulfoton, ethion, malathion, parathion, and phorate, have been useful in controlling mites and aphids in corn. Under certain conditions, granular insecticides may be applied for combined control of European corn borers and corn rootworm larvae.

Control Equipment

Corn growers use both standard and special equipment in applying insecticides. For preplanting applications or applications when corn is young, ordinary ground sprayers and granule applicators (Fig. 10:6) are adequate, but after corn has grown tall, high-clearance machines (Fig. 12:3) or airplanes are necessary. Growers may hire airplane applicators to apply insecticides at any stage of corn development.

To control soil insects, we may apply insecticides either broadcast or in bands along the rows. Broadcast treatment, the application of insecticide evenly over the entire field with immediate disking into the soil, protects the main root mass of surface-planted corn. Broadcast treatments may be made with sprays, granules, or insecticide-fertilizer mixtures or solutions. For effective protection of listed corn, band treatment is necessary because the seed is planted in a deep furrow. Growers also employ the band treatment method on surface-planted corn.

Proper placement of insecticides in the row has been achieved with a variety of devices. Planter- or lister-mounted spray or granule applicators disperse the insecticide in a 5 to 7-inch band behind the planter shoes and in front of the covering disks or press wheels (Fig. 8:14). A further development of this technique is the addition of an incorporation wheel, which works the insecticide into the top inch or more of soil. Another treatment has been to combine insecticides with starter fertilizers and to apply the mixture with a split-boot attachment that puts the material to one side of and below the seed—a position that permits good use of fertilizer but is less efficient for

Figure 10:6. Granular insecticide being applied for control of European corn borer. *Courtesy Gandy Company, Owatonna, Minn.*

insecticides. For this reason an additional attachment is often employed for separate application of the insecticidal sprays or granules. Applying liquid starter fertilizer-insecticide solutions with special attachments is still another method that growers use to place insecticide in the soil.

Coating seeds with insecticide effectively controls some soil insects such as the seedcorn maggot and seedcorn beetle. Seed-treating machines are the most efficient devices for applying chemicals to seed, but more often growers mix insecticidal dusts with corn seed in the planter box.

REPRESENTATIVE CORN PESTS

For our detailed study of important corn pests, we have chosen the chinch bug, the corn earworm, the European corn borer, and the three species of corn rootworms.

Chinch Bug. *Blissus leucopterus leucopterus* (Say) [Hemiptera:Lygaeidae]

The chinch bug, a widely distributed native insect of North America, injures small grains, sorghum, and corn, primarily in the Midwest and Southwest. The states most seriously affected are Illinois, Indiana, Iowa, Kansas, Missouri, Ohio, Nebraska, Oklahoma, and Texas. The first report of its being an injurious pest came in 1785, when it destroyed much of the wheat in North Carolina. Since that time chinch bugs have appeared recurrently in destructive numbers, usually during periods of drought. One of the worst outbreaks took place in 1874, when grain crops suffered an estimated loss of $100 million. In 1934, another year of serious infestation, the chinch bug caused an estimated loss of $27.5 million to the corn crop and $28 million to wheat, barley, rye, and oats. In some areas, as in Oklahoma, chinch bugs are present in more or less injurious numbers every year.

The kind of crop attacked and the extent of injury may vary depending on the generation of chinch bug doing the damage. Normally two generations develop in the Midwest and three in the Southwest. The first generation is injurious to small grains, but as these crops mature, the bugs migrate to corn and sorghums. All generations injure the latter crops, although damage by the first generation is the most severe because the plants are young and tender at the time of attack.

Chinch bugs destroy plants principally by withdrawing enormous quantities of plant juices (Fig. 10:7). Young plants are highly susceptible, sustaining severe, often fatal, injury within a few days. Older and tougher plants are better able to withstand attack, but they too become weak and stunted, yield less, and frequently lodge. Fifty-five bugs per square foot cause severe injury to small grains, whereas 500 are usually necessary for serious injury to corn. Part of the damage to plants results from exudation of sap through the wounds produced by the feeding chinch bugs. Wounds also facilitate the entrance of pathogenic organisms. Chinch bugs do not seem to inject toxic substances into plants—there is little evidence of cell injury in tissues directly pierced—but stylet sheath secretion may clog conductive tissue of the plant resulting in a yellowing or reddening of the affected region.

In addition to destroying small grains, corn, and sorghum, chinch bugs injure other members of the grass family, both cultivated and native.

Description. Eggs are laid on host plants behind the lower leaf sheaths, on roots, or in ground nearby. Eggs are 0.8 mm long; they are white at first, but gradually become reddish. Newly hatched nymphs are pale yellow, but they soon become red except for the first two abdominal segments, legs, and antennae, which remain pale yellow. Although the second, third, and fourth nymphal instars become darker red after each molt, they retain the pale yellow band on the forward part of the abdomen. The fifth and final nymphal instar is a black and gray nymph with a conspicuous white spot on the back between the wing pads. The adult chinch bug is black with conspicuous white forewings, each of which has a black triangular spot at the middle of the outer margin. The adult is about $\frac{1}{6}$ in. long (Fig. 10:8).

Life history. Chinch bugs overwinter as adults, sheltered principally in tufts of bunch and clump-forming grasses. Some may hibernate under leaves or

Figure 10:7. A mass of chinch bugs sucking the sap from the stalk and leaves of a corn plant. *Courtesy USDA.*

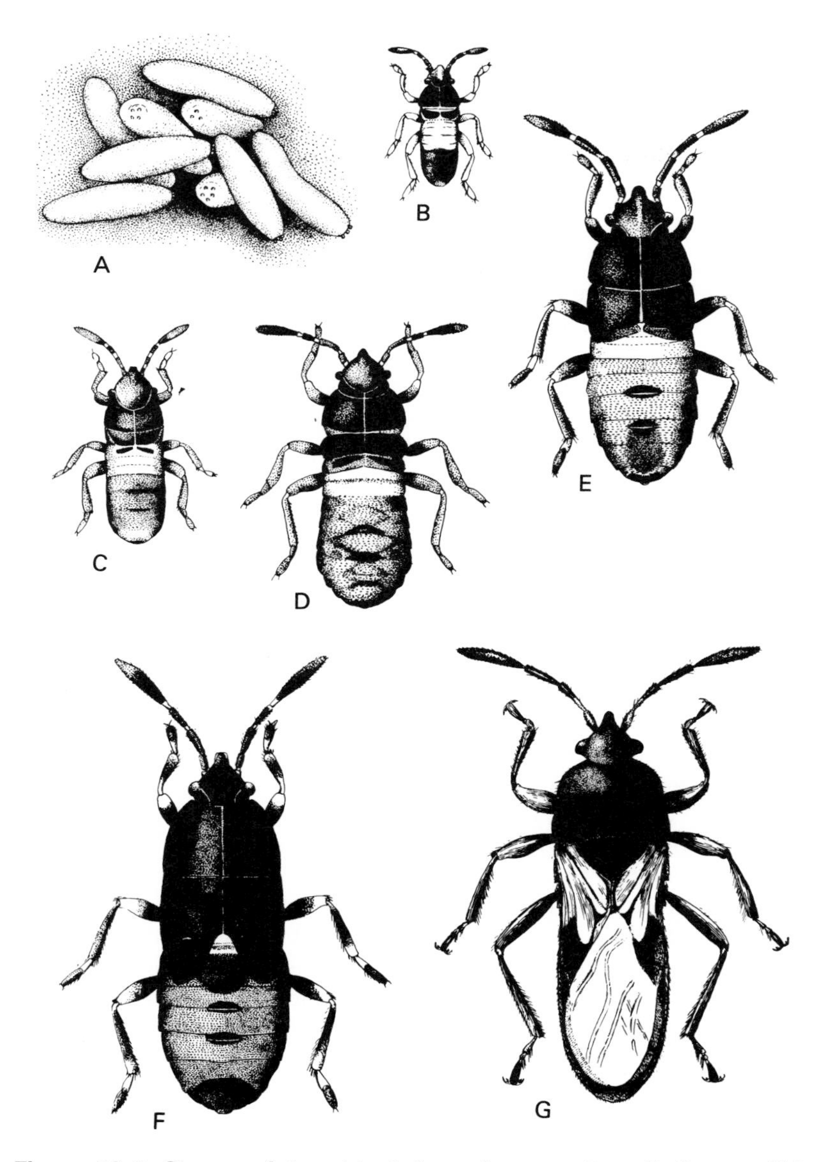

Figure 10:8. Stages of the chinch bug. A, eggs; B to F, first to fifth nymphal instars; G, adult. *Courtesy USDA.*

litter at the borders of woodlands, under hedges, in fence rows, or in fields of sorghum stubble. They remain inactive during the winter unless spells of abnormally warm weather induce them to stir. Migration from winter quarters usually begins in early spring after one or two sunny days when temperatures rise to 70°F or higher. The bugs, preferring the thinner parts of fields, fly to small grains. When cool weather delays migration, they may fly directly to corn or sorghum. Soon after reaching the fields of grain the bugs mate. Females then begin to lay eggs, placing them behind the lower leaf sheaths, on roots, or in ground close to the host plant. They deposit eggs at the rate of 15 to 20 a day over a period of two to three weeks.

Nymphs hatch from the eggs in one to two weeks and begin to feed on the small grain crop, piercing stems and leaves and sucking up plant juices. The bugs pass through five nymphal instars in about 30 to 40 days. When small grains mature before the adult stage of the insect is reached, the bugs must migrate on foot to green food in order to survive. During afternoons of sunny days hordes of young bugs march into fields of corn, sorghum, or other succulent grasses. In the Southwest, where chinch bugs become adult before small grains mature, migration is by flight.

Adult chinch bugs do much flying about and dispersing. Eventually they settle down, and the females start laying eggs on corn or sorghum plants. These eggs hatch into the bugs of the second generation, which become adult in late summer or early fall. From August through October adults gradually leave the grain fields and fly to their wintering quarters. In the Southwest a third generation develops before the bugs enter hibernation.

Size of chinch bug populations, and hence severity of damage, are largely determined by weather. During dry years in the Midwest, infestations build up gradually to a peak. When rainfall becomes normal again, populations drop off. Rainfall may affect chinch bugs in several ways. Heavy rains during the hatching period beat the young nymphs into the mud, killing them, and splashing mud covers the eggs and prevents hatching. The white fungus disease caused by *Beauveria globulifera* (Speg.) spreads rapidly among chinch bugs during periods of warm, damp weather.

In the Southwest, weather affects chinch bugs differently. There, rainfall must be adequate, $2\frac{1}{2}$ in. or more, from July 10 to August 20 to keep green, succulent food available for the third-generation bugs. If rainfall is light, sorghums dry out after ripening, and no second growth develops. Lack of food then prevents the third-generation bugs from maturing and causes a reduction of the population for the following year.

Besides adverse weather, natural enemies destroy chinch bugs. The most important is the white fungus, but chinch bugs have another enemy in the tiny wasp, *Eumicrosoma beneficum* Gahan. The female wasp lays her egg in the chinch bug egg. Upon hatching, the wasp larva feeds on the internal contents of the host egg. In certain localities this wasp parasitizes from 30 to 50 per cent of chinch bug eggs.

Control. Cultural practices, erection of barriers against migrating nymphs, and application of insecticides control chinch bugs. There are several cultural practices that are helpful. Growers can minimize chinch bug injury by reducing either the acreage of small grains or the acreage of corn and sorghum, whichever is the less important in their farming operation. This practice limits the size of chinch bug populations by removing essential food from a part of the life cycle.

Growing nongrass crops like legumes, flax, rape, or buckwheat cuts losses, because chinch bugs rarely feed on any plant outside the grass family. If growing nongrass crops does not fit into the agriculture of a region, farmers may elect to plant resistant grains. Oats are more resistant than wheat or rye, and these latter are more resistant than barley. Spring barley is highly susceptible to injury and should not be planted when there is danger of a heavy infestation of chinch bugs.

Some varieties of corn, such as Hays Golden, a variety adapted to the Southwest, are resistant. Hybrid corn varieties suffer less injury than their inbred parents, whereas open-pollinated varieties vary widely in susceptibility and resistance. Among sorghums, most sorgo and Kafir varieties are resistant, whereas most milo and feterita varieties are susceptible.

Planting legumes among small grains and corn aids in control by producing shade and dampness, which chinch bugs avoid. For the same reason, good cultural practices that stimulate dense, vigorous stands reduce injury. In areas where the European corn borer is not a problem, early planting of corn gets the plants past the most susceptible stage before the chinch bugs attack. Early planting of sorghum is likewise beneficial, but in some areas this practice may lead to poorer stands and lower yields.

For many years growers in the Midwest erected barriers to prevent migrating nymphs from entering fields of corn. Barriers of dusty trenches, creosoted dirt ridges, and creosoted paper fences are now obsolete and have been replaced largely with chemical barriers. To construct the latter barrier, apply the insecticide as a spray to the ground in strips 4 rods wide between adjoining fields of small grains and of corn. Bugs crossing the strip contact the insecticide and die. To stop nymphs from skirting the ends, apply a strip of insecticide 2 rods wide and 8 to 10 rods long at right angles to both ends of the main barrier. Insecticide barriers may have to be renewed during heavy migrations or after rain. Construct barriers when small grains begin to mature and dry and before migration starts. In the Southwest, barriers are not successful, because bugs reach the flying stage before small grains mature.

When entire fields of corn are already infested with chinch bugs, spray $1\frac{1}{4}$ lb of carbaryl or 2 lb of toxaphene per acre, using crop nozzles and directing the spray to base of plants. Use at least 20 gal water per acre. Treatment of entire fields provides effective but more costly control, and certain materials impose feeding restrictions on the stover.

Corn Earworm. *Heliothis zea* (Boddie) [Lepidoptera:Noctuidae]

The corn earworm, a native insect with wide distribution, is the most serious insect pest of sweet corn in North America. It ranges from southern Canada through the rest of North America to as far south as Montevideo, Uruguay. It also inhabits the Hawaiian Islands, apparently having been introduced from North America.

The favored host plant of the corn earworm is sweet corn, but it feeds on and damages field corn and tomatoes too. Further, it is an important pest of cotton, tobacco, various legumes, and vegetables in the South. Depending on the crop it attacks, this insect may be called the corn earworm, the cotton bollworm, or the tomato fruitworm.

The corn earworm damages corn in several ways. Larvae, which hatch from eggs laid on the leaves and stalks of young early corn, migrate to the whorl and feed on the tender, folded leaves. When the damaged leaves unfold, they appear ragged. Growers refer to this as ragworm or budworm injury. Usually budworm injury does not reduce yields greatly, but occasionally it may stunt the plant, which then produces little grain.

Earworms do their greatest damage when the ears begin to silk. Larvae desert all other parts of the plant to feed on the silks, and the moths prefer to lay their eggs there. The larvae work their way through the mass of silk to the tip of the ear, where they feed on the kernels (Fig. 10:9). Occasionally young or partially grown larvae make side entry into the ears. Sweet corn growers lose heavily because the damaged portion must be clipped from the ears. Trimming of ears not only takes time but also reduces quality and lowers value of the corn.

Figure 10:9. Nearly full-grown corn earworm devouring silks and soft kernels of corn. Note the dark brown frass. *Courtesy USDA.*

In field corn or seed corn, the larvae burrow under the kernels when they begin to harden and feed on the softer germ. Later, during husking and handling, the injured kernels drop from the ear. Corn earworms damage corn indirectly by opening entrances for molds and insect pests such as grain beetles, sap beetles, and weevils. Ears with thick growths of mold are unsafe to use for livestock feed, especially for horses. In the United States, agriculturalists have estimated corn losses of $75 million to $140 million annually caused by corn earworms.

Description. Corn earworm eggs have the appearance of tiny, flattened spheres. They are prominently ribbed and approximately 1.2 mm in diameter. When deposited they are light yellow, but they soon darken, and at hatching time they are dusky brown. Newly hatched larvae are about 1.5 mm long, nearly white, with shiny black heads and legs. They grow rapidly and become variously colored, ranging from pink, green, or yellow to almost black. Many are conspicuously striped. Often down the side there is a pale stripe edged above with a dark one, and down the middle of the back there is a dark stripe divided by a narrow white line. Full-grown larvae are robust and $1\frac{1}{2}$ to 2 in. long. Pupae are about $\frac{3}{4}$ in. long, first green in color and later brown. Moths are about $\frac{3}{4}$ in. long with a wingspread of $1\frac{1}{2}$ in. They vary in color from dusty yellow, olive green, or gray to dark reddish brown (Fig. 10:10).

Life history. The corn earworm passes the winter in the pupal stage, protected in the soil. In Canada and in the northern part of the United States, pupae are unable to survive the winter. Infestations in these areas arise from moths flying in from southern overwintering grounds. Available data indicate that the northern overwintering limit of the species in North America corresponds to the area of the last spring freeze before April 30. In coastal areas such climatic conditions extend as far north as the state of Maine and Washington and the province of British Columbia.

The corn earworm has several generations annually in all but the most northern part of its range. In southern Florida and southern Texas, moths remain active the year round and produce as many as seven generations annually. In the Corn Belt there are usually two or three generations a year.

Moths emerge from overwintering pupae in spring at a time when early corn is in the seedling stage. Female moths lay the eggs of the first generation on young corn, vetch, alfalfa, and on other available host plants. Duration of life stages and of complete life cycles of the several generations varies with prevailing temperatures and available host plants. Incubation of eggs takes from two to eight days. Upon hatching, young larvae feed on the empty egg shells. Soon afterwards they begin feeding on plant tissues. Molting five times, they develop to full-grown larvae in two to four weeks. Those inhabiting corn ears gnaw their way through the husks and drop to the ground, where they bore into the soil to depths of 1 to 9 in. and construct cells. In order that the moth may reach the surface easily, the larvae construct a smooth passage to within $\frac{1}{2}$ in. of the surface. Then they return to their cells, where they transform to pupae. In summer the pupal period lasts from two to three

Figure 10:10. Life stages of the corn earworm. A, eggs attached to leaf of corn plant; B, larvae; C, pupa in soil; D, adult. *A, C, and D courtesy USDA; B, courtesy Kan. Agr. Exp. Sta.*

weeks. Under favorable conditions a life cycle may be completed in 30 days.

Emerging moths mate, and the females, as young as a day old, begin to lay eggs. They deposit eggs singly but usually place many on an individual plant before seeking another host. The moths are strong fliers and may fly far from the fields in which they developed as larvae. Thus they infest other areas and invade the North. They are most active in the evening and feed heavily on the nectar of flowers. Females live about 12 days and each may deposit from 350 to 3000 eggs. Fecundity as well as rate of larval growth and size of the insect depend upon which species of plant the larva feeds. Moths reared from larvae fed on corn ears are larger and produce twice as many eggs as those reared on any other host. Less favorable host plants include cotton, soybean, lima beans, alfalfa, and tomato.

Although females may lay many eggs on the silks of a single ear, only a few larvae survive to full growth in an individual ear because of their cannibalistic habits. Whenever two earworms come together, they fight until one or both are fatally injured.

Earworms are present in damaging numbers every year, but in some years populations are unusually heavy and the damage is correspondingly greater. Favorable weather and a favorable succession of host plants foster outbreaks of corn earworms. Natural limiting factors include cannibalism of larvae, predators, parasites, and disease. The wasp, *Trichogramma minutum* Riley, parasitizes the egg; a fly, *Winthemia quadripustulata* (Fabr.), parasitizes the larvae. A bug, *Orius insidiosus* (Say), feeds on corn earworm eggs. At least 21 species of birds feed on the larvae, and moles feed on the pupae. Wet weather promotes disease among the larvae and wet soil among the pupae.

Control. The methods used by growers to control corn earworms vary depending on whether they grow sweet corn, seed corn, or field corn and on whether they farm in the North or in the South. Application of insecticides is currently practiced only on sweet corn.

For control of earworms in field corn or seed corn, cultural methods—good agronomic practices such as fertilizing, rotating crops, and planting adapted varieties—are relied on to reduce damage. Corn planted at the usual time for an area sustains less injury than either early- or late-planted corn. In the South, corn earworms infest both early- and late-planted corn more heavily. In the North, the European corn borer attacks the early corn more seriously, and the corn earworm attacks the late corn.

Planting resistant varieties of corn is an economical way of reducing injury. Plant breeders have found that resistance in corn is associated with tighter husks, harder kernels, less attractiveness for oviposition, and lower food value for larvae. Generally lower concentrations of amino acids occur in the resistant lines and higher concentrations in the susceptible lines. Research has indicated that a susceptible dent corn single cross (MP317×MP319) showed 22 per cent reducing sugars as compared with 15 per cent in a resistant single cross (F44×F6). Studies support the possibility of the presence of either a feeding inhibitor, a feeding deterrent, a growth inhibitor, or a combination of these substances in the silks of resistant corn.

Southern dent corns, evolved in regions where corn earworms are a constant feature of the environment, show the highest type of resistance. Resistant varieties of field corn include Dixie 18, Dixie 11, Georgia 281, Louisiana 521, Texas 30, Texas 11W, and Coker 811. The exotic corn strain Zapalote Chico is extremely resistant to corn earworm.

Variations in susceptibility and resistance are also present among varieties of sweet corn. In stands untreated with insecticides some sweet corn varieties may have as high as 90 per cent marketable ears, whereas others may have as low as 10 per cent or less. La2W×14S, Country Gentlemen, and related types are among the more resistant varieties. Iona, Aristogold, Seneca Scout, Seneca Chief and others exhibited considerable resistance in southern tests. The sweet corn hybrid 471-U6×81-1 is resistant in Georgia. Plant breeders are working constantly on transferring earworm resistance to new and better varieties.

Growers of sweet corn rely chiefly on insecticides to control earworms. They apply carbaryl, methomyl, parathion, or parathion-carbaryl combination with high-clearance sprayers, dusters, or airplanes. A common spray recommendation for ground application suggests applying $1\frac{1}{2}$ lb of actual carbaryl per acre in 20 to 25 gal water at pressures of 100 to 200 psi. Four flat fan nozzles per row are adjusted so that the spray hits the ears. Some growers add white mineral oil to the spray at a rate of $1\frac{3}{4}$ gal/acre. Mineral oils should be applied to corn with caution for several reasons. Certain oils cause injury to the ears; mineral oils may cause foliage and ear injury during hot dry weather; some varieties are highly susceptible to oil injury.

Several applications are necessary for satisfactory control: In the South, where the problem is severe, growers may apply as many as 17 treatments at daily intervals, making two of the treatments before silks first appear. They may even make earlier applications when whorls are heavily infested. Growers in the North manage with fewer treatments. Under conditions of light infestation, they make the first application when 25 per cent of the top ears show silk, the second three days later, and successive applications at four-day intervals, continuing the treatments until several days before harvest. Under conditions of heavy infestation, growers may begin treatment when silks first appear and apply successive applications at two-day intervals until harvest.

Growers may also control corn earworms with insecticidal dusts. They use high-clearance dusters to apply 5 per cent carbaryl dust at the rate of 30 lb/acre or employ airplanes to apply 10 per cent at 15 lb/acre. Where control of corn earworms and other insects in the whorl is needed, granular carbaryl is a convenient and effective formulation to apply. By employing the inexpensive granule applicators on the whorl stage corn, growers of multiple plantings are able to release the high-clearance machines for treatment of maturing fields.

European Corn Borer. *Ostrinia nubilalis* (Hübner) [Lepidoptera:Pyralidae]

The European corn borer, the most destructive pest of corn in America, is an introduced insect, having been discovered in 1917 on sweet corn near Boston. The borer is believed to have entered this country about eight years earlier in broom corn imported from Italy or Hungary. Entomologists quickly recognized the serious threat that this pest presented to the corn industry and attempted to prevent its spread. Despite stringent foreign and domestic quarantines and the application of other control measures, the borer invaded new areas. It has since extended its range from the Atlantic to the Rocky Mountain states and from Canada to the southern states.

During its early period of expansion, the European corn borer had only one generation each year. In the 1930s a second generation began to appear in some areas, and as time passed, the capacity of the insect to produce more than one generation annually became widespread. Now in North America the second generation constitutes an important part of European corn borer populations. Within the Corn Belt, several biotypes have developed.

First-generation borers greatly affect yields because they attack corn plants in an early stage of development. The borers begin feeding in the whorl on the leaf surface, but later they bore into the stalk, where their tunneling destroys food channels. The latter injury is a major cause of reduction in yield as it weakens the plant and starves the ears. The tunneling of second-generation borers in older plants does not affect yields greatly, but it does mechanically weaken the plants so that much stalk breaking and ear dropping occur. Borings in the shanks lead to chaffy ears and nubbins, and invasions of the cob may make sweet corn unsalable (Fig. 10:11).

A serious secondary effect is the opening of the stalk and ears to both saprophytic and pathogenic fungi and bacteria and their distribution inside the plant. Rots developing in individual plants may cause greater weakening of the stalks and more dropping of ears than do the borers themselves. A survey in Minnesota revealed that 80 to 90 per cent of borer tunnels were infected with rots. Thus, damage often attributed to borers may be due in part to damage caused by stalk- and ear-rotting fungi.

Annual losses due to European corn borer vary with the intensity of infestations. In the United States from 1968 to 1974, losses of corn have

Figure 10:11. European corn borer damage. A, ears attacked directly, nubbins and light chaffy ears; normal ear on right from an uninfested field; B, stalk rot developing around borer tunnel. Note borer in left stalk. *Courtesy University of Minnesota, Minneapolis.*

averaged more than $210 million annually. The European corn borer infests over 200 kinds of plants. Besides corn, it attacks sorghums, soybeans, millet, buckwheat, oats, barley, potatoes, beans, and many large-stemmed flowers and weeds.

Description. Adult females of the European corn borer lay their eggs on the underside of corn leaves usually in clusters of 14 to 20. The eggs overlap one another like fish scales. Individual clusters measure $\frac{1}{8}$ to $\frac{3}{16}$ in. wide and $\frac{1}{4}$ to $\frac{3}{8}$ in. long (Fig. 10:13A and B). The egg is nearly flat and about 1 mm in diameter. It is white when first laid but turns yellow, and just before hatching the black head of the larva shows through the shell. The newly hatched larva, approximately 1.5 mm long, has a black head and a pale yellow body bearing several rows of brown or black spots. The full-grown larva is about one inch long, gray to light brown or pink, and faintly spotted on the dorsal surface. The underside of the body is cream colored and unmarked. The pupa is about $\frac{1}{2}$ to $\frac{3}{4}$ in. long and light to dark reddish brown. The adult female is a moth with a robust body and a wing spread of about 1 in. Its general color varies from a pale yellow to light brown. The male moth is slightly smaller, more slender bodied, and darker than the female (Fig. 10:12).

Life history. The European corn borer passes the winter as a full-grown larva inside its tunnel in stubble, in a stalk, in an ear of corn, in a weed, or in

Figure 10:12. Adult moths of the European corn borer on a corn leaf. The female, larger and lighter, is near an egg mass. The male is smaller and darker. *Courtesy Illinois Natural History Survey.*

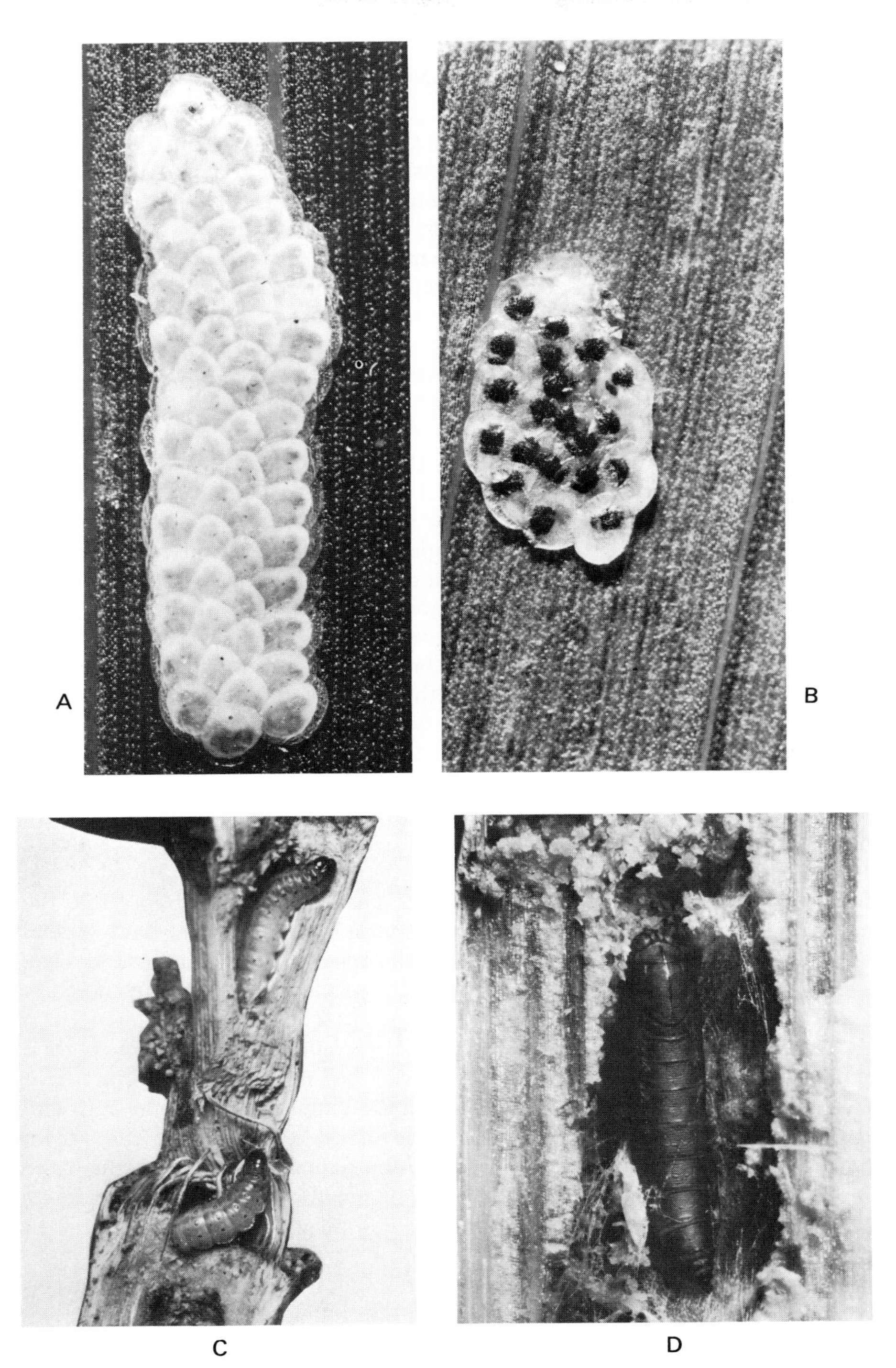

Figure 10:13. Stages in the life of the European corn borer. A, fresh egg mass; B, egg mass in "black head" stage; C, two larvae exposed in ear shank; D, pupa exposed in stalk. *Courtesy USDA.*

other protective plant material. In May or June, before changing to the pupal stage within the plant, the borer cuts a circular exit hole for the escape of the future moth. It then closes the hole with a thin webbing of silk, returns to the tunnel, and usually spins a flimsy cocoon before it pupates. After the pupal stage which lasts from 10 to 15 days, the moths appear in late spring and early summer. Within 24 hours after emergence, mating takes place, usually at dusk, when European corn borer moths are most active. During the day they remain hidden under the leaves of corn or in weeds growing in and around the fields. Three days after mating the females begin to lay eggs. In the evening they fly from plant to plant frequently selecting the tallest upon which to lay their eggs. On the under surface of the leaves, they usually deposit eggs in clusters of 14 to 20 but may deposit as few as 1 or as many as 162. During their lifetime of 6 to 24 days individual females may lay up to 1900 eggs, although the average is about 400.

The eggs hatch in 3 to 12 days depending on temperature. The young larvae wander for a while on the leaf and feed slightly on the surface. Soon they migrate down into the growing whorl, where they live as young larvae. The borers have five or six instars and become full grown in about 25 to 35 days. The part of the plant upon which borers feed is influenced by their age and generation and also by the stage of the corn plant. Hatching when corn is in the whorl stage, first-generation larvae feed in the whorl on the leaf surface. Later, as third and fourth instar larvae, they feed heavily on the sheath and midrib, and around the collar. The fifth and sixth instars bore extensively in the stalk. Larvae of the second generation, hatching when corn plants are at their maximum height, feed on plant tissues near the leaf axil where pollen grains often accumulate or they seek shelter and feed in between the husks and around the silk of the ear. They begin to bore into the stalks at an earlier age than do first-generation larvae, and they tunnel extensively in stalk, shanks, and ears. Many larvae that hatch do not survive. The highest survival occurs when eggs hatch at the time corn is in the early green tassel stage to the midsilk stage.

In midsummer as first-generation larvae become full grown, some enter a state of diapause and eventually overwinter, others pupate and emerge as moths in August. The latter, the first-generation adults, produce the eggs that hatch into second-generation larvae. The larvae that become full grown before cold weather sets in are able to hibernate and survive through the winter. The development of all stages is influenced by seasonal temperatures and may be either advanced or retarded depending on the earliness or lateness of the growing season.

Recent laboratory studies have indicated that the induction of diapause is controlled mainly by temperature and photoperiod acting upon the last larval instar. At 65°F and 9½ to 14 hours of light per day, 95 per cent of the larvae entered diapause. Both greater and lesser amounts of light reduced the number entering diapause. Larvae kept at a temperature of 75°F could not be induced to diapause at any photoperiod.

Biological control. Because the European corn borer was introduced from abroad, it had few parasites and predators to contend with in its new home. In 1919 entomologists began searching for its natural enemies in Europe and in the Orient. They imported 24 species, and of these 21 were released over the borer-infested areas of this country. Four of the parasites have succeeded in reducing infestations. The most abundant and most widely effective is *Lydella thompsoni* Herting, a fly that looks much like the common house fly. Live maggots, deposited by female flies in vicinity of host larvae, penetrate the bodies of the pests, feed internally, and cause their death. The other three parasites are small wasps that also attack the borer stage.

A five-year study in several midwestern Corn Belt states indicates that predators play an important part in population fluctuation of the corn borer at some locations during certain years. Predators, however, cannot be depended upon year after year, or in any given year, to play a significant part in controlling a population at a specific location. The predators included eight species of ladybird beetles, one chrysopid, one syrphid, one anthocorid, mites, and perhaps ground beetles. Trials with *Bacillus thuringiensis*, gamma irradiation, and chemosterilization have also shown promise in the control of the European corn borer.

Cultural control. Corn growers can reduce the amount of borer injury by employing several cultural methods: (1) selecting resistant or tolerant varieties, (2) correctly timing the planting, and (3) destroying overwintering borers by plowing and by sanitation.

1. Plant breeders have developed inbred lines of field corn with a good degree of resistance to the borer. These are used to produce hybrids with even greater resistance through the cumulative effect of an undetermined number of resistant factors. In hybrids that contain three resistant lines, 50 per cent fewer borers survive than in hybrids made up of three susceptible lines. Certain hybrids tolerate borer infestation, for they produce high yields in spite of being infested. They stand up well and retain their ears, permitting efficient mechanical harvesting. More than 24 inbreds and about the same number of hybrids have been released in Corn Belt states as resistant to the European corn borer. All have been grown, but because of secret pedigrees of commercial hybrid corn, it has been impossible to link breeding accomplishments for European corn borer resistance to increases in yield by borer-resistant hybrids. In a recurrent selection program, plant breeders have shown that two cycles of selection were sufficient to shift the frequencies of resistant genes to a high level in all varieties. Research has indicated the importance of the aglucone DIMBOA (2,4-dihydroxy-7-methoxybenzoxazin-3-one) as an active agent in the resistance of maize to the borer.

Some inbred lines contributing resistance to hybrids are A392, B2, B14, B15, B30, B52, L317, Oh7, Oh40B, Oh41, Oh43, Oh45, Oh51A, Oh561, R4, R61, W10, W22, W23, and WR3. The number of resistant lines in a hybrid determines its level of resistance. For effective resistance, a hybrid should contain at least three.

2. Planting corn at the time that gives best yields for an area keeps the injurious effects of borers at a minimum. Early planting results in a heavier infestation of first-generation borers, which reduce yields; late planting results in increased infestations of second-generation borers, which cause stalk breakage and ear dropping. Late planting, in itself, may reduce yields and increase moisture content at harvest.

3. Entomologists have found that plowing under corn stubble and residues reduces the number of hibernating larvae by as much as 99 per cent. Plowing is done preferably in fall but may be done in spring before the moths emerge. Borers are able to crawl to the surface, but finding no plant fragments for shelter, they succumb to adverse weather and predators. Buried moths cannot make their way to the soil surface.

In actual practice, control of the European corn borer by plowing and sanitation methods has been disappointing. In Ontario, where the government enforced cleanup practices, the population of this insect did not differ from that in Michigan just across the Detroit River, where these practices were not enforced.

Chemical control. Because cultural methods alone do not afford sufficient protection of field corn from heavy infestations of European corn borer, chemical methods have been developed and are now recommended. Although DDT has been the most widely used insecticide in the past for control of European corn borers, its registration was canceled; for a time other chemicals such as endrin and toxaphene were used extensively. Now, because of restrictions, endrin can no longer be used, and toxaphene is being replaced by such materials as carbaryl, diazinon, EPN, and other less persistent insecticides.

Proper timing of applications is a critical factor in achieving adequate control and differs with the area in which corn is grown. Various procedures have been devised to determine correct timing. For control of first-generation borers, one method consists of determining the number of plants showing feeding injury in the whorl. When 75 per cent of plants in a field show the characteristic holes in leaves, it is time to apply insecticide. A heavily infested field or a valuable seed crop may require a second treatment seven days later.

Field experiments have shown the possibility of controlling European corn borers and corn rootworms with a single well-timed application of insecticide granules of diazinon, parathion, or phorate. The success of such a program will depend highly upon proper timing and synchronizing of the application with the life cycles of the two insects.

When heavy populations of first-generation moths develop and reproduce, a treatment to control second-generation borers may be necessary in order to prevent stalk breaking and ear dropping. The level of infestation requiring treatment is approximately 100 egg masses or more per 100 plants, and the suggested time for applying the treatment is when eggs begin to hatch.

Growers treat field corn with insecticidal granules, sprays, or dusts. Granules are most efficient against first-generation borers because they roll down into the whorl of growing leaves where the young borers congregate. Dust is the least efficient formulation. Insecticides that have been recommended are carbaryl, carbofuran, diazinon, fonofos, EPN, methomyl, parathion, phorate and others.

Insecticidal control of borers attacking sweet corn is essential to obtain marketable ears. The rates and methods of applying insecticides are the same as for field corn, but the number of treatments is usually increased. Adequate control against borers of either first or second generations may require up to four applications spaced at intervals of five days. It is advisable to begin treating when eggs start to hatch.

Northern Corn Rootworm. *Diabrotica longicornis* (Say)

Western Corn Rootworm. *Diabrotica virgifera* LeConte

Southern Corn Rootworm. *Diabrotica undecimpunctata howardi* Barber
[Coleoptera:Chrysomelidae]

Three species of beetles, whose larval stages are called corn rootworms, are major pests of corn. The economic importance of each species varies in different regions of the United States. Although the northern corn rootworm was at one time the most serious in the Corn Belt states, it is being replaced rapidly by the western corn rootworm. Once primarily a pest of corn in Colorado, Kansas, and Nebraska, the western corn rootworm is now moving east and north into most of the Corn Belt states. The southern corn rootworm is found mainly in corn in the South and Midwest. In Canada the southern corn rootworm is widely distributed in the eastern provinces, whereas the northern corn rootworm is found chiefly in southwestern Ontario.

Rootworms damage corn by feeding on and tunneling inside the roots. The injury reduces the amount of food available to the plant for growth and ear development and consequently lowers yield. Tests have indicated yield reductions of 10 to 30 per cent where corn was picked by hand and weighed. Rootworms may destroy completely both the main roots and the brace roots touching the soil. Affected plants readily lodge during wind and rain storms and are difficult to harvest. In addition to direct injury, rootworms transmit bacterial wilt or Stewart's disease of corn and make wounds which allow the entry of rot organisms. The southern corn rootworm not only bores into roots but also enters the stalk just above the roots. Here it eats out the crown of young plants and kills the bud. In the South this species often destroys 25 per cent of the stand when plants are 6 to 18 in. high. Although the larvae cause most of the injury to corn, the adults feed upon every part of the plant above ground. Their feeding on newly emerging silks prior to pollination is most serious. This injury can result in sparsely filled ears.

Description. The egg of the northern corn rootworm is oval, 0.5 mm long, pale yellow, and sculptured with hexagonal pits. The larva is a slender, white or pale yellow grub which grows to a length of $\frac{4}{10}$ in. (Fig. 10:14). The pupa is

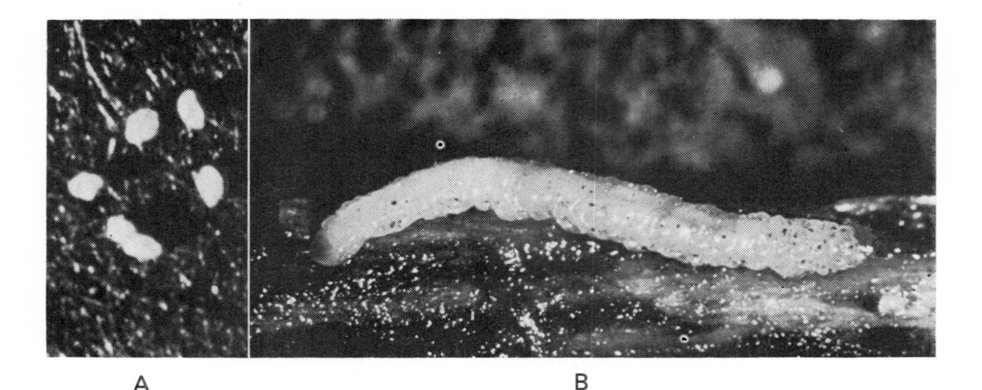

Figure 10:14. Eggs (A) and larva (B) of the northern corn rootworm. *A, courtesy University of Nebraska, Lincoln; B, courtesy Illinois Natural History Survey.*

white and fragile. The adult beetle is $\frac{1}{6}$ to $\frac{1}{4}$ in. long and is uniform green to yellowish green in color (Fig. 10:15).

The egg, larva, and pupa of the western and southern corn rootworm are similar in appearance to the same stages of the northern corn rootworm. The adults, however, are easily distinguishable: the western species is yellowish green with three dark stripes or a large dark area on the elytra, whereas the southern species is yellowish green with 11 black spots on the elytra.

Life history. The northern and the western corn rootworm have similar life histories. There is a single generation annually, and the species overwinter in the egg stage. Eggs hatch in May and June, and the larvae move about in the soil until they find the corn roots upon which they feed. The western corn

Figure 10:15. A, adult stage of northern corn rootworm; B, southern corn rootworm; C and D, western corn rootworm. *Courtesy Iowa State University, Ames.*

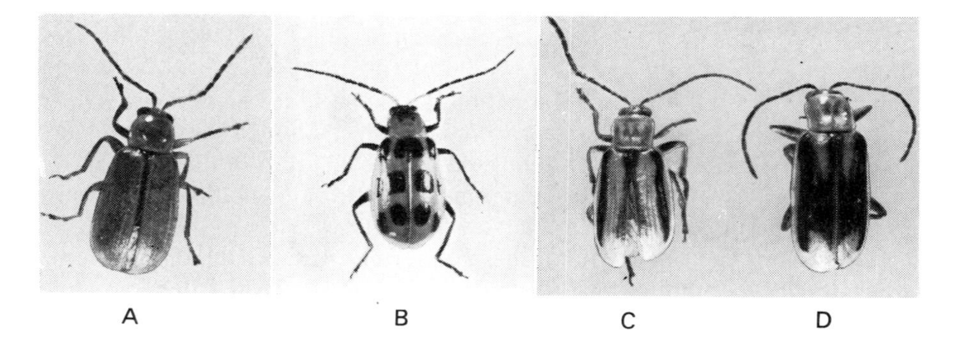

rootworm larvae prefer corn, but they survive and develop on green and yellow foxtail grass, Minter wheat, Omugi barley, and Oahe intermediate wheatgrass.

Rootworms become full grown during July and pupate in cells in the soil. The adult beetles emerge in late July and August. In corn fields they feed principally upon corn pollen and silks, but if they leave corn fields, they consume the pollen of many kinds of plants. The beetles are active insects and fly away rapidly when disturbed. They often congregate in large numbers in the whorls, in axils of leaves, on tassels, or on ears. After mating the females lay clusters of eggs in the soil at depths of $\frac{1}{2}$ to 2 in. usually among the brace roots of corn plants. In many irrigated corn fields in Nebraska, they deposit eggs in the furrows between the rows. The main period of egg laying extends from late July through September.

Control. Growers use both cultural and chemical methods to control rootworms. Cultural practices useful against the northern and western corn rootworms include crop rotation and application of nitrogen fertilizers. A general rule for rotation is to grow corn for one or two years then to follow with a legume or small grain crop for two years. An exception to the rule is rotating corn with oats or certain varieties of wheat or barley because this practice allows rootworms to multiply and injure corn. In the North, rotation is ineffective against the southern corn rootworm because the adults migrate in from the South and lay eggs after the corn is up and because the species has two generations a year.

Applying extra nitrogen helps overcome the injury of rootworms. Whether the beneficial effect results from the remaining roots being able to supply the plant's requirement or whether the additional nitrogen stimulates regeneration of roots is still unknown. If corn is under irrigation, timely application of water also helps to surmount rootworm injury.

In the South, growers use three cultural practices to control the southern corn rootworm. They plow and disk early, at least 30 days before planting corn, which kills many rootworms and discourages further egg laying. They also plant early when rootworms are least active, and plant heavier than normal to ensure a good stand.

Progress has been made in breeding for host plant resistance. Inbreds showing considerable tolerance or resistance include N38A, Ind. 38-11, HD2187, SD10, C.I.38B, B55, Oh05, A251, MO22, H51, MO12, A297, and B57. The following plant introduction materials have also shown promise: PI 177606, PI 177645, PI 214288, PI 239099, PI 257625 and PI 303923. The component most responsible for good performance appears to be tolerance—the ability to withstand and repair damage after larval feeding. Tolerance has been found to be closely associated with root regeneration.

Applying recommended insecticides to the soil before or at time of planting controls all species of rootworms effectively. Growers have a choice of either broadcasting the insecticide over the field or applying it in the rows.

In the broadcast method one applies sprays, granules, or insecticide-fertilizer mixtures uniformly over the field before planting and immediately disks the material into the top 3 or 4 in. of soil. Delaying the disking operation only four hours may reduce control by 50 per cent. Another way of broadcasting insecticides, and one that is becoming quite popular, is to dispense them in fertilizer solutions. For this method, specially formulated emulsifiable concentrates are added to liquid fertilizers. Broadcast applications are more effective in protecting surface-planted corn than listed corn.

In the second method, row treatment, a narrow band of spray or granules is applied when seeding. The insecticide flows through tubes and emerges behind the planter shoes and in front of the covering disks or press wheels. The insecticide thus comes to lie in the soil near the seed and will be in the area of the root zone to protect the roots. By employing a starter-fertilizer attachment at planting time, one may also row treat with insecticide-fertilizer mixtures. The same insecticides and rates of actual toxicant are effective.

After ten years of extensive use of aldrin and heptachlor in Nebraska and Kansas for control of western corn rootworm, the insect developed a high degree of resistance to these insecticides. Following the first evidence of resistance in Nebraska in 1960, resistant western corn rootworms spread rapidly and have since appeared in practically all the northcentral Corn Belt states. The northern corn rootworm developed varying degrees of resistance to aldrin and heptachlor in localized areas of several Corn Belt states. Because use of these materials has now been canceled, they are no longer recommended.

Because of the rise of resistance to the chlorinated hydrocarbons, many substitute insecticides—both organophosphorus and carbamate—were developed and applied for corn rootworm control. Methods of application were also changed. Insecticides are now applied primarily in the granular form and generally as a row treatment rather than broadcast.

The following insecticides have entirely replaced aldrin and heptachlor for corn rootworm control: carbofuran, chlorpyrifos, fensulfothion, diazinon, fonofos, ethoprop, phorate, and terbufos. Most of these are applied at 1 lb or less of active ingredient per acre and as granules in a 5 to 7-in. band over the row ahead of the press wheels at planting time. Pounds of granules per acre are generally based on 40-in. rows, and dosage rates on narrower rows should be increased, approximately 10 per cent for every 4-in. decrease in row width. More recently recommendations are often given in ounces of granules per 1000 ft of row. Evidence indicates that in certain areas, western corn rootworm larvae are developing resistance to some of the organophosphorus insecticides.

When necessary, corn rootworm adults can be controlled by treating with carbaryl, diazinon, malathion, or parathion. For control of complexes of soil insects that attack corn, growers may broadcast 2–4 lb of diazinon, fonofos, or other recommended insecticides. Various feeding restrictions are imposed on the ensilage and stover following soil treatment with some insecticides.

Because restrictions may vary in different states and change from year to year, corn growers should consult with their county agricultural agent or agricultural college on current recommendations.

SELECTED REFERENCES

Blanchard, R. A., and W. A. Douglas, *The Corn Earworm as an Enemy of Field Corn in the Eastern States*, USDA Farmers' Bul. 1651, 1953.

Branson, Terry F., and Eldon E. Ortman, "Host Range of Larvae of the Western Corn Rootworm," *J. Econ. Entomol.*, 60(1):201–3 (1967).

Brindley, T. A., A. N. Sparks, W. B. Showers, and W. D. Guthrie, "Recent Research Advances on the European Corn Borer in North America," *Ann. Rev. Entomol.*, 20:221–39, 1975.

Burkhardt, C. C., *Corn Insects*, Kan. Agr. Exp. Sta. Bul. 373a, 1955.

Chiang, H. C., "Bionomics of the Northern and Western Corn Rootworms," *Ann. Rev. Entomol.*, 18:47–72, 1973.

Chiang, H. C., J. L. Jarvis, C. C. Burkhardt, M. L. Fairchild, G. T. Weekman, and C. A. Triplehorn, *Populations of European Corn Borer,* Ostrinia nubilalis *(Hbn.) in Field Corn,* Zea mays *(L.,)*, N. Cent. Reg. Pub. 129, Mo. Agr. Exp. Sta. Bul. 776, 1961.

Chiang, H. C., A. J. Keaster, and G. L. Reed, "Differences in Ecological Responses of Three Biotypes of *Ostrinia nubilalis* from the North Central United States," *Ann. Entomol. Soc. Amer.*, 61(1): 140–6 (1968).

Davis, John J., *The Corn Root Aphid and Methods of Controlling It*, USDA Farmers' Bul. 891, 1949.

Dicke, F. F., and M. T. Jenkins, *Susceptibility of Certain Strains of Field Corn in Hybrid Combinations to Damage by Corn Earworms*, USDA Tech. Bul. 898, 1945.

Douglas, W. A., and R. C. Eckhardt, *Dent Corn Inbreds and Hybrids Resistant to the Corn Earworm in the South*, USDA Tech. Bul. 1160, 1957.

Everett, T. R., H. C. Chiang, and E. T. Hibbs, *Some Factors Influencing Populations of European Corn Borer* Pyrausta nubilalis *(Hbn.) in the North Central States*, Minn. Agr. Exp. Sta. Tech. Bul. 229, 1958.

Fitzgerald, Paul J., and Eldon E. Ortman, "Two-year Performance of Inbreds and Their Single Crosses Grown Under Corn Rootworm Infestation," *Proc. N. Cent. Br.–E.S.A.*, 20:46–7 (1965).

Guthrie, W. D., F. F. Dicke, and C. R. Neiswander, *Leaf and Sheath Feeding Resistance to the European Corn Borer in Eight Inbred Lines of Dent Corn*, Ohio Agr. Exp. Sta. Bul. 860, 1960.

Henderson, C. A., and Frank M. Davis, *The Southwestern Corn Borer and Its Control*, Miss. Agr. Exp. Sta. Bul. 773, 1969.

Hill, R. E., A. N. Sparks, C. C. Burkhardt, H. C. Chiang, M. L. Fairchild, and W. D. Guthrie, *European Corn Borer,* Ostrinia nubilalis *(Hbn.) Populations in Field Corn,* Zea mays *(L.) in the North Central United States*, N. Cent. Reg. Pub. 175, Nebr. Agr. Exp. Sta. Res. Bul. 225, 1967.

Huber, L. L., C. R. Neiswander, and R. M. Salter, *The European Corn Borer and Its Environment*, Ohio Agr. Exp. Sta. Bul. 429, 1928.

Kirk, V. M., *Corn Insects in South Carolina*, So. Car. Agr. Exp. Sta. Bul. 478, 1960.

Klun, J. A., C. L. Tipton, and T. A. Brindley, "2,4-Dihydroxy-7-methoxy-1,4-benzoxazin-3-one (DIMBOA), an Active Agent in the Resistance of Maize to the European Corn Borer," *J. Econ. Entomol.*, 60(6):1529–33 (1967).

Knapp, J. L., F. G. Maxwell, and W. A. Douglas, "Possible Mechanisms of Resistance of Dent Corn to the Corn Earworm," *J. Econ. Entomol.*, 60(1):33–6 (1967).

Luginbill, Philip, *Habits and Control of the Fall Armyworm*, USDA Bul. 1990, 1950.

Luginbill, Philip, Sr., and T. R. Chamberlin, *Control of Common White Grubs in Cereal and Forage Crops*, USDA Farmers' Bul. 1798, 1953.

Painter, R. H., *Insect Resistance in Crop Plants* (New York: Macmillan, 1951).

Penny, L. H., Gene E. Scott, and W. D. Guthrie, "Recurrent Selection for European Corn Borer Resistance in Maize," *Crop Science*, 7:407–9 (1967).

Robert, Alice L., *Bacterial Wilt and Stewarts Leaf Blight of Corn*, USDA Farmers' Bul. 2092, 1955.

Rolston, L. H., *The Southwestern Corn Borer in Arkansas*, Ark. Agr. Exp. Sta. Bul. 553, 1955.

———, C. R. Neiswander, K. D. Arbuthnot, and G. T. York, *Parasites of the European Corn Borer in Ohio*, Ohio Agr. Exp. Sta. Res. Bul. 819, 1958.

Satterthwait, A. F., *How to Control Billbugs Destructive to Cereal and Forage Crops*, USDA Farmers' Bul. 1003, 1932.

Sifuentes, J. A., and R. H. Painter, "Inheritance of Resistance to Western Corn Rootworm Adults in Field Corn," *J. Econ. Entomol.*, 57:475–7 (1964).

Sparks, A. N., H. C. Chiang, C. C. Burkhardt, M. L. Fairchild, and G. T. Weekman, "Evaluation of the Influence of Predation on Corn Borer Populations," *J. Econ. Entomol.*, 59(1):104–7 (1966).

Sparks, A. N., C. A. Triplehorn, H. C. Chiang, W. D. Guthrie, and T. A. Brindley, *Some Factors Influencing Populations of the European Corn Borer*, Ostrinia nubilalis (*Hubner*) *in the North Central States*, N. Cent. Reg. Res. Pub. 180, Ia. Agr. Exp. Sta. Res. Bul. 559, 1967.

Stoner, Warren N., and A. J. Ullstrup, *Corn Stunt Disease*, Miss. Agr. Exp. Sta. Info. Sheet 844, 1964.

USDA, *Grasshopper Control*, USDA Farmers' Bul. 2193,1975.

USDA, *The European Corn Borer*: *How to Control It*, USDA Farmers' Bul. 2190, 1968.

USDA, *Corn Earworm in Sweet Corn*: *How to Control It*, USDA Leaflet 411, 1967.

USDA, *The Southern Corn Rootworm*: *How to Control It*, USDA Leaflet 391, 1961.

USDA, *Chinch Bugs*: *How to Control Them*, USDA Leaflet 364, 1958.

Wendell, J. and W. W. Copeland, "Distribution and Abundance of the Corn Earworm in the United States," *USDA Cooperative Economic Insect Report* 21:71–6 (1971).

Wressell, Harry B., and Marcel Hudon, *Common Insects of Corn in Eastern Canada*, Canada Dept. of Agr. Publ. 945, 1968.

chapter 11 / **B. AUSTIN HAWS**

INSECTS OF LEGUMES

Cultivated leguminous plants may be grouped for convenience into three categories: (1) forage and pasture plants such as alfalfa, clovers, and vetches; (2) human food plants such as peas, beans, and peanuts; (3) miscellaneous plants that yield medicines, dyes, oil, insecticides, and timber. In this chapter we shall be concerned mainly with the first category.

Pests of human food legumes are discussed in Chapter 13, "Vegetable Crop Insects." Because soybeans may be classified in various ways and are not discussed in Chapter 13, they are included here. Soybeans have become an increasingly important crop, widely utilized as food for humans and livestock and as prime material for industry. Except for insecticides the third category has not been included in this text.

Forage legumes are attractive to many insects. Approximately 500 different insects have been found in alfalfa, for example. There are others yet to be found, and new ones undoubtedly will appear in the future. Fortunately the majority of these insects are not injurious. Many are parasites, predators, visitors, or pollinators. They may be neutral in their relationship with the plants, or they may feed on nectar, pollen, or other plant substances without injuring the alfalfa.

THE PESTS

Among the important insect pests of forage legumes are the alfalfa weevil, lygus bugs, alfalfa seed chalcid, and meadow spittlebug. These four insects will be discussed in detail later in this chapter. The insects most injurious to legumes belong to five orders: (1) the Orthoptera (grasshoppers and crickets); (2) the Hemiptera (plant bugs and lygus bugs); (3) the Homoptera (leafhoppers, aphids, and spittlebugs); (4) the Coleoptera (weevils and beetles); and (5) the Lepidoptera (larvae of butterflies and moths; the adults rarely are injurious).

Grasshoppers and Crickets

Grasshoppers frequently do great damage to alfalfa and other legumes. With their chewing mouthparts, they eat leaves and other plant parts. Four of the most common grasshoppers causing damage to alfalfa are the **differential grasshopper**, **migratory grasshopper**, **redlegged grasshopper**, and **twostriped grasshopper** (Fig. 9:7). The black **field cricket** relishes the soft seeds of legumes.

Plant Bugs and Lygus Bugs

Plant bugs and lygus bugs have piercing-sucking mouthparts. They feed by sucking up plant juices; often the areas of the plant where they feed become yellowish, whitish, and necrotic. If one sweeps alfalfa with an insect net (Fig. 11:8) to sample the insects, these plant bugs are often the most abundant insects caught in the sample. They include the **lygus bugs**, **tarnished plant bug**, **alfalfa plant bug**, and **rapid plant bug**. The plant bugs feed on many different kinds of plants, alfalfa, grasses, vegetables, fruits, cotton, and weeds. They are likely to be found on most of the important forage legumes. They feed on the seed of many crops as well as the foliage. Young seeds attacked by plant bugs may be destroyed or be damaged so they no longer retain the capacity to germinate.

Leafhoppers and Aphids

Leafhoppers (Fig. 11:1) also have piercing-sucking mouthparts. Frequently they are overlooked in legume fields because they take off rapidly when disturbed and because they are not easily seen because of their small size and pale green color. It is not uncommon to catch 300 to 500 or more leafhoppers in a single sweep with an insect net. In eastern United States, west to Colorado and Wyoming, the **potato leafhopper** injures alfalfa and some clovers in an unusual way. In sucking fluid from vascular tissues, the insects inject toxic secretions into the plant. These secretions result in spotting or stippling and a pink or reddish discoloration of leaves beginning at the rib. When injury is severe, the foliage may wilt and exhibit a condition commonly known as **yellows**. The small wedge-shaped, pale green insects do not hibernate or overwinter in northern areas of the United States, but they breed continuously throughout the winter in the gulf states and migrate northward early in the spring. The **clover leafhopper** is commonly found on clovers and sometimes on alfalfa. Leafhoppers vary considerably in size, color, and other characteristics. Telling them apart is usually difficult and the majority can be identified only by a few specialists.

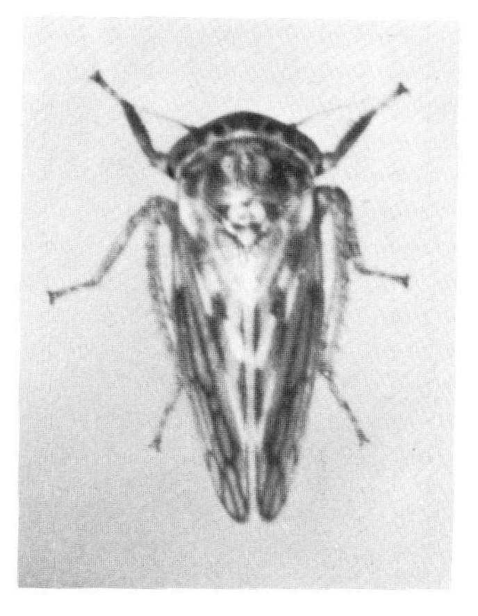

Figure 11:1. A leafhopper, an insect found abundantly in legume forage crops. Species shown here is *Aceratagallia arida* Oman.

Aphids (Fig. 11:2) are common pests of clovers and alfalfa. The **pea aphid** is found in tremendous numbers. Some species of aphids can be present in large numbers without apparent injury to the plants, but with others, such as the spotted alfalfa aphid, a relatively few aphids injure plants causing plants to wilt and become stunted and discolored, decreasing yields substantially. Severe infestations of aphids have destroyed stands of seedlings and even old stands of alfalfa. The **spotted alfalfa aphid** was first found in New Mexico in 1954 and soon spread at a phenomenal rate over most of the United States and into Mexico. Some believe this aphid was introduced into North America inadvertently, possibly from the Mediterranean region. It is estimated that damage to alfalfa by the spotted alfalfa aphid from 1954 to 1956 was approximately $81 million. Spotted alfalfa aphids are smaller than pea aphids; they are yellowish green, with four to six rows of small black spots and little spines on their backs, and have spotted wings. Several biotypes or strains have been discovered since the aphid was introduced into the United States. This aphid has been most abundant on alfalfa, but it also infests clovers, sweetclover, sour clover, and berseem. Fortunately, cultural practices, such as early planting of seed, mowing or pasturing alfalfa, and maintaining vigorous alfalfa stands, have been useful in reducing its damage. Several aphid-resistant alfalfas—including Cody, Dawson, Kanaz, Lahonton, Moapa, Sonora, Washoe, and Zia—have been developed.

The **blue aphid**, *Acyrthosiphon kondoi* Shinji, was found in California in 1974. Since that time the aphid has spread to Arizona, Nevada, and Utah and appears to be a new potential threat to alfalfa and other forage legumes. The **clover aphid** is a common pest of red clover and alsike clover. It reduces growth of these clovers, and its excretions of honeydew result in parts of the flower sticking together in clumps.

Figure 11:2. The pea aphid, *Macrosiphum pisi* (Harr.), a common insect in fields of alfalfa and other legumes. Note female on right giving birth to young aphid.

Weevils and Beetles

A large number of weevils are important pests of legumes. One of the larger weevils commonly found is the **clover leaf weevil**, which has been present for many years in the United States but is a native of Europe. Both larvae and adults feed on the foliage of clovers and alfalfa.

Three of the important insects that injure the roots of legume crops are the clover root curculio, the clover root borer, and the alfalfa snout beetle. Adults of the **clover root curculio** feed on leaf edges and are of relatively minor importance; but the larvae feed on tender roots and chew large cavities in the main roots of the plants. This weevil is an important pest of red, sweet, and alsike clovers and of alfalfa in many areas of the United States and southern Canada.

The **clover root borer** is recognized as a major pest of red clover. Most of the life cycle of the clover root borer is spent in the roots of red clover. The adults are cylindrical, black or dark brown beetles about 2.5 mm long. This insect is distributed in the northern part of the United States and eastern

Canada. **Alfalfa snout beetles** injure roots of alfalfa and red clover. They have a two-year life cycle and spend most of their larval stage in the soil.

Sweetclover weevils are gray or brownish weevils, about $\frac{1}{4}$ in. long with a short snout (Fig. 11:3). They are a major pest of sweetclover in many areas of the United States and Canada. Larval feeding on the roots is reported to have a negligible effect on the plants, but the adults often destroy new seedlings and may defoliate larger plants.

Lesser clover leaf weevils are commonly found on red clover and other species of clovers and alfalfa. Adults feed mostly on leaves, but the larvae feed on the stems, buds, and florets. Larval feeding deforms the clover seed heads. **Clover seed weevils** are frequently pests of white Dutch and alsike clovers. Damage by these weevils is caused mostly by the larvae developing inside the seed pods. Adults may also injure the plant stems by their feeding. **Clover head weevils** live in the heads of clovers and have been reported damaging alfalfa. These weevils appear to be spreading from the eastern United States into previously uninfested areas of the West. The adults of these weevils and the alfalfa weevil (Fig. 11:9) look much alike.

The **vetch bruchid** is known mostly in the Atlantic and the seaboard states, from New Jersey south to Georgia. It is an important pest only in seed production. The larvae are found in seed pods and feed on the developing seed.

Several kinds of adult **blister beetles** feed on florets or other plant parts and are considered periodic pests of forage legumes. Adults may be plain black or gray, or have other colors of spots or stripes, and are usually about $\frac{1}{2}$ to $\frac{3}{4}$ in. long.

Butterflies and Moths

Among the important lepidopteran pests of legumes, the **alfalfa caterpillar** in some states is the worst insect pest of hay. Periodically several species of armyworms—the **fall armyworm**, **western yellowstriped armyworm** (Fig. 11:4), **bertha armyworm**, and **beet armyworm**—infest legumes. Cutworms and armyworms feed at night and hide under surface litter or in the soil during the day. Because of this behavior, fields may be injured severely before the cutworms are discovered. The **army cutworm** and **redbacked cutworm** are serious pests early in spring and may retard growth of alfalfa a month or longer.

Occasionally, webworms—the **garden webworm** and the **alfalfa webworm**—damage clover or alfalfa. These yellowish green caterpillars have conspicuous black spots and are about 1 in. long when full grown. Large numbers may completely defoliate a crop and leave considerable webbing on the plants. A leafroller, *Sparganothis xanthoides* Walker, damages birdsfoot trefoil and prevents pollination and seed setting by webbing the terminal floral parts.

White, red, mammoth, and alsike clovers are sometimes attacked by the **clover head caterpillar**. These small hairy caterpillars, about $\frac{1}{4}$ in. long, feed on

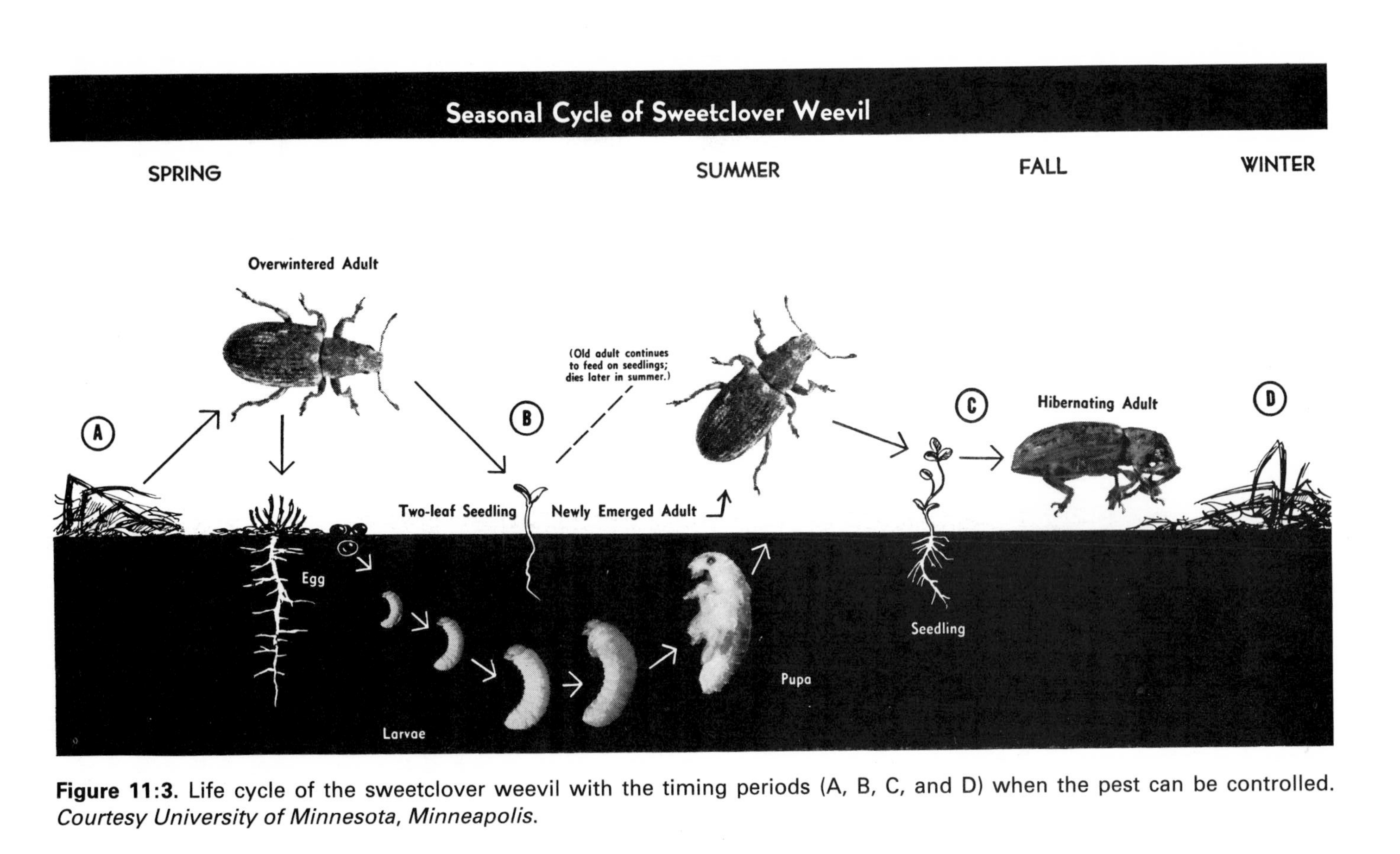

Figure 11:3. Life cycle of the sweetclover weevil with the timing periods (A, B, C, and D) when the pest can be controlled. *Courtesy University of Minnesota, Minneapolis.*

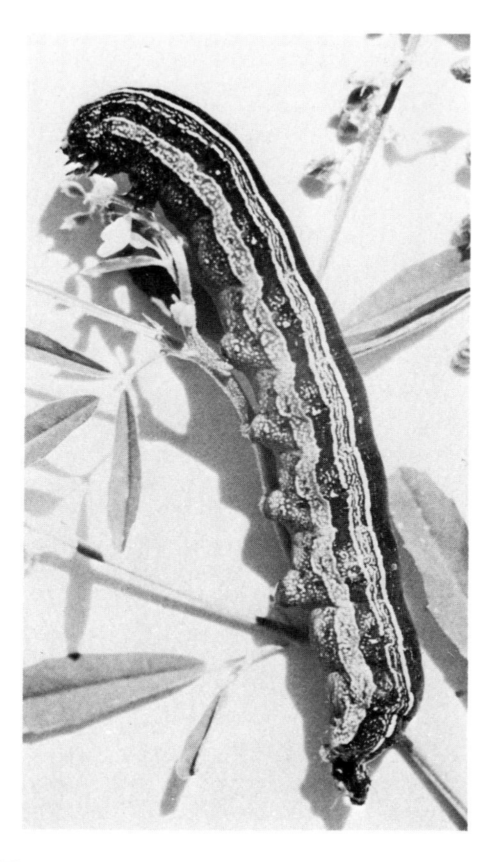

Figure 11:4. The western yellowstriped armyworm, a general feeder that is sometimes a pest of alfalfa and other legumes. *Courtesy William P. Nye, USDA.*

the developing seeds or at the base of the florets. The larvae may attack the leaves of plants when the heads are not present. The **clover bud caterpillar**, a similar species, damages clover in the Pacific Northwest.

Other Insect and Mite Pests of Legumes

At certain stages of their development some of the flies are pests of legumes. For example, the **clover seed midge** is an important pest in England, Europe, New Zealand, and North America. The larvae, which are approximately $\frac{1}{16}$ in. long and deep red or orange in color, destroy the ovaries of developing seeds in red clover; infested heads can be recognized by their uneven bloom. Some of the other general pests of legumes include thrips, ants, and mites (Fig. 11:5).

Figure 11:5. Spider mite damage to alfalfa. Note the yellow spotting of the leaves. *Courtesy William P. Nye, USDA.*

Soybeans have been one of the fastest developing agricultural commodities in our nation's history. In 1975, plantings in the U.S. totaled over 56 million acres, with more than 1.5 billion bushels being produced. Soybeans are attacked by insects from the seedling stage to harvest by many different insect and other pests. More than 1500 species of insects and their relatives have been collected from soybeans, and approximately 40 species have been considered present or potential pests. Some of the most important pests include: green cloverworm, potato leafhopper, larvae of the painted lady butterfly, velvetbean caterpillar, *Heliothis* spp., cabbage looper, Mexican bean beetle, brown stink bug, green stink bug, southern green stink bug, threecornered alfalfa hopper, bean leaf beetle, margined blister beetle, and striped blister beetle.

THE INJURY

Insects injure legumes by: (1) feeding on or in seeds, seedlings, leaves, roots, stems, flowers, and reproductive parts of plants; (2) ovipositing in the plants; (3) injecting toxic or harmful substances; (4) removing cellular or other substances from plants; (5) secreting honeydew or spittle, or spinning webs that interfere with plant development or normal functions; (6) causing plant disease or transmitting agents that cause plant diseases.

There are two main types of insect pest problems on forage legumes: insects may affect the production of forage and hay, or they may be detrimental to seed production and germination. Some insects are pests of both hay and seed.

Seed injury. Insects such as the alfalfa seed chalcid may attack legume seeds as they develop in the field (Fig. 11:11). The female chalcid inserts its ovipositor through the pod and deposits eggs in the soft, developing seeds. The hatched larva destroys the seed by feeding on the internal contents.

Insects with chewing mouthparts, such as grasshoppers, may devour seeds and pods. Other examples include weevil adults, field crickets, blister beetles, and various lepidopterous larvae, such as the western yellowstriped armyworm. After seed is planted in the soil, wireworms and other soil pests may feed on and destroy the seed.

Lygus bugs, plant bugs, stink bugs, and other insects with piercing-sucking mouthparts feed on seeds by inserting their proboscis or "beak" through the pods into the developing seeds. Such feeding on immature seeds often leaves them shriveled, incompletely developed, and nonviable.

The clover seed midge is a fly that lays eggs in young clover heads; the larvae may destroy the developing seed. Clover head weevils and clover seed weevils also attack florets and seeds of clovers.

Some moth larvae, such as the tortricids, tie leaves, flowers, or stems of legumes together and feed or pupate in the "bundle" they have formed. The webbing of bunches of flowers interferes with pollination and seed development.

Seedling injury. New seedlings are often attacked by insects as soon as the tiny plants emerge from the soil. Chewing insects such as grasshoppers, flea beetles, or weevils may consume the small plants. Insects with piercing-sucking mouthparts, such as aphids, may kill or debilitate new seedlings by piercing the leaf surface, injecting toxins or disease agents, or removing plant fluids.

Leaf injury. Insects may injure plant leaves in various ways. Weevils, other beetles, grasshoppers, crickets, and larvae chew out notches (Fig. 11:6), or they may consume entire leaves. Some weevil larvae, such as those of the alfalfa weevil, may not chew sections of the leaves, but eat surface cells and skeletonize the leaves. Leafminers bore between the bottom and top layers of leaf surfaces and leave winding trails in the leaf. Feeding by mites mottles and

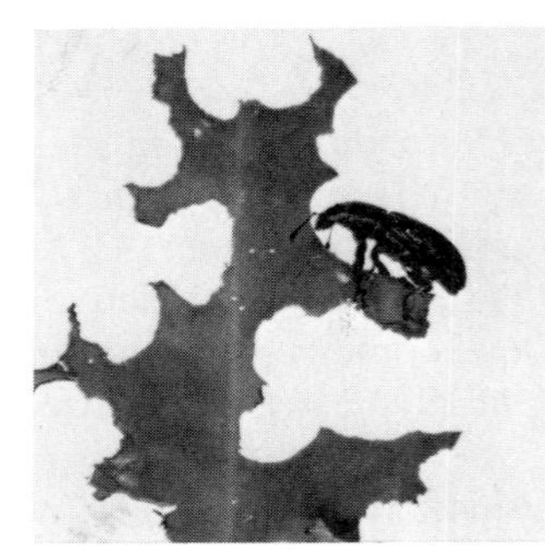

Figure 11:6. Typical notches in leaves made by sweetclover weevil. *Courtesy University of Minnesota, Minneapolis.*

dries the leaves (Fig. 11:5) and eventually turns them brown or other colors. Leafhopper feeding may also discolor leaves.

Some insects, such as the spotted alfalfa aphid, cover plants with large amounts of honeydew, which they excrete. The honeydew makes harvesting of hay difficult. Plants stick together, and the honeydew causes harvesting machines to slow down or break because the sticky parts cannot move freely. Fungi that grow well in the honeydew of aphids may develop so profusely that they degrade or ruin hay.

Root injury. Insects such as the clover root borer chew or bore holes in legume roots or feed on nitrogen nodules attached to the roots. Some of this injury may kill or damage plants directly, or the injury may weaken the plants and facilitate secondary infections of plant diseases. Some aphids feed on the roots of legumes.

Stem injury. Injury to plant stems may include consumption of stems by chewing insects and changes in structure, appearance, and development because of toxins or agents deposited when insects with piercing-sucking mouthparts feed on the plants. Plant breeders need to be aware of physical or physiological changes in plant growth that may be caused by insect feeding so as not to draw faulty conclusions about differences in certain plant characteristics being compared in studies of plant varieties. The differences may be, not genetic, but the result of changes in plant growth patterns or characteristics associated with insect feeding—for example, extra numbers of small leaves, or long side stems with small racemes instead of one healthy, normal raceme in the case of alfalfa.

INSECTS BENEFICIAL TO LEGUMES

Pollinating insects must cross-pollinate legume crops if profitable seed yields are to be realized (Fig. 11:7). In cross-pollinating legumes, various bees, such as alkali bees, leafcutting bees, honey bees, and bumble bees, take the pollen from the flowers of one plant to the flowers of another plant. Wind or rain, or

Figure 11:7. Honey bee tripping alfalfa blossom. *Courtesy William P. Nye, USDA.*

artificial pollinating devices, have not been effective agents of cross-pollination for alfalfas and clovers.

Other beneficial insects in legume fields are the predators and parasites of pest insects. Pollinators, parasites, and predators of insects should be protected when injurious insects are controlled with chemicals. Through reasonable cooperation, this can be done by selecting the appropriate chemicals and applying them at the right time.

INTEGRATED CONTROL AND PEST MANAGEMENT

The problem of insect control in most crops today is something like putting a jigsaw puzzle together and playing a game of chess at the same time. It is necessary to find all the pieces of the puzzle and put them together. You should not "move," as in the chess game, until you consider well the results of your move on all the other pieces in the game or picture.

A grower cannot afford insect control that costs him more than he earns from the crop. Learning to identify all the good and bad insects in a crop, to know their biologies and interactions, and to establish a criteria of how many insects and how much damage represent that critical level when it is or is not economical to control the pest is difficult. In fact, for many pests such information is not available because it is not yet known. Superimposed on all these problems are the effects of climate, the influence of temperature and moisture on the development of insects and their damage and on the crops, markets and costs—and the need to put all this information together to come up with an economical control strategy.

In some states and for some crops organized, interdisciplinary research teams are already working with universities, state and federal agencies, private industry, and farmers to gather information, computerize it, and on the basis of the information put into the system, take advantage of economic models that have been developed to help formulate decisions on pest control.

Because of new laws regulating the purchase and use of pesticides and the complexity and economics of controlling insects, producers of legume hay and seed in certain areas have organized to help each other, or else they are now hiring private companies to supervise programs of pest management. In these programs professional entomologists sample fields regularly to determine what kinds and numbers of beneficial and injurious insects are present, to predict their damage, and to recommend what action, if any, is needed to protect the crop most economically and spare the beneficial insects.

Where pest management services are not available or not economical, it may be necessary for producers to have their own insect nets, collect insects from their fields regularly, and learn through texts and publications of extension services how to prepare a reference collection of pest and beneficial insects and how to identify them and control the major insect pests (Fig. 11:8). In this way producers may develop their own pest management programs.

Integrating chemical control with protection of beneficial insects needs to be a regular consideration in producing legumes, for it is doubtful that the plant world could exist in its present form without the legume plants and the pollinating animals that are required for their survival.

CULTURAL CONTROL

Important and economical control of insect pests is often provided by manipulation of crops or of the insect environment to destroy insects directly or indirectly, or to break insect life cycles or create unfavorable conditions for their reproduction or development. Some cultural methods are economical because the insects may be controlled by adjusting the timing of farm work that has to be done anyhow—for example, planting, irrigating, harvesting, disposing of plant residues, plowing, or cultivating the land.

Control of insects by crop management includes the use of plant varieties known to be well adapted to an area and resistant to the dominant insect pest or pests in the area. Sometimes seed can be planted so that the crop emerges

Figure 11:8. Top, sampling alfalfa with insect net to determine the number and kinds of beneficial and injurious insects present as a guide to control strategy; Bottom, cone of net is pushed up through the hoop and insects carefully identified and counted as they escape or the insects are placed in a plastic bag or other container for counting and evaluation later. *Courtesy William P. Nye, USDA.*

when the injurious pests are absent, dormant, dead, or fewest in numbers, or at a time of the year when the insect pest is confined to a particular host. For example, lygus and some other plant bugs may be controlled partially by cutting hay crops when the eggs are present in the alfalfa stems of the plant. The eggs will be destroyed, and any young nymphs present will be killed by exposure to high temperatures and drought.

Crops may be substituted to avoid damage from a particular pest; in areas where the sweetclover weevil destroys sweetclover and practically nothing else, other legumes that the weevil will not eat may be planted. In areas where the clover root borers have become abundant, grain crops or other plants not fed on by the weevils can be rotated with the legumes. Crop management—by regulating the amount of seed sown so that the density of the plants is unfavorable for injurious insects—has influenced and controlled injurious insects. Proper use of fertilizers and irrigation water have created environments unfavorable to injurious insects and exerted partial or complete control of destructive insects.

Plowing or cultivating legume crops at a particular time—thus removing plant tops or roots, and eliminating an insect's food supply, or burying adult or immature insects or the plant materials containing the insects in the soil—can be an effective method of insect control. Summer fallowing removes food for many soil pests, and the cultivation of the land to control weeds often dries the upper portions of the soil and kills certain insects. Removing volunteer plants and weeds from borderlands, fence rows, or waste land around farm land often helps control insects that overwinter or migrate there. The insect pests develop in the weeds for part of their life, then move into agricultural crops. Flights of moths, leafhoppers, or other insects often end in vegetation along roadsides where the migrant insects feed and develop before they move into major crops. Sweeping such areas regularly with an insect net will show when insects are present and developing so that appropriate controls can be implemented ahead of their movement into crops.

NATURAL AND BIOLOGICAL CONTROL

Entire books are written on natural and biological control, but briefly stated, the fact is that parasites, predators, diseases of insects, and climatological factors are important regulators of legume pest populations. Some entomologists have speculated that these agents of control help keep the plant-feeding insects from overwhelming the rest of the world. The introduction of natural insect enemies has helped control pest insects. For example, the alfalfa weevil is sometimes held to subeconomic levels by an introduced parasite, *Bathyplectes curculionis* (Thomson). The female parasite lays eggs in the alfalfa weevil larvae. Each egg hatches into a parasitic larva, which feeds inside and eventually kills the weevil larva.

Chemical control or other control methods on legume crops need to complement natural and biological control. For example, demeton can be used for effective control for the spotted alfalfa aphid. Properly applied, this insecticide is relatively nontoxic to native predators of the aphids such as lady beetles, nabids, and syrphids.

CHEMICAL CONTROL

Several suggestions can help in using chemicals wisely to solve pest problems of legume crops. First, learn which insect or insects are injuring the crop. It is becoming a common practice for growers to own an insect net and to use it regularly for sampling the insects in their crops. Many growers prepare a reference insect collection of their own to help them identify the major pests and beneficial insects.

Once an insect is identified correctly, it is usually possible to get information about it and recommendations for its control. Such information, however, may not be available for the newer pests. For this discussion we refer to Figure 11:3, which shows in part the kinds of information one might expect to find in the literature about an insect pest and points out several opportunities for cultural, biological, and chemical control. Figure 11:3 shows photographs of the adult and immature insect; its life history; its seasonal cycle as related to crop development; possible times and places during the season and in the life cycle when one can break the life cycle of the insect by removing plant residues (such as removing the plant debris in the fall or spring); cultural practices that will destroy the adults, eggs, or larvae (such as plowing to incorporate the legume tops and roots into the soil as green manure and at the same time killing the larvae and removing the food supply of both the larvae and adults); emergence of new adults from the soil in summer when these adults cannot fly for several weeks (so that one can plan a strategy for starving them by removing their food supply or trapping them by forcing them to walk through thin bands of insecticide without having to apply chemicals to a whole field, resulting in reduced costs and less insecticide on the soil). If it is necessary to apply an insecticide, the panorama in Fig. 11:3 shows when the adults are most vulnerable to the application of insecticides and the times of the year when the materials can best be applied. Information showing how many insects or how much injury is necessary to cause economic losses to a crop and when it is or is not economical to apply insecticide is now available for major pest insects.

Programs where all available methods of control—cultural, biological, chemical, resistant varieties—are combined are often referred to as "integrated control."

Apply only the kinds and amounts of insecticides specified on the insecticide label to control a particular insect. Avoid routine application of insecticides when the target species of insects or the insect problem in a crop is unknown. Such use of insecticides is usually uneconomical, and may even increase insect problems. Follow label directions for all pesticides in determining rates of application, safety conditions, and restrictions controlling the use of each material. Get official recommendations for your area before applying pesticides.

The relative toxicity of most of the better-known insecticides to insect pollinators has been studied. Tests indicate that the following organic phosphates can be applied with relative safety to crops in bloom when the bees are not in the field, dew is not present, and temperatures are not below 65°F: tepp, carbophenothion, naled, phosphamidon, dioxathion, and mevinphos. Certain systemic insecticides—phorate, demeton, disulfoton, and schradan—have only limited toxicity to honey bees. Other insecticides that can be applied safely while bees are not foraging are methoxychlor, toxaphene, and endosulfan. A few of the organic phosphates (such as schradan and ethion) have shown such low toxicity to bees that it has been suggested they can be applied safely to legume crops nearly any time whenever bees are not in the field.

Carbaryl, a carbamate, can be used safely and effectively against many alfalfa insect pests if safeguards are taken to avoid poisoning bees. Dandelions and other plants from which bees collect nectar and pollen should not be in bloom when carbaryl is applied.

In some states research has shown that lesser amounts of certain insecticides provide adequate control of a pest insect and yet permit a more rapid development of predators and parasites that assist in reducing the pest population. Some recommendations of normal rates of application may result in a wipeout of both the pest species and beneficial insects. Differences in temperatures, relative humidity, soil types, and hours of sunshine may affect the toxicity and disintegration of pesticides, so that recommendations are not always transferable from state to state.

Control of insects on hay crops is an acute problem because of insecticide residues that are left after treatment. The following insecticides have been registered for use on **alfalfa hay**. For control of alfalfa weevil larvae: diazinon, carbofuran, azinphosmethyl, phosmet, malathion, methyl parathion, methoxychlor, carbaryl, and methidathion. For various larvae or caterpillars: trichlorfon, methoxychlor, and carbaryl. For grasshoppers: malathion, diazinon, carbaryl, mevinphos, azinphosmethyl, and dimethoate. For aphids: malathion, methyl parathion, diazinon, carbophenothion, mevinphos, disulfoton, and dimethoate.

The following insecticides currently are approved for control of injurious insects on **legume seed crops**. For alfalfa weevil larvae: malathion, methoxychlor, diazinon, methidathion, carbofuran. For lepidopteran larvae, such as cutworms: diazinon, trichlorfon, toxaphene, naled, malathion, and methomyl. For grasshoppers: dimethoate, malathion, and toxaphene.

Because of the importance of pollinating insects, different insecticides are recommended for seed crops prior to bloom and during the blooming period. Insecticides recommended for **prebloom** application include parathion, for control of pea aphids, spotted alfalfa aphids, alfalfa weevil larvae, and tortricid moths; demeton, endosulfan, and mevinphos for pea aphid and spotted alfalfa aphid; phorate and disulfoton as a seed treatment for spotted alfalfa aphid; diazinon and malathion for control of pea aphids, spotted alfalfa aphids, alfalfa weevil larvae, lygus bugs, leafhoppers, thrips, and armyworms. Alfalfa weevil larvae may also be controlled with carbofuran, methidathion, azinphosmethyl and methoxychlor. Carbaryl is used to control armyworms, caterpillars, and grasshoppers.

Included in the insecticides that may be applied **during the blooming** period are tepp, sulfur, naled, mevinphos, trichlorfon, and toxaphene. These should be applied only when bees are not in the field. During the blooming period, tepp, mevinphos, or naled are recommended for control of aphids; mevinphos or trichlorfon for lygus bugs, armyworms, grasshoppers, stink bugs, webworms, thrips, leafhoppers, and leafrollers.

Demeton, parathion, and mevinphos are considered particularly hazardous for the applicator. Special permits are required before they can be purchased or applied in some states. The plant residues of seed crops treated with insecticides (except those registered for use on hay crops) should not be fed to livestock.

For **soybean pest control**, the following have been recommended. For Mexican bean beetle: carbaryl or methomyl. For green cloverworm and corn earworm: methyl parathion, carbaryl, methomyl. For velvetbean caterpillar: methyl parathion, carbaryl, methomyl. For bean leaf beetle and green cloverworm: carbaryl.

CONTROL EQUIPMENT

Low-pressure boom sprayers are commonly used in applying insecticides on legumes (Fig. 8:4). These sprayers are especially effective for control of alfalfa weevil in stubble or in fields where vegetation is short. It is sometimes difficult to get a thorough coverage of vegetation and to cover the undersides of the leaves with low-pressure, boom-type sprayers. Nozzles spaced to give thorough ground coverage are often used on booms 20 to 40 ft long. In general, application of 15 to 50 gal of liquid per acre has proved more satisfactory on lush legume crops than lesser amounts of liquid.

Sprays often give better pest control than dusts (drift less, adhere better) and require less insecticide for equal effectiveness. Generally, sprays do not wash off the plants as rapidly as dust; they can often be applied more evenly and are less hazardous than dusts to man, animals, other crops, or pollinators. Experience has shown that sprays do a better, cheaper job than dusts, but often the decision as to which equipment and materials to use must be based on local conditions and individual situations. One of the disadvantages of ground equipment is the wheel tracks made by the equipment in crops that have been broadcast or drilled. Application of the chemicals by ground rigs is often too slow when time is at a premium and control is urgent.

Aircraft are available for applying sprays, dusts, granular insecticides, or treated seeds. In open areas where the land has a flat or rolling contour, insecticides are easily applied by plane. Use of a plane makes it unnecessary to remodel ditches, make roads, or endure the washboard rides through irrigated fields. Application of a technical insecticide without (or with very little) water, diesel oil, or other diluent, is now possible utilizing the ultralow volume (ULV) formulation. The highly concentrated material permits airplanes to cover large acreages without having to land to reload the chemicals. Helicopters have also proved effective in applying ULV insecticides.

REPRESENTATIVE LEGUME PESTS

Because of space limitations, it is possible to include only a few insects that represent groups of insects, their damage, and a few typical problems. Represented are the chewing insects (alfalfa weevil), sucking insects (lygus bugs), a subtle, less obvious insect (meadow spittlebug), and a seed pest (the alfalfa seed chalcid).

Alfalfa Weevil. *Hypera postica* (Gyllenhal) [Coleoptera:Curculionidae]

The alfalfa weevil was probably imported from southern Europe some time in 1900. It was reported first in Utah in 1904; gradually it infested the other western states. In 1951, the alfalfa weevil was found in Maryland. From there it spread rapidly to other eastern states at a rate of about 50 miles each year. Weevils are now reported present in all states of the continental U.S. and in a few provinces of Canada and some areas of Mexico. A related species, the **Egyptian alfalfa weevil**, *Hypera brunneipennis* Boheman, was first reported in Arizona and southern California in 1939. The Egyptian alfalfa weevil and those found in the eastern and western states have many morphological similarities. Opinions vary, but some evidence suggests that they may represent different strains, biotypes, or subspecies. Certain combinations of the weevils interbreed and reproduce, but the various populations appear to occupy distinct types of habitats. So far, the Egyptian alfalfa weevil is found mostly in the warmer areas and hot valleys of California and Arizona. It has not spread as far or as fast as the western and eastern alfalfa weevils.

The larvae are responsible for the major damage by the alfalfa weevil. They mainly injure the first crop but may also injure the second crop (Fig. 11:9E). Young larvae feed on the stems for a while, then move to the leaf buds at the tops of the plants. They chew cavities in the young alfalfa buds and later feed on the leaves. Their feeding skeletonizes the leaves, so that an infested alfalfa field appears frosted, grayish, or whitish. The injury to buds stunts plant growth, reduces yields and quality of hay and may make fields unfit for seed production.

Alfalfa is usually the only crop injured severely by the alfalfa weevil, but vetch, alsike, zigzag clover, red clover, bur clover, crimson clover, white clover, yellow and white sweetclover, and several other legumes are fed upon

Figure 11:9. The alfalfa weevil. A, eggs; B, larva; C, pupa in cocoon; D, adult laying eggs; E, alfalfa damaged by weevil (right); alfalfa undamaged by weevils on plot sprayed with insecticide (left). *Courtesy William P. Nye, USDA.*

as well. Adult weevils have even been reported feeding on some fruits such as cherries and apricots.

Description. Adult weevils are about $\frac{3}{16}$ in. long; their color varies during their life (Fig. 11:9). They are light brown with a broad, dark brown stripe extending posteriorly from the front of the head along the middle of the back approximately two thirds or three fourths the length of the body when they first emerge. As they grow old, they become gray or brownish black. Eggs are white when first laid, but after a short time they turn black. The larva is yellow, about 1 mm long, with a shiny black head when first hatched. The full-grown larva is legless, about $\frac{3}{8}$ in. long, and green with a brown head and a white stripe down the middle of the back. The larvae tend to curl and bring the tip of the head and the abdomen together when held in the hand. The pupae are tan and are encased in small netlike cocoons spun by the larvae.

Life history. Adult weevils overwinter in plant-debris in or at the edges of fields, in alfalfa crowns, or sometimes in litter some distance from the alfalfa fields. Adults emerged from hibernation feed as the alfalfa grows, and females begin to lay eggs after a few days. Some females of the new generation mature during their first summer and lay eggs which may overwinter. Most females do not become sexually mature and lay eggs until the following spring. Estimates of the number of eggs laid by individual females range from 400 to 1000 during a season. Eggs may be laid in dry litter or dead alfalfa stems in the early spring, but later they are usually placed in cavities chewed by the females in the growing alfalfa stems. The eggs hatch in four days to three weeks, depending on the weather. Eggs laid in the fall may not hatch for more than 170 days.

Each larva passes through four larval instars and takes about 25 days to a month to complete development. A cocoon is spun by the larva before pupation. The cocoon is attached to the alfalfa leaves or to debris on the ground. The pupal stage lasts about 10 days to two weeks. After the adults emerge, they feed on alfalfa for a short time and then enter diapause. Usually there is only one generation of weevils a year, but there is some variation in the time of year that the eggs, larvae, and adults are most abundant and in their rate of development in different areas. Some investigators believe that there are at least two generations per year in certain areas, but present evidence from several states indicates there is only one.

Control. Where possible, use appropriate integrated methods of control, such as cultural practices, protection of beneficial insects, resistant varieties, and chemicals. Follow current recommendations for your area in a weevil control program, deciding when to apply insecticides and how to provide maximum protection for the beneficial insects in your alfalfa.

Early cutting of both the first and the second hay crops is one method of cultural control that has proved fairly effective in reducing weevil populations. If the hay is removed from the fields soon after it is cut and if the ground is hot and dry, many eggs and larvae are killed. Cultivation and dragging fields after the hay has been removed are also recommended.

After first-crop hay is removed, larvae and adults remaining in the field may continue to feed on leaves and shoots of the stubble and prevent the regrowth of the alfalfa. Controlling weevils in stubble and providing water for the plants result in spectacular regrowth (Fig. 11:9E).

Chemical controls for alfalfa weevils are usually aimed at the larvae. Treatment may be applied either before cutting or to the stubble immediately after the first hay crop is removed. Sometimes there is difficulty in obtaining adequate plant coverage when spraying before the hay is cut. Often it is better to cut the hay early than to spray it.

In some areas, when there is an average of 0.5 overwintering weevil per square foot in an alfalfa field, very little damage may be expected the following year. When there is an average of 0.5 to 1.5 per sq ft, there may be some damage, but when the average is 1.5 to 2.5, damage may certainly be expected if weather is favorable. These criteria of potential damage are not applicable everywhere.

Where climate permits weevils to develop early when alfalfa plants are only a few inches tall, a smaller number of weevils will damage the alfalfa more than in regions where the alfalfa is taller when the weevils become plentiful and active. Due to these differences, recommendations for timing the application of insecticides vary considerably from place to place.

The preferred time for applying an insecticide may vary with the season. Fields should be checked about three weeks before the scheduled first cutting date. If obvious damage is seen, the alfalfa should be sprayed before cutting. If little damage is seen, a stubble spray after cutting might be suggested. Under severe outbreak conditions, a precutting spray followed by a stubble spray may be justified. Growers must carefully observe the specified time limits from application of insecticide to harvest.

It is remarkable that all the alfalfa weevils discussed here have been similar in their reaction to individual insecticides. Insecticides registered for weevil control include carbofuran neutralized or acidified as needed with vinegar or buffering agent to a pH of about 5.5 to 6 and applied when alfalfa is 2 to 5 in. high or according to local specifications of damage to buds or number of larvae per sweep or square foot. Early application is least damaging to predators and parasites but may be damaging to bees if early blooming plants such as dandelions are in the field. Methidathion has been effective as a precutting or stubble spray and seems to have minimal toxicity to some beneficial insects. Other insecticides that have specific uses on first-crop hay or stubble or that have certain advantages or disadvantages are: azinphosmethyl, malathion (can be applied up to cutting time), methyl parathion (high toxicity to humans, good for precutting treatment, damaging to beneficial insects), ethyl parathion for stubble spray (highly toxic to humans and damaging to beneficial insects), phosmet (works well in integrated control programs, effective as stubble treatment). Methoxychlor, a chlorinated hydrocarbon that is still approved for alfalfa, can be used effectively in mixtures, for example, with diazinon, malathion, or parathion.

Waiting periods from application to harvest vary from 0 to 21 days depending on the insecticide and its dosage rate. Directions on the insecticide containers should be followed carefully.

Chemicals can be applied by either ground or air. Low-pressure, low-volume, tractor-mounted sprayers capable of developing from 40 to 60 psi pressure, delivering 15 to 50 gal of mixture per acre, are adequate. A combination swather-sprayer may be used at harvest time; this piece of equipment sprays the stubble in a band just under the swath and before the swatch or windrow is dropped. Either regular or ULV applications can be made with aircraft. In general, aerial applications have not been as effective as ground applications.

A weevil parasite, *Bathyplectes curculionis* (Thomson), effective in many areas, was introduced into weevil-infested areas of Utah from Europe just preceding World War I. The parasite deposits an egg inside the weevil larva but does not kill its host until after the weevil larva spins its cocoon. In areas where the parasite is established, 80 or 90 per cent or more of the larvae may be parasitized. The parasite cannot yet be relied upon for complete control of the weevil. *Microctonus* wasps and several other parasites are established widely in the East and in some areas are reported to control the pest effectively. At this writing, **Team** is the only commercially available alfalfa variety that shows some resistance to weevil.

Lygus Bugs

Legume Bug. *Lygus hesperus* Knight

Pale Legume Bug. *Lygus elisus* Van Duzee

Tarnished Plant Bug. *Lygus lineolaris* (Palisot de Beauvois) [Hemiptera:Miridae]

Lygus bugs have been called the most devastating insect pest of the alfalfa seed crop. Their importance as alfalfa pests was discovered first in Utah in 1932, and since that time, many investigators in different areas have confirmed and added to the original observations, which indicated that plant growth, buds, flowers, and seed development are all seriously affected by lygus feeding. Various lygus species feed on a large number of host plants and have proved to be serious pests of the seeds of other plants besides legumes.

In the alfalfa seed-producing areas of Arizona and California, three main species of *Lygus* have been present: the legume bug (the most common species), the pale legume bug, and the tarnished plant bug. In Washington, Oregon, Idaho, Nevada, and Utah the legume bug and the pale legume bug are the two predominant species in alfalfa. The tarnished plant bug is the common eastern species of lygus and has been the most common species found in northern Minnesota. Our discussion of lygus will concern particularly the legume bug and pale legume bug.

Both adult and immature lygus bugs severely injure forage crops and seeds, but the fourth and fifth instar nymphs are more serious pests than the adults. The piercing-sucking mouthparts of lygus enter and leave plants without much immediate, obvious damage, but later plants may develop short internodes, become stemmy, and produce many branches and an unusual number of short racemes (Fig. 11:10C).

When lygus bugs feed on flower buds of alfalfa, the buds turn white and die in two to five days. If lygus are sufficiently numerous in the field, blooming may be completely prevented. If plants in bloom are attacked, lygus feeding may result in flower dropping; if the immature seeds are fed upon, the seeds may shrivel, become discolored, and fail to germinate when planted. Hay production may also be adversely affected by lygus feeding, and the quality of the forage lowered. After lygus are removed from plants and the plants are clipped, the new growth is usually normal, indicating that all of these toxicogenic symptoms are temporary despite the large amount of economic damage done.

Figure 11:10. Lygus bugs and injury. A, nymph; B, adult; C, Lygus-damaged racemes of alfalfa (left) and normal plant (right). *Courtesy William P. Nye, USDA.*

Feeding activity of lygus bugs at night appears to be as great as, or even greater than, during the day. Often during the extremely hot periods of the day the young nymphs will hide under vegetation. They can be seen feeding on the terminal portions of the plant during the cooler parts of the day and in the evening.

Description. Adult lygus (Fig. 11:10) are about $\frac{1}{4}$ in. long. They have four wings that lie more or less flat over the back, and are marked by a distinct V on the back just in front of the wings. Lygus may be light green, various shades of brown, or almost black. The eggs, which are inserted into plant tissues, are slightly curved and approximately 1 mm long.

As the nymphs develop, black spots of various sizes and numbers appear on their backs. They sometimes look similar to aphids, but lygus nymphs usually run about rapidly in contrast to the relatively slow-moving aphids. Lygus nymphs are usually a green or yellowish green color. Newly hatched nymphs are approximately 1.5 mm long, delicate, and not easily seen among a group of insects collected in an insect net.

Life history. The life cycle of lygus bugs comprises three stages: egg, nymph, and adult. In areas where adults hibernate during winter the adults become active the first warm days of spring. They mate soon after they emerge, and the females begin immediately to lay eggs. Egg laying continues throughout the summer and fall. Studies in Utah have indicated that lygus bugs lay most eggs from April through July and that egg production decreases rapidly during the months of August and September. The majority of eggs are laid in the apical 3 in. of alfalfa stems; however, eggs have been found in various plant locations and structures. Eggs are inserted full length into plant tissues with egg caps approximately level with the outside surface.

The length of the egg and nymphal stages varies considerably depending on environmental conditions. In the west, the incubation period for eggs has been found to be from 11 to 19 days, with an average of about 15 days. Lygus nymphs pass through five nymphal instars and require from 13 to 31 days, with an average of 21 days, to complete development and reach the adult stage. Newly emerged females begin to lay eggs in approximately 10 days. Undoubtedly, there is much variation in the life cycle and times of development among the different species of lygus. There are probably four or five generations of lygus a year in certain areas, but the number of generations may vary.

Although adult lygus fly substantial distances from adverse conditions to more favorable ones in summer, several natural factors help to control them. Wet winter seasons with little or no snow cover and alternating mild and low temperatures result in poor survival of hibernating lygus adults. Oviposition, incubation of eggs, and nymphal development are retarded by wet cold spring weather, and these conditions are often fatal to large numbers of lygus nymphs. Lygus nymphs are extremely sensitive to heat and often are unable to survive short walks in the hot soil between plants on hot summer days.

Control. On alfalfa hay or seed, control of lygus is usually aimed at the young nymphs. Nymphs hatch about the time the plants start to develop flower buds. Although a few overwintering lygus adults may be present in the field early in the season and application of insecticide could destroy some of them, the early treatment would not protect the crop from the large number of bugs that sometimes fly in later. Find out when lygus nymphs and adults are present in your area and use the information in planning how to control them.

In deciding when to treat for lygus, sample the alfalfa with a 15 in. diameter sweep net, taking a 180° sweep. Some growers apply insecticide when they find: 2 lygus per sweep in prebloom, 5 per sweep in full bloom, or 8 per sweep in postbloom. Count each nymph as two because they damage alfalfa more than adults. Try to eliminate lygus bugs before the plants bloom and before bees come into the fields. Remember that both wild bees and honey bee pollinators must be protected. If alfalfa weevils, aphids, and other insects need controlling at the same time, investigate the possibility of using a single material or mixture that will control all of them, such as dimethoate or buffered carbofuran on the first crop or trichlorfon mixed with demeton on the second crop. If it becomes absolutely necessary to apply insecticides during bloom, select one known to be least damaging to bees such as trichlorfon or mevinphos, and apply it when bees are not in the field. Get rid of weeds: lygus bugs feed on many of them. Try to cut infested fields of alfalfa at the same time so that lygus cannot migrate from a cut field to one that has not been cut.

The choice of materials for controlling lygus bugs on hay has become more limited, because the use of the longer residual insecticides, such as heptachlor and dieldrin, is no longer permitted. Cutting the hay a little early when the alfalfa is in the bud stage usually will destroy large numbers of eggs and immature lygus. If hay crops become heavily infested with lygus and must be sprayed to save the crop, it is recommended that one of the following insecticides be applied in the bud stage: dimethoate, buffered carbofuran, or methidathion for lygus, aphids, and alfalfa weevils infesting the first crop; dimethoate or methidathion during bloom of the second crop for lygus and aphids; and buffered trichlorfon plus demeton if aphids are present on the first or second crop. Restrictions include not using dimethoate, carbofuran, or methidathion within five days of bloom; methidathion within 10 days of harvesting or feeding alfalfa to livestock; carbofuran or dimethoate within 28 days of harvesting.

For controlling lygus on alfalfa seed crops, a general guide used in many areas is to time the application of insecticides with the bud stage or early bloom to kill newly hatching nymphs. The objective is to destroy lygus before the alfalfa blooms to try to avoid the use of insecticides when bees and other pollinating or beneficial insects are present. If lygus average one or more per sweep, one or a combination of the following insecticides give adequate control: trichlorfon, malathion, diazinon, carbaryl, or dimethoate, applied as sprays as soon as or before the buds appear but not when the crop is blooming. If it is necessary to control lygus when plants are blooming,—if lygus migrate into a field, for example—the use of trichlorfon or mevinphos is less hazardous to bees than insecticides usually recommended for earlier treatments. If nymphs are not present during bloom, some suggest that treatment may be delayed until there is an average of 12 lygus per sweep, presumably because the adults are less damaging per individual than the large nymphs. If spotted alfalfa aphids or mites, as well as lygus, are damaging the crop demeton may be added to the trichlorfon treatment.

Lygus bugs have natural enemies, such as damsel bugs, bigeyed bugs, lady beetles, and ants. Lygus eggs are attacked by chalcid egg parasites.

Alfalfa Seed Chalcid. *Bruchophagus roddi* (Gussakovsky) [Hymenoptera:Eurytomidae]

The alfalfa seed chalcid (Fig. 11:11) is commonly regarded as one of the most important unsolved pest problems of alfalfa seed. It is difficult to indicate accurately what the losses caused by alfalfa seed chalcid are, but observations have shown that from 2 to 85 per cent of the produced alfalfa seed may be infested. Seed losses are commonly estimated to range between 10 and 15 per cent.

The alfalfa seed chalcid appears to be present nearly everywhere that alfalfa seed is grown. It has been reported from Germany, Turkey, Chile, Siberia, and in the major seed-producing areas of the United States. The female chalcid inserts eggs into newly developing seeds by means of a short ovipositor. Seeds are destroyed when the larvae feed on the internal contents.

Description. The adult alfalfa seed chalcid is black, about 2 mm long, and has two pairs of membranous colorless wings (Fig. 11:11). Parts of the legs are yellowish brown. Female chalcids are slightly larger than males, but their antennae are shorter, with joints more closely united; the abdomen of the female is more pointed than that of the male. The eggs are so small that a microscope is needed to see them. One end of the egg is slightly pointed and the other end drawn out into a long tube two or three times longer than the main body of the egg. Mature larvae are white with a pair of brown mandibles. The larvae develop into pupae about 2 mm long.

Life history. The seasonal history of the alfalfa seed chalcid varies greatly at different locations and from year to year. In the intermountain areas of the West, chalcids overwinter as larvae inside alfalfa seed. Infested seed may be found in volunteer plants along roadsides surrounding alfalfa fields, in chaff stacks, or scattered on the ground in alfalfa fields where seed pods have been knocked off plants or blown out of the combines during harvest. It is not uncommon to have large numbers of these overwintered chalcids emerge from clean seed stored over the winter. Infested seeds are about the same size and weight as normal seeds, so that often both are bagged together. Persons handling the seed sometimes get the mistaken idea that the chalcids are infesting or attacking the dried seed in the bags.

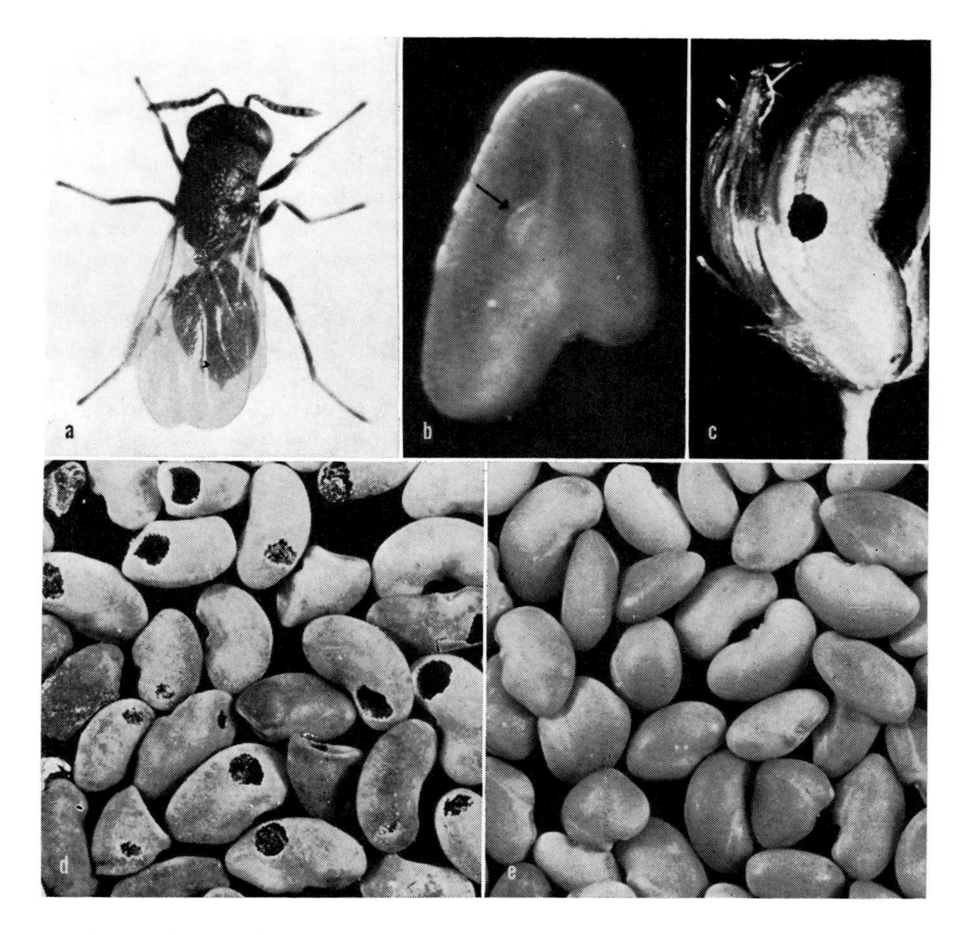

Figure 11:11. The alfalfa seed chalcid. A, adult; B, alfalfa seed containing egg; C, emergence hole; D, chalcid-damaged alfalfa seed; E, noninfested seed. *Courtesy William P. Nye, USDA.*

As temperatures increase in spring, the larvae inside the seeds pupate and develop into adults. When development is complete, the adult escapes by chewing a hole through the seed coat. Changes in temperature influence the growth and activity of immature and adult seed chalcids so that emergence dates of adults in spring may vary considerably. Some studies in the West have indicated that males emerge before females in spring and that they are more numerous than females all season.

Chalcids mate almost immediately after they emerge, and females lay eggs within a few hours if they can find seeds in the right stage of development. If suitable seed for oviposition cannot be found, the females apparently will fly several miles to find seed in which to oviposit. Seeds whose contents are semifluid or jellylike seem to be most suitable for oviposition. The females usually deposit only one egg per seed. The eggs hatch in about 3 to 12 days, depending on the temperature, and the larvae feed on the contents of the seed until larval growth is complete.

If temperature and moisture conditions remain favorable, the larvae may change immediately into the pupal stage, and new adult chalcids emerge from the seeds 5 to 40 days later. If temperatures become low, the larvae may remain inside the seed until more favorable conditions for development return. Under excessively dry conditions the larvae may go into a resting state and remain in dry seeds for one or even two years. The number of generations produced per year varies from almost continuous generations in favorable areas to only two generations a year in some of the colder areas of the United States. The average feeding period for larvae under conditions in northeastern Utah is from 10 to 15 days. Developing from eggs laid the same day in pods on the same alfalfa stem, some larvae pupate and emerge as adults within 10 days to two weeks after they complete their feeding, whereas others hibernate and emerge as adults the following May.

Control. Reports of studies on commercial and experimental fields show that there has been little success in use of insecticides to control the alfalfa seed chalcid. None of the insecticides tested, including some systemic insecticides, has reduced chalcid injury enough to make chemical control economically justifiable. The deposition of eggs within seeds, the constant sources of reinfestation, the development of continuous generations of chalcids, and the presence of insect pollinators that are necessary for seed setting in the field when chalcid adults need to be destroyed have all complicated chemical control of chalcids.

Studies of seed chalcid life history and behavior have resulted in several recommendations for reducing chalcid damage.

1. Grow only either first- or second-crop seed in the same area. If both first- and second-crop seed are grown in the same area, a continuous supply of favorable host plants will be available for the increase of chalcids. Obviously there will be practical difficulties in obtaining the necessary cooperation of all seed growers in an area to establish such a practice and in furnishing enough pollinators to set seed in a large area all at the same time.
2. Grow only one host plant of the seed chalcid in an area. Destroy or prevent the development of volunteer host plants.
3. Manage the seed crops so that the seed ripens uniformly.
4. Feed badly infested seed crops as hay (if not contaminated by insecticides) and remove these crops from the fields as soon as possible.
5. If seed is threshed in a stationary location, eliminate the chaff stacks containing infested seed before the chalcids emerge in the spring. Do not feed chaff contaminated with insecticide to livestock.

6. Reclean harvested seed and destroy the infested seeds.
7. Cultivate fields to bury infested seeds that have fallen to the ground during harvest or have been scattered over the fields by the combine harvesters. There is evidence that burying infested seed to a depth of 2 to $2\frac{1}{2}$ in. will destroy a large percentage of the chalcids; high soil moisture adversely affects their survival.

Ten species of parasites belonging to the superfamily Chalcidoidea are known to attack the larvae or pupae of the alfalfa seed chalcid in the United States. Some studies have shown that often 90 per cent or more of seed chalcids have been parasitized.

The chalcid problem is being investigated by a number of different agencies. One project is a search for plant materials resistant to chalcid infestation. None of the varieties studied thus far is free from chalcid attack, but differences of chalcid infestation suggest that it may be possible to find plant materials with more resistance.

Meadow Spittlebug. *Philaenus spumarius* (Linnaeus) [Homoptera:Cercopidae]

The meadow spittlebug (Fig. 11:12), widely distributed in Europe and North America, is the most important species of spittlebug in the eastern United States. It is frequently a serious pest of legumes (especially alfalfa and red clover) in the northeastern and north central states. Infestations of meadow spittlebugs can be recognized readily by the white, frothy masses of spittle on the plants. Plants may look as though soapy water had been thrown on them.

Most of the injury to plants by meadow spittlebug results from nymphal feeding. The nymphs have piercing-sucking mouthparts, through which they suck juices from the plants. Their feeding on alfalfa results in stunted plants, reduced yields of forage and seed, wilting, difficulty in curing the hay because of dampness from the spittle masses, and rosetting of the terminal plant growth. Plant blossoms and seeds may also be injured. The alfalfa plant internodes are shortened, and the plants are dwarfed. But, unlike the yellows that result from the feeding of potato leafhoppers, spittlebug-infested plants remain green. Clover is stunted by spittlebug feeding, but the rosetting of terminal growth is not as marked.

When densities of spittlebugs ranged from 2 to 26 nymphs per sq ft in New York, control of the pests in certain meadow mixtures resulted in hay increases up to 59 per cent. Infestations of less than an average of six nymphs per stem have appeared to have little effect on alfalfa or red clover hay fields. No carryover effect of spittlebug injury has been observed on later cuttings.

The meadow spittlebug commonly attacks meadow plants, weeds, and garden plants. It will survive on nearly any succulent foliage on which it becomes established, including trees, shrubs, and herbaceous species. The amount of damage seems less than might be expected from the tremendous numbers of spittlebug adults that are often seen on a crop.

Figure 11:12. The meadow spittlebug. A, eggs on stalk; B, nymph; C, adult; D, spittle on legume plant. *A, B and C, courtesy Ohio State University, Columbus, Ohio; D, courtesy Lloyd L. Stitt, University of Nevada, Reno.*

Description. Adult spittlebugs are $\frac{1}{4}$ to $\frac{1}{2}$ in. long and about $\frac{1}{8}$ in. wide. They may be brown or gray, with colors arranged in various designs; the adults have a short, bluntly angular head and prominent eyes located on the sides of the head. They are shaped much like leafhoppers, but the two can be readily told apart by the differences in their hind legs (see p. 128, couplet 3). Spittlebugs are sometimes called "froghoppers" because of their shape, which slightly resembles that of tiny frogs.

The nymphs (Fig. 11:12B) are found in masses of spittle located in various places up and down the stems of the plants they feed on. The first-instar nymphs are orange; as development proceeds, they gradually assume a green color, becoming entirely green in the last nymphal instar. Spittle, a fluid voided from the anus, is a substance possibly secreted by glands. The air bubbles are formed by manipulation of the abdominal segments, which forces air through the ventral "air canal" and into the spittle.

Life history. Meadow spittlebugs have one generation annually. In the fall the females lay eggs on grain stubble, alfalfa stubble, and on stems of many other plants, usually less than 4 in. above ground level. One to 30 eggs per mass may be laid, with an average of 7 being found in Ohio. It is estimated that individual females may lay from 18 to 51 eggs apiece.

The eggs overwinter and hatch the following spring. A critical period in the life of a spittlebug is from the time of hatching until the nymph finds and establishes itself on a suitable host. Apparently many nymphs do not survive this period. At Wooster, Ohio, studies indicate that approximately equal time is spent in each of the five instars (about eight to 10 days each). Although the nymphal period averages five to six weeks, the rate of development may be altered considerably by changes in temperature and moisture.

Nymphs usually rest head downward while feeding, and as the spittle is formed, it covers the insect. This material provides the moist habitat necessary for the survival of the nymph and appears not to be readily removed, even by heavy rains. Each mass of spittle contains one or more nymphs.

Spittle production ceases just prior to the final molt. The foam around the insect dries and forms a chamber in which the molt occurs, usually in June. The new adults crawl or leap out of the mass, leaving a neat hole in the dried spittle ball. Adults may remain in the fields where they developed until the foliage is removed, or fly to other crops that attract them. The direction of adult spittlebug migration appears to be greatly influenced by the wind direction.

In August fully developed eggs begin to appear in ovaries of females; later, in September, females migrate into suitable fields for oviposition. Entomologists have found that intense oviposition usually occurs after the first week in September in Ohio. Females may continue to lay eggs until they die naturally or are killed by frost.

A combination of high humidity and an abundance of good legumes seems to be necessary for the spittlebug to develop as a serious pest. Humidity is an important condition determining survival of spittlebugs in all stages of their existence. The adults are attracted to and survive best on succulent foliage, and an abundance of this succulent foliage must be present to attract the adults for oviposition.

Control. Because climate and cultural practices vary in different areas, so do the recommendations for controlling spittlebugs. Get the latest information from your state, federal, or private agricultural organizations. Application of methoxychlor to infested fields early in September, after the adults have ceased extensive migration has resulted in excellent control of adult spittlebugs and has reduced nymphal infestations the following spring.

Methoxychlor should be applied at least seven days before harvesting or pasturing the treated crop. The criteria for deciding when to apply insecticides to control spittlebugs changes periodically as new pest management principles are developed. Some suggest treatment when plants are 8 to 10 in. tall, which corresponds to the peak hatching period of nymphs in certain areas. Others report that spittlebug infestations are usually more severe on sandy soils than on heavier soils and have suggested treatment on sandy soils when one third of the stems coming up from the crowns have spittle masses on them and when one fifth of those on heavier soils have spittle masses. Others have found that when examination of the stems, during the first warm days of spring, shows an average of more than one spittle mass per stem, treatment is advisable. Find out what these criteria and insecticide treatment recommendations are for your area.

SELECTED REFERENCES

Carner, G. R., M. Shepard, and S. G. Turnipseed, "Seasonal Abundance of Insect Pests of Soybeans," *J. Econ. Entomol.*, 67(4):487, 1974.

Davis, D. W., G. E. Bohart, G. D. Griffin, B. A. Haws, G. F. Knowlton, and W. P. Nye, *Insects and Nematodes Associated with Alfalfa in Utah*, Utah Agr. Exp. Sta. Bul. 494, 1976.

Dickason, E. A., and R. W. Every, *Legume Insects of Oregon*, Oregon Agr. Ext. Ser. Bul. 749, 1955.

Dorsey, C. K., *Experimental Control of Insect Pests of Alfalfa in West Virginia, Other Than the Alfalfa Weevil, 1959–1964,* West Virginia Agr. Exp. Sta. Bul. 549T, 1967.

Haws, B. A., and F. G. Holdaway, *Sweetclover Weevil and Its Control in Minnesota*, Minn. Ext. Folder 190, 1955.

Johansen, C. A., *How to Reduce Poisoning of Bees from Pesticides*, Washington Coop. Ext. Serv. EM 3473 (SR), 1975.

Johansen, C. A., *Insect Management and Control on Crops Grown for Seed*, Pacific Northwest Coop. Ext. Publ. PNW 128 Rev., 1977.

Johansen, C. A., E. C. Klostermeyer, A. H. Retan, and R. R. Madsen, *Integrated Pest Management on Alfalfa Grown for Seed*, Washington Coop. Ext. Serv. EM 3755 (Rev.), 1977.

Kindler, S. D. and J. M. Schalk, "Frequency of Alfalfa Plants with Combined Resistance to the Pea Aphid and Spotted Alfalfa Aphid in Aphid-Resistant Cultivars," *J. Econ. Entomol.*, 68:716–18, 1975.

Klostermeyer, E. C., *Alfalfa Seed Insects*, Washington Agr. Exp. Sta. Bul. 587, 1958.

Koehler, C. S., and S. S. Rosenthal, "Economic Injury Levels of the Egyptian Alfalfa Weevil or the Alfalfa Weevil," *J. Econ. Entomol.*, 68(1):71, 1975.

Lauderdale, R. W., and W. H. Arnett, *Insect Pest Control on Forage Alfalfa*, Nevada Ext. Serv. L40, 4P, 1975.

Miller, M. D., L. G. Jones, V. P. Osterli, and A. D. Reed, *Seed Production of Landino Clover*, California Agr. Ext. Ser. Circular 182, 1951.

Niemczyk, H. D., and G. E. Guyer, *The Distribution, Abundance and Economic Importance of Insects Affecting Red and Mammoth Clover in Michigan*, Michigan Agr. Exp. Sta. Tech. Bul. 293, 1963.

Rockwood, L. P., *The Clover Root Borer*, USDA Bul. 1426, 1926.

Rowell, J. O., and J. C. Smith, *Control Soybean Insects*, Virginia Polytechnic Institute Ext. Pub. 132, 1969.

Ruppel, R. F., *Insect Control in Hay, Forage and Pasture Crops*, Michigan Ext. Bul. E-827, 1975.

Strong, F. E., "Laboratory Studies of the Biology of the Alfalfa Seed Chalcid, *Bruchophagus roddi* Guss. (Hymenoptera:Eurytomidae)," *Hilgardia*, 32(3):229–49, 1962.

Turnipseed, S. G., and M. Kogan, "Soybean Entomology," *Ann. Rev. of Entomol.*, 21:247–82, 1976.

Watkins, T. C., *Clover Leafhopper* (Aceratagallia sanguinolenta Prov.), Cornell Agr. Exp. Sta. Bul. 758, 1941.

Weaver, C. R., and D. R. King, *Meadow Spittlebug*, Ohio Agr. Exp. Sta. Res. Bul. 741, 1954.

Wildermuth, V. L., and F. H. Gates, *Clover Stem-Borer as an Alfalfa Pest*, USDA Bul. 889, 1920.

Wilson, M. C., *Prevent Leafhopper Yellows on Alfalfa*, Indiana Agr. Ext. Serv. Bul. 398, 1953.

Yunus, C. M. and C. A. Johansen, *Bionomics of the Clover Seed Weevil*, Miccotrogus pricirostris (*Fabricius*) *in Southwestern Washington and Adjacent Idaho*, Washington Agr. Exp. Sta. Tech. Bul. 53, 1967.

Zeeman, M. G. and D. W. Davis, "Pesticides and Alfalfa," *Utah Science*, 36(2):64–7, 1975.

chapter 12 / **ROBERT E. PFADT**

INSECT PESTS OF COTTON

THE PESTS

Cotton, with its green, succulent leaves, its large open flowers, its nectaries on every leaf and flower, and its abundance of fruit, attracts and supports a great variety of insects and mites. In the United States over 125 injurious species attack cotton. One of the most destructive of these is the **boll weevil** (Fig. 12:7), a native insect of Mexico or Central America that crossed the border into Texas in 1892. It spread rapidly and is now found in more than 85 per cent of the Cotton Belt.

Other major pests of cotton include the bollworm, tobacco budworm, pink bollworm, cotton fleahopper, lygus bugs, cotton aphid, and several species of spider mites and thrips.

The **bollworm**, *Heliothis zea* (Boddie), also called the corn earworm, feeds on many kinds of cultivated and wild plants (Fig. 10:9). It prefers young corn, but after corn matures the moths oviposit on cotton and other acceptable plants. Usually larvae of the second generation injure corn, whereas larvae of the third and succeeding generations injure cotton. The **tobacco budworm**, *Heliothis virescens* Fabricius, is distributed throughout the cotton belt, but damaging populations generally occur only in Texas and eastward. In recent years this insect has severely damaged late (October) actively growing cotton in the Imperial Valley of California. Although the preferences for host plants of the two cutworms often differ, they are found together in summer infesting and damaging cotton. The budworm is more resistant to insecticides and more difficult to control than the bollworm.

The **pink bollworm** (Fig. 12:9), known as the most serious worldwide pest of cotton, is entrenched in the Southwest. Only stringent control measures have prevented its spread to the entire Cotton Belt. Hibernating larvae of this frail moth entered the United States in 1916 in cotton seeds imported from Mexico.

The **cotton fleahopper** (Fig. 12:10), a widely distributed mirid bug in the United States, inflicts its greatest damage to cotton in Texas and Oklahoma. It feeds on a large variety of weeds in addition to cotton. **Lygus bugs**, also mirids (Fig. 11:10), are most injurious to the cotton crop in New Mexico, Arizona, and California.

The **cotton aphid**, also called the melon aphid, is the most important of several species of plant lice infesting cotton (Fig. 12:1). Enjoying almost worldwide distribution, the aphid is found everywhere in the United States. Today it is mainly a pest of seedling cotton, but formerly, when growers applied calcium arsenate exclusively to control the boll weevil, it often increased to large, damaging populations on older plants. Several other species—the cowpea aphid, green peach aphid, and potato aphid—may infest the foliage of seedling cotton, while several species attack the roots including the corn root aphid and the aphids *Smynthurodes betae* (Westwood) and *Rhopalosiphum rufiabdominalis* (Sasaki).

Twelve or more species of **spider mites** are known to infest cotton in the United States. An important factor that has increased the seriousness of the mite problem on cotton has been the application of chlorinated hydrocarbon insecticides, which have not only killed various cotton pests but also beneficial, predaceous species. To correct this condition growers make suitable miticide applications whenever needed. Among the important mite species are the twospotted spider mite, fourspotted spider mite, Strawberry spider mite, desert spider mite, Pacific spider mite, Schoene spider mite, tumid spider mite, carmine spider mite, and lobed spider mite.

Several species of **thrips**, at least eleven, have been reported attacking cotton in the United States. Although thrips may injure mature plants, they are chiefly pests of seedling cotton. Included in this group of pests are the bean thrips, tobacco thrips, flower thrips, and onion thrips (Fig. 16:12).

Cotton pests of somewhat less importance, which still require control at times, are listed below to show just how numerous and diverse the enemies of this crop are.

Collembola. Three or more species, including the garden springtail and *Entomobrya unostrigata.*

Orthoptera. The differential, migratory, redlegged, twostriped, American, and lubber grasshopper; field cricket; snowy tree cricket.

Isoptera. Termites of undetermined species.

Hemiptera. Rapid plant bug, superb plant bug, ragweed plant bug, and several other mirids; several stink bugs including the conchuela, Say stink bug, southern green stink bug, harlequin bug, brown stink bug, western

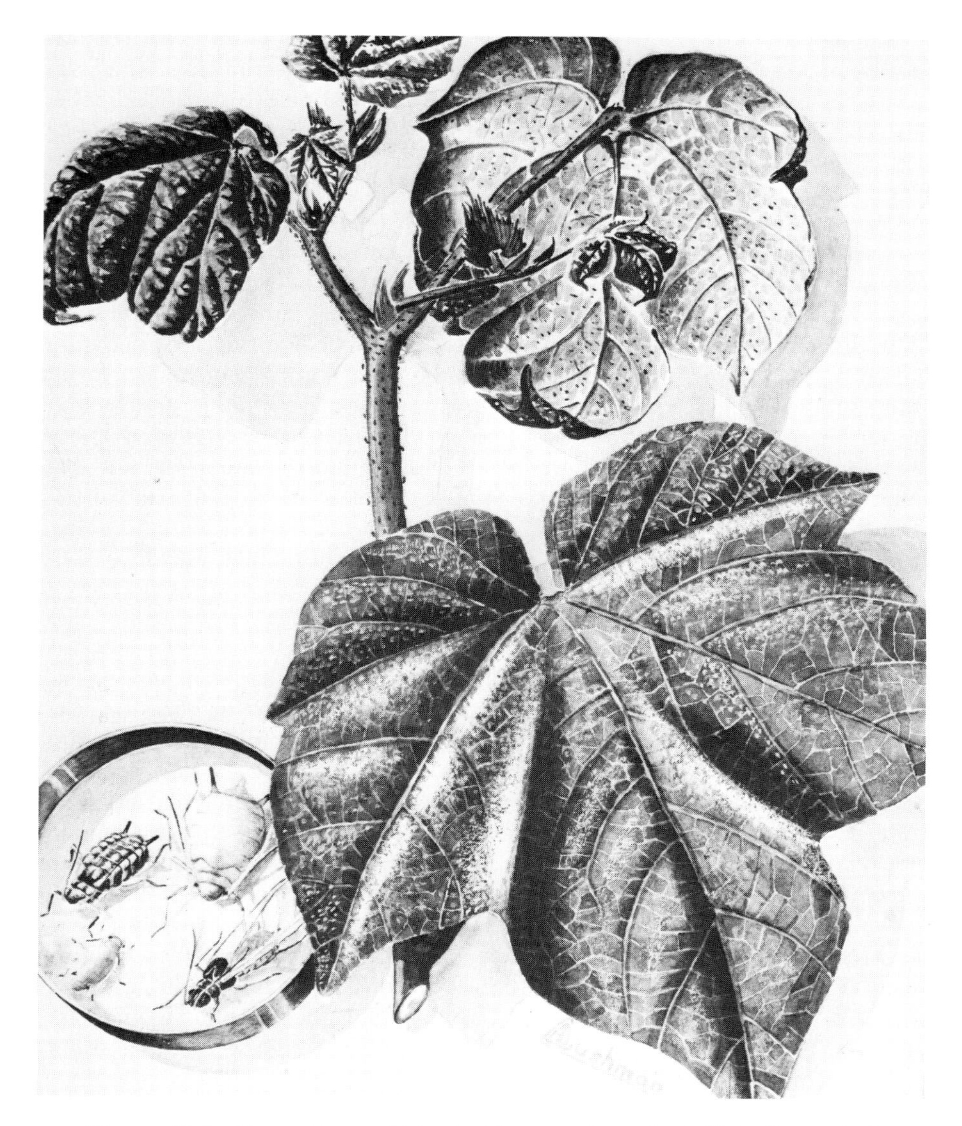

Figure 12:1. The cotton aphid attacking a cotton plant. Note the curled, infested leaves at the top and the honeydew-covered lower leaf. *Courtesy USDA.*

brown stink bug, and redshouldered plant bug; the cotton stainer; false chinch bugs.

Homoptera. The bandedwing whitefly, greenhouse whitefly, sweetpotato whitefly, and *Trialeurodes pergandei*; several leafhoppers, including the potato leafhopper and southern garden leafhopper; several cicadas.

Coleoptera. The sand wireworm and Pacific Coast wireworm; several species of false wireworms; striped blister beetle; whitefringed beetles; several species of *Colaspis*; flea beetles including the palestriped flea beetle, elongate flea beetle, and sweetpotato flea beetle; the spotted cucumber beetle, the cowpea curculio, and several other curculionids; the Fuller rose beetle; the corn silk beetle; white grubs and May beetles, *Phyllophaga* spp., and Japanese beetle; adults of the buprestid, *Psiloptera drummondi* Lap.

Lepidoptera. A large number of noctuids, including the black cutworm, palesided cutworm, variegated cutworm, granulate cutworm, army cutworm, beet armyworm, fall armyworm, yellowstriped armyworm, western yellowstriped armyworm, cabbage looper, cotton leafworm, brown cotton leafworm, leafworms (*Anomis* spp.), and the stalk borer. Representatives of other families include the cotton leafperforator and *Bucculatrix gossypiella*; the garden webworm. European corn borer, and greenhouse leaftier and false celery leaftier; the cotton square borer; the saltmarsh caterpillar and the yellow woollybear; the whitelined sphinx; several species of leafrollers, including *Platynota stultana* and *P. rostrana*, and the barberpole caterpillar.

Diptera. The seedcorn maggot and several serpentine leafminers.

BENEFICIAL INSECTS

Not all insects that live in a cotton field are pests. Many beneficial species of insects, spiders, and mites inhabit cotton fields and prey on or parasitize the harmful cotton-feeders. Because parasites and predators make an important contribution in the control of cotton pests, knowledgeable growers are now making a determined effort to spare the beneficial forms as much as possible from the effects of insecticidal treatment. Common beneficial insects found in cotton fields and in surrounding fields of other crops and wild vegetation from which they migrate to cotton include these predators: insidious flower bug, bigeyed bugs, damsel bugs, lady beetles, green lacewings, ants, wasps, spiders, and predaceous mites. Parasites include many kinds of parasitic flies and tiny wasps. Entomologists have just made a start in mass rearing and releasing certain of these native predators and in managing their habitat to increase numbers. They have also introduced parasites from foreign countries in order to increase natural mortality of pests. Much more research remains to be done before full exploitation of biological control can be realized in the suppression of cotton pests.

THE INJURY

Insect pests not only reduce yields of cotton lint and seed, but also lower their quality. Agricultural scientists have estimated that in the United States insects cause an annual loss of the cotton crop ranging from $200 to $300 million. In addition to these losses, chiefly from reduced yields, growers spend up to $75 million each year for insect control. Partly because of inflation these estimates are now doubled; each year losses to insects exceed $500 million, and expenditures for insecticides amount to $150 million.

Pests attack cotton in the field from the time the seed is planted until the crop is harvested. Those pests that feed on cotton in its early stages of development probably have little effect on quality, but they may affect yields. Insects such as wireworms, false wireworms, and seedcorn maggots may reduce or even prevent the establishment of stands by consuming the seeds and destroying the seedlings.

After the young plants emerge, they are attacked by both chewing and sucking insects. Thrips and the cotton aphid suck juices of seedling cotton and introduce salivary secretions that retard both growth of the plant and production of fruit. Chewing insects—cutworms, beet armyworm, darkling beetles, May beetles, flea beetles, cowpea curculio, and field crickets—feed on the leaves or the tender stems often killing the young plants. Sometimes fields of seedling cotton show symptoms of injury—yellow, weak plants—without obvious evidence of insects being present. This injury may be the work of root-feeding pests such as root aphids.

Once the crop begins to fruit, insects not only reduce yields of cotton but also have adverse effects on its quality. They need not inflict this injury directly on the bolls but may do it indirectly through defoliation, feeding on the roots, or boring in the stem. Notorious defoliators include the cotton leafworm, spider mites, cotton aphid, yellowstriped armyworm, fall armyworm, cabbage looper, and cotton leaf perforator. Insects that bore in the stem are the European corn borer and the stalk borer.

The most serious pests of cotton attack the fruit directly, feeding on squares, blooms, or bolls. Mandibulate insects such as the boll weevil, bollworm, and pink bollworm feed inside the squares and bolls, ruining or devouring lint and seed. Sucking insects, such as the cotton fleahopper, lygus bugs, and stink bugs, suck juices from the squares and cause them to turn brown or black and fall from the plant. Lygus bugs and stink bugs also feed heavily on the bolls, causing young bolls to shed or staining the lint of older bolls.

On open bolls lint may become stained in other ways. The excrement of cotton leafworms may stain the lint. Honeydew from aphids causes gummy lint on which molds frequently grow; the molds stain and weaken the fibers.

Insects are predisposing factors and vectors of several diseases of cotton. In piercing the bolls, stink bugs and cotton stainers introduce a yeastlike fungus, *Nematospora gossypii*, which stains the lint and contributes to the injury known as internal boll disease of cotton or stigmatomycosis. Insects that chew holes through the walls of the bolls open avenues of entry for bacteria and fungi that cause boll rots. Some insects may transport the pathogen as they enter the boll. Thrips injury weakens seedling plants, making them more susceptible to sore shin disease caused by a fungus, *Rhizoctonia solani*. Insects also transmit several virus diseases of cotton. The banded wing whitefly and sweetpotato whitefly transmit the virus of cotton leaf crumple, which affects the crop in the Southwest.

The mite *Siteroptes reniformis* transmits the fungus *Nigrospora oryzae*, which causes Nigrospora lint rot of cotton bolls. An interesting mutual relationship exists between the mite and fungus. The fungus serves as the host plant of the mite; and the mite's contribution consists of each young female transporting two spores of the fungus in an abdominal sac to a partially opened cotton boll. After gaining entry the mite inoculates the cotton boll with the spores.

COTTON PEST MANAGEMENT

Sole use and reliance on insecticides have not provided adequate and durable control of cotton pests. Experience of recent years has dictated a different approach in suppressing pest populations and reducing damage. A new strategy, designated pest management, requires the employment and integration of all possible control methods for effective insect control, including natural, biological, cultural, chemical, mechanical and legal, the objective being to keep pest populations below economic injury levels with minimal adverse side effects to the agroecosystem. For the individual cotton grower insecticides continue to be an important tool. He must integrate other methods available to him with judicious use of these chemicals.

Pest management requires the regular scouting and monitoring of pests and of beneficial species inhabiting each field of cotton and of cotton growth, fruiting, maturity, and damage. In order to protect and let the beneficial insects exert their control of pests for as long as possible, applications of insecticide are withheld from the crop until they are actually needed. Economic thresholds have been established for major cotton pests. Whenever densities of a pest reach the threshold level an appropriate insecticide must be applied quickly to prevent appreciable damage.

Monitoring pests and beneficial insects requires the inspection of a sample of plants in each field on a regular schedule of about once a week during the growing season. The trained grower or scout enters the cotton field and traverses a route that allows him to sample widely. Every 30 to 40 feet he stops and carefully examines a plant, looking at the leaves, terminal buds, squares, and bolls and records his findings of insects or telltale evidence of their presence. For the boll weevil this evidence may be its feeding and egg punctures in squares or for bollworms their feeding damage of squares. Each type of injury is totaled, and then a percentage of plants with the particular kind of injury is calculated. A specified level of attacked plants indicates that the insect population has reached an economic threshold and requires immediate insecticidal control. For example, when boll weevils have injured 10 per cent of the squares, published recommendations advise the grower to make 3 applications of insecticide 4 to 5 days apart. For certain species of pests the insects themselves are counted. To monitor the cotton fleahopper, a scout will count the nymphs and adults on the top 3 to 4 inches of terminals of the main stem of 100 plants as he walks diagonally across a cotton field. When 40 or more fleahoppers are found per 100 terminals, published recommendations advise the grower to treat the cotton field with an insecticide. State colleges of agriculture and agricultural extension services provide training and literature, which are tailored for local conditions, on cotton monitoring, scouting of pests, protection of beneficial insects, and insecticidal control of pests. The basic reason for these regional instructions by states is that the damage potential of an insect species varies over the area of its distribution from insignificant to totally destructive; hence the problem a pest presents and the methods for solution vary with geographical location.

CULTURAL CONTROL

Because cotton pests are difficult to control solely with insecticides, growers must make cultural control methods a regular part of their farm operations. The chief objectives of cultural control are (1) to have an early-maturing crop, so that it escapes peak populations of pests; (2) to harvest the crop early, so that removal of the pests' host plants stops their multiplication; (3) to reduce the numbers of pests that are able to overwinter.

Growers reach these objectives by (1) locating cotton fields as far as possible from other crops that may serve as host plants of cotton insects and as far as possible from favored hibernation quarters of cotton insects, particularly the boll weevil; (2) planting reasonably early; (3) planting healthy seed of a recommended variety for the area; (4) planting during a short period of time within a given area; (5) stimulating rapid growth and maturity through the adoption of recommended agronomic practices; (6) harvesting early; (7) destroying stalks with cutter-shredders immediately after harvest (Fig. 12:2); (8) plowing under crop residues; (9) destroying all stub, volunteer, or abandoned cotton; (10) destroying weed patches, which serve as hibernating quarters for several cotton pests.

Because gin trash may serve as hibernating quarters for boll weevil and pink bollworm, gin-plant sanitation is essential for reducing the overwintering populations. In regulated areas of the pink bollworm, state and federal laws require that gin trash be either burned, sterilized, run through a hammer mill or a special-type fan, or composted.

Agricultural scientists of USDA and state agricultural experiment stations have initiated a vigorous breeding program to develop varieties of cotton that incorporate resistance to insects. Several characters of cotton are responsible for resistance. Nectarless cotton suppresses bollworms, plant bugs, and cotton fleahopper; Frego bract cotton suppresses boll weevil; high gossypol content suppresses bollworms and cotton fleahopper; okra leaf suppresses banded whitefly; smooth leaf suppresses bollworms; hairy leaf suppresses cotton fleahopper and boll weevil; and early maturity allows cotton to escape the peak damage in late season by boll weevil and

Figure 12:2. Cutting and shredding cotton plants after harvest aid in the control of insect pests. *Courtesy Texas A & M University, College Station.*

bollworms. Plant breeders are bringing together compatible characters into commercially acceptable varieties. A new variety developed in Texas combines nectarless, Frego bract, okra leaf, and earliness characteristics plus disease resistance. The newer varieties are increasing yields and reducing costs of production because of better insect control with fewer applications of insecticide.

CHEMICAL CONTROL

To protect it from injurious insects and mites, cotton receives a wider variety and greater quantity of insecticides than any other single crop. Each year cotton growers normally purchase some $150 million worth of insecticides, about 45 per cent of the total used in agriculture. Reasons for applying a variety of insecticides to cotton are that no one material controls all pest species, some materials used alone create problems with bollworms, aphids, and spider mites, and still others have decreased effectiveness because of the development of resistance by several cotton insects.

The insecticides used on cotton include various chlorinated hydrocarbons, organophosphorus, carbamates, several miscellaneous organic compounds, and a few inorganic chemicals. The principal insecticides for controlling important cotton pests are shown in Table 12:1. In addition to these, several compounds developed more recently, mainly organophosphorus, look promising, and some, no doubt, will become widely utilized. No one chemical controls all cotton pests; therefore growers often make a practice of combining two or more insecticides in a single treatment. One popular mixture combines toxaphene and methyl parathion.

Because most insecticides kill beneficial predators and parasites as well as pest insects and thereby contribute to outbreaks of still other pests, informed growers use them only when needed. Correct timing is essential for satisfactory control of cotton insects. Growers apply insecticides early in the season to control infestations of cutworms, beet armyworms, grasshoppers, and aphids. In some localities growers are advised to make early-season treatments for weevil and likewise for thrips, fleahoppers, and plant bugs. In midseason and in late season, growers make treatments to control such serious pests as the boll weevil, bollworm, leafworm, fleahopper, plant bugs, aphids, and spider mites. In outbreak years of the boll weevil they may treat cotton every four or five days, making as many as 20 or more applications during the season. A most promising new way of controlling the boll weevil is treatment of populations before the weevils enter hibernation. The method has been named "the reproduction-diapause boll weevil population control program."

A grave problem in dealing with cotton pests is the development of resistance to insecticides. Resistance in cotton insects first appeared in 1953 in the cotton leafworm. But not until 1955 was the problem fully appreciated, when in some areas it was found that the boll weevil had developed resistance to the chlorinated hydrocarbon insecticides. Entomologists have now found resistance in 25 species of cotton insects and spider mites, and several other species are suspect. Resistance is mainly to the chlorinated hydrocarbon insecticides, but four species of mites, the bollworm, the tobacco budworm, the beet armyworm, and the bandedwing whitefly are resistant to organophosphorus compounds. The bollworm is resistant to the chlorinated hydrocarbons—DDT, TDE, toxaphene, strobane, and endrin—to the carbamate carbaryl, and to the organophosphorus compound methyl parathion. Resistance has been found also in the beet armyworm, cabbage looper, cotton aphid, cotton fleahopper, cotton leafperforator, cotton leafworm, lygus bugs, pink bollworm, saltmarsh caterpillar, southern garden leafhopper, a stink bug species, and several species of thrips. Resistance of the majority of these species is limited to relatively small areas. No species is known to be resistant throughout its entire distribution.

Methods of combating the resistance problem include the use of insecticides having physiological actions different from the ones to which the pests have become resistant, protection of beneficial predators wherever possible, and greater reliance on cultural methods.

Because much valuable research is conducted every year on both cotton pests and cotton culture, agricultural scientists interested in this crop meet in an annual conference to exchange and discuss new findings and new field experiences. The Thirtieth Annual Conference on Cotton Insect Research and Control was held in 1977 in Atlanta, Georgia. Summaries of the deliberations are published each year by the Agricultural Research Service of USDA.

Table 12:1. Recommended Dosages of the Principal Insecticides Used for Control of Cotton Pests

(Pounds Per Acre of Technical Material in a Dust or Emulsion Spray)

Insecticide	Boll Weevil	Bollworm or Tobacco Budworm	Cabbage Looper	Cotton Aphid	Cotton Leaf Perforator	Cotton Leafworm	Cutworms
aldicarb[1]	0.6–1.0			0.3–0.5	2.0		
azinphosmethyl	0.25–0.5			0.25–0.5		0.25–0.5	
carbaryl	1.0–2.0	1.0–2.0				0.5–1.0	
carbophenothion				1.0			
chlordimeform		0.5–1.0			0.5–0.75		
demeton				0.38			
diazinon							
dicrotophos				0.1–0.5			
dimethoate				0.1–0.5			
disulfoton[2]				0.6–1.0			
endosulfan		1.0	1.0				
endrin		0.3–0.6					
EPN	0.5	1.0					
ethion				0.5			
malathion	0.5–2.0			1.25		0.4–1.25	
methamidophos		0.5–1.0	0.5–1.0	0.5–1.0			
methomyl		0.45–0.67			0.45–0.67		
methyl parathion	0.25–1.0	1.0–1.5	0.5–1.0	0.25–0.5		0.12–0.5	
monocrotophos[3]	0.6–1.0	0.6–1.0	0.6–1.0				
naled							
parathion				0.1–0.38		0.12–0.25	
phorate[4]				0.5–1.5			
phosphamidon				0.18–0.5			
toxaphene	2.0–3.0	2.0–4.0				2.0–3.0	2.0–4.0
trichlorfon							
B. thuringiensis			$3.6–8 \times 10^9$ IUs				

Source: *1976 Conference Report on Cotton Insect Research and Control*, ARS, USDA, in cooperation with cotton-growing states.

[1] Granules in furrow at planting.
[2] Granules in furrow at planting. Seed treatment of 0.25 to 0.5 lb AI/cwt of planting seed for control of cotton aphid and thrips.
[3] Dosage to control thrips is for application to 100 lb planting seed.
[4] Granules in furrow at planting. Seed treatment of 1.3 to 1.5 lb AI/cwt of planting seed for control of cotton aphid and thrips.

In the United States study of cotton insects and their control is an important part of the total cotton research program. The primary goal of the program is the maintenance of the competitive position of cotton with synthetic fibers by finding ways to reduce the costs of production and to improve quality. Support for this work comes not only from state and federal government and from industry but from cotton producers themselves. Soon after congress passed the **Cotton Research and Promotion Act** in 1966, growers voted to assess themselves $1 per bale. Receipts go to the recently reorganized Cotton Incorporated, which allocates money for research and promotion. In 1975 the budget for research on cotton insects was $530,000. This amount helped support research projects in several state universities and research institutions.

To keep growers abreast of new developments, the extension service of each cotton state publishes detailed guides every year for control of cotton

Fall Army worm	Cotton Flea-hopper	Garden Webworm	Grass-hoppers	Lygus Bugs	Pink Bollworm	Saltmarsh Caterpillar	Stink Bugs	Thrips
	0.60–1.0			0.6–1.0				0.3–0.5
	0.1–0.25			0.25–0.5	0.5–1.0			0.08–0.4
1.0–2.0	0.5–1.5	1.0–2.0	0.5–1.0	0.5–2.5	2.0–2.5	2.0	1.25–2.5	0.35–1.0
						1.0		
	0.1–0.4			0.25				0.1–0.25
	0.1–0.4			0.5				0.1–0.4
								0.6–1.0
				1.0			1.0	
	0.7–1.25	1.0–2.0	1.0–2.0	0.5–2.5				0.4–1.5
0.25–2.0	0.12–0.5	0.25–0.5	0.25–0.5	0.125–0.5		1.0	0.75–1.5	0.12–0.5
				0.5	0.6–1.0			0.25–1.25
			0.5–0.75					
	0.25–0.5						0.5–1.0	
								0.5–1.5
	0.18–0.5			0.5				0.18–0.5
2.0–3.0	1.0–4.0	3.0–4.0	2.0–4.0	2.0–4.0				1.0–1.5
	0.25–1.0			1.0		1.5	1.0–1.5	

pests. These may be obtained from county agricultural agents and state colleges of agriculture.

CONTROL EQUIPMENT

Beginning in 1916, when calcium arsenate was found effective against several serious cotton pests, growers have applied insecticides to cotton plants mainly as dusts. With the discovery of the chlorinated hydrocarbons and organic phosphates, which are formulated conveniently as emulsifiable concentrates, a trend toward the use of sprays has developed. Sprays are more efficient in covering seedling plants with insecticides, and although both methods are equally effective later in the season, sprays drift less than dusts and can be applied under a wider range of weather conditions.

The types of dusters that growers use depends a great deal on the acreage of cotton they must treat. Hand dusters are practical for treating about 5 acres daily, horse-drawn traction dusters for treating 20 to 30 acres daily, and tractor or airplane dusters for treating larger acreages. Growers usually apply 10 to 20 lb of dust per acre except in the West where they may apply heavier dosages. On ground rigs they adjust nozzles so that there is one nozzle over each row about 10 in. above the plants.

A significant development in cotton-dusting and -spraying equipment is the introduction of self-propelled, high-clearance ground machines (Fig. 12:3). These machines provide efficient coverage in rank cotton without injuring the plants mechanically. They are able to treat 75 acres daily, but they have the disadvantage of all ground rigs in their inability to operate in wet fields.

Instead of dusters some growers employ low-gallonage, low-pressure, boom-type sprayers. Normally from 1 to 8 gal of emulsion spray are applied per acre, although in the Far West up to 15 gal may be applied. One nozzle above each row is used to spray seedling cotton. On larger plants three nozzles per row may be used, one above and one on each side, and in rank growth as many as five or six.

A recent development is the aerial application of ultra-low volumes (ULV) of insecticidal sprays. In the case of malathion this innovation means the application of $\frac{1}{2}$ to 1 pt of the technical insecticide per acre. The advantages of ULV application are several: planes can spray more acres per load; spray droplets, being heavier, drift less, and therefore more hit the target; deposits of the technical insecticide are more residual and hence more effective than deposits of emulsion. Other insecticides that are applied in ultra-low volumes for the control of certain cotton pests are azinphosmethyl, endosulfan, and a mixture of malathion and methyl parathion. Progress is being made in applying several more insecticides in this way and in developing ground equipment for applying ultra-low volumes.

Figure 12:3. A self-propelled, high-clearance, 8-row sprayer treating cotton with insecticide. *Courtesy John Blue Co., Huntsville, Ala.*

Another method of applying insecticides to cotton is to treat the seed. When applied in this way, chlorinated hydrocarbons, such as endrin, provide protection from wireworms, false wireworms, and seedcorn maggots, whereas organophosphorus systemics such as phorate and disulfoton protect seedlings from thrips and cotton aphid.

Granules and fertilizer-insecticide mixtures are useful for control of soil insects such as whitefringed beetles and wireworms. Granule formulations of systemic insecticides placed in the seed furrow at planting are useful in controlling various early season foliage-feeding pests. Systemic insecticides are applied also as sprays in the seed furrow at planting.

REPRESENTATIVE COTTON INSECTS

Although there are numerous species of insects that attack cotton, a limited number of **key pests** are clearly of major importance—the ones that occur every year and must be controlled to avert crop losses. Different groups of key pests attack cotton in the three major producing regions of the United States. In the Southeast the key pests are the boll weevil, bollworm, and tobacco budworm; in the Southwest they are the boll weevil, cotton flea-hopper, bollworm and tobacco budworm; in the far West they are the pink bollworm, lygus bugs, and bollworm. For our study of individual cotton pests, we have chosen the boll weevil, the pink bollworm, the cotton fleahopper, and the black cutworm.

Boll Weevil. *Anthonomus grandis* Boheman [Coleoptera:Curculionidae]

The boll weevil causes more damage to cotton in the United States than any other insect. Native to Mexico or Central America, it was first reported in the United States in the fall of 1894 in Brownsville, Texas. Investigation showed that the boll weevil was established in several southern Texas counties and had probably crossed the Mexican border in 1892. By name and reputation, the boll weevil became known to every southerner; even songs were written about it. In the 1890s Texas Negroes began singing an early blues song, the "Ballad of the Boll Weevil."

From the point of invasion, the boll weevil spread rapidly. It advanced at the rate of 40 to 160 mi a year, mainly by dispersal flights. Entomologists believed that by 1922 it had attained its general distribution in the United States infesting more than 85 per cent of the Cotton Belt (Fig. 12:4). Panic followed the boll weevil. Because Southern agriculture and industry depended almost entirely on one crop—cotton—weevil destruction of one-third to one-half normal production in newly invaded areas bankrupted farmers, merchants, and bankers. Many farmers deserted their places and fled from the ravages of the weevil to other regions.

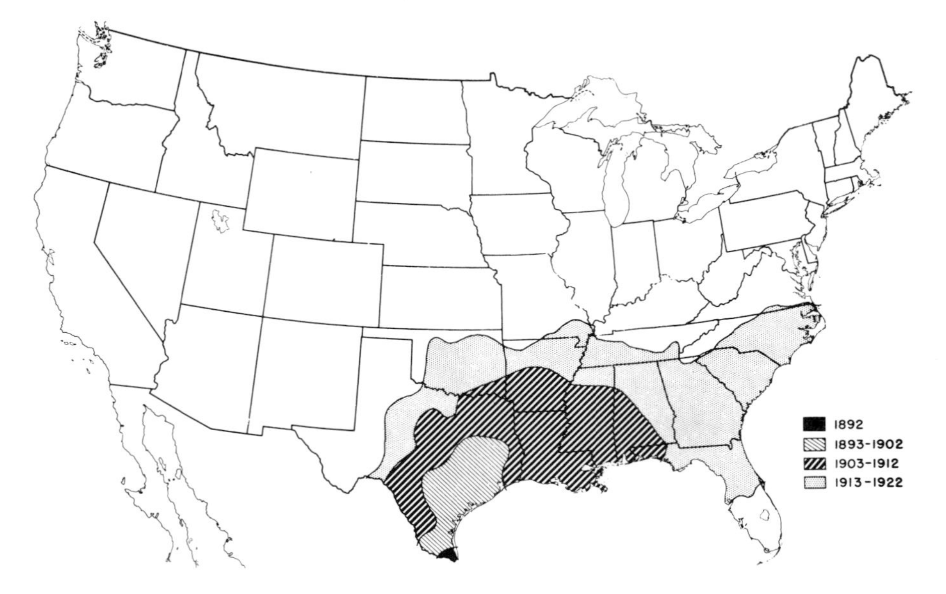

Figure 12:4. Spread of the boll weevil in the United States. It was assumed that the insect had reached the extent of its general distribution by 1922, but actually this was not the case. The boll weevil has continued to spread and adapt to new areas. *Courtesy USDA.*

During recent decades, despite the earlier belief of entomologists, the boll weevil has continued to spread and to adapt itself to new areas. Students of the insect now say that the boll weevil is on the move and predict that it will eventually extend its distribution to every area that produces cultivated cotton.

A recent taxonomic study indicates that there are three closely related forms of weevil attacking cotton: the boll weevil *Anthonomus grandis grandis* Boheman, the thurberia weevil *Anthonomus grandis thurberiae* Pierce, and an intermediate form, the Mexican boll weevil, which may be the common ancestor of the first two. Figure 12:5 shows the distribution of the three forms. Present evidence indicates that the boll weevil and the thurberia weevil spread along diverging lines and followed different host plants.

The boll weevil is a key pest on 52 per cent of the cotton grown in the United States; in 1972 it infested in damaging numbers 7.3 million acres of the 14 million total acres planted. Amount of boll weevil damage varies from year to year because the number of insects fluctuates. From 1909 to 1949 agriculturalists have estimated average loss of cotton and cottonseed of more than $203 million annually. The loss exceeded $500 million in each of five years. In only 16 of the 41 years was the annual loss less than $100 million. In 1973 a reviewer of boll weevil literature concluded that cotton growers continue to lose $200 million or more annually because of this insect.

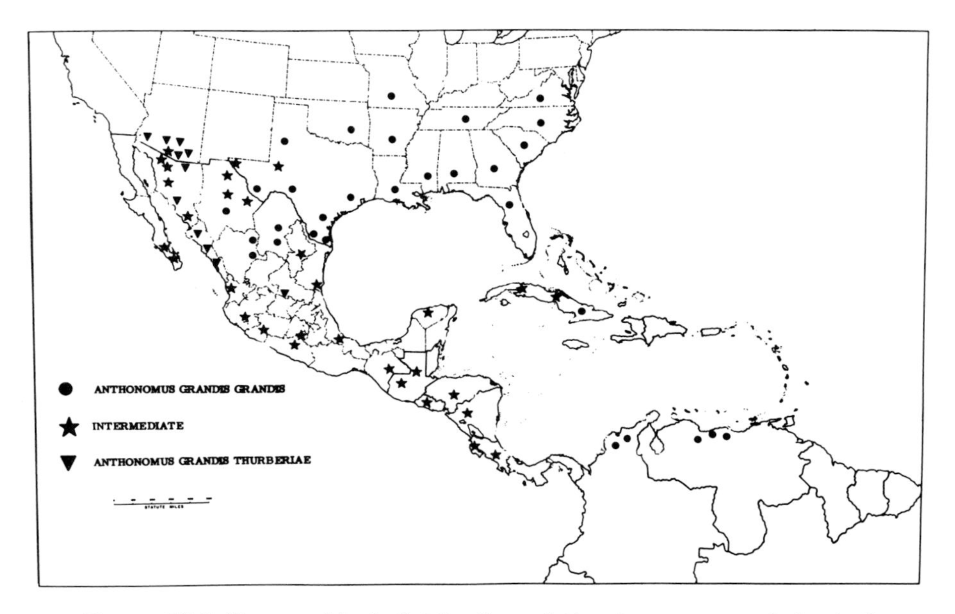

Figure 12:5 Geographical distribution of the three races of the boll weevil based on combined frequency occurrence within the population. *Courtesy Rose Ella Warner, USDA.*

In spring before cotton begins to fruit, boll weevils feed on growing terminals and destroy a few leaves by puncturing the leaf petioles. As soon as flower buds, the so-called "squares", start to form, the weevils feed on them, gouging out small cavities with their long snouts. Feeding on the squares often induces the bracts to open up or "flare," the color to fade to a yellowish green, and the plant to shed the injured buds.

Further injury results when the female weevils lay eggs in the squares, usually one egg to a square. Upon hatching the larva feeds inside the square and quickly causes it to flare and fade in color, and in about three days to drop from the plant. If the square does not drop, it stops growing and becomes hard and dry.

Because heavy numbers can destroy all of the squares, a bad infestation of boll weevils may prevent cotton from blooming. In lighter infestations uninjured squares open up into flowers, which eventually form fruit or bolls. Weevils destroy young bolls as they do squares. Older bolls in which females oviposit usually stay attached to the plant and may continue to grow. A larva developing in a lock will cut, stain, and ruin the lint. If several larvae develop within a boll, as often happens when food is scarce, they destroy the entire boll. Usually the spring infestation is light so that cotton plants produce a good "bottom crop" of bolls. As the season progresses and weevils increase in number, they may severely injure or completely destroy the "top crop."

In 1919 the citizens of Coffee County, Alabama, erected a monument to the boll weevil in the town square of Enterprise, Alabama (Fig. 12:6). They inscribed these words: "In profound appreciation of the boll weevil and what it has done as the herald of prosperity." Their reason for this action was the fact that the boll weevil forced diversification of crops and production of livestock in the South, bringing about a more stable economy.

Another benefit attributable to the boll weevil is the establishment of the federal and state extension services. In 1904 part of an appropriation to fight the boll weevil was made available to the Bureau of Plant Industry to demonstrate to cotton farmers how to control the pest. The demonstrations were so successful that the Farmers' Cooperative Demonstration Work grew into what is now the Agricultural Extension Service.

Cotton (*Gossypium* spp.) is the principal host plant of the boll weevil, but it also develops on certain species of the related genera, *Thespesia, Cienfuegosia,* and *Hampea.* In southern Mexico it severely infests trees of the species *Hampea nutricia.* A rather large number of other malvaceous plants

Figure 12:6. Monument to the boll weevil on town square of Enterprise, Alabama. *Courtesy Enterprise Ledger.*

are used for food only. How the boll weevil became a pest of cotton and the identity of its original host plant are questions that are still without conclusive answers. Until recently a common belief held that the boll weevil did not become a pest of cotton until around 1900. But the discovery of a well-preserved, "intermediate" adult weevil in a cultivated cotton boll found in Mexican archeological excavations places the relationship at least as far back as A.D. 900.

Description. Eggs of the boll weevil are pearly white, elliptical, and approximately 0.8 mm long. The eggs hatch into legless, white grubs with light brown heads. After feeding from 7 to 12 days the larvae become full grown and pupate within the squares or bolls (Fig. 12:7B). The pupa is all white at first but becomes suffused with brown as it develops. After a pupal period of three to six days, the adult boll weevil emerges. Adult weevils range from $\frac{1}{8}$ to $\frac{1}{3}$ in. in length and are reddish brown or grayish in color. The conspicuous snout is about one half as long as the body. A distinctive character is the spur on the inner surface of each front femur.

Life history. The boll weevil passes the winter in the adult stage sheltered underneath ground litter in cotton fields, along fence rows, in the edge of woods, in weed patches, and in other protected places. Many even seek shelter in clumps of Spanish moss hanging from trees. Adult weevils start to come out of hibernation quarters as soon as the weather warms up in spring.

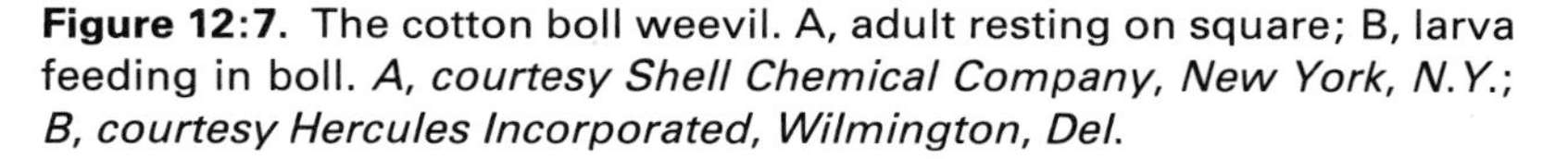
Figure 12:7. The cotton boll weevil. A, adult resting on square; B, larva feeding in boll. *A, courtesy Shell Chemical Company, New York, N.Y.; B, courtesy Hercules Incorporated, Wilmington, Del.*

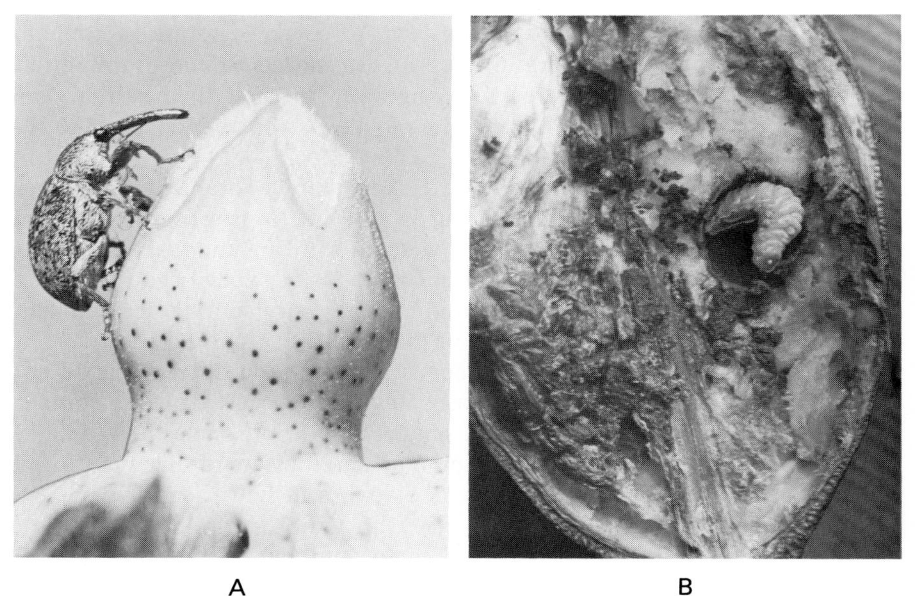

Although the period of emergence may extend from March to the middle of July, the majority of weevils emerge in June. They usually invade cotton fields close to their hibernation site, but some weevils may fly 6 miles in search of host plants. After finding a field of fruiting cotton, they settle down to feed and reproduce.

Weevils that emerge before cotton plants have begun to square feed on the leafbuds and growing terminals and live for only one to two weeks. Those emerging later feed on squares, their principal food, and may live as long as six weeks. To obtain food, weevils usually puncture the tips of squares, where they feed internally on the developing pollen and other structures. After a few days of feeding, the females begin to produce eggs.

Weevils prefer to lay in squares, but they also utilize bolls. They deposit eggs singly into cavities made with their mouthparts. After placing an egg deep in the cavity, they seal the hole with a glue-like substance. More than one egg may be laid in a square or boll; the female apparently is unable to detect an already infested one. In cases of multiple infestation of a square usually only one larva develops successfully; the others die from starvation or cannabilism. Overwintered females produce fewer than 100 eggs, but females of later generations produce up to 300 or more eggs. A female weevil's average rate of reproduction is 5 to 6 eggs a day.

In summer eggs hatch in three or four days, and the larvae, which develop inside the square or boll, feed upon the anthers and pollen or the lint. Influenced by temperature and by the nutritional value of their food, larvae complete development in 7 to 12 days. The larvae then pupate within the squares or bolls, which afford protection for the delicate pupae. This stage lasts from three to six days after which the adult weevils cut their way out. Depending on latitude, there are from two to seven generations annually in the United States. Beginning in late August and continuing through fall, adult boll weevils take to flying and migrating to their hibernation sites. Timing of flights is influenced by densities of weevils and abundance of cotton fruit.

Outbreaks of the boll weevil are linked with favorable weather and adequate food and shelter. Cotton that is allowed to grow and to develop squares in late summer provides the weevils with an abundant supply of breeding sites and nourishing food. Because well-fed weevils go into hibernation with higher levels of accumulated fat, they survive in larger numbers. Apparently it is a case of the survival of the fattest.

For successful hibernation boll weevils first go into a state of diapause, a condition characterized by an increased fat and decreased water content, lower respiratory rate, and atrophied sex organs of both males and females. Although boll weevils of any generation may enter a state of diapause during the growing season, increasing numbers diapause in fall when days become short (less than 12 hours of daylight). The short photoperiod acts upon both the immature and the mature stages of the weevil, but the response may be modified by the quality of the adults' food and by the temperature to which they are exposed. Fewer boll weevils enter dormancy when they continue to experience high temperatures and to consume favorable food.

Mild winters and satisfactory shelter favor survival of the hibernating weevils. Field studies have shown that winter mortality is usually quite high. Nevertheless, significant variations in survival occur as it may range from less than 1 per cent up to 20 per cent.

After weevils emerge in spring, frequent rains, moderate temperatures, and plenty of squares for food and for breeding sites promote high populations. On the other hand, prolonged hot, dry weather is lethal to developing larvae and effectively checks a potential outbreak. Excessive heat and low humidity kill the larvae inside squares, particularly those in fallen squares exposed to the sun.

Although as many as 42 species of parasites attack the boll weevil, they are only of slight value in limiting the population. Parasites normally kill about 6 per cent of boll weevil larvae; in a few localities parasitism may run as high as 20 per cent throughout the season. The commonest parasite is a small wasp, *Bracon mellitor* Say, which in its larval stage feeds on and destroys the weevil larva inside the square or boll. A number of predators attack the boll weevil, including insects, toads, lizards, and birds; a few pathogens also infect this insect.

Recent research of the behavior of the boll weevil in cotton fields and in the laboratory has revealed several interesting facts. One of the most important is the discovery that adult males produce a pheromone which is both a sex and an aggregating attractant. Unmated females are attracted by the pheromone at all times, and both sexes are attracted at special times, such as in spring after the weevils come out of hibernation. A male that finds a field of cotton, perhaps by random flight, will begin to feed and to produce pheromone. Volatilizing into the air and borne by the wind, the pheromone reaches other weevils, which are then attracted to its source. The pheromone content in a male is never greater than 200 nanograms, but the content in the frass passed during one day averages 1,300 nanograms. In 1968 the pheromone was identified as a mixture of four compounds and in the same year was synthesized. The synthetic product was given the name **grandlure**.

Cultural control. Satisfactory control of the boll weevil depends on a program of pest management utilizing chiefly cultural practices and insecticidal treatments. Beneficial cultural practices include (1) planting of recommended varieties; (2) early planting; (3) stimulating rapid growth by thorough preparation of seed bed, by fertilizing, and by frequent shallow cultivation; (4) delaying thinning and leaving more plants per acre; (5) chemical defoliation and desiccation of the cotton plant; (6) early destruction of cotton stalks; (7) destroying favorable hibernation quarters.

Cultural methods have two main objectives. The first and foremost is to hasten development of cotton plants so that they set a crop before the weevil population has increased to ruinous numbers. The second objective is to reduce the overwintering weevil population and thereby minimize the problem for the next year.

As plant breeders improve cotton, a number of superior varieties are replacing the multitude formerly grown. Currently recommended varieties fruit early, mature quickly, have determinate growth, yield well, and produce quality lint. Few include a useful level of resistance to boll weevil.

As soon as danger from frost has passed, early planting of all cotton within an area enables the crop to set bolls before the weevils become numerous. Planting during the earliest optimum period makes early stalk destruction possible.

Stimulating rapid growth and maturity by preparing a good seed bed, fertilizing, and cultivating also helps the crop to escape the large summer population of weevils. A number of cotton pests may be attracted by the increased plant growth and require closer vigilance and at times insecticidal control. Delaying thinning until the plants are several inches tall and leaving the plants thick in the row suppresses vegetative growth and stimulates early fruiting. Much of the cotton in Texas is now planted to a stand; that is, it is planted and not thinned.

Growers apply defoliating chemicals chiefly to aid in the harvest, but they receive other benefits from this practice. The checking of plant growth and the acceleration in the opening of mature bolls depresses the late season buildup of boll weevil and reduces damage of the weevil and several other pests. Because defoliation speeds up cotton harvesting, stalks can be destroyed earlier.

Soon after the harvesting of the cotton crop, early community cutting and shredding of stalks prevents further multiplication of the weevil population and starves and weakens the adults that go into hibernation. Unfortunately this method is not practicable in the more northern areas, where growers are unable to harvest the crop until October and later.

Destroying or reducing the number of favorable hibernating sites such as patches of volunteer or stub cotton, patches of weeds in or near cotton fields, along fence rows, and in waste land decreases the number of weevils that overwinter successfully.

Chemical control. Because larvae develop inside squares and bolls and because adults feed mainly on internal tissues, and migrate extensively, the boll weevil is a difficult insect to control with insecticides. On a single cotton crop during years of heavy infestation farmers may apply as many as 20 insecticidal treatments at three- to five-day intervals. During light years boll weevils may require no treatment in northern parts of the Cotton Belt and perhaps only two or three in southern parts.

From 1919 until 1947 growers made calcium arsenate dust the primary boll weevil insecticide, applying it full strength at rates of 5 to 10 lb/acre. In 1947 they began turning to the newly discovered chlorinated hydrocarbons. In some areas development of resistance to chlorinated hydrocarbons by the boll weevil and cancellation of registration of many chlorinated hydrocarbons have compelled growers to turn to organophosphorus and carbamate insecticides.

Table 12:1 shows the dosage of insecticides that have controlled the boll weevil in one or more areas of the Cotton Belt. Because applying these insecticides singly may create other insect or mite problems, they are often combined with insecticides that will counteract any side effects.

The timing of treatments is important for successful control of the boll weevil but varies with the region, weather, and severity of infestation. Some states recommend controlling boll weevils only during midseason (blooming period) and late season (maturing period), whereas other states must recommend treatment during early season (prebloom period) as well to avoid serious damage.

A new way of utilizing insecticides against the boll weevil consists of continuing their application late in the season in a so-called "reproduction-diapause boll weevil population control program." The objective of this program is to reduce populations of the boll weevil so severely in the fall of the year that the progeny of the few surviving weevils will not build up damaging numbers in next season's crop until the second or third generation. By reducing the overwintering population so that damaging numbers do not come back until August, cotton growers can delay the first application of insecticide and benefit from natural control of bollworm and spider mites. This method reduces the cost of cotton insect control 30 to 50 per cent; but it requires that growers in an area act collectively and that they scout the fields regularly to detect threatening populations of pests before they reach damaging levels.

Other control techniques that have been found effective include planting trap cotton and installing grandlure traps in and around cotton fields. The first technique consists of planting in each field about 4 rows of cotton 2 to 3 weeks ahead of the main crop. To this trap cotton granules of aldicarb are applied in the furrow at planting at 1 lb AI per acre and again as a side dressing at the pinhead square stage at 2 lb AI per acre. Emerging boll weevils concentrate on the trap cotton because of its more advanced growth and are killed by feeding on the plants, which have taken up the aldicarb systemically. Another way of handling trap cotton is to plant 8 to 16 rows ahead of the main crop along field borders and near hibernation sites, then, at the pinhead square stage, to begin a schedule of 5 foliar applications 5 days apart of an approved boll weevil insecticide.

The second technique, trapping with grandlure in spring, reduces the few boll weevils surviving an effective reproduction-diapause program. A control model in which 10 traps are installed per acre suggests a reduction of 75 per cent of the emerging weevils.

Combining and integrating these different control techniques—planting early maturing varieties with degrees of pest resistance, reproduction-diapause applications of insecticide, trap crops, grandlure traps, delayed in-season applications of insecticide, defoliation and desiccation by application of chemical desiccants, and stalk destruction—make an attractive pest management scheme designed to control the boll weevil, bollworm-tobacco budworm, and other cotton pests infesting cotton in the southeast and southwest production areas.

The possibility exists that by imposing the sterile male technique of control on the few survivors of these several control methods, eradication of the boll weevil can be accomplished. This idea was put to the test in the **Pilot Boll Weevil Eradication Experiment** in southern Mississippi and parts of Alabama and Louisiana from 1971 to 1973. Although boll weevils were greatly reduced, doubt remained whether eradication had been accomplished. After improvement of needed technology another eradication experiment has been proposed, this time taking place in North Carolina.

Despite the new and the improved methods of control, we still have not completely solved the problem of the boll weevil; hence many entomologists are continuing to devote full or part time to its study. In 1961 a $1.1 million USDA Boll Weevil Research Laboratory was completed on the campus of Mississippi State University.

Pink Bollworm. *Pectinophora gossypiella* (Saunders) [Lepidoptera:Gelechiidae]

The pink bollworm, a native of Asia (perhaps India), has become a major worldwide pest of cotton. In shipping infested cotton seed, man has spread this injurious species to all the important cotton-growing countries of the world. Once the insect gains entry into a country, it may disperse widely by flight.

Egyptian seed, imported in 1911 into Mexico, carried the pink bollworm to North America. Not long afterwards, in 1917, entomologists found the insect in fields near Hearne, Texas. Apparently it had entered the country in the fall of 1916 in a shipment of cotton seed sent to an oil mill at Hearne from the Laguna district of Mexico.

In spite of stringent quarantines and control measures, the pink bollworm has fanned out in Texas and crossed state lines into New Mexico, Arizona, Oklahoma, Arkansas, Louisiana, and most recently into California, Nevada and Missouri. Unrelated to this infestation, a separate population attacks wild cotton in southern Florida and presents a threat to the cotton crop of the Southeast.

Even though control measures have not yet stopped the pink bollworm's gradual invasion of the Cotton Belt, they have kept injury down to an insignificant amount. The year 1952 was an exception: losses reached almost $35 million in counties of the Coastal Bend Area of Texas. The insect has staged a resurgence in recent years, becoming a severe problem in some of the most productive cotton acreages in the Southwest. Damage in foreign countries has often ranged from 15 per cent to as much as 70 per cent of the crop. In Puerto Rico the pink bollworm forced growers completely out of cotton production.

Pink bollworms damage cotton by feeding inside the squares and bolls. Because moths prefer to lay eggs on the bolls, and the larvae prefer to enter them, squares are infested for only a short period early in the season. Larvae burrow into the squares, where they feed upon the pollen and other embryonic flower parts. Invaded squares do not open normally, because the pink bollworm spins a web of silk around the petals holding them together. This gives the flower its characteristic rosette appearance. Actual loss from square infestation is negligible, but the spring generation developing in squares is important as the progenitor of populations that may severely attack bolls.

Larvae prefer to feed on bolls, usually attacking those one-half to three-fourths grown (Fig. 12:8). Boring into them at any point, larvae may go several routes and feed on several tissues. They may feed in the soft carpel tissues, producing brownish discolored tunnels—commonly called "railroads"—which are characteristic of pink bollworm damage. Or the larvae may completely perforate the carpel wall, entering and feeding on the immature lint. Some larvae go from one lock to another by cutting round or oval holes through the partitions. Eventually all reach the developing seeds on

Figure 12:8. Pink bollworm in cotton. A, nearly full-grown larvae inside cotton seeds; B, exit holes of larvae which have completed their feeding and growth. *Courtesy USDA.*

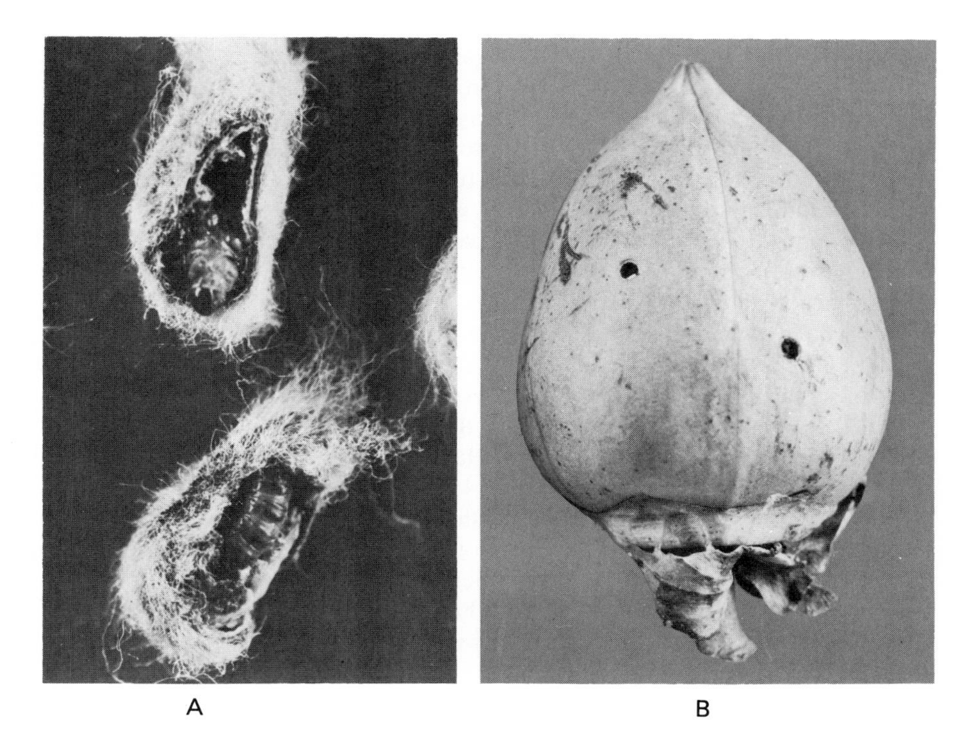

which they finish their feeding. In completing growth each pink bollworm eats out the contents of several seeds. From one to several larvae may infest a single boll, and in heavily infested cotton there may be an average of six or seven larvae per boll.

Because of their tunneling and feeding within bolls and their cutting and staining of fibers, pink bollworms reduce yields and lower the quality of the lint. Damaged fibers, mixed with undamaged lint in the harvesting and ginning process, reduce the average grade of the crop. The insects that feed on seed destroy seed viability and reduce the quantity and quality of oil. The pink bollworm is also instrumental in initiating boll rot, because the exit holes of larvae provide pathways for entry of bacteria and fungi. Boll rot may be extensive and severe in heavily infested fields. Infection by the fungus *Aspergillus flavus* is especially serious because of the production of the poisonous aflatoxins in cottonseed that is used for feedstuffs. Another loss caused by the pink bollworm is cost of control, which runs high because of the number of treatments necessary. In 1967 in Arizona and California the cost often amounted to $40 to $60 per acre.

The pink bollworm is known to propagate on 39 plant species under natural conditions in this country. Okra is probably preferred after cotton and is considered in the same category as cotton in the overwintering of the species and quarantine regulations.

Description. The moths are grayish brown with irregular markings on the forewings. They have a wingspread of about $\frac{1}{2}$ in. The eggs, usually laid in masses on the cotton plant, are elongate oval and 0.5 mm long. They are pearly white with a greenish tint when first deposited but become reddish before hatching. The larvae are creamy white with dark brown heads and thoracic shields. Usually not until the fourth instar do they acquire the pink color that gives them their name. The full-grown larvae measure about $\frac{1}{2}$ in. long. The pupae are brown, pubescent, and about $\frac{3}{8}$ in. long (Fig. 12:9).

Life history. The pink bollworm passes the winter in the larval stage. Starting in spring and extending into summer, the overwintered larvae pupate and about ten days later emerge as moths. Peak numbers usually occur before the first bolls appear.

The moths are seclusive during the day, hiding under trash in the field, in soil crevices at the base of young plants, or in the lower canopy of older plants. At dusk they become active. Through the night they fly from plant to plant feeding on nectar, resting, mating, and laying eggs. The virgin female "calls" for a male nightly between 1:00 and 4:00 A.M. by protruding the terminal segments of her abdomen and releasing a sex pheromone. In 1973 the pheromone was identified as a mixture of four isomers of 7,11-hexadecadien-1-ol acetate. This material is now synthesized and called **gossyplure**.

Moths that emerge before the bolls have formed lay eggs on other parts of the plant. When these eggs hatch the young larvae search out squares, which they enter and feed upon. Earlier emergence of moths before plants have produced squares is called "suicidal" because the larvae are unable to survive.

Figure 12:9. Life stages of the pink bollworm: larva, pupa, adult moth, and cluster of eggs. *Larva, pupa, and adult, courtesy Hercules Incorporated, Wilmington, Del.; eggs, courtesy USDA.*

Females prefer to deposit eggs on green bolls, usually in clusters under the calyx. When bolls are not available they lay their eggs on vegetative parts. During their lifetime of about 15 days, female moths produce from 100 to 200 eggs. The eggs hatch in about five days, and the young larvae bore through the carpel wall of bolls. Cutting into the tissues and throwing the fragments outside, they become completely hidden in 20 to 40 minutes. Much mortality occurs at this time, because only about 10 per cent of the larvae succeed in entering bolls.

Once larvae penetrate the bolls, they are protected from enemies and adverse weather, and unless crowded by other larvae inside the boll, their chances of survival are good. Pink bollworms are antagonistic and often kill one another on contact. Having gained entrance into the boll they feed on the tender tissues, the inner carpel wall, the immature lint, the outer seed coat of immature seeds, and particularly the inside of older seeds. They molt three or four times and become full grown in 10 to 14 days. Then they either cut a round hole through the boll wall and drop to the ground, or they remain in the boll, usually inside a hollowed-out seed. A recent study of larval development showed that 25 per cent of the larvae require five instars to complete development, whereas 75 per cent require only 4 instars.

In spring and summer the majority of larvae are the so-called short-cycle or summer larvae, which spin light cocoons, pupate mainly in the soil, and give rise to succeeding generations. The pink bollworm completes a generation in 25 to 30 days during summer, so that there may be as many as six generations a year.

A few larvae in summer and almost all in fall enter dormancy. They are the long-cycle or diapausing larvae. Spinning a tough cocoon, they remain in the boll, usually inside a hollowed-out seed, or inside two hollowed-out seeds that they tie together. Because the united seeds, called "double seeds," show up after ginning they serve as good indicators of pink bollworm infestation. In warm, humid areas winter survival of resting larvae is greatest in bolls that remain attached to standing plants or that lie on the soil surface. Few survive the winter buried in the soil. In comparatively cold arid areas, minimum winter temperatures of 15°F or lower kill the larvae in bolls on standing cotton.

Several parasites and predators attack pink bollworms. Two common parasites are the small wasps *Bracon mellitor* Say and *Bracon gelechiae* Ashm. Neither species appears to contribute much toward economic control of the pest. Tests of biological control with two introduced parasitic wasps, *Bracon kirkpatricki* and *Chelonus blackburni*, have resulted in parasitization rates as high as 41 per cent. The egg stage of the pink bollworm is the most vulnerable to predation, but unfortunately adequate numbers of effective predators do not inhabit cotton fields. Currently entomologists are studying several insect pathogens, such as a nuclear polyhedrosis virus placed in baits, for biological control of the pink bollworm.

Control. Although insecticides are necessary to control heavy infestations of the pink bollworm in order to prevent crop losses, growers should make cultural practices their first line of defense. The chief objectives of cultural control are to have an early-maturing crop so that it escapes the high populations of pink bollworm that come late in the season, to harvest early to stop pink bollworm reproduction, and to reduce the numbers of overwintering pink bollworm. Mandatory cultural control zones have been established in the regulated areas of Arizona, Arkansas, California, Louisiana, and much of Texas.

Recommended cultural practices include the following:

1. Plant at the optimum time and shorten the planting period. Select a recommended early-maturing variety. Use seeds that have been culled, treated with a fungicide, and tested for germination.
2. Leave as thick a stand of cotton plants as recommended for a particular area and type of soil.
3. Produce the cotton crop in the shortest practicable time. Early-season insect control has proved advantageous in some states. Protection of early fruit from insects will assure an early harvest.
4. Withhold late irrigation and use defoliants or desiccants to hasten opening of bolls.
5. Pick and carry as much of the cotton as possible to the gin because most of the overwintering larvae will be in the seeds in seed cotton and more than 99 per cent will be killed by the modern ginning process.
6. Shred and plow under cotton stalks as soon as possible after harvest. Okra stalks should be shredded and plowed under at the same time. Irrigation in winter further reduces the larvae's chances of survival.
7. In cold arid areas, leave stalks standing until lowest temperatures have occurred in order to obtain a maximum kill of pink bollworms infesting bolls on the stalks. But if a large amount of crop debris such as seed cotton or locks is on the soil surface, high survival of the pest may ensue, so the stalks should be shredded and plowed under as early and as deeply as possible. The flail-type stalk shredder is more efficient than the horizontal rotary type for pink bollworm control—a kill of 85 per cent by flail as compared with 55 per cent by horizontal rotary shredders of the pink bollworms remaining in the field after harvest.

Because late-season cotton is potentially more profitable, many growers are reluctant to follow all of the recommended cultural practices; instead they rely heavily on insecticides to control infestations. In areas affected seriously by the pink bollworm, growers or scouts should begin examining cotton fields in the early fruiting stage for pink bollworm. As soon as bolls develop they should be examined weekly for presence of larvae. One method is to pull 25 bolls in each of four areas of a cotton field (30 to 80 acres in size, proportionately more in larger fields). Bolls are cut or cracked open and examined for presence of larvae. When 15 per cent of the bolls are infested (5 per cent in boll rot areas), a grower is advised to begin insecticidal treatment, making a series of applications six days apart until bolls become hard and harvest is near. Recommended sprays include azinphosmethyl, carbaryl, or monocrotophos (Table 12:1).

In the Imperial Valley of California detection of threshold densities by trapping male moths with hexalure bait is useful in indicating the time to begin insecticidal treatment. When the daily catch of moths averages 3.5 to 4 per trap, a field is treated within 24 hours and then kept on a regular schedule of treatment every 6 to 7 days. Fields treated on the basis of moth counts averaged 4 fewer spray applications than fields on a fully scheduled program of 9 to 12 treatments. Several advantages accrued from this pilot program: equal pink bollworm control with less insecticide and fewer applications, smaller chance of causing outbreaks of other pests, and less likelihood of fostering insecticide-resistant insects.

After the pink bollworm became established in southern California in 1966 fear arose that the pest would invade the San Joaquin Valley to the north, where 12 per cent of the nation's and 90 per cent of California's cotton is grown. A program of exclusion by mass releases of sterile pink bollworms was begun in 1968 and continued in subsequent years. In 1975 more than 150 million moths, irradiated at the APHIS facility in Phoenix, Arizona, were released in the valley. Although wild moths have been caught every year since 1967 and a few larvae found in 1967 and 1970, no reproducing population has been discovered. To detect wild moths of the pink bollworm and to monitor the released sterile ones, thousands of traps baited with gossyplure are placed each year in the 1.2 million acres of cotton grown in this area of California.

Ever since the pink bollworm was first found in the United States, federal and state quarantines and sponsored survey and control programs have slowed down the spread of the pest. The federal quarantine prohibits the entry of cottonseed and cottonseed hulls from most of Mexico and from all other countries. Cotton, cottonseed cake, and meal enter only under restricted conditions.

Quarantines also govern the movement of cotton and cotton products produced in infested areas of the United States. In general the regulations require that cotton, cotton products, and all articles connected with the production of cotton be treated to render them free of pink bollworms before they are moved to nonquarantined areas. Other regulations include deadlines for planting cotton, harvesting and destroying stalks, and processing-plant sanitation. Entomologists and plant breeders are intensifying their efforts to transfer resistant characteristics into commercially acceptable varieties.

Cotton Fleahopper. *Pseudatomoscelis seriatus* (Reuter) [Hemiptera:Miridae]

Several species of mirid bugs, lygus bugs (*Lygus hesperus* and *L. elisus*), the clouded plant bug, tarnished plant bug, ragweed plant bug, superb plant bug, black cotton fleahoppers (*Rhinacloa forticornis* and *Spanogonicus albofasciatus*), and the cotton fleahoppers are serious pests of cotton. Although the injury they cause to cotton is similar, the species vary in importance in different areas of the Cotton Belt.

Widespread in North America, the cotton fleahopper causes greatest damage in Texas, Louisiana, and Oklahoma. It is a native insect whose host plants include many weeds such as goatweed (*Croton*), horsemint (*Monarda*), eveningprimrose (*Oenothera*), horsenettle (*Solanum*), cudweed (*Gnaphal-*

ium), and orach (*Atriplex*). Cotton, the only cultivated plant seriously attacked, appears to be an acquired food plant of this bug.

Described in 1876 by Reuter from specimens collected in Texas, the cotton fleahopper attracted little attention as an economic pest until 1920, when investigators showed that the appearance of a new type of damage to cotton, excessive shedding of minute squares, was associated with the feeding of this insect. The cotton fleahopper is an early-season, occasionally mid-season, cotton pest.

In feeding, cotton fleahoppers prefer to suck juices from the tender growing terminals and squares, causing the tiny squares to turn brown and fall from the plant. They also feed on other parts, inducing swelling and splitting of stems and deformities in leaves. Although injury is of a local nature and is caused by toxic salivary juices injected into the plant by the insect, mass attack brings about abnormal development of the whole plant. Fruiting is retarded, internodes are decreased in length but increased in number, plants either become bushy or tall and whiplike and usually produce only a few bolls near the tops. In Oklahoma, especially the drier sections where growers rely on early crop set, heavy infestations can cause complete loss of a cotton crop.

Description. The cotton fleahopper inserts small, glistening white eggs, about 0.8 mm by 0.2 mm, into the stems of host plants. Before hatching, the eggs become suffused with yellow. First-instar nymphs are pale green, 1 mm long, and have prominent scarlet eyes. Older instars become greener and densely mottled with black spots. The adults are light green, densely spotted black, and approximately $\frac{1}{8}$ in. long (Fig. 12:10).

Life history. The cotton fleahopper passes the winter in the egg stage mainly inside the stems of wild host plants such as goatweed. Overwintering eggs hatch early in spring; populations build up rapidly, because the bugs are able to complete seven or eight generations during a growing season in the South.

Development of large populations of cotton fleahoppers depends upon a favorable sequence of host plants as well as favorable weather. In early spring the bugs feed on early weeds such as cudweed and eveningprimrose. In late spring and early summer they move to goatweed, horsemint, and horsenettle. When host weeds become mature and tough in early summer, adult cotton fleahoppers fly to the succulent, squaring cotton. In late summer and early fall adults of the last generation return to wild hosts, principally goatweed. Of significance in the natural history of the cotton fleahopper is its utilization all through the growing season of goatweed (*Croton*) as a favored host plant. Unlike many of the other favored weeds, this plant starts growing in spring and does not mature until fall.

Depending on temperature and moisture, eggs vary in rate of development. Under optimum conditions eggs may hatch in five days, but on an average they hatch in seven or eight days. The nymphs develop rapidly passing through five instars, in as little as ten days, to reach the adult stage.

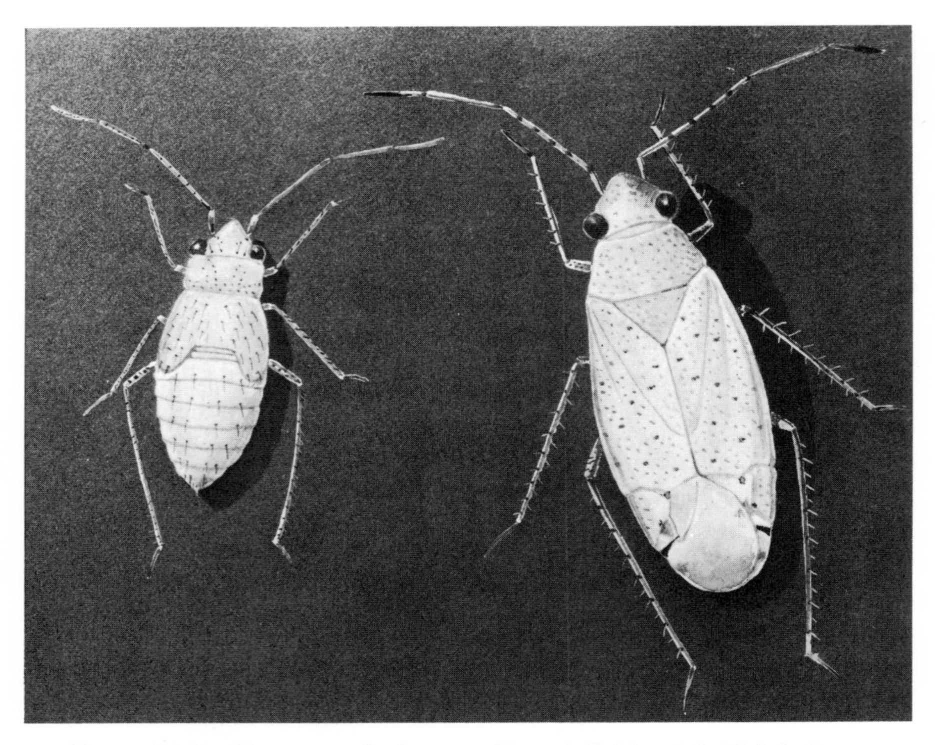

Figure 12:10. The cotton fleahopper. Nymph (left); adult (right). *Courtesy Hercules Incorporated, Wilmington, Del.*

Adult females feed for three or four days before they begin to insert eggs in the stems and other parts of the host plant. They lay the eggs singly by means of a strong swordlike ovipositor. Observations on bugs confined in cages indicate that females live from 14 to 29 days and produce around 21 eggs each. Males have a somewhat shorter lifespan.

Newer studies done in the laboratory differ somewhat from the preceding observations, which were made in the late 1920s. Reared at 85°F under controlled laboratory conditions on cut potato tuber and green bean, immature stages of the cotton fleahopper develop rapidly and survive well; the egg stage requires 9.4 days and the nymphal stage 10.8 days. A constant temperature of 80°F appears optimum for the adults; longevity averages 22.1 days, and females produce an average of 30 eggs each. The species and stage of the host plant significantly affect development and egg production. At 80°F nymphs reared on a preferred host such as flowering *Croton* require 9.6 days instead of the 14 days that nymphs require on potato and bean; females reared on flowering *Croton* average 76 eggs each.

Control. Although growers place their main reliance on insecticides to control damaging infestations of cotton fleahopper, clean culture can be helpful in preventing outbreaks. Destruction of weeds near cotton fields eliminates important host plants, and plowing under old cotton stalks destroys overwintering eggs.

As soon as cotton begins to produce squares, growers or scouts make weekly inspections for density of fleahoppers, which they count by walking diagonally across a cotton field and examining the top 3 or 4 inches of terminals of the main stem of 100 cotton plants. In early season when numbers of nymphal and adult fleahoppers per hundred terminals reach a given number—15 in certain states and up to 50 in others—published recommendations advise the application of insecticide. As plants increase in size and fruits become abundant, cotton can tolerate larger populations. Delaying insecticidal applications for as long as is economically feasible protects beneficial insects and often defers the start of treatments against the bollworm-budworm complex. Yet uncontrolled heavy infestations of the cotton fleahopper delay maturity of the cotton, which then becomes more susceptible to damage by the boll weevil and the bollworm-budworm complex. Dusting or spraying with toxaphene, carbaryl, or one of several organophosphorus insecticides is effective in controlling the cotton fleahopper and curbing its damage (Table 12:1). Make only one application unless more are required to control an infestation; apply additional treatments at 7- to 10-day intervals. Preferably, insecticide applications are terminated before first bloom; fleahoppers do little damage to cotton after blooming begins. This practice allows conservation of predators and parasites. In Texas the cotton fleahopper has become resistant to organochlorine and organophosphorus compounds.

Black Cutworm. *Agrotis ipsilon* (Hufnagel) [Lepidoptera:Noctuidae]

The Noctuidae, containing some 2700 species in the United States and Canada, has many members that are serious crop pests in their larval stages. The larvae, named after their habits, are known commonly as cutworms, armyworms, loopers, semiloopers, leafworms, borers, and fruitworms.

Cutworms in cotton are enemies chiefly of the seedling plants. The pests frequently develop in weeds, in legumes, and in other crops and will attack and often destroy stands of cotton planted in or adjacent to infested land. The various species of cutworms are usually polyphagous in their feeding habits. The black cutworm, for example, seriously attacks not only seedling cotton but also young corn plants, tobacco, and many vegetables. It also finds forage legumes highly palatable.

The black cutworm is a highly destructive species, much out of proportion to the amount of vegetation it consumes. Relatively small populations are able to destroy entire stands of cotton because of the cutworm's habit of eating into the stem of a plant, cutting it down near ground level, and then moving on to repeat the havoc on other plants.

The species is widely distributed throughout the world, occurring in the United States, southern Canada, Mexico, South America, Europe, Asia, Africa, and Australia. Preferring moist or wet soil, the species exhibits a high tolerance to moisture. Outbreaks often arise in river bottoms after floods and in wet muck lands.

Description. The eggs are deposited in groups of 1 to 30 on leaves and stems near the ground. Each egg is sinuously ribbed and subconical in profile; it measures approximately 0.45 mm high and 0.65 mm in diameter. The egg is pale yellow when laid but becomes suffused with brown the second day. The larvae are almost unicolorous dorsally and vary from light gray to nearly black. Full grown, they measure $1\frac{1}{4}$ to 2 in. long. The pupa is naked, brown, and $\frac{6}{8}$ to $\frac{7}{8}$ in. long. The adult moth has a wing expanse of $1\frac{1}{2}$ to 2 in. The forewings are long, narrow, and usually dark; each is characteristically marked with three black dashes. One dash arises from the outer side of the reniform spot and the other two point toward the first one from near the outer margin of the wing (Fig. 12:11).

Life history. The black cutworm normally passes the winter in the pupal stage. In spring the adults emerge and mate, after which the females lay their

Figure 12:11. The black cutworm. A, adult; B, larva or cutworm; C, pupa in its soil cell; D, pupa removed from cell; E, right forewing showing reniform spot and the three black dashes. *A, C, and D, courtesy Can. Dept. Agr.; B, courtesy University of Nebraska, Lincoln; E, original.*

eggs—usually at night. Virgin females emit a calling pheromone from midnight to 3:00 A.M., attracting the males for mating. Peak attractancy and peak ovariole development occur at 4 days of age among laboratory-reared females.

Dissection of females indicates that they have reproductive capacities ranging from around 500 to 2250 eggs each. Preferring low or overflowed wet soil, moths select plants growing in such sites for oviposition. They deposit eggs singly or in small clutches on the leaves and stems of plants. They also deposit eggs on soil in the laboratory and no doubt in the field.

In three to six days the larvae hatch from the eggs, move to the soil, and begin to feed and grow. In laboratory experiments first- and second-instar larvae react photopositively to dim light, but older larvae react photonegatively. Young black cutworms feed first on the leaf surface; later they eat holes in leaves. Only the older larvae develop the cutting habit. Most feeding is done in the evening. Older larvae may drag their food into tunnels that they make in the soil. The black cutworm is noted for its aggressive, pugnacious, and cannibalistic nature. They are exceedingly antisocial, for 1 + 1 = 1.

Although some larvae require seven instars and a few eight, the majority become full grown in six instars. Varying with temperature, the larval period lasts from 25 to 35 days. The full-grown larva forms a cell in the soil in which it pupates. During the summer the pupal period lasts around 12 to 15 days. The entire cycle from egg to adult takes around 45 to 55 days, and a multiple number of generations is produced each year, ranging from two to three in northern states and in Canada to four or more in southern states.

As with most cutworms, a number of parasites, predators, and diseases take their toll, but their importance in limiting outbreaks of the black cutworm is doubtful.

Control. Although growers rely mainly on insecticides to control damaging infestations of cutworms, cultural methods are helpful. Cultural control includes plowing under of vegetation three to six weeks prior to seeding of the cotton crop, destruction of weed host plants, and thorough seed-bed preparation.

For many years poisoned baits have been used to control cutworms, and although baits are highly effective against a few species, the present trend is toward using sprays and dusts. The following insecticides have been recommended as sprays or dusts to control cutworms: toxaphene, endrin, trichlorfon, and carbaryl. These may be applied as a broadcast treatment of ground and plants or as a band treatment in crop rows. Because these insecticides vary in their effectiveness depending on the species of cutworm and on environmental conditions, growers should get specific recommendations from local entomologists or county agents. For control of cutworms in cotton, toxaphene at 2 to 4 lb AI per acre is generally recommended. Two of the most important considerations in preventing destruction of seedling stands are the early discovery of cutworms and then prompt insecticidal treatment.

SELECTED REFERENCES

Adkisson, P. L., et al., *Proceedings U.S.–U.S.S.R. Symposium: The Integrated Control of the Arthropod, Disease and Weed Pests of Cotton, Grain Sorghum and Deciduous Fruit*, Texas Agr. Exp. Sta. MP-1276, 1976.

Adkisson, P. L., L. H. Wilkes, and S. P. Johnson, *Chemical, Cultural, and Mechanical Control of the Pink Bollworm*, Tex. Agr. Exp. Sta. Bul. 920, 1958.

Beck, S. D., et al., *Pest Control: An Assessment of Present and Alternative Technologies Vol. III Cotton Pest Control* (Washington, D.C.: Nat. Acad. Sci., 1975).

Bottrell, D. G., and P. L. Adkisson, "Cotton Insect Pest Management," *Ann. Rev. Entomol.*, 22:451–81, 1977.

Brazzel, J. R., and L. D. Newsom, "Diapause in *Anthonomus grandis Boh*," *J. Econ. Entomol.*, 52:459–62 (1959).

———, and D. F. Martin, *Resistance of Cotton to Pink Bollworm Damage*, Tex. Agr. Exp. Sta. Bul. 843, 1956.

———, J. S. Roussel, C. Lincoln, F. J. Williams, and G. Barnes, *Bollworm and Tobacco Budworm as Cotton Pests in Louisiana and Arkansas*, La. Tech. Bul. 482, 1953.

Bryan, D. E., R. E. Fye, C. G. Jackson, and R. Patana, *Nonchemical Control of Pink Bollworms*, USDA ARS W-39, 1976.

Chapman, A. J., L. W. Noble, O. T. Robertson, and L. C. Fife, *Survival of the Pink Bollworm Under Various Cultural and Climatic Conditions*, USDA Production Res. Rpt. No. 34, 1960.

Cross, W. H., "Biology, Control, and Eradication of the Boll Weevil," *Ann. Rev. Entomol.*, 18:17–46 (1973).

Crumb, S. E., *Tobacco Cutworms*, USDA Tech. Bul. 88, 1929.

Faulkner, L. R., *Hemipterous Insect Pests: Their Occurrence and Distribution in Principal Cotton-Producing Areas of New Mexico*, New Mex. Agr. Exp. Sta. Bul. 372, 1952.

Fenton, F. A., and E. W. Dunnam, *Biology of the Cotton Boll Weevil at Florence, S.C.*, USDA Tech. Bul. 112, 1929.

Folsom, J. W., *Insect Enemies of the Cotton Plant*, USDA Farmers' Bul. 1688, 1932.

Gaines, R. C., *Ecological Investigations of the Boll Weevil*, USDA Tech. Bul. 1208, 1959.

Gaylor, M. J., and F. R. Gilliland, Jr., *The Relative and Seasonal Abundance of Selected Predaceous Arthropods in Alabama Cotton Fields*, Alabama Agr. Exp. Sta. Circ. 227, 1976.

Hunter, R. D., T. F. Leigh, C. Lincoln, B. A. Waddle, and L. A. Bariola, *Evaluation of a Selected Cross-Section of Cottons for Resistance to the Boll Weevil*, Arkansas Agr. Exp. Sta. Bul. 700, 1965.

Isely, D., *The Cotton Aphid*, Ark. Agr. Exp. Sta. Bul. 462, 1946.

Lacewell, R. D., D. G. Bottrell, R. V. Billingsley, D. R. Rummel, and J. L. Larson, *Impact of the Texas High Plains Diapause Boll Weevil Control Program*, Texas Agr. Exp. Sta. MP-1165, 1974.

Loftin, U. C., *Living with the Boll Weevil for Fifty Years*, Ann. Rpt. Smithsonian Institution, pp. 273–91, 1945.

Mangum, C. L., N. W. Earle, and L. D. Newsom, "Photoperiodic Induction of Diapause in the Boll Weevil, *Anthonomus grandis*," *J. Econ. Entomol.*, 61:1125–8 (1968).

National Cotton Council, *Official 1968 Cotton Pest Control Guides* (Memphis: National Cotton Council), 1968.

Newsom, L. D., J. S. Roussel, and C. E. Smith, *The Tobacco Thrips: Its Seasonal History and Status as a Cotton Pest*, La. Agr. Exp. Sta. Bul. 474, 1953.

Noble, L. W., *Fifty Years of Research on the Pink Bollworm in the United States*, USDA Agr. Handbook 357, 1969.
Ohlendorf, W., *Studies of the Pink Bollworm in Mexico*, USDA Tech. Bul. 1374, 1926.
Pack, T. M., and P. Tugwell, *Clouded and Tarnished Plant Bugs on Cotton: A Comparison of Injury Symptoms and Damage on Fruit Parts*, Arkansas Agr. Exp. Sta. Report Series 226, 1976.
Race, S. R., *Importance and Control of Western Flower Thrips, Frankliniella occidentalis, on Seedling Cotton*, New Mexico Agr. Exp. Sta. Bul. 497, 1965.
Reinhard, H. J., *The Cotton Flea Hopper*, Tex. Agr. Sta. Bul. 339, 1926.
Reynolds, H. T., and T. F. Leigh, *The Pink Bollworm a Threat to California Cotton*, California Exp. Sta. Ext. Ser. Circ. 544, 1967.
Roach, S. H., *Heliothis spp. and their Parasites and Diseases on Crops in the Pee Dee Region of South Carolina, 1971–73*, USDA ARS-S-111, 1976.
Snow, J. W., and J. R. Brazzel, *Seasonal Host Activity of the Bollworm and Tobacco Budworm During 1963 in Northeast Mississippi*, Mississippi Agr. Exp. Sta. Bul. 712, 1965.
Sterling, W. L., and P. L. Adkisson, *Seasonal Biology of the Boll Weevil in the High and Rolling Plains of Texas as Compared with Previous Biological Studies of this Insect*, Texas Agr. Exp. Sta. MP-993, 1971.
Tugwell, P., S. C. Young, Jr., B. A. Dumas, and J. R. Phillips, *Plant Bugs in Cotton Importance of Infestation Time, Types of Cotton Injury, and Significance of Wild Hosts near Cotton*, Arkansas Agr. Exp. Sta. Report Series 227, 1976.
USDA, *The Boll Weevil: How to Control It*, USDA Farmers' Bul. 2147, 1969.
USDA, *Lygus Bugs on Cotton: How to Control Them*, USDA Leaflet No. 503, 1969.
USDA, *Twenty-second Annual Conference Report on Cotton Insect Research and Control*, USDA Entomology Research Division, 1969.
USDA, *Controlling the Pink Bollworm on Cotton*, USDA Farmers' Bul. 2207, 1968.
USDA, *Spider Mites on Cotton: How to Control Them*, USDA Leaflet No. 502, 1968.
USDA, *The Bollworm: How to Control It*, USDA Leaflet No. 462, 1960.
USDA, *The Cotton Aphid: How to Control It*, USDA Leaflet No. 467, 1960.
USDA, *The Cotton Fleahopper: How to Control It*, USDA Leaflet No. 475, 1960.
USDA, *The Cotton Leafworm: How to Control It*, USDA Leaflet No. 468, 1960.
van den Bosch, R., and K. S. Hagen, *Predaceous and Parasitic Arthropods in California Cotton Fields*, California Agr. Exp. Sta. Bul. 820, 1966.
Warner, R. E., "Taxonomy of the Subspecies of Anthonomis grandis (Coleoptera:Curculionidae)," *Ann. Entomol. Soc. Amer.*, 59:1073–88 (1966).
Watts, J. G., *A Study of the Biology of the Flower Thrips* Frankliniella tritici *(Fitch) with Special Reference to Cotton*, S.C. Agr. Exp. Sta. Bul. 306, 1936.
Wene, G. P., L. A. Carruth, A. D. Telford, and L. Hopkins, *Descriptions and Habits of Arizona Cotton Insects*, Arizona Ext. Ser. and Agr. Exp. Sta. Bul. A-23, 1965.
Wene, G. P., L. W. Sheets, H. E. Woodruff, I. Pearson, and L. A. Carruth, *Winter Survival of the Pink Bollworm in Arizona*, Arizona Agr. Exp. Sta. Tech. Bul. 170, 1965.
Young, Jr., D. F., *Cotton Insect Control* (Birmingham, Ala.: Oxmoor House, 1969).

chapter 13 / **W. DON FRONK**

VEGETABLE CROP INSECTS

THE PESTS

All vegetables grown in the United States are attacked by one or more insects, and any grower, whether a backyard gardener or a commercial producer, is vitally interested in protecting his crop. Entomologists have estimated that insects cause an annual loss of at least $185,892,000 to vegetable growers in the United States.

Many of the vegetables grown in the United States are of foreign origin. In their native land these plants are attacked by a number of insects. When such pests become introduced accidentally into the United States, controlling factors such as parasites, predators, and diseases are often left behind. Enjoying the lack of enemies, the pests become much more serious than they were in their original home. For example, The European corn borer is of little significance to corn production in Europe, but in the United States it is considered one of the most serious pests of both sweet and field corn. Problems may also arise from native insects when they show a preference for introduced vegetables over original host plants. When first discovered in 1820 by Thomas Say in the grasslands of the West, the Colorado potato beetle fed upon buffalobur, *Solanum rostratum*, and was not a numerous insect. After settlers introduced the potato into the territory, the beetle preferred this cultivated species, *Solanum tuberosum*, thrived on it, and became a serious pest of the crop east of the Rocky Mountains. The beetle appeared in damaging numbers in Indiana about 1868 and reached the Atlantic coast in 1874.

We can divide vegetable insect pests into two groups according to their selection of food plants: (1) those restricted more or less to one species of vegetable or to closely related species, and (2) those feeding on a wide variety of vegetables and often referred to as "general feeders." In the first group are such pests as asparagus beetles, squash vine borer, onion maggot, tomato hornworm, and many species of aphids. The second group contains such pests as cutworms, grasshoppers, white grubs, wireworms, spider mites, blister beetles, vegetable weevil, and lygus bugs.

The importance of vegetable insects will often vary in different sections of North America. Some species inhabit limited regions and for this reason may be only of local importance; others may be widely distributed yet reach injurious numbers only in certain areas favorable for their increase; still others, representing about half the total number of vegetable pests, attack and injure their hosts severely wherever they are grown. Because vegetables belong to several different plant families, their insect enemies are many and diverse, so we shall call attention to only the most important ones.

Insects injurious to solanaceous crops—potato, tomato, eggplant, and pepper—include the Colorado potato beetle (Fig. 4:16A), tomato hornworm, potato leafhopper, potato tuberworm, tomato fruitworm (Fig. 10:10), eggplant lacebug, potato psyllid, pepper weevil, blister beetles, cutworms, flea beetles, wireworms, mites and aphids.

Cruciferous crops—cabbage, broccoli, brussels sprouts, radish, turnip, and mustard—are attacked by the cabbage looper, imported cabbageworm (Fig. 13:10), diamondback moth caterpillar, cabbage webworm, cabbage maggot, seedcorn maggot (Fig. 13:11), harlequin bug (Fig. 4:10B), cabbage aphid, vegetable weevil, flea beetles, and cutworms.

The pests of cucurbits—cucumbers, squash, pumpkin, and various melons—include the striped cucumber beetle (Fig. 13:6), spotted cucumber beetle (Fig. 10:15B), pickleworm, squash vine borer (Fig. 13:1), squash bug (Fig. 4:10A), melon aphid, mites, and cutworms.

Insects that damage beets, spinach, and chard—all members of the family Chenopodiaceae—include the beet webworm, beet armyworm, beet leafhopper (Fig. 13:14), sugarbeet root aphid, sugarbeet root maggot, sugarbeet wireworm, spinach flea beetle, spinach leafminer, spinach carrion beetle, and mites.

Beans are fed upon by the Mexican bean beetle (Fig. 13:8), bean leaf beetle, bean aphid, bean weevil, limabean pod borer, potato leafhopper, seedcorn maggot, and corn earworm. The twospotted spider mite is a serious bean pest, particularly in hot, dry weather. Peas are attacked by the pea aphid, pea weevil, and pea moth.

Onions are injured by the onion maggot, onion thrips, bulb mite, and wireworms. Lettuce is besieged by the cabbage looper, aster leafhopper, cutworms, wireworms, and aphids. Vegetables belonging to the Umbelliferae—carrots, celery,parsnips—are relished by the carrot rust fly (Fig. 4:32), carrot weevil, celery leaftier, celery looper, parsnip webworm, aster leafhop-

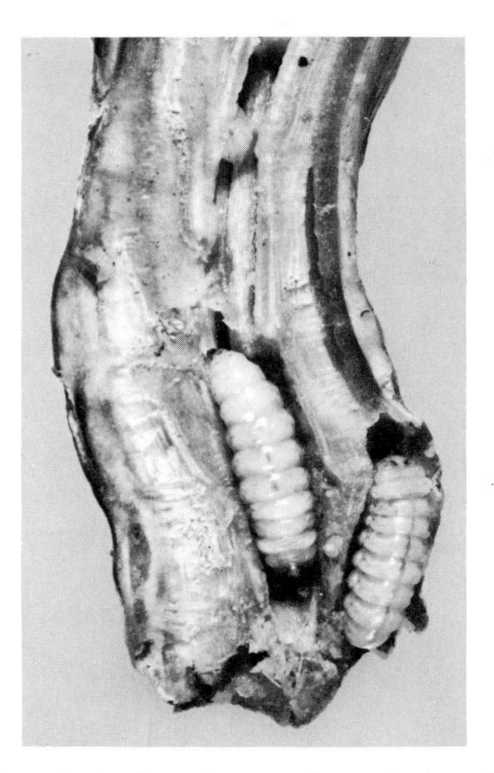

Figure 13:1. Squash vine borer in stem of squash plant. *Courtesy Iowa State University, Ames.*

per, tarnished plant bug (Fig. 4:10C), wireworms, slugs, and aphids. Asparagus is attacked by the asparagus beetle, spotted asparagus beetle, and asparagus miner (Fig. 4:33). Sweet potato suffers from the feeding of the sweetpotato weevil, sweetpotato leaf beetle, sweetpotato flea beetle, sweetpotato or tortoise beetles, and wireworms. Sweet corn is damaged by the corn earworm, fall armyworm, European corn borer (Fig. 10:13), seedcorn maggot, corn flea beetle, corn sap beetles, cutworms, grasshoppers, white grubs (Fig. 13:2), mites, and wireworms.

THE INJURY

Insect injury to vegetables has many forms, the most obvious being the actual consumption of plant parts above ground. Chewing insects like cabbageworms often destroy cabbage heads completely. Colorado potato beetles sometimes strip all of the leaves from potato and eggplant. In satisfying their voracious appetites grasshoppers may eat everything but the stalks of sweet corn.

Figure 13:2. A white grub feeding on corn root. *Courtesy Shell Chemical Company, New York, N.Y.*

Many vegetable pests confine their feeding to particular parts of the plant. The larvae of the cowpea curculio, pea weevil, and bean weevil feed within the green seeds; although corn earworms may feed in the whorl of young corn plants, they feed chiefly on the juicy kernels at the end of the cob; tomato fruitworms eat holes in the tomato fruit; larvae of the pepper weevil feed within the buds and pods of pepper plants. Some insects, such as the squash vine borer, burrow inside of the stems of vegetable plants (Fig. 13:3).

Just as serious, and frequently just as noticeable as the damage of chewing insects, is the damage caused by sucking insects: aphids, leafhoppers, stink bugs, and lygus bugs. Besides withdrawing juices from the plant, thus depressing growth, they may inject toxic salivary secretions that cause disease. By their feeding, potato leafhoppers produce a disease known as **hopperburn**. Conspicuous symptoms in potatoes are the upward curling of the leaves at the tips and sides, triangular necrotic areas at the edges of the leaves, and shortened internodes. Similar symptoms appear in beans, except that the edges of the leaves curl down rather than up.

Lygus bugs feeding on lima beans inject a toxin that causes the blossoms and small pods to shed and young seeds to shrivel. One species of aphid

Figure 13:3. A squash plant dying from attack of the squash vine borer. *Courtesy Iowa State University, Ames.*

causes white spots on celery leaves, another yellowing and chlorosis, and a third curling of the leaflets. Squash bugs feeding on cucurbits inject a toxin that produces **anasa wilt**, the symptoms of which resemble those of bacterial wilt.

A devastating systemic disease of potatoes and tomatoes called **psyllid yellows** results from the injection of toxic salivary juices by nymphs of the potato psyllid. In potatoes the symptoms are a marginal yellowing and upward curling of the leaflets, reddening or purpling of the terminal leaves of pigmented varieties, and stunting of the plant. In young plants infected with the disease, tuberization is suppressed, whereas in older plants the tubers increase in number but remain small and worthless.

Another important way in which insects cause serious injury to vegetables is through their transmission of pathogenic fungi, bacteria, and viruses. A few examples will illustrate this kind of damage. The green stink bug transmits the fungus that causes **yeast spot of lima beans**. The striped and spotted cucumber beetles transmit **bacterial wilt of cucurbits**, and the toothed flea beetle and corn flea beetle transmit **bacterial wilt of sweet corn**. The seedcorn maggot and the onion maggot transmit **bacterial soft rot** of vegetables.

Insects are the most important vectors of virus diseases of plants; and of these vectors aphids transmit the greatest number. They transmit **common bean mosaic, squash mosaic, cucumber mosaic, potato leafroll, onion yellow dwarf**, and many more. But there are other insects important in transmitting plant viruses, such as the beet leafhopper—which transmits **curly top** virus to beets, tomatoes, and beans—and thrips, which transmit the virus of **tomato spotted wilt**. The aster leafhopper transmits the mycoplasmalike organism of **aster yellows** to carrots, lettuce, potatoes, and many other plants.

Vegetables are also bothered by injurious soil insects that eat the seed, cut off roots, shred underground stems, burrow in roots and tubers (Fig. 13:4), and even cut down entire plants. Numbered among the serious soil pests of vegetables are wireworms, whitegrubs, seedcorn maggots, cabbage maggots, corn rootworms, flea beetle larvae, and cutworms.

Insects need not necessarily feed on vegetables to be destructive. One of the chief concerns of the asparagus grower is the presence of asparagus beetle eggs on the spears. These eggs are difficult to remove, and if too many are present, the particular shipment of asparagus may be refused for canning purposes.

CULTURAL CONTROL

Growers cannot produce good yields of high-quality vegetables consistently unless they control the insects that threaten their crops. There are a number of general and specific methods of cultural control that vegetable growers should employ to reduce insect populations and injury. These have an advantage over chemical methods in that they are usually cheaper and leave no poisonous residues. Chapter 6 contains a discussion of the general methods. In the detailed discussion of five representative vegetable insects in this chapter we shall review several specific methods.

CHEMICAL CONTROL

The modern vegetable grower is faced with the problem of producing quality vegetables that are uncontaminated by insects yet bear no harmful residues.

Figure 13:4. Sweet potato damaged by the feeding of a wireworm. *Courtesy Iowa State University, Ames.*

To a large degree modern insecticides have enabled him to meet the challenge. One way in which the grower achieves this goal is by employing materials that are nontoxic to higher animals, such as the botanical insecticides pyrethrum and rotenone. As used in agriculture today, these insecticides constitute no hazard to man or to his livestock. They are effective against a wide range of insect pests, but they possess little residual action and are relatively expensive. A second way in which the grower controls injurious pests and yet avoids toxic residues is by applying short-lived synthetic insecticides. Because insecticidal residues are of great concern, the grower often uses chemicals with less residual effect even though they may be more expensive.

Commonly used foliage insecticides are parathion, carbaryl, malathion, endosulfan, and methoxychlor; commonly used foliage miticides are dicofol, ethion, and carbophenothion. The vegetable grower also employs systemic insecticides—demeton, dimethoate, disulfoton, mevinphos, and oxydemeton-methyl. He still uses a few of the chlorinated hydrocarbon insecticides—methoxychlor, lindane, and endosulfan—as seed dressings or preplant soil treatments and in other ways. Baits are useful in controlling several pests of vegetable crops. The grower may use apple pomace-calcium arsenate bait to control the carrot weevil, wheat bran-trichlorfon bait to control cutworms, and bran-metaldehyde bait to control slugs.

To protect crops of high-acre value from nematodes, wireworms, and other soil pests, the vegetable grower may apply, in fall or spring, two or three weeks before planting, fumigants such as dichloropropene and dichloropropane mixture (D-D), ethylene dibromide (EDB), dibromo-chloropropane (DBCP), or a dichloropropene mixture (DCP). Preplant fumigation in spring, frequently done in areas of light sandy soils, is less desirable than fall fumigation because fumigants that linger in cool wet soils may injure young plants. Because most treatments of soil with fumigants are relatively expensive, the grower should estimate his cost-profit possibilities before applying them. Some fumigants not only control nematodes and insects, but also reduce the population of disease-producing organisms and weed seeds.

Because vegetable growers require a large number of insecticides to handle their many pest problems, entomologists have searched for a mixture of insecticides to simplify the task. One such mixture contains methoxychlor, malathion, and a fungicide. The so-called general-purpose dusts or sprays have found greater acceptance among home gardeners than among commercial growers.

To be sure, the control of insects on commercial plantings and the control of insects in the home vegetable garden present two different problems. The commercial grower, trained to use a wide range of chemicals, has an advantage in having available not only special application equipment to apply insecticides efficiently and effectively, but also special produce-washing equipment to take care of residues of toxicant. Yet in some ways he has greater limitations than the home gardener, for he must consider closely labor costs and the price of various insecticides.

In addition to the usual precautions he must take in using insecticides, the vegetable producer must guard against phytotoxicity and the development of off-flavors in his crop. Phytotoxicity may be hidden or clearly evident. An insecticide may cause a slight reduction in growth and yield, or it may kill plants outright. Cucurbits are extremely susceptible to insecticide injury. On sensitive vegetable varieties it is safer to apply insecticidal sprays prepared from wettable powders than to use emulsifiable concentrates.

Because many pest problems of vegetables occur at about the same time year after year, growers have been able to schedule regular insecticide applications and thereby keep insect infestations at low levels and injury to a minimum. They obtain best results by making the first application before an infestation is apparent. Nevertheless growers need to inspect fields regularly for detection of any pest that may not be susceptible to the treatment being applied. Once aware of the problem growers may make necessary adjustments in the constituents of the spray. Entomologists interested in control of vegetable insects recommend that growers apply two or more insecticides in an alternating schedule to prevent development of insecticide resistance.

Exact methods of chemical control that a grower may use depend on the specific crop and its insect pests. In general we may point out the following practices, employed commonly by vegetable growers:

1. **Preplant field treatment and seed-bed treatment**. A complex of soil pests, cutworms, wireworms, white grubs, root maggots, and others, attacks a variety of vegetable crops—sweet corn, cabbage, cucumber, potato, and tomato. Just before seeding or transplanting, growers spray onto the soil surface an insecticide such as diazinon and then disk the material into the top 3 or 4 in. of soil. Granules may be used in place of spray.
2. **Transplant water treatment**. Growers protect vegetables that are transplanted (e.g., cabbage and tomato) from cutworms, root maggots, and other soil insects by mixing insecticide in the transplant water. Wettable powders are preferable because emulsifiable concentrates may injure roots. One formula suggests 8 oz of diazinon 50 per cent WP in 50 gal of water. At transplanting growers apply to the base of each plant one-half pint of solution either by hand or by mechanical transplanter.
3. **Seed treatment**. Seeds and seedling plants can be protected from certain soil insects such as seedcorn maggots and wireworms if the seeds are dressed with an insecticide, such as diazinon, lindane, or chlorpyrifos, plus a fungicide. Seeds may come treated, or else the grower may apply a commercial seed protectant which includes a fungicide as well as an insecticide.

4. **Seed-furrow or row treatment**. At time of planting insecticides are introduced into the row near the seed by planter attachment or sprayer. Either sprays or granules are useful in making this application. The grower may use diazinon or fonofos to control soil insects, or a systemic insecticide such as phorate or disulfoton to control aphids, leafhoppers, and mites after the plant emerges. A side dress of a systemic insecticide may also be applied in the soil along the row after plants have emerged.
5. **Postseeding or posttransplanting drench**. To protect crops such as cabbage and radishes from root maggots, application of a drench is often necessary. The grower uses straight drop pipes from the boom over the rows and directs a coarse spray to the center of the plant to wet the soil around the stem. A fan tip at the end of each drop pipe distributes the spray in a band 2 to 3 in. wide over the row.
6. **Foliage and fruit treatment**. To control a large number of the insects that attack vegetables, growers spray or dust the foliage of plants, often starting at emergence and continuing at regular intervals until the crop is harvested. They apply surface-active or systemic insecticides depending on the insects to be controlled. When fruits begin to set, they require protection from insects that have a special attraction for them; for example, tomatoes need protection from the tomato fruitworm, and peppers need protection from the European corn borer.
7. **Postharvest treatment**. For several reasons vegetable growers may apply insecticides after harvest. Fruit flies (*Drosophila*) lay eggs on picked tomatoes, particularly in the fresh cracks that result from improper handling. Because the females deposit most eggs in the evening and early morning, growers do not leave picked tomatoes in the field overnight. For additional protection they dust tomatoes with 0.1 per cent synergized pyrethrum. Lettuce growers may dip harvested heads in solutions of pyrethrum to eliminate aphids from their produce. After harvest of asparagus, growers may spray fern growth and seedlings to control the asparagus beetle.
8. **Adjacent vegetation treatment**. Insect pests may breed, hibernate, or rest in vegetation surrounding fields of vegetables. Grasshoppers inhabit margins, ditch bands, roadsides, and weed patches; unless they are destroyed, they eventually migrate into the vegetable crop. Growers spray these sites when grasshoppers are young, lack wings, and are easy to kill. Adults of the onion maggot seek shelter along ditch banks, windbreaks, and other protective vegetation. Growers reduce populations by spraying these sites with an approved insecticide. Spraying brush surrounding asparagus fields in September and October with insecticide prevents the asparagus beetle from successfully hibernating close to its host plant.

CONTROL EQUIPMENT

Although home gardeners find it convenient to use hand dusters or hand sprayers to treat their backyard vegetables, commercial producers with large acreages of vegetables employ power equipment. They utilize power dusters, power sprayers, air sprayers, high-clearance sprayers, and power granule applicators; they may hire a commercial applicator to treat their crops by means of aircraft, either helicopter or fixed wing.

To obtain adequate coverage of plants, growers have traditionally employed high-pressure (400 psi) sprayers delivering 75 to 125 gal of spray per acre. The sprayers are equipped with booms and drop nozzles so arranged that the spray is directed onto the plant from above and from each side. In certain crops the side nozzles are arranged to direct the spray to the under surface of the leaves for the reason that a number of vegetable pests, such as aphids, mites, and cabbage looper, feed and rest on the underside of the leaf.

Because of labor-saving advantages vegetable growers are turning to low-pressure (50–100 psi) and low-volume (20 to 60 gal/acre) sprayers in greater numbers. Satisfactory pest control can be obtained at these lower rates, but the grower must take great care that the sprayer is operating properly to provide satisfactory distribution of the insecticide. He must pay particular attention to nozzle cleanliness and wear, nozzle placement, boom height, tank agitation, and accuracy of pressure gauge.

To speed up pest control operations further some vegetable growers employ large-capacity air sprayers for treatment of row crops. The operation of an air sprayer is even more critical than that of the low-pressure, boom-type sprayer. Wind conditions should be calm, at least below 5 mph, and even then considerable mist moves off without depositing on the plant. Cross drives in fields allow the insecticidal mist to be blown down the rows instead of across them. This arrangement helps to give better coverage in crops such as tomatoes. In some sections growers use lightweight, back-pack, power dusters to treat vegetable crops.

REPRESENTATIVE VEGETABLE INSECTS

The vegetable insects that we have chosen to treat in detail are the striped cucumber beetle, Mexican bean beetle, imported cabbageworm, seedcorn maggot, and beet leafhopper.

Striped Cucumber Beetle. *Acalymma vittatum* (Fabricius) [Coleoptera:Chrysomelidae]

The American Indians cultivated cucurbits long before the arrival of the white man, and it is likely they had trouble with the striped cucumber beetle native to North America. The first published record of this beetle in America was by Fabricius in 1775.

The striped cucumber beetle is found throughout the United States except for the westernmost states, where it is replaced by the western striped cucumber beetle, *Acalymma trivittatum* (Mannerheim). In Canada, the striped cucumber beetle is a common pest of cucurbits in Ontario, Quebec, New Brunswick, and Nova Scotia and an occasional pest in Manitoba and Saskatchewan. It is known to occur in Mexico and Panama and probably could be found in most of the tropical and temperate regions of South America.

Early in the season, even before all the plants are above ground, the beetles fly from their hibernating quarters into cucurbit fields, where they feed on the young plants, frequently killing them. Later the beetles attack the flowers and may prevent a good set of fruit, despite the fact that they may assist in cucurbit pollination. They also feed upon the fruits and gnaw deep pits in the rind, making the produce unfit for market (Fig. 13:5). Of minor importance is the damage caused by the larvae feeding on the roots of the host plant and on the fruits where they come in contact with the soil. During early spring and late fall the beetles feed upon other plants, chiefly members of the Rosaceae, but economic damage seldom results.

The striped cucumber beetle and spotted cucumber beetle (Fig. 10:15B) are vectors of the serious cucurbit disease **bacterial wilt**, which is caused by *Erwinia tracheiphila* (E. F. Smith). Plants infected with the disease wilt in warm weather. As the disease progresses, they become permanently wilted, the leaves dry out, and the plant eventually dies. The causative bacteria overwinter in the bodies of hibernating beetles that introduce the organisms into plants by contaminating feeding wounds with feces. So far as is known, this is the only natural method of infection. Cucumber beetles are also vectors of **squash mosaic** virus.

Description. The adult striped cucumber beetle is oblong, yellowish green in color, $\frac{3}{16}$ to $\frac{1}{4}$ in. long, and marked by three slate-black stripes (Fig. 13:6).

Figure 13:5. Rind of squash fed upon by the adult striped cucumber beetle. *Courtesy Iowa State University, Ames.*

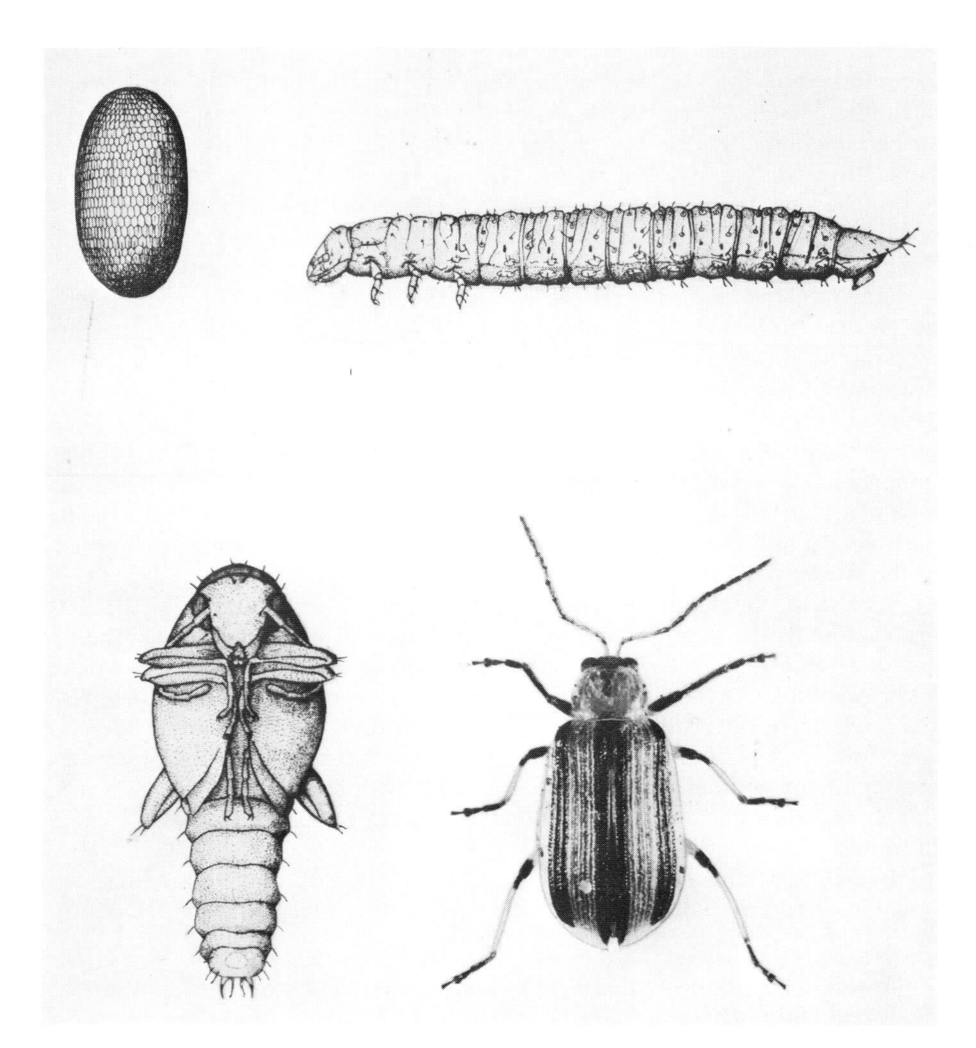

Figure 13:6. Life stages of the striped cucumber beetle. *Egg, larva, and pupa, courtesy University of Arkansas, Fayetteville; adult, courtesy Iowa State University, Ames.*

The head, scutellum, front tibia, tips of middle and hind tibia, and the tarsi are dark. The elytra are covered with very small punctures. The egg is light yellow or orange in color, round to oval in shape, and about 0.6 mm long. The wormlike larva is almost $\frac{3}{8}$ in. long when full grown. The body is white with a dark head and tail plate. It has three pairs of true legs on the thorax and one pair of prolegs on the caudal segment. The whitish-yellow pupa is about $\frac{1}{4}$ in. long.

Life history. Adult beetles leave their hibernating quarters and begin to move about in the spring when the weather has warmed to 65°F or more. If there are no cultivated cucurbits available, the adults feed on the flowers of various plants such as serviceberry, haw, and wild cucumber. As soon as cultivated cucurbits begin to break through the surface of the soil, large numbers of beetles appear suddenly in the field and begin feeding on the seedling plants. Some even crawl into cracks in the soil to reach the sprouting seed.

The beetles soon start to mate and continue to do so throughout the summer. The eggs are laid from 8 to 25 days after the first mating. To deposit her eggs, the female crawls into cracks or depressions about the base of a cucurbit plant, thrusts her abdomen into the soil, and lays her eggs singly or in small clusters. The total number of eggs produced usually varies from 225 to 800, but some females may lay as many as 1514 eggs. In general the beetles that have hibernated through the winter lay fewer eggs than do the females of other generations.

The incubation period varies with external conditions but averages from five to eight days. The larvae, which pass through three instars, spend 15 days feeding on roots and also on stems and fruit that come in contact with the soil. From two to ten days are spent as a prepupa in a small cell formed by the full-grown larva. The pupal period lasts six to seven days. The growth from egg to adult for the first generation requires approximately one month, and the succeeding generations require slightly longer for development.

Dates of emergence of adult beetles from hibernation and the number of generations depend a great deal on temperature. In Arkansas the beetles may become active in March, whereas in Canada the adults leave hibernation in May or June. In the north the beetle has only one generation a year; in southern Texas it produces four generations.

With the beginning of cool fall weather and the maturing of the cucurbit plants, striped cucumber beetles start moving to their winter quarters, usually wooded or bushy areas where the beetles are able to find protection under litter. Hibernating quarters may be as far as a mile from the cucurbit field that produced the beetles.

Control. Recommendations—covering the plants with cloth, dusting with Bordeaux, calcium arsenate, or lime, and the use of trap crops—were common before the discovery of the new organic insecticides. Although there are a number of insecticides that control the beetle, only a few chemicals can be used on cucurbit crops because of their great sensitivity to chemical injury. Application of insecticide is usually recommended as soon as the plants begin to break through the ground. For prevention of bacterial wilt it is advisable to spray at five-day intervals beginning when seedlings emerge or after transplanting and continuing the schedule until vines run. Azinphosmethyl, dimethoate, malathion, parathion, endosulfan, methoxychlor, and rotenone are effective. Sprays prepared from wettable powders are less phytotoxic than sprays prepared from emulsifiable concentrates. Dusts are likewise effective, but plants must be thoroughly covered, using 35 to 40 lb per acre. In humid areas of the South and East, malathion and parathion may cause some foliar burning and should not be applied when plants are wet. Because parathion may injure young plants, it should not be applied before vining.

Research has shown that in addition to spraying foliage, treatment of the seed with a suitable systemic such as dimethoate provides greater protection for the seedling cucurbit which, as a result, grows into a more productive plant.

Mexican Bean Beetle. *Epilachna varivestis* Mulsant [Coleoptera:Coccinellidae]

Etienne Mulsant (1797–1880), a French entomologist, described the Mexican bean beetle in 1850 from specimens sent to him from Mexico. It was first known definitely in the United States in 1883, when it was reported from Colorado. Some entomologists have pointed out that this beetle was known in New Mexico close to the time of the Mexican War of 1846–48, and they conjecture that its introduction into the United States might have resulted from the movement of food for the cavalry. It was not of great importance, however, until it appeared in the eastern part of the United States. In 1918 the beetle arrived in northern Alabama, where it may have entered in a shipment of hay from the West. From this region it dispersed mainly by flight into other parts of the eastern United States at a rate of 100 to 200 miles each year and eventually reached Canada by 1927.

The beetle, a native of Mexico and Guatemala, is distributed from Central America to Canada. In the United States it is generally found east of the Mississippi River and in Arizona, New Mexico, Colorado, Utah, Wyoming, Nebraska, and Idaho. An isolated infestation occurred in Ventura County, California, in 1946, but an energetic control program eliminated it from this state.

Although the Mexican bean beetle is a mandibulate insect, it does not swallow solid food material. Both larva and adult chew off portions of the bean leaf, masticate it, and suck the plant juices. The larvae move backward on the under or upper surface of the leaf with their heads swaying and their mandibles working. In this way they rake the plant tissues into ridges from which they suck the juice. The adults feed somewhat in the same manner. Both larvae and adults usually feed from the under surface of the leaves, raking up tissue from only one side; the membrane on the opposite side remains unbroken. As drying sets in, the unbroken membrane tears apart (Fig. 13:7).

The feeding is confined almost entirely to bush and pole beans (*Phaseolus vulgaris*) and to the lima bean (*P. lunatus*). However, the beetle can live on other plants and has been found feeding on beggarweed, hyacinth bean, cowpea, soybean, peanuts, and alfalfa. It may also feed to a limited extent on adsuki bean, kudzu, some clovers, okra, eggplant, and squash. Left unchecked in spring the Mexican bean beetle develops huge populations of over

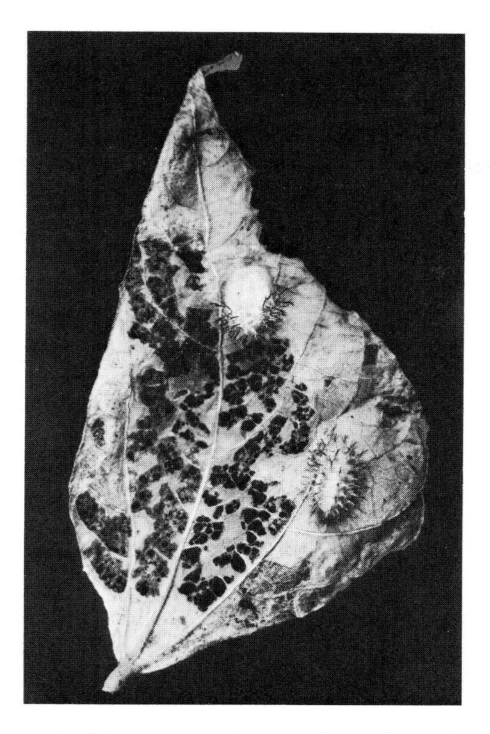

Figure 13:7. Bean leaf injured by the feeding of Mexican bean beetle larvae. *Courtesy Union Carbide Corporation, New York, N.Y.*

two million larvae per acre which destroy the foliage completely, lower the quality, and reduce yields of green beans 55 per cent.

Description. The strongly convex adult beetles are about $\frac{1}{4}$ in. long. The elytra are usually copper red in color and are conspicuously marked with 16 black spots arranged in three rows over the back (Fig. 13:8). The eggs are yellow, about 1.3 mm long, and are laid in clusters on the under surface of bean leaves. Larvae are yellow and covered with dark, branched spines. Full-grown larvae are about $\frac{1}{3}$ in. long. Pupae are yellowish to copper colored and about $\frac{1}{4}$ in. long. The back of the pupa is smooth, but the spine-covered, last larval skin remains attached to the caudal end of the pupa (Fig. 13:8).

Life history. Beetles out of hibernation fly to bean fields about the time the earliest beans are showing their first true leaves. In the South they appear in late March or early April, whereas in Maine they appear in early June.

After a preovipositional period of seven to ten days, females begin to lay batches of 40 to 60 eggs on the underside of bean leaves. Each female may lay an egg mass every two or three days and produce on an average 460 eggs. Some individual females may produce as many as 1669 eggs. In spring eggs hatch in 10 to 14 days; in summer they hatch in 5 or 6 days.

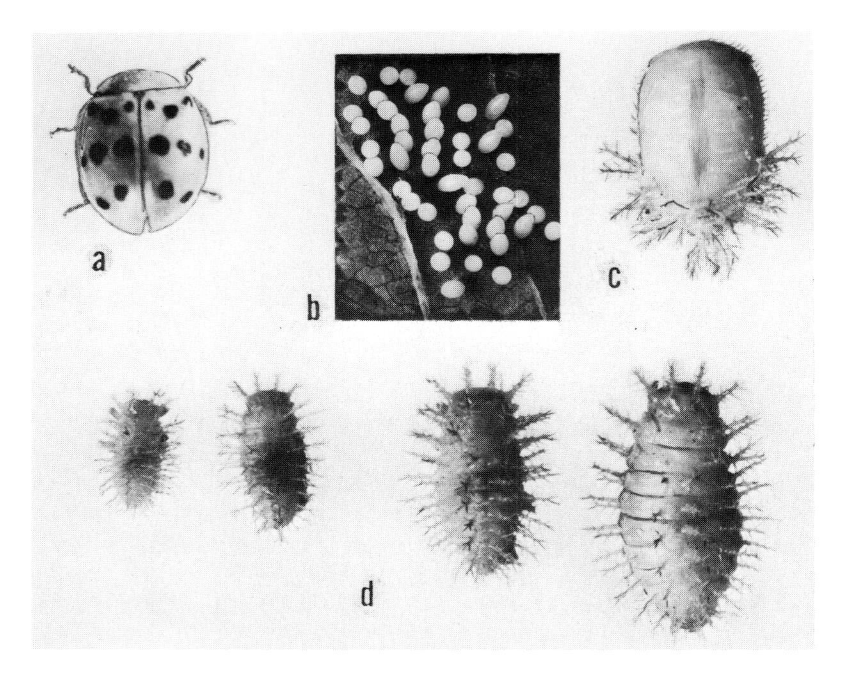

Figure 13:8. Life stages of the Mexican bean beetle. A, adult; B, eggs attached to bean leaf; C, pupa; D, first to fourth instar larvae. *Courtesy Rutgers University, New Brunswick, N.J.*

The larvae, which pass through four instars, require about 35 days to reach the pupal stage in spring and 18 to 20 days in summer. When ready to pupate, each larva fastens its caudal end securely to the underside of a bean leaf or some other convenient object. The pupal period lasts from 7 to 20 days. New females begin laying eggs within 8 to 13 days after emergence. The Mexican bean beetle produces three or four generations in the South and one generation or one and a partial second generation in the North. In the Southwest one generation is the general rule.

With the coming of fall, the adults begin to seek shelter for hibernating. Usually they select the litter in rolling woodlands and show a preference for woods of pine-oak mixture. They hibernate less frequently in the debris along fence rows, stone piles, under rubbish in gardens, and under woodpiles. Ordinarily they do not seek winter quarters more than a mile from the home bean field, and the majority go no farther than one-quarter of a mile. In the West, however, the beetles may fly many miles where they often enter alfalfa fields for hibernation. In the Southwest beetles do not truly hibernate; on warm days they do much moving about.

Weather plays an important part in controlling populations of Mexican bean beetle. Lack of proper cover during the hibernation period greatly increases winter mortality. During the summer unusually high temperatures accompanied by drought reduce populations.

A number of predaceous insects feed on the Mexican bean beetle, but none of them holds populations in check. At times the adult Mexican bean beetle feeds on its own eggs. A tachinid fly, *Paradexodes epilachnae* Aldrich, is an important factor in reducing beetle numbers in Mexico; its introduction into the United States has not been successful. In 1967 a small parasitic wasp, *Pediobius foveolatus* (Crawford), imported from India, was released in several eastern states. In its native country it parasitizes a plant-feeding coccinellid, *Henosepilachna sparsa* (Herbst). Extensive tests in the United States have shown that it readily parasitizes older larvae of the Mexican bean beetle but not the beneficial coccinellid larvae.

Control. For many years the Mexican bean beetle was considered one of the most difficult insects to curb. Even now it is regarded as being a troublesome insect to control. There are two chief reasons for this: (1) the pest is not greatly susceptible to the ordinary insecticides, and (2) larvae and adults live and feed on the undersides of the bean leaves where it is difficult to apply insecticides. Placing a systemic insecticide in the soil at planting time is an effective way of coping with the problem. This method not only controls the Mexican bean beetle as soon as it migrates into the field of seedling bean plants, but also other serious migratory pests such as aphids and leafhoppers. Attachments on the planter dispense granules of insecticide, 10 or 15 per cent disulfoton or phorate at 1.5 lb AI/acre, into the soil 3 in. away from the seed furrow and about 2 in. deep. Care is taken that the insecticide does not contact the seed directly; either compound may cause phytotoxicity.

Mexican bean beetles may also be controlled with thorough foliar applications of one of the following insecticides: azinphosmethyl, dimethoate, malathion, parathion, carbophenothion, carbaryl, methomyl, methoxychlor, and rotenone. For spraying bush varieties, it is necessary to use from 100 to 125 gal of spray per acre at a pressure of 150 to 250 psi. Adjust the sprayer boom so that drop lines fall on either side of the bean row. Point the nozzles upward at an angle of 45° so the spray strikes the underside of the leaves. Direct the third nozzle downward over the center of the row. The ends of the drop lines should be made of rubber hose to prevent their breakage in passing over uneven ground. On young plants lower the boom so that nozzles are about 2 in. above the ground, and as the plants grow higher raise the boom.

Apply dusts at 25 to 35 lb per acre with special nozzles that direct the dust upward, and arrange nozzles to treat both sides of the row. Where there is danger of drift, as on a windy day, control can be improved by using a 10- to 15-ft cloth apron behind the duster so as to hold the dust down around the plants.

It is advisable to apply insecticide as soon as injury is observed or when one cluster of eggs is found about every 6 ft along a row. Examine the underside of leaves for egg clusters. A second treatment of insecticide is necessary seven to ten days after the first to protect new growth. Two applications are usually enough for snap beans, but lima beans may need additional applications.

A community program to control the Mexican bean beetle is advisable, because a few untreated rows of beans can serve as the breeding ground for large numbers of beetles. These insects can then spread to other plantings, including commercial fields, and can destroy large acreages.

There are several farm practices which aid in Mexican bean beetle control. It is wise to burn or otherwise destroy all crop residue soon after the beans have been picked. Plowing refuse at least 6 in. deep in the soil destroys all stages of the beetle. Beans planted 3 to 4 in. apart produce less foliage in the row than beans planted closer, are easier to treat, and as a general rule, are not as heavily infested.

Recent studies have shown that varieties of both snap and lima beans differ in susceptibility to the Mexican bean beetle. Resistant snap bean varieties sustained average leaf damage of 31 per cent compared to 53 per cent for susceptible varieties. The most resistant variety examined was Idaho Refugee with 25 per cent leaf damage. Other resistant varieties included Wade, Logan, Supergreen, Black Valentine, and Refugee U.S. No. 5. Resistant lima bean varieties were Baby Fordhook, Bush Lima, Triumph, Burpee's Bush Lima, Evergreen, and Henderson's Bush. The component of resistance appears to be antibiosis, for female beetles reared on resistant varieties are small and light and produce less than one half as many eggs as females raised on susceptible varieties. A study of the chemical factors underlying selective feeding of the Mexican bean beetle upon beans indicates that variation in sugar content may be the basis for resistance and susceptibility. The higher sugar content of leaves the more susceptible is the variety.

For control of Mexican bean beetle in home gardens commonly recommended insecticides include sprays of malathion, methoxychlor, or carbaryl, or granules of 2 per cent disulfoton.

Imported Cabbageworm. *Pieris rapae* (Linnaeus) [Lepidoptera:Pieridae]

The imported cabbageworm, nearly a worldwide pest of crucifers, was first introduced into North America at Quebec about 1860. Later it entered New York City in 1868 and Charleston, South Carolina and Apalachicola, Florida about 1873. From these points it spread rapidly through the United States and Canada, reaching California by 1883.

When cabbage plants are small, cabbageworms feed primarily on the underside of the developing leaves. Young larvae chew off the lower layers of the leaf leaving the upper layers untouched, but older larvae devour all cell layers and eat big holes into the leaves (Fig. 13:9). When the heads develop, cabbageworms feed on the outer leaves and bore into the centers, making the cabbage unmarketable. Larvae also cause damage by contaminating the heads with their greenish-brown excrement.

Figure 13:9. Cabbage damaged by imported cabbageworms. Note abundant excrement accumulated in axils. *Courtesy University of Florida, Gainesville.*

The imported cabbageworm feeds on all forms of cruciferous plants but prefers cabbage and cauliflower. It frequently damages turnip, kale, collards, radish, mustard, horseradish, and occasionally lettuce. Ornamentals often attacked are nasturtium, mignonette, sweet alyssum, and cleome.

Description. The imported cabbageworm adult, a familiar insect to all growers of cruciferous crops, is a white butterfly with a wing span of about 2 in. in the female and slightly less in the male (Fig. 13:10). The wings bear several black markings. Near the center of the forewing the male bears one black dot, whereas the female bears two. The eggs, laid singly on leaves, are light yellow, spindle-shaped, and about 1 mm long. They have longitudinal ribs reticulated transversely.

When full grown, the larva is about 1 in. in length, velvety green with a faint yellow dorsal stripe and a row of yellow spots along each side in line with the spiracles. The larva crawls slowly over the plant and does not show any of the looping movement that is common to the cabbage looper, another prevalent cabbageworm. The pupa is about $\frac{4}{5}$ in. long, sharply angled, and greenish spotted with black. It is attached usually to the underside of a host leaf by an anal pad and a loop of silk.

Life history. The imported cabbageworm overwinters in the pupal stage on host plants. The first warm days of spring bring out the adults, and they are

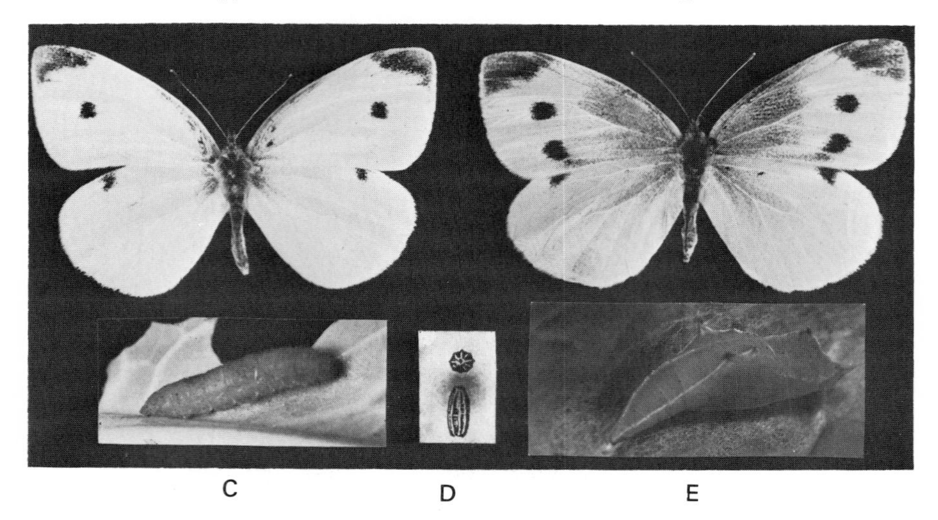

Figure 13:10. Life stages of the imported cabbageworm. A, male butterfly; B, female; C, larva; D, eggs; E, chrysalis or pupa. *Courtesy University of Florida, Gainesville.*

frequently the first insects to be noticed flying about in fields. Adults mate soon after emerging from the chrysalis and begin to lay eggs on their host plants. The eggs, deposited singly and on wild plants if no cultivated crucifers are available, hatch in 4 to 8 days. The larvae, which pass through 5 instars, require 10 to 14 days to reach full growth.

When ready to pupate, the caterpillar spins a carpet of silk over the site chosen, fastens its anal end to the carpet and then spins a silken girdle around its body at the first abdominal segment. After the chrysalis forms, it is held in place by these bonds. The cabbageworm remains in the pupal stage for 7 to 12 days. Laboratory studies have shown that this species enters diapause in the pupal stage when the larvae are exposed to photoperiods of less than 11 hours at a temperature of 20°C but not 24°C.

The generations of cabbageworms greatly overlap, because of the short larval and pupal periods and the long ovipositional period of the females. The time from egg to adult takes 22 to 42 days, and from two to six generations develop annually. In certain parts of Canada the insect is unable to over-winter, and infestations result from a northern migration of butterflies.

Parasites and bacterial and viral diseases play an important role in the natural control of the imported cabbageworm. Among the more important parasites and predators are the hymenopterous parasites *Apanteles glomeratus* (L.) and *Phyrex vulgaris* (Fallén), the wasp predator *Polistes metricus* Say, and ambush bugs.

Control. A complex of cabbageworms requires control on cruciferous crops—cabbage, cauliflower, broccoli, Brussels sprouts, and others. The pests consist of the imported cabbageworm, larvae of the diamondback moth, the cabbage looper, and climbing cutworms. Because of the inevitable presence of these pests, a preventive control program is the most successful way of coping with the problem. Treatment of the crop with insecticidal sprays should begin when plants are young and should be continued at 5- to 7-day intervals. Available for treating cruciferous crops are various insecticides which can be alternated with each other for more effective control and fewer environmental problems; near harvest only the less residual may be applied to reduce minimum days between treatments and harvest. Effective insecticides are endosulfan, methamidophos, mevinphos, parathion, methomyl, chlordimeform, and *Bacillus thuringiensis*.

Growers may apply high-pressure sprays at the rate of about 100 gal/acre and add 2 to 4 oz of spreader-sticker per 100 gal. Some growers use low-pressure sprays and apply around 20 gal of spray per acre. Underleaf spray coverage is required to control newly hatched cabbageworms. With boom-type rigs, growers apply the spray using at least three nozzles per row, directing one downward and one toward each side.

In the home garden malathion, diazinon, carbaryl, and *Bacillus thuringiensis* are approved insecticides for control of cabbageworms.

Studies in North Carolina and Wisconsin have demonstrated that varieties of cabbage vary in susceptibility to insect pests such as the imported cabbageworm, cabbage looper, cabbage maggot, and cabbage aphid. Two chief factors were shown to influence the amount of damage caused by the imported cabbageworm: differences in preference by females for oviposition upon cabbage varieties and differences in larval survival. According to the study, varieties exhibiting resistance include Red Acre, Red Hollander, Mammoth Red Rock, Racine Market, Globe, and Savoy Perfection Drumhead. Significantly, resistant varieties were reported to respond better to chemical control of cabbageworms than susceptible varieties, in that feeding damage to resistant varieties was reduced more by the treatment.

Seedcorn Maggot. *Hylemya platura* (Meigen) [Diptera:Anthomyiidae]

The seedcorn maggot, a widely distributed insect in the temperate regions of the world, was first recorded in North America in 1855 by Asa Fitch, who, in New York State found large numbers of adults feeding upon the flowering heads of wheat during June. The seedcorn maggot is often found associated with the closely related bean seed fly, *Hylemya florilega* (Zetterstedt).

Injury to plants results mainly from the feeding of the maggots on sprouting seed (Fig. 13:11A) or on seedlings. They attack a variety of vegetable crops including beans, peas, corn, cabbage, cauliflower, cucurbits, spinach, potato seed pieces, turnip, radish, and onion. The maggots burrow into the seed, feed on the endosperm, and often leave only a hollow

Figure 13:11. Injury caused by the seedcorn maggot. A, maggot has consumed everything except seed coat of corn seed; B, maggots devouring seed piece of potato. *A, courtesy Shell Chemical Company, New York, N.Y.; B, courtesy USDA.*

shell. By breaking through the tough seed coat, they allow entrance of disease organisms, which cause the seeds to rot. Attacked seeds that manage to germinate often fail to develop true leaves. When seedcorn maggots destroy 75 to 100 per cent of the first pair of unifoliate leaves, they cause a yield reduction of 80 to 99 per cent in snap beans. In some of the larger seeded plants, the maggots may be found feeding in the cotyledons above ground.

Seed pieces of potato are injured by the larvae entering the cut surface and honeycombing the seed piece with their feeding tunnels (Fig. 13:11B). The seedcorn maggot is an important agent in the dissemination and inoculation of **potato blackleg**, which is caused by the bacteria *Erwinia carotovora* (Jones). The disease results in an annual loss of approximately 2 per cent of the potato crop. The seedcorn maggot is also a primary pest of radish and, along with the cabbage maggot, is often found feeding on the fleshy root.

Description. The adult fly is approximately $\frac{1}{5}$ in. long (Fig. 13:12). The gray body has scattered bristles, and the legs are black. The eggs are pearly white, somewhat banana-shaped, and about 1 mm long. The larva is pearly white and when full grown about $\frac{1}{4}$ in. long. The pupa forms inside the last larval skin which hardens to form a protective puparium nearly $\frac{1}{5}$ in. long. It is light reddish brown when first formed, turning darker as the pupa matures.

Life history. Adults and larvae of the seedcorn maggot are most abundant in cool periods of the year—spring and fall. In midsummer populations decrease to their lowest point and adults, chiefly females, survive to carry the species through. The insect remains active during winter in North Carolina, but further north it goes into diapause and overwinters in the pupal stage.

The adult fly emerges from the puparium at night or early in the morning and with the aid of its ptilinum, a bladder-like structure on the head of the fly, works its way through the soil from depths as great as 7 in. The flies feed on

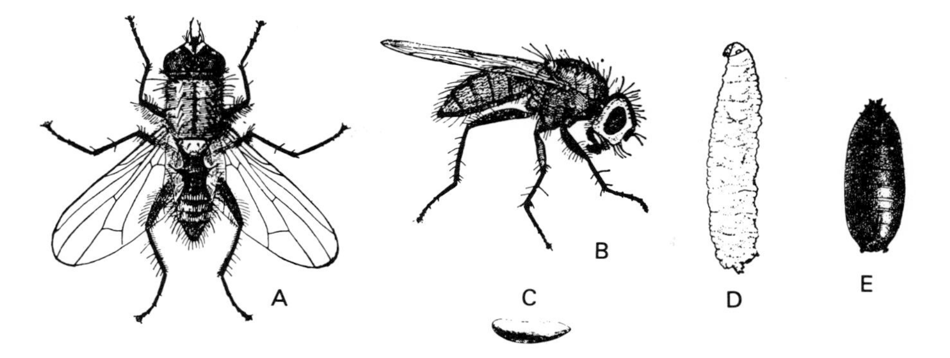

Figure 13:12. Life stages of seedcorn maggot. A, male fly; B, female; C, egg; D, larva; E, puparium. *Courtesy USDA.*

the nectar of flowers and plant juices. During June in the eastern United States, adults are common on the flowering heads of grasses. After a variable preoviposition period of from 5 to 55 days, depending on temperature, females lay eggs singly or in small clusters just beneath the surface of the soil near larval food. Favorite sites are in moist soil containing sprouting or decaying large seeds such as beans, corn, or peas, decaying crop remnants, or organic fertilizers. Research indicates that two seedborne microorganisms—a bacterium *Pseudomonas* sp. and a yeast *Cryptococcus laurentii*—give off odors that attract the ovipositing females. Sterile, germinating seeds by themselves are unattractive. Living for approximately six weeks, the females produce an average of 270 eggs.

Eggs hatch in one to nine days. The larvae, which apparently have three instars, develop rapidly in an abundance of food and at temperatures of 70° to 75°F. Usually in one to three weeks the maggots become full grown and pupate.

The larval skin thickens and becomes darker to form the puparium. Within the puparium the seedcorn maggot passes approximately two days in the prepupal stage and 7 to 26 days in the pupal stage. In the north central states, three to five generations develop each year.

There appears to be a symbiotic relationship between the seedcorn maggot and several species of bacteria, one of which causes blackleg of potato. Larvae deprived of their bacterial flora and reared on sterile media are small and unthrifty, whereas those possessing the usual gut bacteria develop into normal individuals. Other research, however, indicates that the bacteria work on the food source, making it suitable for larval growth, and that fungi may also play a role.

The seedcorn maggot apparently does not have many natural enemies. A parasite that has been reared from this insect is the hymenopteran *Aphaereta auripes* (Provancher). Several predators have been noted, among which are a wasp *Ectemnius stirpicola* (Pack.) and a fly *Scatophaga furcata* Say. Spiders also kill a few seedcorn maggot adults. Fungi belonging to the genus *Empusa* attack the adult flies.

Control. Not until the advent of the chlorinated hydrocarbons was there a satisfactory chemical method for fighting the seedcorn maggot. Now growers can protect vegetable plantings by coating seeds with insecticide. Recommended insecticides for treating seeds of beans, corn, and peas include wettable powders of diazinon, chlorpyrifos (Lorsban), or lindane. Apply these amounts of the actual toxicant per 100 lb of seed: 1 oz diazinon, 1 oz chlorpyrifos, or $\frac{1}{4}$ oz lindane. Lindane may not be as effective as either of the organophosphorus compounds. Recommendations suggest a dry planterbox treatment with wettable powders of diazinon or lindane plus a fungicide, or a slurry treatment of seed with wettable powder of Lorsban 25-SL and fungicide prior to planting. Prepare the slurry by mixing the amounts given on the label of insecticide, fungicide, and a sticker, such as methylcellulose, in water. Apply 12 to 15 fl oz of the slurry to each 100 lb of seed and mix in a rotating blender. Dry the seed immediately after treating and before bagging. Always use a fungicide along with the insecticide in treating seeds. The vegetable grower has also the choice of purchasing commercial seed protectants that combine insecticide and fungicide and have directions for use printed on the label, or of purchasing seed that has been treated by the seed house.

Certain cultural practices tend to lessen populations of the seedcorn maggot. Infestations tend to be lighter on warm, well-drained soils. If organic fertilizer is used, it should be well rotted and plowed into the soil during the fall. Manure should not be added to the soil in spring when the vegetable crop of beans, corn, or peas is to be planted. After a green cover crop is plowed under, it is best to wait about three weeks before planting.

Rapid germination of seed is desirable because the maggot has less time to attack. Growers can hasten germination of seed by working the seed bed thoroughly and planting as shallow as possible. If the spring is wet and cold, seeding should be delayed until warmer weather. Degrees of resistance related to rapid emergence and hard seedcoat have been found in certain lines and varieties of snap beans. A breeding program to exploit these characteristics is desirable.

Because maggots attack sound potato pieces only where the skin is broken or the surface injured, growers can prevent damage by allowing cut seed pieces to heal before planting. If the maggots have infested a stand of vegetables, the best thing to do is disk it under and replant immediately. There is little danger of reinfestation.

Beet Leafhopper. *Circulifer tenellus* (Baker) [Homoptera:Cicadellidae]

The history of the sugar beet industry of the western states is closely tied to the beet leafhopper. The first successful sugar beet factory was erected at Alvarado, California in 1870, and by 1890 the industry was well founded and in full production. In 1898 a plant disease called **curly top** struck the sugar beets, and from then on it occurred sporadically in most of the western beet-growing areas. E. D. Ball, in 1905, suspected the beet leafhopper as being connected with the disease but could offer no experimental proof. In 1910 H. B. Shaw, working on the suggestion of Ball proved that the beet leafhopper was, in fact, the vector of the curly top virus.

The beet leafhopper was probably introduced into North America from the Mediterranean region by early Spanish explorers. It is distributed principally west of the Continental Divide. The beet leafhopper causes damage primarily by acting as a vector of curly top, which attacks not only sugar beets but also spinach, tomatoes, beans, squash, cantaloupe, flax, and several other plants (Fig. 13:13). So far as known, the disease is carried only by this insect, although other species of leafhoppers are the vectors of a similar disease outside North America. Curly top-like diseases are transmitted by *Agalliana ensiger* in Argentina, *Agallia albidula* in Brazil, and *Circulifer opacipennis* in Turkey. Available evidence indicates that all four viruses are closely related; symptoms are very much alike, and the vector-virus relationships are almost identical. A closely related virus, the cause of pseudocurly top disease in Florida, has been found to be transmitted by the treehopper *Micrutalis malleifera* Fowler.

Figure 13:13. Sugar beet plant infected with curly top, a virus disease transmitted by beet leafhopper. *Courtesy USDA.*

Symptoms of the disease vary more or less with the kind of host attacked. In beets the earliest signs are a clearing of the veins and inward rolling of the margins of young leaves. As the disease progresses, curling of the leaves increases, veins swell, and papillae develop on the underside of the leaves. Leaves become dull, dark green, distorted, thick, and brittle. Lateral rootlets die, which induces the beet to develop a mass of hairy and woolly lateral rootlets. Concentric circles of dark tissue alternate with light tissue in cross sections of the root. Plants infected with the disease suffer a general stunting and frequently die.

Curly top has forced sugar beet factories in many places in the West to close and to move to other locations not so seriously affected. Growers generally abandon badly infected fields. In 1944 the beet leafhopper caused an estimated loss of $3,676,000 to vegetable crops in the intermountain region.

The beet leafhopper has been found to transmit *Spiroplasma citri*, the cause of citrus stubborn disease. Because mycoplasma infection also occurs in the mustard, *Sisymbrium Irio*, a favorite host plant, this insect may be an important vector of the disease in citrus, which serves as an occasional food plant of the leafhopper.

Description. The adult beet leafhopper, nearly $\frac{1}{4}$ in. long, is slender and tapers posteriorly, with the widest point of the body just behind the metathorax (Fig. 13:14). Color of adults varies from gray to greenish yellow. Darker individuals have blackish or brownish markings on the front wings, head, and thorax. The egg, 0.06 mm long, is barely visible to the naked eye. It is long, slender, slightly curved, and greenish white, later changing to a lemon yellow. As the egg develops, eye spots appear at the anterior end. The nymphs resemble the parents but are smaller and have no wings. When first hatched, they are white in color but soon darken. Older nymphs are usually spotted with red and brown.

Life history. In northern regions of its distribution, as in Washington and Idaho, the beet leafhopper develops three generations annually; in southern regions, as in California and Arizona, it develops five or more. Populations of the beet leafhopper have two ways of perpetuating themselves. One is to live on desert vegetation in winter, migrate in spring to summer hosts in cultivated areas, and then back again to the desert in fall. Another and somewhat simpler way is to survive on crops and various weeds in the same cultivated area.

In the first case the species overwinters in the adult stage chiefly on overgrazed rangeland that has been invaded by wild mustards, the principal winter host plants of the insect. Whenever temperatures are favorably high during winter, the adults become active and feed. In the Southwest reproduction continues during every month except December and January.

Females usually begin to deposit eggs in quantity when mustards start growing in spring. They insert eggs inside leaves and stems of host plants in rows of two to five and produce an average of 300 to 400 eggs each. Eggs laid during cool, spring weather take as long as 40 days to hatch, but eggs laid in

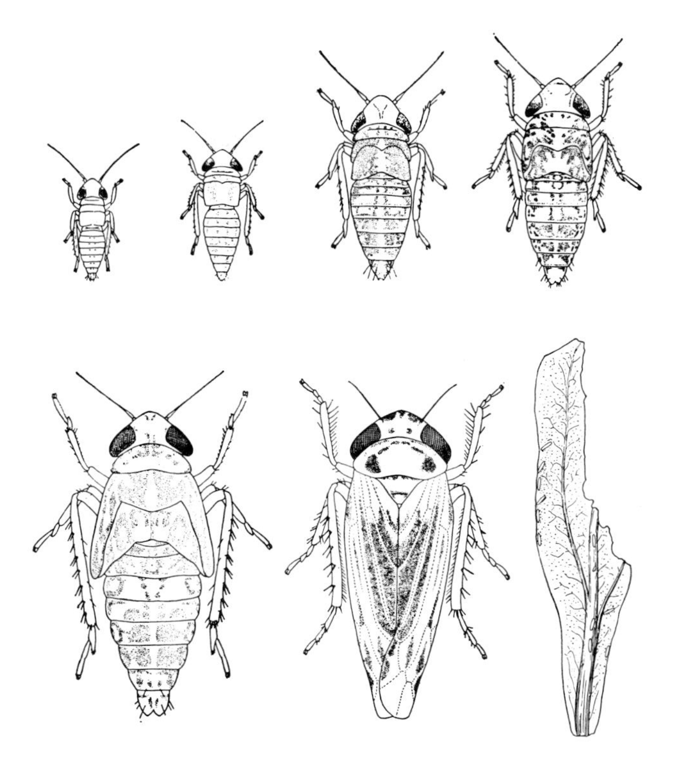

Figure 13:14. Life stages of the beet leafhopper. The five nymphal instars, the adult, and the eggs embedded in tissue of sugar beet leaf. *Courtesy USDA.*

summer may hatch in 5 days. The young nymphs immediately begin feeding on host plants by inserting their beaks and sucking juices. Molting five times, they develop to the adult stage in three to seven weeks. In spring a complete life cycle takes from six to ten weeks; in summer it may take only four weeks. The generations overlap each other, so that all stages are present at any time during the growing season.

The first or spring generation develops on weed hosts, chiefly mustards. About the time the insects reach the adult stage the weeds mature and become dry, inducing the leafhoppers to migrate. Flying with the wind, they infest summer host plants in their path; the infestation of crop plants such as sugar beets, cucurbits, beans, and tomatoes is incidental to their general movement. They do not seek out these crops actively but alight on any vegetation, whether a favorable host plant, such as sugar beets or an unfavorable one, such as tomatoes, beans, or cucurbits.

Russian thistle is the most important summer weed host in the western states. Mixed stands of Russian thistle and mustards are especially favorable for supporting large populations, because the leafhoppers can overwinter and reproduce their spring and summer generations in the same area. Of the cultivated plants, beets are the only important breeding host.

With the maturing and drying of summer host plants in fall, leafhoppers migrate back to their winter hosts. Until the annuals germinate, they may have to feed on plants unfavorable to their survival—perennials such as saltbush, sagebrush, or mesquite—with a resulting heavy mortality. On the other hand, high leafhopper populations enter the winter season if the winter hosts germinate before the summer hosts dry or are killed by frost.

In studying the biology of the beet leafhopper, entomologsts have found six main breeding areas in which the insects have an abundance of food plants and favorable weather conditions (Fig. 13:15). Leafhoppers from any one breeding ground infest the same cultivated areas year after year.

The second way in which populations perpetuate themselves has been discovered in the Imperial Valley of California. In this irrigated region of intense agriculture, sugar beets are planted in fall and harvested in late spring and early summer. The beet crop plus weeds, such as goosefoot (*Chenopodium murale*), mustard (*Sisymbrium Irio*), and rough pigweed (*Amaranthus Palmeri*), furnishes the leafhoppers with a succession of favorable hosts to permit year-round survival. The weed hosts present during summer are particularly important in maintaining the population between beet crops.

A number of parasites and predators attack the beet leafhopper. Three minute wasps—*Polynema eutettixi* Girault, *Abbella subflava* Girault, and *Anagrus giraulti* Crawford—parasitize the eggs. Members of the wasp genus *Gonatopus* act as both predators and parasites of nymphs and adults. The bigeyed fly, *Tömösváryella subnitens* (Cress.), is probably the most important parasite of the nymph and adult. *Geocoris pallens* Stål, a predaceous hemipteran, destroys the active stages of the leafhopper. Spiders, lizards, and birds often prey on leafhoppers.

Research in California on parasites of the beet leafhopper's egg suggests that they may be an important factor in preventing and controlling outbreaks, particularly those originating in the cultivated areas in the Imperial Valley.

Control. Because the beet leafhopper is migratory and is able to transmit curly top virus in a few minutes of feeding, it exerts great pressure on applications of insecticide. Various insecticides are available to control the beet leafhopper on vegetable crops; the exact compounds chosen depend on the particular crop needing protection. For example, table beets may be protected by weekly foliar applications of carbaryl or parathion. Tomatoes may be protected by sidedressing with granules of disulfoton or phorate at planting time; the effectiveness of the systemic lasts about six weeks. When tomato plants are still young, weekly foliar applications of parathion may be started and continued until two weeks before harvest.

Another approach to control is the development of plants resistant to curly top. Several varieties of sugar beets have been developed that are resistant and are well adapted to the beet leafhopper area. Resistant varieties of Great Northern and Pinto beans have been produced by the Idaho

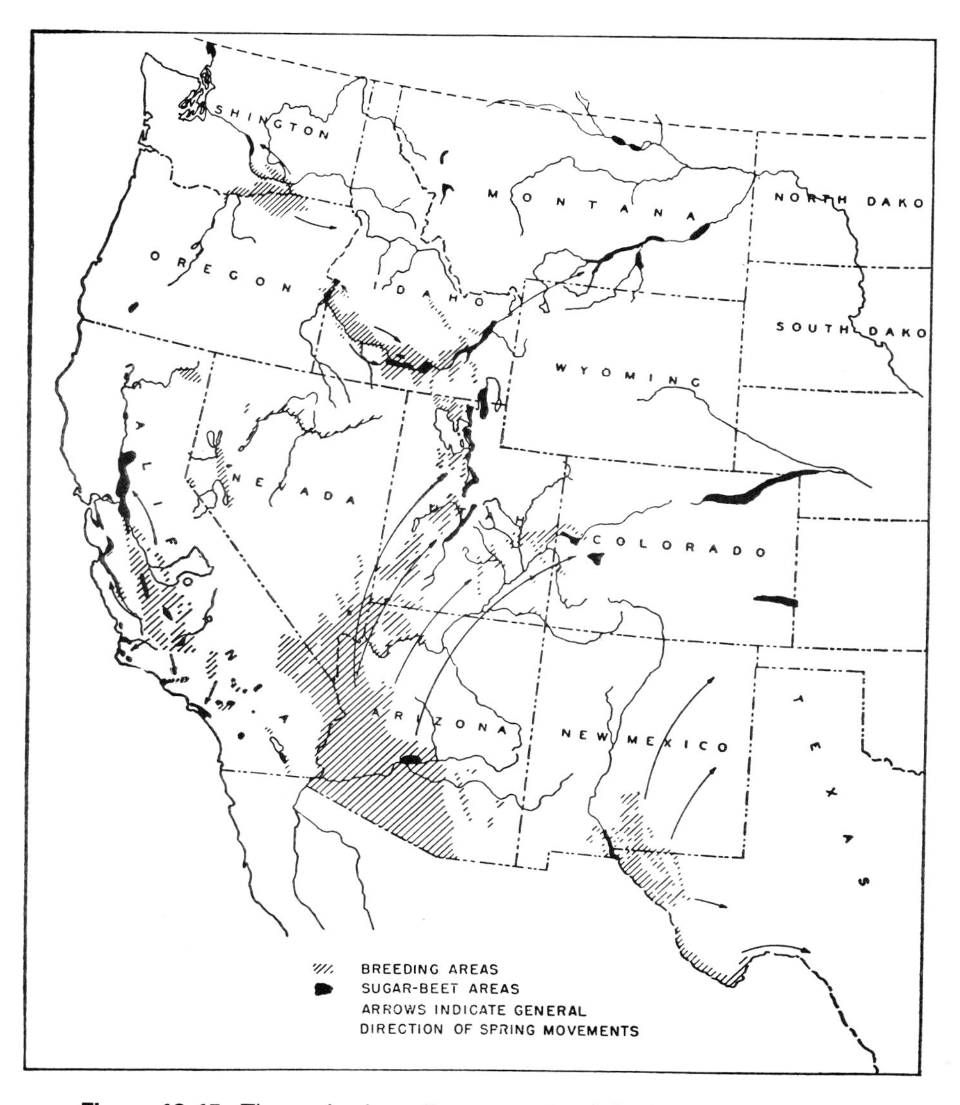

Figure 13:15. The major breeding grounds of the beet leafhopper and the sugar beet areas affected by them. *Courtesy USDA.*

Experiment Station. Work is in progress on breeding resistant tomatoes, but no completely resistant variety has been developed yet. Certain varieties of bean, squash, and pumpkin are naturally resistant to curly top.

In general early-planted sugar beets are less injured by curly top, but often other factors make early planting an unsatisfactory practice. In the coastal area of California beets planted after the spring movement of leafhoppers make a successful crop. It is wise to plant tomatoes as late as possible so as to miss the spring migration of leafhoppers. In Utah double-hill planting of tomatoes has given some protection from curly top; in New Mexico spacing tomatoes 1 ft apart in the row rather than 3 ft has resulted in less curly top and greater production.

Because direct control in commercial plantings has proven difficult, entomologists have attempted to control weed host plants before the leafhoppers migrate depriving them of breeding sites; or the weeds may be sprayed with insecticides to kill the leafhoppers before they have a chance to migrate to cultivated plants.

The hosts carrying the leafhoppers through the winter and during the spring are mostly annuals that develop as weeds on overgrazed areas. One method of control is to limit grazing so that nonhost grasses come back to replace the weeds. Russian thistle, the chief weed host that supports the summer and fall population of leafhoppers, grows principally on abandoned land and on overgrazed areas. Because leafhoppers tend to congregate on this weed for breeding, elimination of Russian thistle by hoeing, cultivating, or dragging a railroad rail over the plant has been suggested for controlling the insect.

Direct control of leafhoppers before they migrate in spring is helpful in preventing transmission of curly top. After survey crews determine and map the areas supporting large populations of leafhoppers, control personnel come in with trucks and planes and cover the area with insecticide.

In the Imperial Valley of California growers of sugar beets have combated the beet leafhopper successfully through a well-organized control program. They have agreed upon a beet-free period during August, the eradication of weed hosts during July and August, and the spraying of weeds along roads with insecticide.

SELECTED REFERENCES

Auclair, J. L., "Life History, Effects of Temperature and Relative Humidity, and Distribution of the Mexican Bean Beetle, *Epilachna varivestis* Mulsant (Coleoptera: Coccinellidae) in Quebec, with a Review of the Pertinent Literature in North America," *Ann. Soc. Entomol. Quebec*, 5:19–44 (1959).

Beckham, C. M., *Biology and Control of the Vegetable Weevil in Georgia*, Ga. Agr. Exp. Sta. Tech. Bul. 2, 1953.

Blickenstaff, C. C., *Sugarbeet Insects: How to Control Them*, USDA Farmers' Bul. 2219, 1976.

Brindley, T. A., J. C. Chamberlin, and Ralph Schopp, *The Pea Weevil and Methods for Its Control*, USDA Farmers' Bul. 1971, 1958.

Canerday, T. D., and J. D. Dilbeck, *The Pickleworm: Its Control on Cucurbits in Alabama*, Alabama Agr. Exp. Sta. Bul. 381, 1968.

Cannon, F. M., *Control of the Colorado Potato Beetle in Canada*, Can. Dept. Agr. Publ. 1071, 1960.

———, *Control of the Potato Flea Beetle in Eastern Canada*, Can. Dept. Agr. Publ. 1072, 1960.

Cockerham, K. L., O. T. Deen, M. B. Christian, and L. D. Newsom, *The Biology of the Sweet Potato Weevil*, La. Agr. Exp. Sta. Tech. Bul. 483, 1954.

Cook, W. C., *Ecology of the Pea Aphid in the Blue Mountain Area of Eastern Washington and Oregon*, USDA Tech. Bul. 1287, 1963.

———, *The Beet Leafhopper*, USDA Farmers' Bul. 1886, 1941.

———, *Life History, Host Plants and Migrations of the Beet Leafhopper in the Western United States*, USDA Tech. Bul. 1365, 1967.

Crosby, C. R., and M. D. Leonard, *Manual of Vegetable-garden Insects* (New York: Macmillan, 1918).

Cuthbert, F. P., Jr., *Insects Affecting Sweetpotatoes*, USDA Agr. Handbook 329, 1967.

———, W. J. Reid, Jr., and A. Day, *The Southern Potato Wireworm: How to Control It on Irish Potatoes*, USDA Leaflet 534, 1965.

Drake, C. J., and H. M. Harris, *Asparagus Insects in Iowa*, Ia. Agr. Exp. Circ. 134, 1932.

Dudley, J. E., and T. E. Bronson, *The Pea Aphid on Peas and Methods for Its Control*, USDA Farmers' Bul. 45, 1952.

Elmore, J. C., and R. E. Campbell, *The Pepper Weevil*, USDA Leaflet 226, 1956.

Fenton, F. A., and A. Hartzell, *Bionomics and Control of the Potato Leafhopper*, Ia. Agr. Exp. Sta. Res. Bul. 78, 1923.

Folsom, D., G. W. Simpson, and R. Bonde, *Maine Potato Diseases, Insects, and Injuries*, Me. Agr. Exp. Sta. Bul. 469, 1949.

Hall, C. V., and R. H. Painter, *Insect Resistance in Cucurbita*, Kansas Agr. Exp. Sta. Tech. Bul. 156, 1968.

Harris, E. H., Jr., and W. A. Rawlins, *Carrot Rust Fly and Its Control*, Cornell Univ. Agr. Exp. Sta. Bul. 946, 1959.

Hayslip, N. C., W. G. Genung, E. G. Kelsheimer, and J. W. Wilson, *Insects Attacking Cabbage and Other Crucifers in Florida*, Fla. Agr. Exp. Sta. Bul. 534, 1953.

Henderson, C. F., *Overwintering, Spring Emergence and Host Synchronization of Two Egg Parasites of the Beet Leafhopper in Southern Idaho*, USDA Circ. 967. 1955.

———, *Parasitization of the Beet Leafhopper in Relation to Its Dissemination in Southern Idaho*, USDA Circ. 968, 1955.

Hills, O. A., *Insects Affecting Sugarbeets Grown for Seed*, USDA Agr. Handbook 253, 1963.

Hills, O. A., and E. H. Taylor, *Cantaloup Insects in the Southwest: How to Control Them*, USDA Leaflet 389, 1967.

Kelsheimer, E. G., N. C. Hayslip, and J. W. Wilson, *Control of Budworms, Earworms and Other Insects Attacking Sweet Corn and Green Corn in Florida*, Fla. Agr. Exp. Sta. Bul. 466, 1950.

Landis, B. J., and J. A. Onsager, *Wireworms on Irrigated Lands in the West: How to Control Them*, USDA Farmers' Bul. 2220, 1967.

Linn, M. B., and J. M. Wright, *Tomato Diseases and Insect Pests: Identification and Control*, Ill. Agr. Exten. Serv. 683, 1951.

Michelbacher, A. E., W. W. Middlekauff, and N. B. Akesson, *Caterpillars Destructive to Tomato*, Calif. Agr. Exp. Sta. Bul. 707, 1948.

Michelbacher, A. E., W. W. Middelkauff, O. G. Bacon, and J. E. Smith, *Controlling Melon Insects and Spider Mites*, Calif. Agr. Exp. Sta. Bul. 749, 1955.

Miller, L. A., and R. J. McClanahan, "Life-History of the Seed-Corn Maggot, *Hylemya cilicrura* (Rond) and of *H. liturata* (Mg.) (Diptera:Anthomyiidae) in Southwestern Ontario," *Can. Ent.*, 92:210–21 (1960).

Peay, W. E., *Sugarbeet Insects: How to Control Them*, USDA Farmers' Bul. 2219, 1968.

Pepper, B. B., *The Carrot Weevil,* Listronotus latiusculus (*Bohe*) *in New Jersey and Its Control*, N.J. Agr. Exp. Sta. Bul. 693, 1942.

Pletsch, D. J., *The Potato Psyllid: Its Biology and Control*, Mont. Agr. Exp. Sta. Tech. Bul. 446, 1947.

Prescott, H. W., and M. M. Reeher, *Pea Leaf Weevil: An Introduced Pest of Legumes in the Pacific Northwest*, USDA Tech. Bul. 1233, 1961.
Read, D. C., *Control of Root Maggots in Rutabagas, Cabbages, and Related Plants in the Maritime Provinces*, Canada Dept. Agri. Publ. 1075, 1960.
Reed, L. B., and R. E. Webb, *Insects and Diseases of Vegetables in the Home Garden*, USDA Home and Garden Bul. 46, 1967.
Reid, W. J., Jr., *Control of Caterpillars on Commercial Cabbage and Other Cole Crops in the South*, USDA Farmers' Bul. 2099, 1968.
———, *Biology of the Seed-Corn Maggot in the Coastal Plain of the South Atlantic States*, USDA Tech. Bul. 723, 1940.
Reid, W. J., Jr., and F. P. Cuthbert, Jr., *Aphids on Leafy Vegetables: How to Control Them*, USDA Farmers' Bul. 2148, 1977.
———, *The Pickleworm: How to Control It on Cucumber, Squash, Cantaloup, and Other Cucurbits*, USDA Leaflet 455, 1966.
———, *Control of Caterpillars on Commercial Cabbage and Other Cole Crops in the South*, USDA Farmers' Bul. 2099, 1960.
Shands, W. A., and B. J. Landis, *Controlling Potato Insects*, USDA Farmers' Bul. 2168, 1970.
——, *Potato Insects: Their Biology and Biological and Cultural Control*, USDA Agr. Handbook 264, 1964.
Simpson, G. W., and W. A. Shands, *Progress on Some Important Insect and Disease Problems of Irish Potato Production in Maine*, Me. Agr. Exp. Sta. Bul. 470, 1949.
Smith, F. F., *Control of Insect Pests of Greenhouse Vegetables*, USDA, Agr. Handbook 142, 1959.
Stone, M. W. *Biology and Control of the Lima-Bean Pod Borer in Southern California*, USDA Tech. Bul. 1321, 1965.
Thomas, C. A., *Mushroom Insects: Their Biology and Control*, Pa. Agr. Exp. Sta. Bul. 419, 1942.
USDA, *Cabbage Insects: How to Control Them in the Home Garden*, USDA Home and Garden Bul. 44, 1969.
USDA, *Controlling White-Fringed Beetles*, USDA Leaflet 550, 1969.
USDA, *Controlling the Mexican Bean Beetle*, USDA Leaflet 548, 1977.
USDA, *The Corn Earworm on Sweet Corn: How to Control It*, USDA Leaflet 411, 1967.
USDA, *The Potato Leafhopper: How to Control It*, USDA Leaflet 521, 1963.
USDA, *The Mexican Bean Beetle in the East and Its Control*, USDA Bul. 1624, 1960.
USDA, *The Onion Thrips: How to Control It*, USDA Bul. 372, 1960.
USDA, *The Sweetpotato Weevil: How to Control It*, USDA Bul. 431, 1960.
Wallis, R. L., *Ecological Studies on the Potato Psyllid as a Pest of Potatoes*, USDA Tech. Bul. 1107, 1955.
Westcott, C., *The Gardener's Bug Book* (Garden City, N.Y.: Doubleday, 1973).
Wilcox, J., and A. F. Howland, *The Tomato Fruitworm: How to Control It*, USDA Leaflet 367, 1968.
Wilcox, J., A. F. Howland, and R. E. Campbell, *Investigations of the Tomato Fruitworm, Its Seasonal History, and Methods of Control*, USDA Tech. B. 1147, 1956.
Wilcox, J., and F. H. Shirck, *The Onion Thrips: How to Control It*, USDA Leaflet 372, 1960.
Wilson, G. F., *The Detection and Control of Garden Pests* (London: Crosby Lockwood and Sons Ltd., 1949).
Wilson, J. W., and N. C. Hayslip, *Insects Attacking Celery in Florida*, Fla. Agr. Exp. Sta. Bul. 486, 1951.

chapter 14 / **CARL JOHANSEN**

INSECT PESTS OF TREE FRUITS

THE PESTS

A variety of insects and mites live on fruit trees—apple, pear, peach, apricot, cherry, prune, and plum. One well-known entomologist has estimated that some 400 species of insects infest the apple, although only 25 of these are of economic importance. Other tree fruits are also fed upon by many different kinds of insects, but the number of major pests are only a fraction of the total. For example, peach in Ohio has 4 major pests and about 36 minor ones, and pear in California has 10 major pests and 20 minor ones.

Orchard insects vary in their specificity for feeding on the different varieties of fruit. Some, like the pear psylla, attack only a single kind of host, but the majority are more general feeders and attack more than one or even all, as does the San Jose scale.

Two orders, the Lepidoptera and the Homoptera, contain the major number of species of tree fruit insects. Among the Lepidoptera, the most familiar species is the **codling moth**—the well-known "worm in the apple" (Fig. 14:11). Related caterpillars that cause damage to tree fruits include the **oriental fruit moth** (Fig. 4:18C), **cherry fruitworm,** and **eyespotted bud moth**.

Leafrollers such as the **fruittree leafroller** and **redbanded leafroller** are common pests in the orchards of North America (Fig. 14:1). Two major pests of peach, the **peachtree borer** (Figs. 14:8 and 4:18A) and the **lesser peachtree borer**, are the larvae of clearwing moths. The **peach twig borer**, a species belonging to another moth family, attacks terminal growth and fruit of peach (Fig. 4:18B).

Figure 14:1. Redbanded leafroller. A, adult and egg mass; B, larva. *Courtesy Michigan State University, East Lansing.*

In the order Homoptera, scale insects and aphids are notorious tree fruit pests. The most important scale insect is the **San Jose scale** (Fig. 14:17), but other highly destructive species include the **European fruit lecanium** (Fig. 14:2), **oystershell scale,** and **scurfy scale**. Major species of **aphids** include the apple aphid, woolly apple aphid, rosy apple aphid, apple grain aphid, black cherry aphid, rusty plum aphid, mealy plum aphid, leaf curl plum aphid, thistle aphid, green peach aphid, and black peach aphid. Other important Homoptera on fruit include several species of **leafhoppers**, which transmit virus diseases especially among stone fruits, and the **pear psylla** (Fig. 14:3), a pest only of pears, found in most orchard regions of America.

Various sucking bugs (Hemiptera), such as the **tarnished plant bug** (Fig. 4:10C), **apple red bug**, and a number of **stink bugs**, cause pitting or scabbing of apples and pears or "catfacing" of peaches and apricots.

Figure 14:2. European fruit lecanium, *Lecanium corni* Bouché. A, Immature scales; B, mature scales. *A, courtesy University of California, Berkeley; B, courtesy Ohio Agr. Res. Dev. Center, Wooster.*

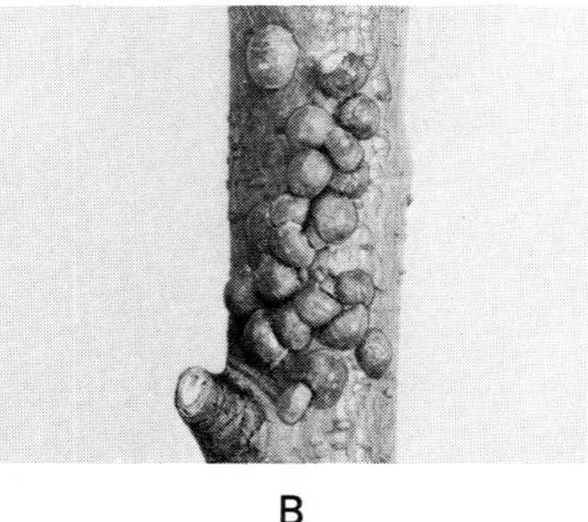

A B

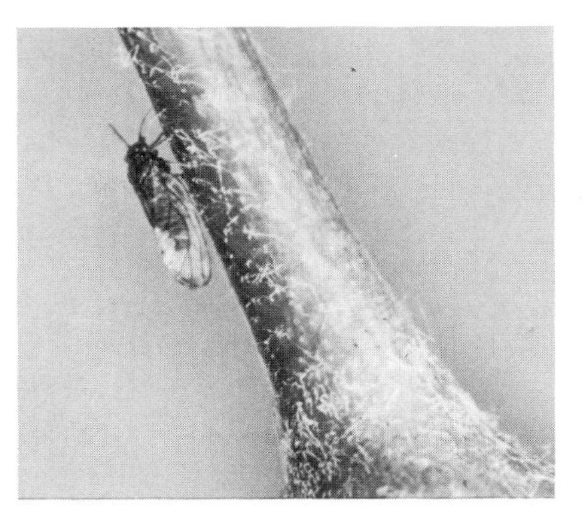

Figure 14:3. Adult pear psylla, a specific pest of pears. *Courtesy Michigan State University, East Lansing.*

Among the Diptera, several species of fruit flies feed internally on the fruit. The **apple maggot** and the **cherry fruit fly** (Fig. 14:19) are distributed in the eastern half of the nation, the **western cherry fruit fly** in the West, and the **black cherry fruit fly** is dispersed widely throughout North America.

A few species of beetles are major orchard pests. The **plum curculio**, found east of the Rocky Mountains, damages all types of fruit (Fig. 14:4). Various species of **roundheaded wood borers** and **shothole borers** injure the trunks and larger branches of orchard trees.

Since World War II, mites have become one of the greatest concerns of fruit growers. **Spider mites** are particularly damaging; **blister mites** and **rust mites** are also serious pests.

THE INJURY

Fruit Damage

Although various insects and mites feed upon all parts of orchard trees, many of the most serious pests feed upon the fruit. Because of consumer attitudes, even small blemishes or feeding scars reduce the value of fruit. If one insect is discovered in samples from a truckload of fruit, the whole shipment is subject to rejection by the processor or fresh fruit merchandiser. Therefore, special attention is given to those pests that feed directly upon or otherwise affect the fruit.

Certain insects damage fruits by tunneling directly into them. Codling moth larvae, cherry fruit fly maggots, oriental fruit moth larvae, peach twig borer, and cherry fruitworm cause this type of injury. Sometimes maggots or caterpillars remain as a contaminant in processed fruits.

Scabs, blemishes, russeting and other disfigurations of the surface of the fruits are often caused by insects. Through their feeding on the fruit, bugs—such as the apple red bug, boxelder bug, and stink bugs—produce

Figure 14:4. The plum curculio, a general fruit pest. *Courtesy Pennsylvania State University, University Park.*

blemishes on apples and pears (Fig. 14:5A). The tarnished plant bug and other species of lygus bugs, as well as certain stink bugs, often cause misshapen or "catfaced" peaches and apricots (Fig. 14:5B). The pearleaf blister mite forms black spots on pear fruits as well as on the foliage, whereas the appleleaf blister mite causes brown blisters on apple foliage. Pitting, stunting, and deforming of apples are caused by the attack of rosy apple aphids. Scabby areas are left by the feeding of climbing cutworms, green fruitworms, and leafroller larvae. Heavy feeding by adult plum curculios produces warty and misshapen fruits. The Japanese beetle eats holes in developing peaches in eastern orchards (Fig. 17:6).

Fruit drop of peaches is caused by the feeding of the green peach aphid upon the blossoms. Feeding of pear midge larvae within developing fruits causes stunting, deforming, and dropping of pears. Fruits attacked by the plum curculio usually drop soon after the eggs hatch and young begin to feed. Tunneling near the skin by the apple maggot causes the fruits to drop prematurely.

Honeydew produced by aphids, scale insects, and others feeding on various parts of the tree may soil the fruit. Honeydew causes blemishes, and a black, sooty fungus, which often grows in the honeydew, causes further disfiguring.

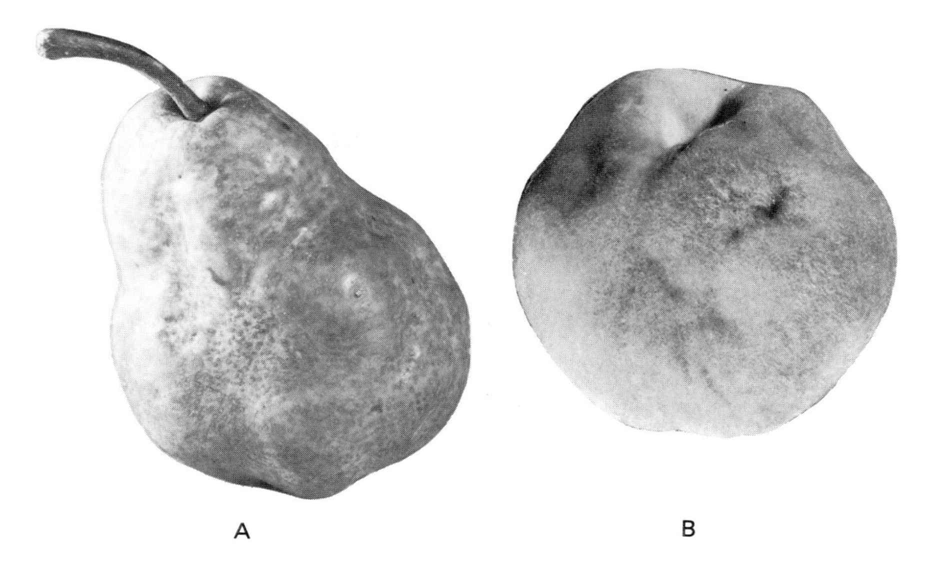

Figure 14:5. Fruit injured by insects. A, stink bug damaged pear. Depressed areas resulted from bugs' feeding on developing fruit; pear is corky under skin where bugs have fed. B, cat-faced peach caused by feeding of tarnished plant bug. *A, courtesy University of California, Berkeley; B, courtesy Ohio Agr. Res. Dev. Center, Wooster.*

Foliage Damage

Insects and mites may damage the foliage by sucking up the cell contents of leaves, by injecting toxic substances, or by devouring leaf tissues. Leafhoppers, mites, and lace bugs remove the chlorophyll along with other cell contents and cause a white speckling of the foliage. If the injury is severe, the leaves usually turn brown and drop. Rust mites cause a general discoloration of the foliage. One species, the peach silver mite, often imparts a silvery appearance to the leaves. Pear and appleleaf blister mites form black and brown blisters on pear and apple foliage, respectively.

Most tree fruit aphids cause a curling of the foliage, a symptom that is particularly characteristic of the rosy apple aphid, the black cherry aphid, the green peach aphid, and the leaf curl plum aphid. Aphids concentrate on the tender growth of the terminals causing a killing back or stunting of these portions.

Foliage pests with chewing mouthparts, such as various leafminers, pearslugs, cankerworms, and leafrollers, consume leaf tissues and may seriously defoliate trees (Fig. 14:6). By their damage of leaves and impairment of leaf function, foliage pests indirectly affect the fruit, which may fail to develop properly and attain normal size.

Figure 14:6. Pearslug skeletonizes the leaves of pear and cherry trees. *Courtesy USDA.*

Trunk, Limb, and Twig Damage

Many pests of fruit trees feed on the surface of twigs, limbs, and trunk or bore into these structures. A variety of scale insects attach themselves on the surface, suck plant juices, kill branches, and weaken trees. Certain insects, such as the buffalo treehopper, the periodical cicada, and tree crickets lay their eggs in young branches and twigs and thereby injure or kill these parts.

Insects, such as the roundheaded appletree borer, sinuate peartree borer, and shothole borer, feed under the bark of the trunks and larger limbs (Fig. 14:7). The larvae of these beetles may weaken and even girdle the trees. Larvae of clearwing moths burrow beneath the bark and are particularly destructive to young trees (Fig. 14:8).

Root Damage

Only a few pests are found on the roots of fruit trees. The woolly apple aphid and the black peach aphid feed on roots as well as top growth and are especially damaging to nursery stock. White grubs sometimes cause minor root injury. In the Pacific Northwest, several species of rain beetles, *Pleocoma* spp., severely injure the roots of orchard trees in local areas.

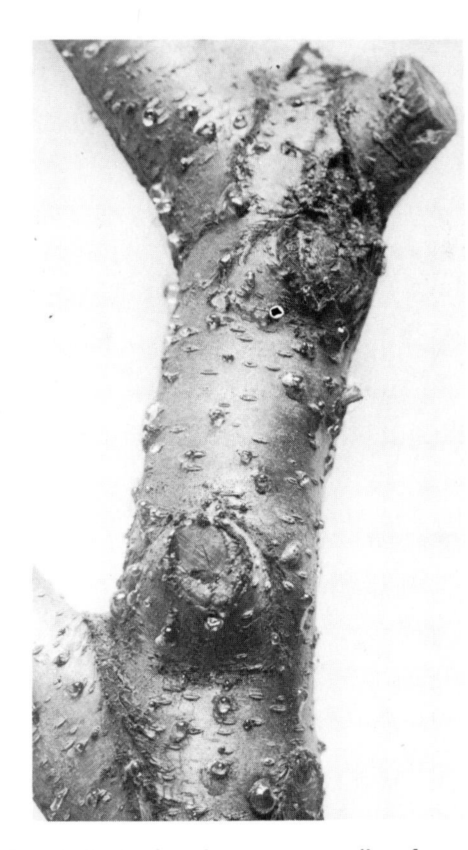

Figure 14:7. Peach limb showing gum exuding from injuries of shot-hole borer, *Scolytus rugulosus* (Ratzeburg). *Courtesy Ohio Agr. Res. Dev. Center, Wooster.*

Disease Transmission

Several serious virus diseases of stone fruits are transmitted by insects. In the northeastern states, peach yellows and little peach are transmitted by the plum leafhopper and the meadow spittlebug. In the West, the geminate leafhopper and certain other species transmit the little cherry-western X disease complex to both cherries and peaches. Workers in California have demonstrated recently that eriophyid mites transmit peach mosaic and cherry mottle leaf virus diseases in the West. Phony peach, peach ringspot, and peach yellow bud mosaic are transmitted by leafhoppers, an aphid, and a nematode, respectively, whereas at least four different virus diseases of plum and yellow mosaic of stone fruits are transmitted variously by the leaf curl plum aphid, the green peach aphid, and the hop aphid. Pear decline virus is transmitted by the pear psylla in California. Similar symptoms are caused by the toxic effects of the psyllid salivary secretions.

Figure 14:8. Young peach tree infested with peachtree borer. *Courtesy Ohio Agr. Res. Dev. Center, Wooster.*

CONTROL

Orchardists cannot rely solely on cultural methods to control insects. Modern consumer demands for top-quality processed fruits have forced increasingly greater use of insecticides in order to obtain nearly perfect control. Furthermore, because tree fruits have a high unit value, use of chemical controls is often economically sound where it might not be on other types of crops. But there are helpful cultural practices as well.

Cultural Control

Use of nonlegume cover crops is suggested in some areas in order to reduce certain insect pest populations. This practice is particularly useful in peach and apricot orchards, because catfacing insects are attracted to legumes both for feeding and for overwintering. Leafhoppers, which transmit stone-fruit virus diseases, also prefer legume hosts.

Several cultural methods are effective in lessening damage from shothole borers. Removal and burning of badly infested trees, together with all prunings, will reduce the possibility of further damage. Keeping trees in good condition, including watering and fertilizing when necessary, will protect them from both shothole borers and flatheaded borers. Removal of wild host plants near the orchards will help reduce shothole or roundheaded borer populations in the area.

Cold storage and controlled atmosphere (CA) storage of fruits stop damage from insects, and continued storage kills many species. CA storage involves the reduction of oxygen, increase of carbon dioxide, and replacement of a major portion of the storage atmosphere with nitrogen. CA storage kills

San Jose scale, apple maggot, and plum curculio more quickly than common cold storage.

The plum and apple curculios winter in plant debris or rubbish on the ground or in other sheltered places. Keeping the orchards and surrounding lands free from such material aids in the control of these pests.

Northern Spy apple rootstocks are used by nurserymen in some areas because of their resistance to the woolly apple aphid. This variety's resistance to the attack of the aphid is caused by its adverse effect upon the insects growth and survival.

Biological Control

Biological and integrated systems have provided excellent control of a number of orchard pests. Naturally occurring predators and parasites are particularly effective against aphids, scale insects, and spider mites. Unfortunately these beneficial forms are often reduced by insecticide programs.

Use of a eulophid parasite, *Aphelinus mali*, to control the woolly apple aphid has been an outstanding worldwide success. The parasite was obtained first in the northeastern states and eastern Canada, the native home of the aphid. The parasite quickly controlled the aphid in commercial orchards of the Pacific Northwest after its introduction in the 1920s and 1930s. *Aphelinus* has continued to be effective, except for temporary setbacks associated with the use of new synthetic organic insecticides.

Other biocontrol agents that have been rated as at least partial successes against tree fruit pests include a chalcidoid parasite, *Blastothrix sericea*, on European fruit lecanium; a predaceous mite, *Hemisarcoptes malus*, on oystershell scale; a wasplike parasite, *Alloptropa utilis*, on apple mealybug; other hymenopterous parasites on Comstock mealybug; a eulophid parasite, *Prospaltella perniciosi*, on San Jose scale; and a braconid parasite, *Macrocentrus ancylivorus*, on oriental fruit moth.

Integrated control methods have been particularly adaptable to orchard situations throughout much of the world. The most common type of program involves the use of carefully selected and timed minimal dosages of insecticides against the major insect pests. Under these conditions, many natural enemies are preserved to control the plant-feeding mites.

Chemical Control

Consumer demands for top-quality fruits, free from blemishes or contaminations of any kind, have forced orchardists to rely more and more on chemical control measures. Because fruit trees are invariably attacked every year by a host of pests, growers regularly apply a series of treatments timed with seasonal development of the trees. These so-called **spray programs** or **spray schedules** for fruit trees have become very complex in recent years and vary from crop to crop and from region to region. Many states publish spray schedules annually as guides for growers to follow, or to modify as actual insect and disease conditions in the orchard warrant.

Specific insect problems are complicated by the necessity of applying fungicides, chemical thinning sprays, hormone sprays, minor-element deficiency sprays, and fertilizers. When several materials are applied in combination, plant injury or other evidences of incompatibility may result. Insecticidal residues remaining on fruits at harvest time must be controlled carefully so as not to exceed the tolerances set by the Environmental Protection Agency. Insecticide poisoning of honey bees has made it increasingly difficult to obtain pollination services in certain areas.

The standard dormant or delayed dormant sprays of oil and lime-sulfur are applied in many areas against spider mites, rust mites, pear and apple leaf blister mites, scale insects, aphid eggs, fruittree leafroller eggs, and pear psylla.

Although it has been replaced largely by newer materials, lead arsenate is still used against apple maggot, various caterpillars and cutworms, plum curculio, roundheaded appletree borer, and codling moth in certain localities. Lead arsenate has the potential advantage for newer integrated programs of not being highly hazardous to most parasites and predators.

Several of the botanical insecticides are employed to control tree fruit pests. Nicotine is used against aphids, leafhoppers, and sucking bugs; rotenone is used against cherry fruit flies and pear psylla. Rotenone or pyrethrum is sometimes injected into the burrows of roundheaded wood borers in eastern orchards. Ryania aroused considerable interest throughout the U.S. and Canada as a promising control for codling moth because it is relatively harmless to the natural enemies of both the codling moth and spider mites. However, it has proven effective only in the northeastern United States and adjacent Canadian localities, where the codling moth has only one to two generations per season and is controlled partially by parasites.

Until about 1960, the organochlorine insecticides were exploited against a wide variety of tree fruit insects. Although DDT often proved to be a major factor in the production of heavy mite populations, it was one of the most frequently used insecticides in the orchard. It controlled apple maggot, codling moth, Japanese beetle, leafrollers, cutworms, caterpillars, oriental fruit moth, peachtree borer, shothole borer, pearslug, pear midge, pear thrips, certain scale insects, sucking bugs, and leafhoppers. In 1969, only a few minor orchard uses of DDT were being recommended by either federal or state agencies. These specific uses were discontinued in 1973. Methoxychlor is still used against cherry fruit flies, cherry fruitworm, plum curculio, Japanese beetle, and apple maggot. The specific use of TDE against two leafrollers of the genus *Argyrotaenia*, the redbanded leafroller in the East and the orange tortrix in the West, is no longer registered.

Perthane has several specialized orchard uses, either alone or in mixtures with superior oils against the pear psylla and used singly against cherry fruit flies. Endosulfan is a cyclodiene material which does not have as many undesirable qualities as related organochlorine compounds. It is used against aphids, peachtree borers, peach twig borer, rust and blister mites, sucking bugs, and leafrollers. Lindane and dieldrin were once used against several orchard pests. However, these materials have been removed from such uses in recent years.

A carbamate insecticide, carbaryl, came into orchard use in 1959. It causes outbreaks of spider mites by killing their predators and also is hazardous to pollinating bees. Nevertheless, it is still used in some localities against apple maggot, codling moth, redbanded leafroller, apple rust mite, and oriental fruit moth. Carbaryl is also used as an apple fruit-thinning spray at 10 to 25 days past full bloom, especially on the Delicious varieties. Special precautions must be taken not to disrupt integrated control programs by killing predator mites and rust mites.

Organophosphorus compounds are utilized widely in orchard insect control. Parathion is effective against a greater variety of pests than any other material in use at the present time. It controls aphids, mites, scale insects, pear psylla, sucking bugs, leafrollers, codling moths, cutworms and caterpillars, plum curculios, oriental fruit moths, cherry fruit flies, peachtree borers, shothole borers, and treehoppers (except where resistant strains of the pests have appeared). A delayed dormant spray of superior-type oil plus parathion, ethion, carbophenothion, or diazinon is used widely against mites, scale insects, and pear psylla in the Pacific Northwest.

One of the most useful new organophosphorus compounds is azinphosmethyl. It rivals parathion in the number of effective orchard uses; however, it does not control aphids. A notable characteristic for a phosphorus material is its unusually long residual effect. Other organophosphorus compounds employed currently against tree fruit insects include malathion, diazinon, EPN, demeton, ethion, carbophenothion, phosphamidon, dimethoate, stirofos, phosmet, and phosalone.

The so-called "specific miticides"—dicofol, chlorobenzilate, chloropropylate, tetradifon, and ovex—are effective in controlling spider mites and rust mites. Each of these materials has limitations and is most useful when applied to certain kinds of fruit trees for control of certain species of mites. Nonorganochlorine specific miticides in use at the present time include dinocap, binapacryl, oxythioquinox, propargite, chlordimeform, and cyhexatin.

Fumigants provide effective control of various tree borers. Paradichlorobenzene is placed in a trench in the soil around the base of the tree to kill the peachtree borer. It is mixed with cottonseed oil and painted on the infested areas for control of lesser peachtree borers in the South. Ethylene dichloride or propylene dichloride emulsions are also used in soil trenches for control of the peachtree borer.

Control Equipment

Since 1948, there has been a rapid change from dilute to semiconcentrate and concentrate sprays for orchard pest control. The method consists of reducing the normal gallonage of spray material per tree and compensating by increasing the concentration of chemicals in the spray. Thus a "2X" concentration means one-half the normal gallonage applied per tree and double the amount of chemicals per 100 gal of spray mixture. By 1954, 80 per cent of the tree fruits in New York were being sprayed at some degree of concentration by air blast sprayers; 25 per cent of the tree fruit acreage in Michigan was being treated with spray concentrations averaging 150 to 300 gal/acre; 90 per cent in northern California was by air blast; 75 per cent of the orchards in eastern Washington were being treated at 200 to 300 gal/acre; and 90 per cent of the orchards in British Columbia were treated with concentrate sprays as low as 60 gal/acre. A survey conducted in 1969 indicated little change, with growers in most commercial orchard areas of the United States applying about 400 gal/acre, a few still applying 800 gal, and 25 per cent averaging about 100 gal. A large proportion of Nova Scotia and British Columbia orchardists are using 50–60 gal/acre, some as little as 20 gal. Air blast sprayers, conventional sprayers with blower attachments, and mist blower concentrate sprayers are the main types of equipment utilized in commercial tree fruit operations (Fig. 8:6). Hand gun application of dilute sprays (Fig. 14:9) is still the mainstay in Utah. Most cherry orchards in The Dalles and Willamette Valley areas of Oregon are treated by means of power dusters.

Figure 14:9. Grower spraying his small orchard with hand gun. *Courtesy Hudson Mfg. Co., Chicago, Ill.*

Aerial application of orchard sprays or dusts is desirable under certain conditions. Orchardists with limited acreages often would rather hire aerial applicators than buy expensive equipment. Growers who own ground equipment may turn to aerial applications near harvest time when the tree branches are bending down with fruit and might be damaged by a spray machine. Aerial sprays are also useful in irrigated orchards in early spring or late fall when water may not be readily available. Nozzles emitting a narrow droplet size spectrum through sintered-metal sleeves (units 20 and 40 μ in diameter have proved best to date) are providing greater penetration of orchard trees and increasing the use of aerial application.

The use of concentrate and semiconcentrate sprays has developed as a consequence of several factors. A major problem of the modern fruit grower is his labor costs. Even when he pays high wages, the orchardist is not sure of obtaining efficient and experienced help. One man with an air blast sprayer can treat an orchard about twice as fast as three men with a conventional high-pressure machine and hand guns.

Potent insecticides such as azinphosmethyl have replaced lead arsenate and oil for many orchard pest problems. Although an experienced man with a hand gun can put a more complete spray cover on the trees than can be accomplished with an air blast sprayer, maximum coverage and deposit of sprays are no longer required. It has also been shown that the use of semiconcentrate and concentrate methods of application results in a saving of up to 20 per cent in spray materials. Growers no longer need to apply sprays until there is "runoff' from the foliage to be sure of adequate coverage. However, careful attention must be paid to the pruning of trees to obtain best results with concentrate applications.

REPRESENTATIVE TREE FRUIT PESTS

For detailed study we have chosen the following serious pests of tree fruits: codling moth, spider mites, San Jose scale, and cherry fruit flies.

Codling Moth. *Laspeyresia pomonella* (Linnaeus) [Lepidoptera:Olethreutidae]

The familiar "apple worm" is an introduced pest from southeastern Europe. It was first observed in New England (1750), in Iowa (1860), in Utah (1870), and in Washington (1880). The increasing destructiveness of this pest closely paralleled the development of the apple-growing industry in the United States. It occurs wherever apples are grown in the world except Japan and Western Australia.

Codling moths feed upon apple, pear, English walnut, quince, crab apple, hawthorn, and wild apple, and they infrequently attack various stone fruits. California investigators have shown that strains of codling moth adapted to English walnut do not readily transfer back to apple. The larva often enters the fruit through the calyx end. Sometimes it will tunnel in where two fruits are touching or a leaf touches a fruit. The larva eats its way into the center of the apple and feeds upon the seeds and the core. Later, it tunnels back out and leaves the fruit. **Stings** are shallow blemishes on the surface of the fruit, usually caused when a newly hatched larva has taken a few bites and then died from the effects of an insecticide.

Damage by the codling moth reached sizable proportions by 1880—ranging from nearly total loss in many areas down to 10 to 20 per cent crop destruction in the extreme northern apple and pear producing regions. About this time, arsenical insecticides were found effective, and for many years thorough and timely spray applications largely prevented codling moth injury. Gradually control became more difficult and spray programs more intensive, until in the 1930s and extending until about 1947, losses from this pest often reached 30 to 50 per cent or more of the crop, despite the heavy spraying. Since 1947, losses caused by codling moth have been kept down to 1 to 5 per cent by use of DDT and newer types of insecticides. Yet the moth is present in all apple regions, and populations build up whenever a farmer is careless in his spray program.

Description. Female codling moths lay their eggs singly on the foliage or fruit. The egg is pearly white when first laid. It is oval in shape and flattened, resembling a minute drop of wax (Fig. 14:10). Newly hatched larvae are semitransparent white with a shiny black head and are about $\frac{1}{16}$ in. long. When the caterpillar is mature it is pinkish white with a brown head and about $\frac{3}{4}$ in. long. The full-grown larva spins a silken cocoon under bark or other suitable shelter.

Figure 14:10. Codling moth eggs. A, eggs laid on apple leaf; B, single egg greatly enlarged. *Courtesy USDA.*

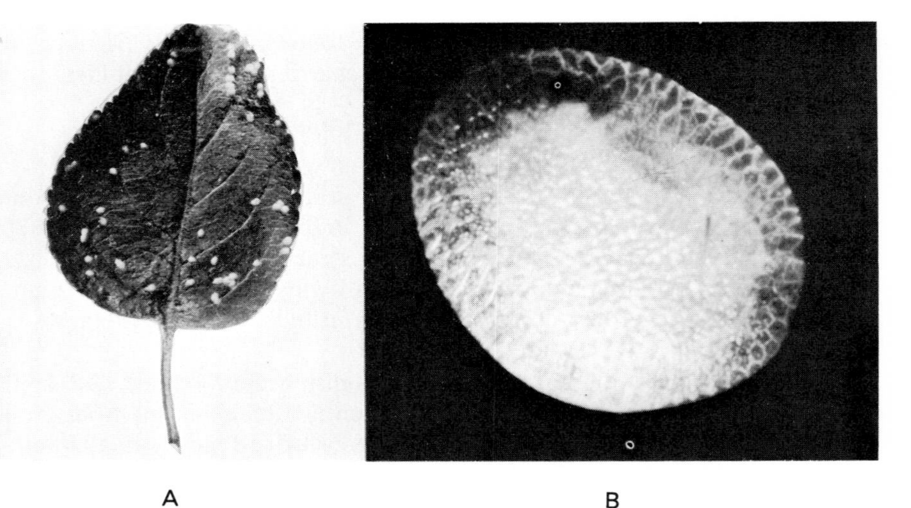

The pupa is about $\frac{1}{2}$ in. long and varies in color from yellow to brown, depending upon age. Adult moths are variable in size, with a wing span of $\frac{3}{4}$ in. or less. The wings are brownish gray with dark bands. Near the tip of each forewing is a dark brown color spot in which are two irregular coppery lines (Fig. 14:11).

Life history. Codling moths overwinter as mature larvae in waterproof cocoons under the bark or in the ground at the base of the tree. In the spring, usually in April or May, the overwintering larva changes into a light brown pupa. The pupal stage may last four to six weeks in the cool weather of early spring. When about to emerge, the pupa works its way toward the surface. After part of its length is out of the cocoon, the shell splits, and the adult moth emerges, leaving the brown pupal shell protruding from the exit hole. Male moths emerge first and often continue emerging for about a week before the females appear.

Moths of the spring brood reach a peak of activity during May or June in most localities (April in southern areas). The females oviposit during late

Figure 14:11. Codling moth. A, larva and pupa in cocoons under bark; B, adult resting on apple leaf. *Courtesy USDA.*

A B

afternoon or evening when the temperature is 60°F or higher. Temperatures during spring influence the extent of egg laying and the development of first-generation caterpillars and are important in determining codling moth activity during the season. However, larval injury depends mostly upon the initial infestation. The greater the population of overwintering larvae, other things being equal, the greater the damage.

Eggs of the first generation usually take 12 to 14 days to hatch because of cool weather. First-generation larvae enter the fruit over a period of five to six weeks. The caterpillars feed in the fruit about three weeks, then leave and search for a place to spin their cocoons. About one fourth of the caterpillars remain in the cocoons until the following spring. Most change to pupae within four to six days, and the pupal period lasts from ten days to two weeks. Larvae enter diapause when the photoperiod (day length) is less than 14.5 hr, and they begin to develop again when it is greater than 16 hr. First-generation moths appear by midsummer and may be present until cool weather in northern areas. The length of the first generation varies from 35 to 50 days in southern areas and 50 to 60 days in northern.

Eggs of the second generation hatch after six to seven days. Second-generation caterpillars appear in mid- to late summer, depending on locality, and attack the fruit for about 6 weeks. In northern states, the second-generation caterpillars leave the fruit and go into winter quarters during August and September or later. A small portion may pupate, emerge, and lay a few eggs. In southern states, the second-generation caterpillars leave the fruit in July and August and give rise to a complete third generation. A fourth and a partial fifth generation may occur in some areas.

Control. During the forty years prior to 1947, lead arsenate was the standard control measure for codling moths. By 1925 the pest was becoming difficult to kill with this insecticide. Strains of codling moth from Washington and Virginia were shown to be resistant to lead arsenate in 1928. Growers in certain regions began applying as many as ten cover sprays in a single season in an attempt to reduce the fruit injury. The apple industry was in serious jeopardy in some areas just before DDT became available for agricultural work in this country. Since 1947, two to four summer applications of DDT gave effective control of the codling moth in most orchard regions. However, resistance to DDT has been reported from all tree fruit areas. Since about 1960, DDT has been replaced largely by other materials, except for home orchard use, and in 1973 it was removed from the remaining agricultural recommendation lists. Substitute materials include methoxychlor, EPN, parathion, diazinon, carbaryl, azinphosmethyl, phosmet, phosalone, and stirofos. Those insecticides that have been most useful in integrated control programs are azinphosmethyl, diazinon, phosmet, phosalone, and ryania.

As in all insect control work, timing is of utmost importance. Codling moth sprays are applied most effectively just before newly hatched larvae attempt to enter the apples. Pheromone trapping to determine moth flight activity combined with daily weather records to determine egg hatch and most effective timing have been used in a computerized program throughout the state of Michigan since 1975.

Ascogaster quadridentata Wesm. is a braconid parasite that helps keep codling moth populations under control in eastern United States and Canada. It was introduced into the Pacific Northwest during the 1920s but apparently is not effective in that region.

Spider Mites.

European Red Mite. *Panonychus ulmi* (Koch)

Twospotted Spider Mite. *Tetranychus urticae* Koch

Schoene Spider Mite. *Tetranychus schoenei* McGregor

McDaniel Spider Mite. *Tetranychus mcdanieli* McGregor

Fourspotted Spider Mite. *Tetranychus canadensis* (McGregor)

Yellow Spider Mite. *Eotetranychus carpini borealis* (Ewing)

Brown Mite. *Bryobia rubriculus* (Sheuten)
[Acarina:Tetranychidae]

Since the development and use of DDT and other synthetic organic insecticides in the orchard, spider mites have become major pests. The fact that DDT kills natural predators of spider mites is the most obvious reason for this sudden change in extent of damage caused by them. However, investigators have recognized that the problem of increase of mite pests on crops has a multiple origin. It is known that indirect effects of DDT and other chemicals upon the mite or its host plant are at least partly responsible. Another causal factor involves certain changes made in the spray program. In the past, an oil was usually added to the lead arsenate sprays which helped to keep mites in check. Oils could not be combined with DDT because such mixtures caused plant injury.

Spider mites were observed damaging tree fruits in California as early as 1854. The European red mite (Fig. 14:12) was first recorded in Oregon in 1911 and was causing injury in the eastern United States and Canada by 1918. Twospotted spider mite injury to orchard trees was noted in New England during the early 1800s. The brown mite was also observed in the east during the 1800s.

Twelve species of spider mites are known to infest deciduous fruit trees in the United States and Canada. Only seven of the species are distributed widely or are of serious consequence at the present time. Several species are usually present in any given orchard, and different species usually predominate at different times throughout the growing season. Orchards in the same area will not necessarily contain the same kinds of spider mites.

The twospotted spider mite, the most common species, is found throughout the country on a wide variety of crops (Fig. 14:13). It is nearly cosmopolitan in distribution. It feeds upon tree fruits, small fruits, vegetables,

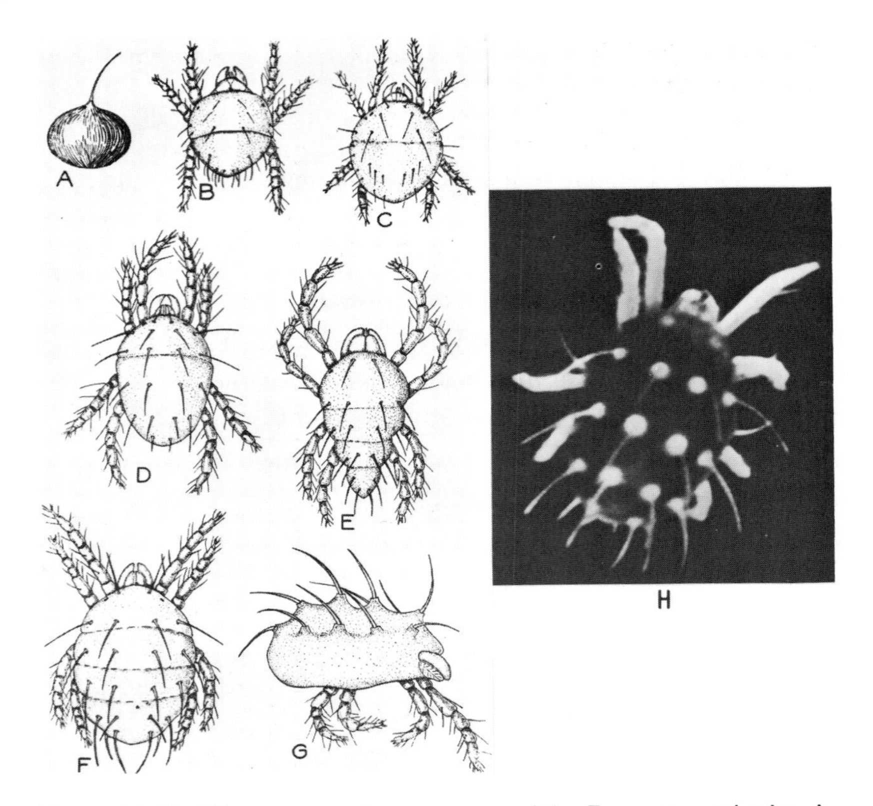

Figure 14:12. Life stages and appearance of the European red mite. A, egg; B, larva; C, protonymph; D, deutonymph; E, adult male; F and G, adult female top and side views; H, photograph of adult female. (*A–G, courtesy USDA; H, photograph by R. R. Kriner, Rutgers University, New Brunswick, N.J.*

forages, and florist and ornamental crops. It is a destructive pest of roses, beans, cotton, hops, strawberries, cucurbits, almonds, and apples.

European red mite (Fig. 14:12) is found throughout most of the United States and southern Canada, as well as in Europe, Russia, Australia, and Tasmania. It attacks all tree fruits as well as other trees, shrubs, and berries. The brown mite (Fig. 14:14), which was formerly thought to be a treefeeding race of the clover mite, is found in all parts of the United States and Canada and in South America, Europe, and Australia. It feeds upon all tree fruits, walnut, almond, and various woody ornamentals.

Schoene spider mites and fourspotted spider mites are found mainly in eastern orchards. They are pests of various crops in the southwestern United States. Schoene spider mites feed on cotton, beans, brambles, black locust and other shade trees, as well as on apple. Fourspotted spider mites occur on apple, plum, cotton, and a number of shade trees.

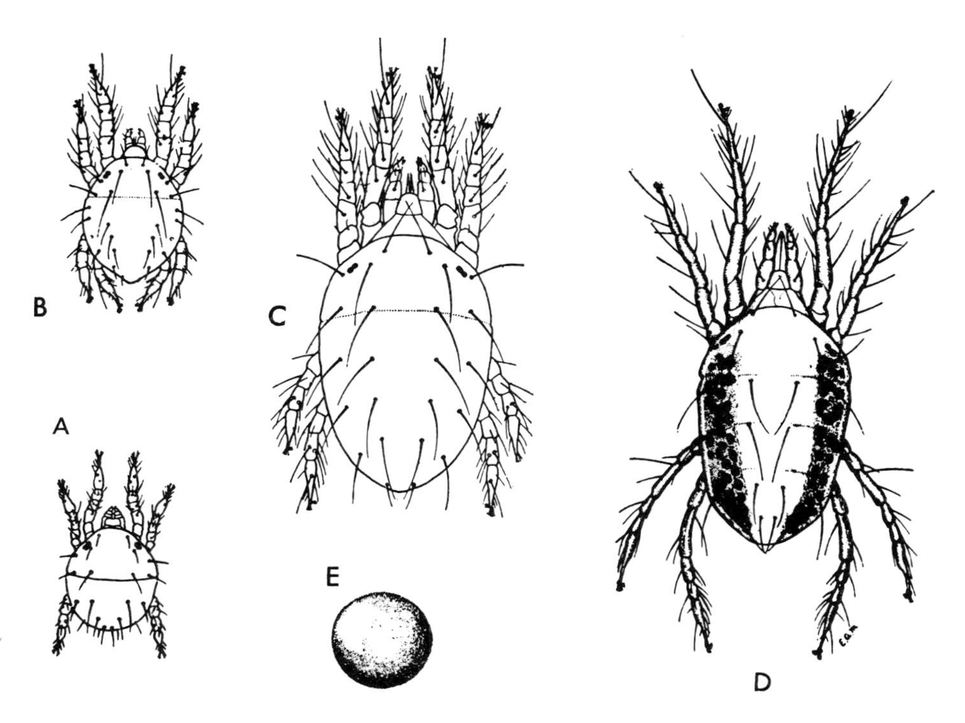

Figure 14:13. Life stages of the twospotted spider mite. A, larva; B, protonymph; C, deutonymph; D, adult female; E, egg. *Courtesy USDA.*

The McDaniel spider mite is a serious pest of tree fruits in the Pacific Northwest and has also been collected in Utah, California, Montana, North Dakota, Michigan and New York. This is the same mite which was formerly called Pacific mite[1] on tree fruits in Washington. McDaniel mites tend to be a major pest in arid area orchards, whereas twospotted spider mites are often found in humid area orchards. Yellow spider mite is damaging to apples, pears, cherries, several berries, and certain ornamentals from British Columbia to Central California. This species was previously known as the Willamette mite[1] in certain orchard areas.

Spider mites cause a general weakening of the trees. Studies have shown that 35 per cent or more of the chlorophyll in apple foliage may be removed by mite feeding. Initial damage to foliage appears as fine white speckling, which extends and turns brownish as the mites continue to feed. Photosynthesis and transpiration of the trees are depressed. In severe infestations, the leaves may become entirely brown and drop from the tree. Severe injury to pear foliage has sometimes been called transpiration burn or leaf scorch.

[1] Pacific mite and Willamette mite are now known to be damaging to tree fruits only in localized areas.

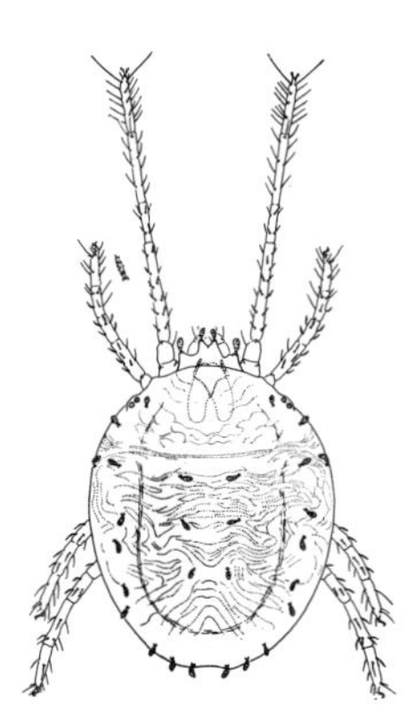

Figure 14:14. The brown mite, *Bryobia rubrioculus* (Sheuten). Note the long front legs. *Courtesy University of California, Berkeley.*

Major portions of pear leaves or entire leaves turn black under these conditions. Injury caused by the brown mite on stone fruits is exceptional in that browning and dropping of foliage does not occur. Damage caused in early summer may cause a light set of fruit the following season. Mite stimulation of late summer or fall regrowth of the tree may result in winter kill of the terminals. Mites may cause a reduction in the size of fruits and also a reduction in the number of fruits produced per tree. They may also cause off-color in fruits and fruit drop.

Description. Spider mites are quite small—the adults being barely visible to the naked eye. The species being discussed vary from about 0.25 to 0.5 mm in length. All of the spider mites go through five stages in their life cycle—egg, six-legged larva, eight-legged protonymph, deutonymph, and adult.

European red mite adults are brick red to reddish brown with prominent white spots at the bases of the dorsal spines (Fig. 14:12H). Nymphal stages are dull green to brown. A light green to straw color is typical of twospotted spider mite adults and nymphs. Actively feeding adults have a large dark spot on each side of the body near the middle. Schoene spider mites and fourspotted spider mites are similar in color, green with two black spots near the front of the body and two near the rear. Active female McDaniel spider mites are translucent green to greenish yellow in color with a triple spot on each side near the middle and another pair of spots near the rear (Fig. 14:15). The large markings of the midregion are quite variable, so the smaller pair of spots near the posterior end of the body are the best distinguishing feature. Yellow spider mites are light pink to pale yellow or greenish in color with two or three pairs of small dusky spots on the body in later life. Adult brown mites usually are reddish on the front portion of the body and greenish brown to dark brown on the rear portion. Spines of the brown mites are short, flattened and lance-shaped (Fig. 14:14). The front legs are unusually long.

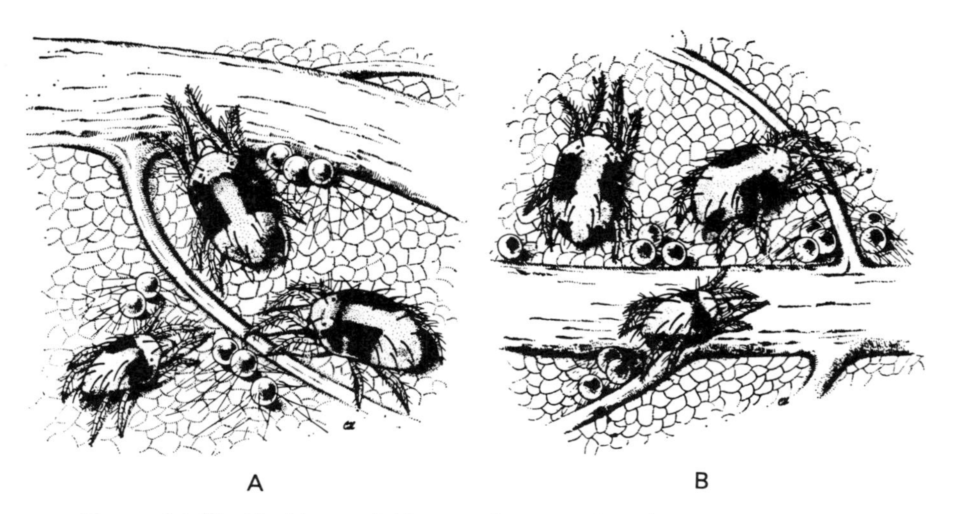

Figure 14:15. Markings of *Tetranychus* species. A, pattern typical of the actively feeding twospotted spider mite (females above right, male below left); B, pattern typical of actively feeding fourspotted spider mite, McDaniel spider mite, and Schoene spider mite (females above, male below.) *Courtesy University of California, Berkeley.*

Eggs of the European red mite are red and slightly flattened on top with a dorsal stipe. This species spins little or no webbing. The eggs of the brown mite are spherical, smooth, and red. Adult females have long front legs, and they spin no webbing. Yellow spider mites lay eggs that are spherical and clear with a fine dorsal stipe. They spin very little webbing. The twospotted, fourspotted, McDaniel, and Schoene spider mites lay spherical, clear to pearly white, smooth eggs; the first two species spin a moderate amount of webbing, and the other two spin a copious amount of webbing (Fig. 16:13).

Life history. Although most spider mites overwinter as bright orange adult females in plant debris on the ground, there are exceptions to this rule. European red mites and brown mites winter as bright red eggs on twigs and limbs. Yellow spider mites hibernate as bright yellow females under the bark. A high proportion of McDaniel spider mites survive the winter in cracks in the bark below the soil level.

Orchard mites usually develop six to eight generations per season, depending on the area. Schoene and twospotted spider mites have nine to ten generations in Virginia. Brown mites have from one and a partial second to five generations per season. European red mites have four generations in northern Europe, five in England and Nova Scotia, six in Ontario and Connecticut, six plus partial seventh and eighth in Quebec and Washington, and nine plus a partial tenth in Virginia. Studies show that McDaniel spider mites normally produce ten to eleven generations per season in Washington and seven to nine in northern Utah.

Hibernating females become active in early spring and migrate to the opening buds. Twospotted spider mites may winter under the bark, move to the cover crop to feed in early spring, and invade the trees again in early summer. McDaniel mites move up the trunk from the ground in early spring and become distributed through the center of the tree. Eggs of the European red mite and the brown spider mite hatch with the first warm weather of spring and the larvae swarm to the young leaves. European red mite tends to populate the peripheral foliage of the tree.

Larvae feed for two or three days, then remain quiescent for about the same length of time. Protonymphs emerge from the split skin of the resting stage. They feed for two to four days; then they pass a quiescent stage of about equal length. Deutonymphs emerge, feed for two to six days, and remain quiescent for an equal number of days. Adults emerge and mate, or else the females reproduce parthenogenetically in which case only males are produced from the eggs. The complete life cycle is from one to three weeks in length, depending upon locality, seasonal variations, and species.

The average number of eggs laid per female varies from 25 for the European red mite to 150 for the McDaniel spider mite. Length of life of adult females is two to three weeks for the brown mite and up to nine weeks for the twospotted spider mite. Apparently the brown mite is entirely parthenogenetic; no males have ever been collected.

Long photoperiods (16 hr) cause first-generation European red mites to become "summer females" that lay eggs destined to hatch in five to ten days. Shorter photoperiods (14 hr) cause late summer generations to become "winter females" that lay mostly diapausing eggs, requiring about 150 days of chilling before hatching.

Brown mites lay their winter eggs early in hot, dry seasons, and the active stages may be greatly reduced by midsummer. European red mite or McDaniel spider mite populations may replace brown mites, building up peaks of infestation by mid- to late summer. Yellow spider mites and twospotted spider mites tend to develop heaviest populations in late summer or fall.

Several types of predators are effective in reducing spider mite populations. Small black lady beetles of the genus *Stethorus* have been cited as the most important natural enemies in several regions. *S. punctum* is the common species in eastern orchards, whereas *S. picipes* is the western species. Phytoseiid mites of the genera *Metaseiulus*, *Amblyseius*, and *Neoseiulus* have been particularly effective against spider mites in orchards where limited spray schedules are used. *Stethorus picipes* does not reproduce in orchards when spider mite densities are low, and it is often killed by integrated program insecticides. Other predators include mirid bugs (*Deraeocoris*, *Campylomma*, and other genera), pirate bugs (*Orius* and *Anthocoris*), and thrips (*Haplothrips*). The stigmaeid mite, *Zetzellia mali*, is an effective predator of spider mites in some localities.

Control. Delayed dormant sprays of lime-sulfur and oil are used to help prevent the development of brown mite and European red mite populations. Insecticides such as parathion, malathion, EPN, TEPP, schradan, and demeton were once effective against orchard mites. However, spider mite strains highly resistant to organophosphorus materials have arisen. Parathion, carbophenothion, diazinon, and ethion are still used in combination with superior-type spray oils as delayed dormant treatments against European red mite and brown mite.

Mites have also become resistant to many of the organochlorine specific acaricides such as Sulphenone, Genite 923, chlorbenside, fenson, and chlorfenethol. Those miticides that are still effective for certain situations include dicofol, ovex, chlorobenzilate, chloropropylate, and tetradifon. Dinocap, binapacryl, oxythioquinox, and cyhexatin are also in current use. All miticides cause phytotoxicity problems associated with certain tree fruit varieties, weather conditions, or spray mixtures. For example, propargite, a sulfite material, cannot be used on pears because of specific injury. Chlordimeform EC in early postbloom sprays is likely to russet pear fruits under slow-drying conditions.

Two sprays, seven to ten days apart, are often applied in mite control. If a prebloom spray of an organophosphorus material plus oil is applied, often at least one summer spray of dicofol, dinocap, binapacryl, ethion, chlorobenzilate, chloropropylate, tetradifon, or cyhexatin will also be required.

Obviously, **integrated control** is the most promising current system for managing the complex spider mite problems on tree fruits. We shall outline the program developed by S. C. Hoyt for use on apples in central Washington as an example of this development. Selective insecticides, selective dosages, selective timing, and altered application techniques are utilized to integrate chemical control of insect pests with natural control of mites. The four principal requirements for practical operation are (1) presence of the predaceous mite *Metaseiulus occidentalis* in the orchard, (2) knowledge of the appearance and habits of plant-feeding and predaceous mites, (3) careful examination of relative numbers of predators and plant-feeding mites, particularly during a period when rapid population changes are occurring, and (4) knowledge of what materials to use, how to use them, and what materials to avoid, in order to conserve predators.

Because European red mite is not controlled by the predator mite during early season, an organophosphorus material plus a superior-type oil are applied during the delayed dormant period. This treatment is a key feature of the integrated program because summer applications for European red mite control would eliminate the apple rust mite and lead to predator mite starvation. High populations of predator mites can develop when their alternate host rust mites are maintained at nondamaging levels. Reduced dosage of azinphosmethyl is applied no later than 21 days after full bloom to control codling moth. The next application is 21 days later, and diazinon, phosmet, or phosalone may be substituted for various reasons. If miticides are required during the summer, cyhexatin and propargite are the preferred materials. They are particularly useful because they improve the balance between predators and plant-feeding mites. Materials that should be avoided include chloropropylate, chlorobenzilate, dicofol, binapacryl, and stirofos. Certain modifications and additions are made to provide control of other insect pests, to allow fruit thinning with carbaryl and to adjust for individual orchard conditions.

Supporting investigations for this program were begun in 1961, when it was found that strains of *Metaseiulus occidentalis* resistant to codling moth control chemicals effectively controlled McDaniel mites in an orchard. By 1966, the integrated program was in commercial use on 9000 acres of apples. It had expanded to 40,000 acres in 1967, and 56,000 acres in 1969. Only about 19,000 acres, however, are being managed correctly for optimum results. The greatest need is for highly competent consulting entomologists who will prepare individual schedules for orchardists.

The benefits of this program include effective management of the insect and mite pests of apple, reduced problems with undesirable side effects of chemicals, increased populations of beneficial predators and parasites that help to control aphids, and reduced operating costs for insecticidal materials (10 to 60 per cent of former expenditures).

San Jose Scale. *Quadraspidiotus perniciosus* (Comstock) [Homoptera: Diaspididae]

San Jose scale, potentially the most destructive orchard pest, is an introduced species, having been brought into California about 1870. By 1873 it had become a serious pest in the San Jose Valley. A "classic" early investigation was the trip by C. L. Marlatt to the Orient in 1901–02 to find the native home of the scale. He found it had originated in the area north and west of Peking on flowering Chinese peach. Meanwhile the insect had been transported inadvertently to New Jersey in 1886 or 1887 with a shipment of Kelsey plums. The two nurseries that received this material shipped nursery stock to various parts of the country, and by 1895 San Jose scale had spread to most of the eastern states. It reached Oregon and Washington by 1882 and Nova Scotia by 1912.

Tremendous damage occurred before control measures were perfected and established. Besides rendering fruit unmarketable, the scales killed twigs and limbs and eventually killed the trees. More than 1000 acres of mature apple trees were killed in southern Illinois in 1922. The insects suck the plant juices from twigs and larger branches and from fruit and foliage. Infested trees show a general decrease in vigor, and terminal twigs usually die (Fig. 14:16). Leaves characteristically remain on infested trees through the winter. This is especially typical of sweet cherry.

San Jose scale attacks all orchard fruits, most small fruits, and many kinds of shade trees and ornamental shrubs. It is most destructive to apple, pear, and sweet cherry; rarely is it found on sour cherry, pecan, or walnut. Records indicate that it is distributed throughout the commercial fruit-growing

Figure 14:16. Peach tree injured by San Jose scale. Several branches have been killed. *Courtesy USDA.*

sections of the United States and Canada. It is also reported from the Orient, Australia, New Zealand, Russia, Germany, Austria, Africa, the Mediterranean Region, and South America.

Description. Female San Jose scales are nearly round and about 2 mm in diameter (Fig. 14:17). The waxy shell is gray with a raised central nipple. The yellowish, sac-like body of the female with its long thread-like sucking mouthparts remains protected under this cover of fibers and wax. Male scales are oval, about 1 mm long, and have a raised dot near the larger end of the shell. When mature the male emerges from the scale as a delicate, yellow, two-winged insect. **Crawlers** or young are borne alive under the female scale. These are very small, yellow nymphs with six well-developed legs and a pair of antennae. Individual scales usually become surrounded by a small reddish spot on the fruit or on the bark of young twigs. On heavily infested trees, the entire surface of the bark is covered with a gray layer of overlapping scales. The twigs appear as if they had been sprinkled with wood ashes when wet. Populations of 150 per sq cm are common on the bark, but they rarely exceed 10 on the fruit or 13 on the leaves.

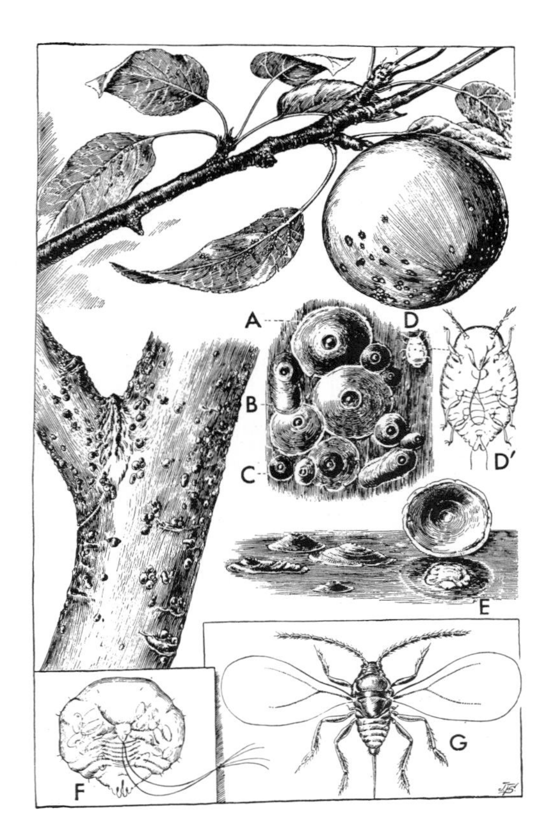

Figure 14:17. San Jose scale. A, adult female scale; B, male scale; C, young scales; D, nymph just hatched; D′, same, much enlarged; E, scale raised, showing body of female beneath; F, body of female much enlarged; G, adult male much enlarged. *Courtesy USDA.*

Life history. Partly grown scales pass the winter upon the tree. Up to 80 per cent of the winter forms may be first nymphal instars. Temperatures of – 15° to – 20°F kill about two thirds of the half-grown scales, and – 25° to – 30°F temperatures kill about 90 per cent or more. In spring, the scales continue their growth and become mature about the time the fruit trees bloom or a little later.

The active males issue from their scales and mate with the females which remain under scales throughout their lives. After being fertilized, females begin producing living young at a rate of nine or ten per day. They reproduce for six weeks or more, each scale bearing from 150 to 500 crawlers.

Young scales may crawl considerable distances during the first few hours of their lives. Often they are carried to other trees by wind, on the feet of birds, on the clothing of orchard workers, or on implements. The first waxy secretion is produced on the dorsal surface of crawlers. Ordinarily they settle down within a few hours on bark or on the leaves or fruit, insert their long, threadlike beaks into plant tissue, and begin feeding.

About three weeks afterwards, they molt or shed their skins and with the old skin lose their legs and antennae. The scales become mere flattened yellow sacks with waxy caps, attached to bark by their sucking mouthparts. As the insect grows, woolly fibers are secreted from the body that mix with a waxy material to continue the formation of the shell. Portions of the shed skins are incorporated into the scales following molts. The female develops through two nymphal instars to the adult; the male develops through four, the last two being called the "propupa" and "pupa."

Two or more generations per season develop in various parts of the country—two in the Pacific Northwest and in the Northeast, three in the Sacramento Valley of California. Overlapping of broods is caused by the long reproduction period of each female and results in all stages being present on the trees throughout the growing season. During the summer each generation is completed in five to seven weeks, depending upon the locality and weather conditions.

Prospaltella perniciosi is a minute wasp parasite of the San Jose scale in the eastern states and as far west as Kansas. Parasitism up to 90 per cent has been reported. Approximately 200,000 *Prospaltella* adults were released in five California counties during 1943 and 1944. The parasite became established and is helping to control the scale in the area. This species, together with the cosmopolitan *Aphytis proclia*, has apparently reduced the scale from a major pest to a minor one in New England.

The twice-stabbed lady beetle is a common scale predator throughout the country but appears to be most effective on the Pacific Coast. *Microweisea misella* is another small lady beetle that preys upon scale insects in both eastern and western orchards.

Control. In North America lime-sulfur was developed primarily as a San Jose scale treatment and was used first in California in 1880. This material is still used as a scalecide, especially as a delayed dormant spray in the Pacific Northwest.

Spray oils are used against the scale throughout the country. A combination spray containing lime-sulphur plus oil is utilized in certain areas of the West. Special grades of oil known as "superior type" are used throughout the country and in combination with ethion, carbophenothion, parathion, or diazinon in the Pacific Northwest. Malathion, azinphosmethyl, diazinon, and parathion are also effective controls for the San Jose scale. A prepink organophosphorus spray has replaced the use of delayed dormant lime-sulfur and oil in some regions. For a number of years, DDT was used as a treatment for scale crawlers, but this recommendation has been discontinued.

Cherry Fruit Fly *Rhagoletis cingulata* (Loew)

Black Cherry Fruit Fly *Rhagoletis fausta* (Osten Sacken)

Western Cherry Fruit Fly *Rhagoletis indifferens* Curran

[Diptera:Tephritidae]

The cherry fruit fly and the black cherry fruit fly are native North American insects. Until recently it was thought that the cherry fruit fly, an eastern species, had been introduced into the West. Investigators in California, however, have shown that the species in the West is indigenous and distinct, the native host being wild bitter cherry, *Prunus emarginata*. The native host of the eastern species—the cherry fruit fly—is wild black cherry, *P. serotina*.

Cherry fruit fly was reported from cultivated cherries in eastern United States and Canada prior to 1900 before its wild host plants were discovered. Western cherry fruit fly was first noted in cultivated cherries in Oregon in 1913 and in Washington in 1916. Although it was present in wild cherries, it did not become an economic pest of cultivated cherries until after 1940 in certain commercial cherry-growing areas of Washington. This unusual lag in adjustment to the domestic host was also noted in California. The maggot was found first in cultivated cherries in 1962 and still is not a pest in the major commerical districts. The fly was discovered in cultivated cherries in the Okanogan Valley of British Columbia in 1968, and control treatments were first applied in 1969.

The black cherry fruit fly was first recorded in British Columbia and later in eastern and midwestern United States. This species is found in all cherry-producing areas but usually is much rarer than the cherry fruit fly or western cherry fruit fly when two species are associated. However, it was the only fruit fly attacking cherries in the Flathead Valley of Montana until 1958. Pin cherry, *Prunus pennsylvanica*, is its primary native host in the East. The insect has been found in bitter cherry, *P. emarginata*, in the West.

Cherry fruit flies mainly infest cultivated and wild cherries. The black cherry fruit fly prefers sour cherries to sweet cherries. Fruit fly maggots burrow to the center of the cherries and feed around the pits. Cherries become somewhat misshapen and undersized. Usually one side of the fruit will become partially decayed and shrunken and closely attached to the pit. Exit holes are sometimes cut in the cherries by the maggots before harvest. Maggots are a contaminant when present in processed fruit.

Description. The cherry fruit fly is about two thirds the size of a house fly, blackish in color, with yellowish head and legs and light green compound eyes (Fig. 14:19). Conspicuous dark bands extend transversely across the wings. The abdomen of the black cherry fruit fly is entirely black, whereas that of the other two species is marked with white bands—four in the female and three in the male.

The egg is white and slightly less than 1 mm in length. The maggot is a dirty white color and about $\frac{1}{4}$ in. long when full grown (Fig. 14:18). The

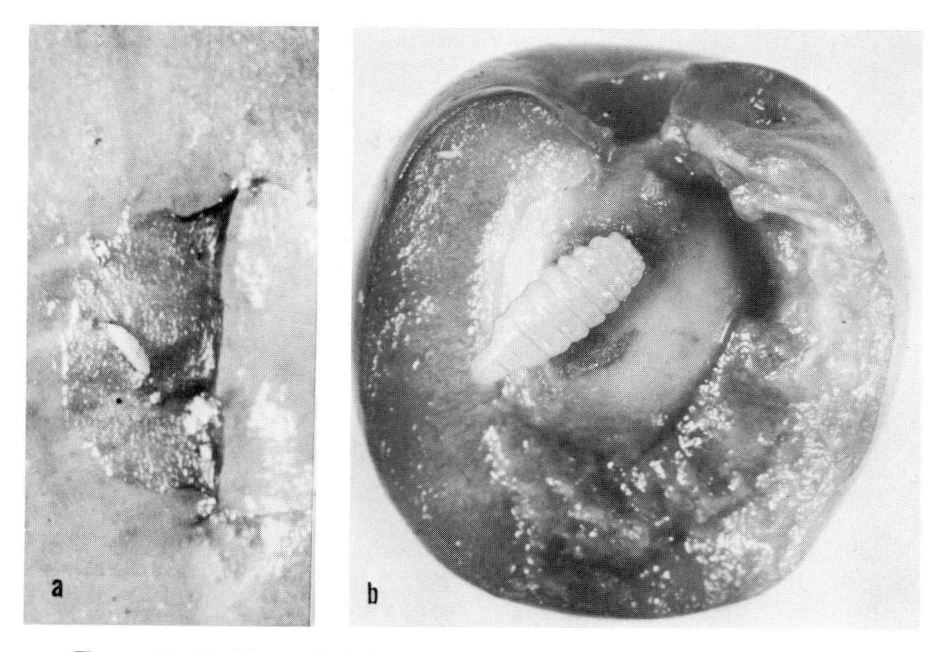

Figure 14:18. Cherry fruit fly. A, egg exposed by removal of flap of skin of cherry; B, full-grown larva. *Courtesy Pennsylvania State University, University Park.*

puparium is about $\frac{1}{7}$ in. long and resembles a grain of wheat. Cherry fruit fly and western cherry fruit fly puparia are reddish brown, whereas those of the black cherry fruit fly are straw colored.

Life history. Cherry fruit flies overwinter as pupae within puparia in the soil. Adult flies begin emerging from the soil in the spring about one month after sweet cherries are in full bloom. Emergence usually begins toward the end of May in the Pacific Northwest and during the first half of June in the Northeast and Midwest.

Depending upon weather conditions, females begin egg laying one week to ten days after emergence. Eggs are usually laid during any daylight hour when it is warm and sunny. The female makes feeding punctures near the bottom of the fruit with her ovipositor and feeds upon the juices that exude. Usually eggs are not deposited in heavily punctured cherries but in nearby fruits. They are inserted just under the skin.

Fruit fly eggs hatch in five to seven days, and the small maggots begin feeding on the pulp of the cherry. Several maggots can develop in a single cherry. It is not uncommon to find five or six mature maggots in individual cherries from heavily infested orchards. The maggots feed for two to three weeks in the fruit. When they are full grown they leave the fruit, drop to the ground, and enter the soil to a depth of 1 to 3 in.

Figure 14:19. Cherry fruit fly. A, pupa; B, adult. *Courtesy Pennsylvania State University, University Park.*

There is one generation per season, except for about 1 per cent of the maggots which pupate and emerge the same season, according to investigations in the Pacific Northwest. The few flies emerging in late summer or fall are lost because there is no suitable host present. Most of the pupae remain in the soil until the following spring.

Infestation of cherries varies considerably with weather conditions in spring and early summer. If cool weather retards the emergence of flies and fly activity after emergence, cherries may escape infestation to a large degree because either too few flies have emerged by harvest time or the female preoviposition period has been extended. Sometimes the season is late, so that cherry blooming, fly emergence, and cherry harvest are all delayed. Then, if high temperatures occur during the last two weeks before harvest, a high potential infestation of the fruit will result.

A number of parasites of the cherry fruit flies have been recorded from both eastern and western regions. Braconid parasites of the genus *Opius* are most common in maggots from wild cherries, while an ichneumonid has been reared from maggots in cultivated cherries. Parasitism only averaged about 2 per cent in an untreated cherry orchard in Washington.

Control. Lead arsenate spray or dust was used widely prior to about 1950; it is quite ineffective against fruit flies, however, and has been replaced by newer and better materials. Methoxychlor is used in two to three applications at ten-day intervals. This material provides fair control of the cherry fruit fly. Parathion and diazinon are the most effective insecticides currently recommended for control of the fruit flies; they also kill the maggots within the fruit. The ideal insecticide for short-season fresh fruit production would (1) leave no visible residues on the cherries, (2) provide a residual action of 10 to 14 days, (3) be safe to mammals, and (4) kill San Jose scale crawlers, spider mites, black cherry aphids, and other pests, as well as cherry fruit flies.

Diazinon has been the best material available for this use for a number of years.

Perthane is the best organochlorine material for cherry fruit fly control. Although it is not as effective as either parathion or diazinon, it presents an extremely low hazard to warm-blooded animals and can be used during harvest. Malathion is recommended especially for use in orchards adjacent to town because of its low toxicity to animals. Aerial application of malathion ULV through a mist dispenser at one pint per acre has provided good control in some areas. Parathion is sometimes applied in postharvest sprays to help reduce the fruit fly population for the following season.

SELECTED REFERENCES

Asquith, D., and R. Colburn, "Integrated Pest Management in Pennsylvania Apple Orchards," *Bull. Entomol. Soc. Am.* 17:89–91, 1971.

Barnes, M. M., and H. F. Madsen, *Insect and Mite Pests of Apple in California*, Calif. Agr. Exp. Sta. Circ. 502, 1961.

Batiste, W. C., "Integrated Control of Codling Moth on Pears in California: A Practical Consideration Where Moth Activity is Under Surveillance, *Env. Entomol.* 1:213–18, 1972.

Boulanger, L. W., *The Effect of European Red Mite Feeding Injury*, Me. Agr. Exp. Sta. Bul. 570, 1958.

Cagle, L. R., *Life History of the Two-spotted Spider Mite*, Va. Agr. Exp. Sta. Tech. Bul. 113, 1949.

———, *Life History of the Spider Mite* Tetranychus schoenei *McG.*, Va. Agr. Exp. Sta. Tech. Bul. 87, 1943.

Chapman, P. J., *Petroleum Oils for the Control of Orchard Pests*, N.Y. Agr. Exp. Sta. Bul. 814, 1967.

Cox, J. A., *The Cherry Fruit Fly in Erie County*, Pa. Agr. Exp. Sta. Bul. 548, 1952.

Croft, B. A., *Integrated Control of Apple Mites*, Mich. Coop. Ext. Serv. Bull. E-825, 1975.

Cutright, C. R., *The European Red Mite in Ohio*, Ohio Agr. Exp. Sta. Res. Bul. 953, 1963.

———, *Insect and Mite Pests of Ohio Apples*, Ohio Agr. Exp. Sta. Res. Bul. 930, 1963.

———, *Codling Moth Biology and Control Investigations*, Ohio Agr. Exp. Sta. Bul. 583, 1937.

Dibble, J. E., H. F. Madsen, G. R. Post, and A. H. Retan, *Concentrate Spraying in Deciduous Fruit Orchards*, Calif. Ag. Ext. Serv. Pub. AXT-131, 1966.

Eyer, J. R., *Ten Years' Experiments with Codling Moth Bait Traps, Light Traps, and Trap Bands*, N.M. Agr. Exp. Sta. (Tech.) Bul. 253, 1937.

Frick, K. E., H. G. Simkover, and H. S. Telford, *Bionomics of the Cherry Fruit Flies in Eastern Washington*, Wash. Agr. Exp. Sta. Tech. Bul 13, 1954.

Garman, P., and J. F. Townsend, *Control of Apple Insects*, Conn. Agr. Exp. Sta. Bul. 552, 1952.

———, *The European Red Mite and Its Control*, Conn. Agr. Exp. Sta. Bul. 418, 1938.

Gentile, A. G., and F. M. Summers, "The Biology of San Jose Scale on Peaches with Special Reference to the Behavior of Males and Juveniles," *Hilgardia*, 27(10):269–85 (1958).

Hoyt, S. C., "Integrated Chemical Control of Insects and Biological Control of Mites on Apples in Washington," *J. Econ. Entomol.*, 62:74–86 (1969).

Hoyt, S. C., and E. C. Burts, "Integrated Control of Fruit Pests," *Ann. Rev. Entomol.* 19:213–52, 1974.
Lathrop, F. H., *Apple Insects in Maine*, Me. Agr. Exp. Sta. Bul. 540, 1955.
Madsen, H. F., and C. V. G. Morgan, "Pome Fruit Pests and their Control," *Ann. Rev. Entomol.*, 15:295–320, 1970.
Madsen, H. F., and L. B. McNelly, *Important Pests of Apricots*, Calif. Agr. Exp. Sta. Bul. 783, 1961.
Madsen, H. F., R. L. Sisson, and R. S. Bethell, *The Pear Psylla in California*, Calif. Agr. Exp. Sta. Circ. 510, 1962.
Madsen, H. F., and M. M. Barnes, *Pests of Pear in California*, Calif. Agr. Exp. Sta. Circ. 478, 1959.
Marlatt, C. L., *The San Jose or Chinese Scale*, USDA Bur. Ent. Bul. 62, 1906.
Marshall, J., *Concentrate Spraying in Deciduous Orchards*, Canada Dept. Agr. Pub. 1020, 1958.
Morgan, N. H., and C. V. G. Anderson, "Life Histories and Habits of the Clover Mite, *Bryobia praetiosa* Koch, and the Brown Mite, *B. arborea* M. & A., in British Columbia (Acarina: Tetranychidae)," *Can. Ent.*, 90:23–42 (1958).
Newcomer, E. J., *Insect Pests of Deciduous Fruits in the West*, USDA Handbook 306, 1966.
Newcomer, E. J., and M. A. Yothers, *Biology of the European Red Mite in the Pacific Northwest*, USDA Tech. Bul. 89, 1929.
Oatman, E. R., *Wisconsin Apple Insects*, Wisc. Agr Exp. Sta. Bul. 548, 1960.
Oatman, E. R., and C. G. Ehlers, *Cherry Insects and Diseases in Wisconsin*, Wisc. Agr. Exp. Sta. Bul. 555, 1962.
Pritchard, A. E., and E. W. Baker, "A Guide to the Spider Mites of Deciduous Fruit Trees," *Hilgardia*, 21:253–86, 1952.
Quist, J. A., *Approaches to Orchard Insect Control*, Colo. Agr. Exp. Sta. Bul. 5175, 1966.
Webster, R. L., *A Ten-Year Study of Codling Moth Activity*, Wash. Agr. Exp. Sta. Bul. 340, 1936.
Westigard, P. H., "Pest Status of Insects and Mites on Pear in Southern Oregon," *J. Econ. Entomol.* 66:227–32, 1973.

chapter 15 / **CARL JOHANSEN**

INSECT PESTS OF SMALL FRUITS

Small fruits are represented by a variety of crops—raspberries, loganberries, blackberries, and other cane berries, strawberries, gooseberries, currants, grapes, blueberries, and cranberries. These plants are grown under widely differing conditions and are attacked by a correspondingly diverse group of insects and mites.

THE PESTS

The greatest number of small fruit pests belong to three orders: Coleoptera, Lepidoptera, and Homoptera. A smaller number belong to the Hemiptera, Thysanoptera, Hymenoptera, Diptera, and Acarina.

Coleoptera

Curculionidae. Weevils are one of the most common types of small fruit pests. Larvae of strawberry root weevils, *Otiorhynchus* spp., feed on the roots of strawberries, raspberries, and other small fruits in all parts of the United States and southern Canada (Figs. 15:8 and 15:9). Root weevils are especially damaging in the Pacific Northwest. Black vine weevil under certain conditions becomes a major pest of grapes. Resulting from changes in control programs, the woods weevil (*Nemocestes*) and obscure root weevil (*Sciopithes*) have become major pests of blueberries, strawberries, cane berries, and certain ornamentals on the Pacific Coast.

The strawberry crown borer, *Tyloderma* (Fig. 15:1), attacks strawberries in the eastern states, especially the Southeast. A closely related species damages strawberries in the Northwest. The strawberry weevil, *Anthonomus*,

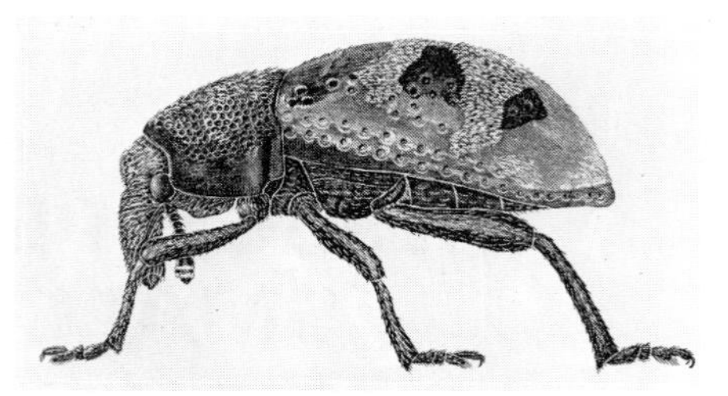

Figure 15:1. Adult of the strawberry crown borer. *Courtesy Ohio Agr. Res. Dev. Center, Wooster.*

is a small native species that occurs in the eastern United States and Canada and in Texas. It attacks strawberries, raspberries, blackberries, and other *Rubus*. Cranberry weevil, a related species, feeds on blueberries and cranberries in the Northeast. Whitefringed beetles damage strawberries in the Southeast. Plum curculio (Fig. 14:4) is a pest of blueberries in the East. Grape curculio is another pest found in the East. Fuller rose beetle is a weevil that attacks small fruits in California.

Byturidae. Eastern raspberry fruitworm and western raspberry fruitworm (Fig. 15:12) are small beetles whose larvae feed upon the receptacles and fruits of raspberries and loganberries.

Chrysomelidae. The grape flea beetle is a pest in the East. Flea beetles of other species attack strawberries throughout the country. The strawberry rootworm, *Paria*, is a small chrysomelid beetle that injures both strawberries and brambles, especially in the East and the Southwest. Cranberry rootworm, a similar species, attacks both blueberries and cranberries. Both eastern and western species of grape rootworm feed on the roots of grape. Grape colaspis is damaging to both grapes and strawberries in the Northeast.

Buprestidae. Rednecked cane borer damages cane berries in the East.

Scarabaeidae. The Japanese beetle attacks blueberries, cane berries, and grapes in the East. The rose chafer is a related species attacking cane berries and grapes. White grubs are general feeders on the crowns and roots of small fruits.

Lepidoptera

Sesiidae. We find that a good number of small fruit pests belong to the order Lepidoptera. Clearwing moth larvae feed upon all types of berries and grapes. For example, larvae of the strawberry crown moth attack blackberries and raspberries as well as the preferred host; the raspberry crown borer damages all cane berries; the currant borer damages both currants and gooseberries; the grape root borer attacks both wild and cultivated grapes.

Tortricidae. The redbanded leafroller (Fig. 14:1) feeds on grapes and berries in the East. The obliquebanded leafroller and the orange tortrix attack cane berries, blueberries, and strawberries in the West. Obliquebanded

Figure 15:2. Larva of the grape berry moth feeding inside a grape. *Courtesy Michigan State University, East Lansing.*

leafrollers are also found in several eastern states now. The garden tortrix and several other species are particularly damaging to strawberries in California. The blueberry leafroller is a common pest of strawberries, blueberries, and cranberries in mideastern states. The yellowheaded fireworm is a serious pest of cranberries.

Olethreutidae. The so-called strawberry leafroller folds over a portion of a leaf and ties it to the main portion flatly; it might be better named "leaf folder." This insect is a common pest in all parts of the country. In the Great Lakes region the grape berry moth is a serious pest, as the larvae (Fig. 15:2) destroy the blossoms and fruit of grapes. The blueberry tip borer is a small caterpillar chiefly injurious in the Northeast. In some localities the cherry fruitworm is a pest of blueberries as well as cherries. The blackheaded fireworm is a major pest of cranberries wherever they are grown in North America.

Gelechiidae. The western strawberry leafroller is injurious to strawberries on the Pacific Coast. The blueberry leaftier is a pest of blueberries in the Southeast. The strawberry crownminer, a pest of strawberries, causes similar injury to that of the crown moth.

Pyralidae. The cranberry fruitworm attacks both cranberries and blueberries wherever these crops are grown. The gooseberry fruitworm attacks gooseberries and currants throughout the northern parts of the United States.

Tischeriidae. In certain seasons, the strawberry leafminer becomes numerous enough in California to cause significant leaf injury.

Homoptera

Cicadellidae. Numerous species of leafhoppers infest small fruits, injuring crops directly by their feeding and indirectly by transmitting plant diseases. Particularly important species are the bluntnosed cranberry leafhopper, bramble leafhopper, grape leafhopper (Fig. 15:14), rose leafhopper, threebanded leafhopper (Fig. 15:14), variegated leafhopper, Virginia creeper leafhopper, western grape leafhopper, and sharpnosed leafhopper. The brambleberry leafhopper is a European species first detected in British Columbia in 1952. It was collected in western Washington in 1968, where it is spreading on both domestic and wild *Rubus*. This species is the vector of *Rubus* stunt virus in Europe.

Aphididae. Aphids effect injury by their feeding and by transmission of plant diseases. Widely distributed and included in this group are the currant aphid, grapevine aphid, raspberry aphid, strawberry aphid, and strawberry root aphid.

Aleyrodidae. Several kinds of whiteflies, including the strawberry whitefly and the grape whitefly, are pests of small fruits.

Phylloxeridae. The grape phylloxera is one of the historical pests of small fruits (Fig. 15:3).

Coccidae. Important soft scales attacking small fruits are the cottony maple scale, the European fruit lecanium (Fig. 14:2), and the wax scale.

Figure 15:3. The grape phylloxera. A, galls on grape root inhabited by root generations of insect; B, female nymph; C, wingless female laying eggs on roots; D, winged female (migrant form). *After Riley.*

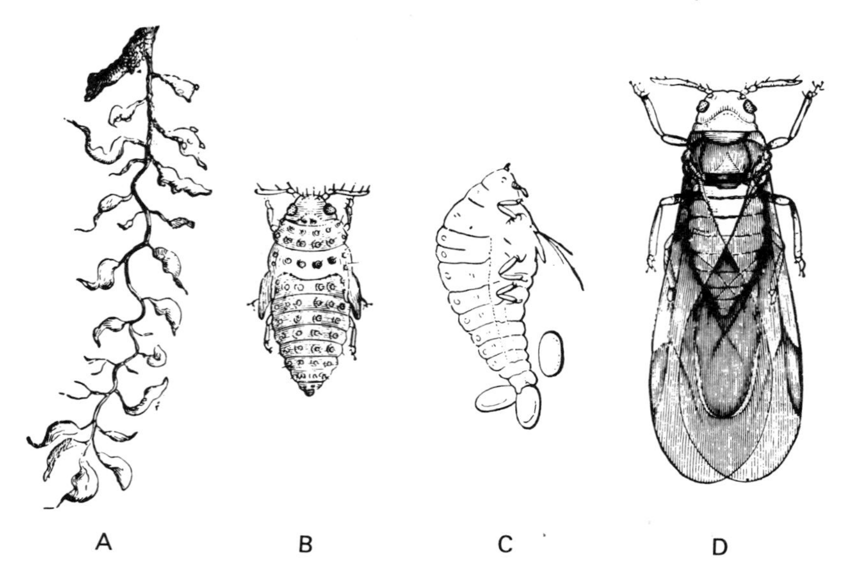

Diaspididae. Important armored scales are the rose scale (Fig. 17:16C) and the San Jose scale (Fig. 14:17).

Pseudococcidae. The grape mealybug sucks juices from canes, stems, and berries of grape vines and secretes honeydew in which sooty mold develops.

Cercopidae. The meadow spittlebug, a serious pest of alfalfa and red clover, attacks all kinds of small fruits.

Hemiptera

Miridae. The tarnished plant bug and several kinds of lygus bugs are pests of small fruits, particularly strawberry and raspberry.

Lygaeidae. The false chinch bug and the strawberry bug attack a variety of small fruits.

Pentatomidae. Several species of stink bugs attack small fruits.

Pyrrhocoridae. The bordered plant bug feeds on many hosts, including cotton and strawberries in the Southwest.

Coreidae. The leaffooted bug is not host specific but is especially common on blueberries in the Southeast.

Diptera

Tephritidae. Several species of fruit flies attack the berries of small fruits. Included in this group of pests are the blueberry maggot and several species of currant fruit flies.

Anthomyiidae. The raspberry cane maggot attacks both raspberries and blackberries.

Cecidomyiidae. The blueberry tip midge damages blueberries in the Northeast.

Hymenoptera

Tenthredinidae. The slugs or larvae of several species of sawfly feed on the foliage of small fruits. The raspberry sawfly attacks raspberry and blackberry, and the imported currantworm attacks currants and gooseberries. The raspberry leaf sawfly is a common species on brambles. Several species of *Empria* are minor pests of strawberries in the northern United States and Canada. The larvae of a *Caliroa* species have been reported on blueberries in North Carolina.

Thysanoptera

Thripidae. Thrips that attack small fruits are the flower thrips, western flower thrips, and blueberry thrips.

Acarina

Eriophyidae. A number of eriophyid mites are serious pests of small fruits. These include the blueberry bud mite, currant bud mite, dryberry mite, grape erineum mite, and redberry mite (Fig. 15:13).

Tetranychidae. Many of the spider mites that infest tree fruits likewise infest small fruits. Injurious pests include the twospotted mite (Fig. 14:13) on strawberry, raspberry, and blackberry; McDaniel and Schoene mites (Fig. 14:15) on raspberry; European red mite (Fig. 14:12) and Pacific and Willamette mites on grape; southern red mite on cranberry; yellow mite on raspberry and blueberry.

Tarsonemidae. Cyclamen mite (Fig. 4:5B) is a major pest of both greenhouse and field-grown strawberries as well as various ornamentals. This pest was discovered first on cranberry in Washington bogs in 1965.

Some general pests of small fruits not listed here are wireworms, cutworms, grasshoppers, garden symphylan (Fig. 16:11), slugs (Fig. 16:16), and snails.

THE INJURY

Fruit Damage

Many of the small fruit insects cause injury by direct feeding on and contamination of the fruits. Both the eastern and the western raspberry fruitworms feed upon the cores and into the drupelets of raspberries and loganberries. Larvae of the currant fruit fly and the blueberry fruit fly are maggots that feed within the fruits. The cranberry fruitworm literally devours the berries. It also feeds on blueberries. Gooseberry fruitworms web together several berry clusters and feed upon the fruits. The grape berry moth is the only pest that causes extensive damage to grape berries (Figs. 15:2 and 15:4).

As a toxic effect of its feeding, the redberry mite causes the red berry condition of blackberries. Dryberry mites cause similar symptoms, often referred to as "dryberry disease" of loganberries and "sun-scald disease" of raspberries.

Tarnished plant bugs and spittlebugs suck the juices of strawberry and raspberry fruit buds, producing small, hard, seedy "nubbins." Strawberry weevils cut the pedicels of the buds, thereby destroying them. Second-brood blackheaded and yellowheaded fireworms usually feed upon the flower buds and the fruits of cranberries.

Honeydew from grape mealybugs will "gum up" the bunches of berries. A black, sooty mold usually develops on the honeydew. Grape juice processors reject fruit that contains honeydew and sooty mold.

The orange tortrix is a unique pest of red raspberries. Its only damage is contamination of the berries by the caterpillars. When the canes are disturbed during harvest, the larvae drop from the foliage and lodge in the depressions of the fruits, where they cannot be washed out. Their direct injury to the foliage is insignificant. Oddly enough, orange tortrix caterpillars feed directly into the cores and drupelets of loganberries, blackberries, boysenberries, and other berries.

Figure 15:4. Grape cluster damaged by larvae of the grape berry moth. Deflated berries injured by third generation larvae, small dried berries by second generation larvae. *Courtesy Pennsylvania State University, University Park.*

Foliage Damage

Insects that consume the foliage of small fruits include various leafroller caterpillars, raspberry sawfly larvae, the imported currantworm, and several species of cranberry fireworms. Sucking pests, such as the strawberry whitefly and grape whitefly, various aphids, spider mites, the cyclamen mite, bramble leafhoppers, and grape leafhoppers, also attack the foliage.

Cane Damage

Canes are tunneled by cane maggots and cane borers. Scale insects infest the stalks or canes of brambles, currants, grapes, blueberries, and cranberries. The currant borer feeds in the center of the canes of currants and gooseberries (Fig. 15:5).

Crown and Root Damage

The crowns and roots of small fruits harbor a variety of pests. Strawberry crowns are attacked by two kinds of caterpillars—the strawberry crown moth and the strawberry crownminer. Certain weevils, such as the strawberry crown borers, feed in the crowns of strawberry plants. Roots are eaten by the strawberry rootworms and strawberry root weevils. Raspberry, loganberry, and blackberry crowns are tunneled by the raspberry crown borer and the strawberry crown moth. Grape roots are destroyed by grape rootworms. A kind of sod webworm, the cranberry girdler, feeds upon the crowns, stems, and roots of cranberries. Only a few sucking insects attack the roots. Of these the strawberry root aphid and the grape phylloxera (Fig. 15:3) are the most injurious.

In addition, a number of general pests feed upon the crowns or roots of small fruits. Cutworms and wireworms are damaging wherever such crops are raised. White grubs, particularly of the genus *Phyllophaga* in the East and the genus *Polyphylla* in the West, cause considerable injury (Fig. 15:6). The garden symphylan (Fig. 16:11) feeds upon the smaller roots and root hairs of many kinds of crops. Although symphylans are mainly soil inhabitants, ripe strawberries touching the soil surface sometimes become infested.

Disease Transmission

Virus diseases represent the major limiting factor in the production of berries in some regions. The yellows disease of strawberries is transmitted by the strawberry aphid, whereas the raspberry aphid is the vector of raspberry mosaic. Certain xylem-feeding leafhoppers, called sharpshooters, and spittlebugs, including the meadow spittlebug, transmit the bacterium of Pierce's disease of grapes. At least 17 leafhoppers and 4 spittlebugs have been incriminated in this disease. Blackberry dwarf and 7 strawberry viruses are transmitted by the strawberry aphid. Strawberry crinkle virus has at least 13 aphid vectors. *Rubus* stunt is transmitted by the newly introduced bramble-berry leafhopper; *Rubus* yellow net is transmitted by two aphids. Aphids

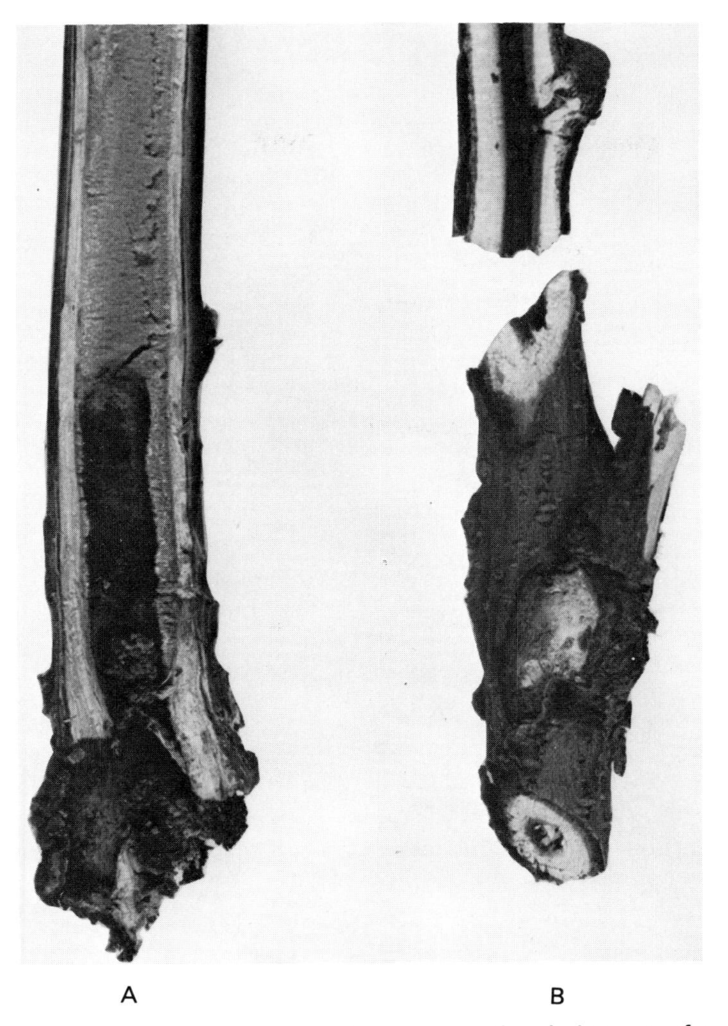

Figure 15:5. Injury by larvae of clearwing moths. A, burrow of raspberry crown borer in blackberry cane; B, burrow of currant borer in currant cane. *Courtesy Washington State University, Pullman.*

transmit a complex of 7 raspberry viruses. Gooseberry, raspberry, and strawberry vein-banding viruses all have aphid vectors. *Vaccinium* false blossom of cranberry is transmitted by the bluntnosed cranberry leafhopper, whereas *Vaccinium* stunt of blueberry is transmitted by the sharpnosed leafhopper.

Two small fruit virus diseases are known to be transferred by noninsect vectors: black currant reversion by the currant bud mite and grape fanleaf by a nematode.

Figure 15:6. A portion of a strawberry row severely attacked by white grubs. *Courtesy Ohio Agr. Res. Dev. Center, Wooster.*

CONTROL

Cultural Control

Because small fruits are shorter-term crops than tree fruits, cultural methods are more useful against the small fruit pests. Some of the more common methods are (1) removal and burning of infested canes or plants; (2) rotation, isolation, and planting in uninfested land; (3) plowing or cultivating to bury or destroy soil insects; (4) use of resistant varieties of plants; (5) flooding of cranberry bogs.

1. **Removal and destruction** of infested canes is a standard practice used against the raspberry crown borer, raspberry cane borers, and the currant borer. Cutting raspberry canes close to the soil or destruction of pruning stubs in the spring removes the pupation site of the raspberry crown borer. Tips of canes infested with raspberry cane maggot are cut several inches below the girdled area. Plants injured by the strawberry crown moth are removed and

burned. Raking and burning of fallen leaves in winter destroys pupae of the grape berry moth.

2. Crop **rotation** helps reduce populations of strawberry rootworm. Establishing new plantings at least 300 yd from old fields lessens the chance of strawberry crown borer infestation. Use only uninfested plants and destroy refuse from old fields after harvest.

Do not establish plantings next to woodlots or allow hedgerows or overgrown fences nearby, because these are overwintering places for strawberry weevil and strawberry rootworm adults. Barriers placed around the perimeters of fields help protect crops from the attacks of strawberry crown borers and strawberry root weevils, flightless pests that migrate by crawling.

3. **Plowing** of vineyards before May 15 buries the overwintering pupae of the grape berry moth. Often a low ridge of soil is thrown under the trellis about 30 to 45 days before harvest so that winter cocoons will be exposed. During the following spring, the ridge is pulled back into the row centers and disked to cover the cocoons. Plowing or cultivation is also utilized to destroy grape rootworm and currant fruit fly pupae.

4. Use of American grape rootstocks highly **resistant** to the attack of the grape phylloxera is a common practice among nurserymen. Desirable European varieties of table and wine grapes are protected from the phylloxera by this method, which has been the most important control practice since about 1850. Lloyd George raspberry is highly resistant to the raspberry aphid. This variety is used in breeding programs to obtain mosaic-escaping qualities in new hybrids, because the raspberry aphid transmits the mosaic virus.

5. **Flooding** is used as a method of control in cranberry bogs against several pests. Bogs are flooded for ten hours during the night or on a cool day about the end of May to kill small caterpillars of the blackheaded fireworm. Winter flooding continued into May reduces the yellowheaded fireworm and certain cutworms. Fall flooding is effective against the cranberry girdler.

Biological Control

Biological control has been utilized much less for small fruit insects than for tree fruit insects, probably because of the size of the supporting industries involved. Natural control of the orange tortrix by the braconid parasite, *Meteorus argyrotaeniae*, can regulate the pest populations below economic levels on red raspberry in Washington. Disruption of the braconid by insecticide applications directed against pests other than the tortrix, however, allows outbreaks of the latter to occur.

The only attempt at biological control of a small fruit pest to date in the United States has been rated as a substantial success. A tachinid parasite (*Sturmia harrisinae*), a braconid parasite (*Apanteles harrisinae*), and four other species were brought into California in 1950–53 to control the western grape leaf skeletonizer. *Apanteles* and *Sturmia* from Arizona quickly attained a high parasitization of up to 75 per cent in the colonized areas. In addition, there is good evidence that these two parasites transmit a granulosis virus disease of the larvae which has helped considerably in reducing the pest.

An integrated control project against the cyclamen mite on strawberries has been partially successful. The predator mite *Amblyseius cucumeris* is effective in reducing the cyclamen mite. The predator cannot breed without cyclamen mites to feed on and is not present in new fields until the pest species becomes established. Release of the predator into young strawberry fields in late summer will greatly reduce cyclamen mite damage the following year. Careful selection of insecticides to be used against other strawberry pests, including acaricides for the twospotted spider mite, will allow this program to succeed. The growers' reluctance to introduce the cyclamen mite purposely along with the predator has precluded the adoption of this technique. It has been used only a few times commercially.

A current integrated program being developed against a complex of pests on grapes in California contains these elements: (1) establishment of three parasitic wasps and use of a microbial insecticide to reduce the grape leaf folder; (2) search for better and more selective acaricides for use against the Pacific and Willamette spider mites, allowing the predator mite, *Metaseiulus occidentalis*, to reduce the pest species; (3) investigation of economic injury levels of the western grape leafhopper to support a modified and less disruptive chemical program; (4) destruction of overwintering larvae of the omnivorous leafroller by plowing and disking and other cultural practices; (5) reduction of the use of disruptive chemicals such as carbaryl, which lead to increased spider mite and grape leafhopper populations.

Chemical Control

Although cultural measures are effective against some small fruit insects, insecticides must be relied upon for control of others. In general, the control programs are less complicated than those for tree fruit pests. Reasons for this are (1) smaller size of plantings, (2) lesser permanence of plantings, (3) fewer major introduced pests, and (4) smaller industry, with less disruption and problems caused by man.

Between 1950 and 1970, soil treatments with the cyclodiene organochlorine materials came into widespread use. This type of application provided excellent control of many soil-inhabiting pests which previously were difficult to control. These persistent materials are currently being phased out because of their potential hazard as environmental contaminants.

As with tree fruits, one of the oldest practices is the use of **lime-sulfur** and/or **oil** in dormant or delayed dormant sprays. These are applied against grape mealybug, San Jose scale, redberry mite, and dryberry mite. Oil (dormant, summer, or superior grades) is used alone against San Jose scale, rose scale, European fruit lecanium, and cottony maple scale.

The **inorganic insecticides** are still used against certain chewing pests—lead arsenate for imported currantworm and redbanded leafroller; calcium arsenate combined with metaldehyde in baits for slug and snail control; sodium fluosilicate in baits for control of *Otiorhynchus* and *Nemocestes* root weevils. The only botanical material still in use against small fruit pests is **rotenone**. It is recommended for control of raspberry fruitworms, rednecked cane borer, gooseberry fruitworm, imported currantworm, and Japanese beetle. Since 1972, TDE and DDT, both **chlorinated hydrocarbons** have been removed from all recommendations; some suitable substitutes are available. Specific pests for which alternate control methods are still sought include the woods weevil and obscure root weevil on blueberries, cane berries, and strawberries; climbing cutworms, grape leaf folder, grape rootworms, and grape flea beetle on grapes. Methoxychlor, with less objectionable properties than DDT, is being used for grape leafhoppers, strawberry weevil, rose chafer, flea beetles, omnivorous leaftier, spittlebugs, and Japanese beetle. Toxaphene is still used for strawberry crown borer control.

Cyclodiene compounds have been particularly useful against soil insects but have been replaced because of their persistent nature. Endosulfan is applied for control of aphids, cyclamen mite, grape leafhoppers, and meadow spittlebug.

Three specific miticides—dicofol, cyhexatin, and tetradifon—are applied for control of spider mites; dicofol is also used against cyclamen mite.

During the past ten years **organophosphorus** compounds have replaced organochlorine materials in many insect control programs of small fruits. Malathion, with low toxicity to warm-blooded animals, is used most often. It is applied for control of aphids, scale insects, leafhoppers, blueberry tip borer, cherry fruitworm, lygus bugs, Japanese beetle, strawberry leafroller, whiteflies, thrips, grape mealybug, currant fruit fly, blackheaded fireworm, cranberry fruitworm, strawberry root weevils, and spider mites (under certain conditions). Parathion controls many of the same pests plus currant borer, raspberry crown borer, grape berry moth, and garden symphylan. Diazinon and azinphosmethyl are each used for about six small fruit pests, whereas demeton and carbophenothion are used mainly against aphids and spider mites. Ethion controls the grape mealybug, and fonofos is one of the best materials to date for use against the garden symphylan.

The **carbamate** insecticide carbaryl has been recommended for control of orange tortrix, redbanded leafroller, blackheaded fireworm, blueberry tip borer, cherry fruitworm, grape berry moth, omnivorous leaftier, strawberry leafroller, Japanese beetle, rose chafer, and stink bugs. It destroys beneficial predators, however, causing outbreaks of spider mites, and it is highly hazardous to pollinating bees. Carbofuran has been registered for use on strawberries against all types of root weevils.

Growers must resort to **fumigants** for control of certain pests. Methyl bromide treatment of strawberry plants for cyclamen mite is conducted in fumigation chambers. In California, this pest is controlled in the field by using machines to cover the strawberry rows with plastic tarpaulins and releasing the methyl bromide underneath. Dichloropropenes or D-D mixture is injected into the soil with a chisel apparatus to control the garden symphylan.

Planting stock is fumigated with methyl bromide to kill the strawberry crown borer.

Control Equipment

Several kinds of insecticide application equipment are suitable for use in small fruit production. Standard high-pressure power sprayers are efficient in various situations. They are equipped with horizontal booms for strawberries and vertical booms for cane berries and grapes. Sometimes a vertical boom is built in an inverted U shape so that it extends over the row of cane berries or grapes with nozzles directed inward from both sides. This method gives good spray coverage. Many growers are now employing low-volume air blast sprayers in their vineyards, and achieving equally good control of pests (Fig. 15:7).

Power dusters are used, with horizontal arrangement of fish-tail outlets for strawberries and vertical arrangement of outlets for cane berries and grapes. Rotary hand dusters and knapsack sprayers are adequate in plantings of a few acres.

Spraying or dusting of cranberry bogs poses a special problem. Some growers use stationary spray plants with long pipelines to carry the material to the ends of the bogs. Hoses and hand guns are attached to the pipeline during the spraying operation. Mobile equipment can be used by carrying long hoses into the bog and spraying with hand guns or multiple-nozzle broom guns. Long horizontal booms that extend out over the planting are sometimes utilized. Many growers apply insecticides through sprinkler systems. Power dusters and airplane equipment have also been used on cranberries.

Soil treatments should always be applied before planting. It is difficult to treat the soil efficiently in an established planting of small fruits. Sprays or dusts are distributed evenly on the soil surface by any suitable method and then thoroughly mixed into the soil to the prescribed depth. Devices that inject the treatment behind a ditching tool or chisel, or in a plow furrow, have also been used.

REPRESENTATIVE PESTS OF SMALL FRUITS

For detailed discussion we have chosen the following insects that are serious pests of small fruits: strawberry root weevils, raspberry fruitworms, redberry mite, and grape leafhoppers.

Strawberry Root Weevil. *Otiorhynchus ovatus* (Linnaeus)

Black Vine Weevil. *Otiorhynchus sulcatus* (Fabricius)

Rough Strawberry Root Weevil *Otiorhynchus rugosostriatus* (Goeze) [Coleoptera:Curculionidae]

All three of the major species of strawberry root weevils were introduced from Europe. The strawberry root weevil was recorded in Massachusetts in

Figure 15:7. Air blast sprayer applying low volume (20 gal/acre) of pesticidal spray to grapes in Erie County, Pa. *Courtesy G. L. Jubb, Jr. Pennsylvania State University, University Park.*

1852 and in the Pacific Northwest by 1904; the black vine weevil was collected in Massachusetts in 1831; the rough strawberry root weevil was found in New York in 1891. They have become established throughout the United States and southern Canada.

Other root weevils are sometimes associated with the three we have listed. In New York, the alfalfa snout beetle, *Otiorhynchus ligustici*, also attacks the strawberry; *O. singularis*, the claycolored weevil, feeds on strawberries in British Columbia; *O. cribricollis* and *O. meridionalis* damage strawberries in California. Two other root weevils have become major pests in Pacific Coast areas during the past 15 years. They are the obscure root weevil (*Sciopithes obscurus*), which is brownish gray with a dark wavy band across the fused elytra, and the woods weevil (*Nemocestes incomptus*), which is brown with lighter markings at the posterior end. Both are about $\frac{1}{4}$ in. long.

The strawberry root weevils attack a wide range of host plants, including all types of small fruits. In addition, the strawberry root weevil and the black vine weevil attack ornamentals and nursery stock. Strawberry root weevil is very damaging to primroses. Black vine weevil is known as "cyclamen bulb borer" or "taxus weevil" in certain areas because of its injury to these ornamentals. Root weevil grubs will also be found among the roots of legumes and grasses and have been known to cause serious damage to grasses grown for seed, peppermint and spearmint, and hops.

Adult *Otiorhynchus* weevils feed upon the foliage or fruits of plants. This injury is usually limited and of minor importance compared to that of the larvae, except for the pruning of grape forms or mature bunches by adults of the black vine weevil. When grass-sod cover crops are used, larger numbers of this weevil may reduce grape yields by as much as $3\frac{1}{2}$ tons per acre. The larvae or grubs of these weevils feed on the roots of host plants. Heavy infestations are capable of killing entire plantings of strawberries. Damaged plants become stunted, the foliage turns red, and the fruit remains small and seedy.

Description. The strawberry root weevil is about $\frac{1}{5}$ in. long. It is usually shiny black, sometimes brown, with rows of small, round punctures on the wing covers. The rough strawberry root weevil is similar except that it is about $\frac{1}{3}$ in. long. Black vine weevil is the largest of the three—about $\frac{2}{5}$ in. long. It is black, but has several yellowish or gray spots on the wing covers (Fig. 15:8). The larvae of strawberry root weevils are white to pinkish in color, with no

Figure 15:8. Three species of root weevils. A, the strawberry root weevil, normal length 5 mm; B, the rough strawberry root weevil, normal length 8 mm; C, black vine weevil, normal length 10 mm. *Courtesy P. M. Eide, Wash. Agr. Exp. Sta.*

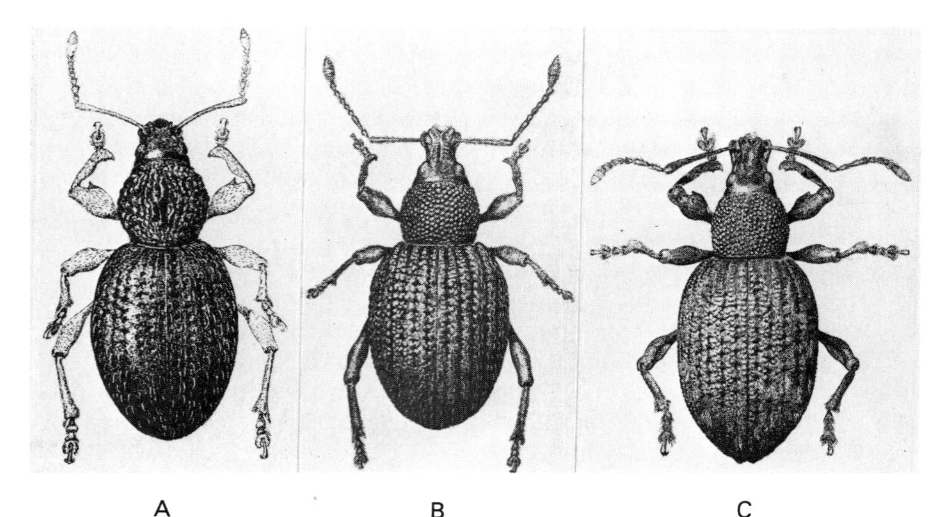

legs, and brown heads. They vary in length, being $\frac{1}{4}$ to $\frac{1}{2}$ in. long when full grown (Fig. 15:9).

Life history. Strawberry root weevils winter as grubs or adults in the soil or plant debris. The larvae become active during the spring and form earthen cells in which they pupate. Adults usually emerge during May or June depending on the locality and the season. Black vine weevils emerge somewhat later than the strawberry root weevil. Rough strawberry root weevils emerge about two weeks later. Adult weevils that pass the winter will usually start emerging in April.

Egg laying begins two to four weeks after emergence and continues for about two months. Eggs are placed on the soil near the plants. All three species are parthenogenetic: only females occur. The strawberry root weevil usually lays 150 to 200 eggs per individual. A black vine weevil has been recorded to lay more than 2000 eggs. The eggs hatch in about ten days, and the larvae burrow down as much as 6 or 8 in. into the soil. They feed upon the fibrous roots and tend to be most destructive in the spring.

Adult weevils feed on foliage and fruit at night. They make two distinct migrations per season—one starting in June during the oviposition period, and the other in fall, possibly for hibernation. In heavily infested areas, the adult beetles may enter houses in sizable numbers and become household pests. All movement is by walking; because the wing covers are fused, they cannot fly. Late-emerging adults may survive through the winter and lay more eggs the following season. Sometimes the adults are able to survive into a third season; this results in overlapping generations.

Control. Baits containing dried fruit or bran plus sodium fluosilicate or calcium arsenate have been used for root weevil control for many years. Apple waste or "pomace" is utilized commonly in commercial bait preparations. An apparatus consisting of a funnel and a metal tube is useful in applying baits to the crowns of strawberry plants.

In recent years, soil insecticides have been proven effective against strawberry root weevils. Aldrin, dieldrin, heptachlor, or chlordane is applied as a preplanting treatment. The insecticide is dusted or sprayed onto the soil surface and then worked in with a disk harrow or rotary tiller to a depth of

Figure 15:9. Larva of the strawberry root weevil. *Courtesy Dept. Agr. British Columbia.*

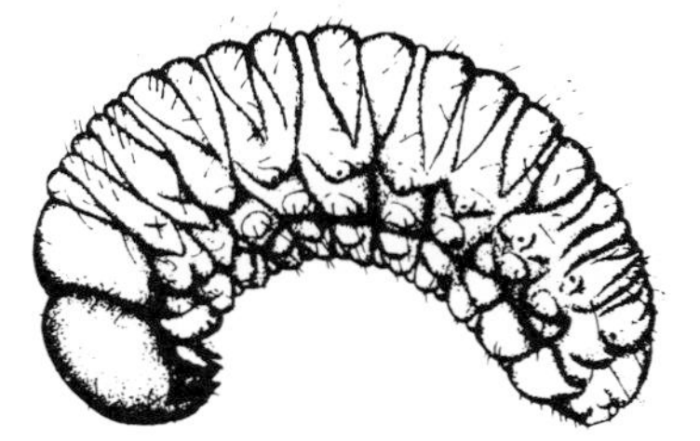

about 4 in. Even on established plantings, dusting with these insecticides has given better control than baits. Recommendations of these persistent cyclodiene materials are being phased out as suitable substitutes become available. Stocks of aldrin and dieldrin were used up in 1975.

In fields where soil insecticides have been used for several seasons, damage from *Nemocestes* or *Sciopithes* weevils may occur, particularly on newly cleared land. In such cases, a return to poison baits has been required. Heavily infested fields may be cleaned up with ethylene dibromide injected into the soil behind a chisel that slices through the center of the strawberry crowns. In recent years, these weevils have been controlled with DDT, but this recommendation has been eliminated. Azinphosmethyl, parathion, and malathion sprays will aid in controlling root weevils, but must be carefully timed to reduce killing of essential pollinators. Carbofuran, which controls all root weevils, has recently been registered for use on strawberries.

Eastern Raspberry Fruitworm. *Byturus rubi* Barber
Western Raspberry Fruitworm. *Byturus bakeri* Barber
[Coleoptera:Byturidae]

The two major raspberry fruitworms in North America are indigenous to this continent. Apparently these closely related species have evolved because the Continental Divide acted as a barrier to keep them separated. Other species of *Byturus* are recorded in this country but are not of economic importance. The eastern raspberry fruitworm is a pest in the eastern United States and Canada, whereas the western raspberry fruitworm is present in Oregon, Washington, Idaho, Montana, and British Columbia.

Raspberry fruitworms attack red raspberry and loganberry, and to a lesser extent black and purple raspberries, blackberry, and strawberry. In the Pacific Northwest, the fruitworm has also been recorded from wild blackcap raspberry, wild dewberry, thimbleberry, and salmonberry.

The overwintered beetles feed on the unfolding leaves, especially those of the small, new canes. As the season progresses, they move up and feed upon foliage of the old canes, flower buds, and open flowers (Fig. 15:10). When the leaves become distended, the damage appears as narrow slits between the secondary ribs. This injury is typical of the beetles and is often used as a reference point for insecticide applications.

Larvae of the fruitworms usually feed upon the receptacles, scoring or tunneling them. Sometimes they feed upon the drupelets, causing misshapen fruits (Fig. 15:11). Adults feeding in fruit buds will destroy them completely or cause the developing fruits to be distorted.

Description. The western raspberry fruitworm beetle is about $\frac{1}{6}$ in. long, and the eastern species is about $\frac{1}{7}$ in. long. They both are light brown in color, the eastern form being somewhat paler than the western. The full-grown larvae are $\frac{1}{4}$ to $\frac{1}{3}$ in. long and white or yellowish in color with light brown areas

Figure 15:10. Damage to leaves and buds of raspberry by adult raspberry fruitworms. *Courtesy Conn. Agr. Exp. Sta.*

on the upper part of each segment. Pupae are $\frac{1}{6}$ to $\frac{1}{5}$ in. long and are white (Fig. 15:12).

Life history. The overwintering adults emerge from the soil in the spring, starting in March and April in the Pacific Northwest and May in New England. Eggs are usually deposited singly on the buds or in the flowers about the time the first blooms appear. Sometimes they are placed on bud pedicels, leaf petioles, or even on a leaf.

The eggs hatch in about one to three weeks. Newly emerged larvae may feed in the blossoms or on the developing fruits for a few days. Some bore through the calyx into the receptacle or between developing drupelets the first day after hatching. The larvae feed in the fleshy receptacles or the drupelets for about four to six weeks.

Mature larvae drop to the ground and enter the soil, burrowing there for several days. They pupate in late summer or fall. Fully colored adult beetles are formed before winter, but they do not leave the pupal cells before the following spring.

Control. The standard control measure for raspberry fruitworms previously was rotenone, applied to kill the beetles before they deposit their eggs. Usually several treatments were required—the first about a week after the first raspberry blooms open, the second about ten days after the first, and sometimes a third about ten days after the second. Diazinon is used now, but it must be applied before bloom to avoid killing bees.

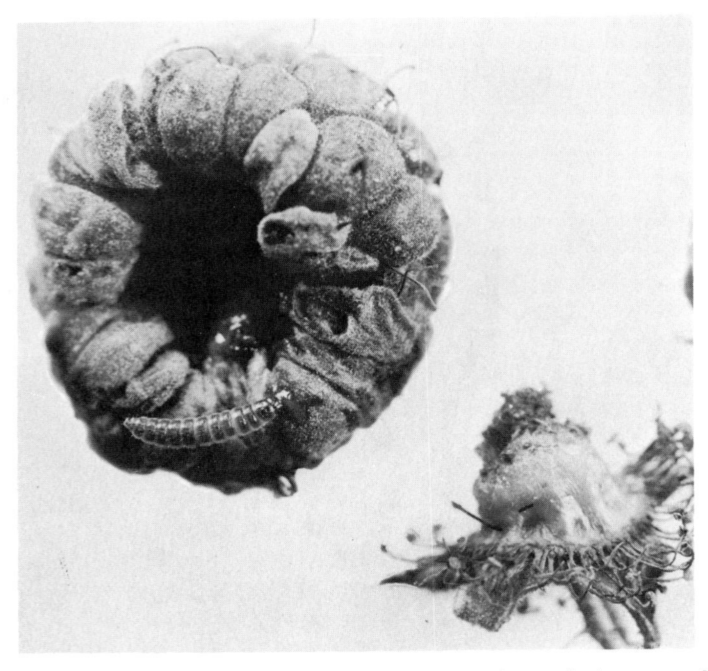

Figure 15:11. Damage to raspberry fruit by raspberry fruitworms. Note larva feeding on drupelet. *Courtesy Conn. Agr. Exp. Sta.*

In eastern Canada, lead arsenate plus Bordeaux mixture or lead arsenate plus hydrated lime are sometimes applied. DDT and TDE have been removed from fruitworm recommendations.

Redberry mite. *Acalitus essigi* (Hassan) [Acarina: Eriophyidae]

The redberry mite was apparently introduced into California from Europe about 1921. When it was discovered in this country, the mite was confused with the dryberry mite, another European species, but it was properly described by Hassan in 1928. It has become established in Oregon, Washington, California, and southern Canada.

At first it appeared that the mite would attack only Himalaya blackberries. Later observations showed that it would feed upon all varieties of blackberries tested, as well as red and black raspberries, wild dewberry, loganberry, and boysenberry. It is primarily a pest of domestic blackberries.

While in the berry, the mites feed near the bases of the drupelets around the core. The fruits grow to full size but do not ripen normally. All or part of the drupelets remain bright red and hard. Berries become hardened and roughened and remain on the plants until the old canes die during winter. **Redberry disease** varies in extent with the numbers of mites present. As many

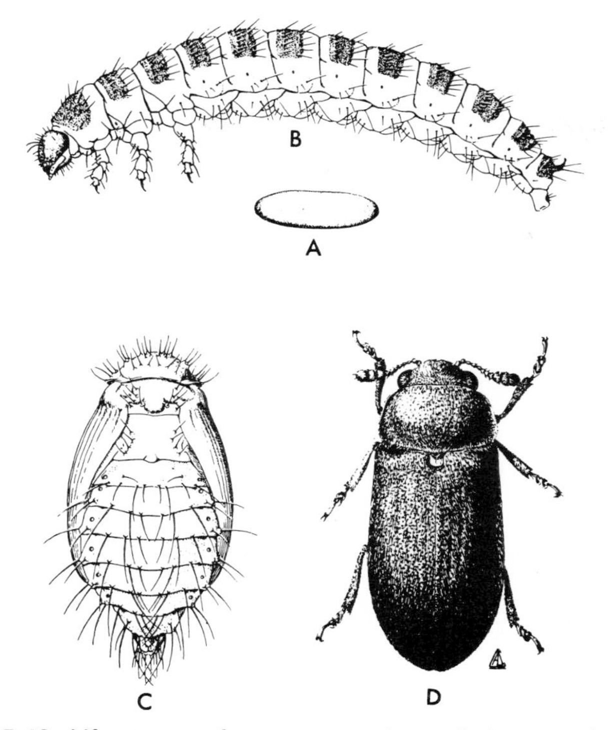

Figure 15:12. Life stages of western raspberry fruitworm. A, egg; B, larva; C, pupa; D, adult. *Courtesy Wash. Agr. Exp. Sta.*

as 683 mites have been recorded in an individual red berry during September. Injury ranges from a single red drupelet in a black berry to a completely red berry.

A closely related species, the dryberry mite, *Phyllocoptes gracilis* (Nalepa), causes so-called **dryberry disease** in loganberries and red raspberries.

Description. The redberry mite is very small, white, and wormshaped. It varies in body length from 0.125 to 0.155 mm. Like all eriophyid mites, this species has only two pairs of legs. The slender, tapered abdomen is marked with 80 fine grooves that encircle the entire body (Fig. 15:13). Although the mite is usually white, winter forms may be yellowish or amber. The male is slightly smaller than the female. Eggs are dull white, spherical, and 0.005 mm in diameter.

Life history. The redberry mite overwinters in the adult stage under the bud scales. Some eggs may be laid during the winter in temperate localities. Reproduction is greatly increased in the spring, when the foliage buds begin developing. Egg laying starts in late February or early March. The first brood is usually active by mid-March in the Pacific Northwest. The mites feed in the buds and sometimes move into the unfolding leaves. As the fruits begin to

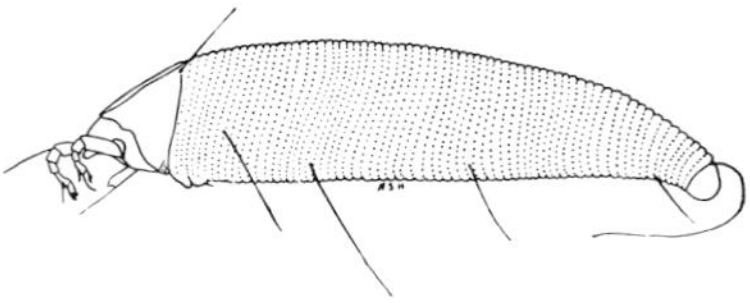

Figure 15:13. Adult of redberry mite, *Acalitus essigi* (Hassan). *Courtesy Ore. Agr. Exp. Sta.*

form, there is a migration from the scales of the winter cane buds to the green berries. It appears that mites are carried from row to row of a planting by air currents. When migrating from the buds, they may be picked up easily by the wind and carried to adjacent rows.

In California, the migration is directly from the cane buds to the developing flowers. In Washington, not all mites leave the scales of the winter cane buds until the following fall. The mite population starts increasing in March and reaches a peak during September. During fall and winter the population decreases.

In each season there are a number of overlapping generations; these are indefinite because of the overlapping of egg and immature stages. The length of the life cycle from egg to adult may be as short as ten days under favorable conditions. During the summer, a heavily infested field may contain an average of several hundred mites per berry. Early in the fall the majority of the mites on the new canes are found in the axis of the cane buds and the compound leaf rather than inside the cane bud. As winter advances, more are found beneath the bud scales.

The mortality from midsummer to the following spring is very high. Two hundred mites per berry in summer may be reduced to ten per berry by next spring. Eggs have been observed as late as December. Some injured fruits retain a faded red or purplish color and hang on into the winter. Mites have been recorded in such berries as late as January.

Control. Lime-sulfur or sulfur sprays have been used against redberry mite in California since 1923. Lime-sulfur is still the best control measure for the mite. It is usually applied in a delayed dormant spray during February or March. Sometimes a second application is required in May. Lime-sulfur is a standard treatment for eriophyid mites, including the pearleaf blister mite, apple rust mite, dryberry mite, and citrus rust mite.

Important natural enemies of the redberry mite are predaceous mites belonging to the family Phytoseiidae. The life history of these predators is closely related to that of their host. They winter as adults under bud scales but remain active, feeding upon redberry mites throughout the winter.

Grape Leafhopper. *Erythroneura comes* (Say)

Western Grape Leafhopper. *Erythroneura elegantula* Osborn

Variegated Leafhopper. *Erythroneura variabilis* Beamer

Threebanded Leafhopper. *Erythroneura tricincta* Fitch
Virginiacreeper Leafhopper. *Erythroneura ziczac* (Walsh)
Other Species. *Erythroneura dolosa* Beamer & Griffith, *E. vitifex* Fitch, *E. vulnerata* Fitch, *E. vitis* Harris, and *E. octonotata* Walsh [Homoptera:Cicadellidae]

Many of the native North American species of *Erythroneura* attack grapes. *E. comes*, *E. tricincta*, *E. vulnerata*, and *E. vitis* are found in the Northeast, the central states, and the Southeast. *E. ziczac* inhabits the northeastern states, central states, and northwestern states. *E. octonotata* is recorded from the central states only. *E. vitifex* is found in the Northeast, Southeast, central states, and the Southwest. *E. variabilis* inhabits only the southwestern states. *E. elegantula* is a major species in the Southwest, Northwest, and certain central states, whereas *E. dolosa* is restricted to the Pacific Northwest.

All species of *Erythroneura* overwinter as adults. Nearly all live on woody plants, especially oaks and grape. Other hosts of various grape leafhoppers are Virginia creeper, currant, gooseberry, raspberry, blackberry, dewberry, apple, plum, cherry, wild cherry, hawthorn, hollyhock, dahlia, Boston ivy, various grasses, and weeds. Such plants as holly, wild strawberry, wild blackberry, dewberry, curled dock, and dandelion may be more important for shelter from wind than as host plants.

The grape leafhopper has been known as an injurious pest of grape foliage since 1825. Major outbreaks have been recorded beginning in the early 1900s. In New York, peaks occurred in 1902, 1911, 1922, and 1938–47. California records indicate that the western grape leafhopper was particularly abundant in 1907–08, 1913–14, and 1929–32.

Two major damaging effects of grape leafhoppers are the lowering of the quality of fruit and the reduction in yield. Removal of chlorophyll and other cell contents causes a white speckling of the foliage. As feeding punctures become more numerous, the light areas become larger, and the leaf appears mottled (Fig. 15:14).

In advanced cases the whole leaf may turn pale, die, turn brown, and then fall. If a large portion of the foliage drops, the bunches of grapes may be sunburned. Lesser damage may cause deformed leaves, low sugar content in berries, poorly maturing canes in fall, and weak growth of vines the following spring. Droplets of honeydew that are exuded by the leafhoppers may disfigure the grapes. Dust may accumulate in these sticky droplets, and sooty molds may also develop on them.

Description. Grape leafhoppers are small, agile insects about $\frac{1}{8}$ in. long. The grape leafhopper adult is pale yellow with red, brown, and black markings (Fig. 15:14). Western grape leafhoppers are pale yellow with reddish and dark brown markings. The variegated leafhopper is usually darker than the above species with larger areas of cloudy brown on the forewings. Threebanded leafhoppers have a pair of dark markings across the thorax and

Figure 15:14. Grape leafhoppers and injury. A, white speckling of leaf due to feeding of leafhoppers; B, threebanded leafhopper (left) and grape leafhopper (right); C, nymphal leafhopper. *Courtesy USDA.*

two more pairs forming color bands across the forewings. Overwintering adults have darker red markings than those of the spring and summer broods.

Grape leafhopper eggs are less than 1 mm long, slightly curved, and white to pale yellow in color. Newly hatched nymphs are very small and nearly white. As they develop, they become more yellowish or brownish. Faint indications of wing pads occur in the second instar and continue to enlarge with each molt until the adult form is reached.

Life history. All of the grape leafhoppers overwinter as dark-colored adults in plant debris. These adults become active and migrate to the grapevines during March and April in California. Farther north, they become active in May. In New York there are two generations per season; in Kentucky there are two and a partial third; in California there are three. Mating takes place on the grapes. Females insert eggs singly under the surface tissue of leaves. Each female may lay up to 100 eggs over a period of one to two months. Eggs of the first brood hatch in 15 to 20 days, depending upon the locality and the species. The newly hatched nymph emerges through a slit in the egg and the leaf. Eggs are inserted mainly into the lower surface of leaves, and leafhoppers prefer to feed upon the lower surface. Development through the five nymphal instars is completed in about 15 to 25 days. First generation adults appear in late May in California and about mid-June in more northern areas.

Mating takes place about two weeks after adults emerge, and egg laying commences about a week later. Second-generation eggs take only 5 to 15 days to hatch. Nymphal stages are completed in about two weeks. A third generation develops during August and September in California. Adults overwinter in old leaves, dead grass, and straw along fence rows, ditches, and in alfalfa fields. There is a greater mortality of adults during winters with long periods of cold, wet weather than during mild winters.

Control. At one time, the standard treatment for grape leafhoppers was DDT, but because of the appearance of DDT resistance in most areas, this use was discontinued. A delayed dormant treatment of parathion plus either superior or dormant oil controls leafhoppers, grape mealybug, and black vine weevil in the Pacific Northwest. Parathion, malathion, or ethion are used in summer sprays, malathion as close as three days before harvest. Malathion, with its very low toxicity to warm-blooded animals, is the preferred material in residential areas. Endosulfan and methoxychlor are alternate chemicals used in some areas.

Current studies of population densities and economic injury levels of grape leafhoppers in California are aimed at modifying the disruptive insecticide program. If low levels of fruit spotting can be tolerated on table grapes, natural enemies of both the leafhoppers and spider mites can be encouraged as a part of the integrated control program.

A number of natural enemies of the grape leafhoppers are recorded in the literature. One that is sometimes present in numbers large enough to provide some control is a wasp egg parasite, *Anagrus epos* Girault. Weather conditions have been shown to be largely responsible for variations in the prevalence of the grape leafhoppers from year to year. The entomogenous fungus, *Entomopthora sphaerosperma*, is often an important natural factor adversely affecting populations of the leafhoppers.

SELECTED REFERENCES

Allen, W. W., *Strawberry Pests in California*, Calif. Agr. Exp. Sta. Circ. 484, 1959.

Andison, H., *Common Strawberry Insects and Their Control*, Canada Dept. Agr. Pub. 990, 1956.

Baker, W. W., S. E. Crumb, B. J. Landis, and J. Wilcox, *Biology and Control of the Western Raspberry Fruitworm in Western Washington*, Wash. Agr. Exp. Sta. Bul. 497, 1947.

Barber, H. S., *Raspberry Fruitworms and Related Species*, USDA Misc. Pub. 468, 1942.

Barnes, M. M., *Grape Pests in Southern California*, Calif. Agr. Exp. Sta. Ext. Ser. Circ. 553, 1970.

Bournier, A., "Grape Insects," *Ann. Rev. Entomol.*, 22:355–76, 1977.

Breakey, E. P., D. H. Brannon, and P. M. Eide, *Controlling Insect Pests of Small Fruits*, Wash. Ext. Bul. 450, 1957.

Chamberlain, G. C., W. L. Putman, and A. T. Bolton, *Diseases and Insect Pests of the Raspberry in Canada*, Can. Dept. Agr. Pub. 880, 1963.
Doutt, R. L., J. Nakata, and F. E. Skinner, "Integrated Pest Control in Grapes," *Calif. Agr.*, 23(4):4–11, 16–17 (1969).
Essig, E. O., *The Blackberry Mite: The Cause of Redberry Disease of the Himalaya Blackberry, and Its Control*, Calif. Agr. Exp. Sta. Bul. 399, 1925.
Franklin, H. J., *Cranberry Insects in Massachusetts*, Mass. Agr. Exp. Sta. Bul. 445, 1948; Parts II–VII, 1950; Suppl. 1952.
Hanson, A. J., *The Blackberry Mite and Its Control*, Wash. Agr. Exp. Sta. Bul. 279, 1933.
Lamiman, J. F., *Control of the Grape Leafhopper in California*, Calif. Ext. Circ. 72, 1933.
McGrew, J. R., and G. W. Still, *Control of Grape Diseases and Insects in the Eastern United States*, USDA Farmers' Bul. 1893, 1968.
Neunzig, H. H., and J. M. Falter, *Insect and Mite Pests of Blueberry in North Carolina*, N.C. Agr. Exp. Sta. Bul. 427, 1966.
Rings, R. W., and R. B. Neiswander, *Insect and Mite Pests of Strawberries in Ohio*, Ohio Agr. Res. Dev. Cen. Res. Bul. 987, 1966.
Smith, L. M., and E. M. Stafford, *Grape Pests in California*, Calif. Agr. Exp. Sta. Circ. 455, 1955.
Stearns, L. A., W. R. Haden, and L. L. Williams, *Grape Leafhopper and Grapeberry Moth Investigations*, Dela. Agr. Exp. Sta. Bul. 198, 1936.
Still, G. W., *Cultural Control of the Grape Berry Moth*, USDA Inf. Bul. 256, 1962.
Still, G. W., and R. W. Rings, *Insect and Mite Pests of Grapes in Ohio*, Ohio Agr. Res. Dev. Cen. Res. Bul. 1060, 1973.
Taschenberg, E. F., and F. Z. Hartzell, *Grape Leafhopper Control—1944 to 1947*, N.Y. (Cornell) Agr. Exp. Sta. Bul. 738, 1949.
USDA, *Strawberry Insects—How to Control Them*, USDA Farmers' Bul. 2184, 1972.
Vaughan, E. K., and R. G. Rosenstiel, *Diseases and Insect Pests of Cane Fruits in Oregon*, Ore. Agr. Exp. Sta. Bul. 418, 1949.
Walden, B. H., *The Raspberry Fruit Worm*, Conn. Agr. Exp. Sta. Bul. 251, 1923.

chapter 16 / **JOHN A. NAEGELE**

INSECT AND OTHER PESTS OF FLORICULTURAL CROPS

The floricultural industry in the United States is a multibillion-dollar enterprise that produces a diverse number of crops grown for flowers and foliage to satisfy the esthetic appetite of modern Americans. Because these crops must be appealing, insect damage need not be greatly destructive to be of economic importance. Any impairment of beauty or form amounts to a reduction in quality. Hence, floricultural crops must be virtually pest- and injury-free in order to command top prices.

THE PESTS

The concentrated units in which floricultural crops are grown—greenhouses, lath houses, and cloth houses—are ideally suited for the buildup of injurious populations of insects, mites, and other pests. The mild, regulated environment provides ideal conditions for the survival and multiplication of a variety of species. Furthermore, when some plants are maintained in one location for any length of time, conditions not only allow for the perpetuation of populations but also set the stage for the development of long-term problems such as insect resistance to insecticides.

Many floricultural pests are indigenous and are normally of only moderate importance on crop plants or native foliage, but when they are introduced to the floricultural crop growing under optimum conditions, the pests assume a major destructive role. The known enemies of floricultural crops exceed 370 species distributed among 93 families and 13 orders. For this reason, the following survey of pests can treat only the most important species.

Lepidoptera

The Lepidoptera contain by far the greatest variety of floricultural pests, but because of new and effective control measures, they are generally no longer the most destructive. At the present time, the most serious pests in the Lepidoptera are found in the following groups: the leafrollers and leaftiers, plume moths, tussock moths, loopers, cutworms, spinx moths, Pierids, and leafminers. Particularly important species in this order are the **orange tortrix, omnivorous leafroller, carnation leafroller, rose budworm, rose leaftier, greenhouse leaftier, plume moths, cabbage looper, variegated cutworm, corn earworm, beet armyworm,** and **European corn borer**. European corn borers feed on the foliage and bore into stems of chrysanthemums, dahlias, gladioli and other plants; once within the stems, they are impossible to control.

Acarina

Mites are the most common and destructive pests on floricultural crops. There are a large number of species; the most important are those called red spiders or **spider mites** (family Tetranychidae). In this family, the **twospotted spider mite** is the primary pest (Fig. 16:14). Generally spider mites are small and unobtrusive, but when present in large numbers, they may web over plants completely and thus call attention to themselves. Mites may also be detected by observing ball-like aggregations composed of many migrating mites at the tips of leaves, or by the presence of feeding damage, which upon close examination appears as minute speckling.

In addition to the twospotted spider mite, there are a number of other spider mites commonly found on floricultural crops: the **clover mite, asparagus spider mite, tumid spider mite, luden spider mite, oxalis spider mite,** and **hydrangea spider mite.**

The **false spider mites** (family Tenuipalpidae), closely related to the true spider mites, cause similar foliage damage. They are distinguishable from the spider mites because they are exceedingly small, red or yellowish, and noticeably flat. The eggs are bright red and are usually seen more easily than the mites themselves. They do not produce silken webs. The most important species in this family is the **privet mite** (Fig. 16:1). This pest is bright red and very easy to spot on its favorite hosts, azalea and fuchsia. It has been widely distributed in America through commercial shipment of azalea plants. A false spider mite commonly found on orchids, particularly phalaenopsis orchids, is the **phalaenopsis mite**. This pest, which is yellowish or reddish in color, forms pits on the upper surface of phalaenopsis leaves.

The **threadfooted mites** (family Tarsonemidae) constitute another group of great importance to floricultural crops. Their feeding characteristically produces deformity of the growing tips and young leaflets of the plants. The most widespread and destructive species of threadfooted mite found in greenhouses throughout the country is the **cyclamen mite**. It is found on many

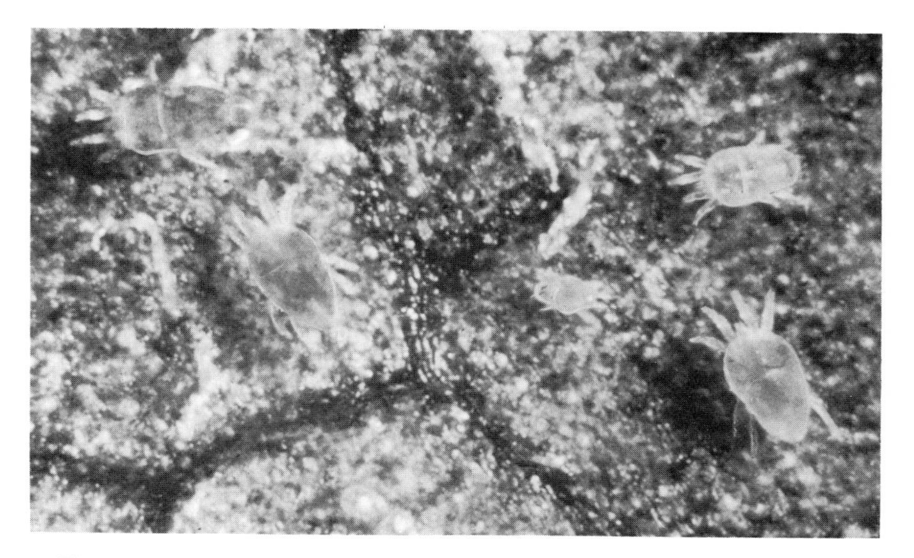

Figure 16:1. The privet mite, adults, nymphs, and larvae. *Courtesy R. N. Jefferson, University of California, Riverside.*

plants and is particularly important in the production of the African violet, gloxinia, and cyclamen.

The **broad mite** is a threadfooted mite similar to the cyclamen mite in appearance and in the injury it produces. Unlike the cyclamen mite, it occurs more generally over the plant rather than just in the developing buds and feeds on the undersurface of the leaves, causing them to curl outward and take on a brittle appearance. The **bulb scale mite** is found principally on the stems, leaves, and bulbs of narcissus. The principal damage is done to the bulbs and bulb scales.

The **louse mites** (family Pyemotidae) are found generally on grasses or grains and are occasionally pestiferous to humans. Floricultural plants are attacked by one species, namely the **grass mite**, which burrows into the carnation bud to feed on the developing flower parts. They carry on their bodies the spores of the fungus responsible for **carnation bud rot** and a similar disease of grasses called **silvertop**. The mites are peculiar in that the abdomen of the female swells enormously to the shape of an egg or ball which contains as many as 200 eggs. These eggs hatch within the female, and the young mites develop to maturity before they emerge from the parent.

Fungus mites (family Acaridae) have been associated generally with the soil or moldy plant tissue, but they can be the primary cause of injury in narcissus and related bulbous crops. The mites are usually found in enormous numbers on decayed or partially decayed bulbs; this has brought about the confusion concerning their importance. A common species is the **bulb mite**.

Among the **blister** or **rust mites** (family Eriophyidae) the only species of importance in floriculture is the **carnation bud mite**, which causes a grassiness of the carnation often confused with boron deficiency.

Homoptera

Numerous pests of floricultural crops are to be found in the order Homoptera: scale insects, mealybugs, whiteflies, and aphids. Because of the boom in green plant production, scales and mealybugs are increasing in importance. Among the soft scales, a common and destructive species is the **brown soft scale** (Fig. 16:2). The adult female is flattened, quite soft, and yellowish or brownish in appearance. This insect produces large amounts of honeydew, which tends to coat the infested plant and make it unsightly. A sooty mold then grows on this honeydew, coating the plants with a brown or black soot. The **hemispherical scale** is commonly found in greenhouses throughout the country. The females are very convex, hard, brown, shiny, and smooth. The hemispherical scale infests a large number of plants and is particularly fond of ferns. The **black scale** is also common on a large number of hosts. It is readily recognized by the raised H-shaped ridges on its back. The **cottony camellia scale** is found on camelias and several other plants under greenhouse conditions.

Numerous species of **armored scales** attack greenhouse plants. Perhaps the most common is the **oleander scale** (Fig. 16:3). It has a circular shield with a central nipple. The **Florida red scale** is found on a large variety of floricultural plants. The **Boisduval scale**, recognized by the cottony mass in which the males are clustered, is common on certain orchids. Often the females become so numerous that the entire stem is encrusted. The armored scales do not secrete honeydew.

Figure 16:2. The brown soft scale on ivy. Note the two groups of scales along two of the veins.

Figure 16:3. The oleander scale, a common armored scale on green house plants.

Mealybugs, close relatives of scale insects, are important pests of floricultural plants. Plants infested with mealybugs are often covered with waxy, white cottony sacks filled with their eggs. The most common species is the **citrus mealybug**, which is found on a wide variety of plants. The **Mexican mealybug** (Fig. 16:4) is destructive to chrysanthemums and other host plants. The **grape mealybug** is an important pest of ferns. The **ground mealybug** is found occasionally on the roots of a large variety of plants, particularly cactus and palms, and is a particularly difficult pest to control.

Figure 16:4. The Mexican mealybug on chrysanthemum leaf.

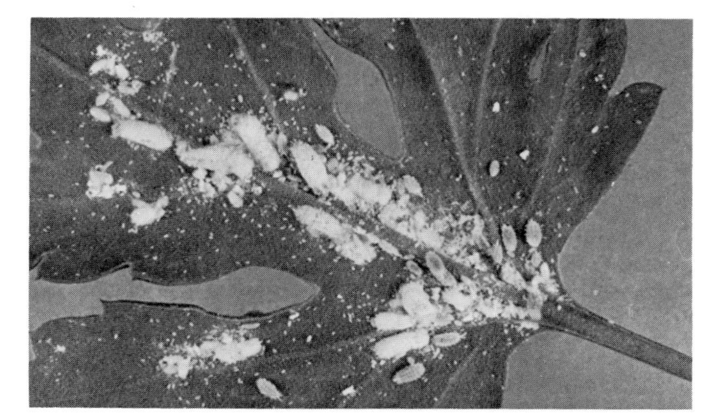

Whiteflies, also related to scale insects, are notorious pests in greenhouses. The principal damage is caused by the larvae that suck sap from the plants and excrete large quantities of honeydew. The **greenhouse whitefly** (Fig. 16:5) is the most common species in greenhouses and infests many kinds of plants. Other destructive species include the **iris whitefly, fern whitefly**, and **azalea whitefly**.

Aphids are common pests of floricultural crops and include a number of important greenhouse species. Some of these attack a wide variety of plants, for example, the **green peach aphid, foxglove aphid, potato aphid, melon aphid** (Fig. 12:1) and **ornate aphid**. Others are more restrictive, such as the **crescentmarked lily aphid, rose aphid, tulip bulb aphid**, and **chrysanthemum aphid**.

Leafhoppers are not usually found in greenhouses, but on outdoor floral plantings they can be serious pests because of their ability to transmit virus diseases and aster yellows.

Thrips are of importance to florists. In greenhouses, various species of thrips occur because they are carried by the wind from native hosts to greenhouse crops through the greenhouse ventilators. On outdoor crops, such

Figure 16:5. The greenhouse whitefly. Adults, white in color, congregated on leaf; shed skins of the "pupae" are gray.

as gladioli or field-grown roses, thrips can be noxious pests, destroying not only flowers such as gladioli, but on newly grafted rose plants, the grafted bud as well. Of special importance to the gladiolus is the **gladiolus thrips**, found wherever these flowers are grown. Within greenhouses the most common species are the **greenhouse thrips** and the **onion thrips**.

Hemiptera

Several Hemiptera are common enemies of floricultural crops, in particular the **tarnished plant bug** and the **fourlined plant bug** (Fig. 16:6). The **azalea lace bug**, the **chrysanthemum lace bug**, and the **rhododendron lace bug** are periodically destructive.

Coleoptera

Beetles are serious pests of floricultural crops. In most cases the adult beetle is responsible for the injury, and such groups as the **cucumber beetles**, the **tortoise beetles**, and **blister beetles** are destructive flower feeders. **Flea beetles** are particularly damaging to the foliage of several crops. Both adults and larvae of the **black vine weevil** may destroy the foliage and roots of several crops. Adult scarabs such as the **Japanese beetle, rose chafer, Asiatic garden beetle**, and **oriental beetle** are generally wholesale skeletonizers of the foliage and flowers in outdoor floricultural production. The larvae of some beetles (e.g., the **threelined potato beetle**) feed directly on the foliage and attack

Figure 16:6. Fourlined plant bug, adult and nymph on chrysanthemum leaf. Note dark, round necrotic spots resulting from feeding of bug.

several kinds of plants. The larvae of this beetle have the peculiar habit of plastering their bodies with their own excrement.

Diptera

Several species of Diptera are destructive to floricultural crops. The **narcissus bulb fly** is particularly important to narcissus production in the Northwest. Leafminers, the larvae of certain small flies, feed between upper and lower leaf surfaces, leaving narrow winding trails or mines. The **columbine leafminer** is common on columbine grown outdoors, while the **serpentine leafminer** and **chrysanthemum leafminer** are pests of greenhouse crops. **Fungus gnats**, small gray flies, may breed in greenhouse soil. The larvae feed on fungi and decomposing organic matter and may injure seedlings and rooted cuttings by feeding on roots and root hairs. They may damage bedding plants, and broods of gnats emerging from potted plants are objectionable in homes and hospitals. Another family of delicate flies, the gall midges, have members that are greenhouse pests, including the **rose midge** and the **chrysanthemum gall midge**.

Hymenoptera

The Hymenoptera are of relatively little importance on floricultural crops. With the exception of several sawfly species on rose and violet, the Hymenopterous pests are no longer important. At one time, the **orchidfly** was the major pest of orchids and often a limiting factor in orchid production. Gall wasps, ants, and stem-nesting bees may cause some concern.

Miscellaneous Orders

Occasionally grasshoppers and cockroaches are pests, but infestations of these insects are controlled easily in greenhouses. In certain areas of the South and West, grasshopper damage to plants grown outdoors can be severe. Springtails, termites, earwigs, sowbugs, millipedes, and snails often associated with the production of floral crops, are sometimes destructive. Of major importance are **slugs** (Fig. 16:16) and the **garden symphylan** (Fig. 16:11).

THE INJURY

The kinds of injury produced by pests of floricultural crops can be summarized in seven categories.

1. Physical removal of plant parts by the **consumption** of external feeders. This type of injury is produced by pests with chewing mouthparts and is usually the most obvious kind. The pests that are responsible for the damage may not be as obvious, for many do their work at night or leave the scene of the crime when feeding is completed. In this kind of injury, generally associated with lepidopterous pests, the plant may be completely eaten or skeletonized by free-feeding larvae, or the leaves and flowers may be first folded and then devoured, as by leafrollers and leaftiers. Cutworms and beet

armyworms create a problem on chrysanthemums and other floral crops by their habit of taking a quick bite and pinching the terminals. Slugs are direct feeders with huge appetites. Their feeding is not confined to one location, for they can easily climb the plant ignoring the green leaf tissue and sample flower parts like practiced gourmets. Some insects burrow into the soil and destroy the underground stems and roots.

2. The **mines** produced by miners in leaf tissue are conspicuous blemishes. The azalea leafminer, whose handiwork is easily visible, is usually not responsible for wholesale destruction of leaves except in the Northwest. The mines of several species of leafminers are unsightly and, if present in large numbers, may reduce quality. Leafminers generally do not constitute an economic threat because they are controlled by insecticides used against other pests. Resistance to many insecticides, however, is developing among leafminers.

3. **Borers** working in the stem, crown, or base of the plant cause serious damage. Generally the first indication of their presence is the wilting or breaking of the infested stem. Usually the injury is not apparent until considerable time and effort have gone into the growing of the crop, and the flowers are ready to harvest. The destruction of shoots that might have borne flowers is another way by which borers reduce profits. Although not considered true borers, narcissus bulb flies produce a somewhat related type of injury; these species burrow into the mature bulb and destroy it. Likewise, the iris borer completely destroys the rhizome by hollowing it.

4. **Discoloration, curling, stunting** of the leaves and flowers, and on occasion necrosis produced by the inconspicuous sucking of equally inconspicuous pests are common types of injury. At the present time the most destructive of the sucking pests are the mites. They are responsible for general discoloration, loss of vigor, and decrease of flower production (Fig. 16:7). Rose bushes severely infested with twospotted spider mite take on a brown, dried appearance, called "hardness" by growers. By its injection of toxic salivary secretions, the tarnished plant bug induces serious malformations in developing flower buds and developing foliage. Defects in the flowers and leaves caused by feeding of thrips have created serious financial problems for many growers (Fig. 16:8). The damage mites produce is confined largely to the flowers, although some species may injure the foliage as well. Flower damage is particularly annoying because the crop is generally ready for market when it is injured, and the damage is not apparent until the crop is harvested.

5. Apart from the general loss of vigor and loss of productiveness, some sucking insects contribute to the general unsalability of floricultural crops by the excretion of sticky **honeydew**. Several aphid and scale species are responsible for this type of injury, which is compounded by the unsightly brown or black sooty mold which grows on the honeydew.

6. The presence of **insect bodies** or insect scales is unsightly and hence responsible for loss of quality in floricultural plants. Whiteflies reduce quality

Figure 16:7. Deformed leaves and buds of cyclamen caused by feeding of the cyclamen mite.

Figure 16:8. Thrips damage to carnations. Note the injured petal edges.

when nymphs attach to leaves and when adult whiteflies flit about the plant. Aggregations of aphids on host plants lower quality because of their unattractive appearance.

7. The **transmission of plant diseases** by insects and mites also contributes to the problems of growers of floricultural crops.

PEST MANAGEMENT

Because insects and diseases are critical factors limiting the health and marketability of greenhouse plants, a planned pest management program is essential for keeping on top of the array of problems affecting floral crops. Regular practices of pest exclusion, isolation and examination of plant introductions, pasteurization of plant media, sanitation, weed control, regulation of the greenhouse environment, and promotion of vigorous plant growth are imperative and must be employed along with adopted methods of chemical control.

Pests gain entrance to the greenhouse in a number of ways. In summer they will enter through open vents and doors. At night greenhouse lights will attract insects, especially moths, which may fly through the same openings or crawl through structural cracks and broken panes. Screening doors and ventilators (Fig. 16:9) and repairing breaks in the structure will stop the entry of many pests. Insects and other pests may be carried into the greenhouse on infested plants, infested equipment, and in infested soils and mulches. Purchasing certified pest-free plants and utilizing pest-free soil and equipment ensures against inadvertent introductions. It is advisable to have an inspection and isolation house for examination—and when necessary for treatment—of incoming plants and other material before taking these into the main growing areas.

Figure 16:9. Screening the side vents of a greenhouse with cheesecloth to exclude thrips. A, erecting the screen; B, completed job.

A

B

Sanitation in the greenhouse involves pasteurizing of planting media with steam, thorough cleaning of carts, potting benches, and tools, and sterilizing used pots with steam or formalin solution. Keeping areas beneath benches dry, clear, and free of equipment and accumulations of trash eliminates potential hiding places for a variety of pests.

During the interval between crops a general greenhouse cleanup is advisable; in addition to the above sanitation practices it should include sterilizing walkways and benches with commercial disinfectant, painting the growing benches with copper naphthenate, and finally fumigation.

In summer, weeds growing outside and around the greenhouse harbor pests that gain entry in devious ways, while all year long weeds growing unchecked inside provide a continuous habitat and source of food for pests. Hence weed control is important for good insect control.

Providing optimum conditions of light, temperature, humidity, and nutrition promotes vigorous growth and results in strong plants that are more resistant to pests and disease pathogens than weak ones.

CHEMICAL CONTROL

Because of the rapidity of their action, their effectiveness, and their general reliability, chemicals are essential means of controlling floricultural pests today. The phenomenal decimation of extremely destructive pests by chemicals has revolutionized control practices in floriculture. Prior to the introduction of modern insecticides, chemical and mechanical methods were largely inadequate. The older methods included the use of nicotine painted on the steam pipes, naphthalene fumigated from hot plates, lead arsenate and paris green sprayed on plants, and forceful streams of water to wash off the pests.

Although the intrinsic properties of the new organic insecticides were largely responsible for their success, the development of the aerosol method of distribution firmly entrenched them as favored agents for the control of floricultural pests. Another advancement came with the discovery of systemic insecticides, which provided an easy way to combat certain sucking pests.

The good results achieved by these compounds, however, were accompanied by problems unique to greenhouse insect control. Parathion was effective for most pests, but it was not very effective against brown soft scale. As a consequence, populations of this species became prevalent in most rose houses. The introduction of other phosphates such as malathion and sulfotepp, however, alleviated the problem. Likewise, DDT was not especially effective against leafrollers, but subsequently other materials such as TDE provided that control.

Organophosphorus and carbamate systemic insecticides are now used in the protection of several crops, particularly lilies and chrysanthemums. Three systemics are recommended currently—demeton, oxydemetonmethyl, and aldicarb; they are useful for the control of aphids, thrips, whiteflies, mites, and several other pests.

Because of the cancellation of many chlorinated hydrocarbon insecticides, few are still available for use in greenhouse floriculture, but those few are needed by the industry. Lindane is recommended for control of fungus gnats, spittlebugs, symphylans, centipedes, and sowbugs; methoxychlor for control of beetles, thrips and caterpillars; and endosulfan for control of aphids, spittlebugs, whiteflies, and cyclamen mites.

The organophosphorus insecticides have found favor both as general insecticides and as acaricides. Applied as sprays, aerosols, vapors, fogs, or smokes, they include the widely-used compounds of malathion, parathion, diazinon, dichlorvos, naled, and trichlorfon, and the special greenhouse compound sulfotepp. For control of caterpillars on chrysanthemums the microbial, *Bacillus thuringiensis* is an effective insecticide. The synthetic pyrethroid, resmethrin, applied as a foliar spray or a nonthermal fog, is especially effective against the greenhouse whitefly.

Because of the serious infestations of floral crops by mites and their resistance to many of the insecticides, specific miticides are extensively employed in controlling these pests. They include the older compounds of dicofol, chlorobenzilate, propargite, and the more recently developed compounds of Pentac and Plictran. Baits, sprays, or dusts of metaldehyde or methiocarb (Mesurol) are successful in controlling slugs and snails.

Because most floricultural crops are subject to serious damage from more than one pest, a single application, chemical, or procedure may not be sufficient. A common approach is to make several applications or to use several chemicals. There are generally two types of control programs: preventive programs and watch-and-wait programs. **Preventive programs** are the preferred approach to insect control in floricultural crops. In this type program, insecticides are applied on a regular schedule regardless of whether pests are present or not. In this way, insecticide applications become part of the routine of growing the plants instead of a special event. The principal strength of the preventive program is that it does not allow insects to develop into large, destructive populations. Such programs also encourage the purchase and installation of adequate equipment for regular applications.

Watch-and-wait programs consist of applying insecticides when an infestation occurs. This type of program requires a minimum of labor and materials; but in order to be effective it must have as a necessary part of the program daily checks on the plants to determine the presence of insects. It is at this point that the program generally breaks down, for observation of pests is generally not a part of the growing routine. Consequently, an infestation generally becomes large, injurious, firmly established, and difficult to control before it is noticed.

In establishing any control program several considerations must be made: the determination of which pests are consistently serious and which pests are only periodically serious. A preventive program can become wasteful if every pest of unpredictable importance is taken into account. What, in fact, happens in successful preventive programs is that pests that appear regularly are anticipated by a prearranged application schedule, and pests that appear irregularly are kept under surveillance. When the irregularly appearing pests are observed in damaging numbers, they are combatted by employing prearranged treatments.

Preventive programs do not consist solely of chemical treatment. The wise use of other control methods, cultural and mechanical, plays an important part. A well-planned program correlates the life history of the pest, the culture of the host, and any peculiar characteristic of the control procedure.

CONTROL EQUIPMENT

There is considerable variation in the way control chemicals are applied to floricultural crops. The method used depends in part on the kind of structure in which the crop is being grown. The use of **hydraulic sprays** has been popular on crops grown out of doors or under lath or cloth. Spraying is also a standard method used within the greenhouse. Depending on availability of formulations and the crop requiring treatment, growers may prepare sprays from emulsifiable concentrates or wettable powders. Emulsions are more likely to be phytotoxic than wettable powder suspensions, but they are easier to keep mixed and do not leave as much visible residue. It is essential to place spray on the undersurface of the foliage as well as on the upper surface. (There is, however, a so-called translaminar effect of nonsystemic acaricides: deposits on the upper leaf surface exert acaricidal action on the mites inhabiting the undersurface.)

There has been an increased interest in the use of small **mist blowers** or mist concentrate machines for applying insecticide in floricultural establishments. The advantage of this type of machine is the virtually invisible residue that results from the application. This invisible residue, of course, means that the chemical residue on the foliage does not detract from the quality of the plant.

A convenient device for the small retail grower is the **Whitmire Aerosol Generator** No. 1200, which weighs about 15 lb, is charged with 1 per cent resmethrin under pressure, and is equipped with screw-on hose and a special nozzle. The grower carries the sprayer and directs a fine spray to the undersurfaces of leaves, holding the spray gun lower than foliage and spraying upward at an angle. Dosage is based on timing; 100 sq ft of plants are sprayed in 5 to 10 seconds.

Although insecticidal **dusts** may be used effectively to control floricultural pests, they have several disadvantages. Dusts leave visible residues that are objectionable at sale time; particles are redistributed when plants are handled either to remove cut flowers or to tie or "pinch" them; and it is difficult to avoid inhaling dust particles under greenhouse conditions.

Space treatment of greenhouses is a common method of pest control. The grower has several ways of putting the insecticide into the air—aerosols, fogs, vapors, smoke generators, or fumigation. The **aerosol method** employs a liquified gas in a container or "bomb" in which insecticide is dissolved with auxiliary solvents. The liquefied gas is discharged through a nozzle and evaporates immediately upon its discharge, leaving small particles of insecticide suspended in the air. These particles fall rapidly, settle on the foliage, and to some extent volatilize, producing a fumigant effect. The aerosol method of applying insecticides in greenhouses has been popular because of its economy of time and labor.

Proper use of aerosols, and other methods of space treatment, demands proper safety equipment. The insecticides employed in aerosols are generally quite toxic to mammals so that precautions must be taken to avoid inhaling them. Furthermore, the propellent gas, often methyl chloride, can be toxic at high concentrations. Aerosoling is generally done with a full-face gas mask and with an appropriate canister attached to the mask. Protective clothing to cover all exposed skin is also necessary. The aerosol method of application can be used only in greenhouses or closed areas because the particles are so small (generally 1 to 10 microns) that the slightest breeze or air current will influence their distribution and produce uneven coverage.

One type of aerosol generator, called the **Hi Fog machine**, has been used extensively on the west coast for the control of floricultural pests. This method of application utilizes an insecticide concentrate under extremely high pressure. The insecticide is released through a small nozzle under pressures of 350–1000 psi, so that there is a fine breakup into small droplets. The resultant particles are not quite as small as liquefied gas aerosol particles, and consequently Hi Fog particles settle more rapidly, producing a heavier deposit on plants. The Hi Fog machine can be used under lath, under cloth, and even in the open field on calm days (Fig. 16:10).

Special apparatus have been developed to dispense **thermal fogs** in which oil-based formulations are injected into heated exhaust or heated air stream. Growers carry or pull the foggers through walks of the greenhouse and direct

Figure 16:10. The Hi Fog aerosol sprayer. A, closeup of sprayer; B, grower treating greenhouse carnations with Hi Fog sprayer.

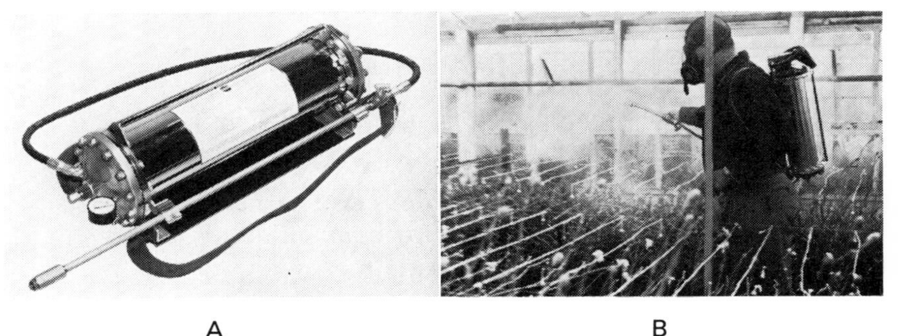

A B

the fog under the benches or toward the ground—not at the plants, which may be injured by hot particles and scales.

A method of space treatment that requires no special equipment is the production of **vapors** by painting insecticide on steam pipes or on hot plates. A slurry of a wettable powder or an emulsifiable concentrate can be used in this way. Suitable compounds include dichlorvos, naled, and nicotine.

Smokes likewise require no special equipment. Formulations of insecticides are purchased as **smoke generators**; when pest control is needed, growers place these devices in walkways and ignite them. The smoke generator burns a low-temperature combustible material in which the insecticide is dissolved. The smoke that is emitted is a mixture of combustion byproducts and insecticide in intimate contact and is distributed throughout the greenhouse by convection currents. The particles, generally less than 1 micron in diameter, are smaller than aerosol particles. Consequently they are much more subject to breezes and convection currents than liquefied gas aerosol particles and have not been effective for treating plants under lath or cloth in the field. The number of insecticides that can be successfully formulated as smokes is limited because insecticides of relatively high volatility and heat stability are required.

Traditional **fumigation** has been employed in the past to control insects in greenhouses. Although the fumigant, calcium cyanide, was widely used at the turn of the century, it had some serious drawbacks. Not only was its mammalian toxicity high, but in addition, it would often cause serious damage to a number of plants; yet it is still a useful fumigant for general cleanup of pests in the interval between crops.

The application of insecticidal dusts, granules, or drenches to the soil is effective in the control of several greenhouse pests. **Dusts** of nonsystemic insecticides may be used on the surface to control pests that crawl over the soil such as sowbugs and slugs, while drenches and granules of nonsystemics may be worked into the soil to control soil-inhabiting pests such as symphylans. **Drenches** and **granules** of systemic insecticides placed in the soil are effective in controlling sucking pests such as aphids, whiteflies, and mites. Special applicators are available for placing measured amounts of granules into the soil of potted plants.

REPRESENTATIVE FLORICULTURAL PESTS

Garden Symphylan. *Scutigerella immaculata* (Newport) [Symphyla:Scutigerellidae]

The garden symphylan is one of the serious subterranean pests that attacks floricultural crops. Symphylans are particularly injurious to seedlings of such plants as tomatoes, snapdragon, and sweet peas. They are also injurious to germinating seeds and root hairs of corn, lettuce, spinach, celery, cucumber, stock, roses, chrysanthemums, asters, gladioli, smilax, lilies, and ferns. In

California, they may seriously injure field asparagus; in Illinois they have caused severe losses, up to 50 per cent, in yield of corn from scattered fields and, in the East, in field plantings of lettuce. Much of the damage done by these pests is to the roots, for symphylans eat off root hairs or chew cavities into larger roots and crowns. On plants such as sweet peas, stems may be hollowed out and new roots destroyed as quickly as they are produced. Surviving plants are stunted and produce poorly in yield and quality.

Symphylans can also be a serious problem to growers of African violets, especially when these plants are grown on a large scale and plunged into soil or moist sand on a bench. Symphylans will then infest the whole bench and every pot on the bench. Whole benches of chrysanthemums and other florist crops have been infested. On occasion the infestation is confined to a relatively small area or several discontinuous areas.

Description. The garden symphylan is a tiny creature, never more than 8 mm long and $\frac{1}{2}$ mm wide (Fig. 16:11). It is usually pure white; the adult has 12 pairs of legs and 14 body segments. Occasionally the alimentary canal is visible through the body wall as a darker streak. Symphylans have long conspicuous antennae composed of about 50 to 60 segments in adults. The egg is about 0.5 mm in diameter and pure white or pearl colored.

Life history. The symphylan adult lays her eggs in groups of 5 to 25 in cavities in the soil. Usually these eggs are deposited deep in the soil, where it is undisturbed and normally moist. In ground beds this is generally in the subsoil. In the greenhouse, the eggs hatch in 10 to 12 days, whereas in the field they take from one to three weeks. Upon hatching, the young symphylan is about 1.6 mm long and looks very much like its parent, except that it has

Figure 16:11. Life stages of the garden symphylan. A, eggs and newly hatched young; B, adult. *A, courtesy University of California; B, courtesy Oregon State University, Corvallis.*

A

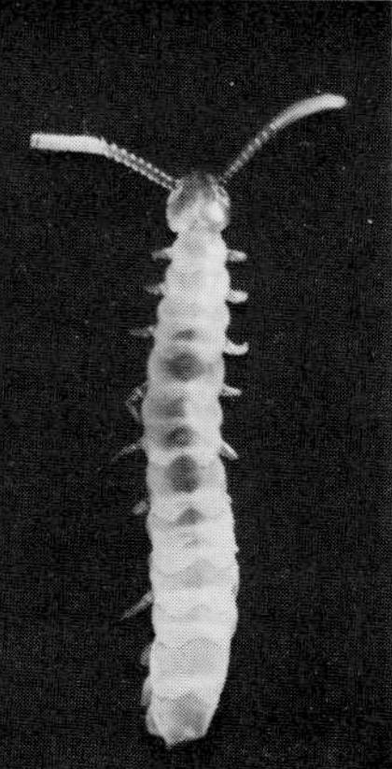

B

only six pairs of legs and six antennal segments. Young symphylans molt six times and add a pair of legs with each molt until reaching adulthood. During the developmental period, symphylans also add antennal and body segments.

As adults they continue to molt for a long period of time, undergoing as many as 50 molts. Symphylans are long lived. Several investigators have reported them as living for four years and longer. Symphylans have silk glands located in the two appendages that project from the rear of their bodies. They can produce fine sticky silken threads, and it has been suggested that these threads help guide the symphylans through the soil.

Symphylans are very active creatures and are quite sensitive to changes in their environment. Although they will often make their way up to the surface to feed on seeds or seedlings, they generally remain under cover of the soil. When disturbed, they run away quickly from the light and escape into the soil. They are delicate animals and apparently are unable to make their own burrows. Instead they depend upon burrows of earthworms, normal air cavities, and holes left by decayed plant material.

Optimum activity of symphylans occurs between 50 to 80°F. They are, however, active and present at soil temperatures as high as 85°F, but do not do well at 90°F and die quickly when the temperature reaches 100°F. They survive outdoor soil temperature as low as 32°F during the winter. Populations in nature have two peaks of accelerated egg deposition, one in spring and one in fall. In the laboratory symphylans exhibit a reproductive periodicity; a two-month period of high egg production occurs followed by a three- to four-month period of low egg production. Symphylan populations increase fastest at a temperature of 75°F.

Control. Control of symphylans has largely centered upon soil sterilization and incorporation of certain chemicals into the soil. Although symphylans have a number of natural enemies such as predatory mites, diseases, the larvae of several species of beetles, and several centipedes, these enemies have never been employed effectively in greenhouses to provide biological control.

Cultural methods have been effective in control. The garden symphylan occurs outdoors in manure piles, leaf mold, and other decayed material; therefore soil obtained from such places should be sterilized before bringing it into the greenhouse. One should also avoid bringing in infested potted plants. Commercial experience has demonstrated that entire beds can be infested by bringing in one or two pots containing the garden symphylan.

Growing plants on raised beds has been suggested as a preventive measure. Symphylans that may infest the walks and area under the raised beds do not crawl up the dry posts from the ground soil to the bed. But raised beds can be contaminated in other ways. Incorporation of cinders or other drainage material in the bottom of the bed provides for rapid movement of symphylans throughout the bed. Yet, in spite of this, drainage material should not be omitted with the intention of thereby controlling symphylans.

A helpful adjunct to soil sterilization practices is the employment of solid-bottom beds. Many floriculturalists who grow their crops in ground beds have not been able to obtain adequate control of the garden symphylan by sterilization methods because the pests will travel down into the soil, to levels below where the heat or chemical can reach them. Putting solid bottoms in ground beds permits sterilization throughout the soil mass.

In addition to these suggestions, cultivation of empty beds between crops has been helpful in that it kills many symphylans and eggs by the mechanical action of turning the soil. Heating empty greenhouses during summer has also been suggested, but the practice is not very effective in ground beds because active stages can always reach cool, moist soil. It has also been suggested that compaction of soil would delay or inhibit rapid spread of symphylans. Steam is probably the earliest and most effective soil sterilant for eliminating symphylans in raised beds. The use of carbon disulphide along with steam is even more effective, but it increases the cost.

Chemicals have also been used to control symphylans in soil. Carbon disulphide, one of the earliest chemicals used, is poured or injected into holes in the top of the bed, and the holes are covered with soil. After application of the fumigant, the soil surface is wetted or covered with a tarpaulin to retain the gas. For best results with carbon disulphide, soil temperatures must be above 60°F and moisture conditions adequate for planting. If the soil is too dry, the gas escapes quickly; if the soil is too wet, it does not penetrate sufficiently. Because the eggs of symphylans are not killed by carbon disulphide, repeated treatment is necessary. There are other sterilant chemicals, such as chloropicrin and methyl bromide, that work well against symphylans.

A regular program of soil sterilization with chemicals can be helpful in keeping symphylan damage to a minimum, because it does take some time for the symphylans to reach injurious numbers again. Consequently, a grower on a regular program of soil sterilization, whether it be with chemicals or with steam, can effectively prevent serious damage from symphylans most of the time.

Beds are often infested with symphylans when the crop is in place, and soil sterilization methods, whether steam or chemical, cannot be used without damage to the plants. In these cases, soil treatment with chemicals such as lindane has been effective. Lindane is probably the most effective chemical and provides long-lasting action against symphylans. Protection up to three to five years has been reported from one application of the material in raised beds as well as ground beds. Lindane can be applied as a dust or as a spray over the surface of the soil and raked in or mixed into the soil so that the symphylans will come in contact with it. Potted plants can be treated with lindane by watering the plants with an emulsion of the insecticide. Lindane has proved so effective that it is now the mainstay of symphylan control in the greenhouse. Tests conducted on outdoor field crops have disclosed several effective organophosphorus insecticides of symphylans—ethoprop, fonofos, and phorate.

Flower Thrips. *Frankliniella tritici* (Fitch) [Thysanoptera:Thripidae]

Because thrips are blown into greenhouses by the wind, a complex of species is involved in the injury caused to floricultural crops. Although composition of the complex varies with location and season, important species include the flower thrips, tobacco thrips, onion thrips, pear thrips, composite thrips, and greenhouse thrips (Fig. 16:12).

Thrips cause damage in several ways. They may injure the flowers directly; with their rasping-sucking mouthparts, they puncture the surface of petals and suck the liberated juices. Their feeding produces streaks and browning of the tips of the petals. On some plants, they induce a stippling and silvering of the leaves. The excrement they leave on foliage creates an unsightly residue. Thrips may also attack and injure bulbs and seeds.

We shall discuss the flower thrips to illustrate the life history and control of this group of insects.

Description. Adults of flower thrips are small insects, about 1 mm long and 0.3 mm wide. They vary in color from brown to yellow. The four immature instars are lighter in color, smaller than the adults, and without wings (Fig. 3:12). The eggs are bean-shaped, 0.2 mm long, and are inserted into plant tissue.

Life history. Dependent on several factors, this pest requires from 7 to 22 days for the completion of one generation. In warmer areas of the country there may be from 12 to 15 generations per year. The females have a preoviposition period of from one to four days or longer depending upon the temperature. The eggs generally hatch in about three days, and the first nymphal instar requires two days. The mean period of the second instar is three days, whereas the third instar, sometimes designated "propupa," requires about one day. The fourth or "pupal" instar requires $2\frac{1}{2}$ days. The total developmental period from egg to adult takes on the average from 11 to 12 days. Under favorable conditions, the short developmental period allows a very rapid increase in numbers. Thrips live on the foliage or in the flowers of the host. The pupa, however, generally resides in the soil. The adult insect is a weak flier but is strong enough to launch into air currents and be carried by winds. Heavy populations of thrips are often carried from southern areas to northern parts of the country.

Control. Control of thrips in greenhouses is complicated by the fact that wind-borne adults continually invade the structures and thereby place tremendous pressure on any chemical used to destroy them. Thrips are blown in through both upper and lower vents, and the intensity of infestation is influenced by location of the greenhouse, prevailing winds, density of thrips outdoors, and location of fields in which they live and

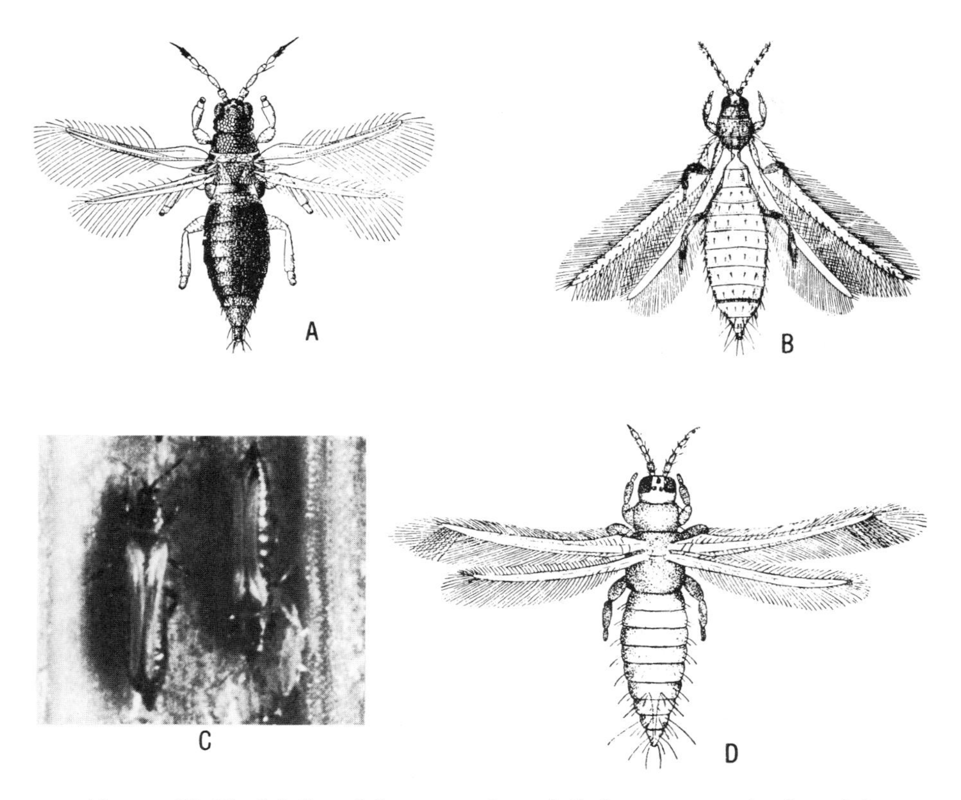

Figure 16:12. Adults of four species of thrips commonly found in greenhouses. A, greenhouse thrips; B, pear thrips; C, onion thrips; D, flower thrips. *A, courtesy USDA*; *B, courtesy Can. Dept. Agr. C, courtesy Hasso von Eickstedt*; *D, courtesy Fla. Agr. Exp. Sta.*

multiply. To prevent this constant invasion growers may cover vents with fine screen or cheesecloth, providing a physical barrier to migrating thrips (Fig. 16:9).

Insecticidal treatment of infested plants within the greenhouse may be accomplished in a variety of ways: foliar sprays, space treatment with aerosols, fogs, vapors, or smokes, or soil treatment with systemic compounds. Effective foliar sprays may be prepared from emulsifiable concentrates of oxydemetonmethyl or from wettable powders of lindane. Spraying of plants may be required at 7- to 10-day intervals until control is achieved. For space treatment dichlorvos, sulfotepp, naled, or nicotine are recommended; treatments may need to be repeated at 5- to 7-day intervals. Granules of aldicarb or drenches of demeton applied to the soil act systemically in controlling thrips. Aldicarb treatment may need to be repeated at 3- to 4-week intervals and demeton at 2- to 3-week intervals.

Twospotted Spider Mite. *Tetranychus urticae* Koch [Acarina:Tetranychidae]

The twospotted spider mite, probably the most serious pest in floriculture today, attacks a wide variety of hosts. Carnations, cymbidium orchids, gardenias, hydrangeas, and roses are greenhouse crops constantly under attack. Roses are especially vulnerable because the cultural conditions of year-round high temperatures provide a favorable habitat for the mite to survive and multiply (Fig. 16:13). The feeding of mites produces a speckled appearance in leaves and flowers. Heavy infestations cause leaves to drop and plants to lose vigor and bear fewer flowers.

Description. The twospotted spider mite (Figs. 14:13 and 16:14) is less than 0.5 mm long. Males are slightly smaller than females. When observed carefully, all four motile stages can be seen to be covered with bristles. The body varies in color from pale orange to dark green and bears a relatively large, dark spot on each side. The eggs are clear to pale green, spherical, and about 0.1 mm in diameter.

Life history. The female twospotted spider mite lays from 3 to 14 eggs per day. The average female usually lays about 70 eggs in her lifetime, but some females will lay as many as 200. Incubation lasts from 2 to 15 days, depending on temperature. Upon hatching from the egg, the first or larval stage has only six legs. Depending on temperature, this stage lasts from $1\frac{1}{2}$ to $4\frac{1}{4}$ days. Larvae

Figure 16:13. Severely infested rose shoot completely webbed by the twospotted spider mite.

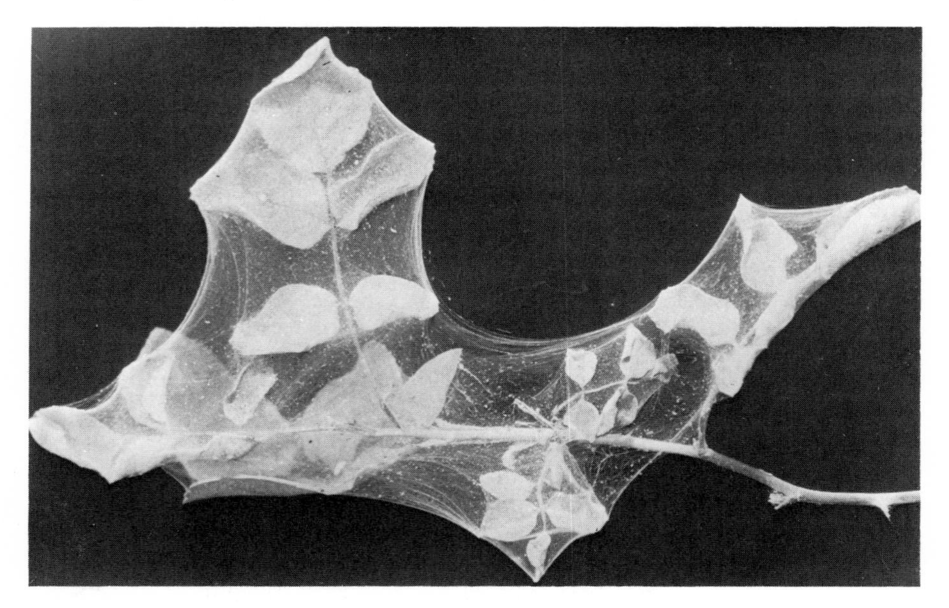

Figure 16:14. Twospotted spider mite, adults and eggs.

go into a quiescent period, at the end of which they shed their skins and emerge as protonymphs. This stage has eight legs and lasts from $1\frac{1}{2}$ to $3\frac{1}{4}$ days. A quiescent period occurs again, at the end of which the skin is cast, and the deutonymphal or second nymphal stage appears. The mite remains in this stage from 1 to 16 days, depending again on temperature, with the male having a shorter deutonymphal stage than the female.

The adult female that emerges after the quiescent period has a preoviposition period of about $1\frac{1}{2}$ days. Females live on the average for three to four weeks and produce eggs throughout this time. Interestingly, the male offspring of any female are haploid, while the female offspring are diploid. Consequently, the sex ratio is affected by the amount of sperm transferred from the male parent to the female parent, because all of the unfertilized eggs will become males. The length of the cycle from egg to egg varies from as little as 5 days to as long as 40 days.

Control. The chemical control of spider mites has been plagued with the pests' development of resistance to acaricides demonstrated to be a complex phenomenon of population genetics. Research conducted to improve control has investigated the usefulness of combinations or alternations of compounds with different modes of action or of different chemical types as well as the value of new and more potent acaricides. Recommended acaricides of different chemical grouping are the following: (1) chlorinated hydrocarbons—dicofol, chlorobenzilate, Pentac; (2) organophosphorus—demeton, oxydemetonmethyl, dichlorvos, naled, parathion, sulfotepp; (3) carbamate—aldicarb; (4) sulfur compound—tetradifon, propargite; and (5) organotin compound—Plictran.

Different methods are used in applying these pesticides. All except dichlorvos, sulfotepp, and aldicarb are useful as foliar sprays. Dichlorvos, parathion, and sulfotepp can be used as aerosols; naled as vapors; and sulfotepp and tetradifon in smoke generators. For production of fogs dichlorvos, dicofol, naled, Pentac, and sulfotepp are suitable. Dusts of 3 per cent tetradifon are available for treating foliage. Granules of aldicarb may be worked into the soil, or drenches of demeton may be applied to the surface. Compounds effective against current strains of resistant mites as well as nonresistant strains are Pentac, Plictran, and aldicarb.

Gray Garden Slug. *Agriolimax reticulatus* (Müller)

Marsh Slug. *Agriolimax laevis* (Müller)

Spotted Garden Slug. *Limax maximus* Linnaeus

Greenhouse Slug. *Milax gagates* (Linnaeus)
[Stylommatophora:Limacidae]

Slugs are common inhabitants of greenhouses and floricultural plantings. Because there are a large number of economic species throughout the United States, we shall discuss slugs as a group instead of concentrating on one species. Slugs, favored by high humidity, are a common pest in greenhouses, but they may also damage crops grown outdoors. They prefer to feed on seedlings and the more succulent parts of plants and devour leaves, stems, or roots. In feeding, young slugs rasp away surface tissues; older slugs make irregular holes (Fig. 16:15). Through commerce, slugs are spread from greenhouse to greenhouse with plants and plant parts. They may also enter greenhouses in soil and flats from the outdoors.

Description. A most vivid description of slugs is that they are snails without shells (Fig. 16:16). Depending upon the species and age, slugs vary in size from $\frac{1}{4}$ in. to 7 in. long. They secrete a characteristic slime, which they leave behind when they locomote. This slime trail is typical and is a good diagnostic characteristic for identifying slug damage. The color of slugs also varies with species, ranging from a dark black brown to a light gray with darker spots. Some have a mantle of lighter color. Also characteristic of slugs are the soft slimy bodies and the extensible eye stalks. The eggs are gelatinous and watery in appearance. They range in size from $\frac{1}{8}$ to $\frac{1}{4}$ in. and vary in shape from round to oval. They are usually colorless but often reflect the color of their surroundings (Fig. 16:17). Baby slugs resemble their parents but may not be colored as fully.

Life history. All slugs lay eggs. Each species requires a different length of time for the development of its eggs and the maturing of its young. Slug eggs

Figure 16:15. Slug injury on gerbera.

are laid in clusters in concealed, moist locations. The number of eggs laid at one sitting by one slug may vary from only 2 to as many as 100, but averages 20 to 30. Apparently young adult slugs lay fewer eggs than older ones.

In many states slug eggs may be found outdoors during any month of the year. Most species overwinter in the egg stage. In spring, the eggs hatch, and the slugs reach maturity and lay eggs by fall. During periods of particularly favorable climatic conditions, the rate at which the eggs and slugs develop may increase to such an extent that eggs are laid in midsummer instead of in fall, thus making a second generation possible. Som‿ species that overwinter as adults may lay eggs any time during the spring or summer. Mating usually takes place from August until mid-October, and eggs are laid from 30 to 40 days after a successful mating. In greenhouses, where climatic conditions are always favorable, egg laying occurs the year around.

Generally eggs are laid on the soil surface but are usually deposited in places of concealment. Mulch, dead leaves, empty flowerpots, plant flats, rocks, and boards all serve as shelter. Particularly preferred are spots beneath clay flowerpots and similar materials, where the nature of the cover keeps the surroundings relatively cool and moist. At least one species, the greenhouse slug, habitually buries its eggs in tunnels beneath the soil surface. Other species may also bury their eggs, especially in the fall.

Little or no development takes place in the overwintering eggs. The minimum temperature at which egg development takes place varies with the species of slug but is in the range of 32 to 40°F. At 40°F as long as 200 days

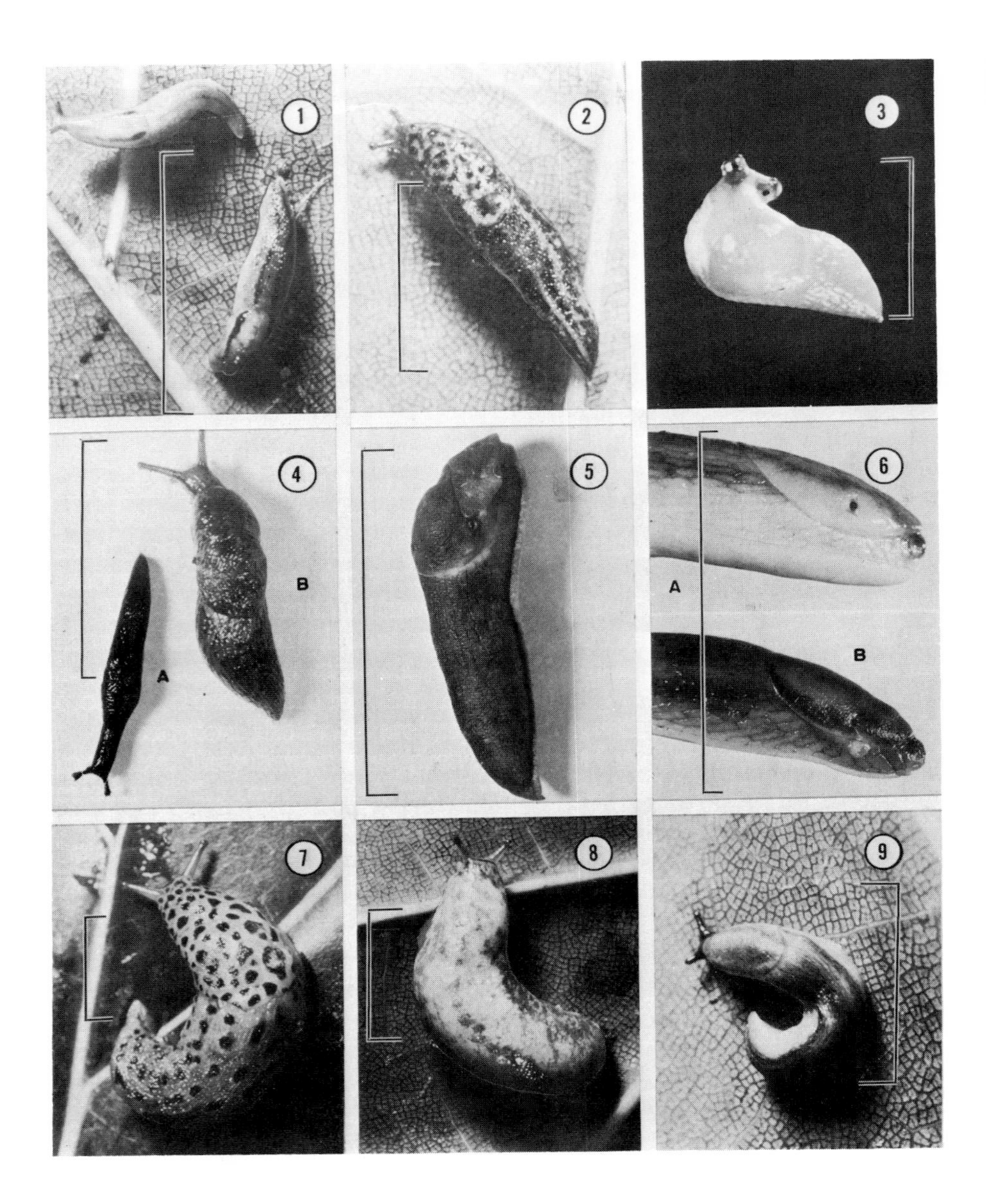

Figure 16:16. Several species of slugs found in greenhouses. 1. *Limax marginatus*; 2. *Limax maximus*, striped form; 3. *Agriolimax reticulatus*, young; 4. (A) *Agriolimax laevis*, (B) *Agriolimax reticulatus*; 5. *Milax gagates*; 6. (A) *Arion circumscriptus*, (B) *Arion subfuscus*; 7. *Limax maximus*, spotted form; 8. *Philomycus carolinianus*, common outdoors but uncommon in greenhouses; 9. *Arion circumscriptus*. Scale in each figure (except figure 3) is equal to 4 cm ($1\frac{1}{2}$ in). The scale in figure 3 equals 5 mm ($\frac{1}{5}$ in.). *Courtesy Cornell University, Ithaca, N.Y.*

Figure 16:17. Egg mass of the spotted garden slug. *Courtesy USDA.*

may be required for the eggs to develop. At higher temperatures, such as those encountered in greenhouses, development is usually completed in ten days to three weeks.

As soon as slugs hatch, they are active and begin to crawl or feed if the temperature and humidity conditions are right. Otherwise they may remain motionless and concealed until nightfall provides suitable conditions for activity. The rate of growth of immature slugs depends on the temperature and humidity of their environment and on the type and amount of food available. Optimum environmental conditions consist of temperatures ranging from 60 to 70°F and high humidity, 85 per cent or greater. Dry conditions usually result in a loss of weight, which is regained rapidly when moist conditions return. Slugs reach full adult size in three months to a year, depending upon environmental conditions. Both large and small slugs can reproduce; under the high-temperature conditions of greenhouses, slugs may lay fertile eggs when they are only six weeks old.

Slugs are hermaphroditic; that is, they can be both males and females at the same time. Usually slugs first develop mature male sexual organs. After the male development occurs, female organs may develop in addition to the male organs; or the male organs may degenerate, leaving a strictly female individual. In some species of slugs, the time of change from male to female is strongly influenced by the amount and nature of available food. Usually cross fertilization occurs; that is, two slugs mate with each other, and both individuals subsequently lay fertile eggs. Self-fertilization can also occur, however, and a single slug raised in isolation may lay fertile eggs.

The rituals associated with the mating behavior of some species of slugs are quite complex. The spotted garden slug mates while suspended several feet above the ground on slime threads; the gray garden slug engages in a "nuptial dance" which lasts for an hour or more before mating.

Very little is known about the maximum length of life of slugs under greenhouse conditions. Outdoors in a temperate climate, slugs usually live less than a year. Overwintering eggs which hatch in the spring provide the nucleus of the new population every year. A few adults of all species, but particularly the spotted garden slug, survive the winter. In greenhouses, many adult slugs live for more than one year.

Control. Formal slug control suggestions were first made during the last decade of the nineteenth century. The discovery, in 1934, of metaldehyde as a slug killer provided one of the most important chemical weapons against slugs. This discovery was the result of an accidental exposure of slugs to the English "Metafuel" (canned heat), which contains metaldehyde as a principal ingredient. This chance observation is all the more remarkable when one considers that several closely related chemicals are of no value whatsoever in controlling slugs. None of the common insecticides or fungicides which have been tested has been of any consistent value in controlling slugs. Application of metaldehyde alone or combined with an arsenical is an effective method available for the control of slugs on floricultural crops.

Although metaldehyde baits are used frequently, most growers who use baits are not satisfied with the results. The baits mold rapidly and must be replaced frequently. For this reason a 15 per cent metaldyhyde dust or the 20 per cent metaldehyde liquid, used according to label directions, is preferred over baits. The 15 per cent metaldehyde dust is preferable to the 20 per cent metaldehyde liquid for slug control. Initial kill from liquid formulations is approximately the same as that achieved with dust; but the dust gives satisfactory residual action for as long as three weeks, whereas the liquid formulation becomes ineffective in a much shorter period of time. Several applications are recommended. Metaldehyde has been used without any phytotoxic effects on a great number of floricultural and vegetable crops.

Metaldehyde has also been used as a dust against snails in orchid houses and elsewhere. The degree of control achieved against snails is much poorer than that against slugs. More experimental work needs to be done on snail control before definite recommendations can be made.

Disturbing reports of metaldehyde resistance in slugs are becoming more frequent. Some commerical formulators have added calcium arsenate or carbaryl to their molluscicides in addition to the metaldehyde. In control experiments several carbamates, such as methiocarb, methomyl, and formetanate, have been shown to be equal or more effective than metaldehyde. A 2 per cent pelletized bran bait of methiocarb (Mesurol), now available, gives faster and greater kills of slugs than the standard metaldehyde bait.

Sanitation is a necessary part of slug control. A cleanup program will minimize slug problems within and around greenhouses. Appropriate cultural control measures are (1) Do not scatter empty flowerpots indiscriminately throughout the greenhouse; avoid underbench pot storage; stack pots on a clean, dry wooden surface on their sides, not on ends. If pots are stored under a bench, keep the pots dry. (2) Stack boards only in a dry area. Store wooden flats in a dry area and turn them on end or keep the open sides down. (3) Remove weeds from the aisles and underneath the benches. Keep areas beneath benches dry and clear. Gravel and cinders aid in underbench drainage and provide unsuitable habitats for slugs. (4) Use cinders, gravel, or sandy soil in benches where potted plants are held. Soil should be present only when plants are grown directly in the benches. (5) Do not bring new plants of any sort into the greenhouse without examining them for insects, slugs, and snails. The number of growers who attribute their troubles to a specific importation of plants, but who do not examine the plants when they first arrive, is surprisingly large. Without this cultural cleanup, money spent for attempted chemical control is largely wasted.

Hand picking, crushing, or drowning slugs is a good idea. The time consumed is amply repaid. Even in the most severely infested ranges, an hour or two spent in a deliberate effort at hand picking will reduce the slug population noticeably. Individual plants of special value can be protected by placing the pots upon boards in a tray of water, by wrapping the stems in cotton batting, or by the use of electrical circuits surrounding the plants. None of these procedures is practical on a large scale.

Many insects, reptiles, amphibians, and birds have been recorded as occasional predators of slugs. None as yet offers an effective and practical means of control. This is an interesting field for future investigation, for it is possible that a tropical species could be introduced that would accept greenhouse conditions but would not be able to spread if introduced accidentally outdoors in areas of low winter temperatures.

SELECTED REFERENCES

Berry, R. E., "Garden Symphylan: Reproduction and Development in the Laboratory," *J. Econ. Entomol.*, 65:1628–32 (1972).

Blauvelt, W. E., *The Internal Morphology of the Common Red Spider Mite* (Tetranychus telarius *Linn.*), Cornell Univ. Agr. Exp. Sta. Memoir 270, 1945.

Compton, C. C., *Greenhouse Pests*, Illinois State Natural History Survey Division, Ent. Series Arc. 12, 1930.

Filinger, G. A., *The Garden Symphylid* Scutigerella immaculata *Newport*, Ohio Agr. Exp. Sta. Bul. 486, Wooster, Ohio, 1931.

Gentile, A. G., and D. T. Scanlon, *Floricultural Insects and Related Pests—Biology and Control, Section I*, Massachusetts Coop. Ext. Ser. Specialty Manual Issue for Commerical Greenhouse Growers, 1976.

Howitt, A. J., "Control of the Garden Symphylid in the Pacific Northwest," *Down to Earth*, 15:6–10, 24 (1960).

Hussey, N. W., W. H. Read, and J. J. Hesling, *The Pests of Protected Cultivation* (New York: American Elsevier, 1969).

Judge, F. D., "Aspects of the Biology of the Gray Garden Slug (*Deroceras reticulatum* Müller)," New York Agr. Exp. Sta., Geneva, *Search Agriculture* 2(19):1–18 (1972).
Karlin, E. J., and J. A. Naegele, "Biology of the Mollusca of Greenhouses in New York State," Cornell Univ. Memoir 372, 1960.
———, *Slugs and Snails in New York State Greenhouses*, Cornell Univ. Bul. 1004, 1958.
———, and ———, "Screening Greenhouses with Insecticide-Impregnated Cloth for Thrips Control," *J. Econ. Ent.*, 50:55–8 (1956).
Lewis, T., *Thrips: Their Biology, Ecology, and Economic Importance*, (New York: Academic Press, 1973).
Michelbacher, A. E., "The Biology of the Garden Centipede, *Scutigerella immaculata*," *Hilgardia*, 11:55–148 (1938).
———, *Chemical Control of the Garden Centipede*, Calif. Agri. Exp. Sta. Bul. 548, 1932.
Miles, H., and M. Miles, *Insect Pests of Glasshouse Crops* (London: Crosby Lockwood and Son, 1948).
McDaniel, E. I., *Insect and Allied Pests of Plants Grown Under Glass*, Mich. Agri. Exp. Sta. Bul. 214, 1931.
McEnroe, W. D., and J. Kot, "Evolution of Organophosphorous Resistance and Fitness in a Hybrid Swarm of the Two-Spotted Spider Mite, *Tetranychus urticae* K.," *Ann. Entomol. Soc. Amer.*, 61:1255–9 (1968).
McEnroe, W. D., and J. A. Naegele, "The Coadaptive Process in an Organophosphorus Resistant Strain of the Two-Spotted Spider Mite, *Tetranychus utricae* K.," *Ann. Entomol. Soc. Amer.*, 61:1055–9 (1968).
Powell, C. C., and R. K. Lindquist, *Insect Mite and Disease Control on Commercial Floral Crops*, Ohio Coop. Ext. Serv. Bul. 538, 1976.
Pritchard, A. E., *Greenhouse Pests and Their Control*, Calif. Agri. Exp. Sta. Bul. 713, 1949.
———, and E. Baker, *Revision of the Spider Mite Family Tetranychidae*, Memoir Series, Vol. 2, Pacific Coast Ent. Soc., San Francisco, Calif., 1955.
Russell, H. M., *The Greenhouse Thrips*, USDA Bureau of Entomology, Circular 151, 1912.
———, *The Greenhouse Thrips*, USDA Bureau of Entomology, Bul. 64, part 6, 1909.
Severin, H. C., *Insect and Other Enemies Harmful to Greenhouse Plants*, in Fifteenth Annual Report of the State Entomologist of South Dakota, Brookings, S. D., 1924.
Smith, F. F., *Controlling Insects on Flowers*, USDA Agr. Info. Bul. 237, 1967.
———, "Spider Mites and Resistance," in Insects, *The Yearbook of Agriculture*, USDA, 1952.
———, T. J. Hennebery, and E. A. Taylor, "How Thrips Get In," *USDA Agr. Research* 7(7):14 (1959).
Thomas, C. A., *The Symphylid or Greenhouse Centipede* Scutigerella immaculata *Newport and Other Pennsylvania Greenhouse Soil Pests*, Pa. Agr. Exp. Sta. Bul. 508, 1949.
Watson, D. L., and J. A. Naegele, "The Influence of Selection Pressure on the Development of Resistance in Populations of *Tetranychus telarius* (L)," *J. Econ. Ent.*, 53:80–4 (1959).
Watts, J. G., *A Study of the Biology of the Flower Thrips*, Frankliniella tritici [*Fitch*], *with Special Reference to Cotton*, S.C. Agr. Exp. Sta. Bul. 306, 1936.
Webb, R. E., *The Narcissus Bulb Fly, How to Prevent its Damage in Home Gardens*, USDA ARS Leaflet 444, 1977.
Weigel, C. A., *The Gladiolus Thrips*, USDA Bul. E-300, 1934.
———, *Insects Injurious to Ornamental Greenhouse Plants*, USDA Bul. 1362, 1925.

chapter 17 / **LELAND R. BROWN**

INSECTS OF ORNAMENTAL SHRUBS, SHADE TREES AND TURF

Man values the plants in this chapter for esthetic reasons. Ornamental plants are pleasing to the eye, and often fragrant; they provide shade from the hot summer's sun and a verdant carpet under our feet; they screen off unsightly scenes or muffle unpleasant screeches and roars of traffic.

Most of us do not realize how many plants we value esthetically. In California alone ornamental shrubs and shade trees grown for sale total over 1600 species in 578 genera, not counting the several kinds of lawn grasses and other ground covers, the multitude of herbaceous flowers, and the large number of races, varieties, or cultivars of these woody ornamental plants.

It is difficult to estimate the worth of ornamental plants. How do you assign a dollar value to beauty? One indication is the value of the 150,000 street and park trees in Riverside, California, valued conservatively at $15,000,000. Nationwide the investment in shade trees is surely in the billions of dollars.

THE PESTS

There are many hundreds of insect and mite pests attacking ornamental plants. The following is a list of a few of the better-known ones; it is arranged in a natural or phylogenetic sequence. Some of these insects are greenhouse pests in the colder states, but in the warmer ones they may be found out of doors the year around. The Homoptera, Coleoptera, and Lepidoptera include the great bulk of the pests.

Orthoptera

Tettigoniidae. Forktailed bush katydid, broadwinged katydid.

Phasmatidae. Walkingsticks.

Isoptera

Termites.

Thysanoptera

Thripidae. Pear thrips, gladiolus thrips, greenhouse thrips, flower thrips.

Phlaeothripidae. Cuban laurel thrips.

Hemiptera

Miridae. Tarnished plant bug, bluegrass plant bug, garden fleahopper, ash plant bug, Pacific ash leaf bug.

Tingidae. Rhododendron lace bug, sycamore lace bug, western sycamore lace bug.

Lygaeidae. Chinch bug, southern chinch bug.

Rhopalidae. Boxelder bug.

Homoptera

Cicadidae. Periodical cicada.

Membracidae. Buffalo treehopper.

Cercopidae. Rhubarb spittlebug, pine spittlebug, acacia spittlebug.

Cicadellidae. Rose leafhopper, whitebanded elm leafhopper, *Draeculacephala minerva*, *Deltacephalus sonorus*, *Erythroneura* spp., *Empoasca* spp.

Psyllidae. Laurel psyllid, boxwood psyllid, acacia psyllid, pear psylla, cottony alder psyllid, hackberry nipplegall maker.

Aleyrodidae. Greenhouse whitefly, crown whitefly.

Aphididae. Green peach aphid, rose aphid, rosy apple aphid, painted maple aphid, pine needle aphid, spirea aphid, viburnum aphid, poplar aphid, sycamore aphid, giant bark aphid.

Eriosomatidae. Woolly apple aphid, woolly ash aphid, woolly alder aphid, elm cockscomb gall aphid, witch hazel cone gall aphid, poplar petiolegall aphid.

Chermidae. Cooley spruce gall aphid, eastern spruce gall aphid, woolly larch aphid, pine bark aphid.

Phylloxeridae. Grape phylloxera, hickory gall aphid.

Coccoidea

Ortheziidae. Greenhouse orthezia.

Margarodidae. Cottonycushion scale, alder scale, sycamore scale.

Diaspididae. San Jose scale, oystershell scale, Bermuda grass scale, Rhodes grass scale, scurfy scale, euonymus scale, rose scale, pine needle scale, California red scale, Forbes scale, Putnam scale, European fruit scale, greedy scale, oleander scale, white peach scale, juniper scale, English walnut scale, olive scale.

Coccidae. Black scale, hemispherical scale, Monterey pine scale, tulip-tree scale, cottony maple scale, European fruit lecanium, Fletcher scale, magnolia scale, terrapin scale, brown soft scale.

Asterolecaniidae. Oak wax scale, oak pit scales.

Pseudococcidae. Citrus mealybug, citrophilus mealybug, grape mealybug, taxus mealybug, longtailed mealybug, coconut mealybug, Mexican mealybug, Comstock mealybug.

Eriococcidae. European elm scale.

Coleoptera

Buprestidae. Flatheaded apple tree borer, Pacific flatheaded borer, oak twig girdler, Pacific oak twig girdler, bronze birch borer, twolined chestnut borer, hemlock borer, California flatheaded borer.

Bostrichidae. California fan palm borer, apple twig borer, spotted limb borer, branch-and-twig borer.

Scarabaeidae. Oriental beetle, Asiatic garden beetle, rose chafer, European chafer, Japanese beetle, green June beetle, tenlined June beetle.

Cerambycidae. Roundheaded apple tree borer, rhododendron stem borer, poplar borer, ribbed pine borer, elm borer, twig girdler, elderborer, locust borer, sugar maple borer, nautical borer, California prionus, twig pruner.

Chrysomelidae. Grape flea beetle, willow leaf beetle, imported willow leaf beetle, cottonwood leaf beetle, locust leafminer, elm leaf beetle, dichondra flea beetle.

Curculionidae. New York weevil, rose curculio, leafrolling weevils, billbugs, Japanese weevil, Asiatic oak weevil, vegetable weevil, poplar-and-willow borer, white fringed beetles, Fuller rose beetle, black vine weevil, cribrate weevil, pales weevil, white pine weevil, nut weevils.

Scolytidae. Engraver beetles, western pine beetle, red turpentine beetle, southern pine beetle, pear blight beetle, native elm bark beetle, smaller European elm bark beetle, shothole borer, hickory bark beetle.

Lepidoptera

Papilionidae. Tiger swallowtail.

Heliconiidae. Gulf fritillary.

Nymphalidae. Mourningcloak butterfly, viceroy.

Hesperiidae. Silverspotted skipper, fiery lawn skipper.

Sphingidae. Catalpa sphinx, whitelined sphinx, achemon sphinx.

Arctiidae. Fall webworm, silverspotted tiger moth, hickory tussock moth, sycamore tussock moth, spotted tussock moth, pale tussock moth, saltmarsh caterpillar.

Noctuidae. Armyworm, granulate cutworm, black cutworm, variegated cutworm, underwings.

Dioptidae. California oakworm.

Notodontidae. Yellownecked caterpillar, walnut caterpillar, redhumped caterpillar, variable oak leaf caterpillar.

Lymantriidae. Gypsy moth, browntail moth, Douglas-fir tussock moth, whitemarked tussock moth, western tussock moth.

Lasiocampidae. Eastern tent caterpillar, forest tent caterpillar, western tent caterpillar.

Geometridae. Spring cankerworm, fall cankerworm, omnivorous looper.

Zygaenidae. Grape leaf skeletonizers.

Pyralidae. Sod webworm, grape leaffolder, lucerne moth, cactus moth, phycitid oak leaftier, genista caterpillar.

Olethreutidae. European pine shoot moth, Nantucket pine tip moth, western pine tip moth, cypress bark moth, cypress leaftier.

Tortricidae. Spruce budworm, fruittree leafroller, oblique banded leafroller, orange tortrix.

Cossidae. Carpenterworm, leopard moth.

Momphidae. Palm leaf skeletonizer, juniper twig girdler, ceanothus gall moth.

Gelechiidae. Lodgepole needleminer, peach twig borer, sycamoreleaf skeletonizer.

Glyphipterygidae. Mimosa webworm.

Sesiidae. Lilac borer, ash borer, maple callus borer, peachtree borer, western sycamore borer, locust clearwing.

Yponomeutidae. Ailanthus webworm, cedar and arborvitae needleminers.

Heliozelidae. Resplendent shield bearer, madrona shield bearer.

Gracillariidae. Lilac leafminer, azalea leafminer, solitary oak leafminer, sycamoreleaf blotch miner, aspenleaf serpentine leafminer.

Psychidae. Bagworm.

Incurvariidae. Maple leafcutter, live oak leafcutter, yucca moths.

Diptera

Tipulidae. Range cranefly.

Cecidomyiidae. Boxwood leafminer, spruce bud midge, rose midge, pinecone willow gall.

Chloropidae. Frit fly.

Agromyzidae. Holly leafminer.

Hymenoptera

Cimbicidae. Elm sawfly.

Diprionidae. Redheaded pine sawfly, European pine sawfly, white pine sawfly.

Tenthredinidae. Larch sawfly, willowleaf galls, birch leafminer, elm leafminer, bristly roseslug, sycamore leafmining sawfly, roseslug, curled rose sawfly, pearslug, imported currantworm, brownheaded ash sawfly.

Siricidae. Pigeon tremex.

Cephidae. Raspberry horntail.

Cynipidae. Oak gall wasps.
Formicidae. Carpenter ants.
Vespidae. Yellowjackets.
Megachilidae. Leafcutting bees.
Anthophoridae. Carpenter bees.

Acarina

Tetranychidae. Southern red mite, clover mite, spruce mite, twospotted spider mite, European red mite.

Eriophyidae. Pearleaf blister mite, Bermudagrass mite, maple bladdergall mite.

THE INJURY

Boring and Girdling

On woody plants, insect boring and girdling can frequently be the most devastating injury the plant must withstand. Often such injury goes undetected until it is too late to save the plant. When scolytid bark beetles and their larvae gnaw out tunnels at the junction of the trunk bark and wood (Fig. 17:1A), xylem and phloem tubes are cut, thus stopping the flow of water and nutrients upward and the flow of sugars downward. If insect tunnels interlace around the trunk, the tree in effect is girdled and will eventually die. The damage can be even more catastrophic if the bark beetle carries a fungus that plugs what vascular tubes are left; such is the case with the deadly Dutch elm disease.

Only slightly less devastating than bark beetles are the flatheaded borers and the larvae of clearwing moths tunneling in trunks, limbs, or twigs. Roundheaded borers (Fig. 17:1B), wood wasp larvae, bostrichids, and carpenterworms may do somewhat similar damage but are just as likely to be found boring in the solid wood, thus weakening the structural strength of the tree. Termites and carpenter ants, by their tunneling, may greatly weaken the strength of wood in trunks and limbs. Frequently weevil larvae, *Otiorhynchus* and *Pantomorus* spp., will girdle shrubs by chewing off most or all the bark and living tissue from the lower stem and larger roots. Rhododendrons and taxus shrubs thus affected wilt and die when the weather warms. Carpenter bees sometimes make rather spectacular tunnels in the wood of trunks, but their borings usually are not extensive enough to be considered very injurious.

The tender terminal twigs of many trees are killed by tunneling of insects such as by the pine tip moths, certain small flatheaded borers, *Agrilus* spp., peach twig borer, and certain scolytid beetles, *Phloeosinus* spp. Sometimes female cicadas and treehoppers injure small twigs by inserting their hard, horny ovipositors. This may weaken the twig or it may be the entrance for disease organisms or the site of sap bleeding.

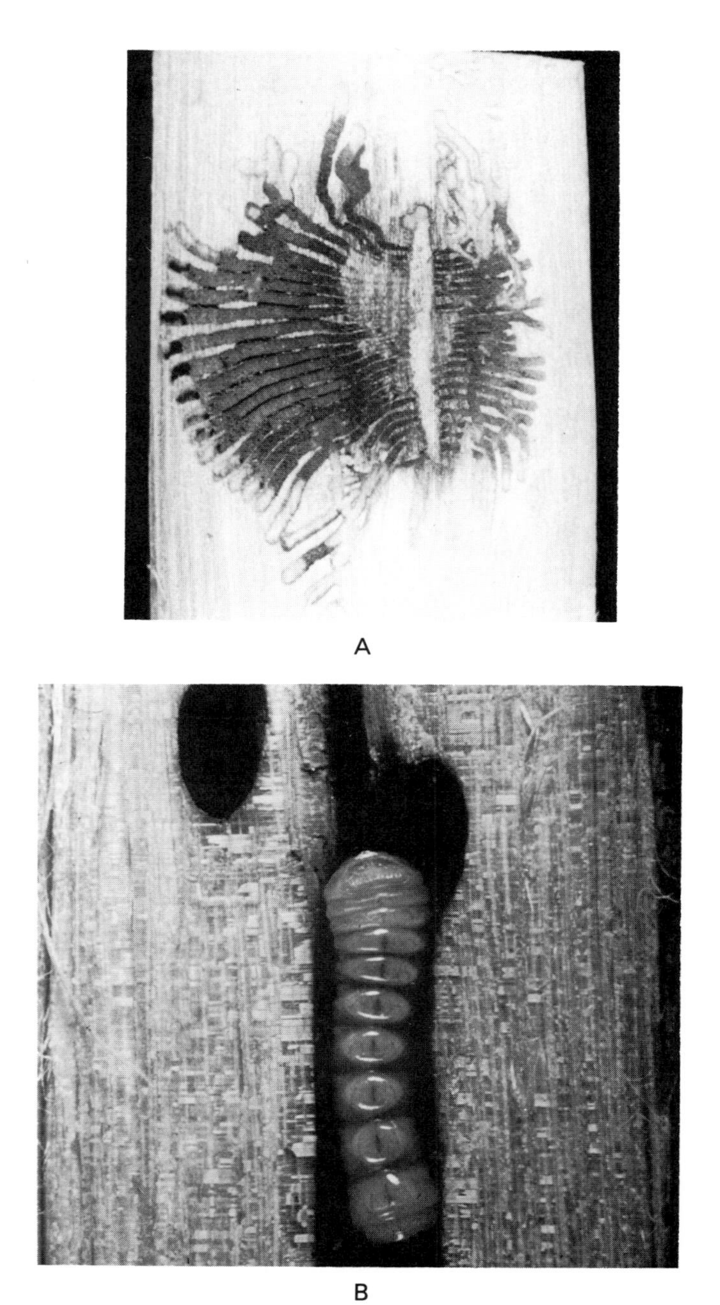

Figure 17:1. Girdling and boring damage by beetles. A, adult and larval engraving patterns of smaller European elm bark beetle; B, locust borer burrowing within the heartwood of locust. *A, courtesy University of California, Riverside; B, courtesy Cornell University, Ithaca, N.Y.*

Root Destruction

Mature shrubs and trees may have some insect feeding on their roots, but usually this is insignificant. On turf grasses, however, this kind of damage is serious and common. Perhaps the worst offenders are the subterranean white grubs, which are larvae of the scarab beetles. Well-known examples are European chafer, Japanese beetle, Asiatic beetle, and *Cyclocephala* and *Phyllophaga* species. Frequently their feeding on and cutting off roots is so severe that the turf can be picked up like a rug. Consumption of roots and crowns of grass plants by billbug weevils is often serious. Vegetable weevil adults and larvae do similar damage. Sod webworms, larvae of the grass moths, *Crambus* spp., are serious threats, particularly on new lawns; so is the fiery lawn skipper. Many species of cutworms feed on grasses and dichondra. Larvae of the dichondra flea beetle feed on the roots and crown, while adults gouge out lines on the leaves. Frit fly larvae are occasionally serious on lawns and golf putting greens.

Leaf and Flower Damage

Injury to leaves and flowers is the most conspicuous insect damage to ornamental shrubs, shade trees, and turf. Leaf feeding can take the form of complete consumption of the leaf, as by gypsy moth larvae, or only the edge of the leaf, leaving an irregular, scalloped margin, as by loopers. If only one surface is fed upon, as by the Japanese beetle, the damage is spoken of as skeletonizing (Fig. 17:2A). Leaf damage may be in the form of mines, as with larvae of certain tenthredinids or gracillariid moths. Sometimes a tree can suffer insect defoliation during one year without serious harm, but if defoliation happens a second or third year the tree may be killed because of a depletion of food reserves. Conifers may be killed after only one defoliation.

An enormous number of insects find the tender plant tissues of leaves and flowers very much to their liking. Every major insect order with chewing mouthparts contributes a share, but the Lepidoptera are the most numerous. The larval stage of Lepidoptera does the damage. Among the tussock moths are several notorious defoliators of trees: gypsy moth, whitemarked tussock moth, and browntail moth. Tent caterpillars have several members that are well-known defoliators: eastern tent caterpillar (Fig. 17:2B), forest tent caterpillar, and western tent caterpillar. The geometrids have the spring cankerworm and the fall cankerworm, as well as many lesser-known leaf-feeding loopers. Of the leafrollers, spruce budworm and fruittree leafroller are serious defoliators of ornamental trees. The woolly bear family contains a number of defoliators, including the fall webworm. One butterfly, the mourning cloak, frequently attacks willow, elm, and poplar; the larvae that feed on elm are called spiny elm caterpillars. There are many other lepidopterous families with one or more lesser known defoliators. One family, the Dioptidae, has only one species in the United States: the California oak moth, which is the most important leaf feeder on oaks in the state. Another

A

Figure 17:2. Leaf damage to hardwood trees by insects. A, adults of the Japanese beetle skeletonizing leaf; B, fully developed nest of the eastern tent caterpillar and partly consumed leaves above. *A, courtesy USDA; B, courtesy Cornell University, Ithaca, N.Y.*

family, Psychidae, has the familiar defoliator of woody ornamentals, the bagworm (Fig. 17:3A). The larvae of several families of tiny moths feed by mining inside leaves; examples are the solitary oak leafminer, azalea leafminer, and aborvitae leafminer.

The beetles are the next most important leaf and flower feeding group, often as adults but sometimes as both adults and larvae. The leaf beetles are important examples of the latter, including the elm leaf beetle and imported willow leaf beetle. The flea beetles, such as dichondra flea beetle, are in the leaf beetle family. Other beetles—such as the scarabs (Japanese beetle, rose chafer), certain weevils (Asiatic oak weevil, black vine weevil, Fuller rose beetle, rose curculio), or blister beetles—may feed on leaves only as adults.

Almost all the primitive Hymenoptera are associated with plants. Many of the common sawflies such as the bristly roseslug, pearslug, and larch sawfly, are leaf feeders. A few are serious leafminers, such as the birch leafminer. Diprionid sawfly larvae, such as the redheaded pine sawfly, feed only on conifer needles (Fig. 17:3B). Some of the higher Hymenoptera may attack plants, as vespid wasps feed voraciously on unopened rose buds and leafcutting bees make neat circular cuts in leaves, not for food but to line their nests.

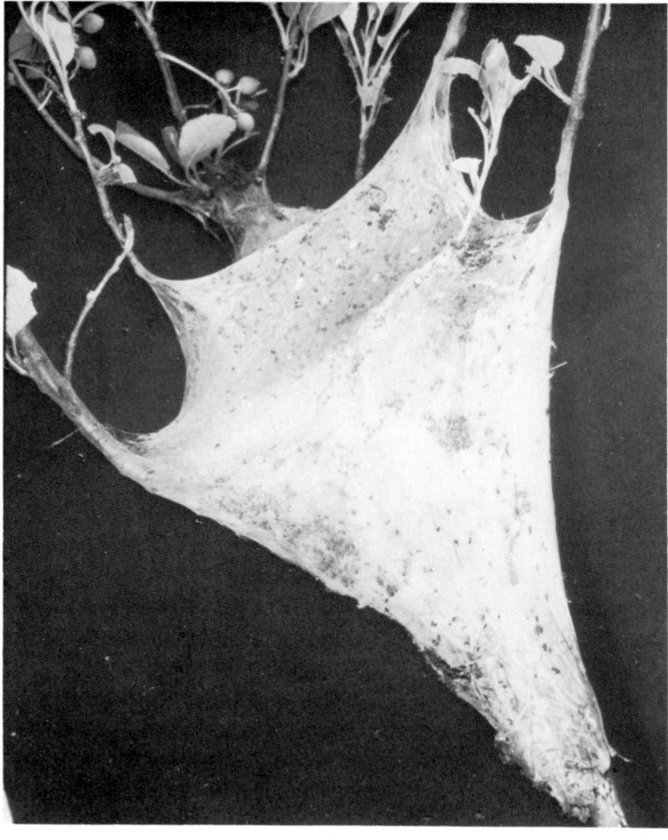

B

Not many Diptera attack woody ornamental plants. An agromyzid that is occasionally serious is holly leafminer. Boxwood leafminer, a cecidomyiid, can be important on boxwood hedges. Some Orthoptera, including grasshoppers and particularly katydids, may harm ornamental shrubs and trees; two injurious katydids are the angularwinged katydid and the forktailed bush katydid.

Sucking Damage

Many injurious insects feed by sucking up sap or liquid cell contents, leaving the cell walls intact. Besides liquid, these insects may ingest such tiny bodies as the chloroplasts (the green coloring matter of leaves). Three orders contain these sucking insects, Hemiptera, Homoptera, and Thysanoptera. The Acarina, or mites, also contain many injurious sucking pests. Frequently infestations of these insects or mites are so heavy that sap loss may amount to many gallons, even on a small tree. The loss of water may severely dehydrate the plant or plant part, particularly under drought conditions. Furthermore

A

Figure 17:3. Leaf damage to evergreen trees by insects. A, nest of the bagworm on cedar and damaged needles; B, redheaded pine sawfly larvae feeding on pine, note the stumps of eaten needles. *Courtesy Cornell University, Ithaca, N.Y.*

the loss of dissolved sugars, amino acids, and other nutrients may be devastating to the plant. It is common for heavy infestations of these insects to cause death of twigs, limbs, or even the whole plant. In many cases the injection of their saliva is toxic to the plant. Insects in this category—especially leafhoppers, aphids, and thrips—often carry the agents of plant diseases, such as the viruses. To rid themselves of excess water and sugars, called honeydew, and to concentrate the protein, homopterous insects have a filter chamber in their gut. The sticky honeydew falling on leaves or other objects below, like automobiles and outside furniture, creates a great nuisance and may affect the paint on these surfaces. A black smut fungus lives in the honeydew on leaves, making the plant appear dirty, as if it were covered with coal dust. Sucking injury may cause enormous distortions or galling of leaves, flowers, fruits, and other terminal growth, especially by aphids and their close relatives.

Gall Damage

Many sucking insects cause the host plant to form unusual plant tissue, called galls, around themselves. Examples are the spruce gall aphid (Fig. 17:4),

B

grape phylloxera, woolly apple aphid, laurel psyllid, and others. But the best-known gall makers are the multitude of cynipid wasps attracted to oak trees; each wasp causes a different-shaped gall to form on a particular part of the oak. Eriophyid mites cause bizarrely shaped galls to form on a variety of plants. Most galls are a curiosity, such as the oak apples, but some, like cork oak cynipid, may injure the tree seriously, killing twigs; others, like the twohorned oak gall, kill leaves. The cynipid wasp, which causes distorted leaf oak gall on white oak, so disrupts the leaf that it cannot function.

CONTROL

Cultural Control

The most valuable cultural control practice available is to maintain woody ornamental plants in a vigorous condition by judicious watering, fertilizing, and pruning. This is especially important in combating or resisting boring insect pests. Such invigorating procedures are used as a matter of course by commercial growers, but homeowners may not be aware how important they are or do not observe their plants critically enough to realize what happens. Almost always in the West and often in the central and eastern states, drought is a problem that makes shrubs and shade trees susceptible to attack by bark beetles and flatheaded borers.

Figure 17:4. Two galls (left and center) of the spruce gall aphid. *Courtesy Cornell University, Ithaca, N.Y.*

A useful practice before planting is to investigate which shrubs and shade trees are resistant or immune to insect attack. Gardening publications, local nurserymen, farm advisors, or county agents may be helpful as to which resistant plants to use in a given locality.

The simple expedient of periodically and systematically hosing off smaller shrubs and trees may prevent infestations of many sucking pests, such as aphids, scale crawlers, and mites—small, fragile creatures that are devastated by the water force and surface tension. After a scale infestation is established, these measures are useless.

Pruning to control insects can be useful, either directly, as pruning out a limb heavily infested with scale, or to open up the plant so that insecticidal sprays can penetrate. Pruning and destroying tips infested by pine tip moths may be almost as effective as insecticidal spraying. If insects infest only new growth (as do cuban laurel thrips, for example), and the shrub or tree has reached its desired size, such new growth can often be pruned away to prevent insect attack. Sometimes plant growth hormones can be sprayed on the plant to prevent new growth from forming.

Biological Control

In theory, biological control agents would appear ideal for the control of pests of ornamental shrubs, shade trees, and turf; but actually there are few outstanding examples of effective control by these agents. It may be, however, that shade trees and shrubs do not have more major pests than they do now because of naturally occurring biological control that goes on all the time.

Control of cottonycushion scale with the Vedalia ladybird beetle from Australia is an outstanding example of biological control initiated by man. This particular control is usually thought of only in connection with citrus orchards, but a great variety of woody ornamentals in California also benefit from the introduction of the Australian insect in 1888. The introduction of *Aphelinus mali* to control woolly apple aphid in apple and pear orchards has also been of great benefit to a large number of rosaceous woody ornamental plants. Certainly notable is the use of milky disease to kill Japanese beetle grubs in turf.

The development and use of biological control agents requires a sophisticated understanding and manipulation of the insects involved. On only the major agricultural food crops have we been able to justify the expense of the necessary research. But many of our most important pests of ornamental plants are of foreign origin such as the three discussed under "Representative Pests" and are therefore much more likely to be controlled by biological agents. Widespread biological control would involve large regional programs that, compared to insecticide use, are not available to the individual homeowner, or to the park or street tree superintendent.

Chemical Control

Food and fiber crops are produced by professional people who must be skilled and knowledgeable in the use of pesticides. Thus many very effective—and very toxic—insecticides may be employed several times a year by the farmer producing such crops. But persons caring for ornamental shrubs, shade trees, and turf are often unfamiliar with insecticides and use them infrequently. Also, ornamental plants are closely associated with people; babies and pets play on turf, for example. For these reasons usually only the "slightly" or "moderately" toxic insecticides are approved for insect and mite problems of ornamental plants. A rule of thumb is that any insecticide as a technical compound having an acute oral mammalian toxicity of less than 50 mg per kg of body weight should not be used on ornamental plants. Thus any insecticide having a skull and cross bones on its package label should not be used, because it is classified as "extremely" or "highly" toxic. But despite these limitations all categories of modern synthetic insecticides are developed for and are being used on ornamental plants, including organophosphorus, organochlorine, and carbamate compounds, as well as the microbials, botanicals, and petroleum oils.

Bacillus thuringiensis, a bacterial that is nontoxic to higher animals, is being recommended increasingly on lepidopterous larvae, including tent caterpillars, webworms, loopers, and leafrollers. Lindane is still used on many sucking pests, some Lepidoptera, and many Coleoptera; it is probably the most effective compound ever invented for boring pests. Endosulfan is useful on sucking pests and some mites. Dicofol and chlorobenzilate, are much used on all categories of plant-feeding mites. Methoxychlor fits into many uses of the now-banned DDT, including controlling elm bark and elm leaf beetles, various armyworms, cankerworms, tent caterpillars, loopers, leafrollers, and borers. It is much safer than DDT and has a much better environmental record. Trichlorfon, the first of several organophosphorus compounds we shall mention, is being used on an increasing number of Lepidoptera, such as pine tip moths, gypsy moth, tent caterpillars, and bagworm. Malathion is one of the few real "work horses" among insecticides, used on woody ornamental plants; exploited for many years, it is effective on a great many pests of most categories, and is among the safest of insecticides to warm-blooded animals, including man, and to the environment. Oxydemetonmethyl and dimethoate are systemic phosphate insecticides effective on most categories of sucking insects and mites and on pine tip moths. Systemic insecticides may be sprayed on the leaves and tender stems, where they can translocate to hidden or burrowed insects: or they may be poured or injected into the soil, where they are absorbed by the roots and translocated to the leaves. Disulfoton, also a systemic, is one of the very few "highly" toxic compounds used occasionally on woody ornamentals; to the small-package trade it is available only as a 2 per cent granular, for soil treatment, in contrast with the 7.5 to 15 per cent granular formulations available to commercial growers. It is used against various mites, some leafminers, and tip moths. Another systemic of interest is dicrotophos. Although it is highly toxic, it has been used extensively as a trunk injection to control the smaller European elm bark beetle; it does control the beetle, but not before the pest has transmitted spores of Dutch elm disease. Diazinon is considerably more toxic than malathion, but is beginning to rival it in the number and variety of ornamental pests it is recommended for, such as aphids, psyllids, scales, webworms, tent caterpillars, bagworm, pine shoot moth, sawflies, and eriophyid mites. One of the sulfonates, tetradifon, is very safe and is used on spider mites and false spider mites. Carbaryl, a carbamate, rivals malathion in the number of woody ornamental pests controlled; almost all categories of Lepidoptera are controlled by it as well as many coleopteran, homopteran, and hymenopteran pests. Carbaryl and malathion are probably the two most commonly used insecticides on woody ornamental plants.

Petroleum spray oils have long been, and continue to be, used against all the scale insects and their sessile relatives on woody ornamental plants, as well as on the overwintering eggs of many insects and mites. Frequently oil is combined with another insecticide, such as malathion or diazinon, to increase its effectiveness. Nurserymen sometimes use such combinations on woody ornamentals as a spring or fall "clean-up" spray, or as a preventive to destroy light infestations before they become damaging. For years, three insecticides of botanical origin—nicotine, pyrethrum, and rotenone—have been available as combinations with a high grade of oil such as Volck. Because of the wide adoption of the synthetic organic insecticides, such combinations as oil–

nicotine, oil–rotenone, and oil–pyrethrum have lost in popularity. Emphasis in recent years on the "environment" and biodegradibility of pesticides, however, has caused renewed interest in these old, reasonably effective oil combinations, especially to the small-package trade in the cities.

Control Equipment

Insecticide application equipment for lawns, shrubs, and shade trees varies from the simplest, least expensive tin "flit gun" types to those complex, expensive high-pressure sprayers suitable for reaching the tallest trees. For small bushes around the home the small, aspirator "flit gun" type sprayer may be adequate. For bigger shrubs and even small trees the liquid displacement types, such as the trombone and stirrup pump bucket sprayers, allow better and quicker spray coverage. A popular and convenient type for home use is the hose-end–type sprayer (Fig. 8:1B), in which the garden hose water pressure continuously dilutes the insecticide, atomizes it, and forces it to where it is needed; the hose-end type may be used on lawns, shrubs, and small trees. The common 1- to 3-gal compressed air tank-type sprayer is also widely used around homes and in commercial nurseries for spot treatments. The 5-gal backpack or knapsack sprayer (Fig. 8:2) is more expensive, and only the more avid home gardener may possess this convenient sprayer; it is commonly used by the commercial nurseryman for anything from spot treatments to rather extensive ones. The same remarks also apply to the still more expensive hand-powered wheelbarrow sprayers.

The least expensive of the power sprayers is the small estate-type sprayer, but even so it may cost as much as several hundred dollars. Thus only the more affluent home owner may possess this very convenient sprayer, which will do all that the preceding hand-powered equipment will do, and will also give finer atomization and better and more even spray coverage of up to moderate-size (15 ft high) trees. The estate-type sprayer is the mainstay of the majority of commercial nurserymen. The low-pressure, low-volume type power sprayer, especially when equipped with a boom, is widely used by those commercial concerns or public agencies who must spray large acreages of turf such as in parks, cemeteries, and golf courses.

For spraying various sizes of ornamental shrubs and trees the high-pressure sprayer is necessary. With a large capacity, reciprocating high pressure pump, and a shade tree gun, it can cover almost any size of shade or ornamental tree. These large-capacity spray rigs, which cost several thousand dollars, are possessed by public parks and street tree and highway departments, as well as by private pest control operators spraying large trees for hire. Also utilized by many of the foregoing public and commercial groups is the air-blast concentrate sprayer, such as Bean's "Roto Mist" unit. These machines are useful only where the truck- or trailer-mounted unit can get under or beside the tree, such as street trees or widely spaced park trees. Although they are very rapid in treatment, these air-blast machines provide a discontinuous type of spray coverage, as contrasted to the continuous film of the slower, conventional sprayer. For this reason air-blast machines are primarily useful for those insects that move around, such as the smaller European elm bark beetle or various moth larvae; they are less effective against sessile insects like the scales.

Many insects affecting turf and other ground covers are controlled by granular formulations of insecticides. Granule applicators, or spreaders, are commonly used by homeowners, and larger-size models (Fig. 10:6) may be drawn behind tractors on golf courses or in parks. These convenient and rapid applicators are of two types: the sharply delineated swath type, such as Scott's, and the broadcast swath type, such as the Cyclone. Certain shade tree insects may be controlled by granular systemic insecticides as applied by spreaders.

The last insecticide applicator to be mentioned is aircraft, either fixed-wing or helicopter. These very expensive, very rapid applicators are used only on such extensive areas of ornamental trees as large parks or forests, examples being the aerial treatments currently being applied for gypsy moth in several northeastern states and the Douglas-fir tussock moth in the Pacific Northwest.

REPRESENTATIVE SHRUB, SHADE TREE, AND TURF PESTS

For detailed study we have chosen the Japanese beetle, a tree and shrub foliage feeder as an adult and a turf root feeder as a larva; the smaller European elm bark beetle, a boring tree insect; the gypsy moth, a serious tree defoliator; and the Coccoidea, the superfamily embracing the sucking pests, the scales and mealybugs.

Japanese Beetle. *Popillia japonica* Newman [Coleoptera:Scarabaeidae]

Although it is an oriental insect, the Japanese beetle was first named and described by Edward Newman, an Englishman, in London in 1841. It first came to America shortly before 1916, when it was found in a plant nursery in August in Riverton, New Jersey, just across the Delaware River from Philadelphia. It is presumed to have been introduced from Japan in soil attached to perennial plants such as iris and azalea. The Japanese beetle evidently liked its new home: in three years it heavily infested over 23 square miles on the New Jersey side of the river. Pennsylvania became infested in 1920 and Delaware in 1929. Today, 60 years after introduction, it infests all states east of the Mississippi River, except possibly Florida, Mississippi, and Wisconsin, and is being trapped with some regularity in eastern Missouri and Minnesota. All or parts of all these states are under quarantine. It heavily infests all states on the eastern seaboard above Florida. It is a continuing threat to all other states. California has had two infestations (Sacramento in 1961; San Diego in 1973), which have necessitated expensive but successful eradication efforts. Because the Japanese beetle is such a vigorous and important pest in this country, it is curious that it is considered only a minor pest in Japan. Possibly in its native home parasites and predators keep it in check and soil and turf conditions limit its activity.

Both the larva and the adult of the Japanese beetle are important pests, each in its own way and each on widely different types of plants. The larva, or white grub, lives in summer just under the surface of the soil and feeds primarily on the roots of grasses. The cutting of the roots kills the grass and loosens the turf so that it can be rolled or lifted like a rug. It is one of the most important pests of pasture grasses and of lawns around homes, in parks, golf courses, and cemeteries. Initial injury by just a few grubs causes a yellowing and slowing of grass growth, but with heavier populations—100 to 500 grubs per square yard—the grass turns brown and dies down so that only the dry earth is seen. Larvae will also damage the roots of many vegetables and woody nursery stock.

The adult is the most conspicuous stage. Very good fliers and essentially omnivorous, the beetles may go to the leaves, flowers, and fruits of any one of over 275 species of woody or herbaceous plants. They skeletonize leaves, consume flowers, devour fruits, or pit and gouge their surface. Shade and ornamental trees attacked include oak, sycamore, maple, linden, ornamental cherry and plum, white birch, larch, poplar, willow, elm, sassafras, chestnut, black locust, and magnolia. Ornamental shrubs and vines attacked include Virginia creeper, rose, althea, tamarisk, flowering cherry, forsythia, spiraea, azaleas, honeysuckle, wisteria, and quince. Flowers attacked include hollyhock, ferns, canna, dahlia, zinnia, geranium, marigold, petunia, iris, sunflower, and peppermint. Fruits attacked include grape, raspberry, apple, cherry, plum and peach. The beetles attack truck crops (beans, asparagus, and cantaloupe) and field crops (corn, soybean, clover, alfalfa, and buckwheat). They also feed on many weeds (smartweed, mallow, primrose, and dock).

Description. The adult Japanese beetle is 8 to 10 mm long, with the female being slightly the larger (Fig. 17:6). From above, the beetle is oval in outline, being broadest (=5mm) just in back of the base of the elytra. The head and prothorax are shining greenish bronze, and the elytra are brownish bronze with green along the sides and midline. Along the sides of the abdomen and at the elytral tips are twelve white spots formed of hair tufts extending outwards. The elytra are short enough to expose two abdominal terga. The elytra are longitudinally striate and punctate. The head and prothorax are minutely punctate. The antennae, typically scarabeiform, are lamellate. The frontal ridge between the eyes is shovellike. The legs are relatively long, with stout claws.

The egg is milky white and is spherical when first laid, with a diameter of 1 mm. Later it becomes elongate-oval, measuring 1 by $1\frac{1}{2}$ mm. The larvae vary from tiny C-shaped grubs to grubs 1 in. long when stretched out at full growth. There are 3 larval instars with head capsule widths of 1.2 mm, 1.9 mm, and 3.1mm. The head capsule is brown, and the powerful pointed

A

Figure 17:5. Life stages of the Japanese beetle. A, eggs; B, larva or grub; C, pattern of larval raster. *Courtesy USDA.*

mandibles are brownish black. The grubs are white, with bluish or blackish cast, especially at the swollen tail end. They appear like most other scarab larvae, except when the raster (pattern of tiny spines on the underside of last abdominal segment) is examined (Fig. 17:5C). Compared to the white grubs likely to be confused with it, only the Japanese beetle grub has a transverse anal slit, together with 2 tiny converging rows of 6 spines each in the center. The pupa is broadly spindle-shaped and has a pale tan color; the developing appendages are largely free of the body (17:6A).

Life history. There is one generation annually (Fig. 17:6C). In the vicinity of New Jersey and New York beetle emergence from the ground extends from mid-June to late October, with a broad peak in late July and early August. Sex ratio is about one to one. On warm, sunny days adults seem attracted to silhouettes against the sky, such as a sunlit tree, and particularly ones giving off fruity odors; they gather there in large, gregarious swarms. After a week or so of leaf skeletonizing and fruit gouging, and of mating, the adult female returns,

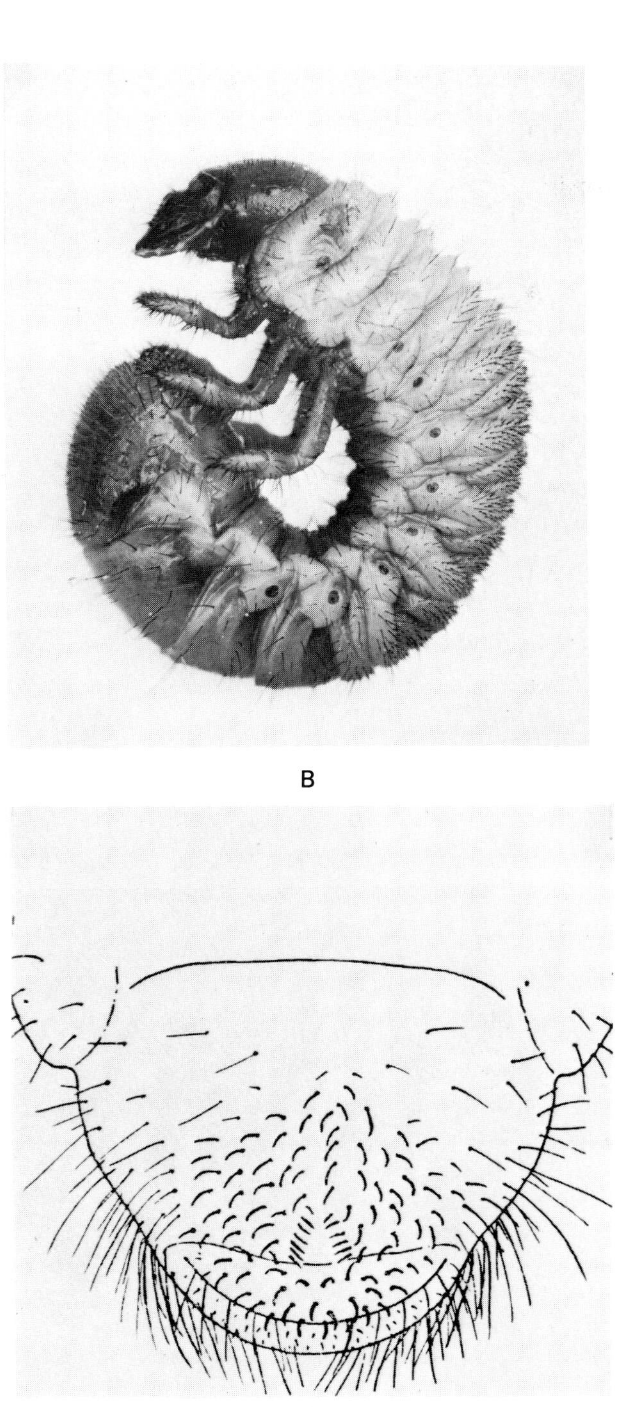

B

C

A
B

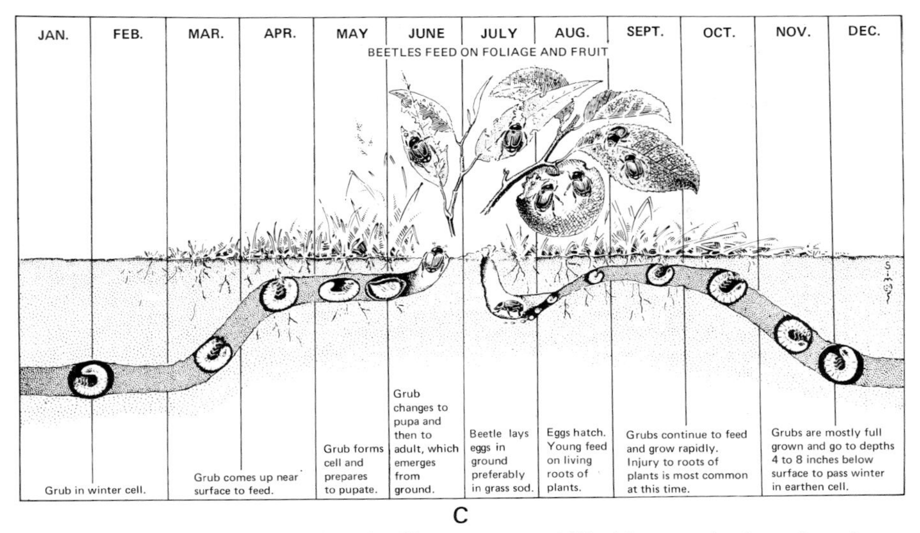

C

Figure 17:6. Japanese beetle life stages and life history. A, dorsal and ventral views of the pupa; B, the adult beetle; C, depiction of the yearly life cycle. *Courtesy USDA.*

preferably to a moist, loamy, sunny, grassy area, burrows 2, 3, or even 5 in. into the soil (of pH 5.3 or less), hollows out a slight chamber, and begins to deposit 3 or 4 eggs singly into the soil. An individual adult will live 3 or 4 weeks; each female lays about 50 eggs. Egg deposition occurs from early July to early October with a peak in mid-August. In about 2 weeks the tiny larvae hatch and begin tunneling upward to just beneath the soil surface. There they feed on the grass roots. By early November all larvae are nearly full grown; they tunnel down 5 to 8 in. into the soil to hibernate during the winter. The next April they become active again and tunnel upwards to within a couple of inches of the soil surface and feed voraciously on grass roots. Beginning in late May, and extending into early August, the grubs hollow a chamber and pupate; pupation peaks in late June to early July. After 2 to 3 weeks, the adults emerge in the chamber, thus completing the annual cycle.

Control. At least 14 different species of insect parasites have been introduced from Japan, but only 5 of these have become established. Of these only two are promising and only in areas of heavy infestation. Attacking the adult Japanese beetle is a tachinid fly, *Hyperecteina cinerea*, which lays 1 to 4 eggs on the beetle's pronotum. On hatching, the parasitic maggot bores into the beetle and feeds on its internal organs, usually killing it within 5 days. Attacking the grub of the Japanese beetle a tiphiid wasp, *Tiphia popilliavora*, bores into the soil, searches for the beetle larva, and lays one egg on the suture between the fifth and sixth abdominal segments. On hatching the wasp larva remains on the outside of the beetle larva, sucking its blood and chewing its internal organs. In 20 or 30 days the grub is killed. Native pentatomids, *Podisus* spp., and wheel bugs, *Arilus* spp., feed on the beetles.

The following birds have been reported feeding on Japanese beetle adults, the first two being quite effective: purple grackle, starling, kingbird, cardinal, meadowlark, catbird, and pheasants. Skunks and moles will feed on the grubs.

Perhaps the most successful biological control measure is a bacterial disease of the grub called milky disease, caused by *Bacillus popilliae.* The larval blood is normally clear and waterlike, but when infected with this disease the blood looks like milk. By culturing the disease in many larvae and grinding them, a milky disease spore powder is obtained. Such a powder is available commercially. It can be applied to the soil, and although it is relatively ineffective the first year, in succeeding years the turf becomes increasingly resistant to grub damage. For 1000 square feet of lawn area, a level teaspoon of spore powder every 10 feet is applied. Applications at 3- or 5-ft intervals will hasten disease establishment. In contrast to applications of chemical insecticides, retreatments are unnecessary.

Application of chemicals to the soil is a common and effective method used against the larvae of the Japanese beetle and other insect pests that live in the soil. For lawns around homes, in parks, and on golf courses, treatments of chlordane are highly effective against the larvae. The effects are rapid and may last for several years. Per 1000 square feet of lawn turf use 4 oz of active ingredient; this is equivalent to 10 lb per acre. Emulsions or suspensions may be sprayed on, or granules may be applied with a lawn spreader, in the spring, summer or fall. After treatment the lawn should be thoroughly watered to wash down the insecticide to the grass roots.

Also effective for soil application are two other chlorinated hydrocarbons, TDE and lindane, and three organophosphorus compounds, diazinon, chlorpyrifos, and trichlorfon. Because of the tightly bound layer, referred to as thatch, of living and dead roots, stems, stolons, and leaves that forms under turf, the less residual organophosphorus compounds are not usually as effective as the more residual chlorinated hydrocarbons. The thatch layer adsorbs the insecticide, preventing its free movement into the soil. Two promising experimental compounds for turf insect control, CGA-12223 and bendiocarb, may soon become available.

Against the adults several insecticides are available for use as sprays, either around the home or for much larger areas. Sprays should be applied when beetles are first noticed. Carbaryl is effective and widely used at the rate of 1 lb of active ingredient per 100 gal of water. But the following are also effective when used at recommended rates: malathion (0.5 lb AI/100); methoxychlor (1.5 lbs AI/100); TDE (1 lb AI/100); lindane (0.25 lb AI/100); and rotenone (2 oz AI/100).

Jet aircraft taking off from runways bordered with infested grass land may scoop up Japanese beetles in their wheelwells and transport them anywhere in the United States. Air passengers landing in major California airports, for example, will notice little yellow cans on top of stakes every few yards along the runway (Fig. 17:7). These are Japanese beetle traps baited with a synthetic

Figure 17:7. Japanese beetle trap baited with a synthetic attractant of phenethyl proprionate and eugenol. Iron stake set into ground holds trap four feet from surface. *Courtesy USDA.*

attractant. They are checked for beetles at frequent intervals. It is probable that recent infestations in California arrived on aircraft.

Smaller European Elm Bark Beetle. *Scolytus multistriatus* (Marsham) [Coleoptera:Scolytidae]

Thomas Marsham described the smaller European elm bark beetle in 1802, in an English entomological journal. It was first found in the United States in Boston, Massachusetts, in 1909, although it may have been here as early as 1904. It is thought to have arrived from Europe in Carpathian elm logs to be used for furniture veneer. By 1938 the beetle had spread as far as western Kentucky. By 1951 it was found in California. By 1969 it was found in all the contiguous states of the continental United States except Florida, North Dakota, Montana, Wyoming, and Arizona. In the last four states it was recorded on or near their borders, so by now it could very likely be in all the states, or at least be coextensive with the distribution of elm hosts.

The **Dutch elm disease**, which is carried by the beetle, was first found in the U.S. in 1930 in Ohio. The disease has spread throughout the Midwest and is found in the West in Colorado, Idaho, and Oregon, and was first detected in California in 1975.

When Marsham described this small beetle, he did not know what plant in Europe it attacked. Since then man has learned that the elm bark beetle is one of the most serious and aggressive pests of weakened elm trees. When it was discovered later that it also carried a deadly parasitic fungus of elm, the agent of Dutch elm disease, little hope remained for survival of this stately, widely planted tree.

The native elm bark beetle, *Hylurgopinus rufipes* (Eichoff), also attacks elms in the U.S. and carries the Dutch elm disease spores. It differs from the European scolytid in appearance, engraving pattern, and in being much less aggressive.

All elms, *Ulmus* spp., are attacked by *Scolytus multistriatus*, but only if they are weakened, as by drought. The adult female feeds first in crotches of tender twigs then makes an entry hole in the trunk or limb bark and bores to the cambium layer. Then, turning parallel to the wood grain, she makes a tunnel along the cambium layer, cutting through both xylem and phloem vessels at the same time. The larvae on hatching burrow perpendicular to the adult tunnel, or around the limb or trunk, and along the cambium layer, cutting also through xylem and phloem vessels. It is this cutting of vessels, particularly the xylem, by adult and larvae that is so injurious to the tree. The flow of water, especially, and nutrients from the roots is blocked, so that the tissue beyond the tunneling becomes desiccated and eventually dies. This is crucial when many beetles attack, and the interlacing larval tunnels in effect girdle the trunk or limb, causing its eventual death. If at the same time the beetle is carrying spores of the Dutch elm disease, *Ceratocystis ulmi* (Buism.) C. Moreau, the spores germinate and the fungus hyphae, or strands clog the xylem and phloem vessels, causing a darkening of these tissues (Fig. 17:8).

Figure 17:8. Cut elm twig showing darkened vascular layer, a symptom of Dutch elm disease. *Courtesy University of California, Riverside.*

Infection of the tree is more likely to occur, however, by initial adult feeding in the young twig crotches. Plugging of the vessels by the fungus thus accentuates the damage by the beetle. In June and July when high summer temperatures prevail normal healthy trees lift through the xylem great quantities of water, which is transpired through the leaves. Seriously beetle-infested and diseased trees cannot do so. It is usually at this time that the tree owner first notices the dried and brown leaves on individual limbs, or "flagging," and realizes that something is seriously wrong. If the bark is peeled back the engraving pattern can be seen on both the wood (Fig. 17:1) and the inner bark surface. Also, inspection of the bark and limb crotches will reveal much reddish brown, powderlike frass.

In driving through New York, Pennsylvania, Ohio, and other northeastern states, one commonly sees dead American elms standing in hedgerows and along fences. In parks or yards only stumps may remain.

Description. The adult *Scolytus multistriatus* is dark reddish brown and stout; it varies from 2.5 to 3.5 mm long (Fig. 17:9). Projecting backwards from the anterior third of the second abdominal sternite is a stout spine. The abdomen between the spine and tips of elytra is concave. The adult has strongly capitate antennae like other scolytids. Any beetle of this description on elm in the United States is almost certain to be this species. The eggs are nearly spherical and pearly white. When fully grown the legless larva is 3 to 4 mm long. It is entirely white except for dark mandibles and brownish head, is thickened in the thoracic area, and may be curved like a white grub. The pupa is entirely white; its short, stubby wing pads are curved around the abdomen. As it matures the elongate, compound eye darkens first, then the mouthparts, then, nearing adult emergence, the entire pupa is dark.

Life history. As many as 80 to 140 eggs are laid and covered with fine frass in niches along the sides of the straight, vertical adult tunnel. Upon hatching, a larva chews an increasingly large tunnel in the cambium region as it bores

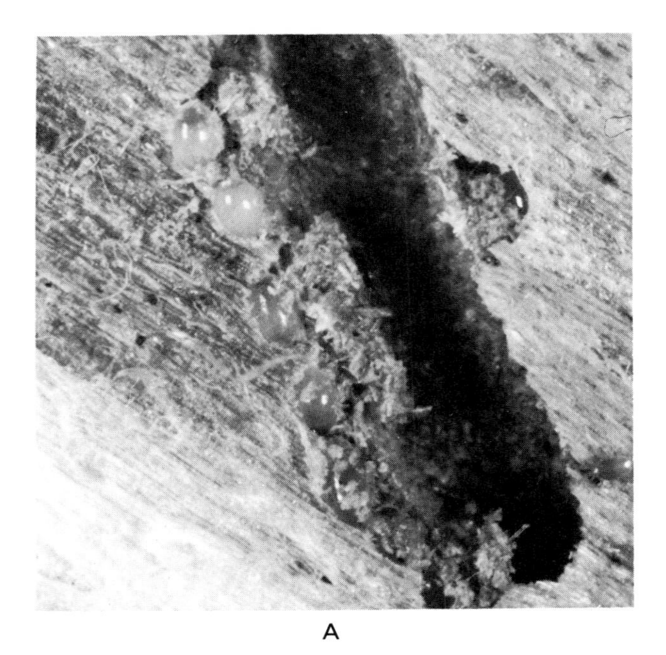

A

Figure 17:9. Life stages of the smaller European elm bark beetle, chief vector of Dutch elm disease. A, eggs; B, larva; C, pupa; D, adult. *Courtesy University of California, Riverside.*

B

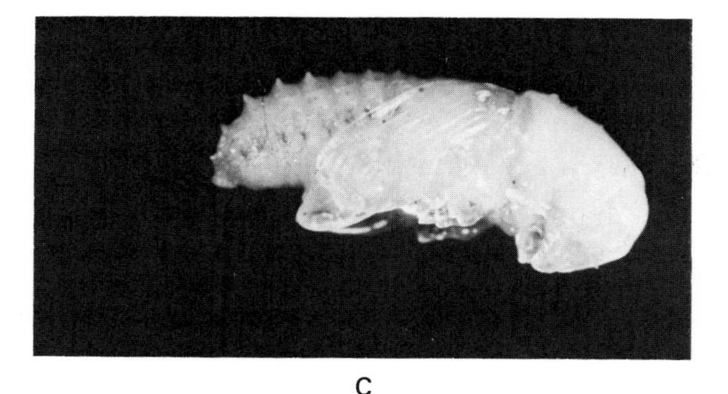

C

D

away from the adult tunnel, leaving dark powdery frass. Nearing pupation it gnaws outward from the cambium layer into the bark and there hollows out a pupal chamber. If the drying bark is sliced into while larvae and pupae are present, they appear superficially like grains of white rice. Adult emergence occurs in the pupal chamber, and after resting a while the adult chews its way out of the bark, leaving a round hole 2 mm diameter. One can easily imagine what the bark looks like when 80 to 140 new adults emerge from one engraving pattern.

The adults then fly to nearby healthy or weakened trees and feed for a few days in crotches of small tender twigs; they infect the tree if they are carrying spores of Dutch elm disease. They are then attracted to the limbs or trunks of weakened trees, or even cut limbs or newly cut firewood, by a distinctive odor emanating from these tree parts. After wandering around on the bark for a while they mate, and the female begins her brood tunnel, which she keeps free of frass by pushing it out the entry hole.

The larvae overwinter in their pupal cells, and when the weather warms in the spring, they pupate. Adults begin to appear on the outside of the bark in May in the New York–New Jersey–Pennsylvania area, and in late April in Kansas. In southern California beetles are first seen in mid-March. One and a partial second generation occur annually in Canada. Two generations and a partial third occur in the northeastern states. Three generations occur in

Kansas and probably the same number in California. In Kansas and California the first brood of beetles peaks in April, the second brood in July, and the third brood in September. In California the second brood is considerably larger than the first and third.

The length of the life cycle—that is, the time from the egg of one generation to that of the next generation—varies depending on temperature, which in turn varies with time of year, and somewhat with latitude and altitude. The greatest percentage of the life cycle is spent in the larval stage.

Parasites and predators. The following parasites and predators have been recorded attacking *Scolytus multistriatus*:

> **Hymenoptera: Braconidae:** *Coeleoides scolyticidae* (Holland), *Dendrosoter protuberans* (France), *Spathius canadensis* (U.S. and Canada).
>
> **Diptera: Lonchaeidae:** *Lonchaea polita* and *L. ciliata* (U.S.).
>
> **Nematodes:** *Rhabditolaimus rickardi* (Holland), *Chiropachys colon* (Italy), and *Parasityonychus scolyi* (Great Britain).

Certain birds (sapsuckers and woodpeckers) also remove great numbers of hatched beetle stages from the bark. To date it is apparent that none of these biological agents will hold bark beetle infestations to acceptable levels.

Control. The most important defense against the smaller European elm bark beetle is to maintain elms in a healthy, vigorous state by judicious irrigation, pruning, and fertilization of the elms. Irrigation is especially important in the more arid western states. In pruning, shaded limbs should be removed as well as any limbs partially broken by wind or with mechanical damage to the bark. The pruned limbs should be destroyed, preferably by fire, because they are very attractive to the beetle. If prunings are to be cut up for firewood, it should be done as soon as they are removed from the tree, and the cut pieces should be sprayed with an effective insecticide. These cultural practices are mandatory if any trees in the area are infected with Dutch elm disease.

Insecticidal sprays are secondary to cultural control measures. Against the beetle DDT was a very good insecticide before it was banned in the United States. Methoxychlor is still available and registered for this use. If a high-pressure hydraulic sprayer is to be used, an emulsifiable concentrate should be mixed with enough water to yield 16 pounds of active methoxychlor per 100 gal of water (8 gallons of a 2 lb/gal concentrate plus 92 gallons of water). If an air-blast mist-sprayer is employed, 6 times the hydraulic sprayer concentration is used (48 gallons of a 2 lb/gal concentrate plus 52 gallons of water) at one sixth the delivery of the hydraulic spray. The hydraulic sprayer should deliver at least 20 gal per minute at 600 lb pressure. The mist blower should be capable of delivering 10,000 cu ft of air per minute. Either of these methoxychlor treatments should be applied just before the leaves appear, and again after ninety days.

Gypsy Moth. *Lymantria dispar* (Linnaeus) [Lepidoptera:Lymantriidae]

Linneaus described the gypsy moth in Sweden over two hundred years ago; it is common throughout Europe, Asia, and North Africa. In 1869 a Professor L. Trouvelot brought gypsy moth egg masses from France into Medford, Massachusetts; his intention was to breed the gypsy moth with the silkworm to overcome a wilt disease of the silkworm. He placed the egg masses on a window ledge, and evidently the wind blew them away. Although he searched diligently he could not find the masses, but he published a notice warning others to be alert for the insect. The incident was forgotten for a few years until numerous shade and forest trees in the area began to be stripped of their leaves by the voracious caterpillars.

By 1891 about 200 square miles north and west of Boston were infested. Despite the fact that the female moths cannot fly, the insect steadily spread until today it infests all of the New England states (except northern Maine and a county or two in northern Vermont and New Hampshire); all of New Jersey; the eastern halves of New York and Pennsylvania; northern Delaware; northeastern Maryland; and a few counties in Canada. All of those areas are under federal quarantine, with Canada cooperating. Furthermore, trapping has revealed male gypsy moths throughout the remainder of New York, Pennsylvania, Delaware, and Maryland, as well as many throughout Virginia and North Carolina and a few in South Carolina, Alabama, Ohio, and Wisconsin. Infestations have started in western Florida and Michigan but fortunately have been eradicated. Even in California gypsy moth egg masses and empty pupal cases have been found many times on furniture and house trailers coming from the New England states, and male gypsy moths have been caught there in two locations in pheromone traps. In 1976 a substantial infestation was detected in California. Humans obviously have been and continue to be the most important factor in spreading gypsy moth over the continent—on packing boxes, crates, freight cars, recreational vehicles, airplanes, nursery stock, lumber, quarry, and other products. Perhaps the second most important factor of the spread is the wind, which transports the tiny, fluffy young larvae for great distances.

In the period from 1891 to 1935, when the gypsy moth was still confined within the New England states, all agencies had spent an estimated $41 million to control the pest. This sum was for control efforts only, and does not include loss of foliage and death and removal of trees. Now it is safe to suggest that we have spent several times this amount in the last 40 years. The infested area has more than doubled in the last 20 years. Through 1972 the gypsy moth has defoliated more than 2 million acres of forest land and killed over 5 million trees (Fig. 17:10).

The leaves of over 500 species of trees and other plants are eaten by gypsy moth caterpillars. The following trees (Group 1) are highly favored by larvae of all instars and will be completely defoliated: alder, all species of oak,

Figure 17:10. Trees defoliated by larvae of the gypsy moth. Note the variation in amount of defoliation among the individual trees. *Courtesy APHIS, USDA.*

gray birch, basswood, willow, river birch, all species of poplar, box elder, hawthorn, and apple. Paper birch and larch (Group 2) are edible by all larval instars but are distinctly less favored than Group 1 and are less likely to be completely defoliated. The following trees (Group 3) are edible by all larval instars but are not favored and defoliation will be only light: all species of maple, yellow birch, black birch, elm, sassafras, all species of hickory, black gum, hornbeam, and black cherry. The following trees (Group 4) are definitely unfavorable in early instars but are highly favored by larger caterpillars, and infestations will not originate in these stands: all species of pine, hemlock, southern white cedar, beech, and spruce. The trees in Group 5—ash, locust, tulip tree, butternut, red cedar, black walnut, dogwood, American holly, balsam, and sycamore—are not favored by any larval instar and are practically immune to attack. As the percentage of favored trees in a stand increases, so does the number of egg masses. A type of cultural control that has been proposed is lumbering out trees favored by the caterpillars (Group 1) and encouraging those trees known to be immune (Group 5). Conifers may be killed by a single defoliation, whereas broadleaf trees may stand two or three defoliations before death.

Description. The female gypsy moth is a dirty, creamy white, with some faint and some dark transverse wing markings (Fig. 17:11). Wingspan varies from 2 to $2\frac{1}{2}$ in. The female's body is stout and heavy, and densely covered with hairs. Her antennae are so slightly bipectinate as to appear filiform. The male is smaller and much darker (Fig. 17:12); his wingspan is about $1\frac{1}{2}$ inches.

Figure 17:11. Gypsy moth females. A, female extending tip of abdomen as she releases sex pheromone to "call" a mate; B, three females laying clusters of eggs. *Courtesy APHIS, USDA.*

A

B

Figure 17:12. Life stages of the gypsy moth. A, male moth showing the plumose antennae, the sensitive olfactory organs that detect sex pheromone secreted and released by the females; B, caterpillar or larva feeding on tree leaf; C, pupa in upper left corner and ovipositing female moths. *Courtesy APHIS, USDA.*

C

His abdomen is much narrower than the female's, and his antennae are plumose. The adult mouthparts and alimentary tract are not developed, which suggests that they do not feed. They lack ocelli.

The egg is globular, whitish, and about 1 mm in diameter. Eggs are laid in a cluster of 400 to 500, covered with the buff hairs from the female moth. Such oval egg clusters look and feel like a piece of soft chamois skin. Length of a cluster may vary from $\frac{1}{2}$ in. to over 2 in., with an average of 1 in.

The fully grown caterpillar varies from 2 to 3 in. in length. It is creamy yellow, finely marked with black, giving it a granulated appearance. It may also look brown or gray over all. Each body segment has 6 tubercles, 1 on each side of a yellowish middorsal line, and 1 on each side of each yellowish midlateral line. The dorsal tubercles on the 5 anterior body segments are tipped in blue, and the remaining 6 tubercles are tipped in red. Each body tubercle is the source of a tuft of many brownish black to yellowish radiating hairs, which on the lateral tubercles are much lighter in color and may be as long as half the body length. The head is dark, with dark yellow lines. The spiracles are creamy yellow. The true legs are dark red. The younger larval instars are generally similar to the mature larvae, with the hairs longer in proportion to the body.

The pupa is enclosed in only a few strands of silk. Male pupae vary in length from $\frac{2}{3}$ in. to almost an inch, whereas females may be as long as $1\frac{1}{3}$ inches. Coloration is from chocolate to dark reddish brown. The pupae are cylindrical, rounded on the front and tapering in the abdomen to a spinelike cremaster having several minute crochets. A few yellowish-brown hairs may occur on the eyes and head and in 10 evenly spaced clumps on each abdominal segment.

Life history. In the vicinity of Connecticut the overwintering egg masses hatch from late April to mid-May (Fig. 17:13). The larva requires 40 days, which may be from late June to mid-July, for completion of its development. The pupal stage requires 10 to 14 days, with pupae being present from late June to late July. Adults may live 6 to 10 days and may be found from early July to late August. Beginning in late July the females deposit their eggs, which remain dormant through the winter to mid-May of the next year, a total of 8 to 9 months. Thus there is a single annual generation of gypsy moth.

The egg masses may be deposited by the female on any part of the tree, especially the lower parts like the trunk or larger limbs, or anything under or near the tree: buildings, fence posts, piles of lumber, packing boxes, piles of rocks, automobiles, housetrailers, railroad cars, and airplanes. With the egg stage lasting so long and the eggs being laid on so many things moved about by man, the rapidity of spread and the necessity for quarantines are easy to understand. The larvae feed voraciously on the leaves during the night and hide during the day in bark crevices, cavities, or other dark places, presumably to escape the sunlight. Caterpillars with 5 instars are destined to be males, and those with 6 will be females. Pupation may be on any part of the tree or on anything under or near the tree.

Parasites and predators. Native birds, such as cuckoos, orioles, robins, catbirds, blue jays, crows, chipping sparrows, chicadees, and vireos, feed upon the caterpillars. In heavy infestations a wilt disease (a nuclear polyhedrosis virus) will decimate large numbers of larvae. Native parasites and predators and those imported from Europe include:

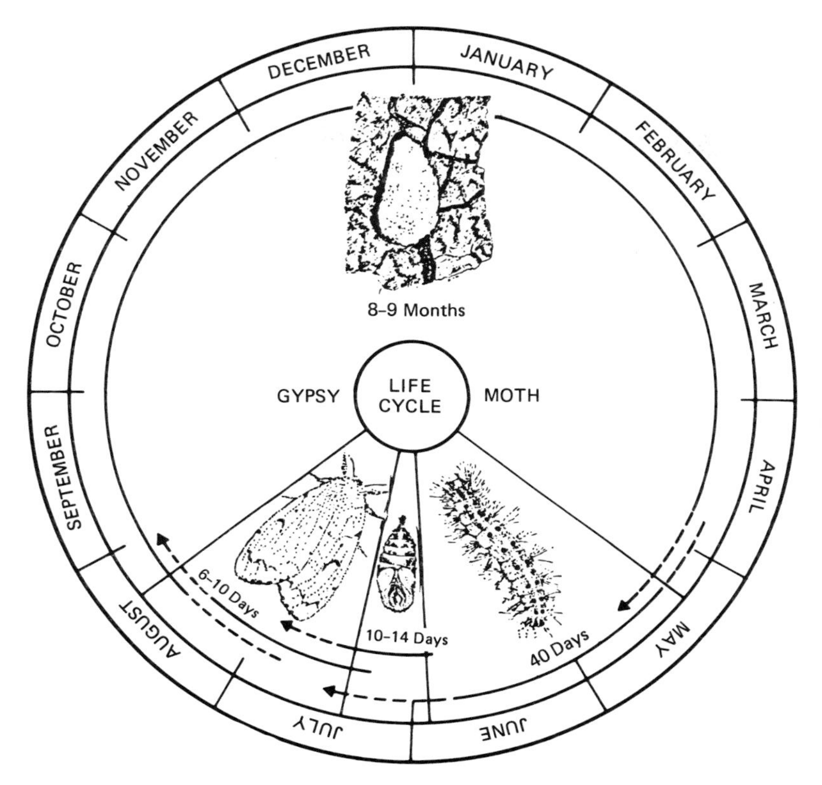

Figure 17:13. Life cycle of the gypsy moth. *Courtesy Connecticut Agricultural Experiment Station.*

Ground beetles: *Calasoma sycophanta, C. calidum, C. frigidum, C. scrutator.*

Tachinid flies: *Compsilura concinnata, Exorista larvarum, Sturmia scutellata.*

Braconid wasps: *Apanteles melanoscelus, A. lacteicolor.*

Ichneumonid wasps: *Coccygomimus pedalis, Itoplectis conquisitor, Theronia atalantae, Campoplex validus, Phobocampe disparis.*

Chalcidoid wasps: *Syntomosphyrum esurus, Ooencyrtus kuwanai, Anastatus disparis, A. bifasciatus, Monodontomerus aereus, Dibrachys cavus, Eupteromalus nidulans, Psychophagus omnivorus, Brachymeria ovata, B. intermedia.*

Soldier (pentatomid) bugs: *Podisus* spp.

None of these parasites and predators is known to keep gypsy moths in check in the United States.

Control. Legal control measures, or quarantines, have already been referred to, as well as the cultural control of planting trees immune to injury. Measures that an individual can use on his home lot property include the long-used daubing of egg masses with a sponge wet with creosote. Because the larvae migrate on trunk and limbs between hiding place and leaves, sticky bands (Tanglefoot or Stik-Em) may catch many, as well as the use of burlap bands for the hiding larvae and their periodic destruction.

Spraying with insecticides has provided the only rapid and effective—although temporary—relief from ravages of the caterpillars. The timing of spray applications coincides with the appearance of the larvae (first to third instar), which is approximately mid-May in the vicinity of Connecticut, and possibly slightly earlier in Pennsylvania and Delaware. Insecticides currently suggested for gypsy moth include *Bacillus thuringiensis*, trichlorfon, stirofos, carbaryl, phosmet, and diflubenzuron. All of these may be applied with a high-pressure hydraulic sprayer, and most of them may be used also with helicopters or fixed-wing aircraft (in 1971 as many as 372,000 acres were sprayed aerially). By air these insecticides may be used as low as 1 qt volume per acre. Persons or agencies intending to spray for gypsy moth should be guided by the label on the insecticide package as well as by county agents or entomologists in their locality.

Scale Insects

[Homoptera: Coccoidea]

The assemblage of animals known as Coccoidea contains specimens so strange and unusual that it is hard to convince students that they really are insects, or even animals. Most do not fit the usual definition, and it is difficult to imagine how these immobile blobs of protoplasm could possibly be injurious. But they really are highly specialized insects of great economic importance. Their sucking of copious quantities of plant sap, by typically enormous, gregarious populations, causes the weakening and death of twigs, limbs, leaves, fruits, and even whole trees. The honeydew excretion of many of the Coccoidea, and the accompanying black smut fungus, detracts from the esthetic appeal of woody ornamental plants, not to mention the objectionable swarms of ants that the honeydew attracts.

Included in the superfamily Coccoidea are at least 15 families (depending on which authority is followed), 8 of which are fairly common in North America; 6 of these have specimens of sufficient economic importance on woody ornamentals to be discussed here. These families are the Margarodidae (giant coccids), the Pseudococcidae (mealybugs), the Eriococcidae, the Coccidae (soft scales, wax scales, tortoise scales), the Asterolecaniidae (the pit scales), and the Diaspididae (armored scales). For accurate identification and taxonomic study coccoid scales must be cleared and mounted on microscope slides.

The females of the Coccoidea, the individuals usually encountered, are wingless, without segmentation in many cases, and frequently without legs or antennae; their long, filamentous, sucking mouthparts often appear to originate in the center of the body, and superficially they may appear like plant bud scales or galls. Females of the armored, pit, and soft scales are sessile—that is, they remain immovably attached to the plant hosts during all of their life after the first instar. Other families (Eriococcidae, Margarodidae) remain attached less of their life, and the mealybugs have usable legs and are mobile all their hatched life, although even they move very slowly.

The first instar of the Coccoidea, called a **crawler**, appears insectlike in that it has legs, antennae, and mouthparts, and is segmented. The tarsus is single-segmented and bears a single claw. The crawler is usually the means of natural dispersal of scale populations, either by crawling about on the plant, or crawling on the legs of birds and hitchhiking to new locations.

Many cases of parthenogenesis exist in the Coccoidea, but where males are known they usually have a single pair of wings of simple venation; they lack mouthparts, and frequently have a pair of long, waxy, filamentous anal appendages. Males have one more instar than the females.

Margarodids have abdominal spiracles and a reduced anal ring, but beyond these common characteristics there is great diversity in the family. Some foreign margarodids are very large—25 mm in diameter—but the sycamore scale in California is one of the smallest of scale insects (Fig. 17:14A), only 0.6 mm long, and looks like an insect egg. The cottonycushion scale is intermediate in size; at one time it was a widespread and very serious pest, but now it is only occasionally important.

The **mealybug** family is characterized by having an anal ring with four or more setae, nine-segmented antennae, dorsal ostioles, and ventral circuli. Mealybugs are covered with powdery wax and usually have lateral and anal wax filaments. The family has many important and well-known pests; shown in Fig. 17:14B, C, D are the citrus mealybug, longtailed mealybug, and taxus mealybug.

The **eriococcids** have an anal ring with four or more setae, but lack both dorsal ostioles and ventral circuli. One species worthy of mention is the European elm scale, which is perhaps the most important sucking pest of elm in the United States (Fig. 17:15A).

The **Asterolecaniidae** has wax glands in the shape of a figure 8, which open on the periphery of the body. One genus affecting oaks, the oak pit scales, is quite important (Fig. 17:15B). After the oak pit scale crawler settles down to feed the oak twig swells around its body; if the scale is removed a pit is left on the twig.

The **coccids** have the posterior end of the body cleft, and the anus is covered by two dorsal plates. This family includes a great number of important woody ornamental pests. Black scale and hemispherical scale infest a large number of host plants, whereas cottony maple scale and irregular pine scale infest only one plant genus (Fig. 17:15 and 16).

The **Diaspididae** have the posterior body segments fused to form a pygidium, which has aggregations of various kinds of wax gland openings and

Figure 17:14. Species of injurious scales and mealybugs. A, sycamore scale; B, citrus mealybug; C, longtailed mealybug; D, taxus mealybug. *Courtesy University of California, Riverside.*

C

D

Figure 17:15. Species of injurious scales. A, European elm scale with attending ants; B, oak pit scale; C, black scale; and D, hemispherical scale. *Courtesy University of California, Riverside.*

D
C

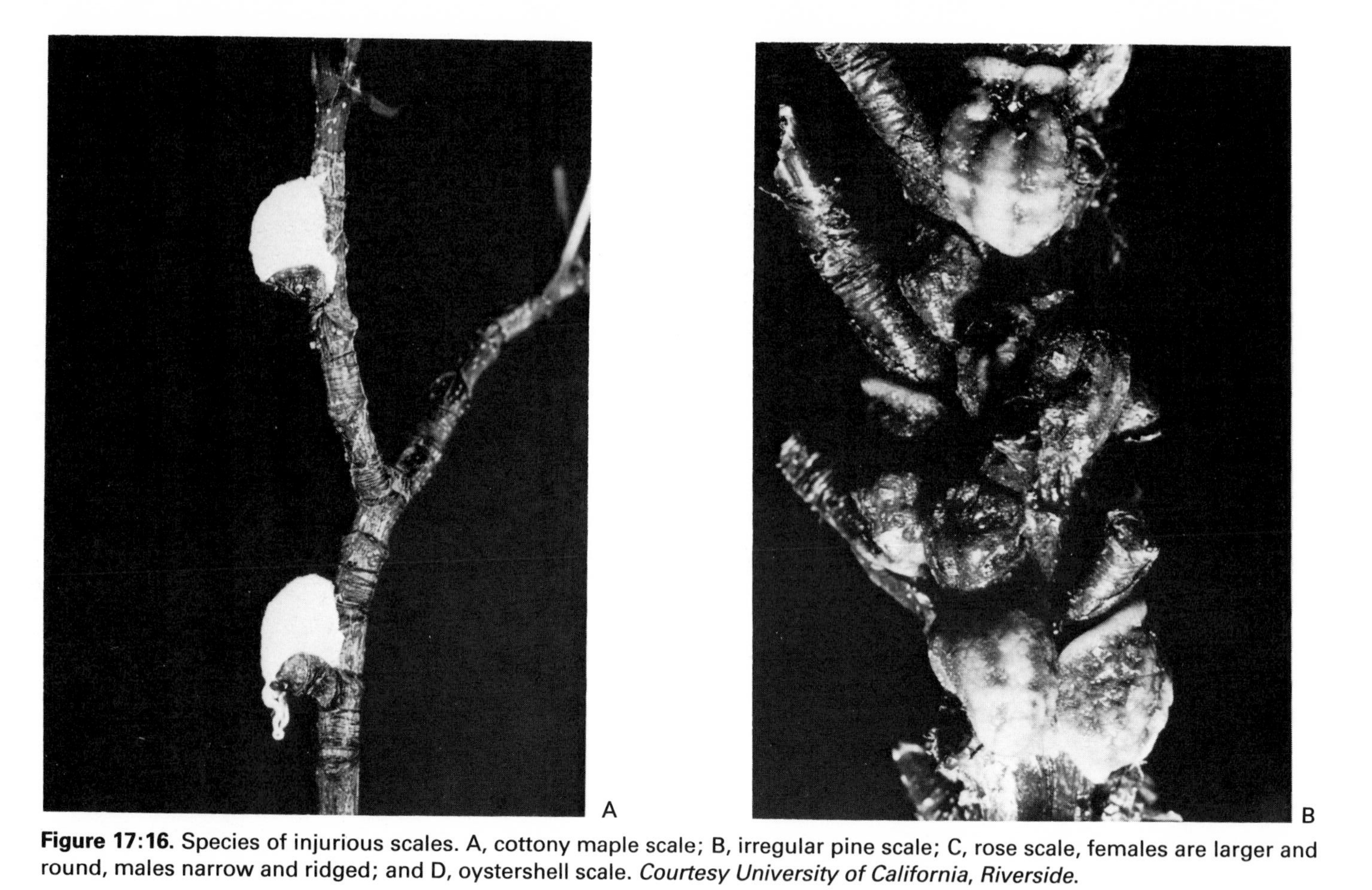

Figure 17:16. Species of injurious scales. A, cottony maple scale; B, irregular pine scale; C, rose scale, females are larger and round, males narrow and ridged; and D, oystershell scale. *Courtesy University of California, Riverside.*

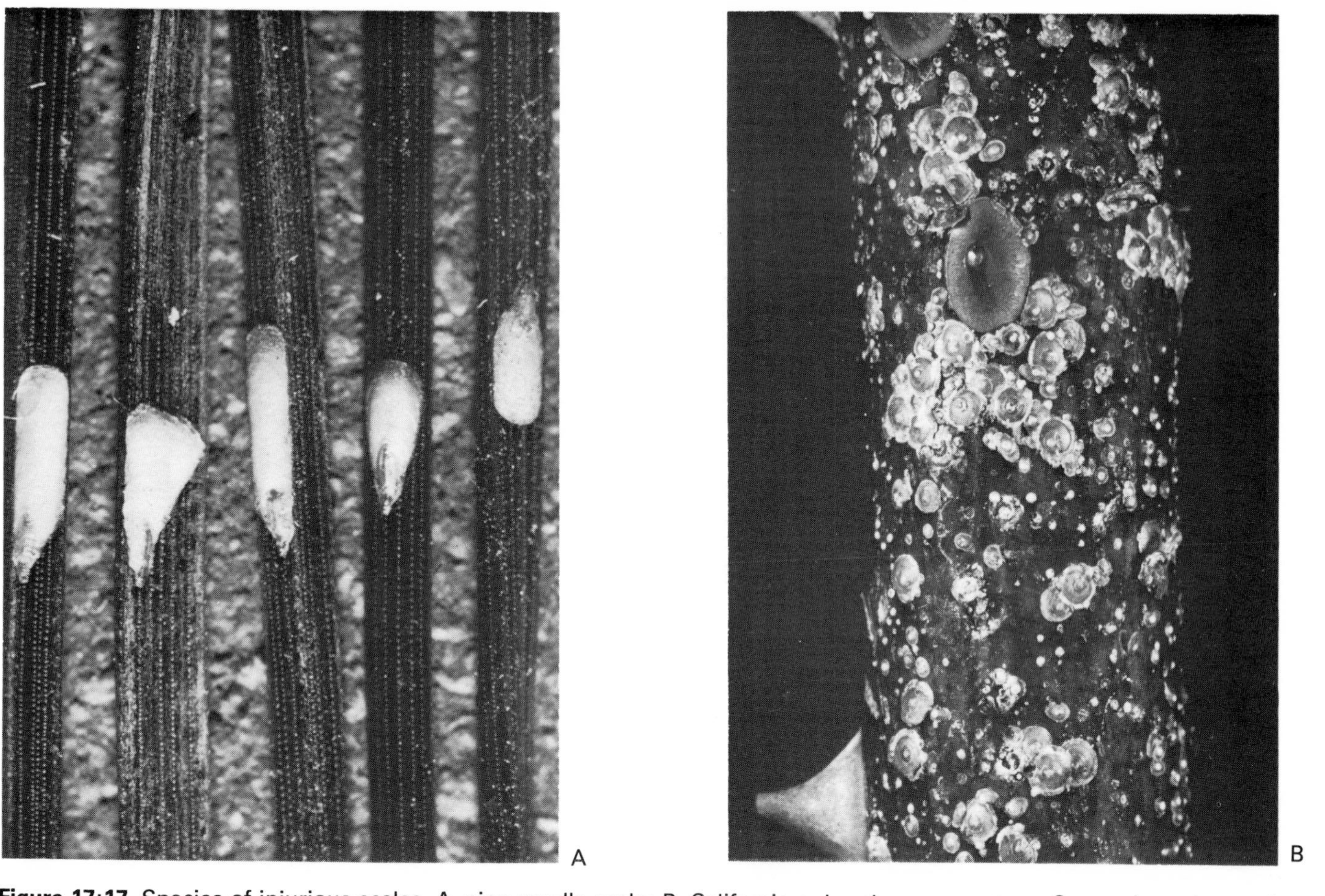

Figure 17:17. Species of injurious scales. A, pine needle scale; B, California red scale on rose stem; C, greedy scale, one has been turned over to show actual body and bottom of armor. *Courtesy University of California, Riverside.*

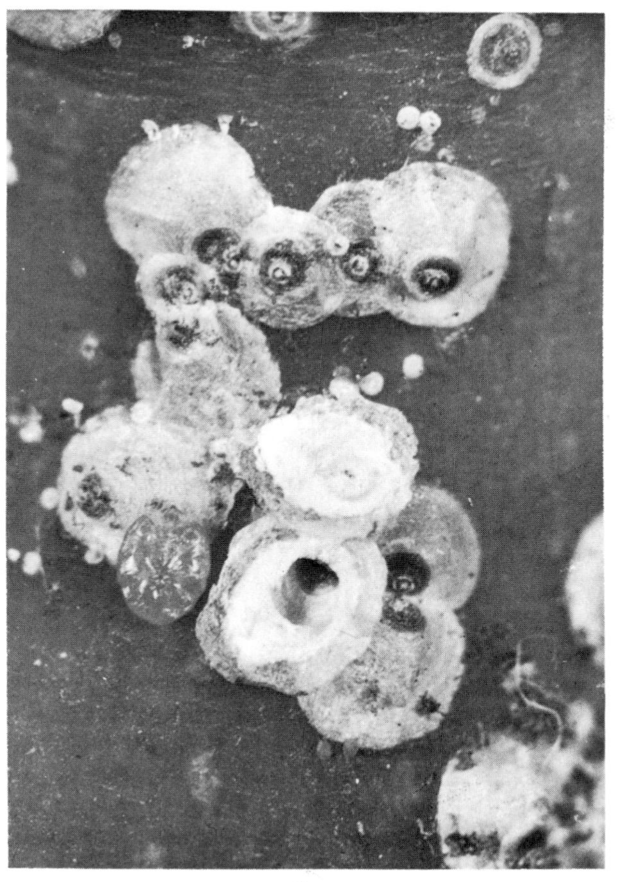

C

a fringe of lobes and spines. Secreted from these wax glands are thin wax strands that form a covering, or armor, over the body. The armored scales have a great number of species affecting woody ornamental plants, many being illustrated in the chapters on tree fruits and floricultural crops. A few that can be illustrated here are rose scale, oystershell scale, pine needle scale, California red scale, and greedy scale (Fig. 17:16C, D and Fig. 17:17).

Control. Biological control of many coccoid species has been attempted; some such attempts have been successful. But control of scales and mealybugs with insecticidal sprays is far more common and successful. Petroleum oil sprays, either in dormant (4 to 5 per cent) or foliage season (1 to 2 per cent), have been used for many years. Frequently a lesser amount of oil may be combined with something else, such as malathion, diazinon, nicotine, rotenone, or pyrethrum. Malathion, diazinon, and dimethoate are examples of organophosphorus insecticides that are commonly used alone in sprays for scale and mealybug control. Much carbaryl is used against the scales, especially the Coccidae. Because ants are attracted to scale honeydew and protect the scales from predators, use of ant sprays around the trunk base will increase the likelihood of scale sprays being successful.

SELECTED REFERENCES

Adams, J. A., and J. G. Matthysse, *The Japanese Beetle*, N.Y. Coll. Agr. Bul. 770, 1952.

Anderson, R. F. *Forest and Shade Tree Entomology* (New York: John Wiley & Sons, Inc., 1960).

Bailey, J. B., and J. E. Swift, *Pesticide Information and Safety Manual.* Calif. Agr. Exp. Sta., 1968.

Baker, W. L., *Eastern Forest Insects*, USDA Forest Service, Misc. Publ. 1175, 1972.

Behre, C. E., and W. L. Baker *Silvicultural Control of the Gypsy Moth*, Mass. Forest and Park Assoc. Bul. 157, 1936.

Blackman, M. W., and W. O. Ellis, *Some Insect Enemies of Shade Trees and Ornamental Shrubs*, N.Y. St. Coll. Forestry Bul. 16, 1916.

Britton, W. E., *The Gypsy Moth*, Conn. Agr. Exp. Sta. Bul. 375, 1935.

Brown, L. R., "Seasonal Development of the Smaller European Elm Bark Beetle in Southern California," *J. Econ. Entomol.* 58(1):176–7, 1965.

———, and C. O. Eads, *A Technical Study of Insects Affecting the Oak Tree in Southern California*, Calif. Agr. Exp. Sta. Bul. 810, 1965.

———, *A Technical Study of Insects Affecting the Sycamore Tree in Southern California*, Calif. Agr. Exp. Sta. Bul. 818, 1965.

———, *A Technical Study of Insects Affecting the Elm Tree in Southern California*, Calif. Agr. Exp. Sta. Bul. 821, 1966.

———, *Insects Affecting Ornamental Conifers in Southern California*, Calif. Agr. Exp. Sta. Bul. 834, 1967.

Craighead, F. C. *Insect Enemies of Eastern Forests*, USDA Misc. Publ. 657, 1950.

Davis, J. J., *The Green Japanese Beetle*, N.J. Dept. Agr. Circ. 30, 1920.

Deal, A. S., R. N. Jefferson, and F. S. Morishita, "Insect and Related Pests of Turfgrass," in *Turfgrass Pests*, Calif. Agr. Exp. Sta. Manual 41, 1971.

Essig, E. O., *Insects and Mites of Western North America* (New York: Macmillan Co., 1958).

———, and W. M. Hoskins, *Insects and Other Pests Attacking Agricultural Crops*, Calif. Agr. Exp. Sta. Circ. 87, 1944.

Felt, E. P., *Insects Affecting Park and Woodland Trees*, N.Y. State Museum Mem. 8 (2 vols.), 1905 and 1906.

———, *Plant Galls and Gall Makers* (Ithaca, N.Y.: Comstock Publ. Co., 1940).

Fernald, C. H., "The Gypsy Moth," in *Report on Insects*, Mass. Agr. Coll. Bul. 19, 1892.

Fleming, W. E., *Integrating Control of the Japanese Beetle—A Historical Review*, ARS USDA Tech. Bul. 1545, 1976.

Herrick, G. W., *Insect Enemies of Shade-Trees* (Ithaca, N.Y.: Comstock Publ. Co., Inc., 1935).

Houser, J. S., *Destructive Insects Affecting Ohio Shade and Forest Trees*, Ohio Agr. Exp. Sta. Bul. 332, 1918.

Johnson, W. T., and H. H. Lyon, *Insects That Feed on Trees and Shrubs* (Ithaca, N.Y.: Cornell Univ. Press, 1976).

Keen, F. P., *Insect Enemies of Western Forests*, USDA Misc. Publ. 273, 1952.
———, "Bark Beetles in Forests," in *Insects, the Yearbook of Agriculture*, 1952.
Ladd, R. L., Jr., *Controlling the Japanese Beetle*, USDA Home and Garden Bul. 159, 1976.
Leonard, D. E., "Recent Developments in Ecology and Control of the Gypsy Moth," Ann. Rev. Entomol., 19:197–229, 1974.
Pirone, P. P., *Diseases and Pests of Ornamental Plants* (New York: Ronald Press Co., 1960).
Rex, E. G. *Facts Pertaining to the Japanese Beetle*. N.J. Dept. Agr. Circ. 180, 1931.
Rose, A. H., and O. H. Lindquist, *Insects of Eastern Pines*, Canadian Forestry Service Publ. 1313, 1973.
Schuh, J., and D. C. Mote, *Insect Pests of Nursery and Ornamental Trees and Shrubs in Oregon*, Oregon Agr. Exp. Sta. Bul. 449, 1948.
Thompson, H. E., S. M. Pady, and R. A. Keen, *Dutch Elm Disease and Its Control in Kansas*, Kansas Agr. Exp. Sta. Bul. 434, 1961.
USDA, *Elm Bark Beetles*. USDA Leaf. 185, 1953.
USDA, *Guidelines for the Use of Insecticides*, USDA Agriculture Handbook 452, 1974.

chapter 18 / **DONALD A. WILBUR AND ROBERT B. MILLS**

STORED GRAIN INSECTS

Storage losses from insect attack are often as great as those sustained by the growing crops. Moreover, losses in growing crops are frequently obvious, whereas losses in stored grain are likely to be insidious. Experienced grain men cannot detect internally infested kernels without employing special techniques. Insect damage to growing crops may be counterbalanced to some degree by partial recovery of the damaged plants or by increased yield from the survivors, but insect damage to stored grain is final and without compensatory adjustments. When the grain is to be used as human food, losses extend beyond that amount actually consumed and must include the effects of contamination from feces, odors, webbing, dead insects, cast skins, and fragments in the manufactured or processed product. Also, the losses must include damage resulting from insect-caused heating and the translocation of moisture, with subsequent molding and caking.

Estimates of losses to the world's supply of stored grain from insect damage range from 5 to 10 per cent of the world's production. In certain tropical and subtropical countries, estimates are much higher. In the United States where storage conditions are more adequate and insecticides are readily available, losses are estimated to be between $300 and $600 million annually. Destruction of food by stored grain insects is a major factor responsible for the low levels of subsistence in many tropical countries. If these losses could be prevented, it would alleviate much of the food shortages in the famine areas of the world. According to one estimate, 130 million people could have lived for one year on the grain destroyed or contaminated by stored grain insects in 1968.

THE PESTS

Original Sources of Stored Grain Insects

Before man began storing grains, the granary weevil, rice weevil (Fig. 18:5), and Angoumois grain moth (Figs. 18:12 and 18:13) were already infesting the seeds of plants, either while the seeds were retained on the plants or after they were stored by rodents, ants, and other seed-harvesting animals. It was a short step for these insects to move to the grain bins of farmers. Other species were originally fungus feeders and scavengers of dead plant and animal materials. Several important species, including the lesser grain borer, cadelle, flat grain beetle, and red and confused flour beetles, were originally wood borers or lived under bark. Some species were normal inhabitants of the nests of other insects and of birds.

Groups of Stored Grain Insects

During recent years several species of stored grain insects, each of which was known to have variable structural and behavioral characteristics, were found to be a complex of two or more distinct species rather than subspecies or strains, as supposed previously. Thus, a confusion exists in relating literature dealing with rice weevil, sawtoothed grain beetle, flat grain beetle, and others to their proper species. It is not surprising that insects of world distribution and with a wide range of feeding habits develop into a complex of species and of strains within species.

The insect species that to varying degrees have become adapted to the environment of commercially stored grains are grouped into the following four categories based on the nature and extent of their association with the grain.

1. Major pests comprise those few species that are particularly well adapted for living in stored grain. The following species, listed alphabetically by families, are generally accepted as being responsible for most of the insect damage in all parts of the world. (Figure 18:1 shows both major and minor beetles.)

Common Name	Scientific Name	Family
Lesser grain borer	*Rhyzopertha dominica* (F.)	Bostrichidae
Sawtoothed grain beetle	*Oryzaephilus surinamensis* (L.)	Cucujidae
Merchant grain beetle	*Oryzaephilus mercator* (Fauvel)	Cucujidae
Flat grain beetle	*Cryptolestes pusillus* (Schönherr)	Cucujidae
Rusty grain beetle	*Cryptolestes ferrugineus* (Stephens)	Cucujidae
Rice weevil	*Sitophilus oryzae* (L.)	Curculionidae

Table continued

Common Name	Scientific Name	Family
Maize weevil	*Sitophilus zeamais* Motschulsky	Curculionidae
Granary weevil	*Sitophilus granarius* (L.)	Curculionidae
Trogoderma complex (including khapra beetle)	*Trogoderma* spp.	Dermestidae
Angoumois grain moth	*Sitotroga cerealella* (Olivier)	Gelechiidae
Cadelle	*Tenebroides mauritanicus* (L.)	Trogositidae
Indian meal moth	*Plodia interpunctella* (Hübner)	Pyralidae
Mediterranean flour moth	*Anagasta kuehniella* (Zeller)	Pyralidae
Confused flour beetle	*Tribolium confusum* duVal	Tenebrionidae
Red flour beetle	*Tribolium castaneum* (Herbst)	Tenebrionidae
Grain and flour mites complex		Tyroglyphidae

2. Minor pests include a larger group of insects and mites that may become abundant locally and are occasionally damaging to grain and milled grain products. Certain of these species approach major pest status when conditions are favorable. The insects in this group either are associated frequently with a particular environment such as high or low moisture or temperature, poor sanitation, out-of-condition grain or product, or else occur within limited geographical areas. The **larger grain borer** is a major pest in Mexico, where it infests corn in the field and continues its development in storage. The **cigarette beetle** is the major pest in tobacco warehouses and occasionally develops into damaging populations in grain and cereal products. **Dark and yellow mealworms** feed in out-of-condition moldy grain and milled grain products. **Hairy fungus beetles** and **foreign grain beetles** may occur in large numbers in stored grain. They are not known to damage grain except by their presence; when they are found, grain handlers are alerted to a high-moisture content in their grain with the probability of present or future damage by molds. The following list includes common minor pests.

Common Name	Scientific Name	Family
Cigarette beetle	*Lasioderma serricorne* (F.)	Anobiidae
Drugstore beetle	*Stegobium paniceum* L.	Anobiidae
Larger grain borer	*Prostephanus truncatus* (Horn)	Bostrichidae
Squarenecked grain beetle	*Cathartus quadricollis* (Guérin-Méneville)	Cucujiidae
Foreign grain beetle	*Ahasverus advena* (Waltl)	Cucujidae
Black carpet beetle complex	*Attagenus* spp.	Dermestidae
Booklouse	*Liposcelis* spp.	Liposcelidae

Table continued

Common Name	Scientific Name	Family
Hairy fungus beetle	*Typhaea stercorea* (L.)	Mycetophagidae
Dried fruit beetle complex	*Carpophilus* spp.	Nitidulidae
Spider beetle complex	*Ptinus* spp.	Ptinidae
Rice moth	*Corcyra cephalonica* (Stainton)	Pyralidae
Meal moth	*Pyralis farinalis* L.	Pyralidae
Longheaded flour beetle	*Latheticus oryzae* Waterhouse	Tenebrionidae
Broadhorned flour beetle	*Gnathocerus cornutus* (F.)	Tenebrionidae
Slenderhorned flour beetle	*Gnathocerus maxillosus* (F.)	Tenebrionidae
Smalleyed flour beetle	*Palorus ratzeburgi* (Wissmann)	Tenebrionidae
Depressed flour beetle	*Palorus subdepressus* (Wollaston)	Tenebrionidae
Dark mealworm	*Tenebrio obscurus* F.	Tenebrionidae
Yellow mealworm	*Tenebrio molitor* L.	Tenebrionidae

3. Incidental pests consist of approximately 150 species of insects and mites. Certain of these are present occasionally in grain; others occur frequently in grain and in large numbers along with major and minor pests. However, they rarely damage grain except by the contamination resulting from their presence. They may serve as a warning that the grain is out of condition and needs attention; thus they are beneficial. The food of many of these species consists of fungi and other microorganisms. The families of Blattidae, Tenebrionidae, Ptinidae, Dermestidae, Bostrichidae, Ostomidae, Cucujidae, Cryptophagidae, Nitidulidae, Pyralidae, and Tineidae provide the more common species among the incidental pests.

4. Parasites and predators comprise an unknown number of insects and mites that prey on the other three groups. Normally they are not a factor in controlling infestations of stored grain insects.

STORED GRAIN AS FOOD AND ENVIRONMENT FOR INSECTS

Food. Stored grains provide all of the essential nutritive requirements for insects capable of chewing into the hard, dry kernels. Yet only a few out of an estimated million described species have been able to adapt to the food and environment provided by stored grain conditioned for safe storage. Frequently the populations of these adapted species become very large. An infestation of only 25 insects in 1 qt of grain is equivalent to 800,000 in 1000 bushels. The biotic potential of adapted species is enormous; by the end of six months without restrictive influences, the progeny of a single pair of rice weevils would attain 675 million adults. In actuality this potential is missed by wide margins in spite of an almost unlimited food supply, a relatively stable physical environment, and a remarkable scarcity of insect parasites and

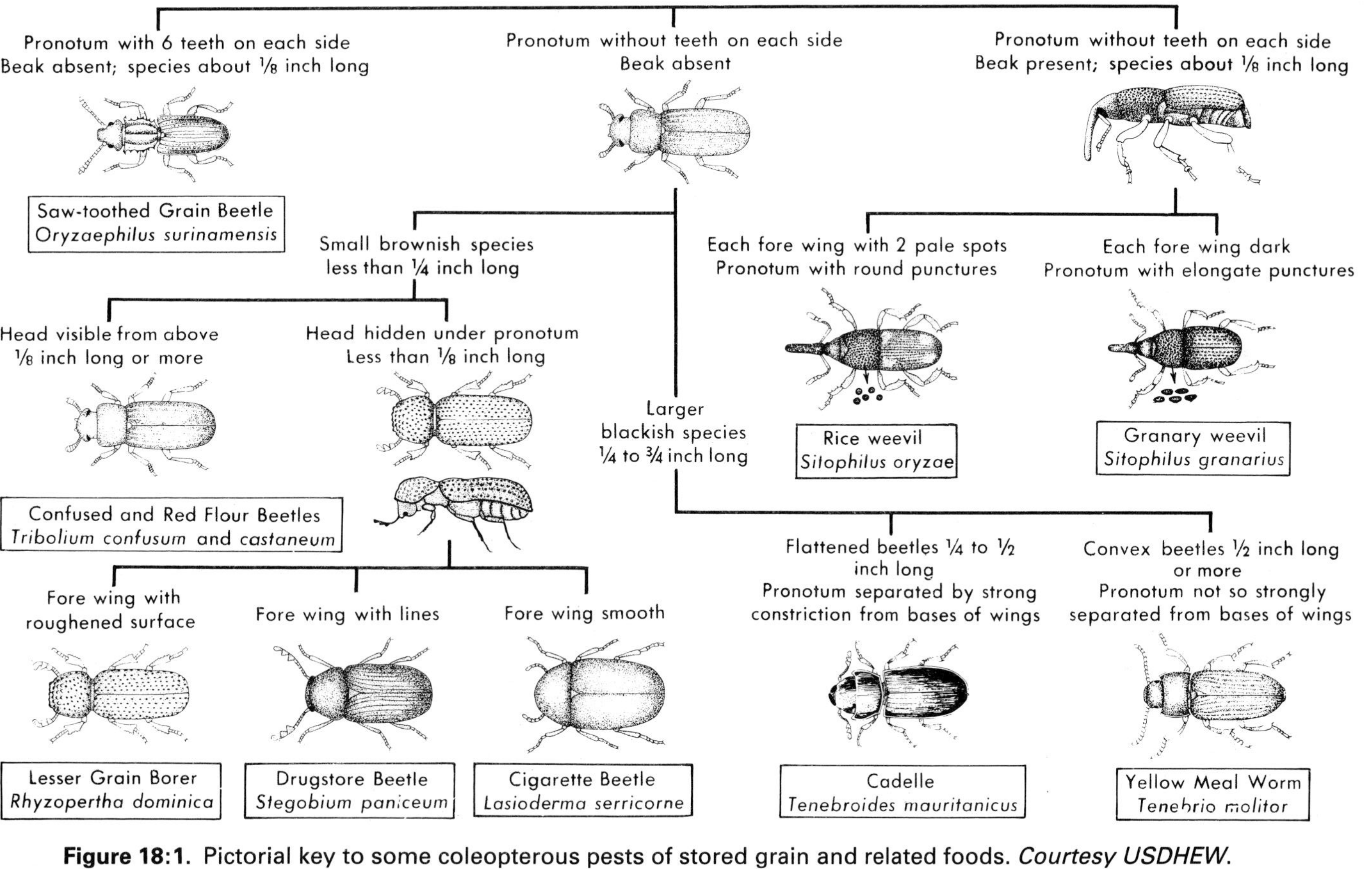

Figure 18:1. Pictorial key to some coleopterous pests of stored grain and related foods. *Courtesy USDHEW.*

predators. The normal feeding and other activities of the insects alter the grain mass by causing heating and subsequent moisture translocation, so that the grain becomes moldy and unsuitable for most forms of insect life. Only certain mold-feeding insects and scavangers are adapted to moldy grain.

Moisture and temperature. The behavior and habits of stored grain insects are closely attuned to the moisture and temperature of their food media. For the most part the major stored grain pests are restricted to a narrow moisture band between 11.5 and 14.5 per cent. A moisture content of about 12.5 per cent favors feeding and reproduction of most major pests. Moistures above 14.5 per cent permit the development of molds and germination when temperatures are favorable. This results in heating, molding, and caking of the grain. Some molds that attack the grain also kill the insects. A few insect species can utilize grain below 11.5 per cent moisture, although their ability to do this varies with the species, the temperature, and the physical condition of the food. If temperatures are increased to approximately 90°F, rice, maize, and granary weevils can reproduce in grain of 10 per cent moisture and can survive on a minimum of 9 per cent moisture. Confused flour beetles cannot maintain a culture on whole kernel wheat of 10 per cent moisture, but on similar grain of 12 and 14 per cent moisture, they maintain vigorous populations. They can survive on such dockage as broken kernels and on flour of 6 per cent moisture if the temperature is favorable. The Mediterranean flour moth can live on a 1 per cent moisture food medium. Kansas farm-stored grain below 11.5 per cent moisture will rarely have damage from weevils, although infestations of sawtoothed grain beetles, cadelles, and lesser grain borers may occur. Under most circumstances dry wheat of 10 per cent or less moisture can be stored indefinitely without serious damage from insects.

Structural and physiological adaptation. Certain structural and physiological adaptations are essential to allow stored grain insects to inhabit such a rigorous low-moisture environment. Their exoskeleton is relatively impermeable to water, preventing the loss of body fluids by evaporation. Moisture is withdrawn from the body wastes that assemble in the hind intestine and is returned through the intestinal wall to the blood, leaving the excrement powder dry. In addition, the **water of metabolism** resulting from the oxidation of fats and carbohydrates to digestible foods is fully utilized.

THE DAMAGE

Direct Damage

Kernel damage. Much of the damage from grain-infesting insects is done directly to the kernels (Fig. 18:2). The rice weevil, maize weevil, granary weevil, lesser grain borer, and Angoumois grain moth consume varying amounts of the endosperm. The germ is consumed by Indian meal moth larvae and by flat and rusty grain beetle larvae. Cadelles, dermestids, and

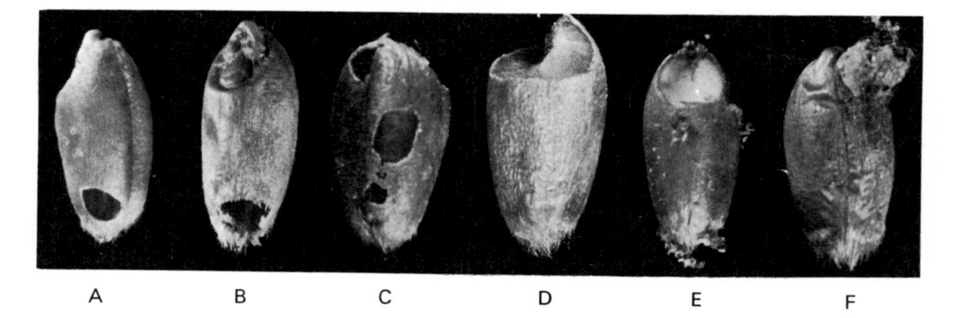

Figure 18:2. Injury to wheat kernels by stored grain insects. A, emergence hole of rice weevil; B, emergence hole of granary weevil; C, emergence hole and adult feeding of lesser grain borer; D, feeding of confused flour beetle on germ; E, feeding of Indian meal moth larva on germ; F, emergence hole and cocoon of Angoumois grain moth. *Courtesy Kansas State University, Manhattan.*

flour beetles first eat the germ and then turn to the endosperm, which may be destroyed completely. The extent of the destruction of germ and endosperm and its accompanying loss of weight, called shrinkage by grain handlers, is determined by such factors as insect species and numbers, type of grain, temperature, moisture, and length of storage.

Contamination. Grain-infesting insects contaminate much more grain than they eat. This contamination results from (1) the presence of living or dead insects (eggs, larvae, pupae, and adults) or parts thereof; (2) cast exoskeletons, egg shells, pupal cases, and cocoons; (3) fecal materials by all species; (4) noxious and persisting odors, especially by lesser grain borers and flour beetles; (5) webbing by Indian meal moth larvae and other moth larvae. Significant amounts of this filth cannot be removed by cleaning processes now available in grain processing plants.

Grain dust. Insect activity inevitably results in dust consisting of flour, excrement, and broken kernels that contribute to the amount of grain dockage determined by official grain graders. Lesser grain borers and cadelles reduce the grain to dust in such exorbitant quantities that a successful fumigation may not be attainable unless the dust is removed.

Damage to wooden structures and to paper and cloth containers. Cadelle larvae, lesser grain borers, and dermestid larvae tunnel into the wooden parts of granaries, warehouses, boxcars, and ships' holds to such an extent that the structures are weakened (Fig. 18:3). Their tunnels provide hiding places for other species of insects and lodging places for nutritious grain dusts and cracked kernels upon which these insects feed. The tunnels prevent thorough cleaning of grain residues from the granaries. In earlier times, when cadelles were abundant in flour mills, they were a scourge to millers because of their

Figure 18:3. Cadelle larval tunnels in wooden grain bin. Note accumulation of wood dust on surface of grain. *Courtesy Kansas State University, Manhattan.*

habit of chewing holes in the bolting cloth screens. Few packages devised for marketing cereal products can be depended on to prevent entry of all stored grain insects.

Indirect Damage

Dry grain heating and moisture migration in storage. These are caused by grain-infesting insects, whose metabolic activities within a grain mass may produce heat up to 106°F even in winter. This warmed air rises to the surface grain, carrying moisture with it. At the surface the air is cooled, and moisture is deposited on the grain, which encourages growth of molds and stimulates seed germination. Caking and spoiling result. **Dry grain heating** develops in grain of less than 15 per cent moisture.

A grain moisture content of approximately 15 per cent or more is usually enough to activate molds and fungi and eventually the germ itself, and thus to produce **wet grain heating**. The microorganisms responsible for wet grain heating usually destroy grain-damaging insects, although certain fungus-eating species thrive.

Lowered germination of seed grains. Germination is impaired or destroyed by heating or high moisture, or by damage to either germ or endosperm.

Distribution of molds and other microorganisms through the grain mass. Grain-infesting insects carry grain-damaging molds to bins of grain and may distribute them throughout the grain mass. If moisture and temperature are favorable for mold development, damage is certain to occur. Disease organisms such as *Salmonella* and the molds that produce toxins such as aflatoxins likewise may be distributed by insects.

Insect fragments in cereal products. Frequently insects inside grain kernels are processed into flour and other cereal products along with the germ and endosperm. Exoskeleton fragments in flour or baked products can be detected readily by simple laboratory techniques. Cereal products containing insect fragments are considered filthy under the articles of the Food, Drug, and Cosmetic Act. The products are liable to seizure, and the manufacturers or processors may be prosecuted. Certainly the presence of insects or of their feces, odors, or silks is repulsive to all consumers and constitutes an esthetic violation.

Field Insects

Field damage resembling damage by stored grain insects. There are several types of insect injury to wheat kernels committed prior to harvest in the central Wheat Belt that superficially resemble injury by stored grain insects. Except in the southeastern states, stored grain insects are rarely found in the fields. Rather, the injury is done by certain pests of the growing crop at the time the kernels are ripening. Frequently holes in the kernels that superficially resemble weevil emergence holes are made by young wheat head armyworms, young cutworms, and cowpea curculios. When these kernels are opened, entrance from the outside is evident because the walls of the hole are parallel and the cavity is free of excrement. Grasshoppers, cutworms, and armyworms may gnaw on the soft developing kernels, leaving scars that resemble feeding of cadelles.

Several species of stink bugs extract the fluids from the developing kernels and inject saliva, causing shriveled kernels that resemble drought damage. This damage does not occur in the southern and central Wheat Belt because the stink bugs are not abundant until after harvest. In the northern Wheat Belt these stink bugs, known as **gluten bugs**, are abundant when the wheat is maturing, and their damage may be extensive. Gluten bug damage is frequently extensive in Eastern Germany, Poland, Russia, and the Near East.

Field infestations. Field infestations in corn by Angoumois grain moth, rice and maize weevils, lesser grain borer, and certain other species are an annual problem in southeastern United States. The area in which damage occurs coincides closely with the area in which the Angoumois grain moth survives normal winters. Combine harvesting has greatly reduced or eliminated field infestations in wheat.

GRAIN STORAGE AND SOURCES OF INSECT INFESTATION

Storage on the Farm

Enormous stocks of grain are stored on the farm either as feed or as market grain. On January 1, 1975, farm storage of cereal grains amounted to 3.6 billion bushels, according to estimates of the USDA Crop Reporting Board. Feed grains are used primarily for livestock consumption on the farm and are retained normally for extended periods; market grains are disposed of through commercial channels and usually are moved from the farm within the year following harvest although longer storage is now more common. There is a close correlation between insect damage and length of storage. Feed grain infestations are more frequent not only because of the longer storage period but because less importance is attached to such infestations and less effort is made to provide adequate sanitation. The extent of grain shrinkage caused by insect feeding is scarcely realized because feed grains are rarely weighed in or out of the granaries, and the filth factors are ignored.

Farmers store market grains on the farm for two reasons: (1) because they choose to do so as part of a planned grain-marketing program or (2) because they must do so when local elevators are full and when boxcars, shipping, or truck drivers are not available. Out of expedience they place the grain in temporary bins in animal shelters, chicken houses, and under haymows, often near infested grains. Grain sanitation procedures are unknown or ignored. When grain storage is a part of a planned-marketing program, most farmers are acquainted with and are prepared for proper grain sanitation procedures and other adequate storage methods.

Important Sources of Insect Infestation in Farm-Stored Grain

Sources of infestation include bins of old grains; stocks of animal feeds, whether home prepared or commerical; granaries and cribs, particularly those with double walls and floor, cracks, and cadelle tunnels; feed and seeds from infested sources; accumulations of waste grain and feed in any of the buildings and in machinery and implements; migration of insects by flight from nearby infested sources; field infestations brought into the bins at harvest (in the South).

An intensive survey of grain storage conditions on central Kansas farms, which are typical of farm storage over a wide area, revealed that accumulations of grain or feed could be found in most farm buildings, including garages, machine sheds, all types of animal shelters and hay barns, and under granaries. During the summer months, with a peak in September, many of these grain and feed accumulations were infested with some or all of the insect species found in the granaries. Farm grain sanitation, therefore, cannot be confined to the granaries but must extend into and around each of the farm buildings.

Commercial and Government Storage

Grain stocks in elevators, commodity credit bins, and interior mills totaled over 2.2 billion bushels on January 1, 1975. For handling and storage of such huge quantities of grain, a system of elevators was developed. Country elevators constructed along railroads received grain from farmers and funneled it directly to terminal elevators. Many old country elevators designed solely for this purpose are still in use. During the great surplus years a considerable number of country elevators expanded their operations to include grain storage and feed manufacture, storage, and sale. Those developments greatly complicated and aggravated the problems of insect sanitation. Elevator management has turned increasingly to grain sanitation specialists, usually a part of the pest control or the agricultural chemicals industries, to advise or to take care of stored grain insect problems.

In older elevators and in those under improper management, insect infestations stem from the elevator premises. Insects reside permanently in tunnels, in cavities of the walls, in dead stocks in the grain-moving equipment, in holes and cracks in the wooden or concrete bins, and in any accumulations of grain. Frequently, infested grain from farm storage is binned directly with uninfested grain—without fumigation. A practice at some elevators is to place newly received infested grain in isolated bins for fumigation before storage. When feed mills are a part of the elevator business and heavily infested grain is run through the mill machinery, the disturbed insects migrate in all directions and particularly to the grain in storage. Frequently infested sweepings and cleanings are sold back to farmers for scratch feed, thus infesting the farm premises.

Grain in commercial channels can become infested from insects present in boxcars during shipment; from July to November nearly all boxcars are infested. The months of highest populations of stored grain insects coincide with the period of greatest grain-shipping activity.

CONTROL

Control Through Management Practices

General concepts of stored grain. If control of insect pests is to be achieved through management practices, farm and commercial grain handlers must approach the problem with the proper attitude. There are four concepts of stored grain, any one of which or any combination of which a grain handler may hold. These are:

1. Stored grain is a commodity to be handled like gravel, baled cotton, or other articles of commerce of similar weights and bulks.
2. Stored grain is money and wealth, or is a means of accumulating money.

3. Stored grain is a mass of living organisms, with properties and behavior characteristics of other living organisms.
4. Stored grain is primarily food for man.

An objective of grain storage management should be to see that grain handlers understand all four of these grain concepts so that they will lend their full support to its proper care and treatment. Grain, originally clean and pure, can be kept that way if grain handlers have the knowledge and desire to do so.

Granary construction. Control of insects in stored grain calls for the application of all known measures. Granaries should be constructed so as to hold the grain securely, exclude moisture, rodents, and birds, and be tight enough to retain fumigant gases. In farm storage the removal of double-walled construction in bins and feed rooms, where such walls are not required for structural reinforcement, will eliminate important lodging places for grain and hiding places for insects. If foundations of bins and granaries are sealed, grain cannot accumulate under the floor by spillage or through the activities of rodents. When new grain is binned, ample head room should be left over the grain for inspection and treatment.

Granary cleanup. Cleanup of the granary by removing accumulations of infested grain and feed from the premises or by fumigation is important. This cleanup is most effective if done in early spring when the insect populations are lowest. Leftover grain should be removed from the bins, and the walls should be swept and vacuumed before new grain is binned. Grain and feed accumulations that are frequently overlooked include empty feed sacks, nutritious dusts created by the feed grinders, seed litter from the haymows, and grain left in truck and wagon beds, in combine and elevator hoppers, and in unused animal self-feeders.

Storage in feed rooms and animal shelters. Market grains should be stored apart from feed rooms and feed bins because animal feed supports all grain-infesting insects. Feed rooms are difficult to clean and cannot be treated adequately with residual sprays. Market grains should not be stored in buildings that shelter animals and hay. The heat from the animals and from their manure may prevent the grain from cooling and thus enable insects to remain active throughout the winter. Mangers, feed boxes, and feed troughs are continuously infested with insects that can move directly to grain stored nearby. Large bulks of hay alongside or over grain bins insulate the insects in bins from the winter's cold.

Importance of dry grain. The drier the grain, the less it will deteriorate during storage as a result of insect activity. When storing grain of 11.5 per cent or more moisture, treatment with a grain protectant or one or more fumigations each year will be required in some parts of the country. The most serious insect problems would be avoided if harvest were delayed until the grain is dry. When this is not possible, it may be necessary to utilize commercial driers, as is done with rice and hybrid corn. Mistakes made by harvest crews that result in high-moisture grain include harvesting too early or too late in the day when atmospheric moisture makes the grain tough, and neglecting to cut around unripe patches in the field. When green weeds or insects such as grasshoppers and cutworms are harvested with the grain, the grain may require screening to avoid a moisture problem.

Chemical Control

Residual insecticides for granaries. After granaries and bins have been swept out and the old grain disposed of in preparation for the reception of new grain, many stored grain insects usually remain in double walls, in cracks and crevices, and in cadelle tunnels. These insects cannot be reached by brooms or vacuums. If an approved residual insecticide is sprayed on the walls and floor about three weeks before the new grain is binned, many of the insects will emerge from their hiding places, walk over the treated areas, and receive a lethal amount of the chemical. Several quarts of dead cadelles were destroyed in this manner in treated wooden farm bins in Kansas. Residual sprays alone have prevented infestations from developing in many bins of new wheat where the chief source for reinfestation was from within the granary itself.

Residual insecticides for treatment of grain bins include (1) a solution of pyrethrins synergized by piperonyl butoxide and/or MGK 264, (2) a spray of methoxychlor as a wettable powder or emulsifiable concentrate, and (3) a spray of malathion applied as an oilbase or emulsifiable concentrate. Rates and dosages for all EPA-approved insecticides are specified on the label. Any of these insecticides may be removed from the approved list at any time and other residual insecticides may be added.

Grain protectants. Grain protectants are formulations of chemicals having residual toxic or repellent action, or both, that are applied directly to grain to prevent damage by grain-infesting insects. Various inert powders have been applied to grain for many years, but at present their use is limited to seeds and to grain for animal feeds. Modern grain protectants have been in use since 1950. They are applied either as sprays or as powders to uninfested grain to prevent infestation. Application to farm-stored grain is made by hand or by small sprayers at some convenient place between the combine and the bin (Fig. 18:4). Mechanical applicators are used in treating grain in commercial storage as the grain is being turned. Thorough, good housekeeping measures should accompany the use of protectants.

Grain protectants have certain advantages over fumigants:

1. Grain can be treated while uninfested and infestations thereby avoided.
2. Protectants are effective when the grain is stored in bins too loose to be fumigated. Much of the grain in emergency storage can be treated effectively only by protectants.

Figure 18:4. Farmer applying wheat protectant to grain as it is transferred from combine hopper to truck. *Courtesy Kansas State University, Manhattan.*

3. Protectants are less dangerous than most fumigants.
4. Generally one application at harvest time is adequate for an entire year.
5. Protectants do not affect germination adversely.

Protectants currently approved for use include:

1. Pyrethrins synergized by piperonyl butoxide or MGK 264, relatively safe insecticide formulations when applied according to label instructions.
2. Premium grade malathion, which may be applied either as a powder or a spray. "Premium grade" is stressed, because the agricultural grade may impart a disagreeable odor to the grain.
3. Diatomaceous earth, mixed at times with the grain at the grain stream; on occasions it is applied only on the grain surface. It should be noted that diatomaceous earth may cause a reduction in the test weight of the grain and that it may have an abrasive effect on the machinery that moves the grain.

Fumigation. Grain fumigants are chemicals used alone or in combination; in their vapor phase, they are toxic to grain-damaging insects. The ideal fumigant would be highly toxic to insects, not harmful to warmblooded animals, inexpensive, vaporized rapidly, applied easily, nonflammable, nonexplosive, harmless to the grain, highly penetrating, and effective at all temperatures. Fumigants in use today fall short of these qualifications on one or more counts.

Chemicals have been used as grain fumigants since 1854, when carbon disulphide was first applied to kill granary weevils. The use of hydrocyanic acid began in 1886, chloropicrin in 1907, ethylene oxide and the 3:1 mixture of ethylene dichloride and carbon tetrachloride in 1927, methyl bromide in 1932, and aluminum phosphide later in the 1930s. Many chemicals are now available for fumigation of grains, but only a few are in common use. Few new grain fumigants have been developed during the past 25 years, but new formulations have been made by combining two or more of the basic chemicals. All commonly used fumigants fall in the "restricted use" category.

Grain handlers and farmers depend on fumigants to solve their stored grain insect problems because when properly applied fumigants rapidly kill all stages of insect life. In most cases fumigants do not require moving or turning grain already in storage, although on occasions fumigants are applied during moving operations.

The limitations of fumigants should be borne in mind. Once the toxic concentration of gas is dispelled, fumigants have no continuing effects. Fumigants are useful only when the structure that holds the grain is sufficiently tight so that a concentration of gas can be maintained long enough to be lethal. Most fumigants are anesthetics. When the fumigant vapors are not concentrated sufficiently to be lethal, the insects may recover completely.

Types of fumigants. There are three physical types of fumigants used for treatment of stored grain.

1. Gaseous fumigants, such as methyl bromide and hydrocyanic acid, are chemicals that are in a liquid state when under pressure or low temperatures but become gases upon release of the pressure or at normal room temperatures. The gaseous types of fumigants are especially adapted for grain fumigation using the closed recirculation type of forced distribution.

2. Liquid fumigants include those chemicals that are liquids under normal temperatures but vaporize rapidly when exposed to air at room temperatures or higher. Although there are many different formulations of liquid grain fumigants, the two basic mixtures (by volume) are the 3:1 mixture of 3 parts ethylene dichloride and 1 part carbon tetrachloride, and the 80:20 mixture of 4 parts carbon tetrachloride and 1 part carbon disulphide. Other formulations are modifications of the basic mixtures with other chemicals (additives) that are included to provide additional toxicants, irritants, and warning gases.

3. Solid fumigants include certain pellets, tablets, and granules that release a toxic gas on exposure to atmospheric moisture. Generally they are applied to the grain as it is being binned, although tablets and pellets can be inserted into binned grain through hollow tubes. During recent years world-wide acceptance has been given to a solid fumigant, aluminum phosphide, which releases the toxin phosphine (PH_3). Aluminum phosphide is available as pellets, tablets, and granules. A tablet will release one gram of phosphine (PH_3), whereas the smaller pellet will produce 0.2 gram of phosphine. Aluminum phosphide is now used extensively in the United States in fumigating grain, space, flour, and processed cereals and tobacco.

Methods of application of fumigants. There are several effective means of applying fumigants, depending on the fumigant used, the type and size of the grain storage facilities, the availability of grain handling equipment, and the type and condition of the grain. In fumigations of all kinds it is necessary to hold an adequate concentration of the fumigant within the grain mass long enough to effect an insect kill; consequently, careful sealing of the structure or grain mass is essential. Because the fumigant is effective only as long as it can be retained at the required concentration, the two most important factors become (1) containing the gas and (2) using the proper dosage. In practical terms, if the structure will not hold the gas or if the dosage is insufficient, the fumigation will fail. The most important recent developments in fumigating grain have been in methods of application, including the use of aeration systems designed to circulate air through the grain mass, to distribute or to recirculate fumigant vapors, and subsequently to remove them. These developments have resulted in greater efficiency, greater economy, and greater safety both to the grain and to the fumigator.

The following methods are used in applying fumigants:

1. Mixing the fumigant with the grain as the bin is being filled. Liquid and solid fumigants can be applied to the grain stream by hand or by automatic applicator. Turning the grain from one bin to another is necessary if grain already in storage is to be treated, except for liquid fumigants, which can be applied to the surface grain, and aluminum phosphide tablets and pellets, which can be probed into the grain through hollow tubes. In applying aluminum phosphide the easiest and most effective method is to apply it to the grain stream at the recommended rate.

Aluminum phosphide can also be used effectively in treating grain in farm storage or in flat storage when turning or moving the grain is impossible or uneconomical. The fumigant is applied to the grain surface and the grain is covered with a poly tarp. The pellets or tablets can also be probed into the grain mass with the aid of one of several types of probing devices or tubes.

2. Gravity penetration of the vapors of liquid fumigants into the grain mass. The surface grain is leveled, and the liquid fumigant is sprayed uniformly over the surface grain in a coarse, wet spray. The heavier-than-air vapors penetrate the grain mass and replace the air in the spaces between kernels. This treatment is popular in the fumigation of farm-stored grains and is used widely with commercially stored grains. Uneven distribution may occur when hot spots or accumulations of dockage are in the grain mass because the vapors tend to flow around such obstructions. Longer exposures and higher dosages must be employed to overcome these difficulties. Occasionally this treatment results in damage to grain by lowering germination and by increasing deposits of harmful residues and objectionable odors.

3. Forced distribution of fumigant vapors throughout the grain by any mechanical or physical force other than gravity, such as motor-driven fans. Forced distribution has been employed in Europe for many years. Improvements in the techniques of gas sampling and of gas analysis have stimulated research in this country on the problems of forced distribution. Adaptations are in use in some large grain storage sites because they permit the use of reduced dosages, cause quick and uniform distribution of the vapors, and provide a means for their removal. Most of the installations for forced distribution of fumigants are adaptations of existing aeration systems, following tests to determine how much air is being moved and where it is going. The air flow pattern in a grain mass can be determined by measuring the static pressures exerted on the air as it is moved through the grain. Static pressure readings indicate the total amount of air being moved per minute of fan operations. From this information, the time required for the fan to produce a single change of air within the bin can be determined. There are two general types of forced distribution.

1. Single pass distribution uses an aeration system without modification. The fan operates long enough to produce one complete change of air within the grain mass. This is adequate to replace the air with the vapors that have resulted from the application of fumigant to the surface of the grain. If desired, the fan can be reversed and the heavy vapors lifted to the surface again, after which gravity penetration takes over.

2. Closed recirculation distribution requires a return duct incorporated in the aeration system. A fumigant in the vapor stage is introduced into the aeration system at any place and is recirculated through the grain for a period equal to one or more air changes.

Fumigant dosage rates. Despite technical developments in materials and methods, grain fumigation remains more of an art than a science. Developments in the use of gas analyzers and air-moving equipment to grain fumigations may hasten the science of fumigation. The art and science of fumigation lie in the proper appraisal of conditions at time of treatment. These conditions have so many variable factors associated with the storage buildings, the grain to be treated, the insects involved, and the chemicals available for use that standard recommendations are only approximations. The following list of storage conditions together with their demands on fumigant dosages suggests the considerations that should be given these variables.

Storage Conditions	Amount of Fumigant Required	
	Less	More
Bin structure	Steel or concrete	Wood
Bin condition	Tight	Loose
Dockage and chaff	Little	Much
Percentage moisture	Low	High
Condition of grain	Normal	Heating
Extent of grain surface to volume	Smaller	Greater
Depth of grain	Deep	Shallow
Compaction of grain	Loose	Tight

Precautions in use of fumigants. All chemicals used in grain fumigation are poisonous to humans and animals. Fumigation should be attempted only by persons fully acquainted with the type of fumigant to be used and the method for its application. Fumigation is not for ignorant, careless, or foolhardy operators. The operator should take all necessary precautions and follow directions on fumigant and gas-mask labels. In addition, fumigators should recognize that many grain fumigants are anesthetic in their effects on man and animals. After the first breath the sensory nerves may be paralyzed to varying degrees and the sense of smell reduced to such an extent that the fumigator may be unaware that he is breathing fumigants.

In handling and applying all fumigants the operator should wear an approved full-face gas mask equipped with a canister labeled for the fumigant to be used. The used canister or one of unknown capacity should be replaced with a fresh one as specified in label directions for the fumigant. Canisters are designed to protect fumigators from light concentrations of gas only; they quickly become useless in heavy gas concentrations. It is sound practice to adopt the "buddy" system and to fumigate grain only when help is on hand for emergencies. Usually masks are not worn when applying aluminum phosphide because of the slow evolution of the toxic gas, phosphine, with its pungent odor. A mask should always be available to the applicator, however, in case of emergency.

Other Methods of Control

Stored-product insects have long been controlled by manipulation of temperature, such as refrigerated storage of processed foods or forced aeration using cool ambient air to lower grain temperature below that which permits activity of insects. Alteration of the atmosphere in storage facilities—including airtight storage and use of various lethal combinations of CO_2, N_2, and O_2—has received considerable study in recent years.

Physical force is commonly used in impact machines through which grain and grain products flow in flour mills. Resistant packages, including insecticide treated ones, are an important means of preventing or retarding infestations.

Electromagnetic energy has not been used extensively in stored-product insect control, but it is being further investigated and may be developed for broader usage. Types of electromagnetic energy considered are ionizing radiation (gamma and X rays), infrared, ultraviolet, radio frequency, and microwaves. Sound waves of various frequencies have been investigated.

Certain insect pathogens, such as bacteria and viruses, show promise for controlling some of the stored-product insects. These along with insect growth regulators and sex pheromones are under study as potential control agents, which, if effective and approved, will reduce the amount of traditional chemical insecticides required.

In the foreseeable future fumigants and other insecticides will be needed to protect grain and other stored products, but investigations should continue in order to develop alternate methods of control to supplement or replace chemicals. The integrated approach to insect control in stored grain and other stored products—combinations of sanitation, temperature and moisture adjustment, mechanical and physical methods, and chemicals—has been used for a long time. More study and refinement, however, are necessary before the "ideal" program is developed.

REPRESENTATIVE STORED GRAIN INSECTS

The following major pests have been selected for detailed study: granary weevil, rice and maize weevils, lesser grain borer, confused and red flour beetles, cadelle, and Angoumois grain moth.

Granary Weevil. *Sitophilus granarius* (L.)

Rice Weevil. (Black Weevil, Lesser Rice Weevil) *Sitophilus oryzae* (L.)
Maize Weevil. (Corn Weevil, Larger Rice Weevil) *Sitophilus zeamais* Motschulsky
[Coleoptera:Curculionidae]

Weevils were present in grain that was stored in the tombs of ancient Egypt. Today, weevils are the most destructive insect pests of grain in farm and commercial storage in many parts of the world.

Granary, rice, and maize weevils are worldwide in distribution. Granary weevils appear to be better adapted to northern areas. All species can live throughout the bulk of stored grain no matter how deep the bins. Generally, rice and maize weevils are better adapted to southern regions, where the maize weevils fly to fields of corn and oviposit on the developing kernels. They also fly to shocked, stacked, or cribbed grain. In more northern regions, rice weevils infest wheat in commercial and farm storage. Maize weevils are primary pests of corn; rice weevils are primary pests of wheat.

Description. Granary weevils are larger than rice and maize weevils when they develop on the same host grain. However, when rice weevils, normally the smallest of the three species, are reared on corn, they may become as large as granary weevils reared on wheat. Adult weevils are approximately $\frac{1}{8}$ in. long and have the head capsule prolonged into a slender snout with chewing mouthparts located at its tip (Fig. 18:5). They may be distinguished by the characteristics listed in the following table.

Granary Weevil	Rice and Maize Weevils
Larger (when reared on the same food)	Smaller (when reared on the same food)
Uniformly shiny black or reddish brown	Black or reddish brown with 4 yellowish spots, 2 on each elytron
Larger, loosely compacted oval pits on pronotum	Smaller, closely compacted rounded pits on pronotum
Without functional wings	With functional wings

For many years differences in size and behavior of rice weevils were noted, but they were considered to be slight and to represent subspecies or strains. In 1961, G. Kuschel worked out the synonymy of the *Sitophilus oryzae* complex and determined that the smaller rice weevil, which had been given the species name of *Sitophilus sasakii* (Tak.), was actually the *Curculio oryza(e)* of Linnaeus. The larger species was found to be *Sitophilus zeamais* Motschulsky. *Sitophilus oryzae* (L.) is now properly known as the rice weevil, and *S. zeamais* is known as the maize weevil. These two species can be separated according to Kuschel's key.

1. Upper surface of aedeagus convex, without two longitudinal impressions. Microsculpture of prothorax and elytra (with high magnification) more alutaceous, dorsal surface consequently duller . . . . *S. oryzae* (L.)
2. Upper surface of aedeagus flattened, with two distinct longitudinal impressions. Microsculpture of prothorax and elytra less alutaceous, dorsal surface consequently more shining . . . . *S. zeamais* Motschulsky.

The sexes of granary, rice, and maize weevils can be determined by the following characteristics. (1) Snouts of males are shorter, thicker, and rougher than snouts of females. (2) In addition, when observed in profile, males have a downward curve to the tip of the abdomen, whereas the female abdomen extends straight backward.

Life history. The life history and habits of the three species are quite similar. Under optimum laboratory conditions of 85°F and 14 per cent moisture, rice and maize weevils require from 30 to 45 days to develop from egg to

Figure 18:5. Adult of the rice weevil, *Sitophilus oryzae* (L.). *Courtesy Kansas State University, Manhattan.*

emerged adult; granary weevils generally require from 35 to 50 days.

Rice and maize weevils have a longevity of two to five months; granary weevils have a longer life. Females may deposit as many as 5 eggs per day with a total of about 400. The female selects a site and chews a narrow cylindrical hole into the kernel. On wheat kernels the favorite oviposition sites are in two narrow bands, a band margining the germ and a band at the brush end. When the hole in the kernel is completed, a whitish egg is deposited in the inner end by the ovipositor. As the ovipositor is withdrawn, the hole is filled to the surface of the kernel with a gelatinous material, known as the egg plug, which protects the egg from desiccation and to some extent from predation.

The eggs hatch into white, footless, wrinkled larvae that mature during four larval instars (Fig. 18:6). After completing the excavation within the kernel, the mature larva constructs a pupal chamber by secreting a fluid

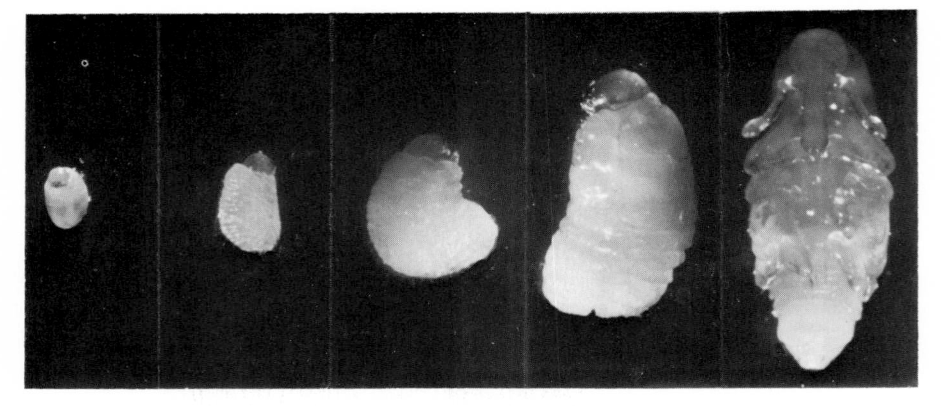

Figure 18:6. Rice weevil, the four larval instars and the pupa (right). *Courtesy Kansas State University, Manhattan.*

against the tunnel wall. This gives the walls a smooth, firm texture. On the day before pupating, the larva elongates into a prepupa. Adults remain in the kernel for several days before chewing an escape hole.

A granary weevil larva consumes approximately 55 per cent (by weight) of the interior of a wheat kernel. Rice or maize weevil larvae usually live on one side of the crease of a wheat kernel; a larva consumes about 25 per cent of the kernel (Fig. 18:7). Two rice or maize weevils may occupy a single wheat kernel if each remains on its own side of the crease. If one larva invades the area occupied by another, however, a fight to the death of one results. Only one granary weevil can develop in a kernel of wheat.

A granary weevil adult chews its way out of the kernel, leaving a large irregular hole in the bran coat by contrast with the roundness and neatness of rice and maize weevil emergence holes (Fig. 18:2A and B). Kernels with weevil emergence holes are called **weevil-cut kernels** by the grain trade, even though the holes may have been made by Angoumois grain moths or lesser

Figure 18:7. Radiograph of rice weevil infesting wheat kernel (dark object top center in each kernel); pupa (left); larva (right). *Courtesy Kansas State University, Manhattan.*

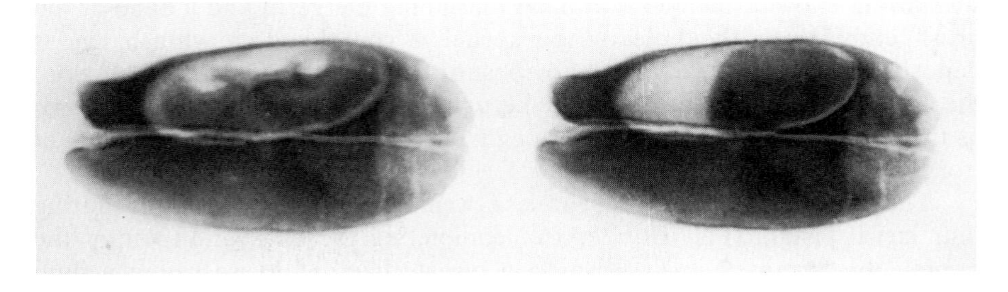

grain borers as well as by weevils. Samples of wheat taken from mills in different parts of the United States have indicated that, for every kernel exhibiting an emergence hole, there are likely to be four additional internally infested kernels. Practical tests in mills have established that when there are 7 infested kernels in 100 g of wheat (approximately 3500 kernels), the loss in flour amounts to several cents per hundredweight.

Determination of internally infested kernels. It is important that grain buyers for milling companies know the full extent of the internal infestation by weevils. The following techniques are the most useful for this purpose:

1. X-ray grain inspection units have been designed to obtain radiographs (negatives) of grain. All internal stages from egg to preemerged adults can be detected on these radiographs or photographic prints made from them. X-ray units are used to some extent by grain-storing and grain-processing industries and to a much greater extent by inspection and research agencies.

2. Acid fuchsin or gentian violet stains have an affinity for the gelatinous egg plugs, so that when treated properly, these plugs show a highly contrasting cherry red or violet spot against an unstained bran coat. The more commonly used acid fuchsin stain is prepared by adding 0.5 g of acid fuchsin to a mixture of 50 cc of glacial acetic acid and 950 cc of distilled water. Grain, wetted by soaking for two minutes in warm water, is covered with the dye solution for about one to three minutes, after which the dye is strained off and the grain washed under the tap to remove excess dye. A hand lens or low-powered, binocular microscope is useful in examining the sample for the stained egg plugs.

3. A cracking-flotation method has been used widely by flour mill technicians. A sample of grain is cracked or ground coarsely so that the insects are released and can be floated to the surface when gasoline or mineral oil is added to the sample.

4. A specific gravity or flotation method provides a rough test that is particularly useful to country elevator operators. A quantity of sodium silicate or water glass diluted with an equal amount of water will result in a specific gravity of approximately 1.16 to 1.19. When a sample of wheat is added to this liquid and stirred, most of the infested kernels will float whereas the uninfested kernels will sink.

Lesser Grain Borer. *Rhyzopertha dominica* (F.) [Coleoptera:Bostrichidae]

Lesser grain borers have become increasingly common and destructive in wheat during recent years. Both adults and larvae are voracious feeders. In heavy infestations the kernels are so completely consumed that only the bran covering remains (Fig. 18:2C).

Description. The adult is a small, cylindrical beetle less than $\frac{1}{8}$ in. long (Fig. 18:8). The head is positioned downward underneath the pronotum and is invisible from above; thus the beetle appears to be structured in two parts. Dorsally, the anterior third of the pronotum bears several tubercles.

Figure 18:8. Adult of the lesser grain borer. *Courtesy Kansas State University, Manhattan.*

Life history. Each female may lay 300 to 400 eggs among the kernels of grain. The newly-hatched larvae normally chew into the kernels, in which they develop to adults. During the approximately 30 days' development (optimum conditions) from oviposition to adult, the larva pushes out through the entry hole a large amount of odoriferous dust composed of feces and flour. The odor is easily identifiable. Accumulated and compacted dust in the infested areas may prevent penetration by fumigants, especially the liquid fumigants.

Broken kernels and nutritious dust in the grain enhance the ability of the species to develop populations. Populations can be maintained in grain with 9 to 10 per cent moisture content if the temperature is above 80°F. Optimum temperature for reproduction is higher than for many of the stored-product insects (about 93°F).

Confused Flour Beetle. *Tribolium confusum* duVal

Red Flour Beetle. *Tribolium castaneum* (Herbst) [Coleoptera:Tenebrionidae]

Flour beetles are cosmopolitan insects that have been associated with grain stored by man at least since early Egyptian times. In nature they lived under bark, where they were semipredators feeding on both living and dead materials. Today they are found in stored grains of many kinds and in the innumerable products manufactured from grain. They are known to millers as "bran bugs" because they were thought, erroneously, to feed only on flour and broken kernels. They will feed on whole kernels if the moisture content is 12 per cent or higher. In addition to grain they attack dried fruits, nuts, spices, and numerous other stored food products.

About 85 per cent of all insects in flour mills in the mid-1930s were flour beetles that populated all available dead flour stock present in the milling machinery, turning it to a dirty gray. At that time many of the millers were convinced that these beetles and their larvae developed spontaneously from the flour. Consequently they were slow to adopt the fumigation and good housekeeping procedures that are necessary if flour beetles are to be eliminated. When disturbed, the beetles secrete a vile-smelling liquid into flour. Modern sanitation procedures practiced by millers and bakers have largely corrected this situation. Flour beetles together with sawtoothed grain beetles and Indian meal moths are the most serious food pests in grocery stores and home kitchens.

Flour beetles contaminate cereal products by their feces and cast-off exoskeletons, by the vile odors that they leave, and by discoloration of white cereal products to a dirty gray.

Flour beetles are without peer as laboratory tools for researchers interested in studying animal genetics and population ecology and behavior. They reproduce abundantly and complete their life cycle in about one month.

Description. The two species closely resemble one another in appearance, behavior, and life cycle. They are flat, reddish brown, and approximately $\frac{1}{7}$ in. long (Fig. 18:9). They may be distinguished by the following characteristics:

Confused Flour Beetle	Red Flour Beetle
Larger	Smaller
Antennae gradually enlarged toward the tip	Antennae with the last three segments abruptly enlarged
Eyes smaller	Eyes larger
On underside of head, the width of each eye is about $\frac{1}{3}$ the distance separating them	On underside of head, the width of each eye is equal to the distance separating them
Wings not functional	Wings functional

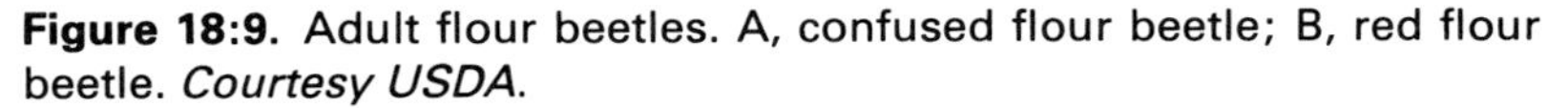

Figure 18:9. Adult flour beetles. A, confused flour beetle; B, red flour beetle. *Courtesy USDA.*

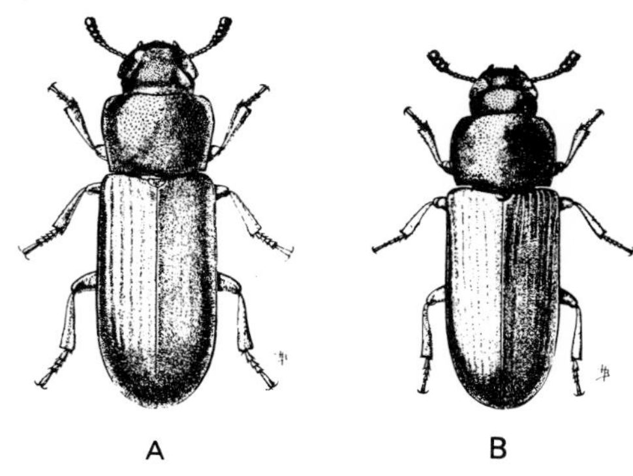

The sexes of the confused flour beetle can be distinguished by the patterns created by the keels and grooves of the apical halves of the elytra. The sixth strial groove of females curves with the curvature of the apex of the wing and unites with the third strial groove. This never occurs on males.

The eggs are small and white. Each egg, when laid, is covered with a fluid to which food particles stick, making the egg difficult to see. Eggs hatch into yellowish white larvae that bear small, dark, forked processes on the tip of the abdomen (Fig. 18:10).

Life history. Each female may deposit from 400 to 500 white eggs among the food particles; frequently eggs are pushed through the mesh of sacks containing cereal products. A 5XX-mesh bolting cloth will remove all flour beetle eggs from flour. When the food medium becomes overpopulated, the populations become self-limiting by cannibalism; the larvae eat the eggs, and the adults eat eggs and pupae.

The number of larval instars varies between 5 and 12. On a suitable food more than one half of both species will have seven or eight instars. The time required to develop from egg to adult depends upon the moisture and temperature. As with most grain-damaging insects within the limits required by the species, when the humidity and the temperature are higher, the developmental period is shorter. Under optimum conditions of environment and food, the developmental period is approximately 30 days. As they mature, the larvae come to the surface of their food medium and transform into naked pupae.

Cadelle. *Tenebroides mauritanicus* (L.) [Coleoptera:Trogositidae]

The cadelle is frequently the first of the stored grain insects to infest newly harvested wheat stored in wooden farm bins in the central Wheat Belt; also, it is one of the most destructive. Generally cadelle larvae are known to the farmers as "flour worms" because of the large amount of grain dust left in the bins after cadelle infestations. This dust is particularly noticeable as the grain

Figure 18:10. Larvae of the confused flour beetle. *Courtesy USDA.*

is moved from the storage bins to market. Formerly cadelles were known to millers as "bolting cloth beetles" because of their habits of eating holes in the bolting cloth; today these insects rarely occur in flour mills. Cadelles are found in all parts of the world where grain and grain products are transported or stored. In nature they are believed to have lived under bark as predators.

Description. The adults are shiny black beetles about $\frac{1}{3}$ in. long with the body divided into two units. The head and prothorax are distinct from the rest of the body. Full-grown larvae are about twice as long as the adults. They are the largest insects likely to be encountered in grain in suitable storage condition. The males can be distinguished by numerous punctures on the underside of the abdomen, some of which are very fine. There are fewer punctures in the same region of females, and they are always coarse. Larvae are highly contrasting black and white; the black areas are the head, parts of the dorsal surface of each thoracic segment, and a plate with two projections on the tip of the abdomen (Fig. 18:11).

Life history. The females insert white, slender, elongate eggs in cracks and crevices. Under favorable conditions they are prodigious egg producers; more than 3500 have been reported from a single female. Oviposition continues

Figure 18:11. The cadelle, larva and adult. *Courtesy Kansas State University, Manhattan.*

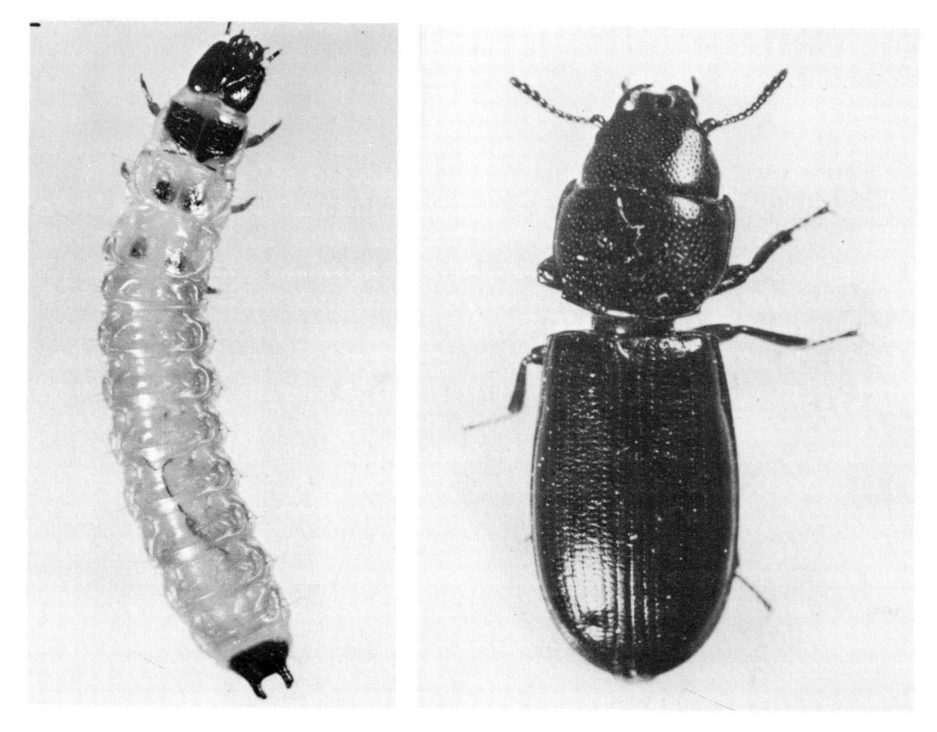

throughout the summer, and larvae of all sizes may occur in the grain until the larvae migrate to their hibernation sites. The larvae tunnel into the wooden parts of bins, boxcars, and ships' holds to hibernate and to pupate. Pupation takes place in early spring; beetles emerge in late spring and in early summer. Beetles are exceptionally long lived, some surviving for more than three years. Only full-grown larvae and the adults normally survive the winters.

The tunnels constructed by cadelle larvae frequently damage storage structures and provide lodging places for grain and grain dust and hiding places for other grain-damaging insects.

Angoumois Grain Moth. *Sitotroga cerealella* (Olivier) [Lepidoptera:Gelechiidae]

Since it was first reported in this country in 1728, the Angoumois grain moth has been the most destructive Lepidoptera of stored grains in the United States. Its larvae feed inside the kernels, weevil fashion, and consume germ and endosperm. Greatest damage is done in the southeastern states, but the moth extends into southeastern Kansas and across southern Illinois.

Description. The adults are buff to grayish or yellowish brown moths with an average wingspread of $\frac{1}{2}$ in. (Fig. 18:12). The moths range in weight from 0.7 mg to 11.2 mg depending on the larval food. Each forewing has two or three tiny dark spots, and the apical tip of each hindwing is narrowed and pointing—like an accusing finger. The hindwings are grayish white and are heavily margined with long hairs. The eggs are 0.6 mm long. The larvae are white with a yellowish brown head and reach a length of about $\frac{1}{5}$ in. when full grown (Fig. 18:13).

Figure 18:12. Angoumois grain moth reared on corn. Female (left); male (right). *Courtesy Kansas State University, Manhattan.*

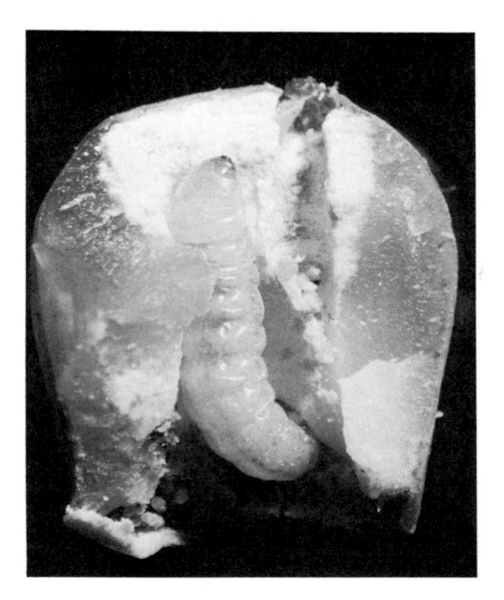

Figure 18:13. Larva of the Angoumois grain moth feeding in a kernel of corn. *Courtesy Kansas State University, Manhattan.*

Life history. Eggs are deposited outside of the kernels. On ears of corn they are likely to be pushed deep between the kernels; on wheat in the field, shock, or stack, eggs are placed anywhere on the head; on shelled corn and wheat in the bin, they are laid among kernels at the surface or within a few inches of the surface.

A newly hatched caterpillar may construct an entrance cocoon on the outside of the kernel before eating an entrance hole into it or may enter the kernel at the soft stem attachment. Cracks in the pericarp over the germ of many wheat kernels provide easy entry for the larvae. Usually there is only one caterpillar per wheat kernel, but there may be more than one per corn kernel. When infestations occur in sorghum grain with kernels too small for complete larval development, a larva may tie several kernels together into a ball with silk and live inside the cluster but outside the kernels. There are from four to eight or more instars, depending largely on whether the young larva fed initially on germ as well as endosperm or whether it was confined to endosperm for some time. Larvae with a normal germ-endosperm diet from the start have four or possibly five instars. Larvae that do not feed on germ for a considerable time may have six or more instars. The length of the larval-pupal period ranges from 23 to 63 days and averages 38 days in wheat, 35 days in corn, and 33 days in sorghum. Insects with longer developmental periods spend disproportionately more time as earlier instars than the insects with shorter developmental periods. Before pupating the mature larva extends the cavity to the inner surface of the seed coat and then spins a tough silk lining for the cavity and prepares a thin escape hatch. Viewed from the outside, this escape hatch in a kernel of corn looks like a glass window. Upon completion of the pupal stage the moth easily pushes its way out of the kernel through the escape hatch, frequently dragging the cocoon along with it (Fig. 18:2F).

The moths largely spend the winter as mature larvae that pupate in the spring and emerge in late May or early June. There are from two to four generations in a year depending on environmental conditions. The moths are sufficiently strong fliers so that they can infest wheat and corn in the field, shock, or stack. The harvesting of wheat by combine has largely eliminated the Angoumois grain moth as a wheat pest, but it is still one of the most important pests of stored corn in those regions in which it can survive the winter.

SELECTED REFERENCES

Birch, L. C., "The Biotic Potential of the Small Strain of *Calandra oryzae* and *Rhizopertha dominica*." *J. An. Ecol.* 14(2):125–7 (1945).

———, "Influence of Temperature on Development of the Different Stages of *Calandra oryzae* L. and *Rhizopertha dominica* Fab." *Aust. J. Exp. Biol. and Med. Sci.* 23:29–35 (1945).

Bond, E. J., "Chemical Control of Stored Grain Insects and Mites," in *Grain Storage: Part of a System*, R. N. Sinha and W. E. Muir, editors, pp. 137–79. The Avi Pub. Co., Inc., 1973.

Boudreaux, H. B., "The Identity of *Sitophilus oryzae*." *Annals of the Entomol. Soc. Amer.* 62(1):169–72 (1969).

Cotton, R. T., *Pests of Stored Grain and Grain Products* (Minneapolis, Minn.: Burgess, 1963).

Cotton, R. T., and D. A. Wilbur, "Insects," in *Storage of Cereal Grains and Their Products*, C. M. Christensen, editor. Monograph Series, Vol. V, revised, pp. 193–231 (Amer. Assoc. Cereal Chem., Inc., 1974).

Cotton, R. T., H. H. Walkden, G. D. White, and D. A. Wilbur, *Causes of Outbreaks of Stored-Grain Insects*, Kan. Agr. Exp. Sta. Bul. 416 (Revision of North Central Regional Publ. 35), 1960.

Cotton, R. T., and N. E. Good, *Annotated Lists of the Insects and Mites Associated with Stored Grain and Cereal Products, and of Their Arthropod Parasites and Predators*, USDA Misc. Publ. 258, 1937.

Daniels, N. E., "Damage and Reproduction by the Flour Beetles, *Tribolium confusum* and *T. castaneum*, in Wheat at Three Moisture Contents," *J. Econ. Entomol.*, 49(2):244–7 (1956).

Fraenkel, G., and M. Blewett, "The Natural Foods and Food Requirements of Several Species of Stored Products Insects," *Trans. Roy. Entomol. Soc.* (London), 93:457–90 (1943).

Frankenfeld, J. C., *Staining Methods for Detecting Weevil Infestation in Grain*, USDA Bur. Entomol. and Pl. Quar. Circ. ET-256, 1948.

Good, N. E., *The Flour Beetles of the Genus Tribolium*, USDA, Tech. Bul. 498, 1936.

Gray, H. E., "The Biology of Flour Beetles," *Milling Prod.* 13(12):7, 18–22 (1948).

Harein, P. K., and E. de las Casas, "Chemical Control of Stored-Grain Insects and Associated Micro- and Macro-Organisms," in *Storage of Cereal Grains and Their Products*, C. M. Christenson, editor, Monograph Series, Vol. V, revised, pp. 232–91 (Amer. Assoc. Cereal Chem., Inc., 1974).

Howe, R. W., "The Biology of the Rice Weevil, *Calandra oryzae* (L.)," *Ann. Appl. Biol.*, 39(2):168–80 (1952).
Kirkpatrick, R. L., and D. A. Wilbur, "The Development and Habits of the Granary Weevil, *Sitophilus granarius*, Within a Kernel of Wheat," *J. Econ. Entomol.*, 58(5):979–85 (1965).
Kuschel, G., "On Problems of Synonymy in the *Sitophilus oryzae* Complex," *Annals and Magazine of Natural History*, 4(40):241–4 (1961).
Linsley, E. G., "Natural Sources, Habitats, and Reservoirs of Insects Associated with Stored Foods Products," *Hilgardia*, 16(4):187–224 (1944).
Mills, R. B., "Early germ feeding and larval development of the Angoumois grain moth," *J. Econ. Entomol.*, 58(2):220–3 (1965).
Mills, R. B., and D. A. Wilbur, "Radiographic Studies of Angoumois Grain Moth Development in Wheat, Corn, and Sorghum Kernels," *J. Econ. Entomol.*, 60(3):671–7 (1967).
Milner, M., "New Methods to Detect and Eliminate Insect-Infested Grain," *Advances in Food Research*, 8:111–31 (1958).
Milner, M., M. R. Lee, and R. Katz, "Application of X-Ray Technique to the Detection of Internal Infestation in Grain," *J. Econ. Entomol.*, 43:933–5 (1950).
Parkin, E. A., "Stored Product Entomology," *Ann. Rev. of Entomol.*, 1:223–40 (1956).
Potter, C., "The Biology and Distribution of *Rhizopertha dominica* (Fab.)," *Trans. Roy. Entomol. Soc.* 83(4):449–82 (1935).
Richards, O. W., "Observations on Grain-Weevils, *Calandra* (Col. Curculionidae). I. General Biology and Oviposition," *Proc. Zool. Soc.* (*London*), 117:1–43 (1947).
Robinson, W., *Low Temperature and Moisture as Factors in the Ecology of the Rice Weevil,* Sitophilus oryza *L., and the Granary Weevil,* Sitophilus granarius (*L.*), Minn. Agr. Exp. Sta. Tech. Bul. 41, 1926.
Sharifi, S., and R. B. Mills, "Developmental Activities and Behavior of the Rice Weevil Inside Wheat Kernels," *J. Econ. Entomol.* 64(5):1114–8 (1971).
Simmons, P., and G. W. Ellington, *Life History of the Angoumois Grain Moth in Maryland*, USDA Tech. Bul. 351, 1933.
Soderstrom, E. L., and D. A. Wilbur, "Biological Variations in Three Geographical Populations of the Rice Weevil Complex," *J. Kans. Entomol. Soc.*, 39(1):32–41 (1966).
USDA, *Stored Grain Pests*, Stored Products Insects Section, AMS, USDA Farmers' Bul. 1260, 1962.
Walker, D. W., *Population Fluctuations and Control of Stored Grain Insects*, Wash. Agr. Exp. Sta. Tech. Bul. 31, 1960.
Wilbur, D. A., Jr., "The federal environmental pesticide control act and its effect on the food industry," MBAA Tech. Quarterly 12(2):103–6, 1975.
Wilbur, D. A., and G. Halazon, *Pests of Farm-Stored Wheat and Their Control*, Kan. Agr. Exp. Sta. Bul. 481, 1965.
———, and L. O. Warren, "Grain Sanitation on Kansas Farms," *Proc. Tenth Int. Cong. of Entomol.*, 4:29–31 [1956 (1958)].

SELECTED JOURNALS

Journal of Stored Products Research. Published quarterly by Pergamon Press, Ltd., Oxford, England.
Tropical Stored Products Information. Bulletin of The Tropical Stored Products Centre, Slough, England.

chapter 19 / **JOHN V. OSMUN**

HOUSEHOLD INSECTS

In our urban communities today there are diverse environments that are attractive to insects, including lawns, flowers, shrubs, parks, industrial complexes, and dwellings. Pests that infest dwellings are commonly referred to as household insects, and they are the ones of particular concern here.

From the day that man first started providing shelter for his family, he was accompanied by insect pests in the household. Early biblical references frequently mention ants, flies, and moths as part of man's abode; one of the benefits of a desirable hereafter was indicated by Matthew, who suggested that proper preparation should be for heaven where moths do not consume. As people have improved their homes, they have unwittingly made them increasingly favorable environments for insects. In addition, food is kept in kitchens and pantries of most homes. Ideal conditions for many insects exist today in many households, and although there may be profound differences among the environments, all of them bear the similarity of an interdependence with the daily activities of man.

Household insects are a part of the total complex of pests that are of direct concern to man and his immediate environment. We speak of this branch of entomology collectively as urban and industrial entomology. Many of the pests described in Chapters 18 and 22 of this book are included. Household insects are also the focal point of one of the principal business phases of entomology, the work engaged in by commercial pest control operators. Urban and industrial pest control is important in helping maintain the health and well-being of mankind and in protecting his food products.

THE PESTS

At first consideration, the insects that one might think of as household pests are flies and mosquitoes. These pests, however, are incidental to the house, having been attracted by food odors or the presence of man. Generally the two groups do not breed or complete life cycles in structures, unless abnormal situations exist. For this reason, mosquitoes and house flies are considered elsewhere in this book as insects affecting man. We are concerned here with the insects that commonly utilize our homes and buildings as their preferred environments.

The most common insects in this category are the cockroaches. Although at least nine species of roaches are found in buildings, four of them predominate: **German cockroach** (Fig. 19:6), **oriental cockroach** or "waterbug" (Fig. 19:7), **American cockroach** (Fig. 19:8), and **brownbanded cockroach** (Fig. 19:9). The first three names suggest origins from countries of the world; it is more likely the nomenclature arose at various times in history as a slander to a neighboring country. These pests are widely distributed in the temperate areas of the world.

Clothes moths are commonly associated with the destruction of woolen clothing and rugs, but of equal or sometimes greater importance in this category of household pests are the several species of carpet beetles (Fig. 19:12), especially the **black carpet beetle**. Of the two clothes moths, the **webbing clothes moth** is more commonly encountered than the **casemaking clothes moth** (Fig. 19:2).

Many species of ants invade the household, and several species are able to nest and complete their life cycles completely within a structure without contact with the ground. As a group, the ants are numerous and highly successful. Those capable of developing within a structure include the **black carpenter ant**, **Pharaoh ant**, and the **odorous house ant**. Other forms that normally nest in the foundation soil and range through dwellings include the **pavement ant**, field ants, **thief ant**, and in the southern and western parts of the United States, the **Argentine ant** (Fig. 19:13). The much-publicized imported fire ant in the South is primarily a pest of agricultural land.

Closely related to the ants are bees and wasps, which are occasional household pests. One of these, the **carpenter bee**, attacks various wood members of houses such as window sills, siding, and exposed rafters. This bee is not in search of food as it enters the wood but is tunneling for nesting purposes. Because of its ominous appearance, the carpenter bee is more of a nuisance than a serious destroyer of wood (Fig. 19:15).

Throughout our country, one of the most feared insects is the termite (Fig. 19:17). As common as these insects are, most people know very little about them, and perhaps it is the element of the unknown that causes such great anxiety. Historically the relatively ancient and socially organized termite has been considered a useful organism. It serves its place in the balance of nature by reducing fallen timbers and dead roots in the ground. Only because man has placed the wood of his dwellings in situations accessible to termites have these insects become destructive. **Subterranean termites**, which normally colonize in the ground and move upward into buildings, predominate and are reported from 49 states. **Drywood termites**, which can colonize in homes without soil or moisture connection, are generally limited to the Gulf states, Texas, Arizona, California, and Hawaii. The less important dampwood termite has a similar distribution.

Powderpost beetles cause wood damage that is sometimes confused with the work of termites. Larvae of these small beetles may tunnel through furniture, baseboards, and joists as they feed. Their activity becomes most noticeable when the adult beetles emerge, leaving small holes in the surface and often telltale piles of sawdust (Fig. 19:20).

There are many insects and related arthropods of lesser importance in households, such as bed bugs, food-infesting kitchen pests, spiders, book lice, earwigs, clover mites, and silverfish (Fig. 19:1). The presence of any one of these reflects an ecological condition favorable for that particular pest.

IMPACT

In considering the economic importance of household insects, one must take into consideration more than just injury or damage. Monetary significance is attached to insects as contaminators of food and as factors in the esthetic well-being of people. No one likes insects in his food or crawling about in his home.

Cockroaches are important in all of these categories. Through their ability to chew, they will attack book bindings, other sized paper, and cloth, and will devour the nutrient material of a gummed surface. They are odorous, obnoxious pests, and habitually they leave trails of fecal droppings. The German cockroach is often rightfully associated with unsanitary conditions and poor disposal of dirt and waste. In addition, at least this species is associated with transmission of disease (e.g., gastroenteritis). The oriental cockroach can be found the year around in homes, but in the northern climates it increases in numbers as individuals move in from the outside during the fall for protection from the cold. American cockroaches prefer a warm, moist environment, which accounts for infestations in steam rooms, moist basements, and in regions where the temperature is high, in sewers and areas under homes. Because of the large amount of interregional moving of people during World War II, the brownbanded roach, once an inhabitant of the South, is now scattered throughout the country. This insect is not as gregarious as the German cockroach it resembles, and it is found widely dispersed in household cupboards and drawers, behind pictures, in electric clocks and in furniture.

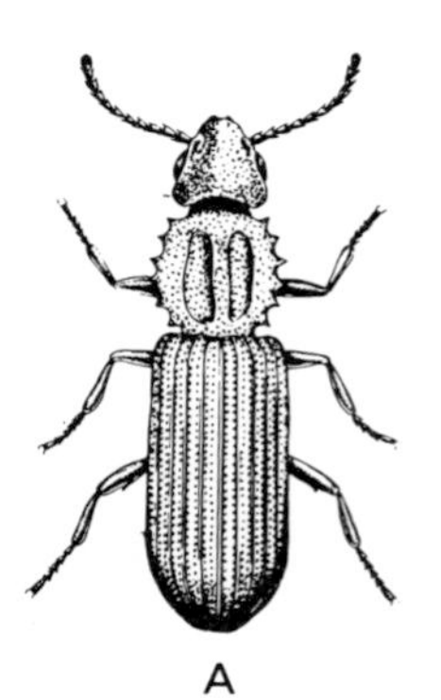
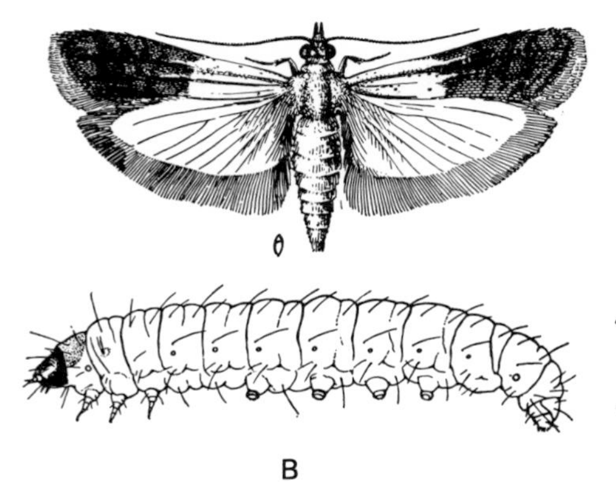
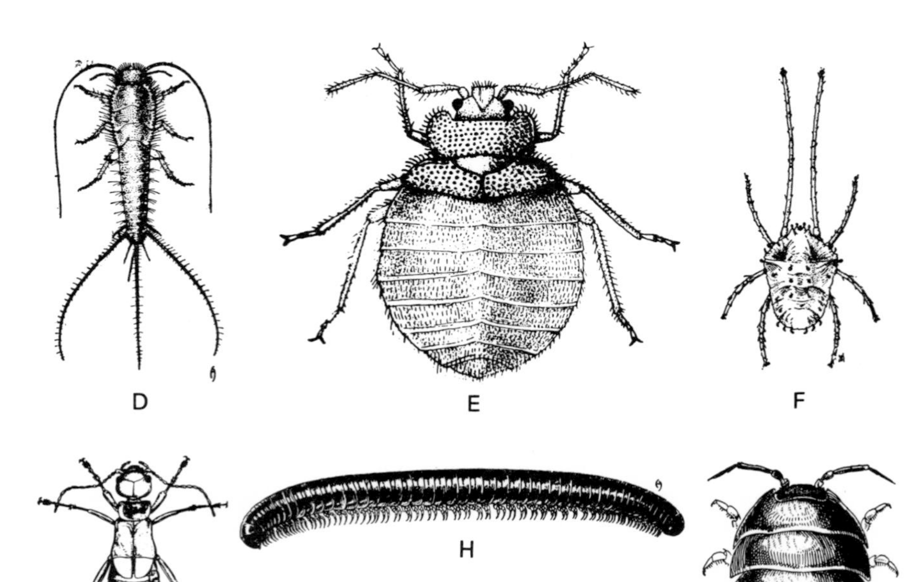

Figure 19:1. Lesser household pests (not to scale). Pantry-pests: A, sawtoothed grain beetle; B, Indian meal moth and larva; C, booklouse. Pests of various locations: D, silverfish; E, bed bug; F, clover mite; G, earwig; H, millipede; I, centipede; J, pillbug. (*A and G, courtesy Washington State University, Pullman*; *F, courtesy USDA*; *B–E and H–J, courtesy University of California, Berkeley.*

Nearly everyone has at one time or another experienced the misfortune of having woolens attacked by clothes moths; this situation is summarized nicely in verse:

> Friends may criticize my clothes
> And all the furnishings I buy
> But moths and carpet beetles like them
> Equally as well as I.

These keratin-loving insects possess a keratinase enzyme that permits protein disintegration and nutrient consumption of wool fibers (Fig. 19:2). Soiled wool is especially attractive to developing larvae, the only stage which feeds.

Figure 19:2. The two species of clothes moths and their damage. A, stages of the webbing clothes moth: pupal case, adult, and larva on badly damaged woolen cloth; B, larva of the casemaking clothes moth feeding on woolen carpet. *A, courtesy of Clemson University, Clemson, S.C.; B, courtesy Edward S. Ross.*

A

Figure 19:2B. Larva of case making clothes moth feeding on woolen carpet.

The adult moths are active at night, moving readily from closet and drawers to furniture, rugs, feather pillows, fur, and even pianos, in order to lay their eggs. Although synthetic fibers are not digested, damage frequently occurs to mixtures of wool and synthetic fibers. Carpet beetle larvae are even more destructive because they move extensively while feeding and because they have a wider range of food, including cereals and grains.

From pants to pantry, ants can be important in a home; when their activities conflict with those of man, they are a problem. Food is a normal attractant to these insects, and the resulting contamination can be considerable. Soiled spots on clothing are also attacked, resulting in damage. In addition, ants are annoying because of their presence and because of their ability to bite and insert irritating formic acid into the wound. Some ants, such as the fire ants, have retained the ability to sting. One group of ants, commonly called carpenter ants, is of economic importance because of their destruction of wood. These ants do not derive nourishment from wood but frequently hollow out supporting timbers in homes and other structures to provide protected nesting sites for colony development (Fig. 19:3).

Termites are injurious because of their ability to penetrate wood and wood products and to devour cellulose. The principal economic damage done by these insects is the weakening of structural timbers in homes and buildings (Fig. 19:4). The activity of a strong colony can be considerable, causing severe damage to foundation plates, joists, wall supports, and flooring. Man seemingly builds his homes for termites in that he places wood members of his dwelling in situations accessible to them. Subterranean termites, those that colonize in the ground and extend their activities from such focal points, can attack wood in buildings by finding protected passage from the soil to the source of food or by making connecting earthen tubes. Because warm temperature accelerates development, extensive damage occurs when there are extended periods of warmth. For many years, the predominantly high incidence of infestation occurred in the southern states. Today, however, infestations are widespread. Homes are better constructed; basements are well heated, many are wood paneled; foundations are hollow, permitting

Figure 19:3. Piece of 2″×4″ wood, excavated by carpenter ants to provide nesting place. *Courtesy Purdue University, Lafayett, Ind.*

undetected passage of termites from soil to wood; slab homes are usually equipped with floor heating, which makes the soil beneath an ideal environment for termite development. Combine this "better living" with the cryptobiotic nature of the termite, and it is little wonder that undetected damage can occur to harass the homeowner.

Although old pieces of furniture are sometimes considered better antiques because of the small holes in their surfaces, normally wood with many holes in it reflects an undesirable infestation of powderpost beetles (Fig. 19:20). These small insects attack a wide variety of processed wood such as implement handles, building joists, and furniture. They damage wood by making extensive tunnels within the wood as the larvae feed, a process that may continue for months or even years before the adults emerge.

Figure 19:4. Wooden beam damaged by subterranean termites. Damage was not visible until torn apart. *Courtesy Purdue University, Lafayette, Ind.*

CONTROL

Control Through Sanitation and Building Management

Although household insects are in general difficult to eliminate by means other than chemicals, some of them can be kept to a minimum by good housekeeping practices. Cockroaches and ants either enter a building by their own volition, or they are carried in with commodities. Their chance for survival, however, may depend on the degree of sanitation or household management that exists there. Included in good management would be rapid and proper disposal of wastes; general cleanliness in the kitchen areas; protected storage of food; clean, well-ventilated basements; a minimum of concealed areas suitable for hiding; and storage free of superfluous accumulations. In the case of fabric pests, seasonal airing of woolens, washing or dry cleaning before storage, and thorough vacuuming of rugs, baseboards, heating ducts, and closet space are good preventive practices. Even when infestations of household pests do occur, such procedures will greatly enhance the effectiveness of chemical control.

In the case of termites, certain measures can be taken that will decrease the chance of termite attack but certainly will not guarantee it. During construction, efforts should be made to prevent the accumulation of scrap lumber in the soil around a building or in crawl spaces beneath it. Structural wood should not touch the soil or be in protected proximity to it. Proper drainage of the soil around buildings will decrease the possibility of moist environments attractive to termites.

Chemical Control

The use of chemicals to control household insects is widely accepted. A great variety of materials is needed, and it is necessary to employ safety in their use. The inhabitants of the household, whether pets, children, or adults, are living organisms just as are insects. Insecticides can be hazardous if not properly used. The safest materials, yet the most rapid in toxic action in insects, are pyrethrum and related synthetic pyrethroids. If residual action is desired, certain organophosphorus insecticides, such as malathion, diazinon, and chlorpyrifos are available.

Household insecticides are formulated commonly in solutions of refined, nonstaining petroleum oil. Some emulsifiable concentrates are available for dilution with water, and there are several dust formulations on the market. Wettable powder suspensions are not recommended for use in household insect control.

Control with chemicals is only as good as the quality of application. Any of these insecticides must be applied thoroughly and, based on the habits of each particular pest, placed where most needed. The operating procedure of diagnosis, prescription, and proper application should be borne in mind. In many cases, especially with termite control, the householder himself will not be competent to utilize insecticides effectively against household insects. Fortunately there are several thousand reliable pest control operators in the country who are specially trained and certified to serve the public.

Control Equipment

The best single piece of equipment for household insect control is a 1-gal compressed air sprayer equipped with a precision nozzle that produces a nondrip, fan-shaped spray pattern (Fig. 19:5). This equipment is convenient for the application of small quantities of insecticide in selected areas, and at the same time it can be used for fairly liberal spraying where needed. Most packaging devices, including pressurized dispensers and "handy" plunger sprayers, are limited in their usefulness and should be relied upon only where light application is desirable. Standard equipment of a commercial pest control operator for termite control includes highly specialized power equipment, modified nozzle assemblies, and drilling devices.

Precautions

The introduction of an insect toxicant into a household automatically implies the possession or use of a household chemical which might be hazardous. Children are often present, and human error can lead to difficulties. Although household insecticides can be used safely by following label directions, it is always best to consider each container as a potential hazard and store it in a safe place. Replace caps securely and keep dusts away from food supplies

Figure 19:5. A one-gallon compressed air sprayer for applying insecticides in the home. Items apt to be contaminated are removed before spraying. *Courtesy Purdue University, Lafayette, Ind.*

where they could be mistaken for flour or seasoning. Storage away from cold is also desirable because cold temperatures cause precipitation of the active ingredient.

When applying household insecticides, avoid contamination of foods and food preparation surfaces. In addition, do not apply materials where infants will be apt to crawl across treated surfaces. Prolonged and repeated exposure to aerosol mists should be considered hazardous. Environmental contamination can result from improper disposal of unused household insecticides. Never flush them down a drain but dispose of them according to the label or recommendations of specialists in your community. Use pesticides as needed, but do so with respect and caution; always read and follow the label.

REPRESENTATIVE HOUSEHOLD INSECTS

German Cockroach. *Blattella germanica* (Linnaeus) [Orthoptera: Blattellidae]

The German cockroach is one of our most gregarious household pests and is distributed widely in the temperate regions of the world. This insect usually breeds and lives in buildings generation after generation without influence of seasons. In apartment buildings, German cockroaches move readily from floor to floor following pipe lines or moving through wall voids. Because they generally carry their egg capsules until nearly hatching time, the spread of these pests is facilitated. The species is commonly introduced into buildings by means of infested food cartons, laundry, and various containers. It is the undesirable presence of the German cockroach that is important, and to a lesser extent, damage that it could cause. Food and dishes become contaminated by the excreta and by various secretions. The German cockroach tends to congregate under tables, in and between cupboards, in wall cracks and behind wall hangings, inside beer and soft drink cartons, and in similar secluded places. In recent years, an apparent change in its habits has been noted in areas where considerable insecticides have been used and where resistance of this species to chlorinated hydrocarbons occurs. The effect has been one of scattering, and individuals may be found almost anywhere in homes or other buildings. It is also increasingly common to find the German cockroach outdoors under siding, in worm holes, around garbage cans, and other protected places. As a result of this change in behavior, control is more complicated.

Description. The German cockroach is pale brown in color with two parallel, dark stripes on the pronotum (Fig. 19:6). When full grown, these

Figure 19:6. German cockroach life stages. A–D, nymphal instars; E, male adult; F, female adult carrying egg case; G, deposited egg case; H, female adult with wings spread. *Courtesy USDA.*

insects are $\frac{1}{2}$ to $\frac{5}{8}$ in. long, oval in appearance, and possess fully developed wings covering the abdomen; their antennae are longer than the body. The wings are functional but are usually used only for downward flight. The nymphs, which are without wings, resemble small editions of the adults; the visual impression is one of an insect with a light area in the center of the thorax. The egg capsules or oothecae are slender and light brown.

Life history. Each egg capsule contains up to 48 eggs. The female carries the capsule throughout most of the three-week incubation period and produces about five capsules during her life. Growth is by simple metamorphosis and takes an average of three months. Provided maximum hatching and optimum developmental conditions prevail, it is theoretically possible for one female to be the progenitor of more than 30,000 offspring in one year. The species is, from the standpoint of population development, the most efficient of the cockroaches. The productivity is an important consideration in understanding the development of chemical resistance in this insect.

Oriental Cockroach. *Blatta orientalis* Linnaeus [Orthoptera:Blattidae]

The Oriental cockroach is more commonly called a "waterbug," a term freely used to avoid the apparent stigma attached to having cockroaches in a household. This insect is found in homes and other buildings where dark, protected areas, food, and water are available. Basements most frequently provide such environments, but these insects will also migrate to higher levels. In most regions of the country, oriental cockroaches are common outdoors in warm weather and may become unusually abundant around trash and garbage. Although they are more sluggish and less wary than other cockroaches, they migrate readily and, especially in the fall of the year, find their way into buildings through drains, wall openings, and open doors. These insects are annoying and may do considerable damage to book bindings and starched materials.

Description. The oriental cockroach is shiny black, although occasionally the wings of the male are a mahogany color. The female is oval and has only wing stubs; its length is $1\frac{1}{4}$ in. Males have developed wings partially covering the abdomen, are more slender, and about 1 in. long. Egg capsules are very dark and bulky in appearance. Nymphs of both sexes resemble miniature adult females (Fig. 19:7).

Life history. The oriental cockroach has retained seasonal influence in its development cycle. The egg capsules, containing a maximum of 16 eggs, are dropped as soon as formed and often are glued to objects. Incubation may take two months. Because nymphs mature in the spring, the period of development may be just short of a year to more than two years.

American Cockroach. *Periplaneta americana* (Linnaeus) [Orthoptera:Blattidae]

Humidity and high temperatures favor the development of the American

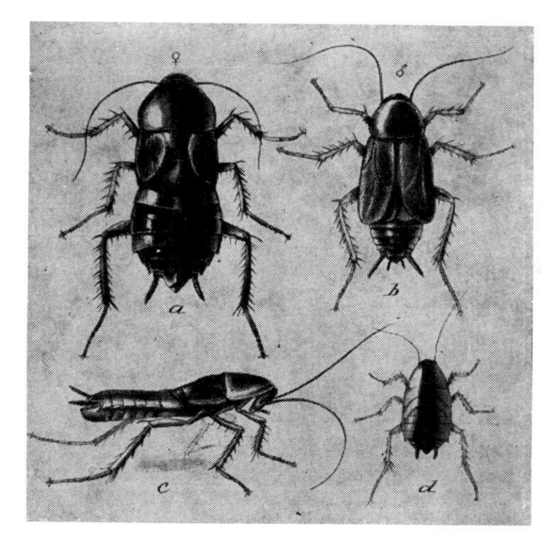

Figure 19:7. Oriental cockroach. A, adult female; B, adult male; C and D, nymphs. *Courtesy Purdue University, Lafayette, Ind.*

cockroach. It is less of a household pest than the other three considered in this chapter, but it frequents the steam tunnels of apartment buildings and works upward from such focal environments. In the southern states this species and others of the same genus are more common as household pests, frequently migrating in from the outside where climatic conditions favor their existence. These insects are capable of moderate flight. They are economically important for the same reasons as oriental cockroaches.

Description. The American cockroach is reddish brown except for a light brown margin around the side and back edges of the pronotum (Fig. 19:8). Males and females both of which are winged, are large insects, about $1\frac{1}{2}$ in. in length, and bear long, active antennae. The nymphs are wingless but otherwise resemble the adults except for size. The dark brown or blackish egg capsule resembles that of the oriental species.

Life history. As many as 16 nymphs can hatch from an ootheca, incubation taking less than two months to complete. Like the other cockroaches discussed, development of the nymphs is influenced by temperature, and may take as little as 285 days or as long as a year and a half. Adults are long lived, and females may produce as many as fifty egg capsules in a lifetime, dropping them in protected places soon after formation.

Brownbanded Cockroach. *Supella longipalpa* (Fabricius) [Orthoptera: Blattellidae]

The influence of war played a very great part in the present distribution of the brownbanded cockroach. Confined principally to tropical regions and the southern states before World War II, this species hitchhiked in household goods of military personnel to establish itself successfully in heated homes in

Figure 19:8. American cockroach adult, actual size. *Courtesy Purdue University, Lafayette, Ind.*

the North. This cockroach is not strictly gregarious. Adults, nymphs, and well-attached egg capsules are found in such secluded places as stationery drawers, inside furniture, beneath loose wall paper and pictures, behind moldings and kitchen utensils, in closets, and not infrequently in electric clocks. Their food is varied, and they have the annoying habit of removing glue from stamps and envelopes.

Description. This dimorphic cockroach is about the size of the small German cockroach but bears a light tan band across the base of the wings and a broken band across the center of the wings. The male has functional wings covering the tip of its slender abdomen; the female is broad with stubby wings. The name *brownbanded* is especially appropriate for the nymphs, on which the banding on the abdomen is prominent in the absence of wings (Fig. 19:9). Egg capsules are small, with sides nearly straight and parallel, and tan in color (Fig. 19:10).

Life history. Capsules of this species may have as many as 18 eggs each, and hatching occurs in a little over two months. (Note below the relationship of this to control practices.) The development period for the nymphs allows just two generations a year.

Cockroach Control

Even though these four insects are all cockroaches, specific knowledge of the species and the habits and habitats of each is essential for effective control. This information has always been important because of the distinct differences among our common species (Fig. 19:11), but it is even more

Figure 19:9. Nymphs of the brownbanded cockroach, natural size. *Courtesy Purdue University, Lafayette, Ind.*

important now because of the insecticide resistance problem present with the German cockroach.

Except for the German cockroach, the several species may be killed readily with chlordane provided the material is applied properly. It is used at 2 per cent in either an oil solution or an emulsion.

The chlorinated hydrocarbon–resistant German cockroach, as well as other species of cockroaches, can be controlled with such materials as 0.5 per cent diazinon, 0.5 per cent chlorpyrifos, or 1 per cent propoxur. Provided application is thorough, success can be achieved using the silica gel desiccant dusts, which have a dehydrating action on the cockroaches.

Figure 19:10. Egg capsules of cockroaches. A, brownbanded; B, woods; C, American; D, German; E, oriental. *Courtesy Purdue University, Lafayette, Ind.*

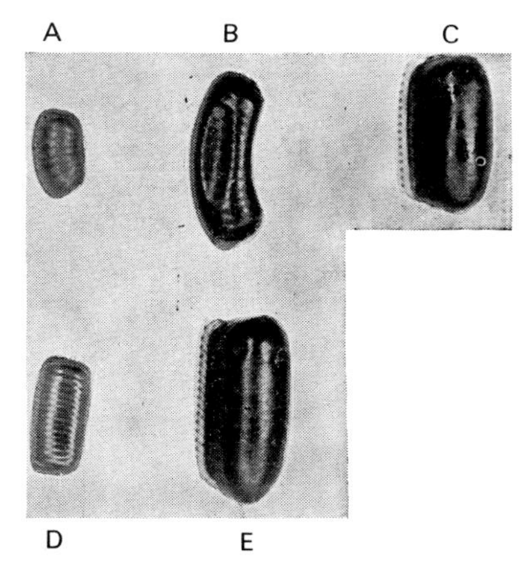

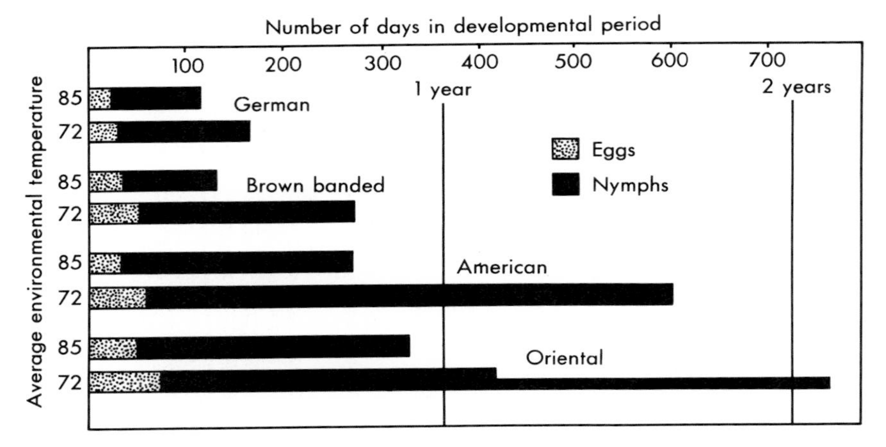

Figure 19:11. The duration of developmental periods in the life cycle of four species of cockroaches. Note effect of temperature and the great variation between species. *Based on Gould and Deay.*

Liquid applications should be made with a reasonably coarse, droplet spray applied in such a way that a uniform residual will remain. If reinfestations are a problem, no longer than two weeks' protection should be expected without reapplication. Dusts can be used only where moisture is not a factor and where they will not be unsightly. Whatever the material, effort must be made to apply it where the particular cockroaches concerned will be most apt to contact it for a rather prolonged period of time.

Certain special uses of insecticides should be noted. In food processing and serving establishments, the use of pesticides is restricted to certain locations to prevent contamination of food. Using special equipment, application is restricted to cracks and crevices where cockroaches are most likely to hide. Dichlorvos, because of its vapor toxicity, continues to be the best control of well-scattered, brownbanded cockroaches, but reapplication is often necessary in two months to kill nymphs hatched from capsules. Pyrethrum, applied as an aerosol, is very useful in locating hiding places of all roaches because it stimulates almost immediate activity which will guide one to the localities where longer-lasting insecticides should be used.

Black Carpet Beetle. *Attagenus megatoma* (Fabricius)

Carpet Beetle. *Anthrenus scrophulariae* (Linnaeus)

Varied Carpet Beetle. *Anthrenus verbasci* (Linnaeus)
[Coleoptera:Dermestidae]

The larvae of carpet beetles attack keratin-containing materials such as wool, fur, hair, feathers, and occasionally skins. The black carpet beetle is the most persistent and devastating species and thus is considered here in detail. In addition to the above materials, the larvae can feed on cereals, grain, dead insects, dust, and even animal droppings. They are very active and may feed on a variety of materials during their development. They prefer darkness and may be found almost anywhere in a home when not feeding. Damage by the black carpet beetle can be considerable, and frequently woolens are left with many irregular holes.

Description. The adult black carpet beetles are coal-black, shiny and hard-shelled, oval in shape, and only $\frac{1}{8}$ to $\frac{3}{16}$ in. in length (Fig. 19:12). Larvae

Figure 19:12. Three species of carpet beetles, adults and larvae. A, black carpet beetle, *Attagenus megatoma* (F.); B, carpet beetle, *Anthrenus scrophulariae* (L.); C, varied carpet beetle, *Anthrenus verbasci* (L.). *A and B, courtesy Conn. Agr. Exp. Sta.; C, courtesy University of California, Berkeley.*

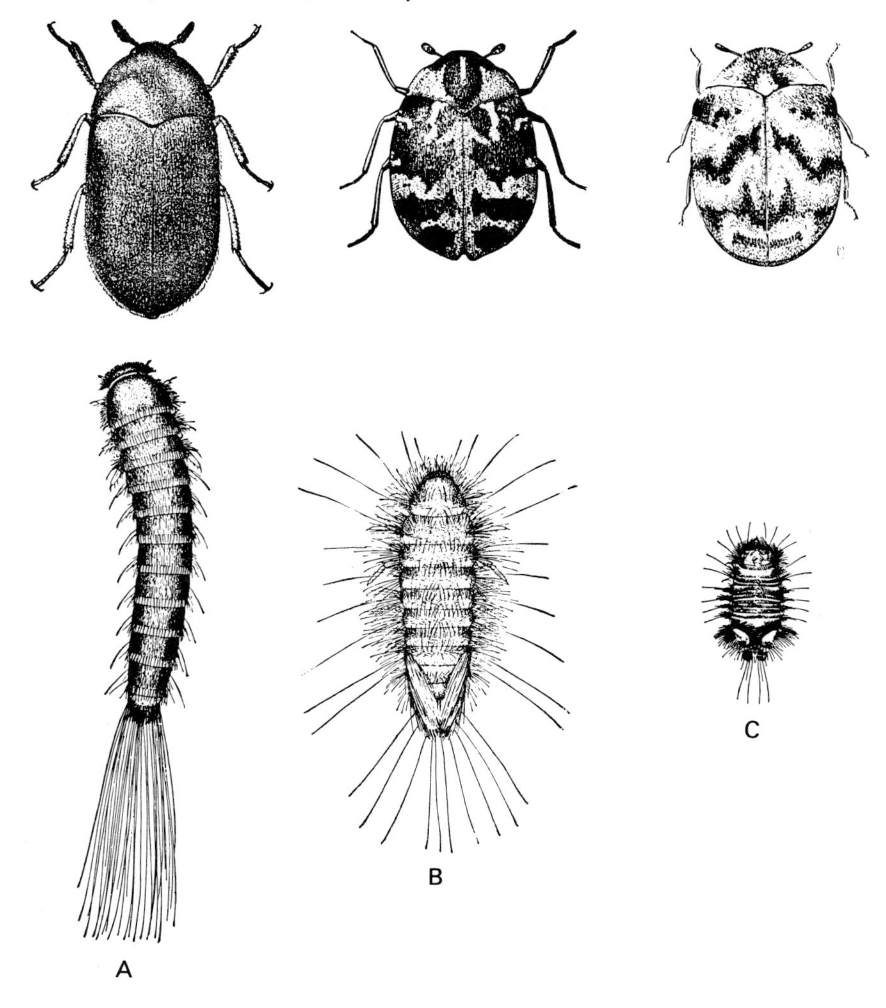

are quite tiny when they hatch but have the same distinctive elongated carrot shape and the long brush of tail bristles of the larger larvae. Body color ranges from yellow-brown to dark brown.

Life History. Adults are frequently found outdoors visiting flowers. Eggs are laid either indoors or out. A female black carpet beetle lays approximately 50 eggs over a period of several weeks, and then dies. Eggs may be deposited in accumulations of lint in air ducts, beneath baseboards, and other similar places; they hatch in a week. Development can occur under a wide range of temperature and humidity conditions. The larval stage varies in length depending on environment and nutrition, so that development may take only 9 months or as long as 3 years. The larvae, which can survive near starvation conditions, molt a variable number of times from as few as 5 to as many as 10 or more. Full grown larvae are wanderers, and thus pupation may occur almost anywhere in a house. The pupal stage lasts one to two weeks.

Control. Preventive measures must be the key to fabric protection, because once the "hole" is made, the damage is done. There are three steps involved in control: (1) sanitation and preventive chemical measures, (2) control of existing infestations, and (3) fabric protection.

Much can be done to prevent trouble by means of household cleanliness: thorough and frequent cleaning with a vacuum cleaner; brushing, airing, and dry cleaning susceptible clothing; avoiding situations predisposed to infestation, such as wall-to-wall wool carpeting and prolonged storage of discarded garments, bedding, and old rugs. Remember that a clean environment is not conducive to carpet beetle activity. Several approaches can be taken to exploit insecticides as preventive treatments. The most common one is the use of paradichlorobenzene (PDB) as a repellent and continuous fumigant in storage. Woolens to be stored should be interspaced with crystals of this material placed on clean paper as the fabrics are packed very tightly into trunks or boxes. Application is at the rate of 1 lb per 30 cu ft of space. Cedar closets and most cedar chests are seldom effective. When wool carpets are to be placed in a room the following preventive measures are recommended: thoroughly spray baseboards with a registered persistent insecticide, and apply a long-lasting insecticidal dust to the underside of the carpets. These measures will afford protection for years. Fur storage in cold vaults is a worthwhile preventive measure.

Once an infestation is established in a household, control becomes a difficult problem. Insecticides cannot be used randomly without regard for the nature of the article to be treated. Registered chemicals can be sprayed on some woolens if care is taken. For control of pests in drawers, closets, wall voids, and air vents, 0.25 per cent lindane or 1 per cent propoxur can be useful. Professional assistance is usually desirable if extensive control measures are necessary.

So-called "mothproofing" or fabric protection is done more readily during the manufacture of fabrics than after they become household goods. Certain organic chemicals related to dyes can be introduced as an integral and essentially permanent part of the basic threads of the fabrics. Dry cleaning establishments also offer fabric proofing; some firms use a closed system employing dieldrin.

Black Carpenter Ant. *Camponotus pennsylvanicus* (De Geer) [Hymenoptera:Formicidae]

Probably because of its large size, usually black color, and roving habits, the black carpenter ant is one of the best known of the ants invading households. This particular species is a wood-nesting form, its colonies occurring in logs, stumps, trees, posts, and the wooden structures of buildings, including beams, sills, rafters, and hollow wooden doors. When wooden members of a house are hollowed out for nesting purposes, serious damage is likely to result. Evidence of nesting is the presence of small or large piles of coarse sawdust thrown out during excavation. Whether the carpenter ant nests inside or outside, it readily enters homes in search of food, and thus is a source of annoyance and sometimes damage to fabrics and paper products that are heavily sized or soiled with food. The presence of these ants in buildings can be involuntary also because they are often introduced in firewood stored in basements or by the hearth. The workers of the carpenter ant are carnivorous and predaceous; they also attend honeydew-excreting insects such as aphids and scales, which provide a source of converted carbohydrates. They are known to feed on moist portions of meat, bread, fruits, and other foods found in a household. Like other ants, this species will bite vigorously when molested. Members of the genus *Camponotus* are well distributed throughout the United States.

Description. The stout workers of the black carpenter ant vary in size from $\frac{1}{4}$ to $\frac{1}{2}$ in. long; this polymorphic condition is related to the age and development of the colony, not to the individual. The winged females may reach a length of nearly 1 in. Like other ants, they are readily distinguished from termites by the presence of a distinct head, thorax, slender waist, large abdomen, and clear wings of unequal length. The prominent gaster has a silky pubescence, and the 12-segmented antennae bear no enlarged terminal segments (Fig. 19:13).

Life history. Colonies of the carpenter ant vary in size from a queen and a few workers to organizations containing several friendly queens and as many as 5000 workers. There are, as with other ants, normally three distinct castes: males, females, and workers (female, but seldom sexually functional). After mating, the winged male dies, leaving the female to develop a colony. She is long lived and propagates the colony through many years of growth. Ants have four developmental stages: egg, larva, pupa, and adult (Fig. 19:14). The egg is nearly microscopic in size and hatches in about the same time in warm weather, but this stage is greatly prolonged in winter. The pupae, enclosed in cocoons which most people mistake for "ant eggs," require three weeks to develop. The winged, sexual forms are produced only by mature colonies and emerge commonly in July. Two generations of workers are possible in a year.

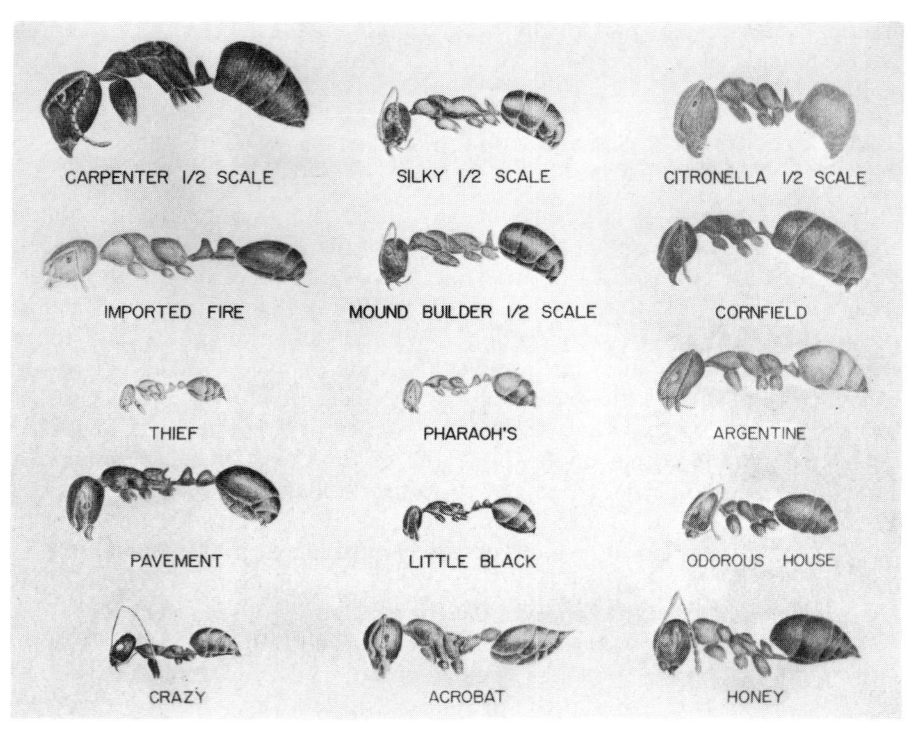

Figure 19:13. Side views of a selected group of common ants (enlarged). Antennal clubs and number and shape of nodes located on abdominal petiole are characters used in identification. *Courtesy Wallerstein, Purdue University, Lafayette, Ind.*

Figure 19:14. Life stages of an ant. A, eggs; B, larva; C, pupa; and D, adult. *Courtesy Purdue University, Lafayette, Ind.*

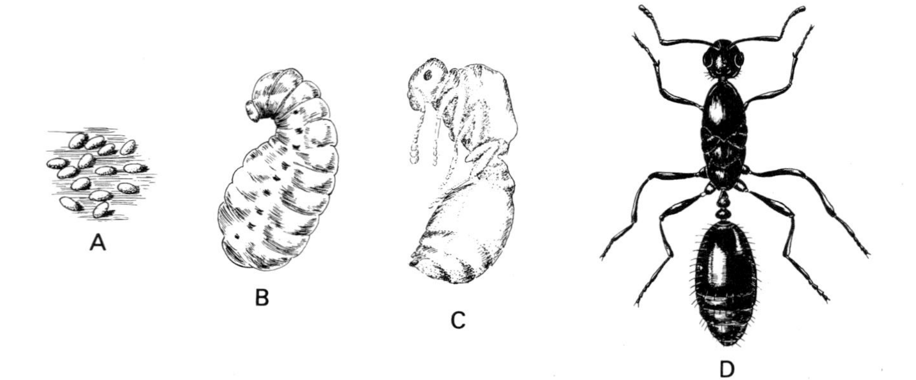

Control. Keen observation is essential as the first step in controlling carpenter ants. Scattered individual ants are seen frequently, and thus other clues must be found. For example, one should look for sources of moisture where wood is present—in siding, under eaves, around windows, in porch areas; look for piles of sawdust, small openings in the floor, and damaged timbers. Check nesting sites such as firewood, door sills, hollow doors, porch posts, and hollow trees.

The next step is to eliminate as many problem areas as possible by taking away moisture sources, food, and in some cases the nesting areas themselves. When the nest is located, introduce a suitable insecticide. It may be necessary to drill into the nest area in several locations. Although liquid formulations are frequently used, sometimes a dust is more effective and will leave no stains. Sometimes the nest simply cannot be located. In such instances, a more general treatment of the premises is necessary. This approach includes spraying outside foundation surfaces, the base of trees, foraging areas, baseboards, door and window frames, and even subflooring. Usually effective insecticides include 0.5 per cent lindane, 1 per cent propoxur, 1 per cent diazinon, and 0.5 per cent chlorpyrifos. Sprays may often be supplemented advantageously with poisoned baits, but the latter alone are seldom satisfactory.

Carpenter Bee. *Xylocopa virginica* (Linnaeus) [Hymenoptera:Anthophoridae]

A bumble bee eating the wood in a house? At first glance, someone seeing a large black bee entering a hole in the side of a building might think so. Because bumble bees do not bore in wood, the conclusion has to be that the insect is a carpenter bee, which merely resembles a bumble bee. Normally the carpenter bee bores a hole $\frac{1}{2}$ in. in diameter into a piece of wood. The tunnel continues inward briefly and then turns at right angles and continues with the grain of the wood. Typically, a tube extends 10 to 18 in. in length. These nesting tunnels (Fig. 19:15) are frequently fashioned in windowsills, supporting wood in garages, car port ceilings, roof overhangs, beams and the like. Carpenter bees will sometimes weaken timber, but principally they are considered undesirable because of their constant activity and the excrement they leave beneath the tunnel opening. The female bees can sting but seldom do.

Description. Although resembling a bumble bee, the carpenter bee, which is also robust and 1 in. long, differs from it by having a broad head with brown eyes, bare, metallic blue-black abdomen, and a thorax covered with fine yellowish hair (Fig. 19:15).

Life History. Both males and females hibernate in northern climates. The nesting tube is divided into several cells, each provisioned with a pollen mass and partitioned from the adjacent one with chips of wood cemented together. Each cell in the nest is occupied by a single larva. There is one generation per year.

Control. If one feels obliged to eliminate the bees, the steps are to kill the insects, prevent reuse of the tunnel, and prevent wood decay. A successful

A

Figure 19:15. Carpenter bee. A, female resting on plant in field where it collects pollen and nectar to provision the nest for its young; B, female entering its nest through opening made in siding board; C, longitudinal section through board showing nesting tunnel with empty brood cells. *A, courtesy Edward M. Barrows; B and C, courtesy Purdue University, Lafayette, Indiana.*

method consists of using a liquid formulation of 1 per cent propoxur together with 5 per cent pentachlorophenol. The former provides kill, and the latter prevents wood deterioration. Caulking of the holes following treatment is recommended.

Eastern Subterranean Termite. *Reticulitermes flavipes* (Kollar)

Western Subterranean Termite. *Reticulitermes hesperus* Banks

Arid Land Subterranean Termite. *Reticulitermes tibialis* Banks [Isoptera:Rhinotermitidae]

The subterranean termites of the genus *Reticulitermes*, which inhabit every state except Alaska, are the most destructive termites in this country (Fig. 19:16). The eastern subterranean termite is discussed as representative. Because they are cryptobiotic in nature and confine their destructive efforts to the inside of wood, they are seldom seen in the worker form. Swarms of termites

B

C

are a familiar sight, however, and serve as indicators of the fact that an active colony exists nearby. Swarming occurs commonly in the early months of the year. It is characterized by the presence of black insects, $\frac{1}{5}$ in. long, with four opaque, white wings of equal length. These swarmers are the primary reproductive forms. They are positively phototactic and are seen around windows or outside at the base of walls. Swarming in itself does not mean that a building is necessarily infested; it simply indicates that a termite colony is somewhere in the vicinity. Unlike the swarmers in appearance, termites in a colony are small, white, wingless insects (Fig. 19:17). Only the workers are destructive.

Damage occurs to almost anything containing cellulose, including rolls and piles of paper, the binding of rugs, cotton cloth, books, adobe wall material containing straw binding, dead roots in the ground, and, of course, timbers found in buildings. Termites will attack most woods, oak being as susceptible as

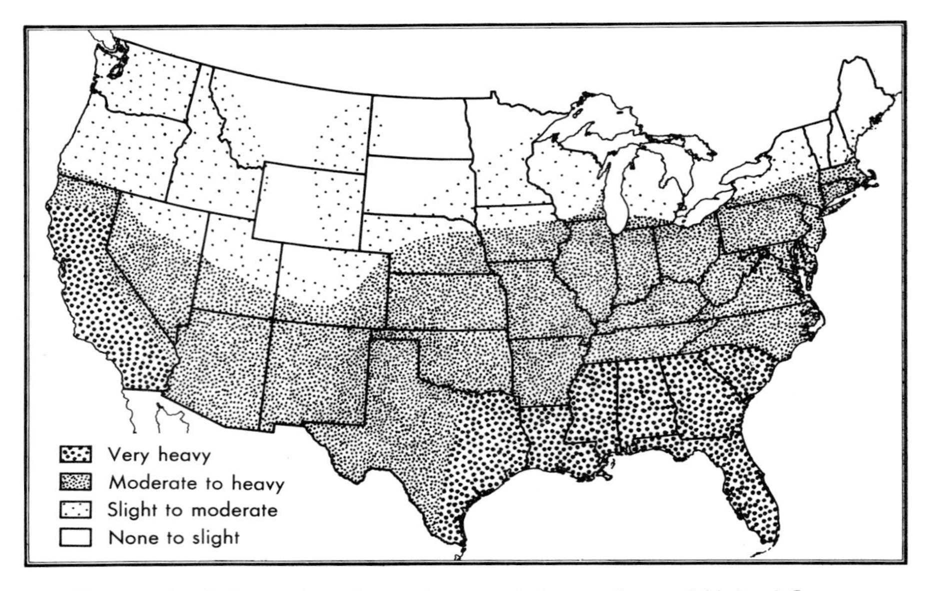

Figure 19:16. Intensity of termite attack in continental United States. Based on a Building Research Advisory Report.

pine; teak, foundation-grade redwood, and juniper are least apt to be attacked. In addition considerable incidental damage is caused to noncellulose products through which termites may chew in search of food and shelter. One of the best examples is the penetration of underground, rubber-insulated cables, an intrusion that results in serious damage. Termites are capable of handling cellulose as part of their diet because of the presence of protozoa and other microorganisms in their gut which reduce cellulose to digestible carbohydrates. In addition, a balance of other nutrients is necessary; these are supplied in part by the microflora present in the wood and soil.

Moisture is an essential part of the environment of subterranean termites. These insects avoid light, a behavior associated with avoidance of dehydrating air rather than light itself. Humid conditions are maintained by occupying moist soil and traveling to food sources through protected passages such as the earthen tubes which termites construct so skillfully (Fig. 19:18).

In northern climates, a great deal of colony movement occurs as the temperature changes. Winters are spent below the frost line unless the termites have access to a heated building. In warm weather the insects move nearer the ground surface, and in some instances they can be located entirely within the wood of a building.

Description. Termites are characterized by colonial existence and the caste system. In the subterranean termites of the genus *Reticulitermes* there are three forms: workers, soldiers, and reproductives. Drywood termites, which inhabit wood without connection to the ground or availability of particular moisture

Figure 19:17. Active colony of subterranean termites. A few slightly larger reproductive forms can be seen among the workers. *Courtesy Purdue University, Lafayette, Ind.*

supplies, differ in caste development from the subterranean form in that no worker caste exists. Labor is performed by the nymphs of the reproductive caste. A normal colony of subterranean termites contains thousands of workers, which perform the functions of excavating wood and soil, building tubes, feeding the young and more helpless soldiers and reproductives, and regulating the environment. Members of the worker caste are small and creamy white (Fig. 19:19A). The soldiers, few in number and about $\frac{1}{4}$ in. long, are morphologically modified with large, dark, muscle-encased heads bearing saber-like mandibles (Fig. 19:19B). Their function is to protect the colony from attack.

The reproductives usually consist of more than one form. The primary (macropterous) form has been described above as the one that swarms (Fig. 19:19C). The sole function of the primary reproductive is to establish new colonies, a feat seldom accomplished because of the hazards of life and general vulnerability of this form. At the end of swarming, the wings are shed, and the

Figure 19:18. Earthen tubes of subterranean termites in semiexcavated area beneath house. *Courtesy R. R. Heaton.*

dealated insects seek mates and a suitable wood niche for initiation of the colony. Natural enemies and unsuitable environment usually preclude success. The secondary (brachypterous) forms are the backbone of the reproductive strength of a colony. Often several hundred males and females are present, the females producing a few eggs per day. These forms can develop new and thriving subcolonies in the event of the loss of the primary pair or when colonies are fragmented. Tertiary (apterous) forms are sometimes present; their importance as reproductives seems to be minor. Colony size of the subterranean form may be as few as a dozen or as many as 100,000.

Life history. The termites are close relatives of the cockroaches, and like the cockroach, development from egg to adult is by simple metamorphosis. The eggs are tiny, white, and kidney-shaped. The nymphs, resembling miniature adults, look alike during the first two instars. Following this early development, they can be separated into the large-headed, early stages of the sterile forms (leading to workers and soldiers), and the small-headed reproductives. Workers and soldiers take one year for development; primary reproductives

Figure 19:19. Eastern subterranean termite. A, worker; B, soldier; C, adult, winged, primary reproductive termite, characteristic of the form which swarms; identifying characters are black body, no abdominal constriction, and wings of equal length. *Courtesy Purdue University, Lafayette, Ind.*

require two years. All stages of development can usually be found in a colony whether it is located in wood or in the ground. Once established, colonies exhibit strength because of individual longevity, cryptobiotic habits, continuous reproductive capacity, ceaseless industry, highly sensitive responses, and unusual adaptability.

Control. Because of the unusual ability of subterranean termites to enter homes and other structures and because control has to be complete to be effective, termite control has evolved into an art with definite standards. Basically, it involves separating or insulating the colony in the ground from its sources of food in a structure. There are four basic steps to control, which are modified only insofar as particular infestations and circumstance may require. The steps are removal of wood debris and mechanical alteration, soil treating, foundation treating, and wood treating.

Before one undertakes control, several questions should be considered. Must the building be treated? It should be treated if active infestations are present or if an owner wants the premise protected against future attack. Must it be treated immediately? Immediate treatment is seldom necessary; there is always time to investigate the situation thoroughly as to need and the particular pest control operator desired. Must the whole foundation area be treated? It is generally desirable to have a complete treatment if a chance exists for termite reentry in various new places. If partial treatments are to be satisfactory, they usually should include all four steps of control. Can termite control be undertaken by the average homeowner? Treatment of buildings by untrained individuals is seldom successful because of the complexity of most situations. Professional help is usually advisable.

The first step in control corresponds to cultural control in agricultural entomology. Because of the close relationship between the incidence of termite infestation and an optimum moisture content of the soil, drainage and proper water disposal are important. Closely related to these is the improvement of ventilation beneath structures. Improvement of such conditions through installation of foundation ventilators will decrease humidity and dry both the wood and the soil. Vapor barriers on the soil surface beneath buildings are also helpful. All miscellaneous wood should be removed from areas near the building. Mechanical alterations include measures that may be as extensive as raising the house on its foundation or simply breaking soil contact of wooden posts erected through a concrete basement floor.

The success of modern subterranean termite control lies in the proper application of *persistent* chemicals in the soil. The procedure is safe and free of environmental contamination. Protection is afforded for periods of five to twenty years.

Suitable chemicals for termite control include emulsions of chlordane at 1 per cent, dieldrin at 0.5 per cent, aldrin at 0.5 per cent, and heptachlor at 0.5 per cent. Pentachlorophenol at 5 per cent is excellent for wood treatment because it deters decay; it is seldom used in the soil because it is less effective than the cyclodienes.

Soil treatments for eradication or as a protective barrier must be applied thoroughly. Around foundations, the normal procedure is to trench the soil and flood or rod-inject the formulation into the soil. The rate for shallow foundations is 2 gal/5 lineal ft; for deep foundations, 4 gal/5 lineal ft. Beneath slab porches and slab floors which should be excavated or drilled, the rate is a minimum of 1 gal/10 sq ft. Special equipment is needed for subslab treatments and for drilling and treating block wall voids.

The general practice is to drill infested wood and treat with a protective chemical. Pentachlorophenol is especially useful for this purpose. It is often good practice to drill and pressure treat unexposed ends of joints and wood supports.

There are no short cuts to subterranean termite control. Each step is a part of the complete procedure and must be done thoroughly.

Pretreatment of buildings. The best time to assure protection of a building from termite attack is at the time of construction. A relatively unsuccessful method has been the use of metal shields or barriers installed on the top of the foundations. Shields are only as good as the quality of installation; unfortunately this is seldom satisfactory and therefore they provide only a false sense of security. Another deterrent is the use of pressure-treated wood (impregnated with a preserving chemical such as pentachlorophenol or copper naphthenate) in critical areas such as the plates and joists. This procedure is helpful and should be encouraged, but it cannot be expected to be a protection against attack elsewhere in the building. Soil treatment with a persistent, protective insecticide expected to last for years is the method of choice. The method consists of treating the soil just before pouring concrete slabs and of thoroughly treating the soil completely around the outside of foundations. The insecticides and the rate of application are those given under the corrective procedures discussed above. Commercial firms employing these methods provide a guarantee for a number of years.

Powderpost Beetles

Lyctus planicollis LeConte [Coleoptera:Lyctidae]

Dinoderus minutus (Fabricius) [Coleoptera:Bostrichidae]

Anobium punctatum (De Geer) [Coleoptera:Anobiidae]

Small beetles that feed in seasoned wood in their larval stage and in doing so produce fine, powderlike particles of frass are known commonly as powderpost beetles. This term applies to three closely related families. There are differences among them in behavior and nutritional requirements; the lyctid beetles are, in the strictest use of the term, the true powderpost beetles (Fig. 19:21). As a group of insects, the powderpost beetles are second only to termites in their destruction of wood. Articles such as ax handles, furniture, joists, rafters, and paneling are examples of wood that may be literally riddled by these insects.

Numerous small holes, resembling those made by birdshot, perforate the surface of infested wood (Fig. 19:20). These holes are made by the emerging beetles, which are seldom more than $\frac{1}{8}$ in. in length. Infestations may be present in wood from three months to more than three years, depending upon the species involved, environmental conditions, and type of wood attacked.

Control. Small infested articles can be heated in an oven to destroy the larvae, but other than that, insecticides are currently the only practical means of control. The objective in using an insecticide is to penetrate the powderpost beetle galleries to kill the larvae and when necessary to leave a lasting deposit on the surface which will deter adults and prevent reinfestations. The number of insecticides available for this purpose is limited because of the curtailment of the use of persistent chemicals. Materials of choice include 0.5 per cent lindane and 5 per cent pentachlorophenol. The first is an excellent insecticide; the latter is a wood preservative with some insecticidal properties, but it is rather odorous. Most favorable results with both materials are obtained with oil-base formulations, which provide maximum penetration when sprayed on the surface or applied with a paint brush. Another approach to the use of chemicals in powderpost beetle control is fumigation, which is sometimes employed for individual pieces of furniture or entire buildings when the infestation is severe enough to warrant it.

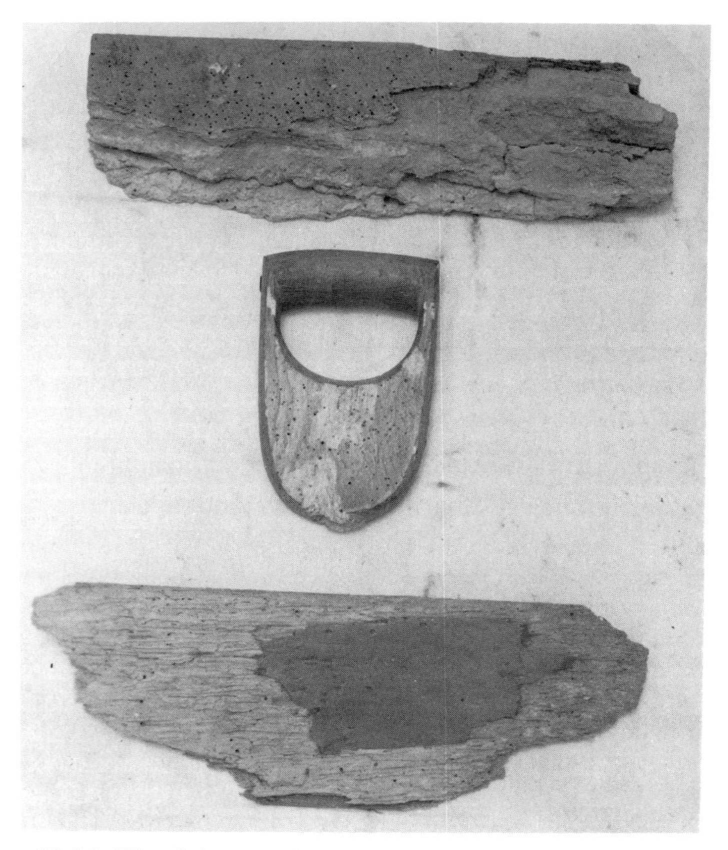

Figure 19:20. Wood damaged by powderpost beetles. A, exit holes of adults of Anobiidae; B, of Lyctidae; C, of Bostrichidae. *Courtesy Purdue University, Lafayette, Ind.*

Figure 19:21. Adult of the southern lyctus beetle, *Lyctus planicollis* LeConte. *Courtesy USDA.*

SELECTED REFERENCES

Ebeling, Walter, *Urban Entomology* (Berkeley: University of California, Division of Agriculture Sciences, 1975).

Gould, G. E., and H. O. Deay, *The Biology of Six Species of Cockroaches Which Inhabit Buildings*, Purdue University, AES Bul. 451, 1940.

Krishna, K., and F. M. Weesner, *Biology of Termites* (New York: Academic Press, 1969).

Mallis, A., *Handbook of Pest Control* (New York: MacNair-Dorland, 1969).

Osmun, J. V., "Recognition of Insect Damage," *Pest Control*, 23:1, 4, 7, 10 (1955).

Osmun, J. V., and W. L. Butts, "Pest Control," *Ann. Rev. Entomol.*, 11:515–48 (1966).

Smith, V. K., Jr., and H. R. Johnston, *Eastern Subterranean Termite*, USDA Forest Pest Leaflet 68, 1970.

Snyder, T. E., *Our Enemy the Termite* (Ithaca, N.Y.: Comstock, 1948).

Truman, L. C., G. W. Bennett, and W. L. Butts, *Scientific Guide to Pest Control Operations* (Cleveland: Harvest Publishing Co., 1976).

USDA, *Controlling Household Pests*, USDA Home and Garden Bul. 96, 1976.

Wheeler, W. M., *Ants, Their Structure, Development, and Behavior* (New York: Columbia Univ. Press, 1926).

chapter 20 / **ROBERT E. PFADT**

LIVESTOCK INSECTS AND RELATED PESTS

Livestock—cattle, swine, sheep, and horses—are harassed, weakened, and sometimes killed by the assault of many kinds of arthropod parasites. The attack is maintained the year round; some pests such as lice and mites are worse during winter, whereas other pests such as flies and mosquitoes are injurious only during the summer.

THE PESTS

Just as was the case among the crop insects we have studied, livestock insects include host-specific members, pests that attack only a single class of host, and general or nonspecific members, pests that attack not only all kinds of domestic stock but wild animals as well. The true flies, members of the order Diptera, contribute the greatest number of species to the group of general pests. Included are the cosmopolitan **house fly** and **stable fly** (Fig. 20:1A) and the introduced **horn fly** (Fig. 20:11). Native species that attack livestock are many kinds of **mosquitoes**, **black flies** (Fig. 22:2), **horse flies** (Fig. 20:1B), several **blow flies**, and the **screwworm** (Fig. 20:12). A recently introduced cattle pest, the **face fly**, *Musca autumnalis* DeG. (Fig. 20:5), has rapidly extended its range in North America. It was first observed in 1952 in Nova Scotia.

Throughout the world various ticks parasitize livestock. In North America at least nine species are important pests: the **cattle tick, southern cattle tick, lone star tick, Gulf Coast tick, blacklegged tick, winter tick** (Fig. 20:16), **Rocky Mountain wood tick, tropical horse tick**, and **ear tick** (Fig. 20:2). None of these has a general distribution over the whole of North

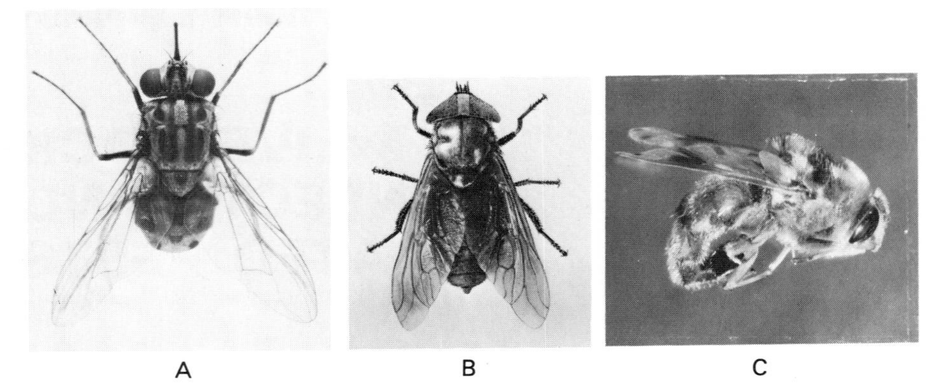

Figure 20:1. A, adult female of the stable fly, *Stomoxys calcitrans* (L.), a cosmopolitan species of blood-sucking fly; B, black horse fly, *Tabanus atratus* F.; C, horse bot fly, *Gasterophilus intestinalis* (De G.), the larval stage is an internal parasite of horses, mules, and donkeys. *Courtesy USDA.*

America, but all inhabit large sections of the continent. The cattle tick, which once infested 15 southern states, has been eradicated from the area by a government-sponsored program of cattle dipping and quarantine. The species is still abundant in many areas of Mexico, and occasionally it invades and breaks out in small numbers at border points in the United States.

Among the more serious and widespread pests are those that attack specific classes of livestock. These parasites, of foreign origin, came to America along with their hosts; because of their importance, we have listed them in Table 20:1.

Figure 20:2. The ear tick, *Otobius megnini* (Dugès). A, female; B, male. Indigenous to the Southwest but now distributed more widely. *Courtesy USDA.*

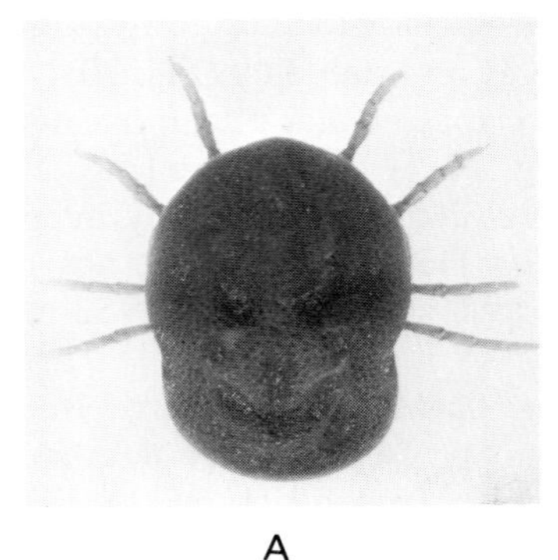

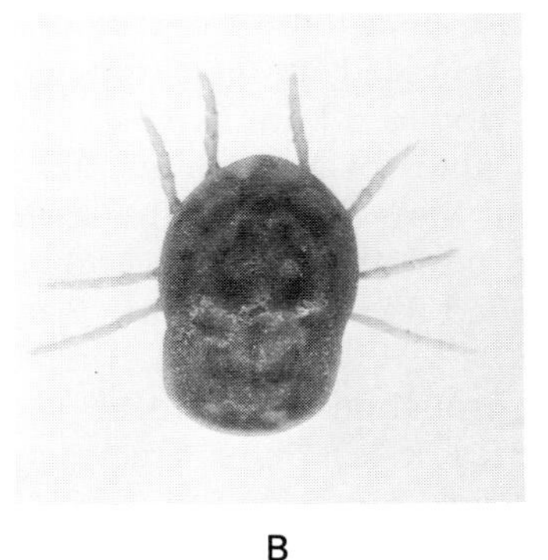

A B

Table 20:1. Insect and Mite Parasites of Livestock in America

Cattle	
Lice	Cattle biting louse, *Bovicola bovis* (Linnaeus) (Fig. 20:8D) Little blue cattle louse, *Solenopotes capillatus* Enderlein (Fig. 20:8C) Longnosed cattle louse, *Linognathus vituli* (Linnaeus) (Fig. 20:8B) Shortnosed cattle louse, *Haematopinus eurysternus* (Nitzsch) (Fig. 20:8A) Cattle tail louse, *Haematopinus quadripertusus* Fahrenholz
Grubs	Common cattle grub, *Hypoderma lineatum* (de Villers) (Fig. 20:10) Northern cattle grub, *Hypoderma bovis* (Linnaeus)
Mites	Chorioptic mange mite, *Chorioptes bovis* (Hering) Scab mite, *Psoroptes equi* (Raspail) (Fig. 4:6B) Itch mite, *Sarcoptes scabiei* (De Geer) Cattle ear mite, *Raillietia auris,* (Leidy) Cattle follice mite, *Demodex bovis* Stiles Cattle itch mite, *Psorergates bos* Johnston

Swine	
Lice	Hog louse, *Haematopinus suis* (Linnaeus) (Fig. 20:17)
Mites	Itch mite, *Sarcoptes scabiei* (De Geer) (Fig. 20:19) Hog follicle mite, *Demodex phylloides* Csokor

Sheep	
Lice	Sheep biting louse, *Bovicola ovis* (Schrank) African sheep louse, *Linognathus africanus* Kellog and Paine Sheep foot louse, *Linognathus pedalis* (Osborn) Goat sucking louse, *Linognathus stenopsis* (Burmeister)
Ked	Sheep ked, *Melophagus ovinus* Linnaeus (Fig. 20:13)
Bot	Sheep bot fly, *Oestrus ovis* Linnaeus (Fig. 4:31D)
Mites	Sheep scab mite, *Psoroptes ovis* (Hering) (Fig. 4:6B) Chorioptic mange mite, *Chorioptes bovis* (Hering) Itch mite, *Sarcoptes scabiei* (De Geer) Sheep itch mite, *Psorergates ovis* Wormersley Sheep follicle mite, *Demodex ovis* Railliet

Horse	
Lice	Horse biting louse, *Bovicola equi* (Denny) Horse sucking louse, *Haematopinus asini* (Linnaeus)
Bots	Horse bot fly, *Gasterophilus intestinalis* (De Geer) (Fig. 20:1C) Nose bot fly, *Gasterophilus haemorrhoidalis* (Linnaeus) Throat bot fly, *Gasterophilus nasalis* (Linnaeus)
Mites	Scab mite, *Psoroptes equi* (Raspail) Chorioptic mange mite, *Chorioptes bovis* (Hering) Itch mite, *Sarcoptes scabiei* (De Geer) Horse follicle mite, *Demodex equi* Railliet

THE INJURY

Livestock are sensitive to the attacks of insect pests and react by general unthriftiness, stunted growth, and sometimes death. The constant **annoyance and irritation** of biting flies make livestock restless and induce them to bunch up during the day when ordinarily they would be grazing. In attempting to lay eggs, female heel flies (adults of cattle grubs) and horse bot flies cause stock to flee in terror and whole herds to stampede. Cattle grubs migrate in body tissues, make holes in the skin of the back, and spoil meat underlying the cysts in which they live. Pests such as the horn fly, lice, and ticks withdraw large amounts of blood, producing **anemia** in their hosts (Fig. 20:3). Screwworms infest wounds and feed on living tissues, often causing death (Fig. 20:4). Parasitic mites cause contagious skin diseases known as **mange or scab** in which the hair coat is lost and the skin becomes thickened, wrinkled, and covered with gray or yellow scabs (Fig. 20:18).

Transmission of livestock diseases by insects results in untold losses to livestock growers. Certain flies such as horse flies, stable flies, blow flies, house flies, and face flies (Fig. 20:5) transmit the bacilli of anthrax, infectious keratitis, mastitis, and "swollen joints" and the virus of hog cholera and of equine infectious anemia. Mosquitoes transmit the viruses of equine

Figure 20:3. A cow heavily infested with the shortnosed cattle louse. Note the languor and the dark, dirty hair coat on normally white areas of the Hereford. This animal died one night of anemia due to gross infestation by the lice. *Courtesy USDA.*

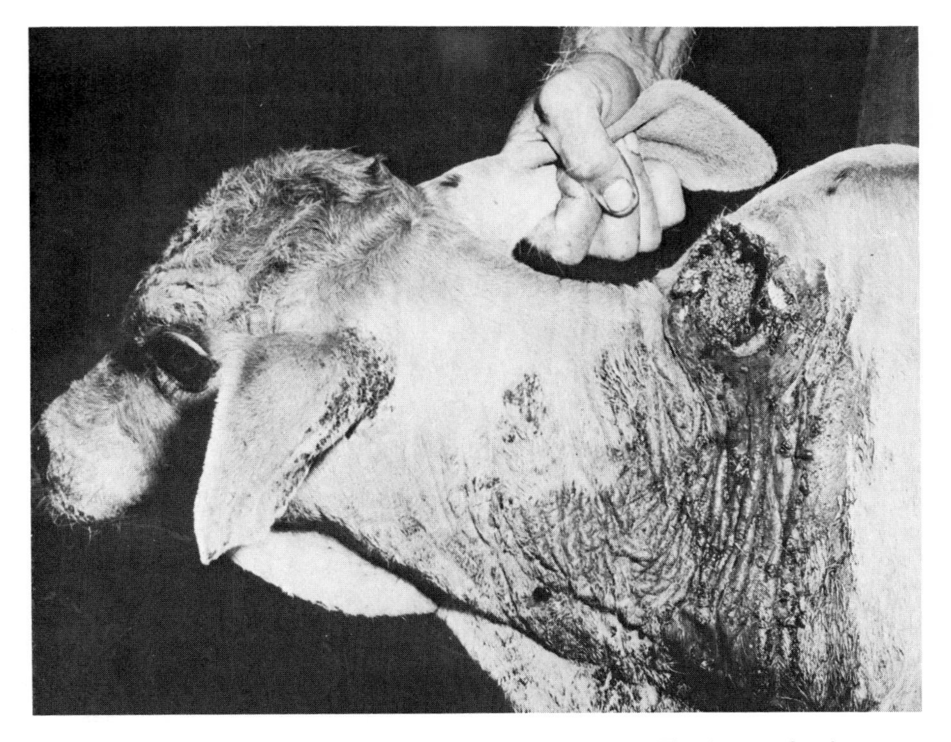

Figure 20:4. A wound infested with screwworms. The larvae feed on living flesh and will kill the host unless treated. *Courtesy USDA.*

encephalites, diseases that cause high mortality among horses. The tiny biting midge, *Culicoides variipennis* (Coquillett), widely distributed in North America, transmits the virus of bluetongue disease among sheep and cattle and the virus of epizootic hemorrhagic disease among cattle. Hog lice may carry swine pox, eperythrozoonosis, and possibly other infectious diseases of swine. In their feeding, the Rocky Mountain wood tick and several other ticks inject toxins into the blood stream of the host, causing a disease known as tick paralysis. Ticks transmit the various microorganisms that live in the red blood cells of animals, causing anaplasmosis, piroplasmosis, and other livestock diseases.

Cattle tick fever or bovine piroplasmosis is a serious and often fatal disease in which the red blood cells are destroyed by a protozoan known as *Babesia bigemina* (Smith & Kilbourne). In the southern states the organisms were found to be transmitted solely by the cattle tick, *Boophilus annulatus* (Say). This discovery by Theobald Smith and Fred L. Kilbourne in 1889 marked a milestone in medicine; it led not only to eradication of cattle tick fever from the United States by eliminating the tick through systematic dipping, but also firmly established the idea that certain other livestock

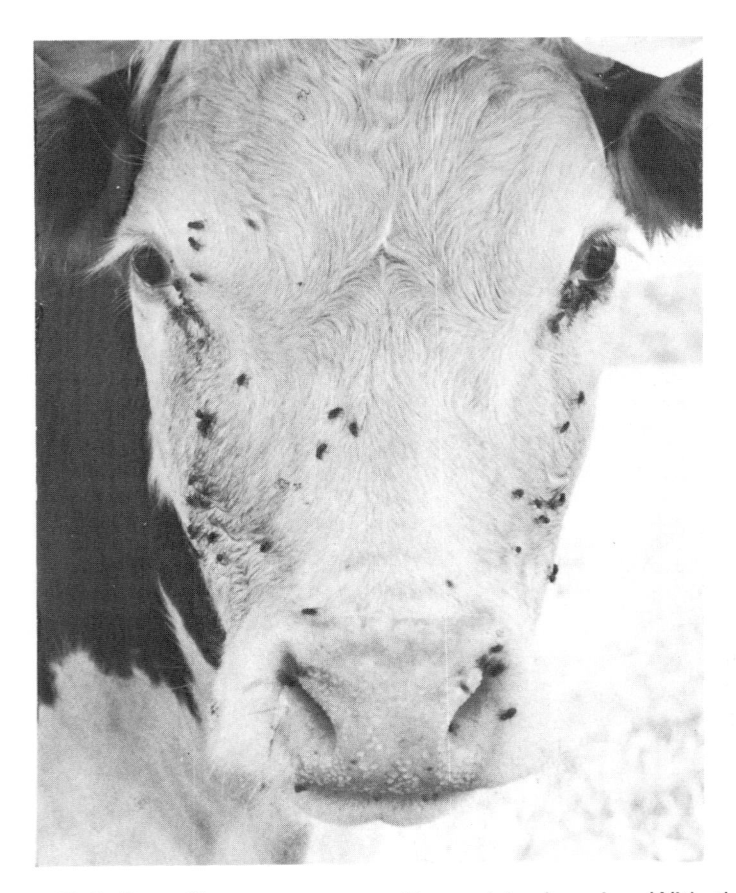

Figure 20:5. Face flies are a new cattle pest in America. With their sponging mouthparts they feed on secretions of eyes, nose, and mouth and on blood oozing from wounds. Evidence indicates that the fly transmits the causative bacterium, *Moraxella bovis* (Hauduroy), of pinkeye or infectious bovine keratitis. *Courtesy Ohio Agr. Res. Dev. Center, Wooster.*

diseases as well as certain human diseases such as malaria, yellow fever, and typhus are transmitted by insects.

Another important way in which insects cause injury to livestock is by serving as **intermediate hosts for parasitic worms**. The gullet worm, *Gongylonema pulchrum* Molin, of cattle, sheep, swine, and wild ruminants requires certain dung beetles and cockroaches for the intermediate stage. Livestock become infested by swallowing the beetles. Dung beetles also serve as intermediate hosts of three species of stomach worms of swine; white grubs serve as the intermediate host of the thorn-headed worm of swine. The broad tapeworms, *Moniezia expansa* (Rudolphi) and *M. benedeni* (Moniez), which infest sheep, cattle, and several other ruminants, require oribatid mites for intermediate hosts. The face fly serves as the intermediate host and vector of the eyeworm *Thelazia lacrymalis* (Gurlt), a nematode that infects the eyes of horses and donkeys.

CONTROL

Control Through Management

The initial defense against insects is always the elimination of breeding sources wherever possible and the avoidance of conditions that predispose animals to being attacked. Many of the flies that torment livestock develop in accumulations of manure, wet straw, spilled feeds, carrion, and in drainage water. Hence, a program of farm sanitation can stop much fly propagation. Proper disposal or treatment of manure is essential to prevent house flies from becoming so abundant that insecticidal applications cannot control them. Stockmen should take care in disposing of wet straw because it is a major source of stable fly development. In stacking straw, it is advisable to select a high, dry location.

Spilled feeds that become soaked with water serve as ideal breeding sites for both house flies and stable flies. Careful handling of feeds not only prevents the propagation of noxious flies but also eliminates wastage. If feeds are accidentally spilled, sweep them up and return them to bins. Dispose of spilled feeds that have become wet or scatter them thinly on the ground to dry them out. Because dead animals are not only potential sources of disease but also breeding sites for blow flies, haul away carcasses to reducing plants, or bury them deeply or burn them completely. Drain or fill low places where water collects and mosquitoes and horse flies breed.

Providing deep dark sheds, so that stock can retreat from the attack of flies during the height of fly activity, is of much value. Groves of trees in pastures can likewise serve as retreats.

To protect livestock from injury by screwworms and blow flies, a program of scheduling births before the fly season starts and of wound prevention and treatment during the fly season is advisable. Stockmen can prevent many accidental cuts and wounds by eliminating excessive amounts of barbed wire and by removing projecting nails and jagged boards from fences, corrals, and buildings.

Raising smooth, open-faced breeds of sheep aids in controlling insects, particularly wool maggots and screwworms. Because wrinkled sheep are difficult to shear, they receive more skin cuts that are attractive to blow flies. Blow flies are also attracted by the feces and urine caught in the wrinkles of the breech. In Australia, where the problem of wool maggots is very serious, the Mules operation—named after its discoverer, J. W. H. Mules—is usually performed on ewe lambs. The operation consists of removing loose skin from each buttock with hand shears. Upon healing of the wound the skin stretches from the vulva, drawing the wool away from the midline and from the path of urine and dung.

Keeping stock in good condition through adequate nutrition makes them less susceptible to some pests and appears to hold populations of mites, lice, and sheep keds at low levels. Of great importance in preventing the spread of parasites and disease is the temporary isolation and treatment of all new animals before introducing them into the herd.

The natural resistance of cattle to insects and ticks is of practical interest to both entomologists and animal scientists. In Australia, breeding has been carried out to amplify (by selection) the resistance of cattle to the southern cattle tick. Zebu cattle show the greatest amount of natural resistance, but it also occurs in the European breeds. By employing a special technique to determine resistance level, Australian workers have found that the Hereford breed has a resistance range of 54 to 90 per cent, the Shorthorn 50 to 96 per cent, and the Brahman Hereford 97 to 99.9 per cent. This trait is inherited, and research indicates that it may be desirable to take it into account in a breeding program along with the usual traits considered.

Recent studies in Louisiana have shown that Brahman cattle are less attractive to the horn fly than several European breeds and carry fewer flies. In studies of the effect of nutrition upon densities of cattle lice, investigators have found large within-treatment variations, which indicate that a wide range of susceptibility to lice exists among cattle. These and other observations of parasitism in livestock suggest that natural resistance is an area of biology needing more research and one possible of exploitation.

Chemical Control

In spite of good management practices, parasites often establish themselves on livestock and cause injury. We must then consider treatment with insecticides; these are utilized not only as cures for established infestations but also as preventive treatments in the control of regularly appearing pests. The use of chemicals on livestock is complicated by restrictions imposed on breeders because of residues which accumulate in fat and are secreted in milk. Yet many insecticides are available to stockmen for protecting their animals if precautions are taken. Greatest limitations occur on dairy animals because few insecticides have tolerances allowable in milk.

The insecticides recommended for livestock belong to several chemical groups. Chlorinated hydrocarbons such as lindane, methoxychlor, and toxaphene are used widely to control lice, various flies, keds, mites, and ticks.

In recent years a number of organophosphorus compounds have been developed for use on livestock. A distinct advantage of certain of these is their capacity to act not only externally but also systemically within the body of the animal, eliminating internal parasites such as cattle grubs and horse bots. The organophosphorus compounds used in controlling livestock pests include the systemic insecticides coumaphos, crufomate, famphur, fenthion, phosmet, ronnel, and trichlorfon and the nonsystemic insecticides crotoxyphos, diazinon, dichlorvos, dioxathion, malathion, and stirofos.

Among the botanical insecticides pyrethrum is still used in considerable amounts, particularly on dairy animals. Several other chemicals, including allethrin, organic thiocyanates (Lethane and Thanite), synergists (piperonyl butoxide and MGK 264), and repellents (butoxy polypropylene glycol) are also applied directly to livestock.

Control Equipment

Stockmen may apply insecticides to animals in the form of dips, sprays, washes, pour-ons, spot-ons, wipe-ons, ear bands, drenches, dusts, ointments, and feed and mineral additives; they may also use self-applicating devices. The most common methods are dipping and spraying. To dip cattle or sheep, growers use permanent concrete vats, which vary in length from 30 to 100 ft. The stock enter at one end of the vat, swim through an insecticidal bath, and leave at the opposite end. Dipping requires a fairly large crew of men to force the animals into the dip and to attend them as they swim through. Men with dipping forks along each side of the vat dunk the heads of animals and prevent accidents (Fig. 20:6). Also available for dipping are portable steel or wooden vats 10 to 20 ft long and plunge cage vats. Dipping is a thorough and rapid method for treating animals with insecticides.

Because of their mobility and adaptability, sprayers are employed widely to treat animals with the insecticides. Confining animals in crowd-pens or chutes, growers apply sprays with spray guns at pressures varying from lows of 50 to 75 psi to highs of 200 to 400 psi. A special spraying device for livestock is the so-called spray-dip or box-spray machine into which cattle are driven one at a time and held until completely soaked. The spray-dip machine is an effective though somewhat expensive and sometimes troublesome piece of equipment. Dairymen, who find it convenient to treat animals when they are stanchioned, frequently use electric mist sprayers to apply insecticides.

Stockmen apply dusts by hand, with hand dusters, or with power dusters. The power dusting of sheep to control keds has gained favor among owners of large range flocks. Dusting after shearing is effective, rapid, and less hazardous than dipping or spraying because there is no wetting of the animals and no rough handling (Fig. 20:14).

Washes and ointments are convenient for treating animals infected with screwworms or other blow fly maggots. Drenches that are used generally for control of parasitic worms may likewise control internal insect parasites such as horse bots and sheep nose bots.

A self-applicating device, the back rubber, has become quite popular for control of flies on cattle. Several designs are available, but all are based on the principle that cattle like to rub themselves against posts and other objects. Burlap or canvas is wrapped around barbed wire or chains strung between deeply set posts and saturated with insecticide. When cattle rub against the treated materials, small amounts of the insecticide adhere to the hair and skin.

Figure 20:6. Dipping cattle in an acaricidal bath to control mange. *Courtesy The Denver Post.*

A more recent development in simple self-application devices is the dust bag (Fig. 20:7).

Other self-applicating devices used to control flies on cattle, particularly dairy cattle, are the treadle sprayer, the lever sprayer, and the automatic photoelectric sprayer. These devices, placed at barn exits or at gates, dispense small amounts of concentrated spray when they are actuated by the passage of an animal.

Figure 20:7. Cattle using insecticide dust bags at entrance to salt station. *Courtesy University of Idaho, Moscow.*

Certain pests of livestock such as the house fly and stable fly may be controlled by applying residual sprays to walls and ceilings of barns, fences, and other resting sites.

Precautions

Because spraying or dipping excites livestock, causing them to mill about, one should make corrals and chutes safe by eliminating protruding nails and boards and by picking up hazardous objects lying on the ground. One can also prevent serious injuries by handling animals as gently as possible, never driving them with clubs or metal bars. Dairy cows are less frightened if they are treated while stanchioned. Because horses are sensitive to the hissing of sprayers and are particularly skittish about treatment, the most successful method is to use a low-pressure, almost noiseless, compressed air sprayer adjusted to dispense a coarse spray or a ready-to-use wipe-on.

Even though animals are reasonably tolerant to wetting during cold weather, it is advisable to select a day that is warm and sunshiny for spraying or dipping. When nights are cold, stop control operations early enough to allow animals to dry off before sunset. Keep swine out of sunlight until they are completely dry to prevent skin blistering. Although there is usually no hazard to the health of livestock in applying recommended dosages of insecticide, very young animals (up to three months old) may be susceptible to poisoning by some chemicals. Always read the precaution statement on label before treating.

Avoid contamination of milking utensils, water tanks, feed troughs, feeds, and vegetation when applying insecticides to livestock or around the barn. In disposing of excess spray or dip do not allow the formation of pools from which animals may drink toxic doses, and avoid runoff into streams.

REPRESENTATIVE LIVESTOCK PESTS

For our detailed study of livestock pests we have chosen cattle lice, cattle grubs, the horn fly, the screwworm, the sheep ked, the winter tick, and the hog itch mite.

Cattle Lice

Shortnosed Cattle Louse. *Haematopinus eurysternus* (Nitzsch)

Longnosed Cattle Louse. *Linognathus vituli* (Linnaeus)

Little Blue Cattle Louse. *Solenopotes capillatus* Enderlein

Cattle Tail Louse. *Haematopinus quadripertusus* Fahrenholz [Anoplura:Haematopinidae]

Cattle Biting Louse. *Bovicola bovis* (Linnaeus) [Mallophaga:Trichodectidae]

Lice are important pests of cattle in all parts of the world. In North America four species of sucking lice and one biting louse do considerable damage, chiefly during winter and spring. Because cattle lice have preferences for particular breeds and ages of cattle, all four kinds rarely coexist on one animal. The longnosed cattle louse more often parasitizes dairy breeds, especially young animals. The shortnosed cattle louse more commonly infests mature animals of the beef breeds. The cattle biting louse occurs on animals of all ages but is more frequent on Angus and Holsteins. The little blue cattle louse appears to have no special preferences as it infests young and old alike of all breeds. The cattle tail louse, a tropical insect, prefers European over zebu breeds and infests all ages of cattle.

Intense irritation set up by lice causes cattle to rub and scratch. Among grossly infested animals large areas of skin become raw, bruised, and denuded of hair because of constant rubbing against posts and fences. The irritation, accompanying nervousness, and loss of blood, mean that the cattle neither thrive nor gain weight normally. Although the exact effects on milk production are still unknown, observation indicates that heavy lice infestations cause substantial decreases in yields.

When sucking lice become numerous, anemia may develop. Heavy infestations of the shortnosed cattle louse have been found to reduce the red blood cells to one fourth the normal number and to bring about the death of infested animals. Entomologists studying this problem in Alberta have observed that mature cattle heavily infested with the shortnosed cattle louse show typical symptoms of anemia—unthriftiness, lack of vigor, and extreme paleness of eyelids, conjunctivae, muzzle, and udder. Cattle in this condition that are moved short distances of only 100 to 300 yd become exhausted and die. Advanced stages of lice-induced anemia cause abortion. The researchers found that only in the severely infested animals were weight gains affected; steers free of lice gained 0.41 lb per animal per day more than a severely infested group. After cattle are treated for lice, the red blood cell and hemoglobin content of the blood begins immediately to return to normal; animals recover completely from anemia in about one month.

Cattle biting lice cause a skin reaction that results in the loosening and falling out of hair. They also produce lesions resembling those of scab or mange. Underneath the scabs colonies of lice live on the raw skin. As yet cattle lice have not been found to transmit any disease.

Description. Cattle lice are small, wingless insects that as adults (Fig. 20:8) range in size from about 2 mm long, the size of the little blue cattle louse and the cattle biting louse, to 3.5 mm long, the size of the shortnosed cattle louse. The cattle tail louse may reach a length of 4.5 mm. Except for smaller size the nymphs look much like the adults. The species can be distinguished from each other by the shape of their heads and by color. The biting louse is reddish brown; the abdomen of the shortnosed louse and tail louse is gray-brown and that of the little blue louse and the longnosed louse dark blue. Eggs of the shortnosed louse are opaque white to brown and 1 mm long; eggs of the longnosed louse are dark blue and 0.95 mm long; eggs of the little blue louse are dark blue and 0.76 mm long; eggs of the cattle biting louse are translucent, colorless to light brown, and 0.64 mm long; eggs of the cattle tail louse

Figure 20:8. Four common species of cattle lice. A, shortnosed cattle louse; B, longnosed cattle louse; C, little blue cattle louse; D, cattle biting louse. *Courtesy Cornell University, Ithaca, N.Y.*

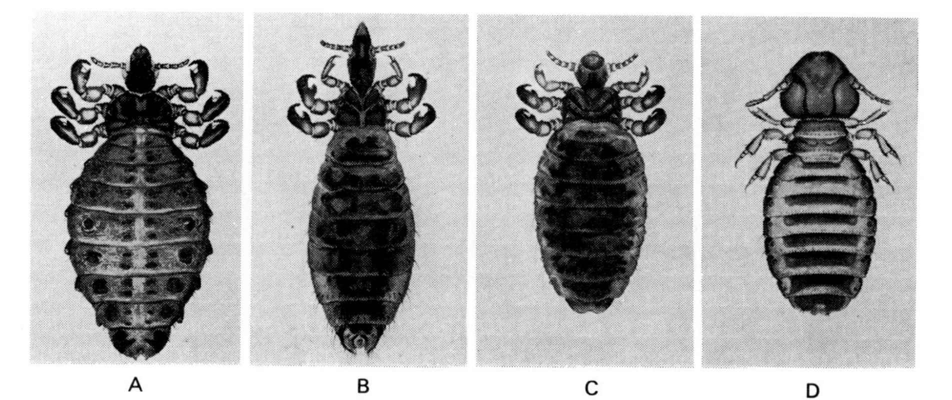

are tan to brown and 1.3 mm long and are found mainly on the hairs of the tail brush. The eggs are also distinguishable by their shape and by the way they are attached to the hairs (Fig. 20:9).

Life history. The life histories of cattle lice are generally similar, with those of the shortnosed louse and the cattle biting louse best known. All cattle lice are obligate parasites; that is, they must remain on their hosts to survive. If eggs or lice fall off with shedding hair, they succumb in a few days. Cattle lice are host specific; they are never found on other animals, and lice from other animals do not live on cattle.

Female lice glue their eggs close to the skin on hairs. Sometimes a single hair will bear two, three, or more eggs. The shortnosed cattle louse, cattle tail louse, and the cattle biting louse have preferred oviposition areas where females congregate and lay clusters of eggs. Summaries of the life cycle of three species are as follow:

	Longnosed Cattle Louse	Shortnosed Cattle Louse	Cattle Biting Louse
Period	Average days	Average days	Average days
Incubation	8	12	8
1st nymphal instar	7	4	7
2nd nymphal instar	4	4	5.5
3rd nymphal instar	4	4	6
Preoviposition	2	4	3
Egg to ovipositing adult	25	28	29.5

Adult males of the shortnosed cattle louse live up to 10 days, whereas adult females live up to 16 days. During their lifetimes females lay 30 to 35

Figure 20:9. The eggs of five common species of cattle lice. A, shortnosed cattle louse egg; B, cattle tail louse egg; C, longnosed cattle louse egg; D, little blue cattle louse egg, note the characteristic bend of the hair at the point of attachment; E, cattle biting louse egg.

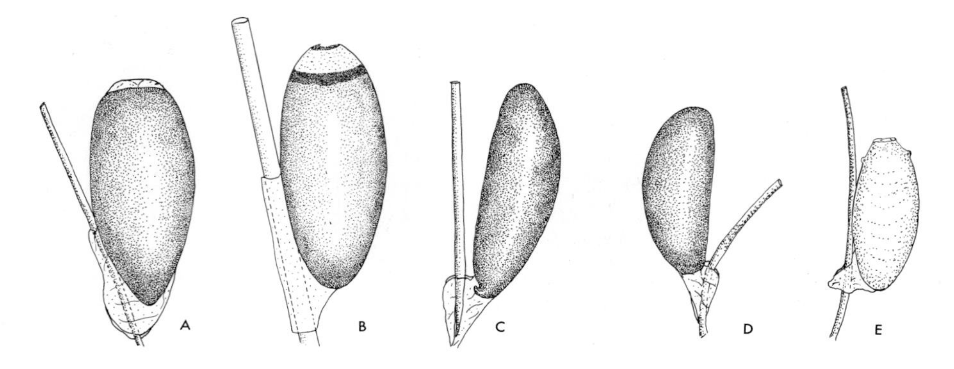

eggs each, depositing 2 to 3 a day. To obtain food, sucking lice pierce the skin and suck blood, whereas biting lice scrape material off the surface of the skin and from the base of hairs. Striking features of cattle biting louse biology are the scarcity of males and parthenogenesis as the normal method of reproduction. A recent study indicates the existence of antagonism between the cattle biting louse and the longnosed louse. When either species was the sole infesting louse, it occupied most body regions; but when both species were present simultaneously and in large numbers on the same host, the cattle biting louse occupied the upper body regions and the longnosed louse the lower. Adults of the cattle tail louse prefer to inhabit the tail and to oviposit on the long hairs. Newly hatched nymphs, however, migrate from the tail to areas of the body such as the vulva. Older nymphs infest other areas such as the neck, shoulder, and back. From these positions the third instar may attach to horn flies and be flown to a distant host.

Cattle lice populations fluctuate with the seasons. They are lowest in summer, begin building up in fall, and reach their peak in winter and early spring. These variations have been found to be related to temperature. In summer it becomes too hot in the hair coat for lice to stay in favored regions of the host and to maintain normal reproduction. During the hot months many animals probably become entirely free of lice, while a few highly louse-susceptible animals called **carriers** perpetuate the infestation. The cattle tail louse is an exception; it reaches greatest abundance in late summer and early fall and is scarce in winter. Recent studies have shown that other factors affect density of cattle lice, including level of nutrition of the host, self-grooming, and shedding of hair in spring.

Control. Because nearly all herds develop infestations of lice, it is best to treat routinely in fall of the year before the onset of cold weather. Treatment then prevents heavy winter infestations. Stock growers usually either dip or spray cattle, but if numbers are small they may apply dusts by hand.

To control lice on dairy cattle, dairymen may treat with crotoxyphos, crotoxyphos plus dichlorvos (Ciovap), coumaphos, ronnel, or stirofos. These insecticides are applied to milk cows in various ways—as sprays of water emulsions, concentrated mist sprays of oil solutions, self-treatment backrubbers charged with oil solutions, or self-treatment dust bags. Each insecticide has restrictions on the ways it can be applied.

Dairymen use power sprayers or compressed air sprayers to apply emulsions of 0.25 per cent crotoxyphos, 0.25 per cent Ciovap, or 0.03 per cent coumaphos at a rate of 1 gal per cow, or they may use electric mist foggers to apply a fine mist of ready-to-use oil solution of Ciovap at a rate of 3.5 oz per cow. They may apply dusts of 1 per cent coumaphos or 3 per cent stirofos with a shaker can and rub the dust into the hair coat at a rate of 2 oz per cow or 3 per cent crotoxyphos at a rate of 3 to 4 oz per cow.

With any of these insecticidal formulations, the whole animal requires treatment and usually two applications 10 to 14 days apart. It is convenient to treat dairy cows while they are stanchioned, but care must be taken to prevent contamination of milk, milking utensils, feed, and water with any insecticide.

Dairymen may employ self-treatment devices to control cattle lice. In fall or winter they may place dust bags of 1 per cent coumaphos or 3 per cent stirofos in loafing sheds or in the exit from the milking parlor. Or they may charge backrubbers located in loafing areas of cows with 1 per cent crotoxyphos, Ciovap, or ronnel.

Pour-ons of systemic insecticides such as ronnel, crufomate, trichlorfon, famphur, or fenthion provide fair to good control of cattle lice. Such treatment must be done before the cutoff date for cattle grub control to avoid adverse side effects. Backrubbers or dust bags charged with recommended insecticides are effective for control of lice when placed where cattle will use them.

Several additional insecticides as well as the above materials may be applied to beef cattle. Recommended sprays include 0.06 per cent coumaphos, 0.15 per cent dioxathion, 0.5 per cent malathion, methoxychlor, or toxaphene, 0.25 per cent ronnel, or 0.375 per cent crufomate or stirofos. Dips include 0.06 per cent coumaphos, 0.15 per cent dioxathion, 0.5 per cent toxaphene, or 0.25 per cent ronnel or crufomate. Recommended dusts for beef cattle are 4 or 5 per cent malathion, 10 per cent methoxychlor or 5 per cent toxaphene. If the stockgrower selects a systemic such as coumaphos or crufomate, he may increase the spray concentration as recommended to control cattle grubs and may thereby solve two of his pest problems with a single treatment.

Generally two treatments two to three weeks apart are needed to control lice infestations satisfactorily in the North; one treatment with a residual insecticide usually suffices in the South.

The cattle tail louse, prevalent in the Gulf Coast states, is harder to control than the other species of cattle lice. Because of the peculiar life history of this louse, the appropriate time to make treatments is from early spring to fall. The most successful method is continuous use of dust bags or backrubbers, but three sprays at three-week intervals are likewise effective.

Cattle Grubs

Common Cattle Grub. *Hypoderma lineatum* (de Villers)

Northern Cattle Grub. *Hypoderma bovis* (Linneaus)
[Diptera: Oestridae]

Cattle grubs and their adult stage, heel flies, are among the most destructive insects attacking cattle. They are distributed principally in the northern hemisphere, where they occur in North America, Europe, northern Africa, and Asia. In North America the common cattle grub is found in Canada, the United States, and in parts of Mexico. The northern cattle grub is somewhat less widely distributed, being found in Canada and in all the United States except the southernmost areas.

Unlike most flies attacking cattle, grubs are injurious not only in the fly or adult stage, but also in the grub or larval stage. For several months the larvae migrate through the body, damaging tissues, until they eventually reach the back, where they perforate the skin. In this location they are enclosed in cysts which are often invaded by bacteria that produce pockets of pus. Heavily infested range animals are weakened. According to some studies, infested feedlot cattle do not realize their full potential of weight gain. A test with dairy cattle indicates that the presence of numerous grubs reduces milk flow.

When infested cattle are slaughtered, packers incur direct losses because grubs spoil the best part of both carcasses and hides. Because yellowish gelatinous areas develop in the meat near grub cysts, carcasses must be trimmed, decreasing yield and lowering quality. Hides with five or more grub holes may be discounted $0.01 per lb.

Probably the greatest loss to the livestock grower and to the milk producer results from the attack of the adult flies. When the females lay their eggs on cattle, they induce an uncontrollable fear that causes the animals to run wildly with tails high in the air seeking shelter in water holes or in the shade of sheds or trees. During the months that heel flies are active, cattle graze less and consequently fail to put on flesh normally and to produce as much milk. The annual loss due to cattle grubs in the United States has been estimated at about $100 million and some estimates place the loss as high as $300 million. The discrepancy in estimates points to the fact that there is yet much to be learned about cattle grub injury and losses.

Description. The eggs of both species are elongate and pale yellow, with a smooth, shining surface. They are 0.7 to 0.8 mm long. The eggs of the common heel fly are laid in rows on the hair of cattle, whereas the eggs of the northern fly are laid singly. There are three larval instars. Young larvae are transparent white and 0.5 to 0.7 mm long. The last larval instar changes color with age, first becoming yellow, then brown, and finally black. It reaches a length of around $\frac{3}{4}$ to 1 in. In this instar the larvae of the two species are easily separated by the shape of the posterior stigmal plates. Those of the common grub are somewhat flat, but those of the northern grub are funnel-shaped. The puparium is black and around $\frac{3}{4}$ in. long. The shape of the stigmal plates remains the same, allowing identification of the pupae. The adult flies are hairy, beelike insects measuring about $\frac{1}{2}$ in. long. They are marked distinctively by transverse bands of black and yellow. The species can be distinguished by the band of hairs covering the end of the abdomen. In the common heel fly, the hairs are orange yellow, and the band is 3 mm or less wide; in the northern heel fly the hairs are lemon yellow, and the band is 5 mm wide. In all stages the northern species is larger than the common (Fig. 20:10).

Life history. Although the life histories of the two species of grubs are similar, there are important differences; one is that all stages of the northern cattle grub appear from two to ten weeks later in the season than do those of the common cattle grub.

Adult flies of the common species emerge in spring. On mild sunny days the males aggregate in small groups at favorite sites, resting and waiting on

A B C D

Figure 20:10. Life stages of the common cattle grub. A, adult or heel fly; B, eggs attached to hair; C, nearly full-grown larva; D, pupa. *A, B, and D, courtesy USDA; C, courtesy Dow Chemical Company, Midland, Mich.*

the ground for females to fly by. When a female approaches, males rise into the air to meet her. The successful suitor clings to the female, and both drop to the ground to complete the act of mating. The mated female deposits up to 500 or 600 eggs on the hairs of the legs and other lower parts of cattle. Several eggs are laid at a time, one immediately next to another in a straight row. When a cow lies down, the common heel fly sometimes approaches so stealthily that many eggs are laid on the hairs of the udder, escutcheon, or sides without causing any disturbance.

The northern heel fly is bolder and more vicious in its attack. It deposits one egg at a time and often chases an animal about the pasture, striking it repeatedly on the thighs and rump. Northern females deposit the majority of their eggs on the shoulders, back, and tail head of cattle and a few on the sides and legs. Because heel flies neither sting nor bite, there has been much conjecture but little study of why cattle are frightened by them.

The eggs of both species hatch in three to five days, after which the tiny white larvae immediately crawl down the hair and, usually near the hatching site, bore through the skin to the connective tissue. From this point, common grubs migrate through connective tissue, chiefly to the esophagus where they remain for several months. Then they resume migration until they reach their subdermal positions in the back, where they produce tumorous swellings. Migrating by a different route, northern grubs accumulate in the spinal canal before reaching the back. The period from hatching to arrival in the back lasts from seven to eight months for both species.

When larvae reach the back they make holes in the skin, through which they obtain air. Cysts form about the grubs, but the holes are kept open. Most growth of grubs occurs in the cysts, as the entire migration is made by the small first instar larvae. Recent research has shown that growth of first instar larvae results from percutaneous absorption of nutrients, little or no food being ingested, and that growth of third instar larvae depends almost entirely on oral ingestion. The second instar larvae take in food by both routes.

Full development of common grubs in the back requires an average of about 55 days, whereas that of northern grubs an average of about 70 days. The first grubs to arrive in the animals' backs spend approximately three weeks longer in the cysts than do those arriving toward the end of the season. The variations are correlated with the different amounts of heat absorbed from the sun by grubs in their subdermal positions.

Upon attaining full growth, grubs crawl from the cysts and drop to the ground, where they pupate in litter. Depending on temperature the pupal period lasts from 15 to 75 days. The majority of flies emerge in spring on bright sunny mornings. Their mouthparts are nonfunctional; life is sustained and eggs are nurtured from stored food acquired during the larval stage. Within an hour after emergence females may mate and start to lay eggs. The life span of individual adults is believed to be short. Aggregating males marked with paint have been resighted as long as 8 days after release. The complete life cycle requires about a year.

Densities of grubs in the same herd often vary widely from year to year. These variations have been found to be related to weather conditions during the pupal and adult stages.

Control. One of the most exciting developments in veterinary entomology has been the discovery that certain phosphate insecticides act systemically and are effective in controlling internal parasites. Research of systemic insecticides began in 1943, but it was not until 13 years later, in 1956, that the first practical compound, ronnel, was developed. Since that time several new animal systemics have become available, including crufomate (Ruelene), coumaphos (Co-Ral), trichlorfon (Neguvon), famphur (Warbex), fenthion (Tiguvon), and phosmet (Prolate); all six, as well as ronnel, are currently recommended for control of cattle grubs.

Cattlemen have a choice of methods by which to apply these insecticides: as sprays, dips, pour-ons, spot-ons, and mineral or feed additives. One treatment kills the grubs before they reach the back, thereby preventing damage to carcass and hide. This damage was not prevented by the old, standard treatment with rotenone, which could not be used effectively until after the grubs had arrived in the back and had cut holes in the hide. Moreover, to achieve adequate control with rotenone, stockmen had to make from two to four spray treatments at regular intervals during winter.

Systemic insecticides are formulated in several ways to suit the method of application and to conform with characteristics of the compound. For use in water sprays, coumaphos is formulated as a 25 per cent wettable powder or

11.6 per cent emulsifiable concentrate, crufomate as a 25 per cent emulsifiable concentrate, and phosmet as a 11.6 per cent emulsifiable concentrate. Depending on the insecticide, spray concentrations range from 0.25 to 0.5 per cent. For achieving effective control with sprays one must wet animals to the skin, not just the hair, because the insecticide has to be absorbed through the skin to act systemically. Spray pressures of 250 to 350 psi are necessary, and 1 gal of spray will usually cover a mature animal.

Dipping vats may be charged with 0.25 per cent coumaphos by diluting the 25 per cent wettable powder formulation or with 0.25 per cent crufomate by diluting the 35.7 per cent emulsifiable concentrate. Dipping wets an animal thoroughly, but special attention must be given to the mixing of the wettable powder with the water so that large amounts do not sink and rest at the bottom of the vat.

Pour-ons come either ready to use as oil solutions or as emulsifiable concentrates or soluble powders for dilution with water. They are applied with a dipper to the backline of cattle in graded amounts depending on the weight of an animal and the systemic used. Cattlemen find pour-ons a convenient method of treatment when they have several compatible operations to perform on their animals. Even more rapid than the pour-on is the new spot-on method, in which metered doses of 20 per cent fenthion are applied to the backs of cattle by means of an automatic refilling syringe.

Cattlemen, particularly livestock feeders, may prefer controlling cattle grubs by mixing either Trolene (a purified grade of ronnel) or famphur with the feed. The mixture is given to the animals daily over a period of 7 or 14 days with Trolene and 10 days with famphur, the number of days depending on size of daily dose utilized. Mineral blocks or granules containing 5.5 per cent Trolene are obtainable ready to use. They are fed continuously for not less than 75 days at a rate of 0.25 lb per 100 lb body weight each month.

Because of the residue problem in milk, no insecticide is approved for control of cattle grubs infesting lactating dairy cows. Grubs are usually fewer in mature cows and may be squeezed out carefully by hand.

Because systemic insecticides enter the bodies of animals and are highly active internally, greater care must be taken with them than the nonsystemic insecticides. Stockgrowers should follow these precautions:

1. Administer only the recommended dosages given on the label.
2. Do not double-dose with another systemic; for example, if you are feeding ronnel, do not use a pour-on of coumaphos.
3. Do not treat sick animals.
4. Observe minimum days from last application to slaughter or freshening of dairy animals.
5. Do not use on lactating dairy animals.
6. Treat after heel fly activity has ceased but in northern states not after November 1 and in southern states not after October 1.

The reason for cutoff dates in treating animals with systemics is because complications may arise when grubs die in sensitive positions within the host. Two syndromes of the host-parasite reaction have been observed depending on the species of cattle grub. If treatment is made when larvae of the common species reside in the tissues around the esophagus, death of the parasites may cause severe inflammation in the area. The affected animal has difficulty in swallowing. Symptoms ranging from frothy salivation to severe bloat may occur because of the constriction in the esophagus, which prevents swallowing of saliva and belching of rumen gases. Treatment of this reaction is often quite simple. Animals experiencing bloat improve quickly when they are forced to exercise, because the movement aids eructation.

If treatment is made when larvae of the northern cattle grub have migrated into the region of the spinal cord, dead grubs may cause an inflammation which results in weakness or paralysis in the hindquarters of affected animals. Most reactions are transitory and disappear within 96 hr. In severe cases of both types of host-parasite reaction a veterinarian should be called.

For some time agriculturalists have thought that death of grubs caused allergic reactions in cattle; however, only rarely have true allergic-type reactions been observed. The symptoms in these cases are swollen eyelids, congested conjunctivae, and urticaria.

A major objective of current research is to find methods and materials that will control several insect pests by one and the same treatment. Recommendations are available for spraying or dipping cattle that will control both cattle lice and cattle grubs. Promising experimental data indicate that dust bags charged with systemic insecticide and made available to cattle for self-treatment will control flies in summer and also cattle grubs, which at this time of year are penetrating the skin and entering cattle. Many veterinary entomologists have the worthy goal of developing techniques—systemic insecticides, sterile-male technique, sex attractants, and others—that will enable stockmen to eradicate grubs from circumscribed geographical areas.

Horn Fly. *Haematobia irritans* (Linnaeus) [Diptera: Muscidae]

Probably brought from southern France to the United States, the horn fly was first collected in 1887 at Camden, New Jersey. It spread rapidly in North America and is now one of the most abundant and persistent of cattle pests. It is not uncommon for a herd of cattle to average 4000 to 5000 flies per animal; indeed, as many as 10,000 to 20,000 may infest individual bulls. Horn flies pierce the skin and feed on the blood intermittently. They irritate and annoy cattle, provoking them to fight back by tail switching, head throwing, licking, and kicking. The flies escape this harassment from the host by clustering around the base of the horns and low on the hind legs.

Several tests conducted in different parts of the United States and in Canada have demonstrated increased gains of 15 to 55 lb by beef cattle accruing from horn fly control during a single season. Calves in treated herds gained as much as 75 lb additional weight. Tests have also indicated that milk production increases when horn flies are controlled.

Description. The adult horn fly is a small insect approximately 4 mm long and dark gray in color, with bayonetlike piercing mouthparts. The eggs are usually reddish brown, but a small percentage are tan, yellow, or white; they are 1.2 mm long. The maggots are white, 1.5 mm long when newly hatched and 10 mm long when full grown. The puparium is barrel-shaped, dark brown, and averages 3.3 mm long (Fig. 20:11).

Life history. Horn flies appear with the first warm days in spring and are present on cattle until cold weather in fall. They cluster in large numbers on animals, particularly resting on the backs, or if the day is hot or rainy they congregate on the belly.

Females leave the host long enough—usually six to eight minutes—to deposit one to fourteen eggs in freshly voided cow dung, mainly under the dropping on blades of grass and litter. Males and nongravid females likewise alight on droppings, insert their proboscises, and apparently feed. Horn flies have a strong capacity to disperse and to seek out hosts; they move principally at night, and females move at a greater rate than males. An individual fly is able to migrate a distance of at least five miles. Horn flies are active insects. They often make short, rapid flights from one part of the animal to another, fly to the ground and back to the host, and dart from one animal to another. They take many small blood meals day and night. The females, each consuming an average of 17 mg of blood per day, feed around 38 times during a 24-hour period, a single feeding taking about 4 minutes. The males feed less often and consume somewhat less.

Figure 20:11. Life stages of the horn fly. A, egg; B, larva; C, pupa; D, adult. *Courtesy USDA.*

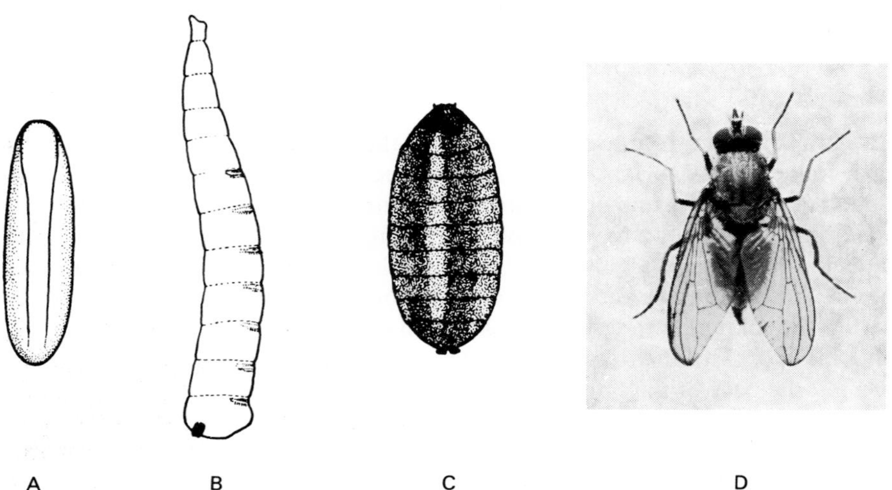

Eggs hatch in 16 to 24 hours; the small maggots crawl into the dung, where they feed and develop. Many arthropods in addition to horn flies develop in cattle droppings. A study in central Texas has shown that because of predation and competition for food and space within a cow pat, only 7 horn flies emerge per pat, whereas without this natural control 10 times as many emerge. The larval stage lasts four to five days. On becoming full grown, larvae pupate either in the dung or in the soil beneath, whichever offers the more favorable moisture conditions. Soils containing less than 0.25 per cent of moisture, or more than 14.5 per cent, inhibit development of pupae. For highest survival soil holding 7 per cent moisture is optimum. After five to seven days in the pupal stage, horn flies transform to adults and in two or three hours begin feeding on the blood of cattle. Mating takes place, usually on the host, as early as the first day after emergence; oviposition may begin on the third day. Females normally mate only once and produce 350 to 400 eggs during their lifetime of approximately three weeks. When cold weather arrives, horn flies hibernate in the soil chiefly as pupae, but in the deep South they continue to breed and develop the year round.

The cessation of development is related to a shorter photoperiod, that is, decreasing length of daylight. In the North pupae begin entering diapause as early as the fourth week of July and in increasing percentages as the season progresses, until 100 per cent enter in September. In the South diapause comes later; pupae in Mississippi begin entering diapause in September, reaching a peak in October, and by early November almost all have entered diapause.

Apparently diapause in the pupae is determined in the unlaid egg through the influence of photoperiod upon the female parent. Recent investigations indicate that the egg likewise receives a diapause message through a response of the female parent to the host's blood. More pupae enter hibernation when the parent feeds on cows deprived of ultraviolet radiation.

In the North, populations reach a peak in summer. In the South they reach a peak in spring, decline during summer due to hot dry weather, then reach a second peak in fall. Horn flies prefer dark-colored animals over light-colored ones, and half-grown or mature animals over calves.

Control. Because the horn fly rests on cattle much of the time, it is highly susceptible to insecticides applied to the host. The stockgrower has a choice of several insecticides and methods of application for controlling this pest. He may treat beef cattle with coumaphos, crotoxyphos, crufomate, dioxathion, malathion, methoxychlor, phosmet, ronnel, stirofos, or toxaphene. Because these insecticides are applied in different ways, he will normally select the method of application based on his management practices, available equipment, and personal preferences and then select a suitable insecticide. Application methods include spraying, dipping, dusting, ULV aerial spraying, backrubber, and dust bag.

If spraying is the method chosen, the beef producer may apply one to two quarts of water-base spray per head, less for calves, and direct the spray around the head, neck, withers, backline, and upper flank area. Complete coverage of the animal is unnecessary because of the movement of flies from one part of the body to another. Sprays of 0.5 per cent carbaryl, malathion, methoxychlor, ronnel, or toxaphene, 0.06 per cent coumaphos, or 0.15 per cent dioxathion are effective. An important consideration in selecting an insecticide is its cost per head.

A beef raiser with large numbers of cattle may prefer to control horn flies by dipping. An advantage of this method besides its thoroughness and rapidity is that it can be used to control several pests at the same time. By choosing a systemic insecticide such as coumaphos, he can control the horn fly, lice, and cattle grubs. If a farmer has only a small number of beef cattle, he may find it convenient to hand dust each animal with 50 per cent methoxychlor, 5 per cent malathion or toxaphene, or 1 per cent coumaphos. The dust is sprinkled on the back and rubbed into the hair coat. Hand dusting of confined animals takes 30 sec per head.

For several reasons self-treatment methods are gaining in favor among beef producers. There is no need to round up cattle and no necessity to spray or dip, which excites livestock and may cause some weight loss, and relatively little labor is required. The backrubber is a self-treatment device developed between 1946 to 1952 by entomologists of the South Dakota Agricultural Experiment Station. It consists of four strands of barbed wire covered and wrapped in burlap and is strung between two posts approximately 20 ft apart in areas where cattle loaf. The burlap is saturated with 1 gal of fuel oil solution containing one of the following: 1 per cent crotoxyphos, coumaphos, or ronnel, 1.25 per cent Ravap (stirophos plus dichlorvos), 1.5 per cent dioxathion, 2 per cent malathion, or 5 per cent methoxychlor or toxaphene. Several other types of backrubbers have been developed some of which are available commercially. Backrubbers are effective if they are located properly and tended regularly. They need to be examined every two weeks and resaturated every three to six weeks.

Developed in the early 1960s by entomologists of several southeastern states, dust bags containing insecticide for self-application by cattle have proved to be one of the simplest and most effective methods of controlling the horn fly (Fig. 20:7). The bags are made of burlap and contain inner liners of various kinds to regulate dust flow and to prevent wetting of the dust by rain. They are suspended from chains or wooden cross pieces by hooks or by baling wire fastened to three grommets in each bag. The bags are usually placed so that animals are forced to use them to get water or salt or pass from one pasture to another. Each bag is charged with 5 to 25 lb of one of the following insecticides: 1 per cent dust of coumaphos, 5 per cent malathion, 10 per cent methoxychlor, or 3 per cent crotoxyphos or stirofos. Dust bags provide some control of the face fly.

Because milk must be protected from objectionable and illegal residues of insecticide, compounds to control horn flies infesting lactating dairy cattle are limited to crotoxyphos, coumaphos, dichlorvos, stirofos, Lethane or

Thanite, pyrethrins plus synergist, malathion, and methoxychlor. Methods of applying these vary. Dusts of 1 percent coumaphos, 3 per cent crotoxyphos, 4 or 5 per cent malathion, or 50 per cent methoxychlor may be applied by hand to the backs of cows every two or three weeks. Oil solutions of crotoxyphos, dichlorvos, Lethane 384, and pyrethrins may be applied daily as mist sprays with hand, electric, or automatic sprayers. Dairymen place automatic sprayers in doorways or gates so that as the herd passes through, each cow actuates the sprayer and receives a small metered dose of insecticide. Engineers and entomologists of USDA have recently developed an automatic sprayer that delivers ultra-low volumes (1 to 5 ml) of insecticide upon dairy cattle. The advantage of ultra-low volumes is that such applications minimize insecticide residues in meat and milk and still provide adequate protection from horn flies.

Dust bags charged with 3 per cent crotoxyphos or stirofos or 1 per cent coumaphos and back rubbers saturated with 1 per cent crotoxyphos, coumaphos, or ronnel may be used to control horn flies on dairy cattle.

Because approved insecticides and methods of application change rapidly, dairymen must keep constantly in touch with their state department of agriculture or agricultural extension service for current recommendations and regulations. In doing so they will not only keep abreast of latest recommendations, but they will also be the first to learn of breakthroughs in controlling stubborn pests such as face flies, horse flies, and mosquitoes.

Screwworm. *Cochliomyia hominivorax* (Coquerel) [Diptera:Calliphoridae]

One of the most serious pests of livestock in subtropical and tropical America is the screwworm, so named because of its resemblance to a wood screw. It is the larval stage of a blow fly that confines its egg laying to wounds on warmblooded animals. Of the domestic animals, cattle, sheep, goats, and hogs suffer most. When the adult flies are active, any minor cuts, as well as serious breaks in the skin of an animal, invite egg laying. Upon hatching the larvae tear into healthy tissues and create a characteristically foul-smelling, bleeding wound. The odor attracts other flies, which lay eggs and increase the severity of the infestation. Because of repeated egg deposition screwworms of different sizes and ages infest untreated wounds (Fig. 20:4).

Infested animals appear nervous and make frantic attempts to scratch and lick the wound. They often leave the herd to hide away in secluded places. Most untreated infestations terminate in the death of animals. A grown steer may die within ten days of infestation. Death is caused directly by the maggots themselves or indirectly by the female flies introducing pathogenic organisms. The causative bacterium of the disease "swollen joints," *Streptococcus pyogenes*, is carried mechanically to the navel of calves by egg-laying screwworm flies, while the feeding activities of the larvae appear to facilitate the infection.

Before control of screwworms was achieved by the several cooperative programs of screwworm eradication, domestic livestock losses were estimated at $20 million annually in the southeastern United States and $50 to $100 million in the Southwest. During the severe outbreak of screwworms in 1935, 1.2 million cases of infestation and 180,000 deaths were recorded among domestic animals.

During favorable years in the past, screwworms have widely populated the southern half of the United States, which they invaded from their overwintering ranges in southern parts of Florida, Texas, New Mexico, Arizona, California, and Mexico. From late spring to early fall the adult flies spread northward by flight to states as far away as Colorado, Kansas, Missouri, Kentucky, and Virginia, individuals migrating as much as 180 miles in a week or two. Epidemiological evidence indicates that a female fly can migrate up to 300 miles during her lifetime. Nearly all outbreaks north of these states originate from maggot-infested animals shipped from the South and not from migration of flies.

Description. Eggs of the screwworm are deposited in oval, shinglelike clusters at the edges of wounds or on their dry surfaces. Eggs are 1 mm long, elongate, and white or cream colored. Maggots are creamy white and vary in length from 1.2 mm at hatching to 16 mm ($\frac{2}{3}$ in.) at full growth. The puparium is dark brown and 10 mm long. The fly, somewhat larger than the house fly, is 8 to 10 mm long. It has a blue to bluish green body and a reddish orange to brown head (Fig. 20:12).

Life history. Under favorable temperatures the species develops and reproduces the year round. During cool weather it survives in the pupal stage, which may last two months. If cool weather (average daily temperatures lower than 54°F) prevails longer than two months, the pupae die in the soil. For this reason screwworms are usually able to overwinter only in the most southern parts of the United States and in countries farther south.

Weather likewise affects the size of populations. Screwworms increase rapidly during warm, humid periods but very slowly during cold or extremely hot, dry periods. The overwintering population is likely to survive in greater numbers and extend farther north when winters are mild and moist.

In nature screwworms develop solely in wounds of live animals; they cannot exist in carrion. Adult flies feed on the juices of manure and the exudations of wounds. Gravid female flies are attracted to wounds, where they deposit 200 to 500 eggs cemented together in shinglelike clusters. Each female produces several clusters at four-day intervals and may lay as many as 3000 eggs. The eggs hatch in about 16 hours into tiny white maggots which invade the wound, feed close together, and form a pocket in the living flesh. Consuming chiefly live flesh and wound fluids, they pass through three stadia in four to nine days to become full-grown larvae. These drop to the ground and burrow into the soil to pupate. Under Texas conditions the pupal stage lasts from 7 days in summer to 54 days in winter. The adult screwworm flies, which exhibit several circadian rhythms, emerge mainly at dawn. Within two

Figure 20:12. Life stages of the screwworm. A, cluster of eggs; B, larva; C, pupa; D, adult. *Courtesy USDA.*

to five days after emerging, flies are ready to mate. During her lifetime a female fly usually copulates only once with a male. In summer the average life cycle of about 20 to 25 days allows several generations to develop annually.

Control. Measures to combat the screwworm have changed drastically consequent to the successful eradication of the species, first in the Southeast in 1959 and then in the Southwest in 1966. The control method employed in these operations was the sterile-male technique, one of the brilliant achievements of applied entomology during this century. The method, first conceived by E. F. Knipling in 1937, consists essentially of rearing vast numbers of the insect in the laboratory, sexually sterilizing them in the pupal stage by exposing them to cobalt-60, and then releasing the adults from aircraft into the wild population. All of this is done in a cooperative program supported by livestock growers and the state and federal governments. Outnumbering their wild brothers, the sterilized males compete successfully in mating with the wild females, which then produce only infertile eggs.

Although eradication of the pest has been achieved in the United States, screwworm flies emigrating from Mexico continue to invade the Southwest and precipitate spotty infestations. To reduce the chances of invasion, a screwworm barrier zone has been established through the release of sterile flies every year along the 2000-mile border between the two countries. Other measures are necessary to form the barrier; the U.S. Veterinary Services examines animals presented for import at the border. Only animals free of screwworms are permitted to enter. The U.S. government likewise maintains a patrol at the international border to prevent infested animals from straying into the United States from Mexico. Entomologists of both countries have proposed that steps be taken to eradicate the fly in Mexico and establish a much shorter screwworm barrier at the 140-mile wide Tehuantepec Isthmus. In 1972 the United States and Mexico signed an agreement to accomplish these objectives and appointed a ten-member commission, The Joint Mexico–United States Commission for the Eradication of Screwworms. The Commission succeeded in its promotion of a sterile fly production plant in Tuxtla Gutierrez, Mexico and of staffing the facility. The plant began production of sterile screwworm flies in September 1976 and by the summer of 1977 it was furnishing in excess of 200 million sterile flies a week for distribution over various sites in Mexico and Baja California.

Practical experience has shown that flies do penetrate the lengthy barrier, get into the freed areas of the country, and start small foci of infestations. Consequently Veterinary Services, APHIS and state extension services recommend that livestock growers in the Southwest follow these measures: (1) inspect livestock regularly for screwworm and other maggot infestations; (2) submit for identification larvae found in wounds; (3) practice wound prevention and treatment; (4) refrain from moving screwworm infested animals from the premises.

Stockmen combat screwworms in three ways: (1) by preventing wounds; (2) by preventing infestation through wound treatment; (3) by treating infested wounds. Gentle handling of livestock prevents many needless wounds. Do not drive sheep, cattle, and other livestock with whips, sticks, or biting dogs. Holding corrals should not have protruding nails and broken boards on which animals can snag themselves. If possible limit operations such as dehorning, branding, castration, shearing, and lamb docking to months when screwworm flies are inactive. If operations are necessary at other times, treat the wounds with an approved insecticidal preparation of coumaphos, ronnel, trichlorfon, or lindane. Coumaphos comes ready to use as a 5 per cent dust in a squeeze bottle or as 3 per cent pressurized spray in a can. Ronnel 2.5 per cent and trichlorfon 3 per cent are also available as pressurized sprays ready to use. Apply a thorough coat of dust or spray immediately to the wound and about an inch around it.

Because of the reduced risk of screwworm infestation, but in spite of recommendations to schedule births in months other than late summer and early fall, some Southern stockmen are now calving the year round. All young born during the normal screwworm season should receive special attention. Treat the navel of a newborn calf by tying off the umbilical cord and cutting the surplus away. Then apply iodine and follow with an application of an approved insecticidal remedy. It is also advisable to treat the vulva of the dam with screwworm remedy before and after she gives birth.

Some wounds are difficult to prevent. The bites of insects such as horse flies, stable flies, and ticks produce breaks in the skin that are attractive to screwworm flies. Equally attractive wounds can result from animals brushing against thorny shrubs or barbed wire. The feeding of sheep and goats on prickly pear cactus causes mouth injuries that lure flies. Moreover, warts, pink eye, and cancer eye may attract egg-laying females. Treat all wounds at all seasons in the deep South; treatment of wounds may be restricted to the normal screwworm season in northern areas. Give uninfested wounds a light coating, but work the material well into infested wounds and paint the surrounding area. Give deep pockets of maggots special attention. A light application to drainage areas below wounds on sheep and goats is helpful in preventing future attacks of the fly. Infestations in beef cattle may also be treated successfully by spraying animals with 0.5 per cent ronnel or 0.25 per cent coumaphos or dipping with 0.125 per cent coumaphos.

Beef cattle may be given a preventive treatment by wetting the whole animal to the skin with a high-pressure spray of 0.25 per cent coumaphos. It is advisable to spray or dip sheep out of the shearing pens with 0.125 per cent coumaphos or 0.5 per cent ronnel. The treament not only protects shearing wounds from screwworms but also controls keds and lice. Wounds of lactating dairy cows may be treated with 5 per cent coumaphos dust or with 5 per cent ronnel dust or 2.5 per cent ronnel pressurized spray.

Sheep Ked. *Melophagus ovinus* (Linnaeus) [Diptera:Hippoboscidae]

The sheep ked, often called sheep tick by American wool growers, is a worldwide pest of sheep. It obtains nourishment by piercing the skin of its host and sucking blood. In feeding it causes irritation and annoyance, prompting the infested sheep to rub, bite, and scratch. Sheep may also roll to the ground in relieving irritation, particularly in spring. Some animals get on their backs and are unable to right themselves. Unless they are found and are helped to their feet, their attempt to relieve irritation ends fatally. In a test conducted in South Africa, entomologists found that keds did not cause any apparent damage to sheep fed an adequate and nutritious diet, but that keds caused a 20 per cent mortality among poorly fed sheep. Experiments in America on the effect of keds upon weight of lambs have yielded conflicting results. Some experiments have shown that keds do not depress weight gains of lambs fed fattening rations, whereas others indicate that keds lower carcass weights an average of 2.4 lb per lamb and cause a reduction of 1.6 lb of clean wool per fleece.

Sheep grazed throughout the year on pasture or range acquire heavy burdens of keds during winter and early spring. Although as yet we are not sure, damage should be greatest at these times because of the consumption of large amounts of blood by the keds and the irritation caused by the bites. In spring injury may occur after transfer of large numbers of keds from ewes to newborn lambs.

It has been discovered recently that ked bites are the cause of a defect in sheepskins called **cockle**. The defect consists of pimplelike blemishes of varying size and elevation in sheepskins. The cockles cannot be completely flattened out or covered with dyes. In the United States alone it is estimated that this defect is responsible for annual losses of about $4 million. If sheep are freed of the parasite, they spontaneously recover from this effect of ked bites. In controlled experiments ARS veterinarians have demonstrated that the sheep ked can transmit the virus of **bluetongue** disease from infected sheep to healthy ones.

Description. Because eggs and larvae develop entirely within the uterus of the female ked, only two stages, the pupal and the adult, are visible on the sheep. Adult keds are large (about $\frac{1}{4}$ in. long) wingless parasites. They have brownish heads, thorax, and legs and a grayish abdomen. The puparia are dark red, barrel-shaped, and about $\frac{1}{8}$ in. long (Fig. 20:13).

Life history. The sheep ked is a permanent external parasite of sheep; that is, it lives upon the host for its entire life. If by accident a ked falls from its host, it does not survive longer than four or five days. Adult keds feed upon the sheep's blood which they obtain by piercing the skin.

Adult females are able to mate 24 hours after they emerge from the puparium, but fertilization of eggs does not occur until the fifth or sixth day. Following pupal emergence at least 12 to 14 days elapse before a young female deposits her first larva. She glues this, as well as subsequent offspring, to several wool fibers. Only one larva develops at a time. During the female's life of 100 to 130 days, she produces around 10 to 15 young, giving birth every eight or nine days. Male keds do not mate before they are 10 to 11 days

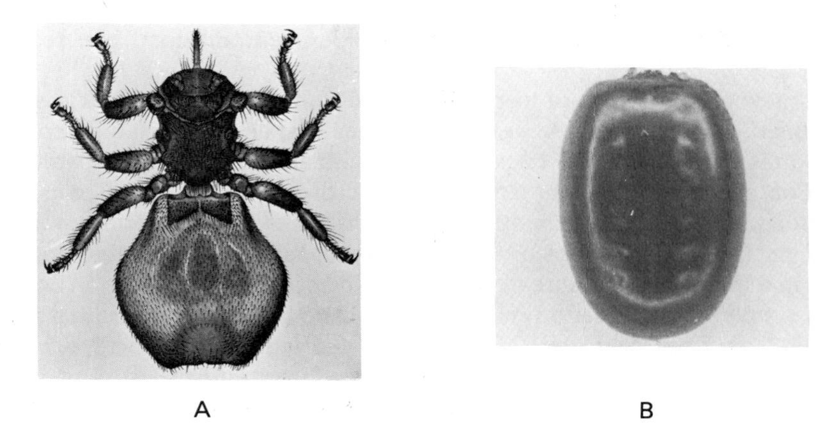

Figure 20:13. The sheep ked, often called sheep tick by ranchers. A, adult; B, puparium. *A, courtesy Cornell University, Ithaca, N.Y.; B, courtesy University of Wyoming, Laramie.*

old. They have a shorter life than females, living for about 80 days. At birth the larva is a light cream color, but within 12 hours it transforms to the red puparium. The pupal stage lasts from 18 to 30 days and averages 22 days.

There appears to be a wide range of susceptibility among sheep to infestation. Lambs and sheep in poor condition bear the heaviest numbers. As many as 1000 adults may parasitize an old weak ewe. Research conducted in Canada indicates that sheep react to increasing numbers of keds by the development of resistance. The acquired resistance is apparently caused by a constriction of blood vessels in the skin that cuts off the flow of capillary blood and starves the keds.

Keds seem to prefer the neck, shoulder, and crutch regions of the sheep, but during warm weather many infest the belly. Few are ever found in the dense wool of the back. Size of populations varies with the seasons. Keds are most abundant during the winter and early spring and lowest during summer. In spring many of the young keds migrate from the ewes to the lambs.

Control. Treating all sheep in a range band or a farm flock with an insecticidal dip, spray, or dust controls sheep keds. Dipping, the most widely recommended practice, is generally the surest way to eradicate an infestation. Several insecticides are approved for dipping sheep: 0.125 per cent coumaphos, 0.15 per cent dioxathion, 0.25 per cent ronnel, 0.5 per cent malathion, 0.025 per cent lindane, 0.25 per cent toxaphene, or 0.5 per cent methoxychlor.

Wool growers often spray sheep to control keds, but spraying is not as effective as dipping because of the difficulty of thoroughly and completely wetting all sheep. The concentrations of the insecticide are sometimes doubled in sprays over those recommended for dips. In addition to the previously mentioned insecticides, high-pressure sprays of 0.03 per cent diazinon and low-pressure sprays of 0.06 per cent diazinon are recommended.

A new, simple way of controlling keds on small farm flocks is the "sprinkler can" method. A wettable powder suspension of 0.06 per cent diazinon is sprinkled on the backs of sheep with an ordinary garden watering can. For convenience about 25 head are crowded into a small pen; the operator walks through them and dispenses 6 gal of the insecticide over the heads, necks, tops, and sides of the sheep. The operation is repeated until all sheep in a flock have been treated.

A simple way to treat a few sheep is to hand dust. Recommended dusts include 4 or 5 per cent malathion, 0.5 per cent coumaphos, or 2 per cent diazinon. One to two ounces of the dust are rubbed into the wool over the entire body. It takes approximately one minute to hand dust a sheep thoroughly.

Power dusting of large range bands of sheep after shearing in spring has proved successful in controlling keds (Fig. 20:14). An effective dust is 2 per cent diazinon. One thorough dusting of a flock eradicates an infestation. Because the label of diazinon sheep dust has been allowed to lapse, the currently used material is 0.5 per cent coumaphos, which provides over 90 per cent reduction of population but not eradication. The dusting method has the advantage of being both rapid and safe: 2000 to 3000 sheep can be dusted in an hour without ever exposing them in a wet condition to inclement weather.

Winter Tick. *Dermacentor albipictus* (Packard) [Acarina:Ixodidae]

The winter tick is widespread and indigenous to North America. It is distributed throughout Canada, the northern United States, the Rocky Mountain states, and southward through Texas into Mexico. A geographical race of the winter tick, *nigrolineatus*, inhabits the southeastern states. Although cattle are attacked seriously, the winter tick is chiefly a pest of horses and of big game—moose, elk, and deer (Fig. 20:15). The direct injury of this tick results from its feeding and withdrawing large amounts of blood. After five to six months of severe infestation during the winter, animals become exceedingly emaciated and weak. Colts are especially vulnerable to attack and often die.

Heavy infestations of the winter tick cause a disease in horses called water belly or tick poisoning. Symptoms include a general rundown condition; loss of appetite; long, rough, dull haircoat; and swelling under the jaws, along the throat and brisket, and under the belly. In advanced stages of the disease death results unless the parasites are removed and the horse is given food and special care.

One of the chief factors contributing to mortality among moose and elk is gross infestation by the tick combined with feed shortages in late winter and early spring. The winter tick is a vector of several diseases of cattle, deer, and

Figure 20:14. Power dusting a large flock of sheep in spring after shearing to control the sheep ked. *Courtesy University of Wyoming, Laramie.*

Figure 20:15. Winter ticks (engorging females) on a yearling elk. Nearly 1400 ticks infested this animal. *Courtesy USDHEW.*

moose. It has been shown to be capable of spreading anaplasmosis among cattle.

Description. The winter tick has four life stages—egg, larva, nymph, and adult. The eggs are laid in masses of several thousand and are coated with a viscid secretion. Eggs are ovoid, smooth, shiny, and yellowish brown. Their longest dimension ranges from 0.45 to 0.5 mm. When first hatched, larvae are pale yellow but soon become dark reddish brown. They are six-legged and when replete with a blood meal measure 1 to 2 mm long. On molting to the nymphal stage, the ticks become eight-legged. Size of engorged nymphs ranges from 3 to 5.4 mm. Newly emerged adults of both sexes range in length from 4 to 6 mm (Fig. 20:16). Females increase their size tremendously as they engorge themselves on blood of the host (Fig. 20:15). They are olive green in

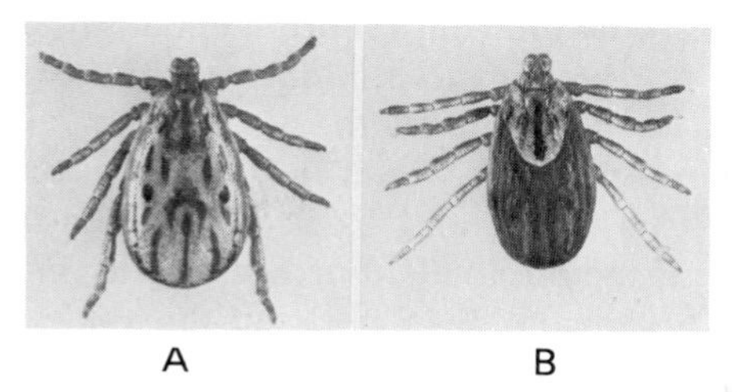

Figure 20:16. Winter ticks, unengorged adults. A, male; B, female. *Courtesy USDHEW.*

color and when replete measure $\frac{1}{3}$ to $\frac{2}{3}$ in. long. Males are reddish brown with pale yellow markings on the dorsum. They neither engorge nor increase in size like the females.

Life history. The life cycle of the winter tick is interesting because of its peculiar habit of feeding during the winter months rather than in spring and summer, as do nearly all other ticks of the same family. It is seldom found on hosts in fall earlier than September or in spring later than June. In fall larval ticks coming out of summer dormancy begin to appear on the ends of grasses and twigs. There they hang head downward waiting for a host to pass by. The larvae are highly tolerant of snow and cold, and unless they attach themselves to a host they may remain in position till spring.

The winter tick is a one-host tick: it will complete its larval and nymphal development and become an adult on one and the same host. Larvae, that succeed in obtaining a host, attach to the skin and begin to suck blood. They feed to repletion in about 9 days, remain quiescent for 11 days, then molt to the nymphal stage. In molting a transverse slit in the cuticula appears near the end of the body. The caudal end of the nymph protrudes from the slit, and the nymph backs out, leaving the exuviae hanging to the host by the inserted mouthparts.

Most of the nymphs crawl about for approximately half a day before they settle down close to the site of larval attachment. Although the winter tick is a one-host tick, nymphs while wandering about are able to and often do transfer to another host when animals are in close proximity. After settling down they feed to repletion in about seven days, then remain quiescent for six to eight days before molting to the adult stage.

After attachment adult female ticks become replete in 5 to 14 days. Adult males feed irregularly, crawl about, and mate with the feeding females. Females have been observed to copulate as many as three times and males two or more times. Replete, mated females drop to the ground and, depending on temperature, may begin laying eggs in one or two weeks. Usually egg laying commences in spring, with each female producing about 4000 eggs in a single mass over a period of several weeks. The eggs hatch in four to six weeks into larval ticks, which remain bunched together in a torpid state during summer. The larvae are in a state of diapause induced by the long day length of spring and summer. When the days shorten in fall, they break diapause and become active. This tick belongs to a minority group of arthropods that are active during seasons of short day length.

Control. Stockmen control the winter tick as well as other ticks such as the lone star tick, Gulf Coast tick, and Rocky Mountain wood tick by either dipping or spraying infested animals. For beef cattle one may use sprays or dips of 0.5 per cent toxaphene, 0.15 per cent dioxathion, or 0.125 per cent coumaphos, or sprays of 0.25 per cent crotoxyphos or 0.75 per cent ronnel. For horses one may use sprays or dips of 0.15 per cent dioxathion or sprays of 0.125 per cent coumaphos. Dairymen may treat lactating cows with sprays of 0.25 per cent crotoxyphos or 0.1 per cent pyrethrins plus 1 per cent synergist.

A single treatment for the winter tick is usually sufficient, but if numbers of the pest are large a second treatment six to eight weeks later may be necessary. A second treatment may also be necessary when the first treatment is made early and animals become reinfested.

Hog Itch Mite. *Sarcoptes scabiei suis* (Gerlach) [Acarina:Sarcoptidae]

Swine are not plagued with many different kinds of external parasites. One species of lice, two species of mites, and one or two species of fleas are the only external parasites commonly found on hogs. Nevertheless, two of these, the hog louse (Fig. 20:17) and the hog itch mite, are both common and serious pests in North America as well as in many other parts of the world. The hog itch mite causes **sarcoptic mange**, a serious skin disease. The mites produce sores and intense irritation by their burrowing and feeding in the upper layer of the skin of the host. Small vesicles develop over the burrows, eventually rupture, and allow serum to ooze over the area. As the mites increase, the infected area expands and the skin takes on the dry, scabby appearance that is characteristic of mange (Fig. 20:18).

The irritation set up by the mites induces hogs to rub and scratch and produce raw, bleeding areas. Infested swine neither grow nor gain weight normally, and some actually die of the condition. When hogs with advanced cases are slaughtered, the meat is of inferior quality, requires special handling, and sells for less.

Figure 20:17. The hog louse, *Haematopinus suis* (L.). A, adult; B, egg attached to hair. *Courtesy USDA.*

Figure 20:18. Hog with an advanced case of sarcoptic mange. Note wrinkled, scabby skin of animal. *Courtesy USDA.*

Hog itch mites are spread from one hog to another by direct contact. The hogs' habit of sleeping close together aids the transmission of the disease. Although sarcoptic mange may start on any part of the body, lesions usually appear first on the head around eyes, ears, and nose. The lesions then spread backward over neck, shoulders, back, and sides and in severe cases involve the entire body.

Mange spreads slowly during warm weather when animals are on pasture. At such times lesions and mites can be found on the inner surface of the ear of infected animals. During winter, when animals are confined and their resistance low, sarcoptic mange erupts and spreads rapidly. Young pigs and hogs in poor condition are more susceptible to the disease than thrifty mature animals. Positive diagnosis of mange is made by taking skin scrapings of lesions and finding the mites.

Description. The life stages of the hog itch mite are egg, larva, nymph, and adult male and immature and mature female. Eggs are ovoid in shape and about 0.2 mm long. The motile stages are small, circular, whitish parasites. Larvae have three pairs of legs and are about 0.2 mm long. Nymphs and adults have four pairs of short legs. Adult females are about 0.5 mm long and adult males about 0.4 mm. The skin has many fine striae, often interrupted by scaly areas or by areas of backward-pointing spines (Fig. 20:19).

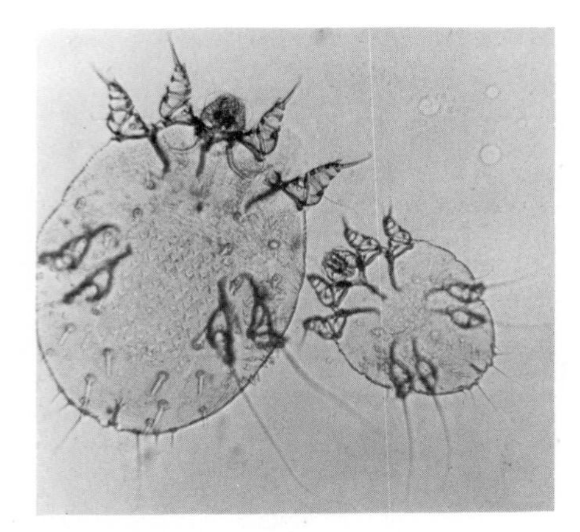

Figure 20:19. Photomicrograph of the hog itch mite, *Sarcoptes scabiei suis* (Gerlach). *Courtesy USDA.*

Life history. The hog itch mite lives its entire life on the hog. Eggs, laid in the burrows by the female mites, hatch in about five days. The larvae leave the burrows and wander over the skin looking for shelter and food. Both larvae and nymphs enter skin follicles of the host. The developmental period from egg to adult is from four to six days long. Although males make short burrows, they spend most of their time on the skin surface searching for unfertilized females. Unfertilized females make short burrows in which they stay for a day or two. Mating probably takes place on the skin surface. To form its burrow, the fertilized female cuts its way into the skin of the host using its mouthparts and legs. Females lay single eggs a few hours after forming tunnels and then at intervals of two to three days for about one or two months stringing them out behind as they expand their burrows. They produce 10 to 25 eggs each before dying.

Control. The modern way of treating for both hog mange and hog lice is by dipping or spraying infested herds with certain of the new synthetic insecticides. A greater number of insecticides are recommended for louse control than for mange control. Dips of 0.06 per cent lindane or 0.6 per cent toxaphene are effective for control of both conditions. Two immersions seven to ten days apart are advisable for treatment of mange. Sprays of 0.06 per cent lindane, 0.6 per cent toxaphene, or 0.5 per cent malathion are also effective if thoroughly applied.

If only lice are a problem, a number of insecticides are recommendable: crotoxyphos, coumaphos, dioxathion, malathion, ronnel, methoxychlor, or toxaphene. A very simple way to control hog lice is to scatter 5 per cent ronnel granules over the floor area used by the animals. Apply granules at the rate of $\frac{1}{2}$ lb over 100 sq ft of bedding.

Treatments are ideally made in early fall so that infestation of fall and spring pig crops is prevented. A farmer should make sure to treat all hogs on his place, as one untreated animal will eventually reinfest the others. Because mange mites can survive for four to six weeks off the host in moist bedding, control suggestions sometimes include the cleaning and disinfecting of shelters and houses with chemicals, steam, or boiling water.

It is advisable to keep treated hogs out of cold, inclement weather until they are completely dry to prevent chilling, and out of sunlight to prevent skin blistering.

SELECTED REFERENCES

Baumhover, A. H., "Eradication of the Screwworm Fly," *J. Amer. Med. Assoc.*, 196:240–8 (1966).

Bishopp, F. C., and W. E. Dove, *The Horse Bots and Their Control*, USDA Farmers' Bul. 1503, 1926.

Bishopp, F. C., E. W. Laake, H. M. Brundrett, and R. W. Wells, *The Cattle Grubs or Ox Warbles, Their Biologies and Suggestions for Control*, USDA Dept. Bul. 1369, 1926.

Bruce, W. G., *The History and Biology of the Horn Fly*, Haematobia irritans (*Linnaeus*); *with Comments on Control.* N. C. Agr. Exp. Sta. Tech. Bul. 157, 1964.

Camin, J. H., and W. M. Rogoff, *Mites Affecting Domesticated Mammals*, So. Dak. Agr. Exp. Sta. Tech. Bul. 10, 1952.

Gregson, J. D., *The Ixodoidea of Canada*, Can. Dept. Agr. Publ. 930, 1956.

Hargett, L. T., and R. L. Goulding, *Studies on the Behavior of the Horn Fly* Haematobia irritans (*Linn.*), Ore. Agr. Exp. Sta. Tech. Bul. 61, 1962.

Haufe, W. O., *Control of Cattle Lice*, Can. Dept. of Agr. Publ. 1006, 1960.

James, M. T., *The Flies That Cause Myiasis in Man*, USDA Misc. Publ. 631, 1947.

Kemper, H. E., *Sheep Scab*, USDA Farmers' Bul. 713, 1952.

———, *The Sheep Tick and Its Eradication*, USDA Farmers' Bul. 2057, 1953.

Kemper, H. E., and H. O. Peterson, *Cattle Lice and How to Eradicate Them*, USDA Farmers' Bul. 909, 1953.

———, *Cattle Scab and Methods of Control and Eradication*, USDA Farmers' Bul. 1017, 1953.

———, *Hog Lice and Hog Mange: Methods of Control and Eradication*, USDA Farmers' Bul. 1085, 1952.

———, *The Spinose Ear Tick and Methods of Treating Infested Animals*, USDA Farmers' Bul. 980, 1953.

Lancaster, J. L., *Cattle Lice*, Ark. Agr. Exp. Sta. Bul. 591, 1957.

Ode, P. E., and J. G. Matthysse, *Bionomics of the Face Fly*, Musca autumnalis *De Geer*, Cornell Univ. Agr. Exp. Sta. Memoir 402, 1967.

Roberts, F. H. S., *Insects Affecting Livestock* (Sydney, Australia: Angus and Robertson, 1952).

Rogoff, W. M., *Cable-Type Backrubbers for Horn Fly Control on Cattle*, So. Dak. Agr. Exp. Sta. Bul. 418, 1952.

Scharff, D. K., *Cattle Grubs—Their Biologies, Their Distribution and Experiments in Their Control*, Mont. Agr. Exp. Sta. Bul. 471, 1950.

Schwardt, H. H., and J. G. Matthysse, *The Sheep Tick*, Melophagus ovinus *L.*: *Materials and Equipment for Its Control*, Cornell Univ. Agr. Exp. Sta. Bul. 844, 1948.

USDA, *Anaplasmosis in Cattle*, USDA Leaflet 437, 1969.

USDA, *Cattle Lice: How to Control Them*, USDA Leaflet 456, 1969.

USDA, *Facts About the Screwworm Barrier Program*, USDA ARS 91-64-1, 1969.
USDA, *Horse Bots: How to Control Them*, USDA Leaflet 450, 1973.
USDA, *How to Control Cattle Grubs*, Leaflet 527, 1972.
USDA, *Portable Dipping Vat for Sheep*, USDA Misc. Publ. 1103, 1968.
USDA, *The Fight Against Cattle Fever Ticks*, USDA PA-475, 1968.
USDA, *Eradicating Sheep Scabies*, USDA PA-458, 1966.
USDA, *Manual on Livestock Ticks*, USDA ARS 91–49, 1965.
USDA, *Wood Ticks: How to Control Them in Infested Places*, USDA Leaflet 387, 1963.
USDA, *Eradicating Cattle Scabies*, USDA PA-471, 1961.
USDA, *Demodectic Mange in Cattle*, USDA Leaflet 438, 1958.
USDA, *Horn Flies on Cattle: How to Control Them*, Leaflet 388, 1968.

chapter 21 / **DEANE P. FURMAN**

POULTRY INSECTS AND RELATED PESTS

Poultry suffer from the attacks of a wide variety of arthropods. These range from temporary nonspecific parasites (e.g., mosquitoes) to pests such as chewing lice, which spend their entire life on the bird and are normally closely restricted to poultry or related wild fowl.

THE PESTS

Chewing lice, belonging to the order Mallophaga, are the most common and widely distributed poultry insects. Nine species of lice are reported from chickens in the United States, with two additional species reported from neighboring regions to the south. Of these, the **chicken body louse** (Fig. 21:1A), **chicken head louse** (Fig. 4:8B), and **shaft louse** (Fig. 21:1B) are the most common and troublesome, with the **fluff louse, wing louse**, and **large chicken louse** of occasional importance. Frequently two or three species are found infesting birds simultaneously.

Of the fleas attacking poultry, the **sticktight flea** is frequently encountered in the southern United States from the Atlantic to the Pacific coast (Fig. 21:9). Two others, the **European chicken flea** and **western chicken flea**, are of occasional importance in North America.

The **fowl tick** is a voracious pest of chickens and turkeys in the southern United States (Fig. 21:12). It is particularly pestiferous in warm, arid regions of the Southwest, where tremendous populations of the pest may build up in a short time during summer.

Of the many species of mites attacking fowl, three species are of primary significance in North America. The most commonly encountered species is

Figure 21:1. Lice infesting chicken. A, the chicken body louse, *Menacanthus stramineus* (Nitzsch), and egg masses on feather; B, the shaft louse, *Menopon gallinae* (L.), as they appear normally on a feather shaft. *Courtesy Steve Moore III, University of Illinois, Urbana.*

the **chicken mite**, which normally spends most of its life hiding in cracks of the poultry housing, only emerging at night to suck blood from birds (Fig. 21:2). The **northern fowl mite**, by contrast, normally spends its entire life on the avian host sucking blood at will (Fig. 21:14). In most parts of the United States it ranks a close second in importance to the chicken mite. The **tropical fowl mite** closely resembles the northern fowl mite in appearance and habits, but it does not normally spend its entire life on the bird. It wanders about on

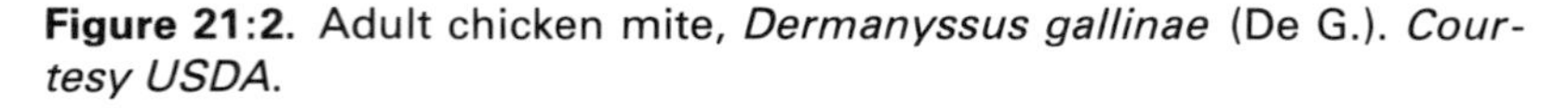

Figure 21:2. Adult chicken mite, *Dermanyssus gallinae* (De G.). *Courtesy USDA.*

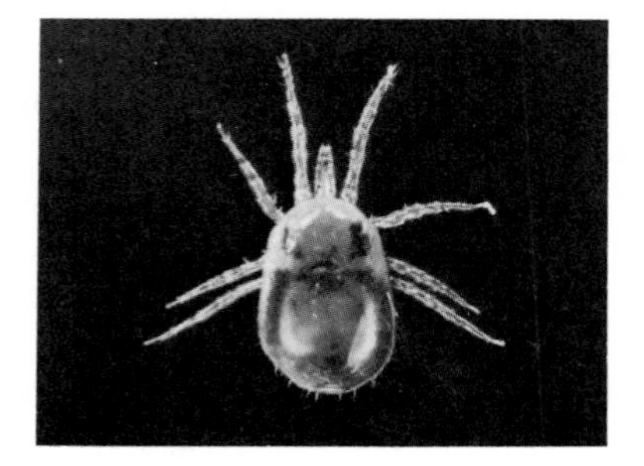

the poultry housing, where it lays its eggs. The feeding stages may attack the birds at any time of the day or night. Less important mites that attack poultry include the **scalyleg mite, depluming mite, airsac mite** (Fig. 21:3), **fowl cyst mite**, and various kinds of **chiggers**, or larvae of Trombiculidae.

The common **bed bug** (Fig. 19:1E) and at least three species of avian bed bugs occasionally attack poultry. These insects have habits similar to that of the fowl tick in that they normally hide in cracks in the housing during the day and attack the birds to suck blood at night.

There are a host of insects that are of varying degrees of economic importance to the poultry industry, but are by no means exclusively pests of poultry. These include blood-sucking reduviid bugs, gnats, and mosquitoes; wound-invading fly maggots, such as the screwworm (Fig. 20:12); house flies and related species which breed in manure; beetles which serve as intermediate hosts of poultry tapeworms.

THE INJURY

If any one feature may be said to be characteristic of successful parasites, it is the insidiousness with which they operate, multiplying frequently to produce tremendous populations before they are detected. During and after the development of heavy infestation of poultry insects and related pests, the damage caused is often difficult to see, because much of it takes place before, or without, accompanying death or marked illness of the birds. The economic

Figure 21:3. Airsac mite, *Cytodites nudus* (Vizioli). Ventral view of male. *From Baker et al., 1956.* A Manual of Parasitic Mites, *by permission National Pest Control Association, Inc.*

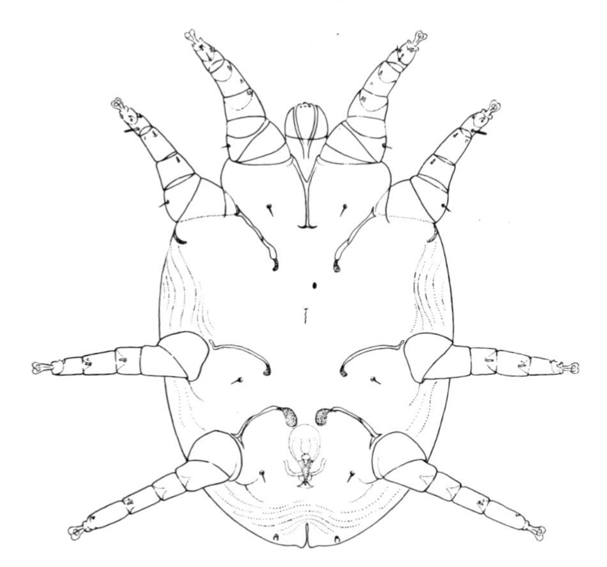

significance of this damage is reflected in such factors as decreased egg production, decreased weight gains, and increased susceptibility to disease agents. An estimate of the average annual cost of external parasites to the poultry industry of the United States was placed at $80 million in 1956.

Poultry lice, with their chewing mouthparts, devour cast-off skin fragments, feathers, and debris adhering to the birds. There is also scratching of the skin by the sharp spines and claws of the insect and, in some cases, by feeding on living skin and young quills. The importance of a few lice lies not so much in the damage they cause, but in their ability to multiply rapidly, with subsequent, serious, accumulative injury. Heavy louse populations produce generalized skin irritation, and scabbing, damaged plumage, and poor condition. Affected birds become restless, lose sleep, and feed poorly. Egg production decreases and weight goes down. Young birds are affected rapidly, becoming droopy and ruffled, developing diarrhea, and often dying.

The blood-sucking poultry mites include the most injurious of external parasites. Their bites are extremely irritating to the birds, and heavily infested birds become scabby, anemic, droopy, and progressively emaciated. Egg production is decreased, and resistance to disease lowered. The fowl tick produces much the same type of damage, but in addition it transmits the causative agent of highly fatal fowl spirochaetosis, a disease now known to occur in North America. On circumstantial evidence the chicken mite has also been suspected of transmitting this pathogen. Piroplasmosis of poultry, caused by *Aegyptianella pullorum*, is transmitted by the fowl tick, but the infection has not been encountered in the United States.

Attacks of fleas and bed bugs, both blood-sucking insects, cause severe irritation and anemia, to which young birds may quickly succumb. The sticktight flea may produce ulcers around the heads of birds which can result in blindness and death (Fig. 21:4).

In the south and central United States, larval trombiculid mites, or chiggers, attach to turkeys that are allowed to range over infested ground during the warm season. Groups of the mites, gathering beneath the wings and on the breast and thighs, irritate the skin and induce the formation of abscesses.

A number of mites with chewing mouthparts invade the skin and even the internal organs of poultry. The scalyleg mite penetrates under the scales of the leg, causing swellings and a rough appearance of the legs and feet due to formation of scales and crusts (Fig. 21:5). Infested birds may be crippled. A closely related form, the depluming mite, burrows into the skin around the feather bases. The resulting intense irritation causes infested birds to pluck out their feathers. The fowl cyst mite invades the loose, subcutaneous tissues of the skin, forming small, caseous or calcified nodules which may reduce the market value of affected fowl. The airsac mite actually invades the deepest portions of the respiratory tract of chickens, but it is evidently well adapted to its host, for little damage has been attributed to it.

Figure 21:4. An infestation of the sticktight flea, *Echidnophaga gallinacea* (Westwood). Note the dark mass of fleas around the eye and on the throat. *Courtesy E. C. Loomis, University of California, Davis.*

Figure 21:5. Legs of a chicken infested with the scalyleg mite, *Knemidokoptes mutans* (R. & L.). *Courtesy USDA.*

The problem created by filth-breeding flies differs from that caused by poultry parasites. Flies do not normally cause direct losses in the form of lower weight gains and decreased egg laying, but they are a nuisance to the poultryman and, most importantly, they create a public health problem for him and his neighbors. Furthermore, three common species of poultry tapeworms reach their avian hosts through house flies. Fly larvae become infected by ingesting tapeworm eggs passed in poultry droppings. Poultry become infected with tapeworms by ingesting larval or adult flies in which young tapeworms have developed to the infective, cysticercoid stage.

Control Through Management

The problem of poultry pest control has been altered considerably in recent years by changing management procedures. In most instances such changes have primarily been directed at something other than poultry pests, and intelligent supervision is necessary to avoid the pitfalls while reaping the benefits associated with use of modern equipment and methods of poultry husbandry.

A few simple precautions will minimize the problems of introduction and control of poultry pests:

1. Isolate replacement birds from the flock for whatever period of time is necessary to free them of those temporary parasites which will leave the birds. For example, in cases of parasitism by fowl tick larvae, a ten-day holding period should free the birds of the pests. The holding crates then may be scalded with steam or hot water, or treated with a suitable insecticide.
2. Treat replacement fowl infested with permanent or semipermanent ectoparasites, such as lice, sticktight fleas, or northern fowl mites, with an insecticide to eradicate the pests before bringing the new birds in contact with the flock.
3. Before partial or complete restocking of the poultry farm, clean housing thoroughly. If pests such as fowl ticks, chicken mites, bed bugs, or fleas are present, treat the housing with a suitable insecticide. Some pests may survive long starvation periods—up to three years for the fowl tick—so that houses that are vacant a few weeks or months may contain hordes of hungry parasites.
4. Guard against introducing pests on used cages or other equipment, or on clothing of humans exposed to infested birds. Even insects such as lice may survive a few days off their hosts.
5. Provide tight, well-constructed housing, free from cracks or loose boards, to aid in the control of those poultry pests that seek shelter in the housing. Easily removable roosts facilitate cleaning and insecticidal treatment. Do not allow poultry to roost in trees or sheds where pests may multiply undetected.

6. In chigger-infested areas, restrict poultry from free range during the months of heavy mite populations. This period will vary from one to two summer months in more northerly areas to practically the entire year in certain subtropical regions.
7. Discourage wild birds from nesting in and around poultry buildings, either by removing their nests as soon as constructed, or by screening out the birds with $\frac{1}{2}$ to 1 in. mesh hardware cloth.

Although house flies and related, filth-breeding flies are by no means restricted to poultry operations, the problems they create on poultry farms have become so critical that they cannot be ignored. The control of such flies by insecticides alone is seldom satisfactory. Effective control requires good sanitation, as represented by the frequent, thorough cleanup and removal of manure, spilled feeds, and other materials that attract flies. Ways to decrease the fly-breeding potential of such materials include the following: (1) open construction of buildings to aid rapid drying of droppings through free air circulation and sunlight; (2) drainage and farm construction to prevent wetting organic waste by leaky watering devices, rain, or ground water; (3) properly maintained, built-up, litter floor, where fly breeding is greatly reduced by rapid turnover of droppings in the dry litter; (4) single-bird, wire-floored cages for laying flocks, rather than group, wire-floored cages; in hot, dry areas, the former procedure results in development of dry cones of manure; (5) compact compost piles with insecticidal treatment of the surface, or use of fly-tight manure boxes or houses; (6) weekly manure distribution in a thin layer on fields; (7) manure removal by a central agency for rapid mechanical composting, or storage in sanitary landfills or pit silos.

Because the life cycles of filth-breeding flies are longer in cool weather than in warm, the weekly disposal period recommended for moist breeding materials may be correspondingly lengthened in cool periods and dispensed with during months of frost or snow.

There is much room for improvement in management practices designed to control house flies. A recent resurgence of interest in the problem has produced some interesting leads. Many large-scale poultry plants now use wire-floored cages constructed so as to permit automatic or semiautomatic removal of manure. Even multilayered banks of cages are provided with dropping boards that are scraped by automatic devices, some as frequently as once every 14 min. Single-layered, wire-bottom cages may be suspended from overhead supports, or back-to-back cages may be supported by a single row of floor-to-cage posts, thus permitting manure scrapers to be drawn down the aisles in uninterrupted fashion (Fig. 21:6).

Biological Control

Research workers, as an alternative, have been investigating the possibilities inherent in biological control, including the introduction into the United States of fly parasites obtained throughout the world. This work is still

Figure 21:6. Wire-bottom cages suspended from over-head supports, a type of construction facilitating waste disposal and fly control. *Courtesy E. C. Loomis, University of California, Davis.*

in the experimental stage, but much information is already available on natural enemies of filth flies in the United States. A type of integrated control has been tried successfully that involves application of insecticides and management of manure so as to favor parasites and predators selectively. Biological control of house fly breeding by means of microorganisms is currently receiving serious attention. Preliminary investigations have demonstrated that application of *Bacillus thuringiensis thuringiensis* spore powder as an additive to chicken feed resulted in 99 per cent reduction of adult fly emergence when 3 g per day were consumed by laying hens.

Chemical Control

Recommendations for use of specific insecticides change from year to year, but most of the biological factors governing our choice remain constant. Thus we may apply contact insecticides to the housing to control fowl ticks, bed bugs, or other pests that rest on or in the housing. Surfaces treated with these materials retain, for weeks or months, a toxic effect for pests contacting them. Such applications of residual insecticides are designed to control existing infestations and to protect against reinfestations.

Materials having a relatively high vapor pressure with accompanying fumigant action on insects may be applied to poultry roosts for control of certain pests remaining permanently on the birds. Thus we may apply nicotine sulfate, in the form of "Black Leaf 40," to roosts for the control of lice, and northern and tropical fowl mites. The warmth of the bodies of roosting fowl increases the fumigant action of the nicotine sulfate. Malathion applied to roosts works in similar fashion. Similarly, recent investigations demonstrate that dichlorvos impregnated in resin strips is effective in controlling poultry lice. The resin strips are applied as leg bands or woven through the meshes of wire-floored cages.

In contrast to the so-called housing and roost treatments, are methods designed for direct application to poultry. The primary objective of such applications is to eliminate those poultry pests which normally spend their life on the birds; a secondary objective is to provide protection from reinfestation from outside sources. This combination of properties is difficult to achieve in view of the fact that many insecticides with residual characteristics may be absorbed through the skin of treated birds, some proving toxic to the birds or resulting in the accumulation of the insecticide in the tissues and even in the eggs. It is to prevent such possibly hazardous residues from reaching the dining room of man that certain of the elaborate safeguards of Public Law 92-516 (Federal Environmental Pesticide Control Act of 1972) have been established.

Properties of some new insecticides have revived interest in the oral route of insecticide application for poultry pest control. This route offers singular advantages over currently used methods; application in the feed would eliminate individual—or even flock—spraying, dusting or dipping, with the attendant drops in egg production. Preliminary experiments reported from California have demonstrated that effective control of the northern fowl mite has been obtained by adding carbaryl or sulfaquinoxaline to chicken feed. No ill effects were produced in the birds.

In applying insecticides around poultry one should follow the precautions given on the insecticide container and avoid contamination of feed and water troughs. Many insecticides are poisonous to man as well as animals if improperly used, so keep containers properly marked and store out of reach of irresponsible individuals.

Control Equipment

Very little special equipment is required in the application of chemicals for control of poultry pests. For small farm flocks and for spot control of pests around larger flocks, a 2 to 5 gal compressed air sprayer is most useful. When used in applying suspension-type sprays, however, the sprayer must be shaken frequently to prevent settling of the insecticide. For larger flocks, a power wheelbarrow sprayer provided with mechanical agitation is most suitable. Either type of sprayer should have an adjustable nozzle to produce a coarse or fine spray. For most operations a power sprayer producing a nozzle pressure of from 50 to 100 psi is satisfactory, but in controlling pests such as fowl ticks in housing, pressures of up to 300 psi are more effective, because they drive the spray into the deep crevices that harbor the pests. Both power and hand sprayers are suitable for treating birds or their housing.

Dusts may be applied directly to poultry by a small, hand-operated, plunger- or puff-type duster provided with an adjustable swivel nozzle and extension tube (Fig. 21:7). These dusters are particularly effective in treating individually caged laying hens. Flock dusting with larger, hand-operated dusters of the rotary or bellows type is also practical, but such equipment fails to provide the penetration of feathers needed by some dusts to assure control of pests. In applying dusts to poultry house litter, the larger hand-operated dusters are useful, but a large shaker can may be used just as effectively.

For applying roost paints or sprays, the simplest device is a paint brush and can. A faster device is a small compressed air sprayer provided with a nozzle producing a flat, fan-shaped spray pattern.

REPRESENTATIVE POULTRY PESTS

For detailed study we have chosen the following four poultry pests: the chicken body louse, the sticktight flea, the fowl tick, and the northern fowl mite.

Chicken Body Louse. *Menacanthus stramineus* (Nitzsch) [Mallophaga:Menoponidae]

The original host of the chicken body louse was probably the wild turkey; the species has been found on this bird and no other wild host. It is now known as a common pest of domestic chickens, as well as turkeys, and is also found on domestic guinea fowls, pea fowls, pheasants, and quail when those birds have been in contact with chickens. The geographic distribution includes North, Central and South America, Hawaii, Europe, East and South Africa, and Australia.

Conflicting opinions appear in the literature concerning the damage caused to poultry by lice. Although the consensus of opinion is to the contrary, various researchers, including some recent ones, believe lice cause no appreciable harm to poultry. It appears, however, that light breeds, such as the White Leghorn, are more susceptible to the effects of louse infestation than some of the heavier breeds, and this may account for the divergence of opinion. Of the several species of poultry lice, the body louse is considered the most harmful to adult birds. In a well-controlled and comprehensive series of tests, investigators in Alabama demonstrated that moderate infestation with the chicken body louse resulted in a 17 per cent decrease in egg production in White Leghorn hens. More heavily infested birds might logically be expected to show an even greater decrease in egg production. In heavy infestations chickens may carry populations well in excess of 8000 lice per individual. The skin of such heavily infested birds becomes irritated and red, with formation of localized scabs and blood clots. In addition to gnawing on the epidermis and feeding on skin fragments, feathers, and debris, the chicken body louse attacks young quills and feeds on blood. Although this louse has been found to be naturally infected with the virus of eastern equine

Figure 21:7. Treating individual birds with a hand duster (two puffs per bird) for control of northern fowl mite. *Courtesy University of California, Berkeley.*

encephalomyelitis, it is not considered to have any epidemiological significance in the transmission of the pathogen.

Description. The adult chicken body louse is a small, yellowish insect 2.8 to 3.3 mm long (Fig. 21:8). The general appearance is that of an insect well adapted for clinging to the skin and feathers of the host. The body is flattened dorsoventrally, with three pairs of short, two-clawed legs projecting horizontally from the thoracic region. The chewing mandibles are located on the ventral surface of the head, invisible from above, and the four-segmented antennae in repose are largely concealed in lateral grooves on the head. Each abdominal segment is provided dorsally with two transverse rows of posteriorly directed setae. Males and females are of similar appearance, but the terminal abdominal segments differ.

The immature stages, or nymphs, resemble the adults in general appearance but are smaller, and the genitalia are undeveloped. The eggs are readily visible to the naked eye and bear characteristic filaments on the anterior half of the shell and on the operculum (Fig. 21:8C).

Life history. The eggs are usually attached to the basal barbs of the feathers (Fig. 21:1A); masses of the eggs may often be seen attached to the small feathers below the vent. The eggs hatch in 4 to 7 days, producing

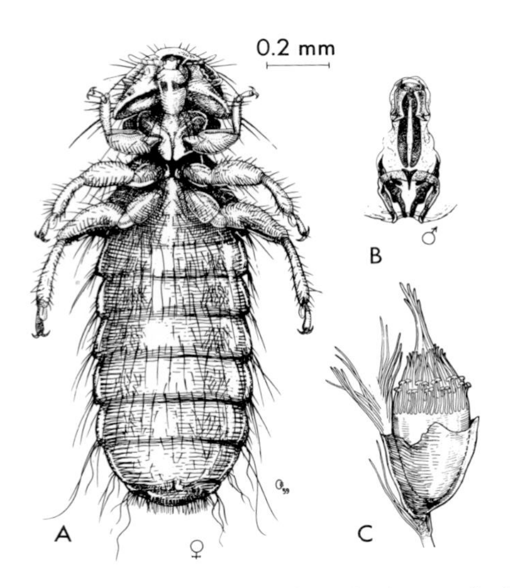

Figure 21:8. Chicken body louse. A, female (ventral view); B, male genitalia; C, egg fastened to feather barb. *Courtesy University of California, Berkeley.*

tiny nymphs. There are three nymphal instars; each instar takes approximately three days to complete its growth. Upon becoming adult the females mate and soon begin to oviposit at a rate of 1 to 4 eggs each day. Females live for about 12 to 14 days and in this time produce an average of 20 eggs each.

Because all stages of the chicken body louse normally remain on or close to the skin of the host, the insect occupies a warm environment despite the season. Consequently large populations may develop very quickly and cause trouble at any time of the year. The louse survives off the host for very short periods, usually less than a day, so transmission from bird to bird primarily follows close contact.

Control. All kinds of poultry lice, and particularly the chicken body louse, are controlled easily with insecticides. Treatment should be made before dense populations of lice develop in order to prevent losses. Birds should be examined frequently by parting the feathers and looking for body lice on the skin around the vent and breast. Other kinds of lice may be found on feathers of the head, body, and wings.

Some of the insecticides used earliest against poultry lice are still considered effective but are not recommended for use today. The major disadvantages of most of them are the residue hazards in treated birds and the amount of time needed for application. This category includes sodium fluoride, sodium fluosilicate dusts, DDT, and other chlorinated hydrocarbon compounds. An early remedy still used occasionally consists in dusting fowl liberally with flowers of sulfur. This provides effective control but occasionally causes severe irritation and blistering of the birds' skin, and likewise causes irritation of the applicator's eyes.

The recent trend in control of poultry lice is that of mass treatment of the flock or application of insecticides to housing, litter, and roosts. These techniques, however, are not new. Roost paints of nicotine sulfate (as Black Leaf 40) have been used for many years to control lice. The material is applied shortly before roosting time at the rate of 1 pt to 200 ft of roost. The fumigant action of the nicotine sulfate kills the lice as the birds roost. A second application seven days following the first is necessary to kill young lice hatching from eggs after the initial treatment. Although the chicken body louse is controlled well by this method, species such as the head louse often survive.

Insecticides for direct application to poultry include water-base sprays of 0.7 per cent malathion, 0.5 per cent carbaryl, or 0.5 per cent stirofos applied at the rate of 1 gal per 100 birds. Turkeys infested with body lice may be treated easily by spraying 0.7 per cent malathion at 60 psi from a row-crop spray boom placed on the ground with nozzles directed upward. Driving the turkeys over the boom and through a curtain of spray treats them rapidly. Dusts for direct application to birds include 4 per cent malathion, 5 per cent carbaryl, or 0.5 per cent coumaphos applied at the rate of 1 lb per 100 birds. The dose for turkeys should be increased proportionately to their size.

Birds held on litter or in range-type pens may be provided with dust bath boxes of about 6 sq ft in area by 1 ft deep, charged with a dust of 4 per cent malathion or 5 per cent carbaryl. Sprays also may be applied directly to deep-litter floors for control of poultry lice. Two per cent malathion sprayed at a rate of 12.5 gal/1000 sq ft of floor space, or 1 per cent carbaryl at the rate of 3.3 gal/1000 sq ft or 0.5 per cent stirofos at the rate of 1 to 2 gal/1000 sq ft are recommended currently.

Specific formulations recommended are subject to change from year to year as new information is obtained. The reader is advised to consult his local state agricultural experiment station or agricultural extension service for current recommendations. Directions to be followed in using insecticides are provided on the manufacturers' labels as specified by law. For example, poultry should not be exposed to carbaryl within 7 days of slaughter. Such directions are designed to protect the safety of the operator, the poultry, and the consumer of the poultry products and should be followed carefully.

The potential of microbial insecticides for control of poultry lice remains to be clarified. Investigators have reported recently that dusts containing spores of *Bacillus thuringiensis* Berliner control lice effectively when applied to chickens.

Sticktight Flea. *Echidnophaga gallinacea* (Westwood) [Siphonaptera:Pulicidae]

Long known as a serious poultry pest in much of subtropical America, the sticktight flea commonly attacks a wide range of hosts. Dogs, cats, rats, and squirrels represent only a few of its mammalian hosts. It also attacks wild birds such as the English sparrow and Brewer's blackbird, and may be disseminated to domestic poultry by such alternative hosts. The geographic range of the flea includes much of the subtropical to tropical regions of the world. In the southern United States, the pest is common, occurring as far north as Oregon on the west coast.

The sticktight flea injures poultry through the semipermanent attachment of female fleas to the skin of the bird, particularly around the eyes, ear lobes, wattles and comb (Fig. 21:4). Irritation by attached fleas produces swelling of the host tissue and often results in ulcerations. Severely infested birds may be blinded, and even light infestations reduce egg production. Young birds may be killed in a short time by heavy infestations.

Description. Adult sticktight fleas are minute, wingless insects about 1 to 1.5 mm long, with laterally compressed bodies of a dark brown color (Fig. 21:9). Of the three pairs of legs the posterior pair are elongated and adapted for jumping. The piercing-sucking mouthparts are prominently placed at the anterior, lower margin of the head. The head, thorax, and abdomen form a compact unit, more or less oval in lateral view. The thoracic segments are greatly reduced, particularly dorsally. The male is readily distinguished from the female by the presence of the coiled, spring-like tendons of the male genital apparatus, visible internally in cleared specimens, and by the characteristic protuberance at the posterior end, representing the protruding genitalia.

The eggs are white and spherical or oval, resembling miniature ping-pong balls. The larvae are slender, white, legless, and wormlike (Fig. 21:10A). They have chewing mouthparts adapted for feeding on solid, particulate materials. Full-grown larvae are 4 mm long. The pupae are of the exarate type; that is, the appendages are free from secondary attachment to the body. In nature the pupa is swathed in a silken cocoon in which are incorporated bits of debris (Fig. 21:10B).

Life history. As with all fleas, only the adults are parasitic, being provided with piercing-sucking mouthparts with which to penetrate skin and suck

Figure 21:9. Sticktight flea adults, female and male. *Courtesy University of California, Berkeley.*

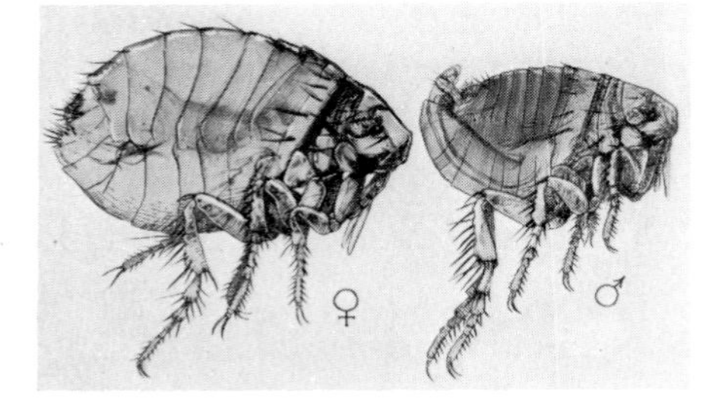

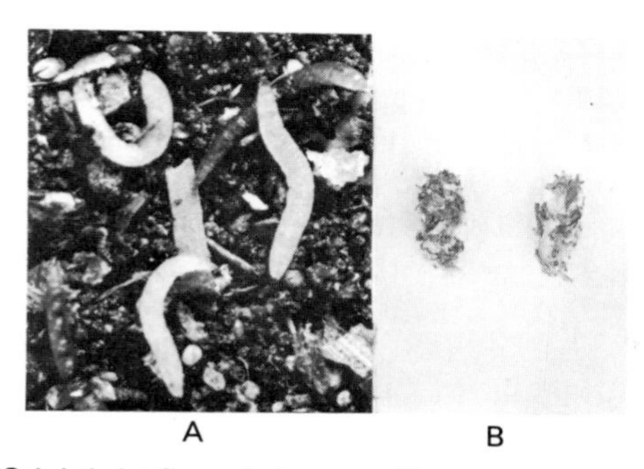

Figure 21:10. Sticktight flea. A, larvae; B, cocoons. *Courtesy USDA.*

blood. Adult sticktight fleas are usually inactive for several days immediately following emergence from the puparium. Following this period they attach to any of a variety of warm-blooded hosts, including poultry. The mouthparts are inserted firmly in the tissues of the host, serving as an attachment organ as well as a food tube through which blood is sucked. Once attached, the fleas may remain in position as long as three weeks, and in this respect they differ from most other fleas, which remain attached for only a few moments at a time. Females will copulate while attached to the host, and egg production commences about a week following their first engorgement. The gravid female lays up to four eggs a day during its feeding period on the host.

The eggs usually drop to the ground unless held by the mass of swollen or ulcerated host tissue. They complete their incubation period in six to eight days at a temperature of 76°F. The wormlike larva is very active and feeds on excrementous matter such as flea feces, solid particles of which it manipulates with its chewing mouthparts. The larvae are found on the floor of infested poultry houses, in poultry nests, and on ground near infested birds. Larvae emerging from eggs that adhere to the bird may feed directly on host tissue, but usually they drop to the ground. After two or three weeks of feeding and developing, the larva becomes full grown and spins a silken cocoon in which the pupal stage is passed. After two to three weeks or more as a pupa, the adult flea emerges. The complete life cycle requires from 30 to 60 days, depending upon the prevailing temperature. Adult fleas may be found in infested debris for five months after all hosts have been removed.

Control. Sparrows, blackbirds, dogs, cats, rats, and, in western areas of the United States, ground squirrels are among the most common hosts introducing sticktight fleas into poultry flocks in the United States. Where possible, measures should be taken to exclude these hosts from the poultry house. Because the immature stages of fleas are passed on the floor, in nests, or on the ground, poultry maintained in wire-floored cages with rollaway-type nests are relatively safe from heavy flea infestations.

Where fleas are breeding, the ground of poultry runs and the floors of houses should be treated thoroughly with insecticide. Flea breeding in soil may be prevented by scattering salt freely about the yards and then thoroughly wetting the soil. Poultry should not be permitted to eat the salt, which is toxic to them. Malathion applied to litter, ground, or floors provides excellent control of flea breeding. It may be applied as a 4 per cent dust at the rate of 5 lb/100 sq ft or as a 0.5 to 1 per cent spray in quantities to wet the treated surfaces. This method of treatment should also provide control of sticktight fleas attached to fowl, without the necessity of individual treatment of the birds. Similarly 5 per cent carbaryl dust may be applied to litter, at the rate of 25 lb/1000 sq ft, or 0.5 per cent coumaphos, at the rate of 50 lb/1000 sq ft. These applications should be repeated once after three to four weeks for malathion or carbaryl or after seven days for coumaphos. Neither malathion nor carbaryl should be applied within seven days of marketing birds.

Where rapid control of sticktight fleas on a few birds is desired, a 2 per cent carbolic ointment or 1 part of sulfur in 5 parts of lard or bland oil may be applied to the masses of attached fleas. Caution must be taken to keep the ointment out of the eyes of the birds.

Fowl Ticks. *Argas* (*Persicargas*) spp. [Acarina:Argasidae]

Until recently it was believed that only one kind of tick, *Argas persicus* (Oken), was a common poultry pest in North America. It was called the fowl tick as well as a variety of other names, such as bluebug, adobe tick, and tampan. As of 1970 we know that this tick actually constitutes a complex of closely related species, including, in the United States, the Persian poultry *Argas*, *A. persicus* (Oken), the North American bird *Argas*, *A. radiatus* Railliet, and Sanchez' western bird *Argas*, *A. sanchezi* Dugès. Because most of the available records on the biology and habits of the fowl tick do not distinguish between these very closely related species they are treated here as a single entity, except where noted otherwise.

The geographic distribution includes Europe, Asia, Africa, America, and Australia; the tick thrives in warm weather and is most troublesome in dry regions. *A. persicus* is an Old World species that appears to be rare in the New World. *A. radiatus* is known from central and southern United States and Mexico. *A. sanchezi* is the common fowl tick of western United States, extending southward into Mexico. Hosts for the fowl tick complex include all species of domestic poultry, in addition to wild birds, such as quail, turkey, dove, vulture, golden crowned sparrow, and owls. On occasion humans may be attacked, particularly if they live close to abandoned roosting or nesting sites of birds.

The fowl tick is a voracious blood sucker in all active stages of its life cycle (Fig. 21:11). Poultry suffering from attacks of ticks show the effects of anemia, becoming pale, weak, emaciated, and diarrheic. The feathers assume a dull, ragged appearance, and egg production is reduced or stopped. Young chicks brought into tick-infested quarters are killed readily, and growth of

Figure 21:11. First instar or seed ticks of the fowl tick attached to skin and feeding on the blood of a chicken. *Courtesy University of California, Berkeley.*

young birds is stunted. Turkeys suffer severely from attacks of the fowl tick.

The fowl tick serves as a vector of *Borrelia anserina*, the causative agent of avian spirochaetosis. This serious poultry disease occurs in the United States as well as in South America, Europe, Africa, India, Java, and Australia. Frequently the disease is transmitted in the absence of ticks, probably through infective bile droppings. The relatively nonpathogenic parasite, *Aegyptianella pullorum*, which has not been reported from America, is also transmissible by the fowl tick.

Description. Mature fowl ticks are flattened dorsoventrally and suboval in outline viewed from above (Fig. 21:12). There is no subdivision of the body into head, thorax, and abdomen as in insects. Females average 8.5 mm long by 5.5 mm wide, and males are slightly smaller. The color varies from light red to dark brown, and the body in general appears leathery and hairless, with more or less distinct cuticular disks arranged around the body margin as well as radially. The piercing-sucking mouthparts are placed anteroventrally and are invisible from above. In common with other argasid ticks, there is very little difference in the gross appearance of the two sexes; the genital aperture in either case is located immediately behind the mouthparts. The eggs of the tick are shiny, reddish brown, and spherical in shape. The larvae are six-legged, active creatures about the size of a pin head before they feed. The two to three nymphal instars each have eight legs; except for the lack of genital

Figure 21:12. Fowl tick adult is ½ in. long and one of the largest poultry parasites. *Courtesy Lorry Dunning University of California, Davis.*

aperture and smaller size each of these instars resembles the adult tick in gross appearance.

Life history. The fowl tick deposits eggs in crevices about the poultry house, or under the bark of trees where these serve as roosting sites for fowl. The female tick produces as many as 900 eggs, laying them in batches numbering up to 180. In the United States, the peak period of egg production is around July. The incubation period of the eggs is from 8 to 11 days at 35°C, but at lower temperatures it may be extended up to three months. The emerging six-legged larvae are ready to feed within a day after hatching, but in the absence of hosts they may survive starvation for several months. Unlike later stages of the tick, the larva remains attached to the avian host for four to nine days, occurring most abundantly on the sparsely feathered parts of the body. During feeding, the larva becomes globose in shape, but following complete engorgement it becomes flattened; at this time it seeks hiding places in crevices near the avian host nest or roost.

Seven to ten days after feeding the larva molts, producing the eight-legged, first nymphal stage. The nymph seeks an avian host at night, feeding to repletion in about two hours. It then hides in crevices for seven to ten days or longer, after which it molts to the second stage nymph; this stage is similar in appearance but larger than the preceding stage. After feeding and resting in a manner similar to that of the first nymphal stage, the second-stage nymph molts to become either a third-stage nymph or the adult male or female tick.

A single mating suffices the female tick for life, but engorgement with blood must precede each successive batch of eggs deposited. Adult ticks normally engorge at night, requiring about one hour on the host. After each feeding they retire to nearby crevices in which to hide. Nymphs and adults may survive starvation up to three years or more.

Control. The general management procedures given earlier should be followed to avoid the introduction of fowl ticks to a poultry flock. It is simpler and less costly to prevent infestation than to eradicate an established population of ticks.

To control fowl ticks with chemicals, it is not necessary to treat the birds. The only ticks normally found on the birds during daylight hours are the larvae, which will drop off within a few days and will be killed by the residual action of acaricides applied to the housing. Before treating the housing, the nest boxes and roosts should be removed from walls to facilitate penetration of sprays into crevices. A coarse, driving acaricidal spray applied under pressures up to 300 psi is recommended for most effective results. Walls, ceilings, roosts, and the outside of nest boxes should be wetted thoroughly with the spray. A 2 per cent carbaryl spray may be used at the rate of 1 gal/1000 sq ft of wooden surface. Malathion or stirofos at 1 per cent concentration may be sprayed at the rate of 5 gal/1000 sq ft.

Control measures effective against the fowl ticks will also eliminate the common chicken mite and bed bugs that hide in cracks of the housing, attacking the birds only for brief periods of feeding.

Northern Fowl Mite. *Ornithonyssus sylviarum* (Canestrini and Fanzago) [Acarina:Macronyssidae]

This tiny mite is a pest of poultry in most temperate regions of the world. In warmer areas the closely related tropical fowl mite usually replaces it.

The northern fowl mite is particularly injurious to domestic chickens and turkeys, to which it may be introduced by a variety of wild bird hosts. The English sparrow and Brewer's blackbird, which frequent poultry yards in the United States, are two common hosts of the mite. Nests of the former, teeming with northern fowl mites, have been observed on rafters and eaves of poultry houses.

As with other serious, external parasites of poultry, infestations of northern fowl mites result in a serious decrease in egg production. An increase of 40 per cent or more in egg production following eradication of infestations is not uncommon. By sucking the blood of avian hosts, the mites produce marked weakening and even death of heavily infested birds. Bloody scabs form around the region of the vent as well as elsewhere on the body. Fluffy feathers around the tail become ragged and unkempt, matted with masses of mite eggs and feces, as well as with scabs (Fig. 21:13).

Heavy infestations of the northern fowl mite are not invariably associated with decreases in egg production. Research is needed on the effects of diet, hormone balance, genetic and other factors on the reaction of poultry to these pests.

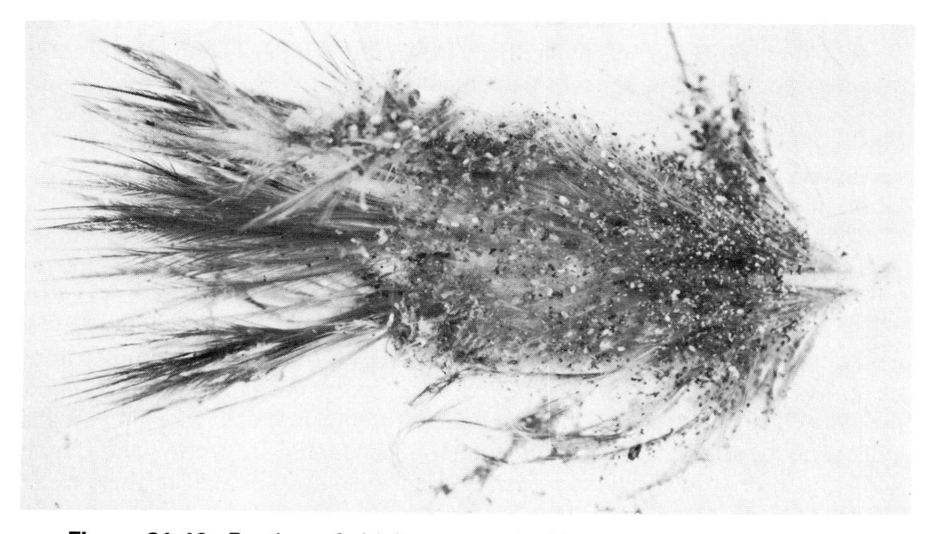

Figure 21:13. Feather of chicken matted with various stages and feces of the northern fowl mite. *Courtesy University of California, Berkeley.*

Viruses of western equine and St. Louis encephalitis as well as Newcastle disease have been recovered from northern fowl mites, but there is no indication that the viruses persist in the mites for any length of time or that they are important in the epidemiology of these diseases.

Description. The gross appearance of the northern fowl mite is similar to that of the tropical fowl mite and the common chicken mite. All are small, active, eight-legged organisms as adults, with a single dorsal plate covering only part of the body. The terminally located mouthparts are borne on a capitulum and are adapted for piercing and sucking by means of slender, paired, retractile chelicerae. Females of the northern fowl mite range from 0.6 to 0.75 mm long and up to 0.5 mm wide (Fig. 21:14). The color varies

Figure 21:14. Northern fowl mite, *Ornithonyssus sylviarum* (C. & F.), dorsal and ventral views of female. *From Baker et al., 1956,* A Manual of Parasitic Mites, *by permission National Pest Control Association, Inc.*

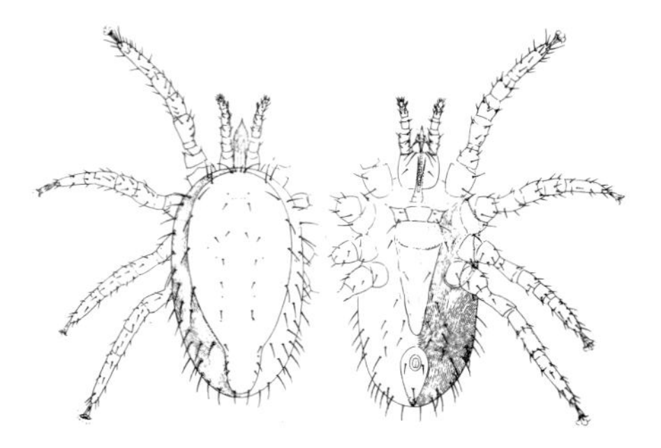

from bright red to black with white flecks, depending on the time elapsed since feeding. The northern fowl mite may be differentiated from the closely related species we have discussed by the presence of a single pair of posterior, dorsal plate setae extending to, or beyond, the tip of the plate. In addition, the northern fowl and chicken mites bear two pairs of sternal plate setae as opposed to three pairs on the tropical fowl mite. The chelicerae of the two fowl mites bear prominent, paired, scissorslike digits, whereas the chelicerae of the chicken mite are long and needlelike, with only rudimentary digits.

Life history. All five stages in the life cycle of the northern fowl mite are normally passed on the feathers and skin of birds. In heavy infestations, however, mites may be found in nests and surrounding structures of the housing. The warmth of newly laid chicken eggs attracts the mites from the feathers of the laying hen, and often the first indication of an infestation noted by the poultryman is the presence of dark red to black mites crawling on the eggs (Fig. 21:15).

The eggs of the northern fowl mite are sticky when laid and usually remain attached to feathers on the bird, often hatching in less than a day. The six-legged larva, a nonfeeding stage, molts to produce an eight-legged protonymph in about one day at moderate temperature. The protonymph feeds two or more times by penetrating the host's skin to suck blood, finally

Figure 21:15. Northern fowl mites and chicken mites crawling on warm, newly laid egg, an indication of heavy mite infestation. *Courtesy University of California, Berkeley.*

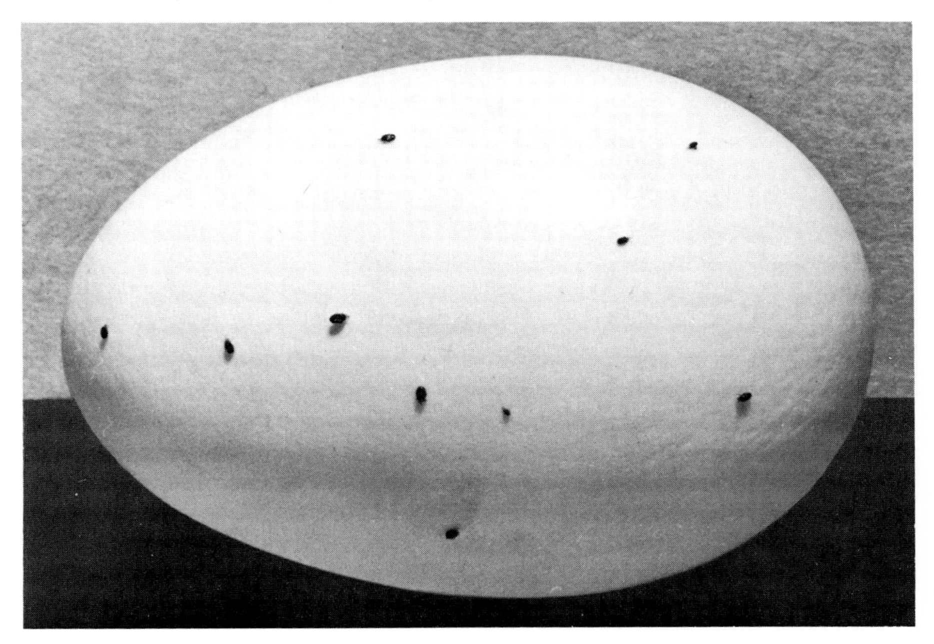

engorging in about one to two days. In another two days it molts to produce the eight-legged, nonfeeding deutonymph, which in turn molts in about one day to produce an adult male or female. Adult mites are blood feeders. The complete life cycle requires less than a week but may be prolonged by unfavorable conditions. Mites removed from the host usually die within ten days, but some may survive up to six weeks.

The northern fowl mite will feed on the mouse and rabbit; hence, rodent-infested premises may serve to prolong an infestation in an otherwise empty poultry house. Although the human being does not appear to be a suitable host, mites will occasionally bite man, causing some itching.

Control. Many materials and techniques of application have been described for control of the northern fowl mite. Several are effective, but their applicability differs, depending upon such variables as prevailing climate, type of housing, and size of the poultry flock.

Northern fowl mite infestations often may be detected in scattered foci in a large poultry plant before a generalized infestation occurs. Usually the mites are noted crawling on newly laid eggs of infested birds. Spot treatment of infested birds and those in nearby cages often is sufficient to check the infestation without need to treat the entire flock.

Insecticide sprays and dusts of malathion, stirofos, and carbaryl are recommended currently. Sprays of 0.7 per cent malathion, 0.5 per cent carbaryl or 0.5 per cent stirofos may be applied directly to birds at the rate of 1 gal per 100 birds, taking care to direct the spray above and below the birds with particular attention to the vent region. Dusts of 4 per cent malathion, 5 per cent carbaryl, or 3 per cent stirofos may be applied by a puff, rotary, or mechanical duster, applying about 0.5 oz per chicken. Turkeys should receive a proportionately larger amount. Sprays may be applied to litter flooring using 2 per cent malathion at the rate of 12.5 gal/1000 sq ft, 1 per cent carbaryl at a rate of 3.3 gal/1000 sq ft, or 0.5 per cent stirofos at a rate of 1 to 2 gal/1000 sq ft of litter. Dust bath boxes may be used as described earlier, using 4 per cent malathion, 5 per cent carbaryl or 3 per cent stirofos. A minimum of 7 days should elapse following any exposure of poultry to carbaryl before birds are slaughtered. Roosts may be painted with 40 per cent nicotine sulfate (Black Leaf 40) using 1 pt per 150 to 200 ft of roosts.

SELECTED REFERENCES

Anderson, J. R., A. S. Deal, E. F. Legner, E. C. Loomis, and M. H. Swanson, *Fly Control on Poultry Ranches*, Univ. Calif. Agric. Ext. Leaflet AXT-72, 1968.

Baker, E. W., T. M. Evans, D. J. Gould, W. B. Hull, and H. L. Keegan, *A Manual of Parasitic Mites of Medical or Economic Importance*, National Pest Control Association, Tech. Publ., 1956.

Bishopp, F. C., and H. P. Wood, *Mites and Lice on Poultry*, USDA Farmers' Bul. 801, 1931.

Briggs, J. D., "Reduction of Adult House-Fly Emergence by the Effects of *Bacillus* spp. on the Development of Immature Forms," *J. Ins. Pathol.*, 2:418–32 (1960).

Cameron, D., "The Northern Fowl Mite (*Liponyssus sylviarum* C. and F. 1877)," *Canad. J. of Research D.*, 16:230–54 (1938).

Chamberlain, R. W., and R. K. Sikes, "Laboratory Rearing Methods for Three Common Species of Bird Mites," *J. Parasitol.*, 36:461–5 (1950).

Crutchfield, C. M., and H. Hixson, "Food Habits of Several Species of Poultry Lice with Special Reference to Blood Consumption," *Florida Entomologist*, 26:63–6 (1943).

Emerson, K. C., "Mallophaga (Chewing Lice) Occurring on the Domestic Chicken," *J. Kans. Ent. Soc.*, 29:63–79 (1956).

Furman, D. P., and W. S. Coates, Jr., "Northern Fowl Mite Control with Malathion," *Poultry Sci.*, 36:252–5, 1957.

Furman, D. P., R. D. Young, and E. P. Catts, "*Hermetia illucens* (Linnaeus) as a Factor in the Natural Control of *Musca domestica* Linnaeus," *J. Econ. Ent.*, 52(5):917–21 (1959).

Gless, E. E., and E. S. Raun, "Insecticidal Control of the Chicken Body Louse on Range Turkeys," *J. Econ. Ent.*, 51:229–32 (1958).

Hoffman, R. A., and R. E. Gingrich, "Dust Containing *Bacillus thuringiensis* for Control of Chicken Body, Shaft and Wing Lice," *J. Econ. Ent.*, 61(1):85–8 (1968).

Kohls, G. M., H. Hoogstraal, C. M. Clifford, and M. N. Kaiser, "The subgenus *Persicargas* (Ixodoidea, Argasidae, Argas). 9. Redescription and New World records of *Argas* (*P.*) *persicus* (Oken), and resurrection, redescription, and records of *A.* (*P.*) *radiatus* Railliet, *A.* (*P.*) *sanchezi* Dugès, and *A.* (*P.*) *miniatus* Koch, New World ticks misidentified as *A.* (*P.*) *persicus*. *An . Ent. Soc. America* 63(2):590–606 (1970).

Kraemer, P., and D. P. Furman, "Systemic Activity of Sevin in Control of *Ornithonyssus sylviarum* (C. and F.)," *J. Econ. Ent.*, 52(1):170–1 (1959).

Legner, E. F., and G. S. Olton, "The Biological Method and Integrated Control of House and Stable Flies in California," *Calif. Agricul.*, 22:2–4 (1968).

Loomis, E. C., "Avian Spirochaetosis in California Turkeys," *Am. J. Vet. Res.*, 14:612–15 (1953).

Loomis, E. C., E. L. Bramhall, J. R. Allen, R. A. Ernst, and L. L. Dunning, "Effects of the Northern Fowl Mite on White Leghorn Chickens," *J. Econ. Ent.*, 63(6):1885–9 (1970).

Matthysse, J. G., "External Parasites of Poultry," in Hofstad et al., eds., *Diseases of Poultry*, pp. 793–843 (Ames, Iowa: Iowa State Press, 1972).

Medley, J. G., and E. Ahrens, "Life History and Bionomics of Two American Species of Fowl Ticks (Ixodoidea, Argasidae, *Argas*) of the Subgenus *Persicargas*," *Ann. Ent. Soc. Amer.*, 63:1591–4 (1970).

Parman, D. C., "Biological Notes on the Hen Flea (*Echidnophaga gallinacea*)," *J. Agric. Res.*, 23:1007–9 (1923).

Reid, W. M., R. L. Linkfield, and G. Lewis, "Limitations of Malathion in Northern Fowl Mite and Louse Control," *Poultry Sci.*, 35:1397–8 (1956).

Reis, J., and P. Nobrega, *Tratado de Doencas das Aves*, Vol. III, Publ. São Paulo, Brazil: Melhoramentos (1956).

Roberts, I. H., and C. L. Smith, *Poultry Lice*, USDA Yrbk. Agric., 1956.

Rodriguez, J. L. Jr., and L. A. Riehl, "Results with Cockerels for Housefly Control in Poultry Droppings," *J. Econ. Ent.*, 52(3):542–3 (1959).

Rodriguez, J. L., Jr., and L. A. Riehl, "The Malathion Dust-Bath for Control of Five Species of Lice on Chicken," *J. Econ. Ent.*, 53(2):328 (1960).

Sikes, R. K., and R. W. Chamberlain, "Laboratory Observations on Three Species of Bird Mites," *J. Parasitol.*, 40:691–7 (1954).

Stockdale, H. J., and E. S. Raun, "Biology of the Chicken Body Louse, *Menacanthus stramineus*," *Ann. Ent. Soc. Amer.*, 58:802–5 (1965).

Warren, D. C., R. Eaton, and H. Smith, "Influence of Infestations of Body Lice on Egg Production in the Hen," *Poultry Sci.*, 27:641–2 (1948).

chapter 22 / **WM. M. ROGOFF**

INSECTS OF MEDICAL IMPORTANCE

Entirely apart from the effects of insects on crops, livestock, timber, and other products are the effects of insects on man himself. Here we leave what has primarily concerned food and fiber production and enter the area of medicine and public health. Many of the insects are the same, and many of the procedures for their control do not differ from those described earlier in this text. What has changed is the point of view and, as a result, the assessment of the relative importance of the species involved. A few mosquitoes on a cow cause considerably less distress to a man than the same number on himself. Disease or death is tolerated at a much lower level in humans than in livestock. The disease-carrying potential of many insects provides the impetus for control efforts that would otherwise not be taken.

THE PESTS

The number of insects and other Arthropods that are of medical importance is very large, but the number of orders involved is relatively small (Fig. 22:1). By far the largest number of these pests fall into the order Diptera; and the arachnid order Acarina is second only to the Diptera in the number of species of medical importance. Relatively small numbers of insects of medical importance belong to the Anoplura, Siphonaptera, Hemiptera, Hymenoptera, Coleoptera, and Orthoptera.

Among the Diptera are the house fly, the various myiasis-producing flies such as blow flies (Fig. 20:12) and bot flies, and the many disease-carrying and discomfort-producing mosquitoes, sand flies, black flies (Fig. 22:2), eye gnats, horse flies, deer flies, and tsetse flies.

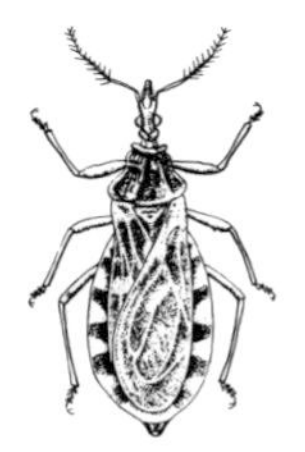

Conenose Bugs
Transmit: Chagas' disease.

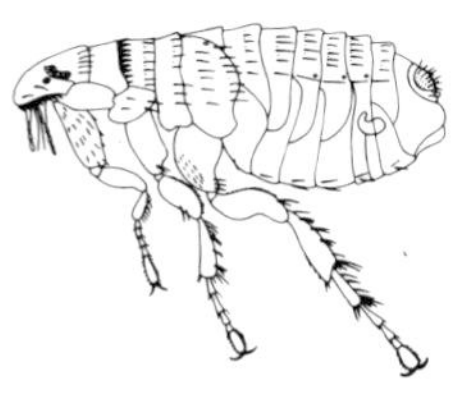

Fleas
Transmit: plague, endemic typhus, dog tapeworm.

Lice
Transmit: relapsing fever, epidemic typhus fever, trench fever.

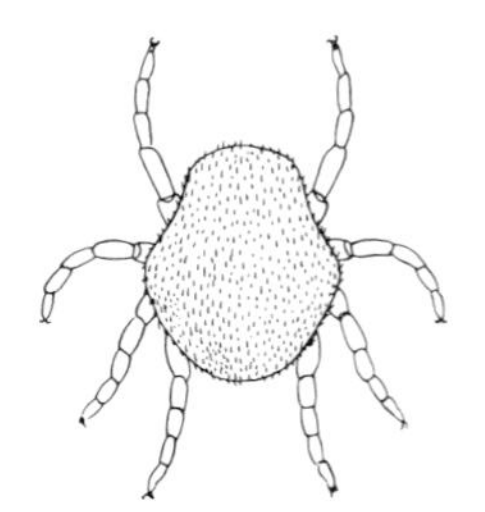

Softbacked Ticks
Transmit: relapsing fever;
Cause: tick paralysis.

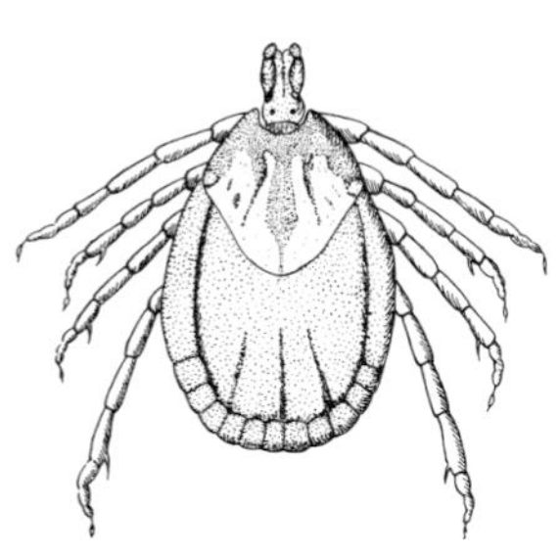

Hardbacked Ticks
Transmit: Rocky Mountain spotted fever, "Q" fever, tularemia, Colorado tick fever;
Cause: tick paralysis.

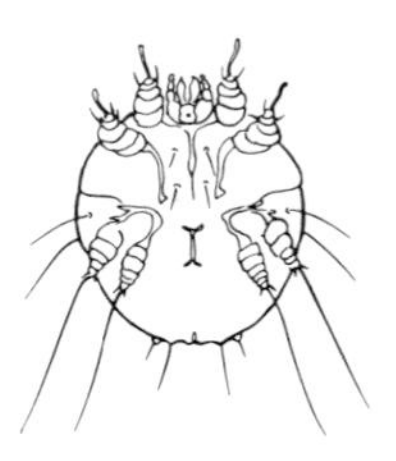

Mites
Transmit: tsutsugamushi (scrub typhus);
Cause: dermatitis.

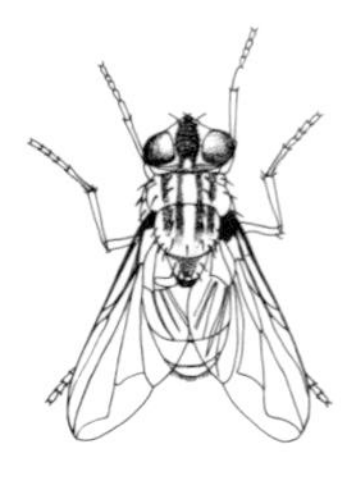

Nonbiting Flies
Transmit: yaws, typhoid fever, dysenteries, cholera, conjunctivitis;
Cause: myiases.

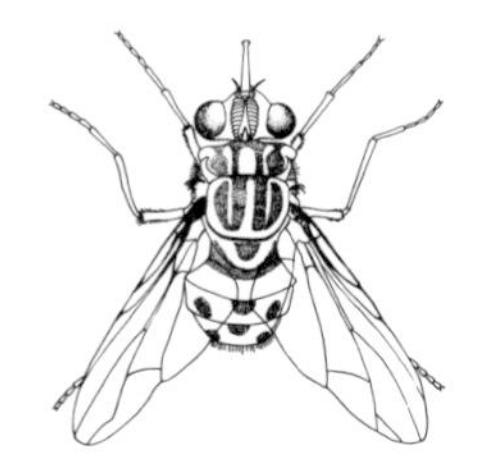

Biting Flies
Transmit: tularemia, sandfly fever, onchocerciasis, African sleeping sickness, kala-azar, bartonellosis.

Mosquitoes
Transmit: malaria, yellow fever, dengue, elephantiasis, encephalitis.

Figure 22:1. Types of arthropods that transmit pathogens of human diseases. *First two rows, courtesy USDHEW, last row, original.*

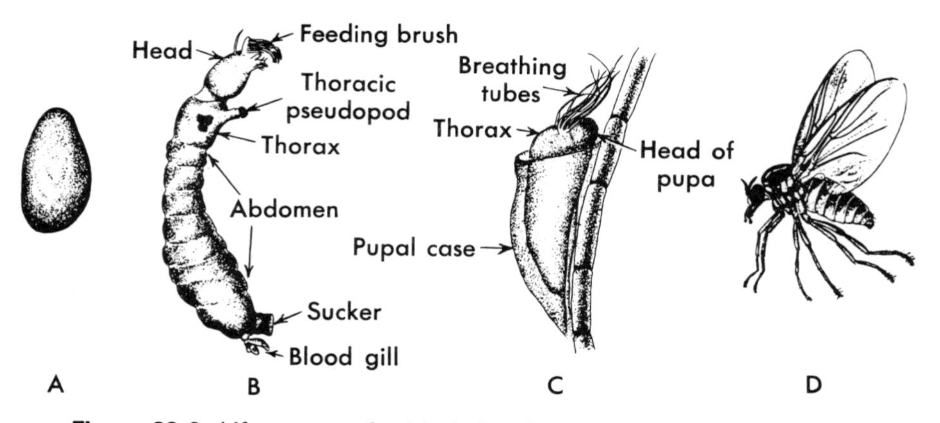

Figure 22:2. Life stages of a black fly, *Simulium*. A, egg; B, larva; C, pupa; D, adult. Species of this genus are intermediate hosts for the filarial worm that causes onchocerciasis in parts of Africa, Mexico, and Central America. *Courtesy U.S. Naval Medical School, Bethesda, Md.*

Although several species of house flies occur in various parts of the world, the common house fly, *Musca domestica* L., is the one most likely to be encountered (Fig. 22:8). Its relationship to transmission of disease-causing organisms is essentially mechanical, but its habits make the insect ideal for transmitting many pathogens of both major and minor importance.

A common medical and veterinary phenomenon is the invasion of body tissues by dipterous larvae of various kinds. Such invasions are generically called **myiasis**. Myiasis-producing species may be obligatory or facultative. **Obligatory myiasis** producers are species whose larvae are found only in living tissues (Fig. 20:4). **Facultative myiasis** forms are those species whose larvae may breed in nonliving organic matter but which occasionally lay eggs or deposit larvae in tissues of man or other animals. **Accidental facultative myiasis** producers are not clearly distinguishable from other facultative forms but usually are considered to include species whose larvae may be ingested by the host along with food, or in a similar unspecific manner. Over 20 species in at least 6 families are obligatory myiasis producers, with larger numbers of species in each of the two facultative categories.

The **torsalo** or human bot fly (Fig. 22:3) is common in the tropical regions of South and Central America and affects wild and domestic mammals in addition to human beings. It has a most unusual method of distribution. The gravid female fly captures various mosquitoes or other arthropods likely to visit a host and deposits one or more eggs on the abdomen of its captive. The arthropod then carries the bot fly egg until it settles upon a mammal. Heat from the body of the host initiates hatching of the egg and subsequent invasion of the host by the larvae.

Mosquitoes (Figs. 22:9 and 22:10) are common throughout the world, from the tropics to the arctic. These insects, by virtue of their blood-sucking

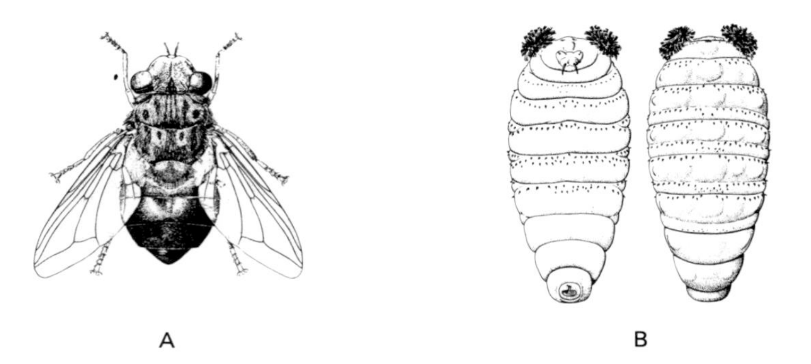

Figure 22:3. The torsalo or human bot fly, *Dermatobia hominis* (Linnaeus, Jr.). A, adult; B, fully developed larvae. *Courtesy USDA.*

habits, enter into disease transmission in a highly specific manner. Among the more important diseases carried by mosquitoes are yellow fever, malaria, dengue, haemorrhagic fever ("breakbone fever"), filariasis, and various types of viral encephalitis. In addition to their importance as carriers of disease, mosquitoes are of concern because of the annoyance they cause in biting.

Two species within the order Anoplura affect man. There are two varieties of one species—*Pediculus humanus humanus* L., the **body louse**, and *P. humanus capitis* De G., the **head louse**—and the other species, *Pthirus pubis* (L.), the **crab louse** (Fig. 22:4). *P. humanus* is an important agent involved in the transmission of trench fever, relapsing fever, epidemic typhus, and to a lesser degree, endemic typhus.

The Hemiptera include several important forms. Bed bugs, whose bites are extremely irritating, are ancient associates of man, found throughout the temperate and tropical regions of the world. Assassin bugs of many genera feed on blood and cause painful wounds. Some species, as in the genus *Triatoma*, are vectors of a flagellate protozoan that causes Chagas' disease.

Figure 22:4. Human lice. A, body louse and head louse are similar in appearance; B, crab louse. *Courtesy USDHEW, Public Health Service.*

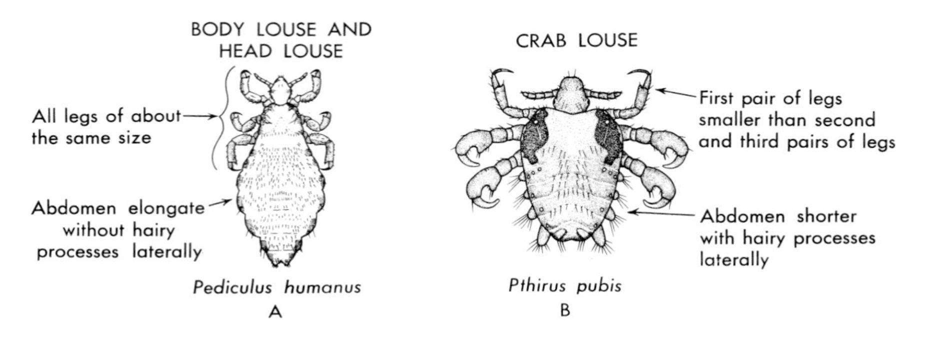

The stings of Hymenoptera are of considerable medical importance. Bees, wasps, and ants, although not known as vectors of animal diseases, can be highly dangerous to some individuals. Anaphylactic (foreign protein) reactions resulting from their stings can cause serious injury or death. Among the Orthoptera, cockroaches are minor, mechanical carriers of disease.

The order Siphonaptera includes a number of medically important species. Among the most important are the **chigoe**, a tropical and subtropical flea that penetrates the skin; the **human flea** (Fig. 22:5); the **cat flea**; and the **oriental rat flea**. The last three fleas are particularly important in the transmission of plague bacilli, and the oriental rat flea is involved additionally in the transmission of endemic typhus virus.

The order Acarina, the mites and ticks, contains a very large number of important and diverse arthropods of medical importance. Many of these attack humans as well as livestock and wild animals. One of the better known tick-borne diseases is Rocky Mountain spotted fever. Of the numerous important Acarina, this chapter will treat only the **chiggers**, all within the mite family Trombiculidae. In many regions chiggers are of great importance as carriers of disease pathogens or as severe sources of irritation. Other arachnids, such as spiders and scorpions, are of occasional importance. The bites of spiders are of a severity sufficient to give rise to the term **arachnidism**. The bite of the **brown recluse spider** (Fig. 22:6) results in local necrosis of the skin and an ugly scar after slow healing of the wound.

THE INJURY

The most significant way in which insects and other arthropods affect the well-being of man is by the transmission of disease-causing organisms. In some instances this transmission is of a mechanical nature, involving only a simple process of contamination. In other instances the disease organism maintains an intimate relationship which may actually require the arthropod in order to complete a part of its life history.

Mechanical transmission. Mechanical methods of disease transmission are typical of the relationships to human health of the house fly and similar

Figure 22:5. Adult of the human flea, *Pulex irritans L. Courtesy USDA.*

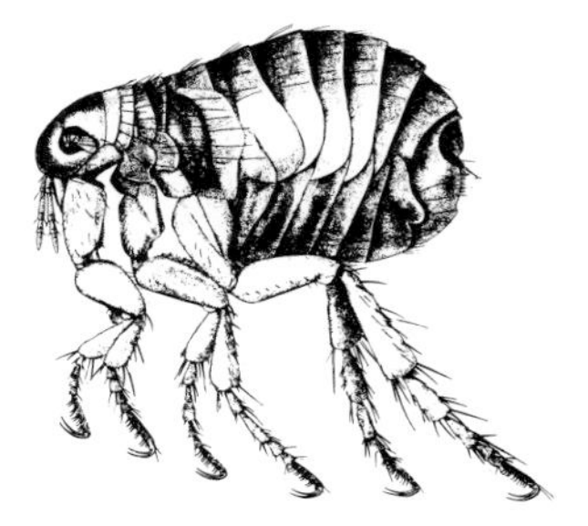

Figure 22:6. Female of the brown recluse spider; note the characteristic violin-like marking on the cephalothorax. *Courtesy University of Missouri, Columbia; photo by Stirling Kyd.*

filth-inhabiting flies, often called synanthropic flies because of their close association with humans. A large number of diseases caused by viruses, bacteria, and protozoa are transmitted by flies from one infested person to another and from contaminated organic matter to humans. In most cases the relationship is based on the habits of the individual fly in exploring a great variety of organic materials and mechanically moving bits of this contamination from one site to another. In some cases the relationship starts with the larva in its breeding medium and continues through pupation, with the adult fly effecting the final transfer. Agents of human disease that are transmitted by flies include those of typhoid, dysentery, trachoma, and to a lesser degree such diseases as leprosy, tularemia, cholera, yaws, tuberculosis, anthrax, trypanosomiasis, leishmaniasis, diphtheria, and poliomyelitis. Various species of flies have been implicated in the transmission of food-poisoning organisms such as *Staphylococcus* and *Salmonella* and of various species of *Streptococcus*. In none of these instances are flies necessary as intermediaries for the normal transmission of the pathogens.

Biological transmission. The mosquito-borne diseases include some that are transmitted mechanically, but include also classical examples of intimate biological relationships between the disease organism, the mosquito, and the human. Among the important mosquito-borne diseases are malaria, yellow fever, dengue, filariasis, and encephalitis. Several of these diseases are actually complexes caused by closely related pathogens. For a clear understanding of the relationship of mosquitoes to disease and biological transmission, we shall examine malaria and yellow fever in some detail.

Malaria, a disease known from antiquity, is distributed widely in tropical, subtropical, and temperate regions. Although it is still of tremendous importance, its range has been reduced greatly in recent years, largely by highly organized efforts of national and international forces. As an example of its former widespread nature, one study, published in 1932, indicated that over 17,750,000 cases had been treated in a single year, and that most probably the actual number of cases would be several hundreds of millions. Malaria, once common, is now a rare disease in the United States and in many other formerly malarious countries. Although it is still of great importance in many parts of the world, intensive international efforts at control have been remarkably successful. The disease is caused by protozoans of the genus *Plasmodium*.

Four species of this genus are known to infect man, although three species are responsible for the majority of cases. *Plasmodium* enters the red blood corpuscles, where the individual parasites grow and reproduce asexually (Fig. 22:7). This process eventually destroys many individual red cells and releases large numbers of infective forms that enter other red cells. The multiplication process continues many times and also involves other tissues associated with the circulatory system (the exerythrocytic phase).

Later, cells of a potentially sexual nature, **gametocytes**, are produced. If, during the time gametocytes are in the blood stream, an appropriate species of *Anopheles* mosquito takes a blood meal, a further cycle takes place in the body of the mosquito. The gametocytes participate in a sexual phase in the alimentary canal of the mosquito and then undergo an asexual multiplication process in cysts in the stomach wall. The resultant parasites migrate to the salivary glands of the mosquito, from which they are introduced into the blood stream of the next victim. Thus, although it is possible to transmit malaria mechanically by moving blood from one person to another, the normal method of transmission involves an active biological participation by the mosquito.

Even though malaria has been known for many years as one of the most important human diseases, details of its nature and transmission have been elucidated only in modern times. The existence of the parasite was discovered by Laveran in 1880. The relationship of mosquitoes to malaria transmission was suggested by Manson in 1894. Within the next four to five years details of the life cycle and experimental transmission of the parasite were

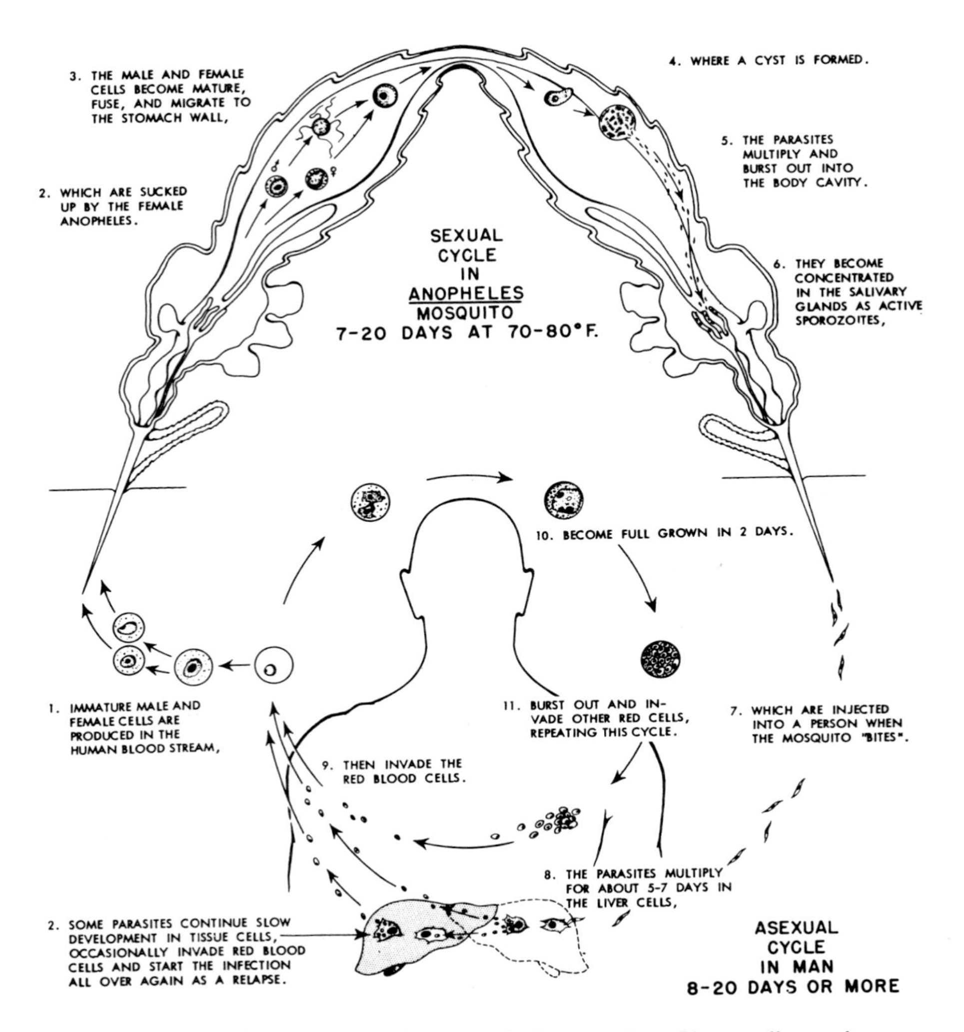

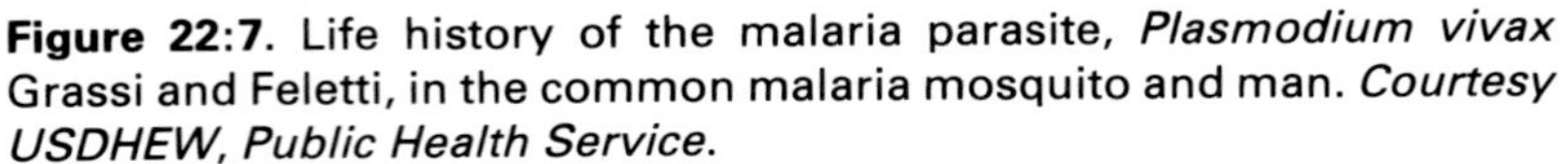

Figure 22:7. Life history of the malaria parasite, *Plasmodium vivax* Grassi and Feletti, in the common malaria mosquito and man. *Courtesy USDHEW, Public Health Service.*

demonstrated by Ross, Grassi, and others. Intelligent efforts at malaria control, therefore, have come only in the twentieth century.

Yellow fever, first described in the seventeenth century, is one of a large number of arthropod-borne viruses, referred to collectively as **arboviruses**. Apparently of African origin, the disease is an acute infectious jaundice caused by a virus which, in the typical urban epidemics, is transmitted from man to man by the yellowfever mosquito *Aedes aegypti*. Severe epidemics were experienced in Philadelphia, Memphis, and New Orleans, but the greater number of epidemics have been seen in tropical and subtropical countries. The mosquito transmission of this disease was investigated by Finlay, starting in 1881 and elucidated in the following twenty years. In the period after 1900, Reed, Carroll, Lazear, and Agramonte, using human volunteers, discovered the essential features necessary for mosquito transmission of the disease. A method of mass vaccination for control of yellow fever was developed in 1932. The combination of appropriate mosquito control measures in the early part of the century, with the mass immunization procedures developed later, has greatly reduced the significance of the classical form of yellow fever.

Insect invasions, bites, and stings. From what has been said, one should not assume that the injury caused by insects, ticks, and mites to the health of man is limited to transmission of infectious agents. An obvious feature is the extreme annoyance caused by arthropods that bite or sting, or that burrow partially or completely into the skin, such as certain mange mites and the chigoe flea. Pestilential arthropods of this nature interfere with the quality of human life over vast areas of the earth.

Insect allergens. Insects and arachnids are involved in a wide variety of allergic diseases that include severe reactions to the stings of bees, wasps, scorpions, and a few species of spiders. Venom, saliva, cuticle, and excreta of many arthropods are allergenic. The effect may be a local reaction of the skin, respiratory or digestive reactions, or general casualties that may result in death.

CONTROL

Control Through Management of Environment

Some arthropod populations may be controlled by environmental management. Most procedures of this nature are aimed at the sites of larval development, but some are directed at adult insects. House fly larvae develop in a wide variety of decaying organic media, some accumulations of which are directly subject to management. Communities located close to areas of cattle, swine, horse, or poultry production or maintenance are frequently subject to infestation from these sources. Sanitation alone rarely provides sufficient control, but excesses of breeding materials greatly complicate the problem of fly control. The routine removal of fly-breeding materials is a necessary prerequisite to control operations. Other large sources of fly-breeding materials that are subject to sanitation efforts are garbage dumps, privies, compost heaps, and other accumulations of organic matter. In situations where dependence is placed on biological control, however, excess depletion of organic matter (such as poultry manure) can be undesirable for maintenance of continuity of predator or parasite populations.

Because water is necessary for the development of the larvae of many insects of medical importance (mosquitoes, for example), major efforts to eliminate or modify such breeding areas are a normal part of control. Filling areas that can be filled or draining appropriate areas are common procedures. Although these are ideal techniques, they have the obvious limitation that most bodies of water are otherwise desirable and cannot be eliminated for purposes of insect control. Advantages can be gained by straightening streams and cleaning and grading shorelines to eliminate quiet pools especially suitable for insect breeding. Mosquitoes generally thrive in water with ample vegetation that affords protection from wave action and predators. Many mosquito control campaigns, such as the project of the Tennessee Valley Authority, utilize vegetation control as the primary technique in reducing mosquito populations. Shoreline vegetation may be controlled by herbicides, by mechanical cutting, or by periodic, controlled fluctuation of lake water levels. In southeast Asia, where stream-breeding species of *Anopheles* are important malaria vectors, periodic flushing of streams is accomplished by the use of automatic siphon dams that release impounded water as it reaches a critical depth.

Management procedures aimed at adult insects are also widely practiced, the most obvious of these being the use of mesh screening on windows and doors. The removal or thinning of vegetation where it serves as adult resting places is also useful, as in mosquito control and in control of mite vectors of scrub typhus.

Biological Control

Biological control is a highly regarded approach, but despite considerable effort extending over many decades, few insects of medical importance are currently subject to adequate management by parasites, predators, or pathogens. Some hymenopterous egg parasites are known to effect partial control of flies. Certain fungi, protozoa, and nematodes, as well as *Hydra* and predaceous beetles, have shown limited promise for fly and mosquito control. Mosquito fish (*Gambusia affinis*) are among the most successful predators and are routinely used in some localities such as rice fields in parts of California.

Chemical Control

Although in many instances management of environment provides a high degree of effectiveness in the control of insects of medical importance, there is a preponderance of situations where the use of chemicals provides the only practical approach. Chemicals particularly help to remedy a problem when too many arthropods are already present.

Chemicals can be used as attractants, repellents, and toxicants. These traditional terms are useful but not precise descriptions of the true mode of action. Attractants and repellents are behavior modifiers. The actual mechanism of operation may be quite different from what the name indicates. For instance ordinary sugar has almost no vapor pressure and hence has no directional influence on a distant fly. Once a fly touches sugar, however, locomotion is inhibited; hence when there are many flies in the vicinity, the number that collect around sugar can give the impression of an actual attraction. The situation is not quite so simple, however, for other chemical and behavioral factors come into play as a result of the fly activity. Repellents, too, may have little or no directional influence at a distance; they, too, may simply induce locomotion or inhibit feeding when a fly or mosquito is in contact or near by.

Some chemicals, of course, do act in a directive manner at a distance. Attractants of this type are of greatest value in connection with trapping for survey purposes. Some highly effective attractants of yellowjackets, however, show promise for the control of these insects by trapping in restricted locations such as parks or campgrounds. Attractants are used widely in baits for fly control. Repellents applied on the person or clothing provide excellent, temporary protection against biting arthropods.

Another approach to behavior-modifying chemicals comes from the insects themselves in the form of pheromones. Sex pheromones have been demonstrated clearly in house flies, stable flies, mosquitoes and ticks. The house fly pheromone **muscalure** is in commerical use as a poison bait additive.

An outgrowth of the highly successful use of radiation-induced sterility for mass releases of insects, as in the classic screwworm control program, is the development of chemical sterilants. These chemosterilants show promise in sterilizing insects in nature which then compete with the fertile forms for mates. Practical use of chemosterilants is dependent on the development of adequate attractants and foolproof trapping and release devices.

By far the greatest use of chemicals for the control of insects of medical importance is in the form of toxicants—insecticides and acaricides. Their use is widespread, and in many situations toxicants are the only effective tools for achieving control—despite a century of research effort directed toward other procedures. Toxicants can be used with precision in highly localized areas, as in bed bug control, or over large areas, as in mosquito control. The relative ease with which toxicants can be adapted successfully to specific insect control problems has resulted in their use on a vast scale. This broad usage has itself brought problems, for by their very nature toxicants can harm nontarget forms as well as the pest species. Their usefulness lies in their capacity to interfere with essential life functions of target pests. This characteristic can seldom provide absolute immunity of nontarget organisms, including man himself. When dealing with insects of public health importance, however, the hazard of using the toxicant must be balanced against the hazard of not using it. Furthermore, one must be realistic in evaluating both hazards. Fortunately there has been and continues to be intensive study of the effects of toxicants on target and nontarget organisms. As new information comes to light, old fears are laid to rest, and new ones are generated. It is a dynamic process with toxicant use changing frequently.

The changing of insecticide recommendations is based in part on new judgments of hazard which in turn are influenced by the remarkable sensitivity of new procedures for residue detection. Changes are also based on the rapid development of resistance on the part of some insect populations. Experience has shown that continuous, aggressive research and development are necessary to maintain an adequate arsenal of toxicants for use against medically deleterious insects.

CONTROL EQUIPMENT

The treatment of adult flies and mosquitoes involves a wide variety of procedures that strongly influence the choice of toxicant. A highly effective method of attack is the space spray, a finely divided mist dispersed throughout the area to be protected. Insects can absorb the insecticide through their cuticle or respiratory system. Space sprays can be applied by hand sprayers, low-pressure bombs, or powered atomizers. The nozzles may be portable or contained in fixed installations that apply insecticide to strategic zones. A large number of special-purpose commerical devices are available for space spray application. Such sprays provide almost immediate relief from current problems but provide relatively small long-term effects. Coupled with adequate screening and sanitation, space sprays can be the method of choice when frequent application is of low hazard and is economically justified. Insecticides used as space sprays must be low-hazard materials because most areas requiring their application are places of human or livestock occupancy.

Fog generators have achieved wide use in protection of outdoor areas from the activity of adult mosquitoes and other blood-sucking Diptera. A wide variety of commercial fog-generating equipment is currently available. Many of these devices were adapted from military smoke-screen generators. Some systems depend on direct heating of oil solutions; some involve secondary heating systems, such as introducing the oil solutions into a stream of steam or into the exhaust system of an internal combustion engine; and some depend on direct mechanical atomization.

Air sprayers, which function by the introduction of small quantities of concentrated insecticide solution into a large-volume air blast, range from small wheelbarrow units to large self-propelled devices for use on land or water. Many of the larger units are turret-mounted to permit ease in directing the discharge. In recent years there has been a tendency to further reduce the amount of solvent used in fluid applications of insecticide to an ultra-low volume (ULV). In community mosquito control projects various equipment and vehicles, including boats and aircraft, apply insecticides to swamps and backwater areas (Fig. 8:8).

REPRESENTATIVE PESTS OF MEDICAL IMPORTANCE

For detailed study we have selected the house fly, several species of mosquitoes, and chiggers.

House Fly. *Musca domestica* Linnaeus [Diptera:Muscidae]

The house fly is a cosmopolitan species. It is associated intimately with humanity, thriving best where people are careless in the disposal of organic wastes. House flies are by no means limited to association with humans, though both house flies and humans tend to maintain high populations in similar general environments. House flies are important mechanical vectors of a great variety of organisms that produce human diseases.

Description. The house fly egg is pearly white, elongate, and approximately 1 mm long (Fig. 22:8). The larva is a typical maggot, blunt at the posterior end and tapering to a point anteriorly. The cuticle is transparent, except toward the end of the larval period when the maggot assumes a creamy or yellowish appearance. When full grown the maggot may be 12 mm or more in length. The puparium, shorter and more robust than the larva, is slightly over 6 mm long and, when mature, is dark brown in color. The adult fly is dark, mottled gray, and 6 to 7 mm long. When retracted the adult mouthparts are inconspicuous, but during feeding they are protruded and capable of acting as an efficient swabbing device.

Life history. The house fly, like other Diptera, undergoes complex metamorphosis. The eggs are laid in considerable numbers in decaying organic matter. A single gravid female fly may lay from 100 to 150 eggs at a time, and up to 600 or more eggs in two months, although in midsummer longevity is thought to be under a month.

Figure 22:8. Life stages of the house fly, *Musca domestica* L. A, egg; B, larva; C, puparium; D, adult. *Courtesy USDA.*

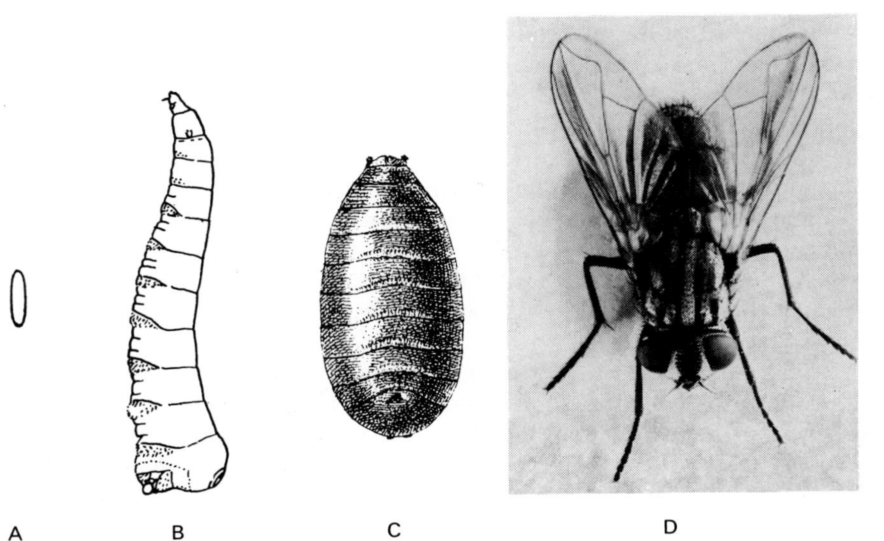

The development of the egg is rapid: the larva hatches within a day in normal summer temperatures. The larva feeds on the organic matter in which it hatches and grows rapidly. It goes through three instars in 5 to 15 days before pupating. The pupa remains within the integument of the last larval instar, which becomes the puparium. The pupal stage lasts from four to ten days. The newly transformed adult has a flexible membrane on the anterior portion of the head. This saclike membrane, called the ptilinum, pushes out from the head as it is filled with fluid. Pulsation of the abdomen causes a corresponding pulsation of the ptilinum, which is used as a tool in forcing open the puparium and in opening a passageway for the newly emerged fly to reach the open air. Two to 4 days after emergence most adults are ready to mate, and within 2 to 3 additional days the gravid female is ready to lay her first batch of eggs.

It is apparent that the reproductive potential of the house fly is very great. Assuming ideal conditions for reproduction, it is interesting to engage in the game of "how many at the end of the season?" Using conservative numbers (but assuming total survival) the offspring of a single gravid female would easily reach 2×10^{20} flies. This enormous number would fill over 2000 cubic miles! Obviously one limiting factor in this sequence is the amount of nutrient material available to developing larvae. In general it can be stated that house flies reproduce to the limit of the available food supply, a fact that emphasizes the importance of sanitation in house fly control programs.

Control. There are three important procedures in house fly control: sanitary, mechanical, and chemical. Sanitary practices by themselves rarely provide adequate control under farm conditions. They are, however, the key to most successful programs. Heavy breeding pressure is very difficult to overcome by attacks on the adults alone. The agriculturist should plan to dispose of fly-breeding materials to as great an extent as practicable.

In some situations mechanical practices such as screening, supplemented by traps or sticky paper, can be valuable. On poultry farms the use of cockerels to search out and eat maggots has proven highly effective in appropriate situations.

Insecticides may be used against larvae or against adults. Larval treatments are employed less frequently than adult treatments because breeding sites may be at some distance and therefore are more difficult to treat. The high organic content of larval sites reduces the effectiveness of many pesticides and, because of the larger numbers of individual insects on which selective pressure operates, larviciding encourages accelerated development of resistance to insecticides.

The development of DDT opened a new era in fly control. It was the first of a series of residual products that provided long-term control. These residues were particularly effective against the house fly for several reasons. The swabbing method of feeding made this fly especially vulnerable to residues. Of at least equal importance, however, is the sensitivity of the fly's tarsi to certain types of toxicants (DDT, for example) that penetrate the cuticle easily and seemingly incapacitate the fly by overstimulating sensory end organs. This last factor is enhanced by the house fly's habit of moving about restlessly on surfaces, thereby coming in contact with a larger amount of residue. The earlier procedures involved overall sprays both in interiors and outside. It was soon recognized that it was wasteful and unnecessary to spray all surfaces. The house fly has a tendency to rest on edges of doors or windows, edges of cracks between boards, pipes, or electric light cords. Anyone can see where house flies have rested during a season by the profusion of black feces spots. This knowledge of fly behavior led to the use of localized applications. Such applications are as effective as over-all sprays and offer substantial savings in material as well as permitting use of less elaborate equipment. Instead of high-pressure, high-gallonage equipment, simple compressed air sprayers are adequate.

Residual applications of DDT in the first few years of use were so outstandingly successful that some people were convinced that the house fly was about to be eradicated. Fly control was one of the largest uses for DDT and other chlorinated hydrocarbons such as lindane, dieldrin, and toxaphene. When widespread resistance to these chlorinated hydrocarbons developed, they were replaced by various organophosphorus compounds such as malathion, diazinon, dichlorvos, dimethoate, crotoxyphos, fenthion, ronnel, naled, stirofos, and trichlorfon. Organophosphorus resistance has also developed to some extent, but residual applications are still generally effective.

Another localized treatment that became widespread again with the use of the organophosphorus materials is the application of bait. One of the first synthetic materials to be used in this connection was tepp, mixed with molasses and water and distributed with a sprinkling can. Diazinon, dichlorvos, malathion, naled, ronnel, and trichlorfon soon replaced the highly toxic tepp. Dry baits, impregnated on sugar or other materials, are also used with great frequency. Baits sprinkled in alleyways, gutters, sills, or on empty sacks require almost no equipment for application. Commerical preparations may contain muscalure, an attractant, in addition to a toxicant.

Pyrethrum, or synthetic pyrethroids, plus a synergist continue to be successful as a space spray in appropriate locations.

Mosquitoes

Anopheles freeborni Aitken

Anopheles quadrimaculatus Say; common malaria mosquito

Anopheles albimanus Wiedemann

Aedes dorsalis (Meigen)

Aedes aegypti (Linnaeus); yellowfever mosquito

Culex pipiens Linnaeus; northern house mosquito

Culex quinquefasciatus Say; southern house mosquito

Culex tarsalis Coquillett [Diptera:Culicidae]

Mosquitoes, like house flies, are known around the world. There are, however, a great many species and genera in this family of insects. Mosquitoes (Fig. 22:9) tend to be relatively independent of human activities, becoming important only as humans live near breeding sites.

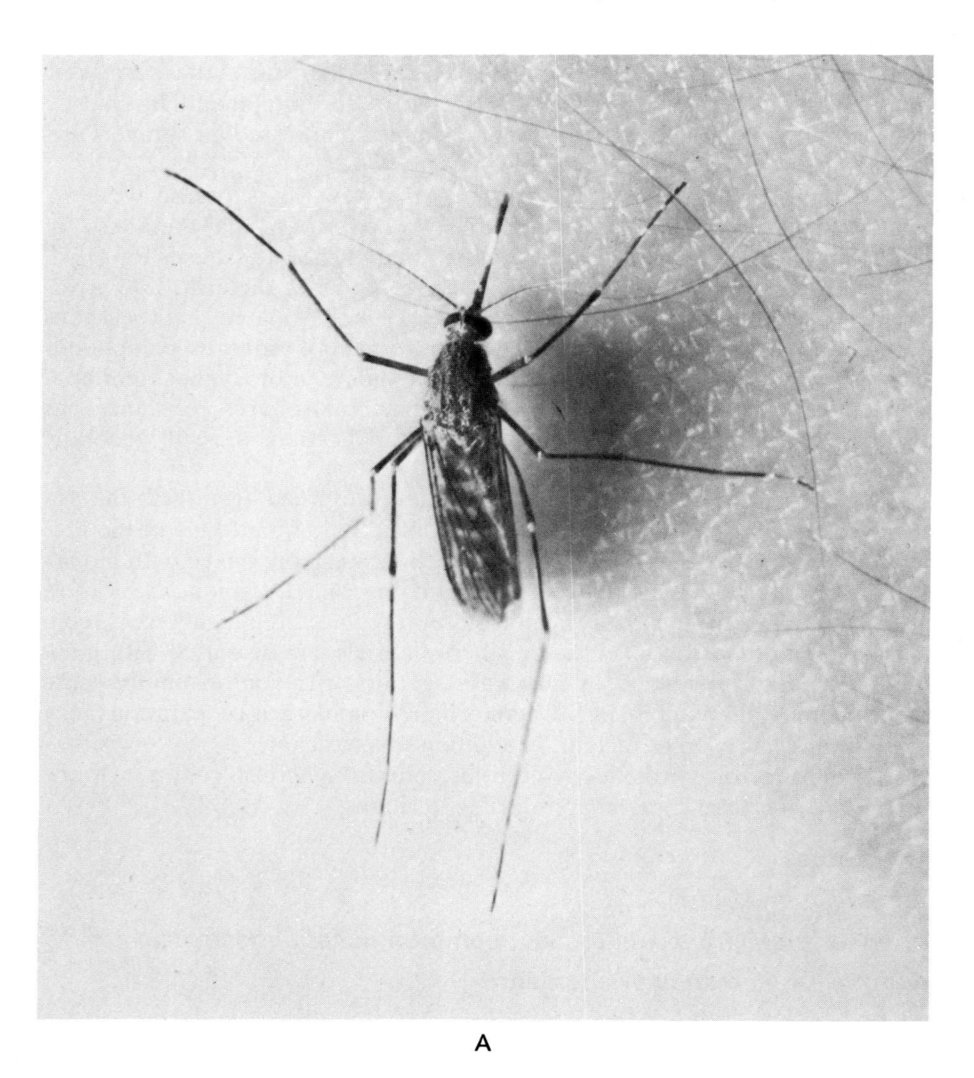

A

Figure 22:9. A, female mosquito, *Culex tarsalis*, feeding on man; B, larva of the yellowfever mosquito. *A, courtesy University of California, Berkeley; B, courtesy USDA.*

The mosquito family, Culicidae, contains about 2500 described species. Many of the species are similar and are grouped into "complexes" below the generic level. Because of the many degrees of variation within the family, taxonomic schemes frequently make use of such subdivisions as subfamily, tribe, genus, subgenus, group, species, and variety. The identification of the many forms is critical because, even though mosquitoes may be barely distinguishable from each other in anatomical detail, their biting habits and other aspects of behavior and physiology may differ enough to affect their importance as vectors of disease.

Of the many genera recognized, *Anopheles*, *Aedes*, and *Culex* are the most important. The discussion that follows is limited to these three genera.

Description. The larval mosquito (Fig. 22:9B) is elongate and has the body clearly differentiated into a head, thorax, and abdomen. The head has a pair of short antennae projecting forward. The eyes are of two types: the eyes of the first instar larva are small spots on each side of the head; as the larva grows, a pair of crescent-shaped compound eyes develops anterior to these original larval eyes. By the fourth instar these crescent-shaped compound eyes (which are the developing compound eyes of the adult) are much larger than the original larval eyes, and are essentially the active optic apparatus of the late larva. The larval mouthparts are of a chewing type, although the actual function varies considerably among different species in relation to dietary habit. The only spiracles of the larva are located dorsally on the eighth abdominal segment. In *Anopheles*, surrounding the spiracles, there is a very short projection which breaks through the surface film. In *Aedes* and *Culex* this projection is greatly elongated and is known as the siphon. The ninth segment has a variety of hairs and a saddlelike plate; it terminates in four projections, the anal gills.

The mosquito pupa (Fig. 22:10) is frequently described as "comma shaped." The main portion of the body contains the fused developing head

B

and thorax of the adult, called the cephalothorax. The functional compound eyes are readily visible. On the dorsal part of the cephalothorax are two respiratory trumpets that connect to the developing thoracic spiracles of the adult. The abdomen, composed of nine segments, is terminated by a pair of conspicuous overlapping plates, the "paddles." In contrast to the relatively inactive pupae of most insects, the mosquito pupa is exceptionally active when disturbed. The abdomen is freely moveable on the cephalothorax, and by utilizing a crayfishlike abdominal motion, the pupa can quickly make an avoidance reaction in response to vibration or shadow.

The adult mosquito (Fig. 22:9A) is a slender, graceful insect, well adapted to independent flight for locating food, a mate, or an oviposition site. The mouthparts are very slender swordlike structures that slide together as a single proboscis. The only mouthparts outside the proboscis are the pair of maxillary palpi; these vary in length and shape with different genera as well as with sex. Both the antennae and the maxillary palpi of male mosquitoes tend to be much more hairy than those of the females, a difference that is easily seen with the naked eye.

Life history. Generally adult mosquitoes lay their eggs on the surface of water or in damp locations. Some species of *Aedes* may lay their eggs out of water but usually in areas that are flooded periodically. The eggs of *Anopheles* and *Aedes* are laid singly, but those of *Culex* are laid in "rafts" of up to several hundred eggs each. These rafts, composed of elongate eggs stacked vertically in a single layer, float on the surface of the water in which they are laid. Mosquito eggs laid on water usually hatch in from two to three days, but the eggs of *Aedes* that are laid in damp locations may either hatch in a similar period or be delayed for periods up to a year or more. (See Fig. 22:10 for mosquito life histories.)

Wherever the eggs are laid, the larvae require water in which to develop. Mosquito larvae, however, are primarily air breathers and are usually found just beneath the surface film, with their spiracles open to the air above. The larvae feed on bacteria, yeasts, protozoans, and particles of organic debris present in water. There are four larval instars that may be passed in as little as three or four days, although more commonly it is a week to ten days. In a few species that can overwinter as larvae, up to half a year or more is passed in the larval stage.

The pupal period lasts roughly from one to four days, depending on temperature as well as the species. At the end of the pupal period air is swallowed, the abdomen straightens, the pupal skin splits, and the adult mosquito extricates itself and stands on the floating old pupal skin or on adjacent vegetation for a short period before flying away.

Ordinarily a few days are necessary after emergence before sexual maturity is attained. Depending on the time of year and on the species, adult mosquitoes may live from a few weeks to over half a year. Many species of mosquitoes have only one generation per year; others have two to five or more, depending on climatic conditions.

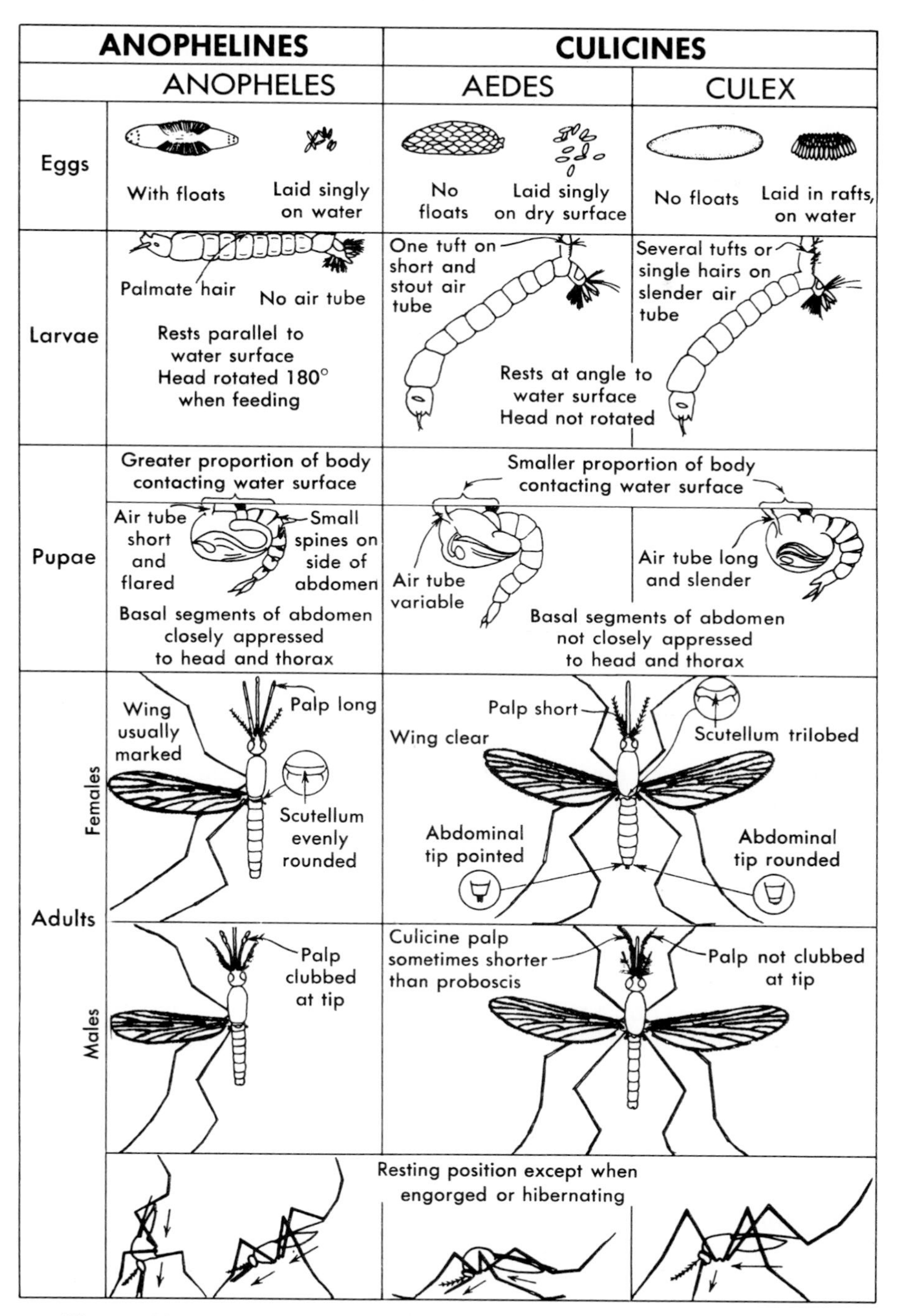

Figure 22:10. Illustration of the life histories of three genera of mosquitoes. *Courtesy USDHEW, Public Health Service.*

Adult mosquitoes on the wing orient to visual stimuli as well as to air movement. Although many species (*Anopheles*, for example) tend to remain localized within $\frac{1}{2}$ mi or so, many others (such as *Aedes*) may range upward of 10 mi or more. Wind may greatly increase these distances.

Male and female mosquitoes exhibit quite different flight patterns: males have a tendency to swarm, females remain solitary. Male swarms tend to orient to dark objects below them and to maintain heights above these signals that are characteristic of the species. Females enter the male swarms, mate, and depart. Many species exhibit highly characteristic flight habits related to time of day or amount of light present. Feeding habits, too, can vary characteristically with the time of day or night.

Among *Anopheles* at least, the female that has just taken a blood meal tends to remain in the immediate vicinity for a significant period of time. In malarious areas this characteristic of resting on walls after feeding has made the routine spraying of houses with residual insecticides so effective in reducing or eliminating malaria transmission.

Control. Activities related to mosquito control are sometimes localized, but more commonly they involve communities, districts, nations, or even cooperating groups of nations. Mosquitoes are controlled by a variety of methods, including the elimination of breeding areas, biological control agents, and many chemical procedures.

Treatment of large areas is costly in both materials and labor. The decision to undertake a major control operation may actually involve little choice. Given some chance at reducing a major disease, such as malaria, a public health agency obviously chooses to make the attempt—witness the worldwide program to control malaria-carrying mosquitoes. Where lesser values are involved, or where the reason for mosquito control is public comfort, decisions must include other considerations.

When DDT first became available, it appeared to be the ideal material for mosquito control because of its efficacy, low cost, and relatively low mammalian toxicity. Although resistant mosquitoes in many areas have forced the use of other insecticides, the important vectors of malaria in much of the world are still susceptible to DDT. In regions where resistance has developed, control agencies have resorted to other organochlorine and organophosphorus insecticides, carbamates, and several miscellaneous chemicals, including insect growth regulators (IGRs).

Application of a residual insecticide such as DDT to the interior of houses in malarious areas is a procedure commonly employed. Such applications, made once or twice a year, have had little effect in reducing total mosquito populations, but have had overwhelming success against those *Anopheles* that once fed on human beings and were then killed before having an opportunity to feed a second time. The result of this breaking of the parasite's life cycle has been a sharp reduction in malaria cases, with the technical possibility of eventually eliminating this disease entirely. The development of resistant strains of mosquitoes may slow the rate of progress, but it is not anticipated that this will prevent the eventual control of malaria. Perhaps a greater barrier to success are public fear of the insecticides themselves and political divisions that inhibit unified control practices at national boundaries. Steadily rising costs of materials and labor have made current efforts increasingly difficult.

Chiggers

Trombicula akamushi (Brumpt)

Trombicula deliensis Walch

Trombicula alfreddugesi (Oudemans) [Acarina:Trombiculidae]

Chiggers cause two types of effect on humans. The first is a direct result of the feeding activities of the larval mite; the second is a result of the transmission of disease organisms. Chigger larvae feed on a wide variety of terrestrial vertebrates, including amphibians, reptiles, birds, and mammals. Man appears to be a host of relatively minor importance to the chigger. The larva attaches itself by means of its chelicerae to folds of skin or to the edges of hair follicles. It secretes a fluid that apparently digests a localized portion of the skin and at the same time seems to induce the underlying tissue to become organized as an elongate tube, frequently as long as the body of the chigger itself. This tube, called the stylostome or histosiphon, is formed over a period of several hours. The chigger remains attached, for a variable length of time, until it is engorged upon the predigested fluid (not blood) that it draws through the stylostome. Between 3 and 24 hr after exposure, the lesion induced by the mite becomes an urticarial papule that itches intensely. The lesions may be from 0.4 to 2 cm in diameter and may persist for several weeks, during which time they may itch intermittently. The resultant scratching may lead to secondary infection.

In the Asiatic Pacific area chiggers are the vectors of **scrub typhus** and are suspected vectors of **hemorrhagic fever**. Ranging from Japan to Singapore, *T. akamushi* is the prime vector of scrub typhus, whereas in Burma, Sumatra, and New Guinea, *T. deliensis* is the major vector of this disease. Scrub typhus, known also as tsutsugamushi disease, is one of the rickettsial diseases, important in Japan and throughout southeast Asia. The disease is frequently fatal and is sufficiently common to be of major importance.

Chiggers, such as *T. alfreddugesi*, are often common on lawns as well as in the wild in areas with a hot, moist climate (Fig. 22:11). They are particularly common in southeastern and south central regions of the United States and on the islands of the Caribbean. Many similar species are found in most of the warmer parts of the world.

Description. Trombiculid mites pass through four life stages during the course of their development (Fig. 22:12). The spherical eggs are small (100 to 200 microns diameter) and for some species are unknown except as dissected from the gravid female adult. Some species may be ovoviviparous. The first

Figure 22:11. A chigger or larva of the mite, *Trombicula alfreddugesi* (Oudemans), the stage that attacks man and other vertebrates. *Courtesy USDHEW, Public Health Service.*

motile, free-living stage is a six-legged larva, from 150 to 300 microns long and red to pale yellow or white depending on the species. It is the larval stage that parasitizes vertebrates. The next stage consists of an eight-legged nymph, 600 to 1000 microns long, hirsute, and resembling the adult. The adult closely resembles the nymph but is much larger.

Life history. Chigger eggs are commonly laid on the ground or in the soil. Some species have been observed to spend from four to ten days in the egg and deutovum stage (the deutovum appears within a secondary chitinous envelope within the egg shell and persists after the egg shell breaks). The larvae move up onto grass or weeds, often in great numbers, and make contact with vertebrate hosts as they brush past. On humans the larvae settle mostly in the groin, under armpits, or where clothing is constricted as at the belt line or under socks. The larvae may remain attached for up to six days, occasionally longer, until engorged, and then drop from the host to the ground. Nymphal and adult stages feed on insect eggs, and perhaps on small arthropods. The nymphal stage apparently lasts several weeks before transformation to the adult. Adults have been kept alive for at least 45 days. Chigger life cycles may last from 2 to 12 months or longer, with from 1 to 3 generations per year in temperate zones and with continuous production in more tropical regions.

Control. Two major approaches in chigger control have been followed. The first is by the use of repellents, applied either directly to the person or by clothing impregnation. The second is an attack on the chigger itself, by increasing the difficulties of contacting a vertebrate host, or by the use of toxicants broadcast on the soil or vegetation harboring the chiggers.

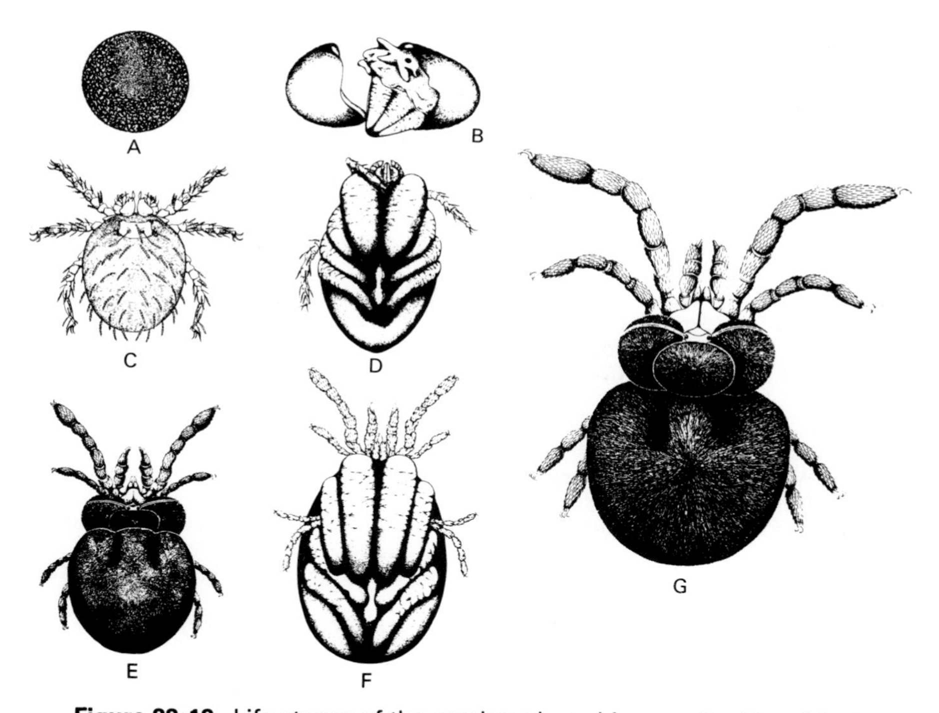

Figure 22:12. Life stages of the scrub typhus chigger mite, *Trombicula akamushi* (Brumpt). A, egg; B, deutovum with ruptured shell exposing developing larva; C, larva, the stage parasitic on vertebrates; D, prenymph, ventral view, showing remnants of larval legs; E, nymph; F, preadult, ventral view, showing remnants of nymphal appendages and developing legs of adult; G, adult. *From T. J. Neal and H. C. Barnett, Walter Reed Army Institute of Research, Wash. D.C.*

Sulfur dust has been widely used for many years, applied directly to the body or broadcast on infested areas. Superior repellents and toxicants have come into use within the last three decades. Repellents include deet, dimethyl phthalate, dibutyl phthalate, and benzyl benzoate; the last two make especially good clothing impregnants, retaining their effectiveness even after several launderings. Liquid repellents are also used as band treatments on cuffs, seams, and tops of socks. Standard repellents appear to act also as toxicants.

Clothing repellents provide excellent protection of the individual, but it is frequently desirable to eliminate the chiggers themselves from campsites or lawns. Several residual insecticides such as chlordane and toxaphene have been used successfully against chiggers. The use of applying toxicants makes the method highly practical for chigger control in limited areas such as pathways and around premises. Where poisons are undesirable, the clearing and removal of vegetation reduces chigger populations by reducing rodent and reptile harborages.

SELECTED REFERENCES

Askew, R. R., *Parasitic Insects* (New York: American Elsevier, 1971).
Baker, E. W., T. M. Evans, D. J. Gould, W. B. Hull, and H. L. Keegan, *A Manual of Parasitic Mites* (New York: National Pest Control Association, 1956).
Bates, M., *The Natural History of Mosquitoes* (New York: Macmillan, 1949).
Brown, A. W. A., *Insecticide Resistance in Arthropods* (Geneva: World Health Organization, 1958).
Carpenter, S. J., and W. J. LaCasse, *Mosquitoes of North America (North of Mexico)* (Berkeley, Calif.: Univ. Calif. Press, 1955).
Christophers, S. R., *Aedes aegypti* (L.), *The Yellow Fever Mosquito. Its Life History, Bionomics and Structure* (London: Cambridge Univ. Press, 1960).
Clements, A. N., *The Physiology of Mosquitoes* (New York: Macmillan, 1963).
Foote, R. H., and D. R. Cook, *Mosquitoes of Medical Importance*, USDA Agr. Handbook No. 152, 1959.
Greenberg, B., *Flies and Disease* (Princeton, N.J.: Princeton, Vol. I, 1971; Vol. II, 1973).
Horsfall, W. R., *Mosquitoes: Their Bionomics and Relation to Disease* (New York: Ronald, 1955).
Glasgow, J. P., *The Distribution and Abundance of Tsetse* (New York: Pergamon, 1963).
James, M. T., and R. F. Harwood, *Herms's Medical Entomology* (New York: Macmillan, 1969).
Leclerq, M., *Entomological Parasitology* (New York: Pergamon, 1969).
Marshall, J. F., *The British Mosquitoes* (London: British Museum, 1938).
Matheson, R., *Handbook of the Mosquitoes of North America* (Ithaca, N.Y.: Comstock, 1944).
Mattingly, P. F., *The Biology of Mosquito-Borne Disease* (New York: American Elsevier, 1969).
Muirhead-Thomson, R., *Ecology of Insect Vector Populations* (New York: Academic Press, 1968).
Oldroyd, H., *The Natural History of Flies* (New York: Norton, 1964).
Roth, L. M., and E. R. Willis, *The Medical and Veterinary Importance of Cockroaches* (Washington, D.C.: Smithsonian Misc. Coll., 1957).
Stone, A., K. L. Knight, and H. Starke, *A Synoptic Catalog of the Mosquitoes of the World*. The Thomas Say Foundation, Vol. 6, Ent. Soc. Amer. Wash, D.C., 1959.
Usinger, R. L., *Monograph of Cimicidae* (College Park, Md.: Entomol. Soc. Amer., 1966).
West, L. S., *The Housefly* (Ithaca, N.Y.: Comstock, 1951).
West, L. S., *An Annotated Bibliography of Musca domestica Linneaus* (London: Dawsons of Pall Mall, 1973).

APPENDIX OF COMMON SCIENTIFIC NAMES[1]

acacia spittlebug* *Clastoptera arizonana* Doering Homoptera:Cercopidae
acacia psyllid* *Psylla uncatoides* (Ferris & Klyver) Homoptera:Psyllidae
achemon sphinx *Eumorpha achemon* (Drury) Lepidoptera:Sphingidae
aedes mosquito* *Aedes* spp. Diptera:Culicidae
African migratory locust* *Locusta migratoria* (Linnaeus) Orthoptera:Acrididae
African sheep louse* *Linognathus africanus* Kellogg & Paine Anoplura:Linognathidae
ailanthus webworm *Atteva punctella* (Cramer) Lepidoptera:Yponomeutidae
airsac mite* *Cytodites nudus* (Vizioli) Acarina:Cytoditidae
alder scale* *Xylococcus betulae* Pergande Homoptera:Margarodidae
alfalfa aphid* *Macrosiphum creelii* Davis Homoptera:Aphididae
alfalfa caterpillar *Colias eurytheme* Boisduval Lepidoptera:Pieridae
alfalfa gall midge *Asphondylia websteri* Felt Diptera:Cecidomyiidae
alfalfa leafcutter bee *Megachile rotundata* (Fabricius) Hymenoptera:Megachilidae
alfalfa looper *Autographa californica* (Speyer) Lepidoptera:Noctuidae
alfalfa plant bug *Adelphocoris lineolatus* (Goeze) Hemiptera:Miridae
alfalfa seed chalcid *Bruchophagus roddi* (Gussakovsky) Hymenoptera:Eurytomidae
alfalfa snout beetle *Otiorhynchus ligustici* (Linnaeus) Coleoptera:Curculionidae
alfalfa webworm *Loxostege commixtalis* (Walker) Lepidoptera:Pyralidae
alfalfa weevil *Hypera postica* (Gyllenhal) Coleoptera:Curculionidae
alfalfa weevil parasite* *Bathyplectes curculionis* (Thomson) Hymenoptera:Ichneumonidae

[1] Common names unmarked by an asterisk (*) are approved by Entomological Society of America ("Common Names of Insects Approved by the Entomological Society of America," November 1975); those marked by an asterisk are common names that have been used in the literature but as yet have not been acted upon or approved by the Society.

alkali bee *Nomia melanderi* Cockerell Hymenoptera:Halictidae
American cockroach *Periplaneta americana* (Linnaeus) Orthoptera:Blattidae
American dog tick *Dermacentor variabilis* (Say) Acarina:Ixodidae
American grasshopper *Schistocerca americana* (Drury) Orthoptera:Acrididae
American syrphid* *Metasyrphus americanus* (Wiedemann) Diptera:Syrphidae
Angoumois grain moth *Sitotroga cerealella* (Olivier) Lepidoptera:Gelechiidae
anopheles mosquito* *Anopheles* spp. Diptera:Culicidae
apple aphid *Aphis pomi* De Geer Homoptera:Aphididae
apple curculio *Tachypterellus quadrigibbus* (Say) Coleoptera:Curculionidae
apple fruitminer *Marmara pomonella* Busck Lepidoptera:Gracillariidae
apple grain aphid *Rhopalosiphum fitchii* (Sanderson) Homoptera:Aphididae
appleleaf blister mite* *Eriophyes mali* Burts Acarina:Eriophyidae
apple leafhopper *Empoasca maligna* (Walsh) Homoptera:Cicadellidae
apple maggot *Rhagoletis pomonella* (Walsh) Diptera:Tephritidae
apple red bug *Lygidea mendax* Reuter Hemiptera:Miridae
apple rust mite *Aculus schlechtendali* (Nalepa) Acarina:Eriophyidae
apple sucker *Psylla mali* (Schmidberger) Homoptera:Psyllidae
apple twig beetle *Hypothenemus obscurus* (Fabricius) Coleoptera:Scolytidae
apple twig borer *Amphicerus bicaudatus* (Say) Coleoptera:Bostrichidae
Argentine ant *Iridomyrmex humilis* (Mayr) Hymenoptera:Formicidae
arid-land subterranean termite* *Reticulitermes tibialis* Banks
Isoptera:Rhinotermitidae
army cutworm *Euxoa auxiliaris* (Grote) Lepidoptera:Noctuidae
armyworm *Pseudaletia unipuncta* (Haworth) Lepidoptera:Noctuidae
ash borer *Podosesia syringae* (Harris) Lepidoptera:Sesiidae
ash plant bug *Tropidosteptes amoenus* Reuter Hemiptera:Miridae
Asiatic garden beetle *Maladera castanea* (Arrow) Coleoptera:Scarabaeidae
Asiatic oak weevil *Cyrtepistomus castaneus* (Roelofs) Coleoptera:Curculionidae
asparagus beetle *Crioceris asparagi* (Linnaeus) Coleoptera:Chrysomelidae
asparagus miner *Ophiomyia simplex* (Loew) Diptera:Agromyzidae
asparagus spider mite* *Schizotetranychus asparagi* (Oudemans)
Acarina:Tetranychidae
aspen serpentine leafminer* *Phyllocnistis populiella* Chambers
Lepidoptera:Gracillariidae
aster leafhopper *Macrosteles fascifrons* (Stål) Homoptera:Cicadellidae
azalea lace bug *Stephanitis pyrioides* Scott Hemiptera:Tingidae
azalea leafminer *Gracillaria azaleella* Brants Lepidoptera:Gracillariidae
azalea whitefly *Pealius azaleae* (Baker & Moles) Homoptera:Aleyrodidae
bagworm *Thyridopteryx ephemeraeformis* (Haworth) Lepidoptera:Psychidae
Banks grass mite *Oligonychus pratensis* (Banks) Acarina:Tetranychidae
barberpole caterpillar *Mimoschinia rufofascialis* (Stephens)
Lepidoptera:Noctuidae
bean aphid *Aphis fabae* Scopoli Homoptera:Aphididae
bean leaf beetle *Cerotoma trifurcata* (Forster) Coleoptera:Chrysomelidae
bean leafskeletonizer *Autoplusia egena* (Guenée) Lepidoptera:Noctuidae
bean seed fly* *Hylemya florilega* (Zetterstedt) Diptera:Anthomyiidae
bean weevil *Acanthoscelides obtectus* (Say) Coleoptera:Bruchidae
bed bug *Cimex lectularius* Linnaeus Hemiptera:Cimicidae
beet armyworm *Spodoptera exigua* (Hübner) Lepidoptera:Noctuidae

beet leafhopper *Circulifer tenellus* (Baker) Homoptera:Cicadellidae
beet webworm *Loxostege sticticalis* (Linnaeus) Lepidoptera:Pyralidae
Bermudagrass mite *Aceria cynodoniensis* Sayed Acarina:Eriophyidae
Bermudagrass scale* *Odonaspis ruthae* Kotinsky Homoptera:Diaspididae
bigeyed bug* *Geocoris punctipes* (Say) Hemiptera:Lygaeidae
bigheaded grasshopper* *Aulocara elliotti* (Thomas) Orthoptera:Acrididae
birch leafminer *Fenusa pusilla* (Lepeletier) Hymenoptera:Tenthredinidae
birdcherry oat aphid* *Rhopalosiphum padi* (Linnaeus) Homoptera:Aphididae
black carpenter ant *Camponotus pennsylvanicus* (De Geer) Hymenoptera:Formicidae
black carpet beetle *Attagenus megatoma* (Fabricius) Coleoptera:Dermestidae
black cherry aphid *Myzus cerasi* (Fabricius) Homoptera:Aphididae
black cherry fruit fly *Rhagoletis fausta* (Osten Sacken) Diptera:Tephritidae
black cutworm *Agrotis ipsilon* (Hufnagel) Lepidoptera:Noctuidae
black grain stem sawfly *Cephus tabidus* (Fabricius) Hymenoptera:Cephidae
blackheaded fireworm *Rhopobota naevana* (Hübner) Lepidoptera:Olethreutidae
black horse fly *Tabanus atratus* Fabricius Diptera:Tabanidae
black lady beetle *Rhizobius ventralis* (Erichson) Coleoptera:Coccinellidae
blacklegged tick *Ixodes scapularis* Say Acarina:Ixodidae
black peach aphid *Brachycaudus persicae* (Passerini) Homoptera:Aphididae
black scale *Saissetia oleae* (Bernard) Homoptera:Coccidae
black vine weevil *Otiorhynchus sulcatus* (Fabricius) Coleoptera:Curculionidae
black widow spider *Latrodectus mactans* (Fabricius) Araneida:Theridiidae
blue aphid* *Acyrthosiphon kondoi* Shinji *Homoptera*:Aphididae
blueberry blossom weevil* *Anthonomus musculus* Say Coleoptera:Curculionidae
blueberry bud mite *Acalitus vaccinii* (Keifer) Acarina:Eriophyidae
blueberry fruit fly* *Rhagoletis pomonella* (Walsh) Diptera:Tephritidae
blueberry leafroller* *Sparganothis sulfureana* Clemens Lepidoptera:Tortricidae
blueberry leaftier* *Aroga trialbamaculella* (Chambers) Lepidoptera:Gelechiidae
blueberry maggot *Rhagoletis mendax* Curran Diptera:Tephritidae
blueberry sawfly* *Caliroa* sp. Hymenoptera:Tenthredinidae
blueberry thrips *Frankliniella vaccinii* Morgan Thysanoptera:Thripidae
blueberry tip borer* *Hendecaneura shawiana* (Kearfott) Lepidoptera:Olethreutidae
blueberry tip midge *Contarinia vaccinii* Felt Diptera:Cecidomyiidae
bluegrass aphid* *Rhopalosiphum poae* Gillette Homoptera:Aphididae
bluegrass plant bug* *Amblytylus nasutus* (Kirschbaum) Hemiptera:Miridae
bluntnosed cranberry leafhopper *Scleroracus vaccinii* (Van Duzee) Homoptera:Cicadellidae
body louse *Pediculus humanus humanus* Linnaeus Anoplura:Pediculidae
Boisduval scale* *Diaspis boisduvalii* Signoret Homoptera:Diaspididae
boll weevil *Anthonomus grandis* Boheman Coleoptera:Curculionidae
bollworm *Heliothis zea* (Boddie) Lepidoptera:Noctuidae
booklice *Liposcelis* spp. Psocoptera:Liposcelidae
bordered plant bug* *Euryophythalmus succinctus* (Linnaeus) Hemiptera:Pyrrhocoridae
boxelder bug *Leptocoris trivittatus* (Say) Hemiptera:Rhopalidae
boxwood leafminer *Monarthropalpus buxi* (Laboulbène) Diptera:Cecidomyiidae
boxwood psyllid *Psylla buxi* (Linnaeus) Homoptera:Psyllidae

brambleberry leafhopper* *Macropsis fuscula* (Zetterstedt) Homoptera:Cicadellidae
bramble leafhopper *Ribautiana tenerrima* (Herrich-Schäffer) Homoptera:Cicadellidae
branch-and-twig borer* *Polycaon confertus* (LeConte) Coleoptera:Bostrichidae
bristly roseslug *Cladius difformis* (Panzer) Hymenoptera:Tenthredinidae
broadbean weevil *Bruchus rufimanus* Boheman Coleoptera:Bruchidae
broadhorned flour beetle *Gnathocerus cornutus* (Fabricius) Coleoptera:Tenebrionidae
broad mite *Polyphagotarsonemus latus* (Banks) Acarina:Tarsonemidae
broadwinged katydid *Microcentrum rhombifolium* (Saussure) Orthoptera:Tettigoniidae
bronze birch borer *Agrilus anxius* Gory Coleoptera:Buprestidae
brownbanded cockroach *Supella longipalpa* (Fabricius) Orthoptera:Blattellidae
brown cotton leafworm *Acontia dacia* Druce Lepidoptera:Noctuidae
brownheaded ash sawfly *Tomostethus multicinctus* (Rohwer) Hymenoptera:Tenthredinidae
brown mite *Bryobia rubrioculus* (Sheuten) Acarina:Tetranychidae
brown recluse spider *Loxosceles reclusa* Gertsch and Mulaik Araneida:Loxoscelidae
brown soft scale *Coccus hesperidum* Linnaeus Homoptera:Coccidae
brown stink bug *Euschistus servus* (Say) Hemiptera:Pentatomidae
browntail moth *Nygmia phaeorrhoea* (Donovan) Lepidoptera:Lymantriidae
brown wheat mite *Petrobia latens* (Müller) Acarina:Tetranychidae
buffalo treehopper *Stictocephala bubalus* (Fabricius) Homoptera:Membracidae
bulb mite *Rhizoglyphus echinopus* (Fumouze & Robin) Acarina:Acaridae
bulb scale mite *Steneotarsonemus laticeps* (Halbert) Acarina:Tarsonemidae
bumble bee* *Bombus americanorum* (Fabricius) Hymenoptera:Apidae
burrowing nematode* *Radopholus similis* (Cobb) Tylenchida:Tylenchidae
cabbage aphid *Brevicoryne brassicae* (Linnaeus) Homoptera:Aphididae
cabbage butterfly* *Pieris rapae* (Linnaeus) Lepidoptera:Pieridae
cabbage looper *Trichoplusia ni* (Hübner) Lepidoptera:Noctuidae
cabbage maggot *Hylemya brassicae* (Bouché) Diptera:Anthomyiidae
cabbage webworm *Hellula rogatalis* (Hulst) Lepidoptera:Pyralidae
cactus moth* *Cactoblastis cactorum* (Berg) Lepidoptera:Pyralidae
cadelle *Tenebroides mauritanicus* (Linnaeus) Coleoptera:Trogositidae
California fan palm borer* *Dinapate wrighti* Horn Coleoptera:Bostrichidae
California flatheaded borer *Melanophila californica* Van Dyke Coleoptera:Buprestidae
California oakworm *Phryganidia californica* Packard Lepidoptera:Dioptidae
California prionus *Prionus californicus* Motschulsky Coleoptera:Cerambycidae
California red scale *Aonidiella aurantii* (Maskell) Homoptera:Diaspididae
California strawberry rootworm* *Paria canella quadrinotata* (Say) Coleoptera:Curculionidae
carmine spider mite *Tetranychus cinnabarinus* (Boisduval) Acarina:Tetranychidae
carnation bud mite* *Aceria paradianthi* Keifer Acarina:Eriophyidae
carnation leafroller* *Platynota stultana* Walsingham Lepidoptera:Tortricidae
carpenter bee *Xylocopa virginica* (Linnaeus) Hymenoptera:Anthophoridae
carpenterworm *Prionoxystus robiniae* (Peck) Lepidoptera:Cossidae
carpet beetle *Anthrenus scrophulariae* (Linnaeus) Coleoptera:Dermestidae

carpet moth *Trichophaga tapetzella* (Linnaeus) Lepidoptera:Tineidae
carrot beetle *Bothynus gibbosus* (De Geer) Coleoptera:Scarabaeidae
carrot rust fly *Psila rosae* (Fabricius) Diptera:Psilidae
carrot weevil *Listronotus oregonensis* (LeConte) Coleoptera;Curculionidae
casemaking clothes moth *Tinea pellionella* Linnaeus Lepidoptera:Tineidae
catalpa sphinx *Ceratomia catalpae* (Boisduval) Lepidoptera:Sphingidae
cattle biting louse *Bovicola bovis* (Linnaeus) Mallophaga:Trichodectidae
cattle follicle mite *Demodex bovis* Stiles Acarina:Demodicidae
cattle itch mite* *Psorergates bos* Johnston Acarina:Cheyletidae
cattle tail louse *Haematopinus quadripertusus* Fahrenholz Anoplura:Hematopinidae
cattle tick *Boophilus annulatus* (Say) Acarina:Ixodidae
ceanothus gall moth* *Periploca ceanothiella* (Cosens) Lepidoptera:Momphidae
cecropia moth *Hyalophora cecropia* (Linnaeus) Lepidoptera:Saturniidae
cedar and arborvitae needleminers* *Argyresthia* spp. Lepidoptera:Yponomeutidae
celery leaftier *Udea rubigalis* (Guenée) Lepidoptera:Pyralidae
celery looper *Anagrapha falcifera* (Kirby) Lepidoptera:Noctuidae
cereal leaf beetle *Oulema melanopus* (Linnaeus) Coleoptera:Chrysomelidae
cherry fruit fly *Rhagoletis cingulata* (Loew) Diptera:Tephritidae
cherry fruit sawfly *Hoplocampa cookei* (Clarke) Hymenoptera:Tenthredinidae
cherry fruitworm *Grapholitha packardi* (Zeller) Lepidoptera:Olethreutidae
chicken body louse *Menacanthus stramineus* (Nitzsch) Mallophaga:Menoponidae
chicken head louse *Cuclotogaster heterographus* (Nitzsch) Mallophaga:Philopteridae
chicken mite *Dermanyssus gallinae* (De Geer) Acarina:Dermanyssidae
chigoe *Tunga penetrans* (Linnaeus) Siphonaptera:Tungidae
chinch bug *Blissus leucopterus leucopterus* (Say) Hemiptera:Lygaeidae
chorioptic mange mite* *Chorioptes bovis* (Hering) Acarina:Psoroptidae
chrysanthemum aphid *Macrosiphoniella sanborni* (Gillette) Homoptera:Aphididae
chrysanthemum gall midge *Rhopalomyia chrysanthemi* (Ahlberg) Diptera:Cecidomyiidae
chrysanthemum lace bug *Corythuca marmorata* (Uhler) Hemiptera:Tingidae
chrysanthemum leafminer *Phytomyza syngenesiae* (Hardy) Diptera:Agromyzidae
cicada killer *Sphecius speciosus* (Drury) Hymenoptera:Sphecidae
cigarette beetle *Lasioderma serricorne* (Fabricius) Coleoptera:Anobiidae
citricola scale *Coccus pseudomagnoliarum* (Kuwana) Homoptera:Coccidae
citrophilus mealybug *Pseudococcus calceolariae* (Maskell) Homoptera:Pseudococcidae
citrus blackfly *Aleurocanthus woglumi* Ashby Homoptera:Aleyrodidae
citrus mealybug *Planococcus citri* (Risso) Homoptera:Pseudococcidae
citrus rust mite *Phyllocoptruta oleivora* (Ashmead) Acarina:Eriophyidae
claycolored weevil* *Otiorhynchus singularis* (Linnaeus) Coleoptera: Curculionidae
clearwinged grasshopper *Camnula pellucida* (Scudder) Orthoptera:Acrididae
clover aphid *Nearctaphis bakeri* (Cowen) Homoptera:Aphididae
clover bud caterpillar* *Grapholitha conversana* Walsingham Lepidoptera:Olethreutidae
clover hayworm *Hypsopygia costalis* (Fabricius) Lepidoptera:Pyralidae

clover head caterpillar *Grapholitha interstinctana* (Clemens) Lepidoptera:Olethreutidae
clover head weevil *Hypera meles* (Fabricius) Coleoptera:Curculionidae
clover leafhopper *Aceratagallia sanguinolenta* (Provancher) Homoptera:Cicadellidae
clover leaf weevil *Hypera punctata* (Fabricius) Coleoptera:Curculionidae
clover mite *Bryobia praetiosa* Koch Acarina:Tetranychidae
clover root borer *Hylastinus obscurus* (Marsham) Coleoptera:Scolytidae
clover root curculio *Sitona hispidulus* (Fabricius) Coleoptera:Curculionidae
clover seed chalcid *Bruchophagus platypterus* (Walker) Hymenoptera:Eurytomidae
clover seed midge *Dasineura leguminicola* (Lintner) Diptera:Cecidomyiidae
clover seed weevil *Miccotrogus picirostris* (Fabricius) Coleoptera:Curculionidae
coconut mealybug *Nipaecoccus nipae* (Maskell) Homoptera:Pseudococcidae
codling moth *Laspeyresia pomonella* (Linnaeus) Lepidoptera:Olethreutidae
Colorado potato beetle *Leptinotarsa decemlineata* (Say) Coleoptera:Chrysomelidae
columbine leafminer *Phytomyza* complex, including *aquilegivora* Spencer and, *columbinae* Sehgal Diptera:Agromyzidae
common cattle grub *Hypoderma lineatum* (de Villers) Diptera:Oestridae
common malaria mosquito *Anopheles quadrimaculatus* Say Diptera:Culicidae
common mud dauber* *Sceliphron caementarium* (Drury) Hymenoptera:Sphecidae
composite thrips *Microcephalothrips abdominalis* (D. L. Crawford) Thysanoptera:Thripidae
Comstock mealybug *Pseudococcus comstocki* (Kuwana) Homoptera:Pseudococcidae
conchuela *Chlorochroa ligata* (Say) Hemiptera:Pentatomidae
confused flour beetle *Tribolium confusum* Jacquelin duVal Coleoptera:Tenebrionidae
convergent lady beetle *Hippodamia convergens* Guérin-Méneville Coleoptera:Coccinellidae
Cooley spruce gall aphid *Adelges cooleyi* (Gillette) Homoptera:Chermidae
corn blotch leafminer *Agromyza parvicornis* Loew Diptera:Agromyzidae
corn earworm *Heliothis zea* (Boddie) Lepidoptera:Noctuidae
cornfield ant *Lasius alienus* (Foerster) Hymenoptera:Formicidae
corn flea beetle *Chaetocnema pulicaria* Melsheimer Coleoptera:Chrysomelidae
corn leaf aphid *Rhopalosiphum maidis* (Fitch) Homoptera:Aphididae
corn leafhopper* *Dalbulus maidis* (De Long and Wolcott) Homoptera:Cicadellidae
corn planthopper *Peregrinus maidis* (Ashmead) Homoptera:Delphacidae
corn root aphid *Aphis maidiradicis* Forbes Homoptera:Aphididae
corn root webworm *Crambus caliginosellus* Clemens Lepidoptera:Pyralidae
corn sap beetle *Carpophilus dimidiatus* (Fabricius) Coleoptera:Nitidulidae
corn silk beetle *Calomicrus brunneus* (Crotch) Coleoptera:Chrysomelidae
corn wireworm* *Melanotus communis* Gyllenhal Coleoptera:Elateridae
cotton aphid *Aphis gossypii* Glover Homoptera:Aphididae
cotton blister mite *Acalitus gossypii* (Banks) Acarina:Eriophyidae
cotton fleahopper *Pseudatomoscelis seriatus* (Reuter) Hemiptera:Miridae
cotton leafperforator *Bucculatrix thurberiella* Busck Lepidoptera:Lyonetiidae

cotton leafworm *Alabama argillacea* (Hübner) Lepidoptera:Noctuidae
cotton square borer *Strymon melinus* (Hübner) Lepidoptera:Lycaenidae
cotton stainer *Dysdercus suturellus* (Herrich-Schäffer) Hemiptera:Pyrrhocoridae
cottonwood leaf beetle *Chrysomela scripta* Fabricius Coleoptera:Chrysomelidae
cottony alder psyllid* *Psylla floccosa* Patch Homoptera:Psyllidae
cottony camellia scale* *Pulvinaria floccifera* Westwood Homoptera:Coccidae
cottonycushion scale *Icerya purchasi* Maskell Homoptera:Margarodidae
cottony maple scale *Pulvinaria innumerabilis* (Rathvon) Homoptera:Coccidae
cottony peach scale *Pulvinaria amygdali* Cockerell Homoptera:Coccidae
cowpea aphid *Aphis craccivora* Koch Homoptera:Aphididae
cowpea curculio *Chalcodermus aeneus* Boheman Coleoptera:Curculionidae
cowpea weevil *Callosobruchus maculatus* (Fabricius) Coleoptera:Bruchidae
crab louse *Pthirus pubis* (Linnaeus) Anoplura:Pediculidae
cranberry fruitworm *Acrobasis vaccinii* Riley Lepidoptera:Pyralidae
cranberry girdler *Chrysoteuchia topiaria* Zeller Lepidoptera:Pyralidae
cranberry rootworm *Rhabdopterus picipes* (Olivier) Coleoptera:Chrysomelidae
cranberry spanworm *Anavitrinella pampinaria* (Guenée) Lepidoptera:Geometridae
cranberry tipworm* *Dasyneura vaccinii* (Smith) Diptera:Cecidomyiidae
cranberry weevil *Anthonomus musculus* Say Coleoptera:Curculionidae
crescentmarked lily aphid *Neomyzus circumflexus* (Buckton) Homoptera:Aphididae
cribrate weevil* *Otiorhynchus cribricollis* Gyllenhal Coleoptera:Curculionidae
crown whitefly* *Aleuroplatus coronatus* (Quaintance) Homoptera:Aleyrodidae
Cuban laurel thrips *Gynaikothrips ficorum* (Marchal) Thysanoptera:Phlaeothripidae
curled rose sawfly *Allantus cinctus* (Linnaeus) Hymenoptera:Tenthredinidae
currant aphid *Cryptomyzus ribis* (Linnaeus) Homoptera:Aphididae
currant borer *Synanthedon tipuliformis* (Clerck) Lepidoptera:Sesiidae
currant bud mite *Cecidophyopsis ribis* (Westwood) Acarina:Eriophyidae
currant fruit fly *Epochra canadensis* (Loew) Diptera:Tephritidae
currant stem girdler *Janus integer* (Norton) Hymenoptera:Cephidae
cyclamen mite *Steneotarsonemus pallidus* (Banks) Acarina:Tarsonemidae
cypress bark moth* *Laspeyresia cupressana* (Kearfott) Lepidoptera:Olethreutidae
cypress leaftier* *Epinotia subviridis* Heinrich Lepidoptera:Olethreutidae
damsel bug* *Nabis ferus* (Linnaeus) Hemiptera:Nabidae
darklipped lacewing* *Chrysopa rufilabris* Burmeister Neuroptera:Chrysopidae
dark mealworm *Tenebrio obscurus* Fabricius Coleoptera:Tenebrionidae
depluming mite *Knemidokoptes gallinae* (Railliet) Acarina:Knemidokoptidae
depressed flour beetle *Palorus subdepressus* (Wollaston) Coleoptera:Tenebrionidae
desert locust* *Schistocerca gregaria* Forskål Orthoptera:Acrididae
desert spider mite *Tetranychus desertorum* Banks Acarina:Tetranychidae
diamondback moth *Plutella xylostella* (Linnaeus) Lepidoptera:Yponomeutidae
dichondra flea beetle* *Chaetocnema repens* McCrea Coleoptera:Chrysomelidae
differential grasshopper *Melanoplus differentialis* (Thomas) Orthoptera:Acrididae
Douglas-fir tussock moth *Orgyia pseudotsugata* (McDunnough) Lepidoptera:Lymantriidae
drone fly *Eristalis tenax* (Linnaeus) Diptera:Syrphidae

drugstore beetle *Stegobium paniceum* (Linnaeus) Coleoptera:Anobiidae
dryberry mite *Phyllocoptes gracilis* (Nalepa) Acarina:Eriophyidae
dusky sap beetle *Carpophilus lugubris* Murray Coleoptera:Nitidulidae
ear tick *Otobius megnini* (Dugès) Acarina:Argasidae
eastern raspberry fruitworm *Byturus rubi* Barber Coleoptera:Byturidae
eastern spruce gall aphid *Adelges abietis* (Linnaeus) Homoptera:Chermidae
eastern subterranean termite *Reticulitermes flavipes* (Kollar) Isoptera:Rhinotermitidae
eastern tent caterpillar *Malacosoma americanum* (Fabricius) Lepidoptera:Lasiocampidae
eggplant lace bug *Gargaphia solani* Heidemann Hemiptera:Tingidae
Egyptian alfalfa weevil *Hypera brunneipennis* Boheman Coleoptera:Curculionidae
elder borer* *Desmocerus palliatus* (Forster) Coleoptera:Cerambycidae
elm borer *Saperda tridentata* Olivier Coleoptera:Cerambycidae
elm cockscombgall aphid *Colopha ulmicola* (Fitch) Homoptera:Eriosomatidae
elm leaf beetle *Pyrrhalta luteola* (Muller) Coleoptera:Chrysomelidae
elm leafminer *Fenusa ulmi* Sundevall Hymenoptera:Tenthredinidae
elm sawfly *Cimbex americana* Leach Hymenoptera:Cimbicidae
elongate flea beetle *Systena elongata* (Fabricius) Coleoptera:Chrysomelidae
English grain aphid *Macrosiphum avenae* (Fabricius) Homoptera:Aphididae
English walnut scale* *Aspidiotus juglansregiae* Comstock Homoptera:Diaspididae
euonymus scale *Unaspis euonymi* (Comstock) Homoptera:Diaspididae
European chafer *Amphimallon majalis* (Razoumowsky) Coleoptera:Scarabaeidae
European chicken flea *Ceratophyllus gallinae* (Schrank) Siphonaptera:Ceratophyllidae
European corn borer *Ostrinia nubilalis* (Hübner) Lepidoptera:Pyralidae
European earwig *Forficula auricularia* Linnaeus Dermaptera:Forficulidae
European elm scale *Gossyparia spuria* (Modeer) Homoptera:Eriococcidae
European fruit lecanium *Lecanium corni* Bouché Homoptera:Coccidae
European fruit scale *Quadraspidiotus ostreaeformis* (Curtis) Homoptera:Diaspididae
European pine sawfly *Neodiprion sertifer* (Geoffroy) Hymenoptera:Diprionidae
European pine shoot moth *Rhyacionia buoliana* (Schiffermüller) Lepidoptera:Olethreutidae
European red mite *Panonychus ulmi* (Koch) Acarina:Tetranychidae
European wheat stem sawfly *Cephus pygmaeus* (Linnaeus) Hymenoptera:Cephidae
eyespotted bud moth *Spilonota ocellana* (Denis & Schiffermüller) Lepidoptera:Olethreutidae
face fly *Musca autumnalis* De Geer Diptera:Muscidae
fall armyworm *Spodoptera frugiperda* (J. E. Smith) Lepidoptera:Noctuidae
fall cankerworm *Alsophila pometaria* (Harris) Lepidoptera:Geometridae
fall webworm *Hyphantria cunea* (Drury) Lepidoptera:Arctiidae
false chinch bug *Nysius ericae* (Schilling) Hemiptera:Lygaeidae
fern scale *Pinnaspis aspidistrae* (Signoret) Homoptera:Diaspididae
fern whitefly* *Aleurotulus nephrolepidis* Quaintance Homoptera:Aleyrodidae
field crickets *Gryllus* spp. (complex of species) Orthoptera:Gryllidae
field crop predator mite* *Amblyseius cucumeris* (Oudemans) Acarina:Phytoseiidae

fiery lawn skipper* *Hylephila phylaeus* Drury Lepidoptera:Hesperiidae
fire ant *Solenopsis geminata* (Fabricius) Hymenoptera:Formicidae
flat grain beetle *Cryptolestes pusillus* (Schönherr) Coleoptera:Cucujidae
flatheaded appletree borer *Chrysobothris femorata* (Olivier)
Coleoptera:Buprestidae
Fletcher scale *Lecanium fletcheri* Cockerell Homoptera:Coccidae
Florida red scale *Chrysomphalus aonidum* (Linnaeus) Homoptera:Diaspididae
flower thrips *Frankliniella tritici* (Fitch) Thysanoptera:Thripidae
fluff louse *Goniocotes gallinae* (De Geer) Mallophaga:Philopteridae
follicle mite *Demodex folliculorum* (Simon) Acarina:Demodicidae
forage looper *Caenurgina erechtea* (Cramer) Lepidoptera:Noctuidae
Forbes scale *Quadraspidiotus forbesi* (Johnson) Homoptera:Diaspididae
foreign grain beetle *Ahasverus advena* (Waltl) Coleoptera:Cucujidae
forest tent caterpillar *Malacosoma disstria* Hübner Lepidoptera:Lasiocampidae
forktailed bush katydid *Scudderia furcata* Brunner von Wattenwyl
Orthoptera:Tettigoniidae
fourlined plant bug *Poecilocapsus lineatus* (Fabricius) Hemiptera:Miridae
fourspotted spider mite *Tetranychus canadensis* (McGregor)
Acarina:Tetranychidae
fowl cyst mite* *Laminosioptes cysticola* (Vizioli) Acarina:Laminosioptidae
fowl ticks *Argas* spp. Acarina:Argasidae
foxglove aphid *Acyrthosiphon solani* (Kaltenbach) Homoptera:Aphididae
frit fly *Oscinella frit* (Linnaeus) Diptera:Chloropidae
fruittree leafroller *Archips argyrospilus* (Walker) Lepidoptera:Tortricidae
Fuller rose beetle *Pantomorus cervinus* (Boheman) Coleoptera:Curculionidae
furniture carpet beetle *Anthrenus flavipes* LeConte Coleoptera:Dermestidae
garden fleahopper *Halticus bractatus* (Say) Hemiptera:Miridae
garden symphylan *Scutigerella immaculata* (Newport) Symphyla:Scutigerellidae
garden tortrix* *Ptycholoma peritana* (Clemens) Lepidoptera:Tortricidae
garden webworm *Loxostege rantalis* (Guenée) Lepidoptera:Pyralidae
geminate leafhopper* *Colladonus geminatus* van Duzee Homoptera:Cicadellidae
genista caterpillar *Tholeria reversalis* (Guenée) Lepidoptera:Pyralidae
German cockroach *Blattella germanica* (Linnaeus) Orthoptera:Blattellidae
giant bark aphid *Longistigma caryae* (Harris) Homoptera:Aphididae
gladiolus thrips *Taeniothrips simplex* (Morison) Thysanoptera:Thripidae
glassy cutworm *Crymodes devastator* (Brace) Lepidoptera:Noctuidae
goat sucking louse *Linognathus stenopsis* (Burmeister) Anoplura:Linognathidae
golden nematode* *Heterodera rostochiensis* (Wollenweber)
Tylenchida:Heteroderidae
gooseberry fruitworm *Zophodia convolutella* (Hübner) Lepidoptera:Pyralidae
grain mite *Acarus siro* Linnaeus Acarina:Acaridae
grain rust mite *Abacarus hystrix* (Nalepa) Acarina:Eriophyidae
granary weevil *Sitophilus granarius* (Linnaeus) Coleoptera:Curculionidae
granulate cutworm *Feltia subterranea* (Fabricius) Lepidoptera:Noctuidae
grape berry moth *Paralobesia viteana* (Clemens) Lepidoptera:Olethreutidae
grape blossom midge *Contarinia johnsoni* Felt Diptera:Cecidomyiidae
grape colaspis *Colaspis brunnea* (Fabricius) Coleoptera:Chrysomelidae
grape curculio *Craponius inaequalis* (Say) Coleoptera:Curculionidae
grape erineum mite *Colomerus vitis* (Pagenstecher) Acarina:Eriophyidae

grape flea beetle *Altica chalybea* (Illiger) Coleoptera:Chrysomelidae
grape leaffolder *Desmia funeralis* (Hübner) Lepidoptera:Pyralidae
grape leafhopper* *Erythroneura comes* (Say) Homoptera:Cicadellidae
grapeleaf skeletonizers *Harrisina* spp. Lepidoptera:Zygaenidae
grape mealybug *Pseudococcus maritimus* (Ehrhorn) Homoptera:Pseudococcidae
grape phylloxera *Daktulosphaira vitifoliae* (Fitch) Homoptera:Phylloxeridae
grape root borer *Vitacea polistiformis* (Harris) Lepidoptera:Sesiidae
grape rootworm *Fidia viticida* Walsh Coleoptera:Chrysomelidae
grape sawfly *Erythraspides vitis* (Harris) Hymenoptera:Tenthredinidae
grape scale *Diaspidiotus uvae* (Comstock) Homoptera:Diaspididae
grape seed chalcid *Evoxysoma vitis* (Saunders) Hymenoptera:Eurytomidae
grape trunk borer *Clytolepus albofasciatus* (Laporte & Gory)
    Coleoptera:Cerambycidae
grapevine aphid *Aphis illinoisensis* Shimer Homoptera:Aphididae
grapevine looper *Lygris diversilineata* (Hübner) Lepidoptera:Geometridae
grape whitefly *Trialeurodes vittata* (Quaintance) Homoptera:Aleyrodidae
grass mite* *Siteroptes graminum* (Reuter) Acarina:Pyemotidae
gray garden slug *Agriolimax reticulatus* (Müller) Stylommatophora:Limacidae
Great Basin wireworm *Ctenicera pruinina* (Horn) Coleoptera:Elateridae
greedy scale *Hemiberlesia rapax* (Comstock) Homoptera:Diaspididae
greenbug *Schizaphis graminum* (Rondani) Homoptera:Aphididae
green cloverworm *Plathypena scabra* (Fabricius) Lepidoptera:Noctuidae
greenfly* *Rhopalosiphum rufomaculatum* (Wilson) Homoptera:Aphididae
green fruitworm *Lithophane antennata* (Walker) Lepidoptera:Noctuidae
greenhouse leaftier *Udea rubigalis* (Guenée) Lepidoptera:Pyralidae
greenhouse orthezia *Orthezia insignis* Browne Homoptera:Ortheziidae
greenhouse slug *Milax gagates* (Linnaeus) Stylommatophora:Limacidae
greenhouse thrips *Heliothrips haemorrhoidalis* (Bouché) Thysanoptera:Thripidae
greenhouse whitefly *Trialeurodes vaporariorum* (Westwood)
    Homoptera:Aleyrodidae
green June beetle *Cotinis nitida* (Linnaeus) Coleoptera:Scarabaeidae
green lacewing* *Chrysopa plorabunda* Fitch Neuroptera:Chrysopidae
green peach aphid *Myzus persicae* (Sulzer) Homoptera:Aphididae
green stink bug *Acrosternum hilare* (Say) Hemiptera:Pentatomidae
ground mealybug *Rhizoecus falcifer* Künckel d'Herculais
    Homoptera:Pseudococcidae
Gulf Coast tick *Amblyomma maculatum* Koch Acarina:Ixodidae
Gulf fritillary* *Agraulis vanillae* (Linnaeus) Lepidoptera:Heliconiidae
gypsy moth *Porthetria dispar* (Linnaeus) Lepidoptera:Lymantriidae
hackberry nipplegall maker *Pachypsylla celtidismamma* (Riley)
    Homoptera:Psyllidae
hairy fungus beetle *Typhaea stercorea* (Linnaeus) Coleoptera:Mycetophagidae
hairy spider beetle *Ptinus villiger* (Reitter) Coleoptera:Ptinidae
Hall scale *Nilotaspis halli* (Green) Homoptera:Diaspididae
harlequin bug *Murgantia histrionica* (Hahn) Hemiptera:Pentatomidae
head louse *Pediculus humanus capitis* De Geer Anoplura:Pediculidae
hemispherical scale *Saissetia coffeae* (Walker) Homoptera:Coccidae
hemlock borer *Melanophila fulvoguttata* (Harris) Coleoptera:Buprestidae
Hessian fly *Mayetiola destructor* (Say) Diptera:Cecidomyiidae

hickory bark beetle *Scolytus quadrispinosus* Say Coleoptera:Scolytidae
hickory gall aphid* *Phylloxera caryaecaulis* (Fitch) Homoptera:Phylloxeridae
hickory tussock moth *Halisidota caryae* (Harris) Lepidoptera:Arctiidae
hog follicle mite *Demodex phylloides* Csokor Acarina:Demodicidae
hog itch mite* *Sarcoptes scabiei suis* (Gerlach) Acarina:Sarcoptidae
hog louse *Haematopinus suis* (Linnaeus) Anoplura:Haematopinidae
holly leafminer *Phytomyza ilicis* Curtis Diptera:Agromyzidae
honey bee *Apis mellifera* Linnaeus Hymenoptera:Apidae
hop aphid *Phorodon humuli* (Schrank) Homoptera:Aphididae
horn fly *Haematobia irritans (Linnaeus) Diptera:Muscidae*
horse biting louse *Bovicola equi* (Denny) Mallophaga:Trichodectidae
horse bot fly *Gasterophilus intestinalis* (DeGeer) Diptera:Gasterophilidae
horse follicle mite *Demodex equi* Railliet Acarina:Demodicidae
horse sucking louse *Haematopinus asini* (Linnaeus) Anoplura:Haematopinidae
house fly *Musca domestica* Linnaeus Diptera:Muscidae
human flea *Pulex irritans* Linnaeus Siphonaptera:Pulicidae
hydrangea spider mite* *Tetranychus hydrangeae* Pritchard & Baker
Acarina:Tetranychidae
imported cabbageworm *Pieris rapae* (Linnaeus) Lepidoptera:Pieridae
imported currantworm *Nematus ribesii* (Scopoli) Hymenoptera:Tenthredinidae
imported fire ant *Solenopsis saevissima richteri* Forel Hymenoptera:Formicidae
imported willow leaf beetle *Plagiodera versicolora* (Laicharting)
Coleoptera:Chrysomelidae
Indian lac insect* *Laccifer lacca* (Kern) Homoptera:Lacciferidae
Indian meal moth *Plodia interpunctella* (Hübner) Lepidoptera:Pyralidae
iris borer *Macronoctua onusta* Grote Lepidoptera:Noctuidae
iris whitefly* *Aleyrodes spiraeoides* Quaintance Homoptera:Aleyrodidae
itch mite *Sarcoptes scabiei* (De Geer) Acarina:Sarcoptidae
Japanese beetle *Popillia japonica* Newman Coleoptera:Scarabaeidae
Japanese weevil* *Pseudocneorhinus bifasciatus* Roelofs Coleoptera:Curculionidae
juniper scale *Carulaspis juniperi* (Bouché) Homoptera:Diaspididae
juniper twig girdler* *Periploca nigra* Hodges Lepidoptera:Momphidae
khapra beetle *Trogoderma granarium* Everts Coleoptera:Dermestidae
Klamath weed beetles* *Chrysolina quadrigemina* (Suffrian) and *C. hyperici*
(Forster) Coleoptera:Chrysomelidae
larch sawfly *Pristiphora erichsonii* (Hartig) Hymenoptera:Tenthredinidae
larder beetle *Dermestes lardarius* Linnaeus Coleoptera:Dermestidae
large chicken louse *Goniodes gigas* (Taschenberg) Mallophaga:Philopteridae
larger cabinet beetle* *Trogoderma inclusum* LeConte Coleoptera:Dermestidae
larger grain borer *Prostephanus truncatus* (Horn) Coleoptera:Bostrichidae
large syrphid* *Scaeva pyrastri* (Linnaeus) Diptera:Syrphidae
laurel psyllid* *Trioza alacris* Flor Homoptera:Psyllidae
leaf curl plum aphid* *Anuraphis helichrys* (Kaltenbach) Homoptera:Aphididae
leaffooted bug *Leptoglossus phyllopus* (Linnaeus) Hemiptera:Coreidae
leafrolling weevils* *Attelabus* spp. Coleoptera:Curculionidae
legume bug* *Lygus hesperus* Knight Hemiptera:Miridae
leopard moth *Zeuzera pyrina* (Linnaeus) Lepidoptera:Cossidae
lesser bulb fly *Eumerus tuberculatus* Rondani Diptera:Syrphidae
lesser clover leaf weevil *Hypera nigrirostris* (Fabricius) Coleoptera:Curculionidae

lesser cornstalk borer *Elasmopalpus lignosellus* (Zeller) Lepidoptera:Pyralidae
lesser grain borer *Rhyzopertha dominica* (Fabricius) Coleoptera:Bostrichidae
lesser mealworm *Alphitobius diaperinus* (Panzer) Coleoptera:Tenebrionidae
lesser peachtree borer *Synanthedon pictipes* (Grote & Robinson) Lepidoptera:Sesiidae
lilac borer *Podosesia syringae* (Harris) Lepidoptera:Sesiidae
lilac leafminer *Caloptilia syringella* (Fabricius) Lepidoptera:Gracillariidae
limabean pod borer *Etiella zinckenella* (Treitschke) Lepidoptera:Pyralidae
little black ant *Monomorium minimum* (Buckley) Hymenoptera:Formicidae
little blue cattle louse* *Solenopotes capillatus* Enderlein Anoplura:Linognathidae
little house fly *Fannia canicularis* (Linnaeus) Diptera:Muscidae
live oak leafcutter* *Vespina quercivora* (Davis) Lepidoptera:Incurvariidae
lobed spider mite* *Tetranychus lobosus* Boudreaux Acarina:Tetranychidae
locust borer *Megacyllene robiniae* (Forster) Coleoptera:Cerambycidae
locust clearwing* *Paranthrene robiniae robiniae* (Hy. Edwards) Lepidoptera:Sesiidae
locust leafminer *Odontota dorsalis* (Thunberg) Coleoptera:Chrysomelidae
lodgepole needleminer *Coleotechnites milleri* (Busck) Lepidoptera:Gelechiidae
lone star tick *Amblyomma americanum* (Linnaeus) Acarina:Ixodidae
longheaded flour beetle *Latheticus oryzae* Waterhouse Coleoptera:Tenebrionidae
longnosed cattle louse *Linognathus vituli* (Linnaeus) Anoplura:Linognathidae
longtailed mealybug *Pseudococcus longispinus* (Targioni-Tozzetti) Homoptera:Pseudococcidae
lubber grasshopper *Brachystola magna* (Girard) Orthoptera:Acrididae
lucerne moth* *Nomophila noctuella* (Dennis & Schiffermüller) Lepidoptera:Pyralidae
luden spider mite* *Tetranychus ludeni* Zacher Acarina:Tetranychidae
lygus bugs* *Lygus* spp. Hemiptera:Miridae
madrona shield bearer* *Coptodisca arbutiella* Busck Lepidoptera:Heliozelidae
magnolia scale *Neolecanium cornuparvum* (Thro) Homoptera:Coccidae
maize billbug *Sphenophorus maidis* Chittenden Coleoptera:Curculionidae
maize weevil *Sitophilus zeamais* Motschulsky Coleoptera:Curculionidae
maple bladdergall mite *Vasates quadripedes* Shimer Acarina:Eriophyidae
maple callus borer *Synanthedon acerni* (Clemens) Lepidoptera:Sesiidae
maple leafcutter *Paraclemensia acerifoliella* (Fitch) Lepidoptera:Incurvariidae
marsh slug *Agriolimax laevis* (Müller) Stylommatophora:Limacidae
McDaniel spider mite *Tetranychus mcdanieli* McGregor Acarina:Tetranychidae
meadow spittlebug *Philaenus spumarius* (Linnaeus) Homoptera:Cercopidae
meal moth *Pyralis farinalis* Linnaeus Lepidoptera:Pyralidae
mealybug destroyer* *Cryptolaemus montrouzieri* Mulsant Coleoptera:Coccinellidae
mealy plum aphid *Hyalopterus pruni* (Geoffroy) Homoptera:Aphididae
Mediterranean flour moth *Anagasta kuehniella* (Zeller) Lepidoptera:Pyralidae
Mediterranean fruit fly *Ceratitis capitata* (Wiedemann) Diptera:Tephritidae
melon aphid *Aphis gossypii* Glover Homoptera:Aphididae
melon fly *Dacus cucurbitae* Coquillett Diptera:Tephritidae
merchant grain beetle *Oryzaephilus mercator* (Fauvel) Coleoptera:Cucujidae
Mexican bean beetle *Epilachna varivestis* Mulsant Coleoptera:Coccinellidae
Mexican fruit fly *Anastrepha ludens* (Loew) Diptera:Tephritidae

Mexican mealybug *Phenacoccus gossypii* Townsend & Cockerell
Homoptera:Pseudococcidae
migratory grasshopper *Melanoplus sanguinipes* (Fabricius) Orthoptera:Acrididae
mimosa webworm *Homadaula anisocentra* Meyrick
Lepidoptera:Glyphipterygidae
minute egg parasite* *Trichogramma minutum* Riley
Hymenoptera:Trichogrammatidae
Monterey pine scale* *Physokermes insignicola* (Craw) Homoptera:Coccidae
Mormon cricket *Anabrus simplex* Haldeman Orthoptera:Tettigoniidae
mourningcloak butterfly *Nymphalis antiopa* (Linnaeus) Lepidoptera:Nymphalidae
Nantucket pine tip moth *Rhyacionia frustrana* (Comstock)
Lepidoptera:Olethreutidae
narcissus bulb fly *Merodon equestris* (Fabricius) Diptera:Syrphidae
native elm bark beetle *Hylurgopinus rufipes* (Eichhoff) Coleoptera:Scolytidae
nautical borer* *Xylotrechus nauticus* (Mannerheim) Coleoptera:Cerambycidae
New York weevil *Ithycerus noveboracensis* (Forster) Coleoptera:Curculionidae
northern cattle grub *Hypoderma bovis* (Linnaeus) Diptera:Oestridae
northern corn rootworm *Diabrotica longicornis* (Say) Coleoptera:Chrysomelidae
northern fowl mite *Ornithonyssus sylviarum* (Canestrini & Fanzago)
Acarina:Macronyssidae
northern house mosquito *Culex pipiens* Linnaeus Diptera:Culicidae
nose bot fly *Gasterophilus haemorrhoidalis* (Linnaeus) Diptera:Gasterophilidae
nut weevils* *Curculio* spp. Coleoptera:Curculionidae
oak gall wasps* Hymenoptera:Cynipidae
oak pit scales* *Asterolecanium* spp. Homoptera:Asterolecaniidae
oak twig girdler* *Agrilus arcuatus* Say Coleoptera:Buprestidae
oak wax scale* *Cerococcus quercus* Comstock Homoptera:Asterolecaniidae
obliquebanded leafroller *Choristoneura rosaceana* (Harris)
Lepidoptera:Tortricidae
obscure root weevil *Sciopithes obscurus* Horn Coleoptera:Curculionidae
odorous house ant *Tapinoma sessile* (Say) Hymenoptera:Formicidae
oleander scale *Aspidiotus nerii* Bouché Homoptera:Diaspididae
olive scale *Parlatoria oleae* (Colvée) Homoptera:Diaspididae
omnivorous leafroller* *Platynota stultana* Walsingham Lepidoptera:Tortricidae
omnivorous leaftier *Cnephasia longana* (Haworth) Lepidoptera:Tortricidae
omnivorous looper *Sabulodes caberata* Guenée Lepidoptera:Geometridae
onion maggot *Hylemya antiqua* (Meigen) Diptera:Anthomyiidae
onion thrips *Thrips tabaci* Lindeman Thysanoptera:Thripidae
orange tortrix *Argyrotaenia citrana* (Fernald) Lepidoptera:Tortricidae
orchidfly *Eurytoma orchidearum* (Westwood) Hymenoptera:Eurytomidae
oriental beetle *Anomala orientalis* Waterhouse Coleoptera:Scarabaeidae
oriental cockroach *Blatta orientalis* Linnaeus Orthoptera:Blattidae
oriental fruit fly *Dacus dorsalis* Hendel Diptera:Tephritidae
oriental fruit moth *Grapholitha molesta* (Busck) Lepidoptera:Olethreutidae
oriental house fly *Musca domestica vicina* Macquart Diptera:Muscidae
oriental rat flea *Xenopsylla cheopis* (Rothschild) Siphonaptera:Pulicidae
ornate aphid* *Myzus ornatus* Laing Homoptera:Aphididae
oxalis spider mite* *Petrobia harti* (Ewing) Acarina:Tetranychidae
oystershell scale *Lepidosaphes ulmi* (Linnaeus) Homoptera:Diaspididae

Pacific ash leaf bug* *Tropidosteptes pacificus* Van Duzee Hemiptera:Miridae
Pacific Coast tick *Dermacentor occidentalis* Marx Acarina:Ixodidae
Pacific Coast wireworm *Limonius canus* LeConte Coleoptera:Elateridae
Pacific flatheaded borer *Chrysobothris mali* Horn Coleoptera:Buprestidae
Pacific oak twig girdler* *Agrilus angelicus* Horn Coleoptera:Buprestidae
Pacific spider mite *Tetranychus pacificus* McGregor Acarina:Tetranychidae
painted leafhopper *Endria inimica* (Say) Homoptera:Cicadellidae
painted maple aphid *Drepanaphis acerifoliae* (Thomas) Homoptera:Aphididae
pale legume bug* *Lygus elisus* Van Duzee Hemiptera:Miridae
palesided cutworm *Agrotis malefida* Guenée Lepidoptera:Noctuidae
palestriped flea beetle *Systena blanda* Melsheimer Coleoptera:Chrysomelidae
pales weevil *Hylobius pales* (Herbst) Coleoptera:Curculionidae
pale tussock moth *Halisidota tessellaris* (J. E. Smith) Lepidoptera:Arctiidae
pale western cutworm *Agrotis orthogonia* Morrison Lepidoptera:Noctuidae
palm leaf skeletonizer *Homaledra sabalella* (Chambers) Lepidoptera:Momphidae
pappataci sand fly* *Phlebotamus papatasi* Scopoli Diptera:Psychodidae
parlatoria date scale *Parlatoria blanchardi* (Targioni-Tozzetti) Homoptera:Diaspididae
parsnip webworm *Depressaria pastinacella* (Duponchel) Lepidoptera:Oecophoridae
pavement ant *Tetramorium caespitum* (Linnaeus) Hymenoptera:Formicidae
pea aphid *Acyrthosiphon pisum* (Harris) Homoptera:Aphididae
peach bark beetle *Phloeotribus liminaris* (Harris) Coleoptera:Scolytidae
peach silver mite *Aculus cornutus* (Banks) Acarina:Eriophyidae
peachtree borer *Sanninoidea exitiosa* (Say) Lepidoptera:Sesiidae
peach twig borer *Anarsia lineatella* Zeller Lepidoptera:Gelechiidae
pea leaf weevil *Sitona lineatus* (Linnaeus) Coleoptera:Curculionidae
pea moth *Laspeyresia nigricana* (Stephens) Lepidoptera:Olethreutidae
pear blight beetle* *Anisandrus dispar* (Fabricius) Coleoptera:Scolytidae
pear lace bug* *Stephanitis pyri* Fabricius Hemiptera:Tingidae
pearleaf blister mite *Eriophyes pyri* (Pagenstecher) Acarina:Eriophyidae
pear midge *Contarinia pyrivora* (Riley) Diptera:Cecidomyiidae
pear psylla *Psylla pyricola* Foerster Homoptera:Psyllidae
pearslug *Caliroa cerasi* (Linnaeus) Hymenoptera:Tenthredinidae
pear thrips *Taeniothrips inconsequens* (Uzel) Thysanoptera:Thripidae
pea weevil *Bruchus pisorum* (Linnaeus) Coleoptera:Bruchidae
pepper maggot *Zonosemata electa* (Say) Diptera:Tephritidae
pepper weevil *Anthonomus eugenii* Cano Coleoptera:Curculionidae
periodical cicada *Magicicada septendecim* (Linnaeus) Homoptera:Cicadidae
phalaenopsis mite* *Tenuipalpus pacificus* Baker Acarina:Tenuipalpidae
Pharaoh ant *Monomorium pharaonis* (Linnaeus) Hymenoptera:Formicidae
phycitid oak leaftier* *Rhodophaea caliginella* (Hulst) Lepidoptera:Pyralidae
pickleworm *Diaphania nitidalis* (Stoll) Lepidoptera:Pyralidae
pigeon tremex *Tremex columba* (Linnaeus) Hymenoptera:Siricidae
pine bark aphid *Pineus strobi* (Hartig) Homoptera:Chermidae
pine cone willow gall* *Rhabdophaga strobiloides* (Osten-Sacken) Diptera:Cecidomyiidae
pine needle aphid* *Essigella californica* (Essig) Homoptera:Aphididae
pine needle scale *Chionaspis pinifoliae* (Fitch) Homoptera:Diaspididae

pine spittlebug *Aphrophora parallela* (Say) Homoptera:Cercopidae
pink bollworm *Pectinophora gossypiella* (Saunders) Lepidoptera:Gelechiidae
plains false wireworm *Eleodes opacus* (Say) Coleoptera:Tenebrionidae
plum curculio *Conotrachelus nenuphar* (Herbst) Coleoptera:Curculionidae
plum gouger *Coccotorus scutellaris* (LeConte) Coleoptera:Curculionidae
plum leafhopper *Macropsis trimaculata* (Fitch) Homoptera:Cicadellidae
plume moths *Platyptilia* spp. Lepidoptera:Pterophoridae
poplar-and-willow borer *Cryptorhynchus lapathi* (Linnaeus) Coleoptera:Curculionidae
poplar aphid* *Chaitophorus* spp. Homoptera:Aphididae
poplar borer *Saperda calcarata* Say Coleoptera:Cerambycidae
poplar petiolegall aphid *Pemphigus populitransversus* Riley Homoptera:Eriosomatidae
potato aphid *Macrosiphum euphorbiae* (Thomas) Homoptera:Aphididae
potato leafhopper *Empoasca fabae* (Harris) Homoptera:Cicadellidae
potato psyllid *Paratrioza cockerelli* (Sulc) Homoptera:Psyllidae
potato stem borer *Hydroecia micacea* (Esper) Lepidoptera:Noctuidae
potato tuberworm *Phthorimaea operculella* (Zeller) Lepidoptera:Gelechiidae
prairie grain wireworm *Ctenicera destructor* (Brown) Coleoptera:Elateridae
privet mite *Brevipalpus obovatus* Donnadieu Acarina:Tenuipalpidae
Puget Sound wireworm *Ctenicera aeripennis aeripennis* (Kirby) Coleoptera:Elateridae
Putnam scale *Diaspidiotus ancylus* (Putnam) Homoptera:Diaspididae
ragweed plant bug *Chlamydatus associatus* (Uhler) Hemiptera:Miridae
rain beetles *Pleocoma* spp. Coleoptera:Scarabaeidae
range crane fly *Tipula simplex* Doane Diptera:Tipulidae
rapid plant bug *Adelphocoris rapidus* (Say) Hemiptera:Miridae
raspberry aphid* *Amphorophora rubi* (Kaltenbach) Homoptera:Aphididae
raspberry cane borer *Oberea bimaculata* (Olivier) Coleoptera:Cerambycidae
raspberry cane maggot *Pegomya rubivora* (Coquillett) Diptera:Anthomyiidae
raspberry crown borer *Pennisetia marginata* (Harris) Lepidoptera:Sesiidae
raspberry horntail* *Hartigia cressonii* (Kirby) Hymenoptera:Cephidae
raspberry leaf sawfly* *Priophorus moris* (Lepeletier) Hymenoptera:Tenthredinidae
raspberry sawfly *Monophadnoides geniculatus* (Hartig) Hymenoptera:Tenthredinidae
redbacked cutworm *Euxoa ochrogaster* (Guenée) Lepidoptera:Noctuidae
redbanded leafroller *Argyrotaenia velutinana* (Walker) Lepidoptera:Tortricidae
redberry mite *Acalitus essigi* (Hassan) Acarina:Eriophyidae
red flour beetle *Tribolium castaneum* (Herbst) Coleoptera:Tenebrionidae
red harvester ant *Pogonomyrmex barbatus* (F. Smith) Hymenoptera:Formicidae
redheaded pine sawfly *Neodiprion lecontei* (Fitch) Hymenoptera:Diprionidae
redhumped caterpillar *Schizura concinna* (J. E. Smith) Lepidoptera:Notodontidae
redlegged grasshopper *Melanoplus femurrubrum* (De Geer) Orthoptera:Acrididae
rednecked cane borer *Agrilus ruficollis* (Fabricius) Coleoptera:Buprestidae
redshouldered plant bug* *Thyanta custator* (Fabricius) Hemiptera:Pentatomidae
redtailed flesh fly* *Sarcophaga haemorrhoidalis* (Fallen) Diptera:Sarcophagidae
red turpentine beetle *Dendroctonus valens* LeConte Coleoptera:Scolytidae
relapsing fever tick *Ornithodoros turicata* (Dugès) Acarina:Argasidae

resplendent shield bearer *Coptodisca splendoriferella* (Clemens) Lepidoptera:Heliozelidae
Rhodesgrass scale *Antonina graminis* (Maskell) Homoptera:Pseudococcidae
rhododendron borer *Synanthedon rhododendri* Beutenmüller Lepidoptera:Sesiidae
rhododendron lace bug *Stephanitis rhododendri* Horvath Hemiptera:Tingidae
rhododendron stem borer* *Oberea myops* Haldeman Coleoptera:Cerambycidae
rhubarb spittlebug* *Aphrophora permutata* Uhler Homoptera:Cercopidae
ribbed pine borer* *Stenocorus inquisitor lineatus* (Olivier) Coleoptera:Cerambycidae
rice moth* *Corcyra cephalonica* Staint Lepidoptera:Pyralidae
rice weevil *Sitophilus oryzae* (Linnaeus) Coleoptera:Curculionidae
Rocky Mountain wood tick *Dermacentor andersoni* Stiles Acarina:Ixodidae
rose aphid *Macrosiphum rosae* (Linnaeus) Homoptera:Aphididae
rose budworm* *Pyrrhia umbra* (Hufnagel) Lepidoptera:Noctuidae
rose chafer *Macrodactylus subspinosus* (Fabricius) Coleoptera:Scarabaeidae
rose curculio *Rhynchites bicolor* (Fabricius) Coleoptera:Curculionidae
rose grass aphid* *Acyrthosiphon dirhodum* (Walker) Homoptera:Aphididae
rose leaf beetle *Nodonota puncticollis* (Say) Coleoptera:Chrysomelidae
rose leafhopper *Edwardsiana rosae* (Linnaeus) Homoptera:Cicadellidae
rose leaftier* *Archips rosaceanus* (Harris) Lepidoptera:Tortricidae
rose midge *Dasineura rhodophaga* (Coquillett) Diptera:Cecidomyiidae
rose scale *Aulacaspis rosae* (Bouché) Homoptera:Diaspididae
roseslug *Endelomyia aethiops* (Fabricius) Hymenoptera:Tenthredinidae
rose stem girdler *Agrilus aurichalceus* Redtenbacher Coleoptera:Buprestidae
rosy apple aphid *Dysaphis plantaginea* (Passerini) Homoptera:Aphididae
rough strawberry root weevil* *Otiorhynchus rugosostriatus* (Goeze) Coleoptera:Curculionidae
roundheaded appletree borer *Saperda candida* Fabricius Coleoptera:Cerambycidae
rusty grain beetle *Cryptolestes ferrugineus* (Stephens) Coleoptera:Cucujidae
rusty plum aphid *Hysteroneura setariae* (Thomas) Homoptera:Aphididae
saltmarsh caterpillar *Estigmene acrea* (Drury) Lepidoptera:Arctiidae
sand wireworm *Horistonotus uhlerii* Horn Coleoptera:Elateridae
San Jose scale *Quadraspidiotus perniciosus* (Comstock) Homoptera:Diaspididae
satin moth *Leucoma salicis* (Linnaeus) Lepidoptera:Lymantriidae
sawtoothed grain beetle *Oryzaephilus surinamensis* (Linnaeus) Coleoptera:Cucujidae
Say stink bug *Chlorochroa sayi* Stål Hemiptera:Pentatomidae
scab mite *Psoroptes equi* (Raspail) Acarina:Psoroptidae
scale insect destroyer* *Microweisea misella* (LeConte) Coleoptera:Coccinellidae
scalyleg mite *Knemidokoptes mutans* (Robin & Lanquetin) Acarina:Knemidokoptidae
Schoene spider mite *Tetranychus schoenei* McGregor Acarina:Tetranychidae
screwworm *Cochliomyia hominivorax* (Coquerel) Diptera:Calliphoridae
scurfy scale *Chionaspis furfura* (Fitch) Homoptera:Diaspididae
secondary screwworm *Cochliomyia macellaria* (Fabricius) Diptera:Calliphoridae
seedcorn beetle *Agonoderus lecontei* Chaudoir Coleoptera:Carabidae
seedcorn maggot *Hylemya platura* (Meigen) Diptera:Anthomyiidae

serpentine leafminer *Liriomyza brassicae* (Riley) Diptera:Agromyzidae
shaft louse *Menopon gallinae* (Linnaeus) Mallophaga:Menoponidae
sharpnosed leafhopper* *Scaphytopius magdalensis* (Provancher) Homoptera:Cicadellidae
sheep biting louse *Bovicola ovis* (Schrank) Mallophaga:Trichodectidae
sheep bot fly *Oestrus ovis* Linnaeus Diptera:Oestridae
sheep follicle mite *Demodex ovis* Railliet Acarina:Demodicidae
sheep foot louse* *Linognathus pedalis* (Osborn) Anoplura:Linognathidae
sheep itch mite* *Psorergates ovis* Womersley Acarina:Cheyletidae
sheep ked *Melophagus ovinus* (Linnaeus) Diptera:Hippoboscidae
shortnosed cattle louse *Haematopinus eurysternus* (Nitzsch) Anoplura:Haematopinidae
shothole borer *Scolytus rugulosus* (Ratzeburg) Coleoptera:Scolytidae
silkworm *Bombyx mori* (Linnaeus) Lepidoptera:Bombycidae
silverfish *Lepisma saccharina* Linnaeus Thysanura:Lepismatidae
silverspotted skipper *Epargyreus clarus* (Cramer) Lepidoptera:Hesperiidae
silverspotted tiger moth *Halisidota argentata* Packard Lepidoptera:Arctiidae
sinuate peartree borer *Agrilus sinuatus* (Olivier) Coleoptera:Buprestidae
slenderhorned flour beetle *Gnathocerus maxillosus* (Fabricius) Coleoptera:Tenebrionidae
smaller European elm bark beetle *Scolytus multistriatus* (Marsham) Coleoptera:Scolytidae
smalleyed flour beetle *Palorus ratzeburgi* (Wissmann) Coleoptera:Tenebrionidae
snapdragon plume moth* *Platyptilia antirrhina* Lange Lepidoptera:Pterophoridae
sod webworms* *Crambus* spp. Lepidoptera:Pyralidae
solitary oak leafminer *Cameraria hamadryadella* (Clemens) Lepidoptera:Gracillariidae
sorghum midge *Contarinia sorghicola* (Coquillett) Diptera:Cecidomyiidae
southern armyworm *Spodoptera eridania* (Cramer) Lepidoptera:Noctuidae
southern cattle tick *Boophilus microplus* (Canestrini) Acarina:Ixodidae
southern chinch bug *Blissus insularis* Barber Hemiptera:Lygaeidae
southern corn rootworm *Diabrotica undecimpunctata howardi* Barber Coleoptera:Chrysomelidae
southern cornstalk borer *Diatraea crambidoides* (Grote) Lepidoptera:Pyralidae
southern fire ant *Solenopsis xyloni* McCook Hymenoptera:Formicidae
southern garden leafhopper *Empoasca solana* DeLong Homoptera:Cicadellidae
southern green stink bug *Nezara viridula* (Linnaeus) Hemiptera:Pentatomidae
southern house mosquito *Culex quinquefasciatus* Say Diptera:Culicidae
southern lyctus beetle *Lyctus planicollis* LeConte Coleoptera:Lyctidae
southern pine beetle *Dendroctonus frontalis* Zimmerman Coleoptera:Scolytidae
southern redlegged grasshopper* *Melanoplus femurrubrum propinquus* Scudder Orthoptera:Acrididae
southern red mite *Oligonychus ilicis* (McGregor) Acarina:Tetranychidae
southwestern corn borer *Diatraea grandiosella* (Dyar) Lepidoptera:Pyralidae
soybean cyst nematode* *Heterodera glycines* Ichinohe Tylenchida:Heteroderidae
spider mite destroyers* *Stethorus picipes* Casey and *S. punctum* (LeConte) Coleoptera:Coccinellidae
spinach flea beetle *Disonycha xanthomelas* (Dalman) Coleoptera:Chrysomelidae
spinach leafminer *Pegomya hyoscyami* (Panzer) Diptera:Anthomyiidae

spirea aphid *Aphis spiraecola* Patch Homoptera:Aphididae
spotted alfalfa aphid *Therioaphis maculata* (Buckton) Homoptera:Aphididae
spotted asparagus beetle *Crioceris duodecimpunctata* (Linnaeus) Coleoptera:Chrysomelidae
spotted cucumber beetle *Diabrotica undecimpunctata howardi* Barber Coleoptera:Chrysomelidae
spotted garden slug *Limax maximus* Linnaeus Stylommatophora:Limacidae
spotted limb borer* *Psoa maculata* (LeConte) Coleoptera:Bostrichidae
spotted tussock moth *Halisidota maculata* (Harris) Lepidoptera:Arctiidae
spring cankerworm *Paleacrita vernata* (Peck) Lepidoptera:Geometridae
spruce beetle *Dendroctonus obesus* (Mannerheim) Coleoptera:Scolytidae
spruce bud midge *Rhabdophaga swainei* Felt Diptera:Cecidomyiidae
spruce budworm *Choristoneura fumiferana* (Clemens) Lepidoptera:Tortricidae
spruce spider mite *Oligonychus ununguis* (Jacobi) Acarina:Tetranychidae
squarenecked grain beetle *Cathartus quadricollis* (Guérin-Méneville) Coleoptera:Cucujidae
squash bug *Anasa tristis* (De Geer) Hemiptera:Coreidae
squash vine borer *Melittia cucurbitae* (Harris) Lepidoptera:Sesiidae
stable fly *Stomoxys calcitrans* (Linnaeus) Diptera:Muscidae
stalk borer *Papaipema nebris* (Guenée) Lepidoptera:Noctuidae
sticktight flea *Echidnophaga gallinacea* (Westwood) Siphonaptera:Pulicidae
strawberry aphid *Chaetosiphon fragaefolii* (Cockerell) Homoptera:Aphididae
strawberry bug* *Myodochus serripes* Oliver Hemiptera:Lygaeidae
strawberry crown borer *Tyloderma fragariae* (Riley) Coleoptera:Curculionidae
strawberry crownminer *Aristotelia fragariae* Busck Lepidoptera:Gelechiidae
strawberry crown moth *Synanthedon bibionipennis* (Boisduval) Lepidoptera:Sesiidae
strawberry flea beetle* *Altica ignita* Illiger Coleoptera:Chrysomelidae
strawberry leaf beetle* *Paria canella* Fabricius Coleoptera:Chrysomelidae
strawberry leafminer* *Tischeria* sp. Lepidoptera:Tischeriidae
strawberry leafroller *Ancylis comptana fragariae* (Walsh & Riley) Lepidoptera:Olethreutidae
strawberry root aphid *Aphis forbesi* Weed Homoptera:Aphididae
strawberry root weevil *Otiorhynchus ovatus* (Linnaeus) Coleoptera:Curculionidae
strawberry rootworm *Paria fragariae* Wilcox Coleoptera:Chrysomelidae
strawberry sawflies* *Empria maculata* (Norton) and *E. ignota* (Norton) Hymenoptera:Tenthredinidae
strawberry spider mite *Tetranychus turkestani* Ugarov and Nikolski Acarina:Tetranychidae
strawberry weevil *Anthonomus signatus* Say Coleoptera:Curculionidae
strawberry whitefly *Trialeurodes packardi* (Morrill) Homoptera:Aleyrodidae
striped blister beetle *Epicauta vittata* (Fabricius) Coleoptera:Meloidae
striped cucumber beetle *Acalymma vittatum* (Fabricius) Coleoptera:Chrysomelidae
striped horse fly *Tabanus lineola* Fabricius Diptera:Tabanidae
sugarbeet crown borer *Hulstia undulatella* (Clemens) Lepidoptera:Pyralidae
sugarbeet root aphid *Pemphigus populivenae* Fitch Homoptera:Aphididae
sugarbeet root maggot *Tetanops myopaeformis* (Röder) Diptera:Otitidae
sugarbeet wireworm *Limonius californicus* (Mannerheim) Coleoptera:Elateridae

sugarcane beetle *Euetheola rugiceps* (LeConte) Coleoptera:Scarabaeidae
sugarcane borer *Diatraea saccharalis* (Fabricius) Lepidoptera:Pyralidae
sugarcane leafroller *Hedylepta accepta* (Butler) Lepidoptera:Pyralidae
sugar maple borer *Glycobius speciosus* (Say) Coleoptera:Cerambycidae
sunflower maggot *Strauzia longipennis* (Wiedemann) Diptera:Tephritidae
superb plant bug *Adelphocoris superbus* (Uhler) Hemiptera:Miridae
sweetclover weevil *Sitona cylindricollis* Fåhraeus Coleoptera:Curculionidae
sweetpotato flea beetle *Chaetocnema confinis* Crotch Coleoptera:Chrysomelidae
sweetpotato hornworm *Agrius cingulatus* (Fabricius) Lepidoptera:Sphingidae
sweetpotato leaf beetle *Typophorus nigritus viridicyaneus* (Crotch) Coleoptera:Chrysomelidae
sweetpotato weevil *Cylas formicarius elegantulus* (Summers) Coleoptera:Curculionidae
sweetpotato whitefly *Bemisia tabaci* (Gennadius) Homoptera:Aleyrodidae
sycamore aphid* *Drepanosiphum platanoides* (Schrank) Homoptera:Aphididae
sycamore scale* *Stomacoccus platani* Ferris Homoptera:Margarodidae
sycamore lace bug *Corythucha ciliata* (Say) Hemiptera:Tingidae
sycamoreleaf blotch miner* *Lithocolletis felinella* (Heinrich) Lepidoptera:Gracillariidae
sycamore leafmining sawfly* *Profenusa platanae* Burks Hymenoptera:Tenthredinidae
sycamoreleaf skeletonizer* *Gelechia desiliens* Meyrick Lepidoptera:Gelechiidae
sycamore tussock moth *Halisidota harrisii* Walsh Lepidoptera:Arctiidae
tarnished plant bug *Lygus lineolaris* (Palisot de Beauvois) Hemiptera:Miridae
taxus mealybug* *Dysmicoccus wistariae* (Green) Homoptera:Pseudococcidae
tenlined June beetle *Polyphylla decemlineata* (Say) Coleoptera:Scarabaeidae
terrapin scale *Lecanium nigrofasciatum* Pergande Homoptera:Coccidae
thief ant *Solenopsis molesta* (Say) Hymenoptera:Formicidae
thistle aphid *Brachycaudus cardui* (Linnaeus) Homoptera:Aphididae
threebanded leafhopper *Erythroneura tricincta* Fitch Homoptera:Cicadellidae
threelined potato beetle *Lema trilineata* (Olivier) Coleoptera:Chrysomelidae
throat bot fly *Gasterophilus nasalis* (Linnaeus) Diptera:Gasterophilidae
tiger swallowtail *Papilio glaucus* Linnaeus Lepidoptera:Papilionidae
tobacco budworm *Heliothis virescens* (Fabricius) Lepidoptera:Noctuidae
tobacco hornworm *Manduca sexta* (Linnaeus) Lepidoptera:Sphingidae
tobacco thrips *Frankliniella fusca* (Hinds) Thysanoptera:Thripidae
tobacco wireworm *Conoderus vespertinus* (Fabricius) Coleoptera:Elateridae
tomato psyllid *Paratrioza cockerelli* (Sulc) Homoptera:Psyllidae
toothed flea beetle *Chaetocnema denticulata* (Illiger) Coleoptera:Chrysomelidae
torsalo *Dermatobia hominis* (Linnaeus, Jr.) Diptera:Cuterebridae
tree fruit predator mite* *Metaseiulus occidentalis* (Nesbitt) Acarina:Phytoseiidae
tropical fowl mite *Ornithonyssus bursa* (Berlese) Acarina:Macronyssidae
tropical horse tick *Anocentor nitens* (Neumann) Acarina:Ixodidae
tuber flea beetle *Epitrix tuberis* Gentner Coleoptera:Chrysomelidae
tulip bulb aphid *Dysaphis tulipae* (Fonscolombe) Homoptera:Aphididae
tuliptree scale *Toumeyella liriodendri* (Gmelin) Homoptera:Coccidae
tumid spider mite *Tetranychus tumidus* Banks Acarina:Tetranychidae
turkey gnat *Simulium meridionale* Riley Diptera:Simuliidae
turnip aphid *Hyadaphis erysimi* (Kaltenbach) Homoptera:Aphididae

turnip maggot *Hylemya floralis* (Fallén) Diptera:Anthomyiidae
twicestabbed lady beetle *Chilocorus stigma* (Say) Coleoptera:Coccinellidae
twig girdler *Oncideres cingulata* (Say) Coleoptera:Cerambycidae
twig pruner *Elaphidionoides villosus* (Fabricius) Coleoptera:Cerambycidae
twolined chestnut borer *Agrilus bilineatus* (Weber) Coleoptera:Buprestidae
twospotted lady beetle *Adalia bipunctata* (Linnaeus) Coleoptera:Coccinellidae
twospotted spider mite *Tetranychus urticae* Koch Acarina:Tetranychidae
twostriped grasshopper *Melanoplus bivittatus* (Say) Orthoptera:Acrididae
underwings* *Catocala* spp. Lepidoptera:Noctuidae
variable oakleaf caterpillar *Heterocampa manteo* (Doubleday) Lepidoptera:Notodontidae
varied carpet beetle *Anthrenus verbasci* (Linnaeus) Coleoptera:Dermestidae
variegated cutworm *Peridroma saucia* (Hübner) Lepidoptera:Noctuidae
variegated leafhopper* *Erythroneura variabilis* Beamer Homoptera:Cicadellidae
vedalia *Rodolia cardinalis* (Mulsant) Coleoptera:Coccinellidae
vegetable weevil *Listroderes costirostris obliquus* (Klug) Coleoptera:Curculionidae
velvetbean caterpillar *Anticarsia gemmatalis* Hübner Lepidoptera:Noctuidae
vetch bruchid *Bruchus brachialis* Fåhraeus Coleoptera:Bruchidae
viburnum aphid *Aphis viburniphila* Patch Homoptera:Aphididae
viceroy *Limenitis archippus* (Cramer) Lepidoptera:Nymphalidae
Virginiacreeper leafhopper *Erythroneura ziczac* (Walsh) Homoptera:Cicadellidae
walkingstick *Diapheromera femorata* (Say) Orthoptera:Phasmatidae
walnut caterpillar *Datana integerrima* Grote & Robinson Lepidoptera:Notodontidae
walnut husk fly *Rhagoletis completa* Cresson Diptera:Tephritidae
wax scales* *Ceroplastes* spp. Homoptera:Coccidae
webbing clothes moth *Tineola bisselliella* (Hummel) Lepidoptera:Tineidae
western bean cutworm *Loxagrotis albicosta* (Smith) Lepidoptera:Noctuidae
western brown stink bug *Euschistus impictiventris* Stål Hemiptera:Pentatomidae
western cherry fruit fly *Rhagoletis indifferens* Curran Diptera:Tephritidae
western chicken flea *Ceratophyllus niger* Fox Siphonaptera:Ceratophyllidae
western corn rootworm *Diabrotica virgifera* LeConte Coleoptera:Chrysomelidae
western flower thrips *Frankliniella occidentalis* (Pergande) Thysanoptera:Thripidae
western grape leafhopper* *Erythroneura elegantula* Osborn Homoptera:Cicadellidae
western grapeleaf skeletonizer *Harrisina brillians* (Barnes & McDunnough) Lepidoptera:Zygaenidae
western grape rootworm *Bromius obscurus* (Linnaeus) Coleoptera:Chrysomelidae
western peachtree borer *Sanninoidea exitiosa* (Say) Lepidoptera:Sesiidae
western pine beetle *Dendroctonus brevicomis* LeConte Coleoptera:Scolytidae
western pine tip moth* *Rhyacionia bushnelli* (Busck) Lepidoptera:Olethreutidae
western raspberry fruitworm *Byturus bakeri* Barber Coleoptera:Byturidae
western strawberry crown borer* *Tyloderma morbillosa* (LeConte) Coleoptera:Curculionidae
western strawberry leafroller* *Compsolechia fragariella* (Busck) Lepidoptera:Gelechiidae
western striped cucumber beetle *Acalymma trivittatum* (Mannerheim) Coleoptera:Chrysomelidae

western subterranean termite *Reticulitermes hesperus* Banks Isoptera:Rhinotermitidae
western sycamore borer* *Ramosia resplendens* (Hy. Edwards) Lepidoptera:Sesiidae
western sycamore lace bug* *Corythuca confraterna* Gibson Hemiptera:Tingidae
western tent caterpillar *Malacosoma californicum* (Packard) Lepidoptera:Lasiocampidae
western tussock moth *Orgyia vetusta* (Boisduval) Lepidoptera:Lymantriidae
western yellowjacket* *Vespula pensylvanica* (Saussure) Hymenoptera:Vespidae
western yellowstriped armyworm *Spodoptera praefica* (Grote) Lepidoptera:Noctuidae
wheat curl mite *Aceria tulipae* (Keifer) Acarina:Eriophyidae
wheat head armyworm *Faronta diffusa* (Walker) Lepidoptera:Noctuidae
wheat jointworm *Harmolita tritici* (Fitch) Hymenoptera:Eurytomidae
wheat midge *Sitodiplosis mosellana* (Géhin) Diptera:Cecidomyiidae
wheat stem billbug* *Calendra parvulus* (Gyllenhal) Coleoptera:Curculionidae
wheat stem maggot *Meromyza americana* Fitch Diptera:Chloropidae
wheat stem sawfly *Cephus cinctus* Norton Hymenoptera:Cephidae
wheat strawworm *Harmolita grandis* (Riley) Hymenoptera:Eurytomidae
wheat wireworm *Agriotes mancus* (Say) Coleoptera:Elateridae
whitebanded elm leafhopper *Scaphoideus luteolus* Van Duzee Homoptera:Cicadellidae
whitefringed beetles *Graphognathus* spp. Coleoptera:Curculionidae
white garden snail *Theba pisana* (Müller) Stylommatophora:Helicidae
whitelined sphinx *Hyles lineata* (Fabricius) Lepidoptera:Sphingidae
whitemarked tussock moth *Orgyia leucostigma* (J. E. Smith) Lepidoptera:Lymantriidae
white oak leafminer* *Lithocolletis hamadryadella* Clemens Lepidoptera:Gracillariidae
white peach scale *Pseudaulacaspis pentagona* (Targioni-Tozzetti) Homoptera:Diaspididae
white pine sawfly *Neodiprion pinetum* (Norton) Hymenoptera:Diprionidae
white pine weevil *Pissodes strobi* (Peck) Coleoptera:Curculionidae
Willamette spider mite* *Eotetranychus willamettei* (McGregor) Acarina:Tetranychidae
willow leaf beetle* *Lina interrupta* (Fabricius) Coleoptera:Chrysomelidae
willow leaf galls* *Euura* spp. Hymenoptera:Tenthredinidae
wing louse *Lipeurus caponis* (Linnaeus) Mallophaga:Philopteridae
winter grain mite *Penthaleus major* (Dugès) Acarina:Eupodidae
winter tick *Dermacentor albipictus* (Packard) Acarina:Ixodidae
witch hazel cone gall aphid* *Hormaphis hamamelidis* (Fitch) Homoptera:Eriosomatidae
woods weevil *Nemocestes incomptus* (Horn) Coleoptera:Curculionidae
woolly alder aphid *Paraprociphilus tessellatus* (Fitch) Homoptera:Eriosomatidae
woolly apple aphid *Eriosoma lanigerum* (Hausmann) Homoptera:Aphididae
woolly apple aphid parasite* *Aphelinus mali* (Haldeman) Hymenoptera:Eulophidae
woolly ash aphid* *Prociphilus fraxinidipetalae* Essig Homoptera:Eriosomatidae
woolly larch aphid* *Adelges laricis* Vallot Homoptera:Chermidae

yellowfever mosquito *Aedes aegypti* (Linnaeus) Diptera:Culicidae
yellowheaded fireworm *Acleris minuta* (Robinson) Lepidoptera:Tortricidae
yellow mealworm *Tenebrio molitor* Linnaeus Coleoptera:Tenebrionidae
yellownecked caterpillar *Datana ministra* (Drury) Lepidoptera:Notodontidae
yellow scale *Aonidiella citrina* (Coquillett) Homoptera:Diaspididae
yellow spider mite *Eotetranychus carpini borealis* (Ewing) Acarina:Tetranychidae
yellowstriped armyworm *Spodoptera ornithogalli* (Guenée) Lepidoptera:Noctuidae
yellow woollybear *Diacrisia virginica* (Fabricius) Lepidoptera:Arctiidae
yucca moth *Tegeticula yuccasella* (Riley) Lepidoptera:Incurvariidae
zebra caterpillar *Ceramica picta* (Harris) Lepidoptera:Noctuidae

GLOSSARY

Acephate (Orthene). A synthetic insecticide, plant systemic; an organophosphate, *O, S*-dimethyl acetylphosphoramidothioate; slightly toxic to mammals, acute oral LD_{50} for rats 866 to 945 mg/kg; effective against lepidopterous larvae, lygus bugs, aphids, leafhoppers, and thrips; product of Chevron.

Abate (trade name). See temephos.

Abdomen. The posterior of the three main body divisions of insects (Fig. 2:1).

Acaraben (trade name). See chlorobenzilate.

Acaralate (trade name). See chloropropylate.

Acaricide. A chemical employed to kill and control mites and ticks.

Acarol (trade name). See bromopropylate.

Acetylcholine. A substance that is present in many parts of the body of animals and is important to the function of nerves.

Acidic. Acid in reaction resulting from excess of hydrogen ions over hydroxyl (OH) ions in solution.

Aerosol. Finely dispersed particles in air, such as smoke or fog.

Aflatoxins. Organic compounds (metabolites) produced by the fungus *Aspergillus flavus*, which are highly toxic and carcinogenic to mammals.

Agroecosystem. The relatively artificial ecosystem in an agricultural field, orchard, or pasture.

AI. Active ingredient, that is, the toxic part of the pesticide which kills pests.

Air sac. A dilated portion of a trachea.

Alate. Winged; having wings.

Aldicarb (Temik). A systemic insecticide, acaricide, and nematicide; a carbamate, 2-methyl-2-(methylthio) propionaldehyde-*O*-(methylcarbamoyl) oxime; highly toxic to mammals, acute oral LD_{50} for rats 1 mg/kg; formulated as granules for soil application to control insects, mites, or nematodes on cotton, ornamentals, sugar beets, and several other crops; product of Union Carbide Corp.

Aldrin (common name). A synthetic insecticide; a chlorinated hydrocarbon of not less than 95 per cent 1,2,3,4,10,10-hexachloro-1,4,4a,5,8,8a-hexahydro-1,4-*endo-exo*-5,8-dimethanonaphthalene; highly toxic to mammals, acute oral LD_{50} for rats 44 mg/kg; phytotoxicity: none when properly formulated, but some crops are sensitive to solvents in certain formulations; product of Shell Chemical Company, New York, N.Y.

Aliphatic. A term applied to the "open chain" or fatty series of hydrocarbons.

Alkaline. Having the reaction of an alkali resulting from an excess of hydroxyl ions (OH) over hydrogen ions in solution; pH greater than 7.0.

Alkaloids. Substances found in plants, many having powerful pharmacologic action, and characterized by content of nitrogen and the property of combining with acids to form salts.

Allethrin. A synthetic insecticide related to the plant derived pyrethrins; slightly toxic to warmblooded animals, acute oral LD_{50} for rats 920 mg/kg.

Alsike clover. A perennial clover, *Trifolium hybridum*, adapted to cool climates with abundant moisture, especially suited for pasture mixtures and also grown for hay.

Altosid (trade name). See methoprene.

Alutaceous. Covered with minute cracks like the human skin.

Ametabola. The insects that develop without metamorphosis, namely the Protura, Thysanura, and Collembola.

Amide. Compound derived from carboxylic acids by replacing the hydroxyl of the $-COOH$ by the amino group, $-NH_2$.

Amine. An organic compound containing nitrogen, derived from ammonia, NH_3, by replacing one or more hydrogen atoms by as many hydrocarbon radicals.

Amino acid. Organic compounds that contain the amino (NH_2) group and the carboxyl ($COOH$) group. Amino acids are the "building stones" of proteins.

Ammonia. A colorless alkaline gas, NH_3, soluble in water.

Anal. Pertaining to the last abdominal segment, which bears the anus.

Anal ring. In coccids, an elevated ringlike structure surrounding the anus.

Anaplasmosis. Infection with *Anaplasma*, a genus of protozoanlike parasites that infest red blood cells.

Anasa wilt. A wilt disease of cucurbits caused solely by the feeding of the squash bug, no parasitic microorganism involved.

Anemic. Deficient in blood quantity or quality.

Angstrom (Å). One hundred millionth of a centimeter, unit used in measuring the length of light waves.

Annulate. Formed in ringlike segments.

Antenna (pl., antennae). Pair of segmented appendages located on the head and usually sensory in function (Fig. 2:1).

Anterior. Front; toward the front.

Anther. Flower part which develops and contains the pollen.

Anthrax. Malignant anthrax, a fatal infectious disease of cattle and sheep caused by a bacterium, *Bacillus anthracis*, and characterized by hard ulcers at point of inoculation and by symptoms of collapse. It also occurs in man.

Antibiosis. An association between two or more organisms that is detrimental to one or more of them.

Anticoagulin. A substance antagonistic to the coagulation of blood.

Anus. The posterior opening of the digestive tract (Fig. 2:19).

Aorta. The anterior, nonchambered, narrow part of the insect heart which opens into the head (Fig. 2:19).

Apterous. Wingless.

Apterygota. A subclass of primitively wingless insects, including the Protura, Thysanura, and Collembola.

Aquatic. Living in water.

Arachnida. A class of arthropods that includes the scorpions, spiders, mites, ticks, and several other groups.

Aromatic. In chemistry, compounds that have a nuclear structure similar to that of benzene.

Arboreal. Living in, on, or among trees.

Arthropoda. A phylum of animals with segmented body, exoskeleton, and jointed legs.

Arthropods. Animals belonging to the phylum Arthropoda.

Aschelminthes. A phylum of invertebrate animals that include roundworms, rotifers, and several other classes.

Aster yellows. A plant disease caused by a mycoplasmalike organism. Disease affects many kinds of wild and cultivated plants and is transmitted by the aster leafhopper. Symptoms are a stunting of plants, sterility, and chlorosis of foliage.

Asymmetrical. Organs or body parts not alike on either side of a dividing line or plane.

Atomization. Process of breaking a liquid into a fine spray.

Atropine. A poisonous, crystalline alkaloid used in medicine; a specific antidote for poisoning by organic phosphate insecticides.

Attapulgite. A magnesium silicate clay mined in Florida and Georgia and used as a dust carrier.

Attractants. Substances that elicit a positive directive response; chemicals having positive attraction for animals such as insects, usually in low concentration and at considerable distances.

Axon. The process of a nerve cell that conducts impulses away from the cell body.

Azinphosethyl (Ethyl Guthion). A synthetic insecticide and acaricide; an organophosphate, *O,O*-diethyl S-4-oxo-1,2,3-benzotriazin-3(4*H*)-ylmethyl phosphorodithioate; highly toxic to mammals; acute oral LD_{50} for rats 7 to 18 mg/kg; product of Chemagro.

Azinphosmethyl (Guthion). A synthetic insecticide and acaricide; an organic phosphate, *O,O*-dimethyl *S*-(4-oxo-1,2,3-benzotriazin-3(4*H*)-ylmethyl) phosphorodithioate; extremely toxic to mammals, acute oral LD_{50} for rats 15 to 25 mg/kg; used on fruit, vegetables, and ornamentals; broad-spectrum insecticide product of Chemagro.

Azobenzene (common and chemical name). a synthetic acaricide, $C_6H_5N{=}NC_6H_5$, used by volatilizing from hot water pipes of greenhouse, or by ignition in a pyrotechnic mixture, or as a spray, either emulsion or suspension; chronic toxicity 1000 ppm in diet kills rats in a few days; phytotoxicity—has injured roses and asparagus fern.

Azodrin (trade name). See monocrotophos.

Bacterial wilts. Plant diseases in which the causative bacteria produce slime that plugs the water-conducting tissue of the invaded plant.

Barley yellow dwarf. A virus disease of cereals, marked by leaves rapidly turning light green and yellow, beginning at the tips; transmitted by certain species of aphids.

Basic. Having the property of ionizing in solution to form hydroxyl ions and of neutralizing acids to form salts.

Baytex (trade name). See fenthion.

Bendiocarb (Ficam W and Garvox). A synthetic insecticide; a carbamate, 2,2-Dimethyl-1,3-benzodioxol-4-yl N-methylcarbamate; moderately toxic to mammals, acute oral LD_{50} for rats 179 mg/kg; effective against coleopteran and dipteran pests of agriculture, effective against household pests and soil pests of turf; product of Fisons Corporation.

Bentonite. A clay composed mainly of silica and aluminum silicate; mined in Mississippi, Wyoming and elsewhere; used as dust diluent and for lining ponds to hold water.

Benzene hexachloride (chemical name). See BHC.

BHC (common name). A synthetic insecticide, a chlorinated hydrocarbon, 1,2,3,4,5,6-hexachlorocyclohexane, mixed isomers and a specified percentage of gamma; slightly more toxic to mammals than DDT, actue oral LD_{50} for rats about 200 mg/kg; phytotoxicity—more toxic than DDT, interferes with germination, suppresses growth, and reduces yields except at low concentration; certain crop plants such as potato absorb crude BHC with consequent tainting of tubers.

Bidrin (trade name). See dicrotophos.

Bilateral symmetry. Similarity of form, one side with the other.

Binapacryl (Morocide). A synthetic acaricide; a nitrophenol derivative, 2-*sec*-butyl-4,6-dinitrophenyl 3-methyl-2-butenoate; moderate mammalian toxicity; acute oral LD_{50} for rats 136 to 225 mg/kg; controls mites on tree fruit and nuts; product of FMC.

Biological control. The control of pests by employing predators, parasites, or disease; the natural enemies are encouraged and disseminated by man.

Bionomics. The study of the habits, breeding, and adaptations of living forms.

Biotype. Groups of insects primarily distinguishable on the basis of interaction with relatively genetically stable varieties or clones of host plants; a strain of an insect species.

Bisexual. Having two sexes distinct and separate; with males and females.

Black spot. A fungus disease of roses caused by *Diplocarpon rosae* and characterized by black spots on the leaves and yellowing and premature dropping of leaves.

Boll. The pod in which the lint and seeds of cotton develop.

Book lung. A respiratory cavity containing a series of leaflike folds.

Boot leaf. The leaf arising from the protective sheath enclosing the young inflorescence of grain.

Bordeaux mixture. Primarily a fungicide but also a repellent to many insects; a popular formula consists of copper sulfate 6 lb, hydrated lime 10 lb, and water 100 gal; ingestion of large quantities may cause fatal gastroenteritis in mammals; toxic to some plants, particularly at low temperatures.

Bot. The larva of certain flies that are parasitic in the body of mammals.

Brachypterous. With short wings that do not cover the abdomen.

Bract. A small leaf at the base of the flower.

Bromopropylate (Acarol). An experimental specific miticide related to DDT; isopropyl 4,4′-dibromobenzilate; slightly toxic to mammals, acute oral LD_{50} for rats 5000 mg/kg; controls mites on fruit and citrus trees; product of Ciba-Geigy Corp.

Brood. In insects, a group of individuals of a given species that have hatched into young or have become adult at approximately the same time, and live together in a defined and limited area.

Bubonic plague. A bacterial disease of rodents and man caused by *Yersinea pestis* and transmitted chiefly by the oriental rat flea; marked by chills, fever, and inflammatory swelling of lymphatic glands.

Bux (trade name). See metalkamate.

Caecum (pl., caeca). A sac or tubelike structure open at only one end (Fig. 2:19).

Calyx. The outer—usually green—leaflike parts of a flower.

Campodeiform larva. A larva shaped like the thysanuran *Campodea*, that is, elongate, flattened, with well-developed legs and antennae, and usually with filaments on end of abdomen, and usually active (Fig. 3:14A).

Capitate. With an apical knoblike enlargement.

Capitulum. Headlike structure of ticks which bears the feeding organs.

Captan (common name). A protective fungicide, particularly for foliage application or seed treatment; chemically, cis-*N*-(trichloromethyl)thio-4-*cyclo*hexene-1,2-dicarboximide; low acute toxicity to warmblooded animals, acute oral LD_{50} for rats 9000 mg/kg; no evidence of phytotoxicity.

Carabiform larva. A larva shaped like the larva of a carabid beetle, that is elongate, flattened, and with well-developed legs; filaments lacking on end of abdomen (Fig. 3:14B).

Carbaryl (Sevin). A synthetic insecticide; a carbamate, 1-naphthyl methylcarbamate; moderately toxic to mammals, acute oral LD_{50} for rats 500 to 700 mg/kg; no evidence of phytotoxicity in use at normal rates; used on many crops; product of Union Carbide.

Carbofuran (Furadan). A synthetic insecticide, plant systemic applied to soil; a carbamate, 2,3-dihydro-2,2-dimethyl-7-benzofuranyl methylcarbamate; highly toxic to mammals; acute oral LD_{50} for rats 11 mg/kg; used to control corn rootworms by incorporating into soil at planting time and other uses; product of FMC Corp.

Carbohydrate. Any of a group of neutral compounds made up of carbon, hydrogen, and oxygen; for example, sugar, starch, cellulose.

Carbon disulfide. Insecticidal fumigant for treatment of stored grain, also a soil fumigant for Japanese and Asiatic beetles, also used to rid horses of bots and roundworms; empirical formula CS_2; highly toxic to mammals and toxic to plants and certain seeds.

Carbon tetrachloride. Insecticidal fumigant for treatment of stored grain; empirical formula CCl_4; MLD for man 3 to 4 cc; vapors may cause acute poisoning at 1000 ppm; phytotoxicity—does not affect germinating qualities of wheat.

Carbophenothion (Trithion). A synthetic insecticide and acaricide; an organic phosphate, *S*-(*p*-chlorophenylthio) methyl *O,O*-diethyl phosphorodithioate; highly toxic to mammals, acute oral LD_{50} for rats 30 mg/kg; slightly injurious to fruit and foliage of certain apple varieties, isolated cases of injury to citrus and to greenhouse roses.

Carnivorous. Preying or feeding on animals.

Carpel. Structural unit of the pistil in which the seed develops.

Carzol SP (trade name). See formetanate.

Caste. A form or type of adult in a social insect, such as among termites and ants (Fig. 19:19).

Caterpillar. The larva of a moth, butterfly, or sawfly (Fig. 4:20).

Catfacing. The injury caused by the feeding of such insects as plant bugs and stink bugs on developing fruit which results in uneven growth and a deformed mature fruit.

Cattle tick fever. A specific infectious disease of the blood of cattle caused by a protozoan, *Babesia bigemina*, which attacks the red blood cells; characterized by fever, anemia, jaundice, and red urine; transmitted by the bite of the cattle tick.

Cement layer. A thin layer on the surface of insect cuticles formed by the hardened secretion of the dermal glands.

Cellulose. An inert carbohydrate, the chief component of the solid framework or woody part of plants.

Cauda. The pointed end of the abdomen in aphids.

Cephalothorax. A body region consisting of head and thoracic segments, as in spiders.

Cercus (pl., cerci). One of a pair of appendages at the end of the abdomen (Fig. 2:1).

Chain. In chemistry, a series of atoms connected by bonds, forming the skeleton of a number of compounds.

Chelicera (pl., chelicerae). The anterior pair of appendages in arachnids (Fig. 2:7).

Chigger. The parasitic larva of a trombiculid mite.

Chitin. A nitrogenous polysaccharide occurring in the cuticle of arthropods and certain other invertebrates. Probably occurs naturally only in chemical combination with protein.

Chlorbenside (common name). A synthetic acaricide, a sulfide; chemically, *p*-chlorobenzyl, *p*-chlorophenyl sulfide; single dose of 3000 mg/kg administered to rats without signs of systemic toxicity; phytotoxicity—none reported, but some harm may be caused to cucurbits; product of Boots Company.

Chlordane. A synthetic insecticide; a chlorinated hydrocarbon, 1,2,4,5,6,7,8,8-octachloro-3a,4,7,7a-tetrahydro-4,7-methanoindane; moderately toxic to mammals, oral LD_{50} for rats about 250 mg/kg; phytotoxicity—high concentrations injurious to some vegetables, residues in soil depress germination; product of Velsicol.

Chlordimeform (Fundal, Galecron). A synthetic insecticide and acaricide; a formamidine, *N*′(4-chloro-*o*-tolyl)-*N,N*-dimethylformamidane; moderately toxic to mammals, acute oral LD_{50} for rats 127 to 352 mg/kg; controls resistant mites and their eggs and various lepidopterous larvae infesting fruit and walnut trees, cruciferous vegetables, and cotton; product of Ciba-Geigy, Nor-am, and Schering AG.

Chlorfenethol (Dimite, Qikron). A synthetic acaricide, a chlorinated hydrocarbon and relative of DDT; chemically, 1,1-bis(*p*-chlorophenyl)ethanol; moderately toxic to mammals, acute oral LD_{50} for rats 500 mg/kg; product of Nippon Soda Co., Japan.

Chlorobenzilate (Acaraben). A synthetic acaricide, a chlorinated hydrocarbon and relative of DDT; chemically, ethyl 4,4′-dichlorobenzilate; slightly toxic to mammals, acute oral LD_{50} for rats 960 mg/kg; product of Ciba-Geigy.

Chloropicrin. Insecticidal fumigant for treatment of stored grain and cereal products and for soil treatment to control insects, nematodes, weeds, fungi; empirical formula CCl_3NO_2; 0.05 oz/1000 cu ft lethal in 10 min to mammals; very toxic to plants when injected into soil.

Chloropropylate (Acaralate). A synthetic acaricide related to DDT; isopropyl 4,4′-dichlorobenzilate; acute oral LD_{50} for rats 5000 mg/kg; controls spider mites on apple and pear trees; product of Ciba-Geigy.

Chlorosis. In plants, yellowness of normally green tissues due to partial failure of chlorophyll to develop or to removal of chlorophyll.

Chlorpyrifos (Dursban, Lorsban). A synthetic insecticide and acaricide; an organophosphate, *O,O*-diethyl *O*-(3,5,6-trichloro-2-pyridyl) phosphorothioate; moderately toxic to mammals, acute oral LD_{50} for male rats 163 mg/kg; broad spectrum insecticide for control of ornamental and turf pests, household pests, and mosquitoes; product of Dow.

Cholinesterase. An enzyme (or enzymes) present in body tissues which hydrolyzes or breaks down acetylcholine.

Chorion. The outer shell or covering of the insect egg.

Chromosomes. At cell division the dark-staining, rod-shaped structures which contain the hereditary units called genes.

Chrysalis. The pupa of a butterfly.

Ciodrin (trade name). See crotoxyphos.

Circadian rhythm. Patterns of behavior recurring regularly during the same time each day.

Circulus. In coccids, a lip on the ventral side of the abdomen between the second and third segments.

Class. A division of the animal kingdom lower than a phylum and higher than an order; for example the class Insecta.

Clavate. Clublike; thickening gradually toward the tip (Fig. 2:3C).

Coarctate pupa. A pupa enclosed in a hardened case formed from the next to the last larval skin; found among higher Diptera (Fig. 3:15C).

Cocoon. A silken case inside which the pupa develops.

Codlelure. Synthetic sex pheromone of the codling moth; (*E,E*)-8,10-dodecadien-1-ol; female scent attractive to male codling moth; used in traps to determine moth flight activity; product of Zoecon.

Collard. A kind of edible kale.

Combine. A farm machine that cuts, threshes, and cleans grain while moving over the field.

Commissure. A bridge connecting any two bodies or structures.

Community. The plants and animals of a given habitat.

Complete metamorphosis. Same as complex metamorphosis. (See definition below and Fig. 3:11.)

Complex metamorphosis. Metamorphosis in which the insect develops by four distinct stages, namely egg, larva, pupa, and adult; the wings (when present) develop internally during the larval stage. (See Fig. 3:11.)

Compound eye. An eye consisting of many individual elements or ommatidia each of which is represented externally by a facet (Fig. 2:29).

Connective. A longitudinal cord of nerve fibers connecting successive ganglia.

Co-Ral (trade name). See coumaphos.

Cornicles. The pair of dorsal tubular processes on the posterior part of the abdomen, as in aphids (Fig. 9:16 and 17).

Corpora allata. A pair of small endocrine glands located just behind the brain (Fig. 3:7).

Cosmopolitan. Occurring throughout most of the world.

Coumaphos (Co-Ral). An animal systemic insecticide: a synthetic organic phosphate, *O*-(3-chloro-4-methyl-2-oxo-2*H*-1-benzopyran-7-yl) *O,O*-diethyl phosphorothioate; moderately toxic to mammals, oral LD_{50} for rats 56 to 230 mg/kg; product of Chemagro.

Counter (trade name). See terbufos.

Coxa (pl., coxae). The basal segment of the leg (Fig. 2:2).

Crawler. The active first instar of a scale insect (Fig. 14:17D).

Cremaster. Terminal spines of the abdomen; the anal hooks by which many pupae suspend themselves.

Crochets. (Pronounced croshays). Hooked spines at tip of the prolegs of lepidopterous larvae.

Crop. The dilated section of the foregut just behind the esophagus (Fig. 2:19).

Crotoxyphos (Ciodrin). A synthetic insecticide; an organophosphate, α-methylbenzyl 3-hydroxycrotonate dimethyl phosphate; moderately toxic to mammals; acute oral LD_{50} for rats 125 mg/kg; controls ectoparasites of livestock; product of Shell.

Cruciferous. Belonging to the mustard family, which includes cabbage, turnip, mustard, radish, and others.

Crufomate (Ruelene). A synthetic insecticide and anthelmintic, an animal systemic; an organic phosphate, 4-*tert*-butyl-2-chlorophenyl methyl methylphosphoramidate; moderately toxic to mammals, acute oral LD_{50} for rats 750 mg/kg; product of Dow.

Cryolite. An inorganic insecticide, sodium fluoaluminate; useful in controlling codling moth, orange tortrix, and several other chewing insects; low acute mammalian toxicity; phytotoxicity—may seriously injure peach trees, corn, or grapes, otherwise little plant injury.

Cryptobiotic. Leading a hidden or concealed life.

Cubé powder. The finely ground roots of certain leguminous trees and shrubs, *Lonchocarpus* spp., which contain rotenone, an insecticidal substance.

Cucurbit. A plant belonging to the gourd family, Cucurbitaceae, such as pumpkin, squash, and cucumber.

Cue-lure. A synthetic attractant of the male melon fly; 4-(*p*-hydroxyphenyl)-2-butanone acetate; developed by USDA; a product of Zoecon.

Cuneus. A small triangular section of the hemelytra at the leading margin and next to the membrane.

Curly top. A virus disease of sugar beets, beans, tomatoes, and other plants transmitted by the beet leafhopper.

Cuticle. The outer noncellular layers of the insect integument secreted by the epidermis (Fig. 2:13).

Cyclic. In chemistry, atoms linked together to form a ring structure.

Cyclodienes. Synthetic insecticides belonging to the cyclodiene group of cyclic hydrocarbons, for example chlordane, heptachlor, aldrin, dieldrin, and endrin.

Cyclorrhaphous Diptera. The group of flies that emerge from the puparium through a circular opening at one end of the puparium. These flies belong to the more advanced families.

Cyhexatin (Plictran). A specific miticide; an organotin, tricyclohexyltin hydroxide; slightly toxic to mammals, acute oral LD_{50} for rats 540 mg/kg; controls plant feeding mites, susceptible and resistant strains, on fruit and citrus trees, almonds, walnuts, and ornamentals; product of Dow.

Cyst. A sac, normal or abnormal, especially one containing a liquid or semisolid.

Cysticercoid. A form of larval tapeworm.

Cytoplasmic polyhedrosis viruses. Insect viruses that are occluded in a polyhedral protein crystal and multiply in the cell cytoplasm; virus particles are nearly spherical; majority occur in larvae of the Lepidoptera.

Dasanit (trade name). See fensulfothion.

D-D (trade name). A soil fumigant for controlling nematodes and garden symphylan; heavy liquid with odor of chloroform; a mixture of 1,3-dichloropropene and 1,2-dichloropropane; moderately toxic to mammals by ingestion or by inhalation, acute oral LD_{50} for rats 140 mg/kg, inhalation LC_{50} for rats 1000 ppm; toxic to plants and germinating seeds.

DDT. A synthetic insecticide; a chlorinated hydrocarbon, 1,1,1-trichloro-2,2-bis(*p*-chlorophenyl)ethane; moderately toxic to mammals, acute oral LD_{50} for rats about 250 mg/kg; phytotoxicity—injures cucurbits, young tomato plants, and beans; product of several companies.

Dealate. Wingless as a result of the insect casting or breaking off its own wings.

Delayed dormant spray. An orchard spray applied during the period from swollen bud to late green tip of bud development; often called "swollen bud" in stone fruit work.

Delnav (trade name). See dioxathion.

Demeton (Systox). A contact and plant systemic insecticide and acaricide; an organic phosphate, mixture of *O,O*-diethyl *O*(and *S*)-2-(ethylthio)ethyl phosphorothioates; extremely toxic to mammals, acute oral LD_{50} for rats 12 mg/kg; little phytotoxicity at recommended dosages, but "pink" and "petal fall" application to McIntosh apples should be avoided; product of Chemagro.

Dengue (pronounced deng'e). A virus disease of man marked by severe pains in head, eyes, muscles, and joints and transmitted by certain mosquitoes.

Density dependent factor. In ecology, a factor that changes in intensity with changes in population density (e.g., lethal effects of a factor intensify as population numbers increase).

Dermatitis. Inflammation of the skin.

Derris powder. The finely ground roots of the leguminous shrub *Derris elliptica*, which contains rotenone, an insecticidal substance.

Deutonymph. The third instar of a mite.

Deutovum. The quiescent, undeveloped larval stage which hatches from the egg of certain mites and from which, after six or seven days, the active, six-legged larva emerges.

Diapause. A state of an animal, such as an insect, in which a reduction of growth processes or maturation occurs which is not necessarily caused by immediate environmental influence, does not depend for its continuance on unsuitable conditions, and is not easily or quickly altered by change to a more favorable environment; frequently related to change in photoperiod. Once the state of diapause comes to an end, normal growth and development are resumed.

Diaphragm. A horizontal membranous partition of the body cavity.

Diazinon. A synthetic insecticide and acaricide; an organic phosphate, *O,O*-diethyl *O*-(2-isopropyl-4-methyl-6-pyrimidyl) phosphorothioate; moderately toxic to mammals, acute oral LD_{50} for rats about 300 mg/kg; phytotoxicity—toxic to Stephanotis and African violets and may cause russeting of certain varieties of apples; product of Ciba-Geigy.

Dibrom (trade name). See naled.

Dibromo-chloropropane (DBCP, Fumazone, Nemagon). A soil fumigant, nematocidal and lethal to garden symphylans; heavy liquid, with odor of chloroform, 1,2-dibromo-3-chloropropane; moderately toxic to mammals; plants sufficiently tolerant that fumigant can be added to irrigation water; product of Dow and of Shell.

Dichlorvos (Vapona). A synthetic insecticide and space fumigant; an organic phosphate, 2,2-dichlorovinyl dimethyl phosphate; moderately toxic to mammals, acute oral LD_{50} for rats 50 to 80 mg/kg; phytotoxicity—none to wide variety of plants at insecticidal concentrations; product of Shell.

Dichloropropenes mixture (DCP, Telone). Soil fumigant, nematocidal and insecticidal fumigant for crop lands; liquid with odor of chloroform, 1,3-dichloropropene and related hydrocarbons; moderately toxic to mammals; phytotoxic; product of Dow.

Dicofol (Kelthane). A synthetic acaricide, a chlorinated hydrocarbon and relative of DDT; chemically, 4,4′-dichloro-*a*-(trichloromethyl)benzhydrol; moderately toxic to warm-blooded animals, acute oral LD_{50} for rats 809 mg/kg; little evidence of phytotoxicity except on eggplant and avocado; product of Rohm & Haas.

Dicrotophos (Bidrin). A synthetic insecticide, plant systemic; an organophosphate, 3-hydroxy-*N,N*-dimethyl-*cis*-crotonamide, dimethyl phosphate; highly toxic to mammals; acute oral LD_{50} for rats 22 mg/kg; controls certain pests of cotton and ornamental trees; product of Shell.

Dieldrin. A highly residual insecticide; a chlorinated hydrocarbon of not less than 85 per cent of 1,2,3,4,10,10-hexachloro-6, 7-epoxy-1,4,4a,5,6,7,8,8a-octahydro-1,4-*endo-exo*-5,8-dimethanonaphthalene; somewhat more toxic to mammals than DDT, acute oral LD_{50} for rats 100 mg/kg; phytotoxicity—none when properly formulated, but some crops are sensitive to solvents in certain formulations; product of Shell.

Differentiation. Increase in visible distinctive morphology.

Diflubenzuron (Dimilin). A synthetic insect growth regulator; 1-(4-chlorophenyl)-3-(2,6-difluorobenzoyl)-urea; relatively nontoxic to mammals, acute oral LD_{50} for rats greater than 10,000 mg/kg; especially effective against foliar feeding lepidopterous insects, registered for control of gypsy moth; product of Philips-Duphar V.B. (Holland) and Thompson-Hayward.

Dimethoate. A synthetic insecticide and acaricide, plant systemic; an organic phosphate, *O,O*-dimethyl S(*N*-methyl-carbamoylmethyl) phosphorodithioate; moderately toxic to mammals, acute oral LD_{50} for rats 250 to 500 mg/kg; slight phytotoxicity to some fruit and field crops; product of American Cyanamid.

Dimilin (trade name). See diflubenzuron.

Dimite (trade name). See chlorfenethol.

Dimorphic. Occurring in two distinct forms.

Dimorphism. A difference in size, form, or color, between individuals of the same species, characterizing two distinct types.

Dinitrocresol. A synthetic insecticide; a nitrophenyl compound, 4,6-dinitro-o-cresol, sodium salt; highly toxic to man, acute oral LD_{50} for rats 30 mg/kg; phytotoxicity—very great, can be used as weed killer; product of several companies.

Dinocap (Karathane). A synthetic insecticide and fungicide; a nitrophenol derivative, 2-(1-methylheptyl)-4,6-dinitrophenyl crotonate; slightly toxic to mammals; acute oral LD_{50} for rats 980 to 1190 mg/kg; controls certain mites and mildew on various vegetables, fruit, and ornamentals; product of Rohm and Haas.

Dioxathion (Delnav). A synthetic insecticide and acaricide; an organic phosphate, *S,S′-p*-dioxane-2,3-diyl *O,O*-diethyl phosphorodithioate cis and trans isomers; moderately toxic to mammals, acute oral LD_{50} for rats 110 mg/kg; not phytotoxic at recommended rates of application; product of Hercules, Incorporated.

Diphtheria. A highly contagious bacterial disease caused by presence of *Corynebacterium diphtheriae*, characterized by fever, heart weakness, anemia, and great prostration; often fatal.

Dipterex (trade name). See trichlorfon.

Disk. A type of plow with a rolling disk bottom.

Disparlure. Synthetic gypsy moth sex pheromone; *cis*-7,8-epoxy-2-methyloctadecane; female scent attractive to male gypsy moth; used in scouting for infestations; product of Zoecon.

Disulfoton (Di-Syston). A synthetic insecticide and acaricide, plant systemic absorbed through the roots; an organic phosphate, *O,O*-diethyl *S*-2-(ethylthio) ethyl phosphorodithioate; extremely toxic to mammals, acute oral LD_{50} for rats 2 to 12 mg/kg; high dosages can injure seed; product of Chemagro.

Di-Syston (trade name). See disulfoton.

Dockage. Foreign material in harvested grain, such as weed seeds, chaff, and dust.

Dormancy. A state of quiescence or inactivity.

Dormant spray. A spray applied to trees in true dormancy, before the buds begin to swell.

Dorsal. Top or uppermost; pertaining to the back or upper surface.

Dorsal ocellus. The simple eye in adult insects and in nymphs and naiads (Fig. 2:29).

Dorsal shield. The scutum or sclerotized plate covering all or most of the dorsal surface in males and the anterior portion in females, nymphs, and larvae of hardbacked ticks (Fig. 20:16).

Dorsum. The back or top side.

Downy mildew. Any plant disease caused by species of fungi in the family Peronosporaceae and characterized by the downy growth on host lesions.

Drone. The male honey bee.

Drupelet. A small drupe; the small individual fleshy fruits that make up the berry, as in blackberries and raspberries.

Dursban (trade name). See chlorpyrifos.

Dyfonate (trade name). See fonofos.

Dylox (trade name). See trichlorfon.

Economic injury level. The lowest pest density that will cause economic damage; or the pest density that causes damage equal to the cost of preventing the damage.

Economic threshold. The pest density at which control measures should be applied to prevent an increasing pest population from reaching the economic injury level.

Ecosystem. The functional ecological system of a given area comprising the community of plant and animal populations and the nonliving environment.

Ectoderm. The outer embryological layer that produces the nervous system, the integument, and several other parts of an insect.

Ectohormone. A substance secreted by an animal to the outside causing a specific reaction, such as determination of physiological development, in a receiving individual of the same species.

Ectoparasite. A parasite that lives on the outside of its host.

Ectothermic. Deriving heat from outside the body; cold-blooded.

Egg pod. A capsule that encloses the egg mass of grasshoppers and is formed through the cementing of soil particles together by secretions of the ovipositing female (Fig. 9:8).

Elateriform larva. A larva with the form of a wireworm; that is, long, slender, heavily sclerotized, with short thoracic legs, and with few body hairs (Fig. 3:14E).

Elytra (sing., elytron). Thickened, horny, or leathery forewings, as in beetles and earwigs (Fig. 9:19D).

Emulsifiable concentrate. A liquid formulation of insecticide which contains an emulsifier so that water may be added to form an emulsion.

Emulsion. A suspension of fine droplets of one liquid in another, such as oil in water. Emulsions are milky in appearance.

Encephalitis. Inflammation of the brain.

Endemic typhus. See Murine typhus fever.

Endocrine. Secreting internally, applied to organs whose function is to secrete into blood or lymph a substance which has an important role in metabolism.

Endocuticle. The innermost layer of the cuticle (Fig. 2:13).

Endosperm. A food storage tissue in seeds.

Endosulfan (Thiodan). A synthetic insecticide and acaricide; a chlorinated hydrocarbon, 6,7,8,9,10,10-hexachloro-1,5,5a,6,9,9a-hexahydro-6,9-methano-2,4,3-benzodioxathiepin 3-oxide; moderately toxic to mammals, acute oral LD_{50} for rats 30 to 110 mg/kg; no phytotoxicity experienced on citrus and deciduous trees and on many row crops; product of FMC Corp.

Endrin (common name). A highly residual insecticide, particularly effective against lepidopterous larvae; a chlorinated hydrocarbon, the *endo-endo* isomer of dieldrin; highly toxic to mammals, acute oral LD_{50} for rats 10–12 mg/kg; phytotoxicity—none when properly formulated, but some crops are sensitive to solvents in certain formulations; product of Shell Chemical Company and Velsicol.

Entex (trade name). See fenthion.

Entoleter. A centrifugal force machine to kill insects infesting grain.

Entomogenous. Growing in or on an insect, for example certain fungi.

Entomopox viruses. Insect viruses that resemble vertebrate pox viruses in having a beaded lipoprotein envelope, a platelike core, and either one or two lateral bodies; they differ in being occluded in large proteinaceous bodies. EPVs have been found in Coleoptera, Lepidoptera and Diptera.

Enzyme. An organic catalyst formed by a living cell.

Eperythrozoonosis. A disease in swine caused by the protozoan, *Eperythrozoa suis*; symptoms are fever and anemia, may be fatal; transmitted by hog louse.

Epicuticle. The thin, nonchitinous, surface layers of the cuticle (Fig. 2:13).

Epidemic typhus. Same as typhus fever.

Epidermis. The cellular layer of the integument that secretes or deposits a comparatively thick cuticle on its outer surface (Fig. 2:13).

Epipharynx. A mouthpart structure on the inner surface of the labrum or clypeus.

Epithelium. The layer of cells that covers a surface or lines a cavity.

EPN. A synthetic insecticide and acaricide; an organic phosphate, *O*-ethyl *O-p*-nitrophenyl phenylphosphonothioate; highly toxic to mammals, acute oral LD_{50} for rats 33 mg/kg; phytotoxicity—may injure McIntosh and related varieties of apples; product of DuPont.

EQ 335. Former screwworm remedy consisting of the following (per cent by weight): lindane 3, pine oil 35, mineral oil 42, emulsifier 10, and silica aerogel 10.

Erinose. Any plant disease in which an abnormal growth of hairs occurs in patches on the leaves, such as is caused by the attack of certain gall mites (genus *Eriophyes*).

Eruciform larva. A caterpillar, a larva with cylindrical body, well-developed head, thoracic legs, and abdominal prolegs (Fig. 3:14C).

Erythrocyte. A red blood corpuscle.

Escutcheon. An area on a cow just above the rear part of the udder and below the vulva.

Esophagus. The narrow part of the alimentary canal immediately posterior to the pharynx and mouth.

Esters. Chemical compounds formed by the elimination of water between a molecule of an alcohol and a molecule of an acid.

Estivate. To enter a dormant state during summer.

Ethers. Organic compounds in which two hydrocarbon radicals are joined through an atom of oxygen.

Ethion. A synthetic insecticide and acaricide; an organic phosphate, *O,O,O′,O′*-tetraethyl *S,S′*-methylene bisphosphorodithioate; moderately toxic to mammals, acute oral LD_{50} for rats 96 mg/kg; phytotoxicity—safe on all crops except that defoliation may result when applied to Wealthy apples; product of FMC corp.

Ethoprop (Mocap). A synthetic insecticide and nematicide; an organophosphate, *O*-ethyl *S,S*-dipropyl phosphorodithioate; moderately toxic to mammals, acute oral LD_{50} for rats 62 mg/kg; a soil insecticide recommended for control of wireworms in tobacco and corn, symphylans, and corn rootworms; product of Mobil Chemical.

Ethylene dibromide (EDB). A soil and commodity fumigant, insecticidal and nematocidal; heavy liquid with odor of chloroform; 1,2-dibromoethane; moderately toxic to mammals; phytotoxic; product of several companies.

Ethylene dichloride. Insecticidal fumigant for use against stored grain insects and peach tree borers; empirical formula $C_2H_4Cl_2$; exposure to 4000 ppm for 1 hr produces serious illness in man.

Ethyl Guthion (trade name). See azinphosethyl.

Eutrophication. The process of enrichment of a body of water in dissolved nutrients either naturally or artificially, as by fertilization.

Exarate pupa. A pupa which has its appendages free and not glued to the body (Fig. 3:15B).

Excretion. The elimination of waste products of metabolism.

Exocuticle. The hard and usually darkened layer of the cuticle lying between endocuticle and epicuticle (Fig. 2:13).

Exoskeleton. Collectively, the external plates of the body wall.

Facet. The external surface of an individual unit (ommatidium) of the compound eye (Fig. 2:29).

Fallow. Of land ordinarily used for crops that is allowed to lie idle during the growing season.

Family. A taxonomic subdivision of an order, suborder, or superfamily that contains a group of related genera, tribes, or subfamilies. Family names end in *-idae*.

Famphur (Warbex). A synthetic insecticide, animal systemic; an organophosphate, *O*-[*p*-(dimethylsulfamoyl)phenyl] *O,O*-dimethyl phosphorothioate; acute oral LD_{50} for rats 35 to 62 mg/kg; controls cattle grubs and lice by pour-on or in feed; product of American Cyanamid.

Fascicle. A small bundle; the bundle of piercing stylets of insects with piercing-sucking mouthparts (Fig. 2:6).

Fat body. An organ in the insect body with multiple functions in metabolism, food storage, and excretion. "Fat body" is a misnomer; protein and glycogen are stored as well as fat.

Femur (pl., femora). The third segment of the insect leg (Fig. 2:2).

Fenbutatin-oxide (Vendex). A specific miticide; an organotin, hexakis (β,β-dimethylphenethyl) distannoxane; slightly toxic to mammals, acute oral LD_{50} for rats 2631 mg/kg; controls mites on apple, pear, and citrus trees and on ornamentals in greenhouse or outdoors; phytotoxic to some varieties of citrus; product of Shell.

Fenson. A synthetic acaricide; a sulfonate, p-chlorophenyl benzene sulfonate; slightly toxic to mammals, acute oral LD_{50} for rats 1560 to 1740 mg/kg; for control of certain mites on pear, peach, and tolerant apple varieties, destroys eggs and young mites; product of Snia Viscosa, Italy.

Fensulfothion (Dasanit). A synthetic insecticide and nematocide; an organophosphate, *O,O*-diethyl *O-p*-[(methylsulfinyl)phenyl] phosphorothioate; highly toxic to mammals; acute oral LD_{50} for rats 2 to 11 mg/kg; controls soil insects and nematodes in corn, onions, tobacco, ornamentals, and turf; product of Chemagro.

Fenthion (Baytex, Tiguvon, Entex, or Queletox). A synthetic insecticide, animal systemic; an organophosphate, *O,O*-dimethyl *O*-[4-(methylthio)-*m*-tolyl] phosphorothioate; moderately toxic to mammals; acute oral LD_{50} for rats 178 to 310 mg/kg; controls adult and larval mosquitoes and cattle grubs; Queletox is formulation for bird control; product of Chemagro.

Ferbam. A protective fungicide used for foliage application; chemically, ferric dimethyldithiocarbamate; of low acute mammalian toxicity, acute oral LD_{50} for rats greater than 17,000 mg/kg; generally nonphytotoxic; product of FMC Corp.

Feterita. A group of grain sorghums.

Filariasis. A disease state caused by infection with filarial roundworms (nematodes of the superfamily Filarioidea).

Filiform. Threadlike; slender and of equal diameter (Fig. 2:3A).

Flax. An annual herbaceous plant, *Linum usitatissimum*, grown for seed or fiber from which linseed oil, linen, and cigarette paper are made.

Flaxseed. The puparium of the Hessian fly.

Floret. A small flower, usually one of a dense cluster.

Flowable. A pesticide formulation in which finely divided particles are suspended in a water or an oil base.

Fluke. A parasitic flatworm of the class Trematoda.

Fonofos (Dyfonate). A synthetic insecticide and acaricide; an organophosphate, *O*-ethyl *S*-phenyl ethylphosphonodithioate; highly toxic to mammals, acute oral LD_{50} for rats 8 to 18 mg/kg; granular formulations much less hazardous than emulsions; a soil insecticide recommended for control of corn rootworms, cutworms, wireworms, and generally safe to plants except when placed directly in contact with seeds of small seeded crops; product of Stauffer Chemical.

Forb. Any herbaceous plant other than grass.

Foregut. The anterior part of the alimentary canal from the mouth to the midgut (Fig. 2:19).

Formetanate (Carzol SP). A synthetic insecticide and acaricide; a carbamate, [3-Dimethylamino-(methylene-iminophenyl)]-*N*-methylcarbamate; highly toxic to mammals; acute oral LD_{50} for rats 24 mg/kg; effective in control of mites, thrips, lygus bugs infesting fruit trees and alfalfa; product of NOR-AM and Schering AG.

Frego bract. A bract that is narrow and twisted, as opposed to the normal flat bracts that envelop the flower bud and boll of cotton.

Fumigant. A substance or mixture of substances which produces gas, vapor, fume, or smoke intended to destroy insect and other pests.

Fundal (trade name). See chlordimeform.

Fungicide. Any substance that kills fungi or inhibits the growth of the spores or hyphae.

Furadan (trade name). See carbofuran.

Galecron (trade name). See chlordimeform.

Gall. An abnormal growth of plant tissues induced by the presence and stimulus of an animal or another plant.

Gametocyte. A sex cell stage of *Plasmodium*, the malarial parasite.

Ganglion. A nerve mass that serves as a center of nervous influence (Fig. 2:19).

Gardona (trade name). See stirofos.

Garvox (trade name). See bendiocarb.

Gastric caeca. The saclike diverticula at the anterior end of the midgut (Fig. 2:19).

Gastroenteritis. Inflammation of the stomach and intestines.

Generation. The group of individuals of a given species that have been reproduced at approximately the same time; the group of individuals of the same genealogical rank.

Geniculate. Elbowed or abruptly bent (Fig. 2:3E).

Genital claspers. Organs of the male genitalia which serve to hold the female during copulation.

Genitalia. The reproductive organs; the external structures which enable the sexes to copulate and the females to deposit eggs, strictly these are the external genitalia.

Genus (pl., genera). A group of closely related species.

Germ. The embryo within a seed.

Glycogen. A carbohydrate synthesized by animals, also called "animal starch."

Gnathosoma. The anterior part of the body of mites and ticks which bears the mouth and mouthparts (Fig. 2:32).

Gonad. The ovary or testis, or the embryonic rudiment of either.

Gossyplure. Synthetic sex pheromone of the female pink bollworm moth; mixture of four isomers of 7,11-hexadecadien-1-ol acetate; attractive to the male pink bollworm moth; used to detect infestations of the pink bollworm and to time insecticidal treatments; product of Zoecon.

Gradual metamorphosis. See simple metamorphosis.

Grandlure. Synthetic aggregating and sex pheromone produced by males of the boll weevil; consists of four compounds, approximately 50 per cent of (+)-*cis*-2-isopropenyl-1-methylcyclobutaneethanol, (*Z*)-3,3-dimethyl-Δ^1, β-cyclohexaneethanol and 50 per cent of (*Z*)- and (*E*)-3,3-dimethyl-Δ^1, α-cyclohexaneacetaldehyde; attractive to virgin females and at times to both sexes; used in scouting infestations and in control programs; product of Zoecon.

Granulosis inclusion viruses. Insect viruses characterized by the presence in large numbers of very small but microscopically discernible granular inclusions in infected cells, and particularly visible in the cytoplasm of the host. The granules consist of proteinaceous material within which the virus particle is located. Granulosis disease occurs in larvae of the Lepidoptera, commonly in cutworms but occasionally in pupae.

Granulosis. Virus disease of insects characterized by the presence of granular inclusions.

Granules. An insecticidal formulation in which the insecticide is impregnated on small particles of clay. Size of particles is expressed in terms of the number of openings per linear inch of two limiting screens. Four common mesh sizes are $\frac{8}{15}$, $\frac{15}{30}$, $\frac{24}{48}$, $\frac{30}{60}$.

Gregarious. Living in groups.

Ground bed. A plat of soil at or near ground level in greenhouses, in contrast to soil in raised benches.

Grub. A scarabaeiform larva; that is a thick-bodied larva with thoracic legs and well-developed head; usually sluggish.

Gular sutures. The lines of division between the gula (throat sclerite) and the genae (cheek sclerites).

Guthion (trade name). See azinophosmethyl.

Halter. A slender, knobbed structure on each side of the metathorax in place of the hind wings.

Harrow. A farm implement used to level the ground and crush clods, to stir the soil, and to prevent and destroy weeds. Five principal kinds are the disk, spike tooth, spring tooth, rotary cross harrow, and soil surgeon.

Heart. The chambered, pulsatile portion of the dorsal blood vessel (Fig. 2:19).

Head. The anterior body region of insects which bears the mouthparts, eyes, and antennae (Fig. 2:1).

Hematophagous. Feeding or subsisting on blood.

Hemelytron (pl., hemelytra). The forewing of Hemiptera in which the basal portion is thickened and the distal portion membranous.

Hemimetabola. Insects with simple metamorphosis, with immature stages aquatic and adults terrestrial; insects of the orders Odonata, Ephemeroptera, and Plecoptera; young are called naiads.

Hemoglobin. Oxygen carrying red pigment.

Hemolymph. The blood plasma or liquid part of the blood; the term is generally synonymous for blood of insects.

Heptachlor. A synthetic insecticide; a chlorinated hydrocarbon, 1,4,5,6,7,8,8-heptachloro-3a,4,7,7a-tetrahydro-4,7-methanoindene; moderately toxic to mammals, acute oral LD_{50} for rats 40 to 188 mg/kg; nontoxic to plants at recommended concentrations; product of Velsicol.

Herbicide. A chemical for killing weeds.

Hermaphroditic. Containing the sex organs of both sexes in one individual.

Heterometabola. Insects with simple metamorphosis, including the Paurometabola and Hemimetabola.

HETP. A synthetic insecticide and acaricide; an organic phosphate mixture of ethyl polyphosphates containing 12 to 20 per cent of tetraethyl pyrophosphate (tepp); extremely toxic to mammals, acute oral LD_{50} for rats 7 mg/kg.

Hexalure. Synthetic pink bollworm moth attractant; *cis*-7-hexadecen-1-ol acetate; synthetic scent attractive to male moths of pink bollworm; used in traps to survey extent of infestations.

Hibernation. Dormancy during the winter.

Hindgut. The posterior part of the alimentary canal between the midgut and the anus (Fig. 2:19).

Histosiphon. Same as stylostome.

Holometabola. The higher insects that have complex metamorphosis.

Homologous. Of organs or parts that exhibit similarity in structure, in position with reference to other parts, and in mode of development, but not necessarily similarity of function.

Honeydew. A sugary liquid discharged from the anus of certain Homoptera.

Hopperburn. A disease of potato, alfalfa, and other plants resulting from the feeding of the potato leafhopper, a toxicogenic insect.

Hormone. A chemical substance formed in some organ of the body, secreted directly into the blood, and carried to another organ or tissue where it produces a specific effect.

Host. The organism in or on which a parasite lives; the plant on which an insect or other arthropod feeds.

Hybrid. The offspring of two plants or animals of different races, varieties, or species.

Hydrocarbons. Compounds that contain carbon and hydrogen only.

Hydrolysis. Chemical reaction in which a compound reacts with water to produce a weak acid, a weak base, or both.

Hypermetamorphosis. A type of complex metamorphosis in which the larval instars in their development assume the form of two or more types of larvae.

Hyperparasite. A parasite whose host is also a parasite.

Hypopharynx. A tonguelike mouthpart arising on the upper surface of the labium; in piercing-sucking insects it may be stylet-shaped and contain the salivary channel.

Hypopus. A nymphal stage in the development of certain mites in which the organism is small and has developed suckers or claspers for grasping insects and thereby effecting dispersal.

Hypostome. In ticks, the median ventral dartlike mouthpart that is immovably attached to the basal part of the capitulum (Fig. 2:7).

Hysterosoma. In mites, the posterior part of the body when there is a demarcation of the body between the second and third pair of legs (Fig. 2:32).

Imago (pl., imagoes or imagines). The adult stage of an insect.

Imidan (trade name). See phosmet.

Insect. A member of the class Insecta.

Insecta. A class of phylum Arthropoda distinguished by adults having three body regions—head, thorax, and abdomen—and by having the thorax three-segmented with each segment bearing a pair of legs.

Insecticide. A toxic chemical substance employed to kill and control insects.

Instar. The form of an insect between successive molts, the first instar being the form between hatching and the first molt.

Integrated control. Control of pests which combines and integrates chemical methods with natural and biological control. Chemical control is applied as necessary and in whatever manner is least disruptive to natural and biological control. Also, a pest management system that in consideration of the associated environment and the population dynamics of the pest species, utilizes all suitable techniques and methods in as compatible a manner as possible and maintains the pest populations at levels below those causing economic injury.

Integument. The covering layers of an animal.

Intermediate host. The host that harbors the immature stages or the asexual stages of a parasite.

Internode. The length of stem between two successive nodes.

Intima. The cuticular membrane lining the tracheae.

Invertebrates. Animals without a spinal column or backbone.

Isomer. Any of two chemical compounds having like constituent atoms, but differing in physical or chemical properties because of differences in arrangement of the atoms.

Johnston's organ. A sense organ located in the second antennal segment of many insects and particularly well developed in male mosquitoes and certain other Diptera.

Joint. In plants, a node; to develop distinct nodes and internodes in a grass stem.

Juvenile hormone. The hormone, secreted by the corpora allata, that maintains the immature form of an insect during early molts.

Karathane (trade name). See dinocap.

Kelthane (trade name). See dicofol.

Keratitis. Inflammation of the cornea of the eyes of man or livestock.

Korlan (trade name). See ronnel.

Keratin. A protein forming the principal matter of hair, horns, and nails.

Labellum. The expanded fleshy tip of the labium.

Labial palpus (pl., labial palpi). One of the pair of sensory appendages (feelerlike and two- to five-segmented) of the insect labium.

Labium. The posterior mouthpart or lower lip of an insect (Fig. 2:1).

Labrum. The anterior mouthpart or upper lip of an insect (Fig. 2:1).

Labrum-epipharynx. A mouthpart composed of the labrum and epipharynx and usually elongate (Fig. 2:6).

Lamellate. Platelike or sheetlike; composed of or covered with thin sheets.

Lannate (trade name). See methomyl.

Larva. The immature insect hatching from the egg and up to the pupal stage in orders with complex metamorphosis; the six-legged first instar of mites and ticks.

Lateral ocellus. The simple eye in holometabolus larvae. Also called stemma (pl., stemmata) (Fig. 3:14C).

Lateral oviduct. In insects, one of the paired lateral ducts of the female genital system connected with the ovary.

LD_{50}. Lethal dose to 50 per cent of the test animals. Usually expressed in terms of milligrams (mg) of toxicant per kilogram (kg) of body weight of the test animal (mg/kg).

Leaf miner. An insect that lives in and feeds upon the leaf cells between the upper and lower surfaces of a leaf.

Leaf sheath. The lower part of the leaf that encloses the stem.

Legal control. Control of pests through the enactment of legislation that enforces control measures or imposes regulations, such as quarantines, to prevent the introduction or spread of pests.

Legume. A member of the plant family Leguminosae, which includes bean, pea, alfalfa, clover, peanut, and many other species.

Leguminous. Having the nature of or bearing a legume.

Leishmaniasis. Any disease caused by infection with species of *Leishmania*, a genus of Protozoa in the class Flagellata.

Leprosy. A chronic, transmissible disease caused by a specific bacterium, *Mycobacterium leprae*.

Lethane 384 (trade name). A synthetic insecticide; a thiocyanate, β-butoxy-β'-thiocyano-diethyl ether; moderately toxic to mammals, acute oral LD_{50} for rats 90 mg/kg; somewhat phytotoxic; product of Rohm & Haas.

Life history. Habits and changes undergone by an organism from the egg stage to its death as an adult.

Lime-sulfur. An insecticide and fungicide made by boiling sulfur and lime together in water, which react to form both soluble and insoluble salts of calcium polysulfide; irritating to eyes, nose, and skin; phytotoxicity—injurious to peach trees and on "sulfur-shy" varieties.

Lindane. A synthetic insecticide; a chlorinated hydrocarbon containing 99 per cent or more of the gamma isomer of BHC; somewhat more toxic to mammals than DDT, acute oral LD_{50} for rats 88 to 125 mg/kg; nonphytotoxic at insecticidal concentrations, can damage plants if used in excess; product of Hooker.

Lint. The fiber surrounding the seed of unginned cotton.

Listed corn. Corn planted in a furrow or trench or below the general level of the ground.

Lister. A plow with double moldboard that heaps soil in both sides of the furrow; it may also be combined with a drill that plants seed in the same operation.

Litter floor. The floor of a poultry house which is composed of straw and shavings or ground corncobs, droppings, and other waste materials and builds up during one laying year.

Lock. A locule or ovary cavity.

Locust. A migratory grasshopper.

Lodge. To throw or beat down, as growing grain.

Looper. A caterpillar with two or more of the ventral prolegs wanting; crawls by looping its body.

Looplure. Synthetic sex pheromone of the female cabbage looper moth; (*Z*)-7-dodecenyl acetate; attractive to male moths of the cabbage looper; product of Zoecon.

Lorsban (trade name). See chlorpyrifos.

Macropterous. Long- or large-winged.

Maggot. A vermiform larva; a larva without legs and without well-developed head capsule (Fig. 3:14G).

Malaria. An infectious febrile disease of man caused by Protozoa of the genus *Plasmodium* which invade the red blood corpuscles and which are transmitted by mosquitoes of the genus *Anopheles.*

Malathion (Cythion). A synthetic insecticide; an organic phosphate, diethyl mercaptosuccinate, *S*-ester with *O,O*-dimethyl phosphorodithioate; slightly toxic to mammals, acute oral LD_{50} for rats 1500 mg/kg; phytotoxicity—tolerated by most plants at concentrations used for insect control; product of American Cyanamid.

Malpighian tubes. Excretory tubes of insects arising from the anterior end of the hindgut and extending into the body cavity (Fig. 2:19).

Mandibles. The anterior paired mouthparts in insects; stout and toothlike in chewing insects, needle- or sword-shaped in piercing-sucking insects (Figs. 2:1, 2:5, and 2:6).

Mange. A group of contagious skin diseases in livestock caused by certain parasitic mites.

Mastitis. Inflammation of the mammary gland.

Maxillae (sing., maxilla). The paired mouthparts behind the mandibles; the second pair of jaws in chewing insects (Figs. 2:1, 2:5, and 2:6).

Mechanical control. Control of pests by mechanical means such as window screens, earth barriers, and so on.

Median oviduct. In insects, the single duct formed by the merging of the paired lateral oviducts; this duct opens posteriorly into a genital chamber or vagina (Fig. 2:19).

Medlure. A synthetic attractant for baiting the Mediterranean fruit fly; *sec*-butyl 4(or 5)-chloro-2-methylcyclohexanecarboxylate.

Melanistic. Characterized by excessive pigmentation or blackening of the integument or tissues.

Membranous. Thin and transparent (in reference to wings); thin and pliable (in reference to integument).

Mesoderm. In insects, the embryological tissue that forms a middle layer and gives rise to muscles, heart, blood cells, fat body, reproductive organ, and others.

Mesothorax. The middle or second segment of the thorax (Fig. 2:1).

Mesurol (trade name). See methiocarb.

Metaldehyde (chemical name). A chemical with slug-killing properties; empirical formula $(CH_3CHO)_4$; moderately toxic to mammals, acute oral LD_{50} for rats 630 mg/kg.

Metalkamate (Bux). A synthetic insecticide; a carbamate, *m*-(1-ethylpropyl)phenyl methylcarbamate mixture (1-4) with *m*-(1-butylethyl)phenyl methylcarbamate; moderately toxic to mammals; acute oral LD_{50} for rats 87 to 170 mg/kg; controls corn rootworms; product of Chevron.

Metamorphosis. Change in form during the development of an insect.

Meta-Systox-R (trade name). See oxydemetonmethyl.

Metathorax. The third or posterior segment of the thorax (Fig. 2:1).

Methamidophos (Monitor, Ortho 9006). A synthetic insecticide; an organophosphate, *O,S*-dimethyl phosphoramidothioate; highly toxic to mammals, acute oral LD_{50} of 75 per cent technical 21 mg/kg; effective in the control of lepidopterous larvae on cotton, potatoes, lettuce, and cruciferous vegetables; product of Bayer, Chemagro, Chevron, and Mobay.

Methidathion (Supracide). A synthetic insecticide and acaricide; an organophosphate, *S*-((2-methoxy-5-oxo-Δ^2-,1,3,4-thiadiazolin-4-yl)methyl) *O,O*-dimethyl phosphorodithioate; highly toxic to mammals, acute oral LD_{50} for rats 25 to 48 mg/kg; effective against alfalfa weevil larvae, spider mites and bollworms of cotton, scales in citrus; product of Ciba-Geigy.

Methiocarb (Mesurol). A synthetic insecticide and acaricide; a carbamate, 4-(methylthio)3,5-xylyl methylcarbamate; moderately toxic to mammals, acute oral LD_{50} for rats 87–130 mg/kg; broad spectrum insecticide and miticide, also effective against slugs and snails; product of Bayer AG, Chemagro, and Mobay.

Methomyl (Lannate and Nudrin). A synthetic insecticide and acaricide; a carbamate, methyl *N*-[(methylcarbamoyl)oxy] thioacetimidate; highly toxic to mammals; acute oral LD_{50} 17 to 24 mg/kg; controls cabbageworms and loopers on cabbage, broccoli, cauliflower, and lettuce; product of DuPont and of Shell Chemical Co.

Methoprene (Altosid). A synthetic insect growth regulator; an acyclic sesquiterpenoid similar to juvenile hormone of cecropia moth, isopropyl (*E,E*)-11-methoxy-3,7,11-trimethyl-2,4-dodecadienoate; relatively nontoxic, acute oral LD_{50} for rats greater than 34,600 mg/kg; especially active against Diptera and Homoptera, used for control of insecticide-resistant mosquitoes; product of Zoecon.

Methoxychlor. A synthetic insecticide; a chlorinated hydrocarbon related to DDT, 1,1,1-trichloro-2,2-bis(*p*-methoxyphenyl) ethane; less toxic to mammals than DDT, acute oral LD_{50} for rats about 6000 mg/kg; generally nonphytotoxic to crop plants; product of DuPont.

Methyl bromide. Insecticidal fumigant for treatment of mills, warehouses, vaults, ships, freight cars; also a soil fumigant; empirical formula, CH_3Br; exposure to 2000 ppm for 1 hr causes serious injury to mammals; toxic to growing plants.

Methyl parathion. A synthetic insecticide and acaricide; an organic phosphate, *O,O*-dimethyl *O-p*-nitrophenyl phosphorothioate, extremely toxic to mammals, acute oral LD_{50} for rats 9 to 25 mg/kg: little phytotoxicity in recommended doses; product of several companies.

Mevinphos (Phosdrin). A synthetic insecticide and acaricide, a plant systemic; an organic phosphate, methyl 3-hydroxy-*alpha*-crotonate, dimethyl phosphate; extremely toxic to mammals, acute oral LD_{50} for male rats alpha isomer 3 mg/kg, beta isomer 46 mg/kg; product of Shell.

MGK 264 (trade name). A synergist for pyrethrum; chemically, *N*-(2-ethylhexyl)-5-norbornene-2,3-dicarboximide; relatively nontoxic to mammals, acute oral LD_{50} for rats 2800 mg/kg; product of McLaughlin Gormley King.

Microflora. Microscopic plant life of an area.

Micron. One-thousandth part of a millimeter.

Micropyle. A minute opening or group of openings into the insect egg through which the spermatozoa enter in fertilization.

Microtrichia. Minute hairs that are formed around cellular filaments and project from the integument.

Midgut. The middle part of the alimentary canal and the main site of digestion and absorption (Fig. 2:19).

Millimeter. One thousandth of a meter, or approximately 0.04 in.

Mineral. An inorganic homogeneous substance; an inorganic foodstuff.

Mite. Any minute invertebrate belonging to the phylum Arthropoda and order Acarina except the ticks (Fig. 4:4).

Miticide. Any poisonous substance used to kill and control mites.

Mitochondria. Small granules or rodlike structures found in cytoplasm of cells after differential staining; in the living cell they are the sites of important cellular enzymes which are concerned with energy transformations.

Mitox (trade name). See chlorbenside, the common name for this chemical.

Mocap (trade name). See ethoprop.

mm. A millimeter.

Mollusca. A phylum of animals containing snails, slugs, clams, oysters, octopus, and others.

Molt. Process of shedding the skin; to shed the skin.

Moniliform. Beadlike, with rounded segments, as in moniliform antennae (Fig. 2:3E).

Monitor (trade name). See methamidophos.

Monitor. To observe or check, especially for a special purpose, as to keep track of crop development and insect infestation.

Monocrotophos (Azodrin). A synthetic insecticide and acaricide, plant systemic; an organophosphate, 3-hydroxy-*N*-methyl-*cis*-crotonamide dimethyl phosphate; highly toxic to mammals; acute oral LD_{50} for rats 21 mg/kg; used on cotton to control boll weevil, bollworms, lygus bugs, cabbage looper, and mites and to control a variety of pests on other crops; product of Shell.

Monophagous. Feeding upon only one kind of food, for example one species or one genus of plants.

Morestan (trade name). See oxythioquinox.

Morocide (trade name). See binapacryl.

Murine typhus fever. A human disease caused by a bacteriumlike microorganism, *Rickettsia mooseri* and transmitted from rats to man by the oriental rat flea. Clinically similar to typhus fever except with milder symptoms.

Muscalure. Synthetic sex pheromone of the female house fly; (*Z*)-9-tricosene; attractive to male house flies; used as an additive in poison baits; product of Zoecon.

Mycetome. In insects, a group of cells that harbor microorganisms.

Myiasis. Infestation of the body by the larvae of flies.

Naiad. An aquatic, gill-bearing nymph.

Naled (Dibrom). A synthetic insecticide and acaricide; an organic phosphate, 1,2-dibromo-2,2-dichloroethyl dimethyl phosphate; moderately toxic to mammals, acute oral LD_{50} for rats 430 mg/kg; phytotoxicity—injury may occur on some varieties of pome and stone fruits; product of Chevron.

Naphthalene (chemical name). A coal tar derivative long used as a fumigant for clothes moth, also employed as a soil fumigant; relatively nontoxic to mammals; very toxic to plants.

Nasutus (pl., nasuti). A type of soldier caste in certain termites; this form bears a median frontal rostrum through which it ejects a defensive fluid; the jaws are small or vestigial.

Natural control. The reduction of pest populations by the forces of nature such as climatic factors, parasites, predators, and disease.

Nectar. The sugary liquid secreted by many flowers.

Nectary. A floral gland that secretes nectar.

Neguvon (trade name). A product prepared for use as an animal medicament; see trichlorfon.

Nematodes. Unsegmented worms with cylindrical, elongate bodies.

Neurone. The entire nerve cell including all its processes.

Neurosecretory. Pertaining to the secretion of hormones by nerve cells.

Newcastle disease. An acute, rapidly spreading respiratory and nervous disease of domestic poultry and other birds caused by a virus and characterized by rales, coughing, sneezing, and nervous manifestations.

Nicotine. A botanical insecticide derived from leaves and stems of the tobacco plant; an alkaloid, *l*-1-methyl-2-(3′-pyridyl)-pyrrolidine; highly toxic to mammals, acute oral LD_{50} for rats 50 to 60 mg/kg; phytotoxicity—safe on most plants.

Nit. The egg of a louse.

Nocturnal. Active at night.

Node. The joint of a stem where a leaf is attached.

Noninclusion viruses. Insect viruses that occur free in tissues as viruses do in higher animals and plants; the viruses are not included in crystals, granules, or other inclusion bodies; as yet only a small number of these viruses have been found in insects; found in larval Diptera, Lepidoptera, Coleoptera, and in larval and adult Hymenoptera.

Notum (pl., nota). In insects the dorsal surface of a body segment.

Nubbin. A small or imperfect ear of corn.

Nuclear polyhedrosis viruses. Insect viruses which are occluded in a polyhedral protein crystal and which multiply in the cell nucleus; virus particles are rod shaped; majority occur in larvae of Lepidoptera.

Nucleus. The spheroid body within a cell that has the major role in controlling and regulating the cell's activities and contains the hereditary units or genes.

Nudrin (trade name). See methomyl.

Nurse cells. Cells that are located in the ovarian tubes of certain insects and that furnish nutriment to the developing eggs.

Nymph. A young insect of a species with simple or no metamorphosis; an eight-legged immature mite or tick.

Obtect pupa. A pupa in which the appendages are closely appressed to the body (Fig. 3:15A).

Ocellus (pl., ocelli). The simple eye of an insect or other arthropod.

Okra. A tall plant of the mallow family with sticky green pods used in soups, stews, and so on.

Omite (trade name). See propargite.

Ommatidium (pl., ommatidia). A single unit or visual section of a compound eye (Fig. 2:29).

Onchocerciasis. Infection with a genus of filarial nematodes, *Onchocerca*. The adults live and reproduce in subcutaneous fibroid nodules; the young, called microfilariae, are carried by the lymph and found chiefly in the skin and eyes; transmitted by certain black flies.

Ootheca (pl., oothecae). An egg case formed by the secretions of accessory genital glands or oviducts, as in cockroaches (Fig. 19:10).

Order. A subdivision of a class or subclass containing a group of related families.

Organophosphates. Organic compounds containing phosphorus; an important group of synthetic insecticides belong to this class of chemicals.

Oribatid mite. A mite belonging to the Oribatei, a large unit of mites containing about 35 families in the suborder Sarcoptiformes.

Orthene (trade name). See acephate.

Ortho 9006 (trade name). See methamidophos.

Osmotic pressure. The maximum pressure which can be developed in a solution which is separated from pure water by a rigid membrane permeable only to water.

Ostiole. External opening of the stink gland in Hemiptera and Homoptera.

Oviduct. The duct leading from the ovary through which the eggs pass (Fig. 2:19).

Oviparae. In aphids, the oviparous females.

Oviparous. Producing eggs that are hatched outside the body of the female.

Oviposition. The act of laying or depositing eggs.

Ovipositor. The tubular or valved structure of the female by means of which the eggs are deposited.

Ovoviviparous. Producing living young by the hatching of the egg while still within the female.

Oxydemetonmethyl (meta-Systox-R). A synthetic insecticide and acaricide, a plant systemic; an organophosphate, *S*-[2-(ethylsulfinyl)ethyl] *O*,*O*-dimethyl phosphorothioate; moderately toxic to mammals; acute oral LD_{50} for rats 65 to 75 mg/kg; systemic foliar spray for control of sucking pests on ornamentals, field, and vegetable crops; product of Chemagro.

Oxythioquinox (Morestan). A synthetic acaricide and fungicide, 6-methyl-2,3-quinoxalinedithiol cyclic *S*,*S*-dithiocarbonate; slightly toxic to mammals; acute oral LD_{50} for rats 3000 mg/kg; used on citrus, most deciduous fruits, and ornamentals to kill mites and their eggs and to control mildew; product of Chemagro.

Paedogenesis. The production of eggs or young by an immature or larval stage of an animal.

Palpus (pl., palpi). A segmented feelerlike process borne by the maxilla or labium.

Paradichlorobenzene. Insecticidal fumigant for control of clothes moth, also soil fumigant for peachtree borer; empirical formula $C_6H_4CL_2$; doses over 300 mg/kg for humans begin to be harmful, acute oral LD_{50} for rats 500 mg/kg; phytotoxicity—seriously injures seed germination.

Parasite. Any animal or plant that lives in or on and at the expense of another organism.

Parathion. A widely used agricultural insecticide and acaricide; an organic phosphate, *O,O*-diethyl *O*-*p*-nitrophenyl phosphorothioate; highly toxic to mammals, oral LD_{50} to rats 4 to 13 mg/kg; phytotoxicity—injurious to certain ornamentals, pears, and McIntosh and related apples under certain weather conditions; product of several companies.

Parenchyma. A plant tissue composed of thin-walled cells which often store food and usually retain the capacity to divide.

Paris green. An arsenical compound which came into use as an insecticide around 1867; chemically, copper acetoarsenite; highly toxic to mammals; toxic to tender plants and fruit trees.

Parthenogenesis. Reproduction by the development of an egg without its being fertilized by a sperm.

Pathogenic. Giving origin to disease.

Paurometabola. Insects with simple metamorphosis and with young and adults living in the same habitat; young are called nymphs.

Pectinate. Comblike; with branches or processes like the teeth of a comb, as in pectinate antennae (Fig. 2:3G).

Pedipalp. The second pair of appendages of an arachnid, used to crush prey.

Pentac (trade name). A specific miticide; a chlorinated hydrocarbon, decachlorobis-2,4-cyclopentadien-1-yl; slightly toxic to mammals, male rat acute oral LD_{50} greater than 3160 mg/kg; controls twospotted spider mite on greenhouse floral crops such as roses, chrysanthemums, and carnations, also on outdoor roses and nursery arborvitae; product of Hooker.

Pentachlorophenol (chemical name). A wood preservative, used to control termites and to protect cut timber from wood-boring insects and from fungal rots; moderately toxic to mammals, acute oral LD_{50} for rats 50 to 140 mg/kg; toxic to plants at point of contact, but not translocated; used as a herbicide.

Perthane (trade name). A synthetic insecticide; a chlorinated hydrocarbon related to DDT; 1,1-dichloro-2,2-bis(*p*-ethylphenyl)ethane; low mammalian toxicity, oral LD_{50} for rats 8170 mg/kg; not phytotoxic; product of Rohm & Haas.

Pesticide. A chemical that is used to poison and control pests, either animal or plant.

Pest management. The control of pest populations by a program that selects and utilizes available control methods so that economic damage is avoided and adverse side effects on the agroecosystem are minimized; or the reduction of pest problems by actions selected after the ecology or "life systems" of the pests are understood and the ecological as well as the economic consequences of these actions have been predicted to be in the best interest of man.

Petiole. A leaf stalk.

Phagocytic. Pertaining to any cell that ingests microorganisms or other cells and substances.

Pharynx. The anterior part of the foregut between the mouth and the esophagus.

Pheromone. See ectohormone.

Phorate (Thimet). A synthetic insecticide and acaricide, a plant systemic with high contact activity; an organic phosphate, *O,O*-diethyl *S*-ethylthiomethyl phosphorodithioate; extremely toxic to mammals, oral LD_{50} for rats 2 to 4 mg/kg; phytotoxicity—seed of wheat, oats, corn, peas, cucumbers, and beans will not tolerate higher dosage than 4 to 8 oz per 100 lb; injures tobacco and apples; product of American Cyanamid.

Phosalone (Zolone). A synthetic insecticide and acaricide, a plant systemic; an organophosphate, *S*-[(6-chloro-2-oxo-3-benzoxazolinyl)methyl] *O,O*-diethyl phosphorodithioate; moderately toxic to mammals; acute oral LD_{50} for rats 125 to 180 mg/kg; controls certain insects and mites on apples, pears, and grapes; product of Rhodia.

Phosdrin (trade name). See mevinphos.

Phosmet (Imidan, Prolate). A synthetic insecticide and acaricide and an animal systemic; an organophosphate, *O,O*-dimethyl *S*-phthalimidomethyl phosphorodithioate, moderately toxic to mammals; acute oral LD_{50} for rats 147 to 299 mg/kg; controls certain insects and mites infesting deciduous fruit trees and several livestock insects; product of Stauffer.

Phosphamidon (Dimecron). A plant systemic insecticide and acaricide; an organic phosphate, 2-chloro-*N,N*-diethyl-3-hydroxycrotonamide, dimethyl phosphate; highly toxic to mammals, acute oral LD_{50} for rats 15 to 33 mg/kg; some phytotoxicity to apples, peaches, and walnuts; product of Chevron.

Phostex (trade name). A synthetic insecticide and acaricide; a phosphorus-containing compound, bis(dialkoxyphosphinothioyl) disulfides (alkyl ratio 25 per cent isopropyl, 75 per cent ethyl); low toxicity to mammals, acute oral LD_{50} for rats 2500 mg/kg; phytotoxicity causes shot-holing of stone fruit foliage, toxic to Wealthy apples.

Phylum (pl., phyla). A major division of the animal kingdom.

Physical control. Control of pests by physical means such as heat, cold, electricity, sound waves, and so on.

Phytophagous. Feeding upon plants.

Phytotoxic. Poisonous to plants.

Pink eye. In cattle an infectious disease of the eyes in which the eye and its protective membranes become inflamed.

Piperonyl butoxide (common name). A synergist for pyrethrum and rotenone; chemically, *a*[2-(2-butoxyethoxy)ethoxy]-4,5-methylenedioxy-2-propyltoluene; relatively nontoxic to mammals, acute oral LD_{50} for rats 7500 to 12,800 mg/kg; product of FMC, McLaughlin Gormley King, and Prentiss Drug.

Pirimicarb (Pirimor). A systemic insecticide; a carbamate, 5,6-dimethyl-2-dimethylamino-4-pyrimidinyl dimethylcarbamate; moderately toxic to mammals, acute oral LD_{50} for rats 147 mg/kg; a fast-killing aphicide by contact, vapor, or systemic action, for use on tree fruits, vegetables, and certain ornamentals; product of Plant Protection Ltd., Great Britain.

Pirimor (trade name). See pirimicarb.

Piroplasmosis. Infection with piroplasma, genus *Babesia*, parasitic protozoans that attack the red blood corpuscles of cattle, dogs, and other animals and cause a high fever, destruction of red blood corpuscles, enlarged spleen, engorged liver, emaciation, and often death; the organism is transmitted by ticks.

Platyform larva. A very flattened larva (Fig. 3:14F).

Platyhelminthes. The phylum containing the flatworms, such as tapeworms and flukes.

Plictran (trade name). See cyhexatin.

Plumose. Featherlike, as in plumose antennae (Fig. 2:3H).

Polar chemical. A chemical compound composed of molecules with a dipole (a pair of equal and opposite charges separated by a small distance), such as water, alcohol, and salt solutions.

Poliomyelitis. Inflammation of the gray matter of the spinal cord; an acute infectious virus disease attended with fever, motor paralysis, and atrophy of groups of muscles.

Pollen. The mass of microspores or male fertilizing elements of flowering plants.

Pollinate. To transfer pollen grains from a stamen to a stigma or ovule of a plant.

Polyembryony. The production of several embryos from a single egg, as in some chalcids.

Polyhedrosis. Virus disease of insects characterized by the presence of polyhedral inclusions.

Polymorphism. The condition of having several forms in the adult stage (e.g., in the honey bee: worker, queen, and drone).

Polyphagous. Feeding on a variety of plants or animals.

Population. A group of individuals of the same species living in a limited and defined area.

Posterior. Hind or rear.

Potassium ammonium selenosulfide. An inorganic acaricide; used as a spray on ornamental plants; highly toxic to mammals.

ppm. Parts per million.

Predaceous. Preying on other animals.

Predator. An animal that attacks and feeds on other animals, usually smaller and weaker than itself.

Preovipositional period. The period between the emergence of an adult female and the start of its egg laying.

Prepupa. The last larval instar after it ceases to feed; often it takes on a distinctive appearance and becomes quiescent.

Presumptive organization. Arrangement of cells in the embryo into groups which in normal development become a particular organ or tissue.

Pretarsus. In insects, the terminal segment of the leg bearing the pretarsal claws.

Primary parasite. A parasite that establishes itself in or upon a free-living host, that is, a host that is not a parasite.

Primary reproductives. See reproductives.

Proctodeal valve. In insects, a valve in the anterior end of the hindgut that serves as an occlusor mechanism.

Prolate (trade name). See phosmet.

Proleg. A fleshy abdominal leg of certain insect larvae.

Propargite (Omite). A synthetic acaricide, 2-(*p-tert*-butylphenoxy) cyclohexyl 2-propynyl sulfite; slightly toxic to mammals; acute oral LD_{50} for rats 2200 mg/kg; used on cotton, fruit and nut trees and certain vegetables to control spider mites; product of UniRoyal.

Propoxur (Baygon). A synthetic insecticide; a carbamate, *o*-isopropoxyphenyl methylcarbamate; moderately toxic to mammals; acute oral LD_{50} for rats 95 to 104 mg/kg; used to control household pests; product of Chemagro.

Propupa. In thrips, the next to the last nymphal instar in which the wing pads are present and the legs are short and thick. Also occurs in male scale insects.

Protein. Any one of a group of nitrogen-containing compounds consisting of a union of amino acids and also containing carbon, hydrogen, oxygen, and frequently sulfur; proteins occur in all animal and vegetable matter and are essential to the diet of animals.

Proterosoma. In mites, the anterior part of the body when there is demarcation of the body between the second and third pair of legs.

Prothoracic gland. One of a pair of endocrine glands located in the prothorax near the prothoracic spiracles.

Prothorax. The first or anterior segment of the thorax (Fig. 2:1).

Protonymph. The second instar of a mite.

Protozoa. The phylum containing the one-celled animals.

Proventriculus. The posterior section of the foregut.

Pseudoscorpions. Small arachnids, seldom over 5 mm long, scorpionlike in general appearance but without sting.

Pseudovipositor. The slender tube to which the posterior part of the abdomen is reduced in the female of certain insects.

psi. Pounds per square inch.

Pterygota. A subclass of insects which are primarily winged but sometimes secondarily wingless.

Ptilinum. In Diptera an organ that can be inflated to a bladderlike structure and thrust out through a frontal suture of the head at the time of emergence from the puparium.

Pubescent. Covered with short fine hairs.

Pupa (pl., pupae). The stage between the larva and the adult in insects with complex metamorphosis, a nonfeeding and usually inactive stage.

Puparium (pl., puparia). In higher Diptera, the thickened, hardened barrellike larval skin within which the pupa is formed.

Pygidium. The tergum of the last segment of the abdomen; in diaspine coccids a sclerotized region ending the abdomen of the adult female.

Pyrethrum (common name). A botanical insecticide derived from the flowers of *Chrysanthemum*, primarily *C. cinerariaefolium*; slightly toxic to warmblooded animals, acute oral LD_{50} for rats 1500 mg/kg; nontoxic to plants.

Pyrophyllite. A mineral used as a dust carrier for insecticides; chemically, $H_2O \cdot Al_2O_3 4SiO_2$.

Queen cell. The special cell in which a queen honey bee develops from egg to the adult stage.

Queletox (trade name). See fenthion.

Qikron (trade name). See chlorofenethol.

Quinones. Organic compounds, benzene derivatives, known to play a role in the hardening and darkening of arthropod cuticle.

Rabon (trade name). See stirofos.

Race. A variety of a species; a subspecies.

Rape. A plant, *Brassica napus* var. *biennis*, belonging to the mustard family and grown chiefly as a pasture crop for forage.

Receptacle. The enlarged end of the flower stalk.

Rectum. In insects, the posterior expanded part of the hindgut, typically pear-shaped (Fig. 2:19).

Relapsing fever. Any one of a group of acute infectious diseases caused by various bacteria of the genus *Borrelia*. The European epidemic fever is caused by *Borrelia recurrentis*, which is transmitted by the human body louse. The disease is marked by alternating periods of fever.

Repellents. Substances that elicit an avoiding reaction.

Reproductives. In termites the caste of kings and queens. They have compound eyes and fully developed wings (before dealation) and are usually heavily pigmented.

Resistance. Of insects to insecticides, the ability of strains of insects to survive normally lethal doses of insecticide, the ability having resulted from selection of tolerant individuals in populations exposed to the toxicant for several generations. Of plants to insect attack, reaction of certain varieties to oppose the nurture of pests or to lack attractancy for them, heritable characteristics that lessen insect damage.

Resmethrin (SBP 1382, Chryson, Synthrin). A synthetic pyrethroid insecticide; (5-Benzyl-3-furyl) methyl-2,2-dimethyl-3-(2-methylpropenyl) cyclopropanecarboxylate; slightly toxic to mammals, acute oral LD_{50} for rats 4240 mg/kg; household, processing plants, greenhouse, ornamental insect control; product of FMC and S. B. Penick.

Rickettsia. An obligate intracellular parasite of arthropods, many types of which are pathogenic for man and other animals. They are thought to be intermediate between bacteria and viruses, because they have features in common with both.

Rocky Mountain spotted fever. A human disease caused by a bacteriumlike microorganism, *Rickettsia rickettsi*, characterized by a rash, fever, headache, backache, and marked malaise; transmitted by the Rocky Mountain wood tick and the American dog tick.

Ronnel (Korlan, Trolene). A synthetic insecticide, an animal systemic and contact insecticide; an organic phosphate, *O,O*-dimethyl *O*-2,4,5-trichlorophenyl phosphorothioate; slightly toxic to mammals, acute oral LD_{50} for rats 1740 mg/kg; product of Dow.

Rotary tiller. A cultivator made up of two gangs of hoe wheels.

Rotenoids. A group of related toxic compounds (rotenone, elliptone, etc.) that are found in certain leguminous plants.

Rotenone (common name). A botanical insecticide; the main toxic constituent in the roots of certain leguminous plants, such as *Derris elliptica* and *Lonchocarpus utilis* and *L. urucu*; moderately toxic to mammals, acute oral LD_{50} for rats 132 mg/kg; nontoxic to plants; product of several companies.

Roundworm. A cylindrical, unsegmented worm tapered toward both ends; a member of the phylum Aschelminthes and the class Nematoda.

Rudimentary. Imperfectly developed.

Ruelene (trade name). See crufomate.

Ryania (common name). A botanical insecticide consisting of the ground stemwood of *Ryania speciosa*; active ingredient is the alkaloid ryanodine; slightly toxic to mammals, acute oral LD_{50} for rats of the ground *Ryania* stems 1200 mg/kg; product of S. B. Penick.

Sabadilla (common name). A botanical insecticide made from the ground seeds of a lily, *Schoenocaulon officinale*; active ingredient is a crude mixture of alkaloids termed veratrine; relatively nontoxic to warm-blooded animals but a nasal irritant. Product of Prentiss Drug.

Salivary glands. Glands that open into the mouth and secrete a fluid with digestive, irritant, or anticoagulatory properties.

Saprophytic. Living on dead or decaying organic matter.

Saturated. In chemistry, having all valences of the constituent atoms of a compound fully satisfied.

SBP 1382 (trade name). See resmethrin.

Scab. A contagious skin disease of animals caused by certain parasitic mites.

Scale. A scale insect; a member of the order Homoptera.

Scarabaeiform larva. A grublike larva, body thick and cylindrical, well-developed head and thoracic legs, and no prolegs, usually sluggish (Fig. 3:14D).

Scavenger. An animal that feeds on dead plants or animals, on decaying organic matter, or on animal wastes.

Schradan (OMPA). A synthetic insecticide and acaricide, a plant systemic; an organic phosphate, octamethylpyrophosphoramide; extremely toxic to mammals, acute oral LD_{50} for rats 9 mg/kg; phytotoxicity—not markedly toxic at insecticidal concentrations, but over 4 lb per acre injurious to some crops; product of Murphy Chemical, Great Britain.

Sclerite. A hardened body wall plate delimited by sutures or membranous areas.

Sclerotization. The hardening and darkening processes in the cuticle (involves the epicuticle and exocuticle).

Scorpion. Any member of the arachnid order Scorpionida; scorpions have an elongated body and a poison sting at the end of abdomen.

Scout. To sample and to observe insects infesting a crop and their damage, also including beneficial insects.

Scrub typhus. A disease prevalent in the Far East caused by a bacteriumlike microorganism, *Rickettsia tsutsugamushi*, which is transmitted by the chigger, *Trombicula akamushi*; the disease is characterized by headache, apathy, general malaise, and fever.

Scutum. In ticks, the sclerotized plate covering all or most of the dorsum in males, and the anterior portion in females, nymphs, and larvae of the Ixodidae.

Sebaceous gland. A gland producing a greasy lubricating substance.

Secondary parasite. A parasite which establishes itself in or upon a host that is a primary parasite.

Segment. A subdivision of the body or of an appendage between joints or areas of flexibility.

Segmentation. The embryological process by which the insect body becomes divided into a series of parts or segments.

Selocide (trade name). An acaricide containing about 30 per cent potassium ammonium selenosulfide.

Semilooper. A caterpillar with one or two pairs of the ventral prolegs wanting; in crawling small loops of the body are formed.

Sessile. Attached and incapable of moving from place to place.

Seta (pl., setae). A slender, hairlike outgrowth of the integument.

Sesamex (common name). A synergist for pyrethrum and allethrin; chemically 2-(2-ethoxyethoxy)ethyl-3,4-(methylenedioxy)phenyl acetal of acetaldehyde; slightly toxic to mammals, acute oral LD_{50} for rats 2000 to 2270 mg/kg; product of Shulton.

Sevin. (trade name). See carbaryl.

Shingling. The placing of shingles or pieces of lumber among crop plantings to become daytime hiding places for slugs, squash bugs, earwigs, and so on, which can then be destroyed.

Silica aerogel. A sorptive dust that is insecticidal, killing insects by desiccation; prepared by treating sodium silicate with sulfuric acid, drying, and then grinding to small particle size.

Simple dorsal eyes. See Dorsal ocellus.

Simple eye. See Ocellus.

Simple lateral eye. See Lateral ocellus.

Simple metamorphosis. Metamorphosis in which the wings (when present) develop externally during the immature stage, and there is no prolonged resting stage preceding the last molt; stages included are the egg, nymphal, and adult. Also called gradual or partial metamorphosis, and paurometabolous development (Fig. 3:9).

Sinus. A recess, cavity, hollow space; an air cavity in a cranial bone.

Skeletal muscle. In insects, a muscle that stretches across the body wall and serves to move one segment on another.

Slug. A relative of snails but having the shell rudimentary or entirely wanting.

Slurry. A thin mixture of water and any of several fine insoluble materials, as clay, derris or cubé powder, and so on.

Small grains. Any cereal having small kernels, as wheat, oats, barley, or rye.

Smear 62. Former screwworm remedy consisting of the following (parts by weight): diphenylamine 3.5, benzol 3.5, turkey red oil 1, and lamp black 2.

Snail. A member of the phylum Mollusca having a single, usually coiled shell and a broad, flat foot.

Social. Living in more or less organized communities of individuals.

Sodium selenate. An inorganic systemic insecticide and acaricide that is applied to soil; empirical formula, Na_2SeO_4; highly toxic to warmblooded animals; limited to treatment of ornamentals; phytotoxic to chrysanthemums at doses about 250 mg/sq ft, carnations stunted by use of selenate for more than 1 year.

Solanaceous. Belonging to the nightshade family, of which potato is a common example.

Soldier. In termites, sterile males or females with large heads and mandibles; they function to protect the colony.

Solitary. Occurring singly or in pairs, not in colonies.

Species. A group of individuals or populations which are similar in structure and physiology and are capable of interbreeding and producing fertile offspring, and which differ in structure and/or physiology from other such groups and normally do not interbreed with them.

Specific miticides. Chemicals that are used to control mites but are relatively nontoxic to insects.

Spermatozoon (pl., spermatozoa). The mature male sexual cell or sperm cell, whose function is the fertilization of the egg.

Spermatheca. The sperm storage receptacle of the female insect.

Spine. A multicellular, thornlike process or outgrowth of the integument not separated from it by a joint.

Spiracle. An external opening of the tracheal system through which diffusion of gases takes place.

Spiracular plate. A platelike sclerite next to or surrounding a spiracle.

Spirochaetosis. Infection with spirochetes, organisms that are regarded as a connecting link between bacteria and protozoa.

Spittle. In insects, a frothy fluid produced by the nymphs of spittlebugs (Cercopidae).

Spur. A spinelike process of the integument connected to the body wall by a joint.

Square. An unopened flower bud of cotton with its subtending involucre bracts.

Stadium (pl., stadia). The time interval between molts in a developing insect.

Stage. A distinct, sharply different period in the development of an insect (e.g., egg stage, larval stage, pupal stage, adult stage); in mites and ticks, each instar.

Stemma (pl., stemmata). The simple eye in holometabolous larvae. Also called lateral ocellus (Fig. 3:14C).

Sterols. Mostly large molecular alcohols, found in plant and animal cells combined with fatty acids and soluble in fat solvents.

Stewart's disease. Bacterial wilt of corn caused by *Bacterium stewartii* and transmitted by the feeding of flea beetles.

Stipe. A small, stalklike structure.

Stirofos (Rabon and Gardona). A synthetic insecticide; an organophosphate, 2-chloro-1-(2,4,5-trichlorophenyl)vinyl dimethyl phosphate; slightly toxic to mammals; acute oral LD_{50} for rats 4000 to 5000 mg/kg; controls corn earworm and fall armyworm on corn seed crops, residual wall spray to control flies on agricultural premises, controls insects on livestock and poultry; product of Shell.

St. Louis encephalitis. A virus disease affecting the brain, producing languor, apathy, lethargy, and death; transmitted by certain mosquitoes and mites.

Stomodeal valve. In insects, the cylindrical or funnel-shaped invagination of the foregut into the midgut.

Stover. Corn stalks used as fodder for animals.

Striate. A virus mosaic disease of wheat transmitted by the painted leafhopper.

Striated muscle. Muscle that is composed of fibers with alternate light and dark bands.

Stubble. The stumps of small grain, corn, and so on, left standing after harvest.

Stylet. A needlelike structure.

Stylostome. The tube formed by the host as a result of the feeding of a chigger; in secreting salivary fluids, the chigger partially digests skin tissues, which induces the host to form a proteinaceous tube walling off the injury.

Subterranean. Living in the ground.

Sulfaquinoxaline. An organic sulfur-containing compound used in veterinary medicine; N′-(2-Quinoxalinyl) sulfanilamide; used in treatment of coccidiosis and several other diseases of intestinal tract of livestock and poultry.

Sulfotepp (common name). A synthetic insecticide and acaricide; an organic phosphate, tetraethyl dithiopyrophosphate; highly toxic to mammals, acute oral LD_{50} for rats 7 to 10 mg/kg; used as greenhouse fumigant; product of Bayer, West Germany.

Sulfoxide (common name). A synergist for pyrethrum; chemically, 1,2-methylenedioxy-4-[2-(octylsulfinyl)propyl]benzene; slightly toxic to warmblooded animals, acute oral LD_{50} for rats 2000 to 2500 mg/kg; nonphytotoxic; product of S. B. Penick.

Superfamily. A group of closely related families; superfamily names end in *-oidea*.

Superior oil. A dormant spray oil of high paraffinic and low aromatic content, characteristics which provide increased plant safety and satisfactory insecticidal action.

Supplementary reproductives. In termites the caste of males and females with short wings, light pigmentation, and small compound eyes. The females lay eggs in the colony, supplementing the work of the queen.

Supracide (trade name). See methidathion.

Suspension. A system of solid particles dispersed in a liquid.

Suture. A linelike external groove in the body wall or a narrow membranous area between sclerites; a line where adjacent parts have united.

Swine pox. A virus disease of swine characterized by small, red skin lesions, weakness, loss of appetite, chills, and fever; transmitted by the hog louse.

Swollen joints. An acute infectious disease, commonly called navel ill, caused by certain bacteria which often gain entrance into the navel soon after birth, becoming septicemic and localizing in the joints.

Symmetry. Similarity of organs or body parts on either side of a dividing line or plane.

Synapse. The region of contact between processes of two adjacent neurons, forming the place where a nervous impulse is transmitted from one neuron to another.

Synergist. A chemical substance that when used with an insecticide, drug, etc., will result in greater total effect than the sum of their indiviual effects.

Systemic insecticide. An insecticide capable of absorption into plant sap or animal blood and lethal to insects feeding on or within the treated host.

Systox (trade name). See demeton.

Talc. Powdered soapstone, anhydrous magnesium silicate.

Tapeworm. A parasitic intestinal flatworm (phylum Platyhelminthes).

Tarsus (pl., tarsi). The foot; the distal part of the insect leg, consisting of from one to five segments.

Tassel. The staminate inflorescence of corn.

TDE (Rhothane). A synthetic insecticide; a chlorinated hydrocarbon related to DDT; dichlorodiphenyl dichloroethane; from $\frac{1}{5}$ to $\frac{1}{10}$ as toxic to mammals as DDT, acute oral LD_{50} for rats 2500 mg/kg; nonphytotoxic at insecticidal concentrations except possibly to cucurbits; product of Rohm & Haas Company.

Tedion (trade name). See tetradifon.

Tegmen (pl., tegmina). The thickened, leathery forewing of an orthopteran.

Temephos (Abate). A synthetic insecticide; an organophosphate, *O,O*-dimethyl phosphorothioate *O,O*-diester with 4,4′-thiodiphenol; slightly toxic to mammals and fish; acute oral LD_{50} for rats 2000 mg/kg; effective mosquito, black fly, and midge larvicide; shows promise on cotton and other crops; product of American Cyanamid.

Temik (trade name). See aldicarb.

Tepp (common name). A synthetic insecticide and acaricide, an organic phosphate, tetraethyl pyrophosphate; extremely toxic to mammals, acute oral LD_{50} for rats 2 mg/kg; toxic to some varieties of tomato and chrysanthemum; blemishes some plants and fruits if applied by aircraft in less than 10 gal water per acre.

Terbufos (Counter). A synthetic insecticide; an organophosphate, *S*-[(2-ethylthio)ethyl)] *O,O*-demethyl phosphorodithioate; highly toxic to mammals, acute oral LD_{50} for male rats 4.5 mg/kg; 15 per cent granular band or in-furrow treatment of soil effective against corn rootworms, seed corn maggots, corn billbugs, corn wireworms, and sugarbeet root maggot; product of American Cyanamid.

Tetradifon (Tedion). A synthetic acaricide, a sulfone; chemically, *p*-chlorophenyl 2,4,5-trichlorophenyl sulfone; relatively nontoxic to mammals, 14,700 mg/kg adminstered to rats without signs of systemic toxicity; nonphytotoxic; product of FMC.

Texas cattle fever. See cattle tick fever.

Thanite (trade name). A synthetic insecticide; a thiocyanate, 82 per cent isobornyl thiocyanatoacetate and 18 per cent other related terpenes; slightly toxic to mammals, acute oral LD_{50} for rat 1603 mg/kg; highly toxic to plants; product of McLaughlin Gormley King.

Thimet (trade name). See phorate.

Thiodan (trade name). See endosulfan.

Thiram (common name). Protective fungicide suitable for foliage application or seed treatment; chemically, bis (dimethylthiocarbamoyl) disulfide; moderately toxic to mammals, acute oral LD_{50} for rats 780 mg/kg; phytotoxicity relatively low when used as directed.

Thrifty. Of livestock, an animal that grows vigorously, efficiently converting its feed to body substance.

Thylate (trade name). See thiram.

Tibia (pl., tibiae). The fourth segment of the leg, between the femur and the tarsus.

Tick. A blood-sucking arachnid parasite of the family Ixodidae or Argasidae (Fig. 4:3).

Tick paralysis. A flaccid, afebrile (without fever), ascending, motor paralysis produced by the attachment of certain species of ticks and believed to be caused by a neurotoxin secreted by the salivary glands of the feeding female tick.

Tiller. An erect shoot arising from the crown of a grass.

Tolerance. The amount of a pesticide that may safely and legally remain as a residue on a food plant or in meat or fat.

Toxaphene. A synthetic insecticide; a chlorinated hydrocarbon, chlorinated camphene containing 67 to 69 per cent of chlorine; moderately toxic to mammals, acute oral LD_{50} for rats 69 mg/kg; toxic to cucurbits, causes off-color in some cured tobaccos; product of Hercules and others.

Toxicogenic. Capable of producing a toxin; said of insects that introduce a toxin into a plant, while feeding.

Toxin. Any of various unstable poisonous compounds produced by some microorganisms and causing certain diseases; any of various similar poisons secreted by plants and animals.

Trachea (pl., tracheae). A tube of the respiratory system in insects.

Trachoma. A contagious form of conjunctivitis, caused by virus and characterized by formation of inflammatory granulations on the inner eyelid.

Trade name. Trade name is a registered trade mark of the company manufacturing the product and is capitalized in this book for identification.

Trench fever. A human disease caused by a bacteriumlike microorganism, *Rickettsia quintana*, nonfatal and characterized by sudden onset of fever, headache, dizziness, and pains in muscles and bones; transmitted by the body louse through its feces.

Trichlorfon (Dipterex, Dylox, Neguvon). A synthetic insecticide; an organic phosphate, dimethyl (2,2,2-trichloro-1-hydroxyethyl) phosphonate; moderately toxic to mammals, acute oral LD_{50} for rats 450 to 500 mg/kg; phytotoxicity—varying degrees of injury to fruit of several apple varieties; product of Chemagro.

Trimedlure. A synthetic attractant for the Mediterranean fruit fly; *tert*-butyl 4 (or 5)-chloro-2-methylcyclohexanecarboxylate.

Trochanter. The second segment of the leg, between coxa and femur.

Trolene (trade name). A refined ronnel for internal medication of livestock. See ronnel.

Trypanosomiasis. The disease caused by the presence in the body of a protozoan parasite of the genus *Trypanosoma*, marked by fever, anemia, and redness of the skin; transmitted by tsetse flies.

Tubercle. A small knoblike or rounded protuberance.

Tularemia. A bacterial disease occurring mainly in rabbits but also in certain rodents, ungulates, carnivores, birds, livestock, and man; caused by *Francisella tularensis* and transmitted by arthropod vectors (ticks, lice, fleas, biting flies) and by contact of skin with infected material; marked by inflammation of lymph glands, headache, chills, and fever.

Typhoid fever. An acute infectious disease caused by a bacterium, *Salmonella typhi*, characterized by continued fever, inflammation of intestine, intestinal ulcers, a rose-spot on the abdomen, and enlarged spleen; food and water-borne but may be transmitted by house flies.

Typhus fever. A human disease caused by a bacteriumlike microorganism, *Rickettsia prowazeki*, and transmitted by the body louse, *Pediculis humanus humanus* L. The disease is characterized by high fever, backache, intense headache, bronchial disturbances, mental confusion, and congested face. Mortality may range from 15 to 75 per cent.

Unsaturated. In chemistry, not having all valences of the constituent atoms of a compound fully satisfied.

Urea. Chief nitrogenous constituent of the urine of mammals and the final product of decomposition of proteins in the body; chemically, NH_2CONH_2.

Uric acid. The chief nitrogenous waste of birds, reptiles and insects; chemically, $C_5H_4N_4O_3$.

Vapona (trade name). See dichlorvos.

Vascular tissues. The fluid conducting tissues of a plant including both xylem (water) and phloem (food) tissues.

Veins. In insects, the riblike tubes that strengthen the wings.

Venation. The arrangement of veins in the wings of insects.

Vendex (trade name). See fenbutatin-oxide.

Ventral. Lower or underneath, pertaining to the underside of the body.

Vermiform larva. A legless wormlike larva without a well developed head (Fig. 3:14G).

Vertebrates. Animals with a spinal column or backbone, such as fishes, birds, mammals, and so on.

Vestigial. Having the nature of a degenerate or atrophied organ, more fully functional in an earlier stage of development of the individual or species.

Virelure. Synthetic sex pheromone of the female tobacco budworm moth; a combination of 2-11-hexadecenal (2-11-HDAL) and Z-9-tetradecenal (Z-9-TDAL); attractive to males of the tobacco budworm; experimentally used in traps and to disrupt mating.

Visceral muscle. A muscle which invests an internal organ.

Vitamin B complex. The group of water-soluble vitamins, including thiamine, riboflavin, nicotinic acid, and others.

Viviparae. Female aphids that bear living young (do not lay eggs).

Viviparous. Bearing living young instead of laying eggs.
Volunteer growth. Plants that spring up from unplanted seed, usually lost before or during harvest.
Warbex (trade name). See famphur.
Western equine encephalitis. A virus disease of horses communicable to man, marked by fever, convulsions, and coma and transmitted by certain species of mosquitoes.
Wettable powder. Insecticidal dusts to which have been added wetting agents as well as insecticide, thus making the product dispersible and suspensible in water.
Wheat streak mosaic. A virus disease of wheat and other grasses marked by yellow streaking of leaves, stunted growth, and reduced seed set, transmitted by the wheat curl mite.
Whorl. An arrangement of organs, such as leaves, in a circle around the stem of a plant.
Wilt. Loss of freshness and drooping of leaves of plants due to inadequate water supply or excessive transpiration or to a vascular disease which interferes with utilization of water or to a toxin produced by an organism.
Wing pads. The undeveloped wings of nymphs and naiads, which appear as two flat structures on each side.
Woollybear. A very hairy caterpillar belonging to the family Arctiidae, the tiger moths.
Workers. In termites, the sterile males and females that perform most of the work of the colony; they are pale and wingless, and usually lack compound eyes; in social Hymenoptera, females with undeveloped reproductive organs that perform the work of the colony.
Yaws. A tropical infectious disease caused by a spirochete, *Treponema pertenue*, marked by raspberrylike excrescences and ulcerations on face, hands, feet, and external genitals; transmitted by certain flies or gnats (*Hippelates*).
Yeast. Unicellular fungi of the family Saccharomycetaceae.
Yellow fever. An acute infectious disease caused by a virus transmitted by certain mosquitoes and marked by fever, jaundice, and albumin and globulin in the urine.
Yellows. A plant disease characterized by yellowing and stunting of the affected plant; caused by a fungus, virus, or insect toxin.
Zineb. A protective fungicide used for foliage application; chemically, zinc ethylene bisdithiocarbamate; of low mammalian toxicity, acute oral LD_{50} for rats greater than 5200 mg/kg; toxic to zinc-sensitive plants.
Zolone (trade name). See phosalone.

SELECTED REFERENCES

Farm Chemicals Handbook 1976 (Willoughby, Ohio: Meister Publishing Company, 1976).
Kenaga, E. E., and C. S. End, *Commerical and Experimental Organic Insecticides*, Entomol. Soc. Am. Special Publ. 74–1, 1974.
Torre-Bueno, J. R., *A Glossary of Entomology* (Brooklyn: Brooklyn Ent. Soc., 1937).
Tullock, G. S., *Torre-Bueno's Glossary of Entomology—Supplement A* (Brooklyn: Brooklyn Ent. Soc., 1960).

INDEX

B

C

D

E

F

G

H

I

J

K

L

M

Q

R

S

T

FUNDAMENTALS OF APPLIED ENTOMOLOGY

with chapters by

LELAND R. BROWN
University of California at Riverside

CHRISTIAN C. BURKHARDT
University of Wyoming

HUAI C. CHIANG
University of Minnesota

W. DON FRONK
Colorado State University

DEANE P. FURMAN
University of California at Berkeley

ROBERT F. HARWOOD
Washington State University

B. AUSTIN HAWS
Utah State University

CARL JOHANSEN
Washington State University

JOHN A. NAEGELE
University of Massachusetts

JOHN V. OSMUN
Purdue University

ROBERT E. PFADT
University of Wyoming

WILLIAM M. ROGOFF
USDA Entomology Research Division

DONALD A. WILBUR and ROBERT B. MILLS
Kansas State University